T0142269

Cognitive Science and Technology

This series aims to publish work at the intersection of Computational Intelligence and Cognitive Science that is truly interdisciplinary and meets the standards and conventions of each of the component disciplines, whilst having the flexibility to explore new methodologies and paradigms. Artificial Intelligence was originally founded by Computer Scientists and Psychologists, and tends to have stagnated with a symbolic focus. Computational Intelligence broke away from AI to explore controversial metaphors ranging from neural models and fuzzy models, to evolutionary models and physical models, but tends to stay at the level of metaphor. Cognitive Science formed as the ability to model theories with Computers provided a unifying mechanism for the formalisation and testing of theories from linguistics, psychology and philosophy, but the disciplinary backgrounds of single discipline Cognitive Scientists tends to keep this mechanism at the level of a loose metaphor. User Centric Systems and Human Factors similarly should inform the development of physical or information systems, but too often remain in the focal domains of sociology and psychology, with the engineers and technologists lacking the human factors skills, and the social scientists lacking the technological skills. The key feature is that volumes must conform to the standards of both hard (Computing & Engineering) and social/health sciences (Linguistics, Psychology, Neurology, Philosophy, etc.). All volumes will be reviewed by experts with formal qualifications on both sides of this divide (and an understanding of and history of collaboration across the interdisciplinary nexus).

Indexed by SCOPUS

Amit Kumar · Gheorghita Ghinea ·
Suresh Merugu · Takako Hashimoto
Editors

Proceedings of the International Conference on Cognitive and Intelligent Computing

ICCIC 2021, Volume 1

Editors
Amit Kumar
BioAxis DNA Research Centre Private Limited
Hyderabad, Andhra Pradesh, India

Suresh Merugu
CMR College of Engineering and Technology
Hyderabad, India

Gheorghita Ghinea
Department of Computer Science
Brunel University
Uxbridge, UK

Takako Hashimoto
Ichikawa, Chiba, Japan

ISSN 2195-3988 ISSN 2195-3996 (electronic)
Cognitive Science and Technology
ISBN 978-981-19-2352-4 ISBN 978-981-19-2350-0 (eBook)
https://doi.org/10.1007/978-981-19-2350-0

This Springer imprint is published by the registered company Springer Nature Singapore Pte Ltd.
The registered company address is: 152 Beach Road, #21-01/04 Gateway East, Singapore 189721, Singapore

Contents

An Extensive Survey of Deep Learning-Based Crop Yield Prediction Models for Precision Agriculture

Srilatha Toomula and Sudha Pelluri

Abstract Precision agriculture, as the trademark of the agriculture 4.0 period, has assured to reform agricultural practices using monitoring and intervention technologies to increase productivity and decrease the environmental impact. Computer vision (CV) and deep learning (DL) models are commonly used as key enablers for precision agriculture. CV technologies utilize digital images for the interpretation and understanding of the world to offer precise, region orient details about the crops and respective surroundings. Today, CV has been widely employed to support precision agriculture processes like crop yield prediction (CYP), crop monitoring, weed control, plant disease detection, weed detection, etc. CYP is a significant process for decision making at the national and regional levels. Several machine learning (ML) and DL-based models have been presented for accurate CYP. Therefore, this paper reviews existing DL-based CYP models developed for precision agriculture. In this view, the major aim of the review is to identify, group, and discuss the existing intelligent agriculture approaches. The existing methods are surveyed based on the underlying techniques, objectives, dataset used, and available datasets. The outcome of the survey pointed out the significance of applying DL models for CYP in precision agriculture.

Keywords Precision agriculture · Computer vision · Deep learning · Crop yield prediction · Image processing

1 Introduction

Crop yield is one of the significant parts of agriculture and has several links with the human community. Yield prediction is highly a difficult process in precision agriculture, which is essential for crop market planning, harvest management, crop insurance, and yield mapping. The crop yield is influenced by several aspects like management practice, crop genotype, and environments [1]. The crop genotype has

S. Toomula (✉) · S. Pelluri
Department of Computer Science and Engineering, Osmania University, Hyderabad, India
e-mail: srilatha.research@gmail.com

A. Kumar et al. (eds.), *Proceedings of the International Conference on Cognitive and Intelligent Computing*, Cognitive Science and Technology,
https://doi.org/10.1007/978-981-19-2350-0_1

been enhanced dramatically for many years by seed companies. Environments which are varying temporally and spatially contain large effects on location to location and year to year differences in crop yield. In these environments, precise yield prediction is more useful for worldwide food production. Import and export decisions in time could be carried out depending upon precise yield prediction. Farmers can make use of the yield prediction for creating financial decisions and knowledgeable management. The efficiency of hybrid crops is predicted in novel and untested positions. But effective crop yield prediction (CYP) is extremely complex because of several composite aspects. For instance, genotype and environmental aspects frequently have interaction with one another that creates yield prediction a difficult process.

Environmental aspects like weather factors frequently contain composite nonlinear effects that are complex for precise estimation. Policy-makers based on the precise prediction for making import and export decisions reinforce the national food security. Seed companies require predicting the efficiency of novel hybrids in several atmospheres to breed for enhanced varieties. Farmers and growers also assist from yield prediction for making financial decisions and informed management. The effect of the genetic marker should be calculated so that it can be subjected to interaction with many fields of management practice and environmental conditions. Several researches have been concentrated on describing the phenotype (like yield) as an environment (E), interaction (G × E), and their explicit function of genotype (G). The most common and direct technique is to assume the additive impacts of E and G and process their interaction as noise [2]. A widespread method for studying the G × E effects is to recognize the interaction and effects of huge environment instead of additional detailed environmental elements. Various researches have been presented for clustering the environment depending upon discovered driver of G × E interactions utilized the location regression and the transferred multiplicative method for G × E interactions analyses by separating environment to equivalent sets.

Burgueño et al. [3] presented a combined method of factor analytic (FA) and linear mixed method for clustering environment and genotype and identify their interaction. They stated that FA method could enhance the prediction up to 6%, while they exit complex G × E patterns in the information. The linear mixed methods have been utilized for the investigation of both interactive and additive impacts of separate genetics and environments. In recent years, machine learning (ML) methods are employed for CYP, involving association rule mining, decision tree, multivariate regression, and artificial neural network (ANN) [4]. An essential characteristic of ML technique is that they process the outcome (i.e. crop yield) as an implied function of the input parameter (environmental and genes elements) that can be a higher complex and nonlinear function.

In recent times, deep learning (DL) methods have been utilized for CYP. Compared to general ML models that have single hidden layer, DL techniques with many hidden layers determine better performance [5]. But the deeper methods are very complex for training process and need more advanced hardware and optimization methods. DL is an extended version of ML, which adds more depth to the models and transforms the data utilizing several functions which enable hierarchical data representation by different stages of abstraction. Feature learning, i.e. automated extraction of features

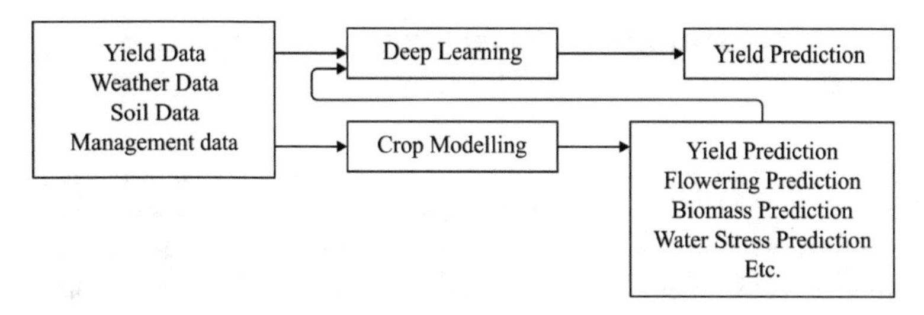

Fig. 1 Process involved in DL-based precision agriculture

from the input data is the major advantage of the DL models [6–9]. IT can resolve complex problems effectively and quickly enabling high parallelization. The complex models applied in DL result in increased classifier accuracy and reduced error in the regression problem, providing large databases. The DL model comprises distinct elements such as convolution, pooling, fully connected (FC), and activation functions.

The hierarchical model and high learning ability of the DL model enable to carry out classification and prediction process. It exhibits flexibility and adaptability to several highly complex issues. Several DL models have been existed in the earlier studies for precision agriculture. An overview of the processes involved in the DL-based precision agriculture is given in Fig. 1. You et al. [6] utilized DL methods like convolution neural networks (CNNs) and recurrent neural networks (RNNs) for predicting soybeans yield in the United States depending upon a series of remote sensor images are captured beforehand the harvest. This method exceeds conventional remote sensing-based methods by 15% based on mean absolute percentage error (MAPE). Russello [7] utilized CNN for CYP depending upon remote sensing images. This technique utilized 3D convolution for including the spatiotemporal feature and exceed other ML techniques [8].

Since several CYP models are available in the literature, there is a need to review the existing DL-based yield prediction models. With this motivation, this survey intends to investigate the works related to the domains of DL and CYP. The existing methods are surveyed based on the underlying techniques, objectives, dataset used, and available datasets. For getting the insights, existing works have been investigated under different aspects. Here, a set of four research questions (RQ) have been stated as follows.

- RQ1—What are the DL models that have been utilized in the existing works of CYP?
- RQ2—What are all the features utilized for CYP by the use of DL models?
- RQ3—What are performance measures and validation models that have been available in the previous works for CYP?
- RQ4—What are challenging issues exist in the area of CYP by DL models?

2 Review of Existing DL-Based Crop Yield Prediction Models

Elavarasan and Vincent [10] developed a new deep recurrent Q-network model for predicting crop yield. The presented technique includes an RNN model over the Q-Learning RL technique. The serially stacked layers of RNN are used along with the Q-learning network designed a crop yield forecasting. The linear map mapped into the RNN outcome to the Q-values. At last, the agent gets an aggregated value for the actions carried out by the minimization of error and maximization of predictive accuracy. Khaki et al. [1] presented a new CNN with RNN model, called CNN-RNN to forecast the crop yield depending upon the ecological data and management practices. The presented CNN-RNN technique with random forest (RF), FC neural network (FCNN), and LASSO is employed for corn and soybean forecast in the US. The presented model can be integrated into the backpropagation technique, which revealed the weather condition, weather predictive accuracy, soil condition, and management practice.

Khaki and Wang [2] developed a new DNN model for the prediction of hybrid corn yield. The DNN model involves a feature selection problem to decrease the dimensionality of the input space with no considerable reduction in the predictive accuracy. An effective dimension decreasing approach: Self organizing map (SOM) is presented together with latent Dirichlet allocation (LDA) in [11]. The SOM approach is the most appropriate dimension decreasing approach for highlighting the self-arranging outline. Later decreasing the measurement, the dimension decreased data are utilized for predicting weather for a satisfactory result. A satisfactory period for a proper crop is ordered with the guideline of DNN classification method. Wang et al. [12] display possible outcomes in forecasting soybean crop yields in Argentina by utilizing DL methods. It also attains reasonable outcomes with a transfer learning method for predicting Brazil soybeans harvest with a small quantity of data. The stimulating for transfer learning is that the achievement of DL method is mainly based on rich ground truth trained data.

You et al. [13] presented an inexpensive, scalable, and accurate technique for CYP utilizing open access remote sensing data. Initially, it declines handcrafted traditional feature, which is utilized in the remote sensing community and introduces a method depending upon the latest illustration of learning concepts. It also establishes a novel dimension reducing method permits for training a long short-term memory (LSTM) or CNN and automatically learn beneficial feature while labelled trained information is insufficient. Lastly, it integrates a Gaussian process component for explicit method of spatiotemporal patterns of the data and additionally enhances the accurateness. In Nevavuori et al. [14], CNN—DL method shows an extraordinary efficiency in image classification process—is employed for building a method to CYP depending upon RGB and NDVI information which is attained from UAV. The impact on several factors of CNNs like tuning of the hyper variable, regularization strategy, network depth, and selection of the training method on the predictive performance is confirmed. The aim of [15] is to calculate the energy of UAV-based multimodal data fusion by utilizing multispectral, thermal sensor, and RGB for estimating soybeans (i.e., Glycine max) grain yield with the DNN architecture. The thermal images, RGB, and multispectral have been gathered by utilizing low-cost multisensor UAV from a testing location in Missouri, USA, and Columbia. Multimodal information, like structure, texture feature, thermal and canopy spectral, was combined and extracted for predicting crop grain yield by utilizing different techniques.

Schwalbert et al. [16] proposed a new method for performing in season ("near real time") soybean yield predictions in southern Brazil by LSTM, NN, weather data, and satellite imagery. The aims of this research are to (i) relate the efficiency of 3 distinct methods like LSTM-NN, RF, and multivariate OLS linear regression to predict soybean yield by utilizing land surface temperature EVI, NDVI, and rainfall as autonomous parameters, and (ii) calculate earlier (at the time of soybean developing period) for predicting yield with moderate accuracy. Khaki et al. [17] presented a novel method is known as YieldNet that uses a new DL architecture which utilizes the transfer learning among soybean and corn yield predictions by allocating the weights of the backbone feature extraction. Furthermore, to assume the multi-target response parameter, it developed a novel loss function. Mathematical outcomes determine that this presented technique precisely predicts the yield from 1 to 4 months earlier the harvest, and is inexpensive for other advanced methods.

Chu and Yu [18] proposed a new end-to-end predictive method that combines two backpropagation neural networks (BPNN) with an independently RNN (IndRNN), called BBI method, which is presented for addressing these problems. In initial phase, BBI method pre-processes the meteorology data and original area. In next phase, the IndRNN and BPNN are utilized for learning deep temporal and spatial features in similar manner. In third phase, additional BPNN integrates two types of deep feature and learn the relations among this rice yields and deep features for making prediction for winter and summer rice yields.

Wang et al. [19] introduced a two branch DL method for predicting winter wheat yield in the major cultivating areas of China at the county level. The initial branch of the method was made depending upon LSTM network with inputs from meteorological and remote sensing data. Additional branch was made by utilizing CNN for modelling static soil features. The method was trained after by utilizing the detrended statistical yield data from 1982 to 2015 and calculated by leave one year out validation. Nevavuori et al. [20] presented the possibility of spatiotemporal DL framework in crop yield time series demonstrating and predictive with RGB time sequence data. By utilizing LSTM and CNN networks as temporal and spatial base frameworks, they trained and developed convolutional LSTM, 3D-CNN, and CNN-LSTM frameworks with full 15-week image frame series from the complete developing period of 2018.

Jiang et al. [21] introduced an LSTM method, which combines meteorology, heterogeneous crop phenology, and remote sensing data for estimating county level corn yields. By combining meteorological indices and heterogeneous phenology based remote sensing, the LSTM method calculates for 76% of yield variances over the Corn Belt, enhanced from 39% of yield variances described by phenology-based meteorological indices. Cao et al. [22] proposed the main winter wheat production areas of China as instance, and it is related to conventional ML technique RF and three DL methods, like LSTM, 1D-CNN, and DNN for predicting crop yield by combining public access data with the Google Earth Engine (GEE) platform, involving satellite, climate, spatial information, and soil properties. Yue et al. [23] presented a data-driven encoder–decoder method, by LSTM and convolutional LSTM that is employed for predicting cumulative precipitation, daily sunshine duration, and average temperature for the future. For testing the efficiency of the convolutional LSTM-based method, in both conventional LSTM and CNN-LSTM encoder–decoder methods are compared.

Feng et al. [24] carried out an in season alfalfa yield prediction by UAV-based hyper-spectral images. In particular, it initially extracts a huge amount of hyper-spectral indices from the original data and accomplished a feature selection for reducing the data dimension. Later, an ensemble ML method was established by integrating three broadly utilized base learners comprise of SVR, K-nearest neighbours (KNN), and RF. This method efficiency was calculated over the research field in Wisconsin.

Khaki et al. [25] presented a new DL technique to count on ear corn kernel in field for collecting real-time information and eventually, enhance problem solving for yield maximization. This DeepCorn approach illustrates that this architecture is powerful in several situations. The DeepCorn calculates the amount of corn kernel in an image of corn ear and predicts the kernel counts depending upon the evaluated density map. DeepCorn utilizes a truncated VGG-16 as a backbone for feature extraction and combines feature mapping from several scales of the network for making it strong towards image scale variation. It accepts a semi-supervised learning method for improving the efficiency of the presented approach. Table 1 shows the comparison of distinct DL-based crop yields prediction models.

Wang et al. [26] are presented for predicting the cotton yield by an enhanced LSTM method that is an artificial RNN framework utilized in the area of DL. The LSTM method has feedback links and integration of distinct gates like forget gate, output gate, and input gate for controlling the required data from storage for prior timestamp information and upgraded from this timestamp inputs. In this research, a UAV imaging system comprising multi-spectral camera of five narrow spectral bands of near infrared (840 $\pm$ 20 nm), red (668 $\pm$ 5 nm), green (560 $\pm$ 10 nm), red edge (717 $\pm$ 5 nm), and blue (475 $\pm$ 10 nm) is utilized for collecting imagery data of cotton in three crucial development phases. The imagery data have been pre-processed for removing calibrate reflectance, background, and registered to produce information-based geo-referenced data. Multivariable aspects of GNDVI, NDVI, canopy temperature, and size have been extracted from UAV multi-spectral images and utilizes as input for the LSTM method. The variables of the LSTM method should be optimum to enhance the efficiency for precise yield calculation.

Table 1 Comparison of different DL-based CYP models

References	Year	Objective	Technique used	Crop type	Performance measures
[10]	2020	Develop a DRL technique for CYP	Deep recurrent Q-network	Paddy	MAE, MSE, RMSE, accuracy
[1]	2020	Design a CNN-RNN model for corn and soybean CYP	CNN-RNN	Corn and soybean	MSE, RMSE, accuracy
[2]	2019	Propose a DL-based CYP model	DNN	Corn	RMSE, accuracy
[11]	2018	Design a weather and CYP technique	Weighted SOM + DNN	Kharif and ragi crops	Sensitivity, specificity, accuracy
[12]	2018	Employ DTL model for CYP using remote sensed data	LSTM	Soybean	RMSE
[13]	2017	Design an inexpensive DL-based CYP using remote sensed data	CNN-LSTM	Soybean	MAE, MAPE
[14]	2019	Introduce a DL model for CYP using NDVI and RGB data from UAV	DCNN	Wheat and malting barley	MAE, MAPE
[15]	2020	Design a UAV-based CYP model using DL and multimodal fusion techniques	DNN, fusion model	Soybean	RMSE, accuracy
[16]	2020	Present a DL-based CYP model using satellite and weather data	LSTM	Soybean	MAE
[17]	2020	Develop a CYP model for multiple crops concurrently	CNN	Corn and soybean	MAE
[18]	2020	Introduce a DL-based fusion model for CYP	BPNN, IndRNN	Rice	MAE, RMSE

(continued)

Table 1 (continued)

References	Year	Objective	Technique used	Crop type	Performance measures
[19]	2020	Employ a DL model for winter yield prediction	LSTM, CNN	Wheat	Accuracy
[20]	2020	Present a CYP model using multitemporal UAV data and spatiotemporal	3D-CNN, LSTM	Nine crops (varieties of wheat, barley, and oats)	MAPE, MAE
[21]	2019	Design a DL-based CYP model to conflating heterogeneous geospatial data	LSTM	Corn	RMSE
[22]	2021	Introduce a scalable and easy model for accurate CYP	DNN, LSTM	Corn	RMSE, accuracy
[23]	2020 ara>	Analyse the growth levels of crops using DL model	LSTM, ConvLSTM	Maize	MAE, RMSE, accuracy
[24]	2020	Design an ensemble model with hyper-spectral images for CYP	Ensemble model	Corn	Scatter plots
[25]	2021	Present a semi-supervised DL model for CYP	DeepCorn	Corn	
[26]	2020	Introduce a DL-based CYP using UAV images	Improved LSTM	Cotton	MSE, RMSE

3 Results and Discussion

This section validates the performance analysis of different CYP models such as deep reinforcement learning (DRL), ANN, gradient boosting (GB), RF, and other DL-based algorithms like Bernoulli deep belief network (BDN), Bayesian artificial neural networks (BAN), rough autoencoders (RAE), and interval deep generative artificial neural networks (IDANN) in terms of accuracy and MAPE.

Table 2 demonstrates the comparative study of reviewed CYP models. Figure 2

Table 2 Comparative analysis of various models in terms of accuracy and MAPE

Models	Accuracy (%)	MAPE (%)
DRL	93.70	17.00
BDN	92.10	20.00
BAN	91.70	27.00
IDANN	91.00	29.00
RAE	90.70	32.00
DL	91.85	28.00
ANN	90.50	38.00
RF	70.70	53.00
GB	81.20	41.00
SOM-KNN	58.99	–
WSOM-KNN	81.23	–
DCNN	–	08.80

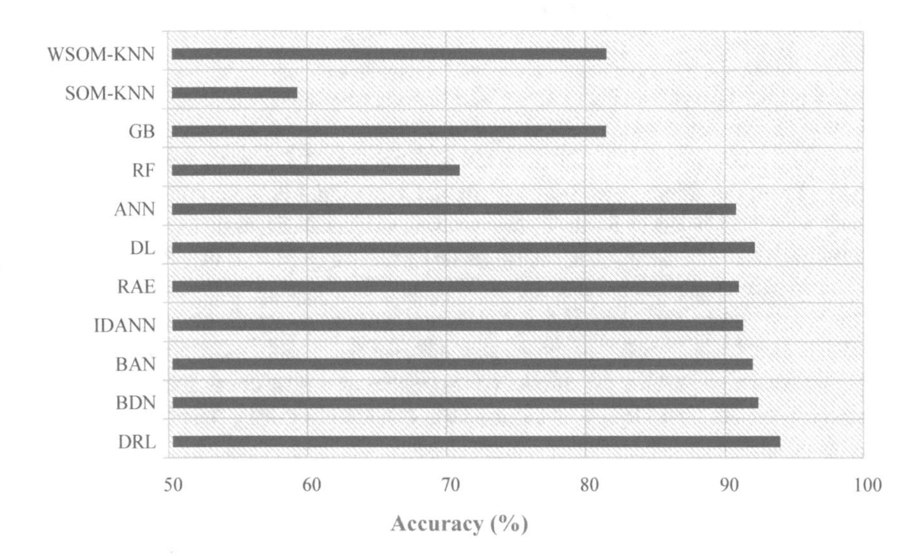

Fig. 2 Accuracy analysis of different CYP models

investigates the results analysis of different CYP models in terms of accuracy. From the figure, it is evident that the SOM-KNN model has accomplished poor results with an accuracy of 58.99%. At the same time, the RF, GB, and WSOM-KNN models have obtained slightly increased performance by offering an accuracy of 70.7%, 81.2%, and 81.23%, respectively. Simultaneously, the ANN and RAE models have reached moderate results with the accuracy of 90.5% and 90.7%, respectively. Concurrently, the IDANN, BAN, and DL models have accomplished reasonable accuracy of 91%, 91.7%, and 91.85%, respectively. Though the BDN model has appeared as a near

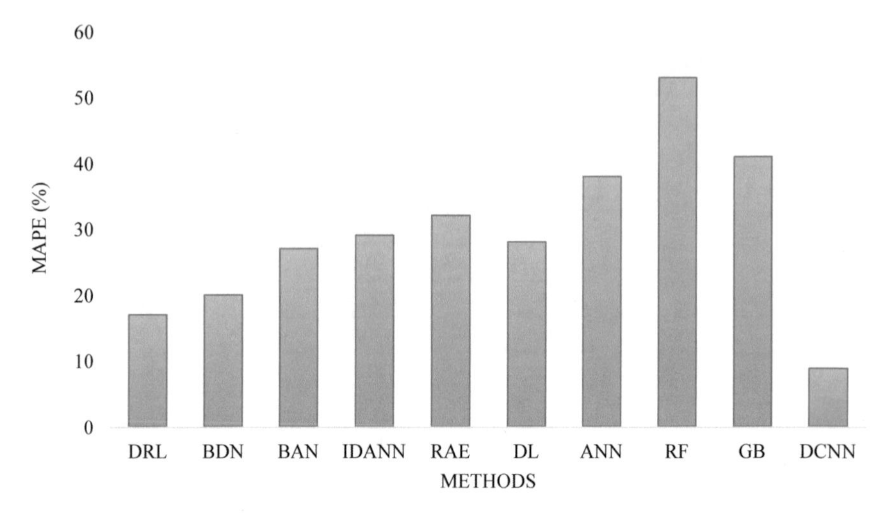

Fig. 3 MAPE analysis of different CYP models

optimal performer with an accuracy of 92.1%, the presented DRL model has offered effective performance with an accuracy of 93.7%.

Figure 3 examines the results analysis of different CYP models with respect to MAPE. From the figure, it is clear that the DCNN model has accomplished worse results with the MAPE of 8.8%. In line with, the DRL, BDN, and BAN techniques have achieved somewhat higher performance by offering a MAPE of 17%, 20%, and 27% correspondingly. Concurrently, the DL and IDANN approaches have attained moderate outcomes with the MAPE of 28% and 29% correspondingly. Followed by, the RAE and ANN techniques have accomplished reasonable MAPE of 32%, and 38%, respectively. But the GB model has appeared as a near better performer with the MAPE of 41%, and the RF technique has offered effective performance with a MAPE of 53%.

4 Conclusion

This survey has aimed to investigate the works related to the domains of DL and CYP in precision agriculture. This study aims to perform a review to identify, group, and discuss the existing intelligent agriculture approaches. The existing methods are surveyed based on the underlying techniques, objectives, dataset used, and available datasets. The outcome of the survey pointed out the significance of applying DL models for CYP in precision agriculture. For getting the insights, existing works have been investigated under different aspects. The outcome of the survey pointed out the significance of applying DL models for CYP in precision agriculture. A detailed results analysis was also performed to highlight the particular characteristics of the reviewed models and a brief comparative study is also made. As a part of future scope, an advanced DL with metaheuristic optimization algorithm-based CYP models will be designed.

References

1. Khaki S, Wang L, Archontoulis SV (2020) A cnn-rnn framework for CYP. Front Plant Sci 10:1750
2. Khaki S, Wang L (2019) CYP using deep neural networks. Front Plant Sci 10:621
3. Burgueño J, Crossa J, Cornelius PL, Yang R-C (2008) Using factor analytic models for joining environments and genotypes without crossover genotype× environment interaction. Crop Sci 48:1291–1305
4. Uthayakumar J, Metawa N, Shankar K, Lakshmanaprabu SK (2018) Intelligent hybrid model for financial crisis rediction using machine learning techniques. Inf Syst e-Bus Manage, pp 1–29
5. Madhan ES, Neelakandan S, Annamalai R (2020) A novel approach for vehicle type classification and speed prediction using deep learning. J Comput Theoret Nano Sci 17(5): 2237–2242
6. You J, Li X, Low M, Lobell D, Ermon S (2017) Deep gaussian process for CYP based on remote sensing data. In: Thirty-first AAAI conference on artificial intelligence, San Francisco, CA, pp 4559–4566
7. Russello H (2018) Convolutional neural networks for CYP using satellite images. IBM Center for Advanced Studies
8. Neelakandan S, Paulraj D (2020) An automated learning model of conventional neural network based sentiment analysis on twitter data. J Comput Theoret Nano Sci 17(5):2230–2236
9. Lakshmanaprabu SK, Mohanty SN, Krishnamoorthy S, Uthayakumar J, Shankar K (2019) Online clinical decision support system using optimal deep neural networks. Appl Soft Comput 81:105487
10. Elavarasan D, Vincent PD (2020) CYP using deep reinforcement learning model for sustainable agrarian applications. IEEE Access 8:86886–86901
11. Mohan P, Patil KK (2018) Deep learning based weighted SOM to forecast weather and crop prediction for agriculture application. Int J Intell Eng Syst 11(4):167–176
12. Wang, A.X., Tran, C., Desai, N., Lobell, D. and Ermon, S., 2018, June. Deep transfer learning for CYP with remote sensing data. In Proceedings of the 1st ACM SIGCAS Conference on Computing and Sustainable Societies (pp. 1–5).
13. You J, Li X, Low M, Lobell D, Ermon S (2017) Deep gaussian process for CYP based on remote sensing data. In: Proceedings of the AAAI conference on artificial intelligence, vol. 1, no 1
14. Nevavuori P, Narra N, Lipping T (2019) CYP with deep convolutional neural networks. Comput Electron Agric 163:104859
15. Maimaitijiang M, Sagan V, Sidike P, Hartling S, Esposito F, Fritschi FB (2020) Soybean yield prediction from UAV using multimodal data fusion and deep learning. Remote Sens Environ 237:111599
16. Schwalbert RA, Amado T, Corassa G, Pott LP, Prasad PV, Ciampitti IA (2020) Satellite-based soybean yield forecast: Integrating machine learning and weather data for improving CYP in southern Brazil. Agric For Meteorol 284:107886
17. Khaki S, Pham H, Wang L (2020) YieldNet: a convolutional neural network for simultaneous corn and soybean yield prediction based on remote sensing data. arXiv preprint arXiv:2012. 03129
18. Chu Z, Yu J (2020) An end-to-end model for rice yield prediction using deep learning fusion. Comput Electron Agric 174:105471

19. Wang X, Huang J, Feng Q, Yin D (2020) Winter wheat yield prediction at county level and uncertainty analysis in main Wheat-Producing regions of China with Deep Learning approaches. Rem Sens 12(11):1744
20. Nevavuori P, Narra N, Linna P, Lipping T (2020) CYP using multitemporal UAV data and spatio-temporal deep learning models. Rem Sens 12(23):4000
21. Jiang H, Hu H, Zhong R, Xu J, Xu J, Huang J, Wang S, Ying Y, Lin T (2020) A deep learning approach to conflating heterogeneous geospatial data for corn yield estimation: a case study of the US Corn Belt at the county level. Glob Change Biol 26(3):1754–1766
22. Cao J, Zhang Z, Luo Y, Zhang L, Zhang J, Li Z, Tao F (2021) Wheat yield predictions at a county and field scale with deep learning, machine learning, and google earth engine. Eur J Agron 123:126204
23. Yue Y, Li JH, Fan LF, Zhang LL, Zhao PF, Zhou Q, Wang N, Wang ZY, Huang L, Dong XH (2020) Prediction of maize growth stages based on deep learning. Comput Electron Agric 172:105351
24. Feng L, Zhang Z, Ma Y, Du Q, Williams P, Drewry J, Luck B (2020) Alfalfa yield prediction using UAV-based hyperspectral imagery and ensemble learning. Rem Sens 12(12):2028
25. Khaki S, Pham H, Han Y, Kuhl A, Kent W, Wang L (2021) Deepcorn: a semi-supervised deep learning method for high-throughput image-based corn kernel counting and yield estimation. Knowle-Based Syst, p 106874
26. Wang S, Feng A, Lou T, Li P, Zhou J (2020) LSTM-based cotton yield prediction system using UAV imagery. In: 2020 ASABE annual international virtual meeting. American Society of Agricultural and Biological Engineers, p 1

Performance Analysis of Routing Protocols for Wireless Sensor Networks

Archana Ratnaparkhi, Radhika Purandare, Arti Bang, Aditya Rajput, and Kaustubh Venurkar

Abstract Many applications are subject to WSN, barely any applications require fast information move with next to no interference and a few applications requires the general throughput rather than the delay. It is entirely dependent on the program's requirements as to which parameters are required for its operation, as each application has its own set of requirements. This paper analyzes routing protocols on the basis of significant parameters such as throughput, delay as well as load on the network. The networks which have been used for experimentation are wireless LAN and ZigBee networks. From this study, we have found that the different protocols have different throughput, delay and network load in same network scale. Like in most of the cases, OLSR provides less jitter, delay and load on network as compared to AODV and DSR. DSR can have the highest of the throughput in large-scale network.

1 Introduction

Selection of suitable optimum route from source to destination is defined by algorithms termed as routing algorithms. There are several difficulties while electing this travel path, based on the network type, nature of channels and its characteristics. The sensor nodes sense the data in a WSN and forwards it in direction of base station which then connects current network with the other networks where different actions

A. Ratnaparkhi (✉) · R. Purandare · A. Bang · A. Rajput · K. Venurkar
Department of Electronics and Telecommunication Engineering, Vishwakarma Institute of Information Technology, Pune, Maharashtra 411048, India
e-mail: archana.ratnaparakhi@viit.ac.in

A. Kumar et al. (eds.), *Proceedings of the International Conference on Cognitive and Intelligent Computing*, Cognitive Science and Technology,
https://doi.org/10.1007/978-981-19-2350-0_2

are taken. The work of creating the route is difficult for a variety of reasons; some of the most significant obstacles are listed below:

- Assignment of universal identification method for the sensor nodes is tedious.
- The observed data flow with variable inputs to the stationary base station is required.
- There is a lot of duplication in the data traffic generated.
- Dissimilarities in motes.

1.1 Routing Protocol Classification

The routing protocols basically form the link for communication between two nodes, and they form the basic unit for sharing information through a network. The routing protocols are divided into four subparts—node-centric routing protocol, source-initiated routing protocol, data-centric routing protocol, destination-initiated routing protocol. The flowchart in Fig. 1 provides a clear idea.

(i) *Node Centric*: In this type, the destination node is symbolized by numerical values and this is not extensively used in WSN as seen in LEACH algorithm.
(ii) *LEACH (Low-Energy Adaptive Clustering Hierarchy)*: Here, the cluster is organized in such a way that energy is equally bifurcated in the respective sensory nodes, and the selection of the cluster head in this type of protocol is based on the process of randomization which also complements its battery life.
(iii) *Data Centric*: Unlike node-centric routing protocol, the main focus of data-centric routing protocol is transmission of information rather than selection of nodes; the example of data-centric routing protocol is the SPIN protocol.
(iv) *SPIN*: The following type of protocol is mainly used for sorting discrepancies like flooding and gossiping. The spin protocol mainly deploys three, messages ADV, REQ, DATA. The representation for spin protocol is given in Fig. 2.

1. *Destination Initiated*:

In this type of protocol, the concept of back tracking is followed where the process of path setup starts from the destination node, i.e., the end node. One of the examples for this type of protocol is directed diffusion (DD).

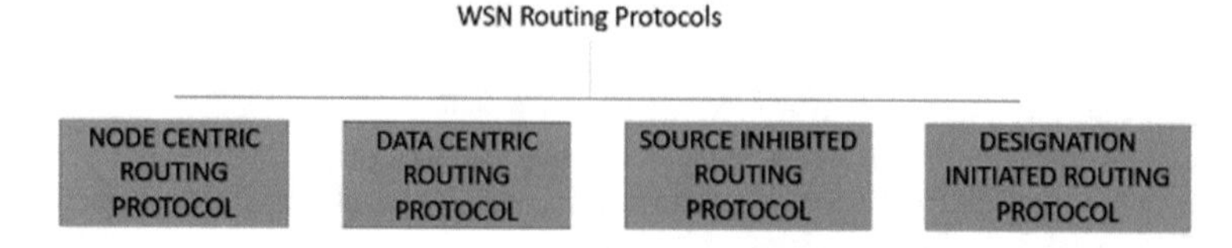

Fig. 1 WSN routing protocols

Fig. 2 SPIN protocol

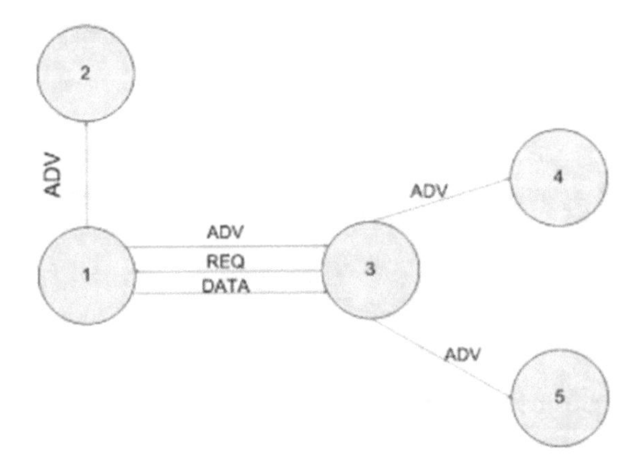

2. *Directed Diffusion*:

It is an energy efficient routing protocol which in turn leads to increasing the overall lifetime of the network. The communication established in this kind is node to node and is data centric, hence, no room for addressing is provided in this protocol.

3. *Source Initiated*:

It is basically an inverse configuration for destination-initiated routing protocol. Here, the source node establishes a signal for availability of data, and correspondingly, the route for data is configured from the source end to destination end, e.g., sensor protocols for information via negotiation (SPIN) protocol.

1.2 Categories of Routing Protocols

There are two techniques for transmitting data in sensor networks, namely flooding and gossiping protocols. There are no routing methods or topology maintenance requirements.

After sensor nodes receive data packets, flooding protocol broadcasts them to all neighbors till following conditions is achieved:

- The packet reaches its destination.
- Maximum number of hops of a packet are achieved [1].

There exists pros and cons for flooding algorithms. Benefits of flooding protocol include its simplicity and trivial implementation. Overlapping and resource blindness can be listed as its disadvantages. The other protocol is gossiping, which is a sophisticated variant of the flooding protocol in which a sensor node receives a data packet and sends it to an arbitrarily picked neighbor, after which these sensing nodes pick another node at random and transfer data to it, and so on. The issue of flooding

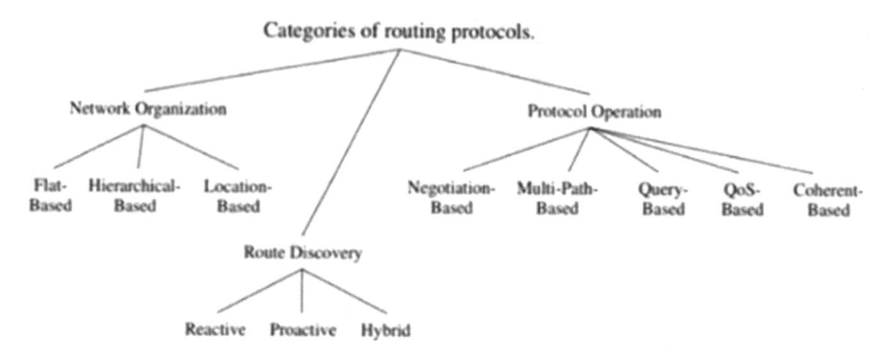

Fig. 3 Types of routing protocols

protocol implosion can now be easily avoided thanks to this development. Routing protocols can be categorized as indicated in Fig. 3.

1.3 Routing Protocols with Route Discovery Method

Reactive protocols: This form of protocol does not keep track of the entire network topology; instead, it is only engaged when a node needs to communicate data to another node.

Proactive protocols: Table-driven protocols, also known as network-wide protocols, keep tabulated information for the entire network by sending network data between the nodes.

Hybrid routing protocols: This has both proactive and reactive benefits while ignoring their drawbacks.

1.4 Network Organization-Based Routing Protocols

Flat Topology: Gradient-Based routing algorithms: Rumor routing, constrained anisotropic diffusion and cougar fall under flat topology protocol. These algorithms treat all nodes the same when all nodes have the same properties and functions. This happens in homogenous networking systems.

Hierarchical-based routing: Mainly for heterogeneous networks, where some of the nodes are more advanced and powerful than other.

Location-based routing (geo-centric): Geographical energy aware routing or SPEED algorithm is examples of location-based routing protocol. This node has ability to determine their current location which improves the routing procedure and provides some extra services to sensor networks.

1.5 *Operation-Based Routing Protocols*

Multipath routing protocols: This protocol gives multiple paths to the data in order to reach the destination. Multipath routing protocols are [2]:

- Multipath and multispeed (MMSPEED)
- Sensor protocols for information via negotiation (SPIN).

Query-based routing: This is a protocol in which initiation is done at receiver. The sensor nodes send data only if there is a generated query by destination node. Sensor protocols for information via negotiation (SPIN) and directed diffusion (DD) are few examples.

QoS-based routing: This protocol when used provide good quality of service as it tries to find a path from source to sink that meets metrics requirements. Examples are sequential assignment routing (SAR) and multipath and multispeed protocols.

Coherent routing: When this protocol is implemented the nodes find a minimum processing on data before transmitting forward.

Negotiation-based routing protocols: Direct diffusion protocol or negotiation-based sensor protocol (SPAN) focuses on keeping the level of redundant data transmission lower. Examples are [2].

2 Methodology

2.1 *Architecture*

Routing in wireless sensor networks (WSNs) is vital in the sector of environment-oriented monitoring. Here, contributions that are made toward routing in WSN are studied. Figure 4 indicates the routing protocol which shows that the routing service framework lies in between application layer and mac layer. Application program interface and link stage table help to communicate in application layer and mac layer, respectively. The packet is first received from mac layer, then it is preprocessed followed by classification in packet classifier after this the packet travels through routing processing, packet scheduler and buffer and reaches user application[3, 4].

3 Design Constraints in Wireless Sensor Networks (WSNs)

Design constraint depends on resources such as energy, storage and bandwidth of processing.

Autonomy: The expectation of a dedicated device controlling radio and routing resources is not suitable for WSN as it can be easy to attack on it. As there is no

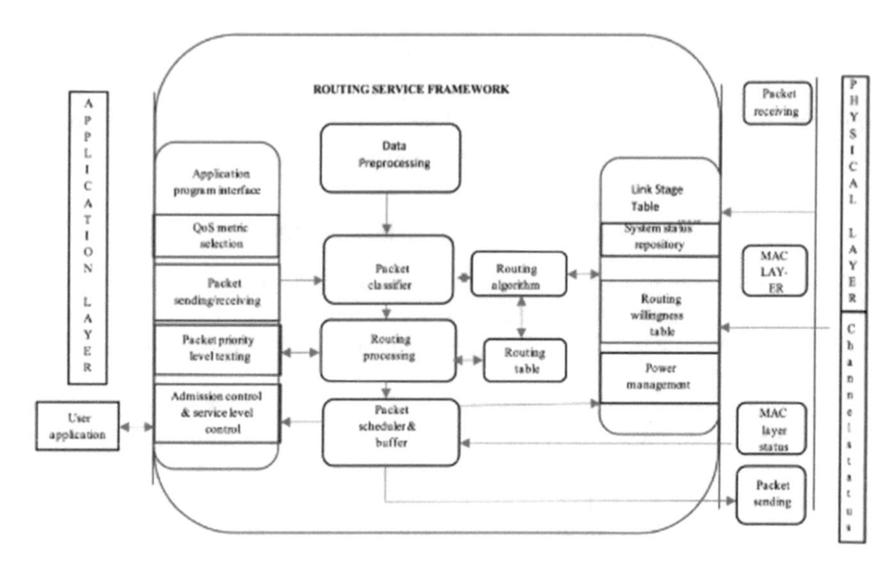

Fig. 4 Architecture of routing protocols in WSN

centralized format to make or change routing decisions or procedures are routed to nodes of the network.

Energy Efficiency: The protocols should long last for the life of the network while maintaining a good level of connectivity to allow communication between nodes. The sensor is mostly randomly placed so it is important factor that the batteries in the sensors cannot be replaced. In some cases, the sensors are not even accessible enough [5, 6].

Scalability: In WSN, there are hundreds of nodes due to this routing protocol should be bound the number of nodes.

Resilience: Mainly sensors are uncertainly terminating the operation due to environmental reasons. It can also cause due to the battery consumption. Due to this, routing protocol should handle this and when the error happens or it fails, so there should be alternative way to recover it.

Device Heterogeneity: Although the majority of WSN civil applications depend on homogeneous nodes, the addition of several types of sensors can provide substantial benefits.

Among others, network scalability, energy draining or bandwidth are potential candidates to take advantage of node heterogeneity. Network characteristics can be improved by using variable nodes that possess different controllers, processors, antennas, power supply or the components use for sensing [7, 8].

Mobility Adaptability: We use WSN in different application so it may be necessary for nodes to deal with their own mobility, as well as the event's movement. As a result, routing protocols should provide adequate support for the situation [9, 10].

4 Performance Analysis of Routing Protocols

The network simulator used here is OPNET Modeler 14.5 to analyze AODV, DSR and OLSR. Routing protocols on WLAN based WSNs are simulated in OPNET Modeler 14.5. These protocols are accordant with WLAN-based WSNs and prior researches stipulate them having better performance (Table 1).

To carry out the performance analysis, we define the three-network metrics stated as: ETE (end-to-end) delay, network load and throughput.

- End-to-end delay is the time taken for a packet to be transmitted across a network from source to destination.
- Network load shows net load in bits per second. When the traffic load exceeds the link capacity, it becomes impossible for a network to handle the traffic. This, in turn, creates congestion in the network.
- Throughput is the rate of data transmission in a specified amount of time from source to destination network node. It is expressed in bytes per second.

We run the simulations for all three routing protocols and observe the following results as seen in Figs. 5, 6 and 7.

The complete extract of the simulations is noted in Table 2. Having performed all the simulations, we can infer that:

- With regard to delay, OLSR is twice as much efficient with respect to other protocols in small-scale network and medium-scale network, whereas AODV surpasses in large-scale networks.
- With regard to network load, OLSR exhibits minimum load for every scale of networks.
- With regard to throughput, AODV gives the best throughput in comparison to rest of the protocols in small-scale network. DSR outperforms other two protocols in medium and large-scale networks.

Table 1 Simulation parameters

Simulation parameters	Values
Time of simulation	120 s
Count of nodes	30 50 70
Rate	10–11 Mbps
Area of simulation	1000 m2
Traffic	FTP (heavy load)
Routing protocols	AODV, DSR and OLSR

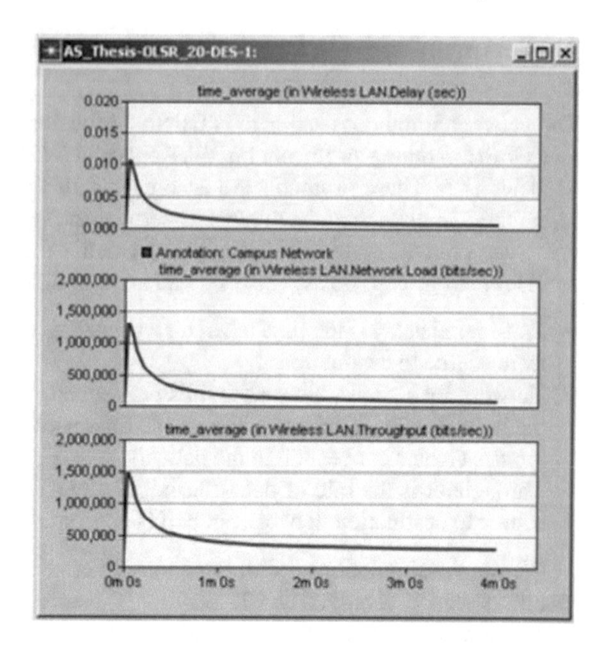

Fig. 5 Simulation results for OLSR

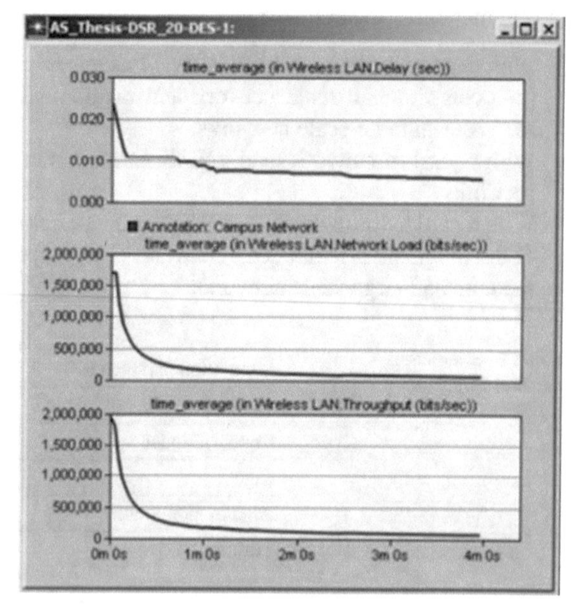

Fig. 6 Simulation results for DSR

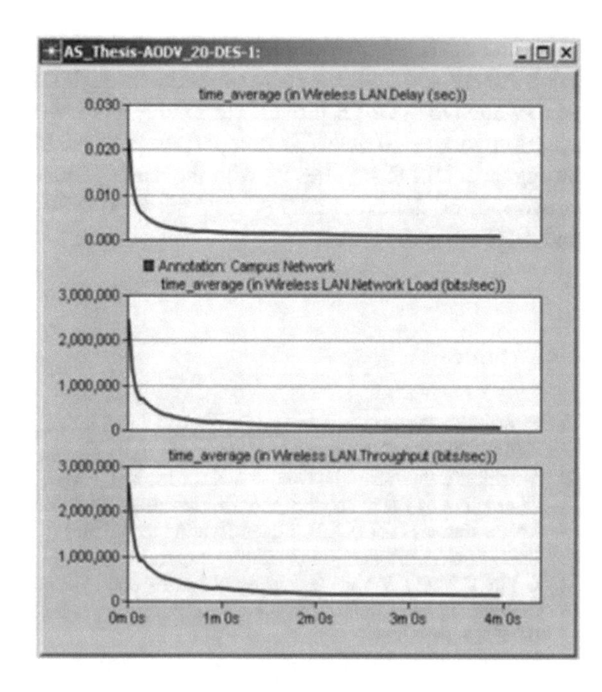

Fig. 7 Simulation results for AODV

Table 2 Extract of simulation

Sensor nodes	Specifications	AODV	DSR	OLSR
30	Delay (s)	0.030	0.034	0.021
	Network load (kbps)	3500	2700	1400
30	Throughput (kbps)	3800	1000	6500
50	Delay (s)	0.036	0.060	0.013
50	Network load (kbps)	3400	3200	2400
50	Throughput (kbps)	3900	4600	3600
70	Delay (s)	0.12	0.17	0.015
70	Network load (kbps)	3600	2600	2400
70	Throughput (kbps)	7200	15,000	12,000

5 Conclusion

Routing protocol serves a critical function in ensuring that communication between trans receiver nodes is uninterrupted and efficient. It all depend on which routing protocol we use to get different performance, service and reliability. On the basis of performance, delay and network load, we examined various protocols. The majority of WSN applications use route discovery-based routing protocols such as AODV, DSR and OLSR. When these three protocols are compared in a small-scale network of 20 nodes, OLSR provides less jitter and less congestion or load than AODV and DSR. AODV and DSR have five times the throughput of OLSR. When compared to AODV and DSR, OLSR provides less delay and lower network burden in a medium scale network of 40 nodes. In comparison with AODV and OLSR, DSR has a high throughput. OLSR supplies us with the same parameters as above in a large-scale network of 60 nodes. In a large-scale network, DSR has the maximum throughput and AODV has the lowest.

References

1. Hassan SR. Performance analysis of ZigBee based wireless sensor networks [thesis]. Lahore: GCU; 2014. Available from: http://library.gcu.edu.pk/theses.htm
2. Mokhlesabadifarahani B, Gunjan VK (2015) EMG signals characterization in three states of contraction by fuzzy network and feature extraction. Springer
3. J.N.A. Karaki, A.N.D.A.E. Kamal Routing techniques in wireless sensor networks: a survey IEEE Wirel. Commun., 11 (6) (2004), pp. 6–28
4. G Yin, G Yang, W Yang, B Zhang, W Jin. An energy-efficient routing algorithm for wireless. In. International Conference on Internet Computing in Science and Engineering (ICICSE'08), IEEE, China; 2008.
5. Kumar, S., Gunjan, V. K., Ansari, M. D., & Pathak, R. (2022). Credit Card Fraud Detection Using Support Vector Machine. In *Proceedings of the 2nd International Conference on Recent Trends in Machine Learning, IoT, Smart Cities and Applications* (pp. 27–37). Springer, Singapore.
6. U. P. Nagane and Dr. A. O. Mulani, "Moving Object Detection and Tracking Using Matlab", Journal of Science and Technology, Volume 6, Special Issue August https://doi.org/10.46243/jst.2021.v6.i04.pp63-66
7. A. O. Mulani1 and G. N. Shinde, "An approach for robust digital image watermarking using DWT-PCA", Journal of Science and Technology, Volume 6, Special Issue 1, August 2021. https://doi.org/10.46243/jst.2021.v6.i04.pp59-62
8. Jadhav M. M., G. H. Chavan and A. O. Mulani, "Machine Learning based Autonomous Fire Combat Turret", Turkish Journal of Computer and Mathematics Education (TURCOMAT), 12(2), 2372-2381, 2021
9. Naik, M. V., Anasari, M. D., Gunjan, V. K., & Kumar, S. (2020). A Comprehensive Study of Sentiment Analysis in Big Data Applications. In Advances in Cybernetics, Cognition, and Machine Learning for Communication Technologies (pp. 333–351). Springer, Singapore.
10. Gunjan, V. K., Kumar, S., Ansari, M. D., & Vijayalata, Y. (2022). Prediction of Agriculture Yields Using Machine Learning Algorithms. In *Proceedings of the 2nd International Conference on Recent Trends in Machine Learning, IoT, Smart Cities and Applications* (pp. 17–26). Springer, Singapore.

LS-TFP: A LSTM-Based Traffic Flow Prediction Method in Intelligent Internet of Things

Nhu-Y Tran-Van, Nhat-Tuan Pham, and Kim-Hung Le

Abstract Intelligent transport system has been emerging as a crucial component of the smart city context, and traffic flow prediction plays an essential role in ITS. Recently, many studies have used algorithms based on time prediction and deep learning. However, their prediction accuracy is insufficient for the significant growth in IoT applications. To overcome this issue, we proposed a novel prediction model, namely the LSTM-based traffic flow prediction (LS-TFP), using the combination of the long short-term memory and recurrent neural network (LSTM-RNN). In our proposal, we stack two LSTM layers to produce a more in-depth model. In addition, as a consequence of the remembering ability of LSTM, the predicted value could achieve high accuracy. Our practical experiments on real datasets show that the LS-TFP accuracy is reached up to 98.1% and outperforms our competitors.

Keywords Traffic flow prediction · Smart city · Intelligent transport system · Long short-term memory

1 Introduction

Intelligent transport system (ITS) is a critical component of the smart city context expected to improve traffic conditions, safety, flexibility, and reduce the effects of significant traffic flow. One of ITS's leading technologies acting as an essential role in several fields is traffic flow prediction. It is currently gaining much attention and achieving many results [1, 2]. Traffic flow information provides continuous and accurate information about traffic conditions and density. As a result, people could

N.-Y. Tran-Van · N.-T. Pham · K.-H. Le (✉)
University of Information Technology, Ho Chi Minh City, Vietnam
e-mail: hunglk@uit.edu.vn

N.-Y. Tran-Van
e-mail: 17521287@gm.uit.edu.vn

N.-T. Pham
e-mail: 17521219@gm.uit.edu.vn

A. Kumar et al. (eds.), *Proceedings of the International Conference on Cognitive and Intelligent Computing*, Cognitive Science and Technology,
https://doi.org/10.1007/978-981-19-2350-0_3

actively choose a more convenient, time-saving, and cost-saving route. It also reduces pressure on traffic infrastructure, traffic jams, exhausted emissions, ensures safer for people in traffic, improves the quality of life, and immensely helps the administration officials have effective strategies for managing and operating the transport system [3].

Over the past decades, there has been a sustained research activity in traffic flow prediction that is a typical time series problem. Processing data are recorded values, real-time values of traffic, and vehicle speed. This information is collected from various sources such as motion detection sensors, cameras, GPS, radars, and social media [4]. Besides, there are massive data generated by intelligent devices in the smart city context. This increases the research community's interest in traffic flow prediction, sensors, and traffic tracking technologies [5]. By using real-time processing and high-accurate requirements, deep learning is considered for solving traffic flow prediction.

The deep learning algorithm requires a large amount of data with many features that make the training and inference process longer and consume massive computational resources (CPU and RAM) [6]. Whereas, LSTM, a special kind of RNN's capable of learning long-term dependencies, could remember the information for long periods. Therefore, LSTM is suitable for problems with time series or sequences. In this paper, we propose a predicting model, namely the LSTM-based traffic flow prediction method (LS-TFP). The LSTM model is used to learn those characteristics of the traffic flow. To improve the processing speed and the traffic flow value's dependence at a location over time, we adaptively select the most characteristic features of the dataset. Along with maintaining information in the long term, LSTM can select relevant information needed to keep or forget during training. As a result, LS-TFP could predict traffic flow with high accuracy. By experimenting on practical datasets, we demonstrate that our proposal's accuracy is recorded at about 98.1% and out-performs other predictive methods (the best accuracy of competitors is 94.1%).

The remainder of this paper is structured as follows. In Sect. 2, we review related works and describe the proposed model. In Sect. 3, we introduce the origin of the dataset used for the experiment and how to process it. The results of the proposal in this dataset are also provided. In Sect. 4, we conclude the whole paper.

2 Related Work and Methodology

In the smart city, the application of technology in traffic flow prediction plays a significant role in making a traffic schedule [7]. For a long time, many research methods about traffic flow prediction were proposed. Early prediction models have been relatively straightforward and easy to deploy, but the performance is insufficient. Therefore, many more complicated and robust models of prediction have arisen. In this section, we introduce related works along with the proposed model.

2.1 Related Work

In the article [8], Zhang was used to forecast traffic flow in a short time by the implementation of deep learning and in particular of the convolution neural network (CNN). Chen et al. [9] used the deep learning model to forecast traffic flow and showed that the RNN has excellent success in estimating traffic flows. Kumar et al. [10] indicated that machine learning models could be enhanced the analysis process in the financial system than the non-machine learning solutions. Kataria et al. [11] proposed a self-healing neural model to predict missed data in the pilot's head-tracking process and proved that the accuracy achieved by their model is better than the back-propagation neural network.

Data on traffic flow are collected every hour and provide an enormous data volume that can not effectively process by prior approaches. Meanwhile, the neural network is capable of processing big data. Many models recommend using a neural network for traffic flow prediction. Lv [12] proposed a prediction model for traffic flow using novel deep learning, which inherently considers spatial and temporal correlations. Fu et al. [13] proposed a model using that recurrent neural network (RNN) including LSTM and GRU gives better performance than ARIMA.

2.2 Overview of LSTM Model

Recurrent neural networks (RNNs) suffer from short-term memory and the vanishing gradient problem [14]. As a result, a consistent structure of RNN, named long short-term memory, was proposed. It has a remember cell that includes three gates to regulate the information flow. These gates allow the LSTM to keep or forget information to optimal time lag for solving time series problems. Figure 1 gives an illustration of the LSTM memory unit.

Let us denoted the input historical traffic flow sequence as $d = (d_1, d_2, \ldots, d_n)$, the hidden state of memory block as $h = (h_1, h_2, \ldots, h_n)$. Below is the detailed operation of the gates.

Forget gate determines what information is kept or discarded. This information is passed through the sigmoid function from the previously hidden state (h_{t-1}) and information from the current state (d_t). The value obtained (f_t) is in the 0 to 1 range. If the value is close to 1, the data are held; otherwise, the data are forgotten if it is close to 1.

$$f_t = \sigma(W_f[h_{t-1}, d_t] + b_f) \tag{1}$$

Input gate, used to decide what information is put into cell state. The previous hidden state (h_{t-1}) and the current state (d_t) are transferred to the sigmoid function. The received results are in the range of 0–1. 1 mean the data are valuable and 0 is not. Besides, put hidden state and current input in the tanh function to regulate the

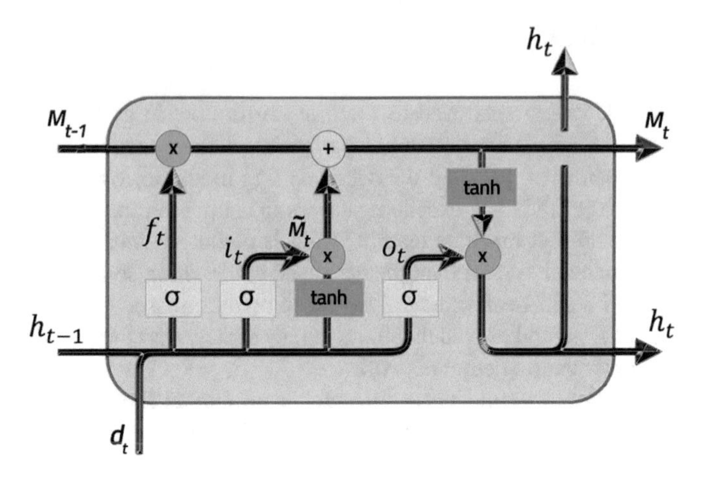

Fig. 1 Structure of LSTM

output in the range $[-1, 1]$. Then, multiply the result of the tanh function $(\widetilde{M}_t)$ by the sigmoid (i_t). And update that result to the cell state.

$$i_t = \sigma(W_i \dot{[} h_{t-1}, d_t] + b_i) \tag{2}$$

$$\widetilde{M}_t = \tanh(W_m \dot{[} h_{t-1}, d_t] + b_m) \tag{3}$$

Now, it is time to update old cell state, $M_t - 1$, to new cellstate M_t. First, multiply the old cell state multiplied by the forget vector (f_t). This has the potential to remove the value in the cell state if the above multiplication result is close to zero. Next, get the result from the input gate plus the cell state to update the new cell state.

$$M_t = f_t * M_{t-1} + i_t * \widetilde{M}_t \tag{4}$$

Output gate, used to determine the next hidden state. First, we run a sigmoid layer to decide what to bring to the output gate. Then, put the newly updated cell state (M_t) into the tanh function. To determine the stored information in hidden state, multiplying the result of the tanh by the sigmoid. New cell state (M_t) and new hidden (h_t) is taken over the next time step.

$$o_t = \sigma(W_o \dot{[} h_{t-1}, d_t] + b_o) \tag{5}$$

$$h_t = o_t * \tanh(M_t) \tag{6}$$

where $\sigma(.)$ is the logistic sigmoid function defined in Eq. (7). $\tanh(x)$ is is a centered logistic sigmoid function with range $[-1, 1]$ defined in Eq. (8).

$$\sigma(x) = \frac{1}{1 + e^{-x}} \tag{7}$$

$$\tanh(x) = \frac{2}{1 + e^{-2x} - 1} \tag{8}$$

2.3 Proposed Model

In this study, we have experimented and compared multiple models with different parameters on the same traffic flow dataset. The result shows that the LSTM model that proposed gives outperformance to other models such as A-LSTM [15], A-GRU [16], A-SAE [17], and B-LSTM [18]. In order to match with traffic flow prediction problems and gain more accuracy, two LSTM layers were stacked to make the model deeper. In this case, each layer can be considered as a hierarchical structure in which the input of the following layer is obtained by the hidden state of the previous layer [19]. In this way, the hidden state could operate independently at a different timescale. According to Fig. 2, the data go through the input layer then continue through two LSTM layers before going into the dropout layer and these layers. Besides, we fine-tuned and optimized the parameters in the model. Thereby, the proposed model is capable of improving accuracy.

In order to build an LSTM model for traffic flow prediction, we need to define precisely the elements in the model, such as the number of input layers, hidden layers, and units in each hidden layer. In this model, we used four hidden layers, and the input layer is configured with the shape is 3×1. We have used two LSTM layers, with the number of units for each of these layers is 64. Besides, we employ a dropout layer because when a layer is fully connected, it may contain most of the parameters, and hence, neurons develop with significant dependency on each other, which reduces each neuron's power over-fitting on the training set by randomly ignoring neurons during training. Finally, the previous layers' results are passed through the dense layer to output the traffic flow prediction value.

The architecture of our proposed model is illustrated Fig. 2. Input data are passed into the input layer, and then, the result is forward propagated to the two LSTM layers. The remaining hidden layers act as regularizing, and the final result is to produce at the output layer. At the final step, the validation and cost values are computed as well as the parameters of weights and biases are adjusted during back-propagation. We applied an Adam optimization, a stochastic gradient descent technique with a learning rate of 0.001, and the number of epochs was 1000, decreasing training error and avoiding local minimum points during the training process. In this model, a total of 49,985 parameters was used, while the first LSTM layer has 16,896 parameters, and the second LSTM layer has 33,024 parameters. The dense layer has a small number of parameters (reported about 65 parameters). This characteristic makes our model more lightweight than competitors. The number of parameters used in this model is shown in Table 1. According to Table 1, we can see that the total parameters

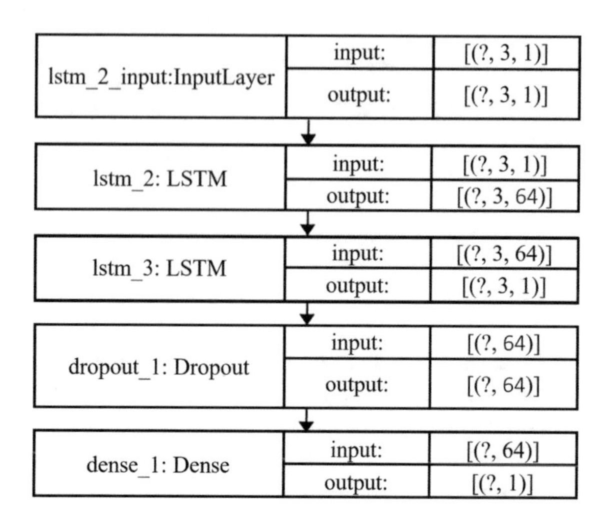

Fig. 2 The proposed model summary

Table 1 The proposed model layers and its parameters

Layer	LSTM1	LSTM2	Dropout	Dense	Total params
Param	16896	33024	0	65	49985

of this model are 49,985, while the first LSTM layer has 16,896 parameters, and the second LSTM layer has 33024 parameters. The dense layer has a small number of parameters (reported about 65 parameters). This characteristic makes our model more lightweight than competitors.

3 Evaluation

In this section, we present the experiential results of our proposal. The evaluation data in this research are gathered from the Caltrans Performance Measurement System (PeMS) [20] based on traffic flow data gathered on a working day in the first quarter of 2016. The first two months' data are used for training, and the other month is used for testing. The original data were collected and aggregated every 5 min at each detector, and the total sample was about 12,000 points over three months.

3.1 Index of Performance

In this experiment, we used different evaluation metrics to evaluate our model performance. Let denote the observed traffic flow f_i, and the predicted traffic flow $\hat{f}_i$, these metrics are defined below:

$$\text{MAE} = \frac{1}{n}\sum_{i=1}^{n} |f_i - \hat{f}_i| \tag{9}$$

$$\text{RMSE} = \left[\frac{1}{n}\sum_{i=1}^{n}(|f_i - \hat{f}_i|)^2\right]^{\frac{1}{2}} \tag{10}$$

$$\text{MAPE} = \frac{1}{n}\sum_{i=1}^{n} |\frac{f_i - \hat{f}_i}{f_i}| \tag{11}$$

$$R^2 = 1 - \frac{\sum_{i=1}^{n}(f_i - \hat{f}_i)^2}{\sum_{i=1}^{n}(f_i - \overline{f}_i)^2} \tag{12}$$

$$\text{Accuracy} = 1 - \frac{|f_i - \hat{f}_i|}{f_i} \tag{13}$$

where n represents the number of data, and $\overline{f_i}$ is the average of all observation value (Fig. 3).

3.2 Results

After performing a comprehensive evaluation on the practical datasets, Fig. 5 shows the proposed model results for traffic flow prediction with an interval time of 60 min. Those data are traffic flow in the workday from March 4, 2016, to March 30, 2016. As shown from the graph, the line representing the predicted value is nearly the same as the line representing the observed value. Compare with 15 min and 30 min traffic flow, the proposed model would be more efficient in the dataset of 60 min traffic flow, as clearly shown in Table 2. All three results are well with high accuracy and low error rates. The good one is 60 min with 98.1%, followed by 30 min and 15 min with 97.1 and 96.5%. In the dataset, the traffic flow values over a short period do not have much variation, and the missing data point represents just a small portion of the entire dataset. As a result, we concentrate more on building models suitable for the 60 min dataset. The following experiments are based on this dataset (Fig. 4).

We compared the performance of our proposed model (LS-TFP) with other models such as A-LSTM, A-GRU, A-SAE, and B-LSTM. The predicted results of the above

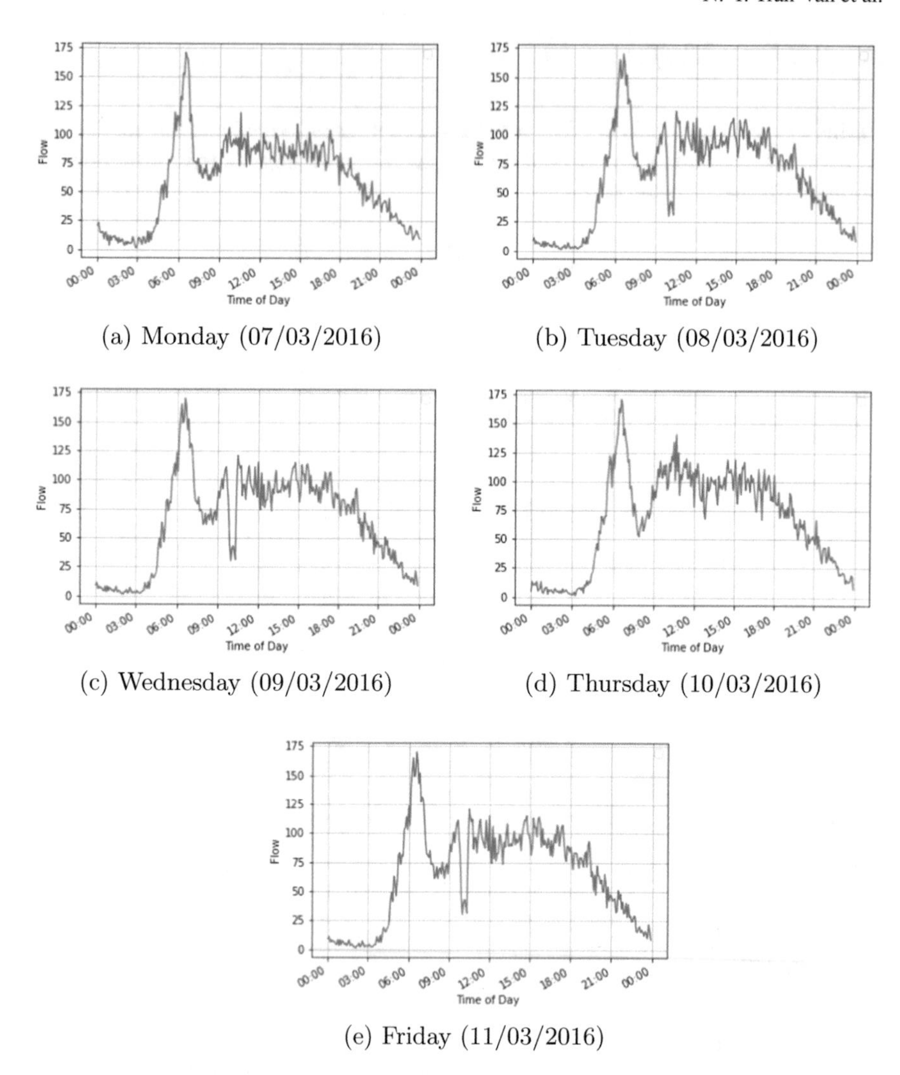

(a) Monday (07/03/2016)

(b) Tuesday (08/03/2016)

(c) Wednesday (09/03/2016)

(d) Thursday (10/03/2016)

(e) Friday (11/03/2016)

Fig. 3 Weekly traffic flow pattern

Table 2 Performance with different time interval

Task interval	MAE	RMSE	MAPE	R^2	Accurary
60 min	3.883	5.251	8.302	0.981	0.981
30 min	4.922	6.809	10.389	0.969	0.971
15 min	5.333	7.565	10.949	0.963	0.965

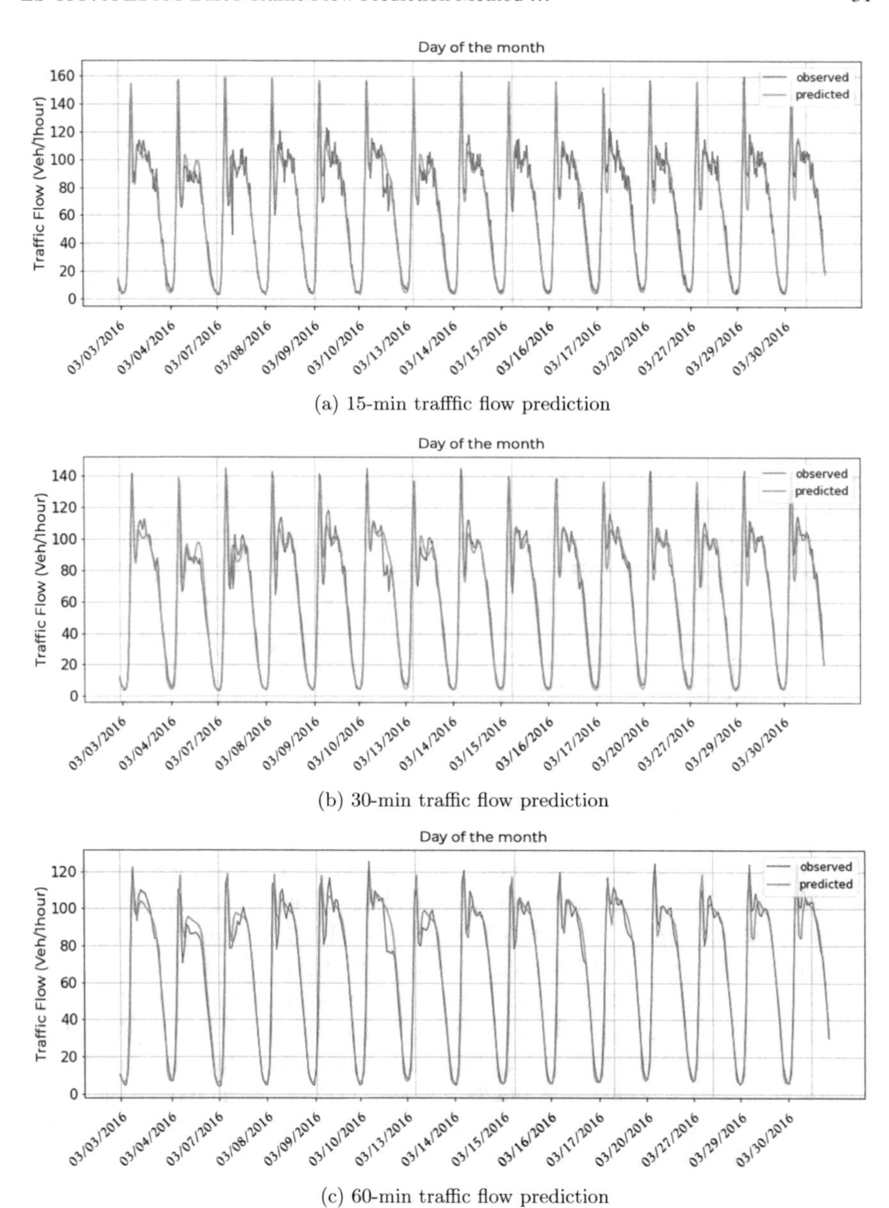

(a) 15-min trafffic flow prediction

(b) 30-min traffic flow prediction

(c) 60-min traffic flow prediction

Fig. 4 Traffic flow prediction with different time interval

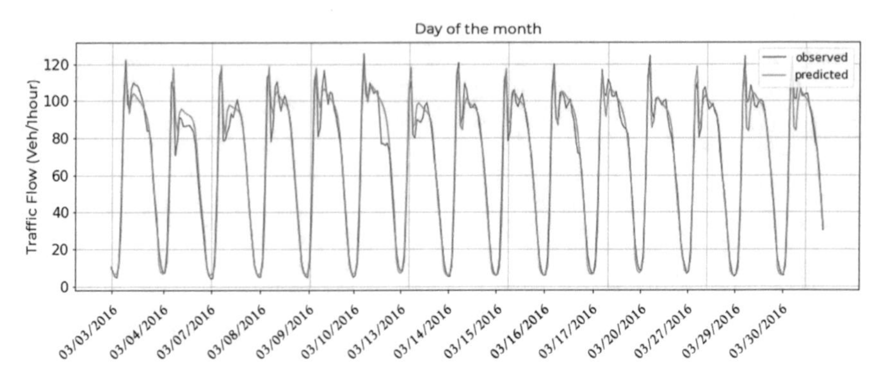

Fig. 5 60 min traffic flow prediction

Table 3 Performance comparison between our proposal and competitors

Metrics	MAE	RMSE	MAPE	R^2	Accurary
LS-TFP	3.883	5.251	8.302	0.981	0.981
A-LSTM	7.131	9.832	17.367	0.941	0.941
A-GRU	7.216	9.901	18.006	0.939	0.939
A-SAE	7.058	9.596	17.801	0.943	0.944
B-LSTM	20.43	27.035	130.459	0.226	0.233

five models are given in Table 3. In all cases, we have used the same set of datasets to make the comparison more objective. The predicted values of LS-TFP are based on testing data with an average 60 min traffic flow rate. Compared to the other four models, LS-TFP has the lowest error rate. Its MAE, RMSE, and MAPE values are reported at about 3.883, 5.252, and 8.3, respectively. LS-TFP is also the model with the highest accuracy at about 98.1%, while with the rest of the models, the highest accuracy is 94.1%. A-LSTM, A-GRU, and A-SAE models use the same method for preprocessing data and algorithms according to the RNN mechanism. Thus, they also produce high accuracy, but the network selection is not suitable and optimized. For B-LSTM, LSTM is also used, but the accuracy is not high on this dataset. As we can see, LS-TFP outperforms other models on the same dataset, which proves the effectiveness of LS-TFP from this experiment.

4 Conclusion

In this paper, we proposed a model called LS-TFP to predict traffic flow. We compared the prediction performance of A-LSTM, A-GRU, A-SAE, and B-LSTM in the same datasets collected from the PeMS. In our practice evaluation, we demonstrate that LS-TFP outperforms competitors. The accuracy of our proposed model is up to

98.1%. It is 4% higher than our best competitor, A-LSTM. In addition, LS-TFP has reduced MAE by about 3% on average compared to the A-LSTM. In the future, we will enhance the proposed algorithm by adding other complex factors like traffic in neighbor observation and weather conditions in neural networks.

Acknowledgements This research is funded by University of Information Technology-Vietnam National University HoChiMinh City under grant number D1-2019-20.

References

1. Hammi B, Khatoun R, Zeadally S, Fayad A, Khoukhi L (2017) Iot technologies for smart cities. IET Networks 7(1):1–13
2. Agarwal PK, Gurjar J, Agarwal AK, Birla R (2015) Application of artificial intelligence for development of intelligent transport system in smart cities. J Traffic Transp Eng 1(1):20–30
3. Sładkowski A, Pamuła W (2016) Intelligent transportation systems-problems and perspectives, vol 303. Springer
4. Xiong Z, Sheng H, Rong W, Cooper DE (2012) Intelligent transportation systems for smart cities: a progress review. Sci China Inf Sci 55(12):2908–2914
5. Vlahogianni EI, Karlaftis MG, Golias JC (2014) Short-term traffic forecasting: where we are and where we're going. Transp Res Part C Emerg Technol 43:3–19
6. Mohammadi M, Al-Fuqaha A, Guizani M, Oh J-S (2017) Semisupervised deep reinforcement learning in support of iot and smart city services. IEEE Internet Things J 5(2):624–635
7. Nagy AM, Simon V (2018) Survey on traffic prediction in smart cities. Perv Mob Comput 50:148–163
8. Zhang W, Yu Y, Qi Y, Shu F, Wang Y (2019) Short-term traffic flow prediction based on spatio-temporal analysis and cnn deep learning. Transportmetrica A Transp Sci 15(2):1688–1711
9. Chen Q, Song X, Yamada H, Shibasaki R (2016) Learning deep representation from big and heterogeneous data for traffic accident inference. In: Thirtieth AAAI conference on artificial intelligence
10. Kumar MR, Gunjan VK (2020) Review of machine learning models for credit scoring analysis. Ingeniería Solidaria 16(1)
11. Kataria A, Ghosh S, Karar V (2020) Data prediction of electromagnetic head tracking using self healing neural model for head-mounted display. Sci Technol 23(4):354–367
12. Lv Y, Duan Y, Kang W, Li Z, Wang F-Y (2014) Traffic flow prediction with big data: a deep learning approach. IEEE Trans Intell Transp Syst 16(2):865–873
13. Fu R, Zhang Z, Li L (2016) Using lstm and gru neural network methods for traffic flow prediction. In: 2016 31st Youth academic annual conference of Chinese association of automation (YAC), pp 324–328. IEEE
14. Hochreiter S (1998) The vanishing gradient problem during learning recurrent neural nets and problem solutions. Int J Uncertainty Fuzziness Knowl-Based Syst 6(02):107–116
15. Xiaochus (2017) Trafficflowprediction. https://github.com/xiaochus/TrafficFlowPrediction
16. Fu LLR, Zhang Z (2017) Using LSTM and GRU neural network methods for traffic flow prediction, pp 324–328. Chinese Association of Automation
17. Lv Y, Duan Y, Kang W, Li Z, Wang FY (2015) Traffic flow prediction with big data: A deep learning approach. IEEE Trans Intell Transp Syst 16(2):865–873
18. Yazdani S (2020) Traffic-flow-prediction. https://github.com/shakibyzn/Traffic-flow-prediction
19. Wang F, Tax DM (2016) Survey on the attention based rnn model and its applications in computer vision. arXiv preprint arXiv:1601.06823
20. Caltrans, performance measurement system (pems) (2020). http://pems.dot.ca.gov/

Prediction of Deformed Shape in Incremental Sheet Forming Processing Feedforward Neural Network

Hai Son Le, Quoc Tuan Pham, Anh Tuan Nguyen, Hoang Son Tran, and Xuan Van Tran

Abstract Numerical simulations of the incremental sheet forming (ISF) process using the finite element method (FEM) provide essential information for designing parts in automotive industries. However, solving numerous high-complexity FEM models during the designing phase requires many resources, leading to an increase in the final product's cost. This study presents a feedforward neural network (FFNN) to predict the deformed shape of an AA1050 sheet subjected to an ISF process. FEM solutions obtained from various vertical step size (Δz) of the forming tool are used to train and validate the FFNN. The model is then used to predict the deformed shape demonstrating by the displacement in the forming depth direction. The norm of the relative errors between the FFNN solution and FEM solution at the last forming step is about 2%. The predictive results illustrate the feasibility and potential of using FFNN as an efficient surrogate model to replace the time-consuming FEM-based ISF process simulation.

Keywords Incremental sheet forming · Finite element method · Aluminum alloys sheet · Plasticity deformation · Feedforward neural network

1 Introduction

In recent decades, incremental sheet forming (ISF), a die-less forming process, has been actively developed to manufacture parts requiring small-batch productions such as prototypes and complicated components [1]. In this process, a rigid tool is programmed to move following designated tool paths and plastically deforms a blank sheet into pre-designed shapes. Several process parameters influence the

H. S. Le · A. T. Nguyen · H. S. Tran · X. V. Tran (✉)
Institute of Strategies Development, Thu Dau Mot University, Binh Duong, Vietnam
e-mail: xuantv@tdmu.edu.vn

Q. T. Pham
Division of Computational Mathematics and Engineering, Faculty of Civil Engineering, Institute for Computational Science, Ton Duc Thang University, Ho Chi Minh City, Vietnam
e-mail: phamquoctuan@tdtu.edu.vn

A. Kumar et al. (eds.), *Proceedings of the International Conference on Cognitive and Intelligent Computing*, Cognitive Science and Technology,
https://doi.org/10.1007/978-981-19-2350-0_4

formability and qualities of the final products manufactured by ISF, for example, tool diameter, tool moving speed, and vertical step size [2]. Notably, an optimization procedure commonly for ISF requires a large number of finite element method (FEM) simulations to achieve the optimized process parameters [3]. That raises the demand for the development of surrogate models, which are able to provide numerical approximations for the complex input–output relationship within an efficient time.

In machine learning (ML)-related publications, for example, Ehsan et al. [4] proposed a multi-network of physics-informed neural network (PINN) to achieve more accurate predictions on various solution fields in linear elastic and nonlinear plastic problems. In addition, Pham et al. [5] adopted a back-propagation neural network (BPNN) to search for Pareto optimal solutions of formability and thickness of AA5052-H32 sheets subjected to an ISF process.

This study aims to develop a surrogate model to predict the displacement fields obtained from FEM solutions of an ISF process. Several FE models are developed to simulate the ISF groove tests performed for AA1050 sheets, which vary the vertical step size (Δz). Based on the simulation results, a feedforward neural network (FFNN) is trained to address the relationship between the imposed Δz and the displacement fields obtained from FE simulations of the ISF process. After validation, a comparison between the FFNN prediction and FEM solutions for the displacement fields of a test performed with an unseen value of Δz is presented. Furthermore, the usefulness of the trained FFNN model is discussed.

2 Data Acquisition and Neural Networks

This section describes the detailed setup configuration of the FEM model used to simulate the ISF groove test. Later, a strategy to achieve the useful data is introduced based on simulated results. Finally, the most suitable architecture of FFNN, as well as the data structure, is discussed.

2.1 Data Acquisition via a Virtual Groove Test

Figure 1a shows the assembly of the groove tests [6]which are developed in Abaqus/Explicit for the aluminum sheet AA1050 with an initial thickness of 0.3 mm [7]. In this test, a rigid tool with a diameter of 10 mm is programmed to move following the tool path presented in Fig. 1b. The blank sheet is modeled by 4 node shell elements with reducing integration (S4R) with a size 1×1 mm^2 of the smallest element. Totally, the sheet is modeled with 2573 nodes and 2577 elements. Whereas, tool and die are modeled by discrete rigid quadrilateral elements (R3D4).

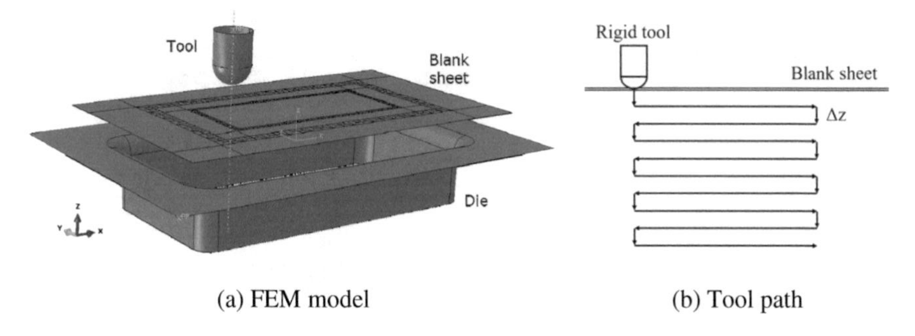

Fig. 1 Finite element model to simulate the groove test

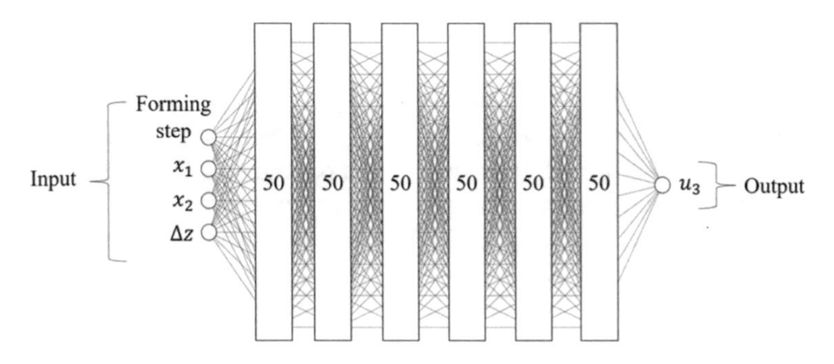

Fig. 2 Proposed FFNN architecture

2.2 *Feedforward Neural Network (FFNN) Architecture*

After a trial–error procedure, the most suitable FFNN architecture for re-building the deformed shape is presented in Fig. 2.There are four essential input features of the developed FFNN, including (i) the forming step, (ii) and (iii) $x - y$ coordinates of the point of interest on the initial blank sheet (labeled as $x_1 - x_2$), and (iv) Δz value. Consequently, the output is the displacement of the considering point in the Z-direction. There are six hidden layers with 50 neurons each. The linear activation is used on the output layer to produce the desired output values.

To generate data used in the FFNN model,2573-dimensional vectors of displacements in the thickness direction of all nodes are recorded according to various values of the forming step Δz (i.e., Δz = 0.5, 0.6, 0.7, 0.8, 0.9, 1.0). Six data samples of different Δz values are split into the training, cross valid, and test sets. In the following work, the developed neural network structure is used to train two models, named FFNN 1 and FFNN 2. The former is trained with three data samples, while the latter is trained with four data samples. After each training epoch, both FFNN models perform the prediction on a cross valid set (i.e., $\Delta z = 0.6$) to record the overfitting phenomena. When the training phase ends, the performance of trained

Table 1 Two dataset structures for training and testing the proposed FFNN

Training	FFNN 1	FFNN2
	$\Delta z = [0.5, 0.8, 1.0]$	$\Delta z = [0.5, 0.8, 0.9, 1.0]$
Cross-validation	$\Delta z = [0.6]$	$\Delta z = [0.6]$
Testing	$\Delta z = [0.7, 0.9]$	$\Delta z = [0.7]$

models for the remaining test set is used to evaluate the goodness of these models. The detailed dataset is reported in Table 1.

3 Results and Discussion

The FFNN models were implemented in Python using TensorFlow [8]. An NVIDIA GeForce GTX 1030 graphics processing unit (GPU) with a 3.6 GHz quad-core CPU and 8 GB RAM are used for data generation and FFNN training. The adaptive moment estimation algorithm [9] is used to update the network weights with a learning rate of 0.001. The batch size is set up at 12,800. The loss function is the mean squared error (MSE) metric defined as:

$$\mathrm{MSE} = \frac{1}{N}\sum_{i=1}^{N}\left(u_i^{\mathrm{FFNN}} - u_i^{\mathrm{FEM}}\right)^2, \tag{1}$$

where u_i^{FFNN} and u_i^{FEM} are the displacements at the node i obtained from the FFNN and FEM model, respectively, and N is the number of nodes containing in the FEM simulation.

Figure 3 illustrates the training and validation errors of the proposed FFNN architecture for the two training models. Generally, by adding one more sample for the training data, the FFNN2 convergence rate is much faster (i.e., epoch = 1993) than that of FFNN 2 (i.e., epoch = 9231).

Table 2 shows the comparison between the calculation time of the FEM and FFNN models. It is seen that both two FFNN models are well-trained within few minutes; meanwhile, the prediction time of the trained models is less than the second. In contrast, the FEM model requires almost 50 min to finalize the results. Although the

Table 2 Computation time of FEM and FFNN models

Model	Computing time (s)	
	Training	Prediction
FFNN 1	513.0	0.373
FFNN 2	198.0	
FEM		2983.0

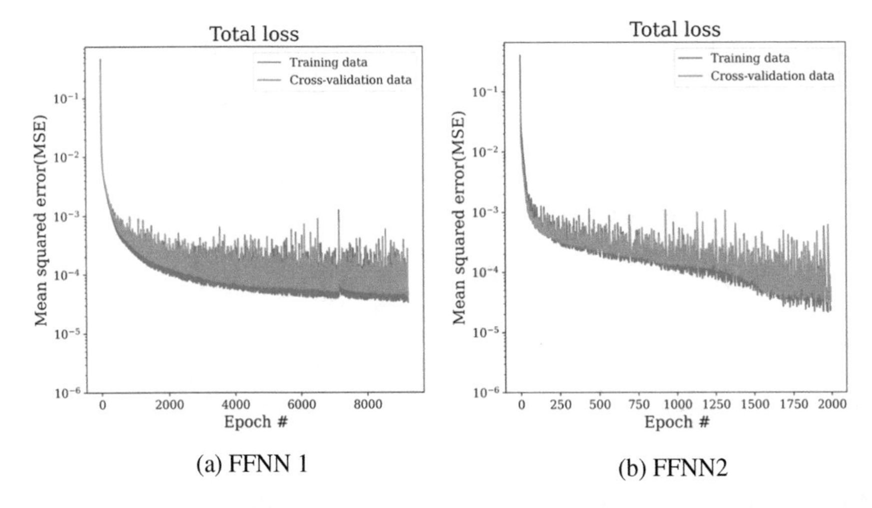

(a) FFNN 1

(b) FFNN2

Fig. 3 Convergence of the proposed **a** FFNN 1 trained with three data samples and **b** FFNN 2 trained with four data samples

efficiency of the trained FFNN models outperforms the conventional FEM model, collaborating the two algorithms may be an efficient way to modeling and analysis the ISF process, particularly.

To evaluate the reliability of the FFNN solutions, the relative error norm metric, which estimates the difference between the FFNN predicted vector solution and those of the FEM model, is calculated as follows at each simulation step:

$$\eta = \frac{\left\| \mathbf{u}_k^{\text{FEM}} - \mathbf{u}_k^{\text{FFNN}} \right\|_2}{\left\| \mathbf{u}_k^{\text{FEM}} \right\|_2} \times 100\%, \quad (2)$$

where $\| \|$ denotes the L^2 norm; $\mathbf{u}_k^{\text{FEM}}$ and $\mathbf{u}_k^{\text{FFNN}}$ are the solution vectors at the forming process step k of the FEM and FFNN models, respectively. Figure 4 shows the norm's evolution based on the cross-validation and test samples. It is indicated that the performance of the FFNN model trained with four samples (i.e., dash lines) is generally better than that of the one trained with three data samples (i.e., solid lines).In practice, the simulation results achieved at the final forming step are the most important during an ISF process. As shown in Fig. 4, in the last forming step, the norm of the FFNN model trained with four data samples is even lower than 2%, which indicates a good approximation.

To illustrate a comprehensive picture regarding the two FFNNs' performance on the test data sample at the final forming step, the absolute error between the FEM solutions and FFNNs' predictions is point-wised in Fig. 5. Comparison between the error distribution exhibited in Fig. 5b, c clarifies that adding more samples to the training dataset increases the accuracy of the FFNN predictions. After all, it is reasonable to conclude that the FFNN expresses great potential and efficiency to

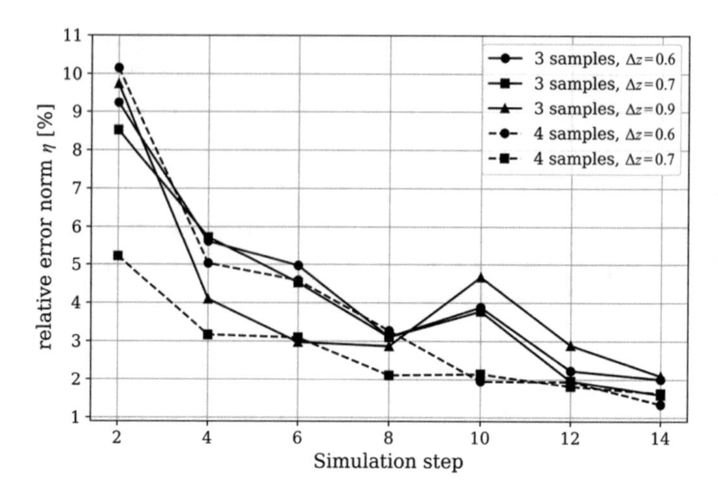

Fig. 4 Relative error norm between the solution vector of FFNN model and FEM model

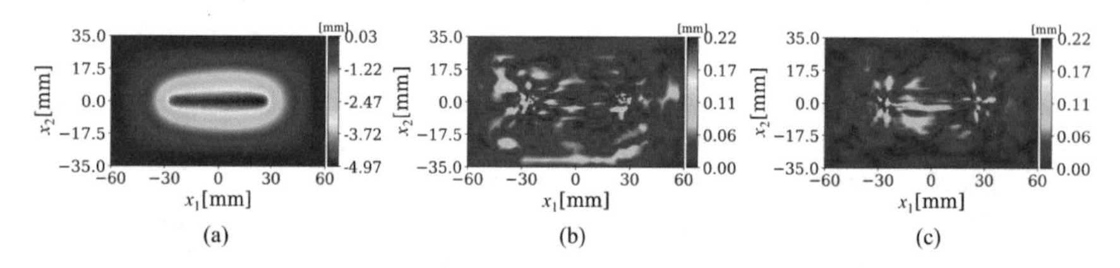

Fig. 5 Forming step number 14: **a** the displacement distribution of the FEM model, **b** error distribution of the FFNN 1 model trained with three samples, **c** error distribution of the FFNN 2 model trained with four samples

reproduce a reliable approximation for the FEM simulations of the considering ISF process.

4 Conclusion

In this work, two simple FFNN model strained with three and four samples for predicting the displacement field in the ISF process are well developed. The validation stage indicates that the model can predict the displacement field of the ISF process at the last forming steps with a relative error norm below 2%. The proposed framework shows the developed FFNN's potential to be a useful surrogate model that can save considerable computing time. From perspective, considering more parameters for the FFNN model, such as the material properties to increase the complexity of high-fidelity solutions, which could be a challenge and the motivation in the further work.

Acknowledgements This work was funded by Vingroup and supported by Vingroup Innovation Foundation (VINIF) under project code VINIF. 2020.DA15.

References

1. Nguyen D-T, Park J-G, Lee H-J, Kim Y-S (2010) Finite element method study of incremental sheet forming for complex shape and its improvement. Proc Inst Mech Eng Part B J Eng Manuf 224(6):913–924. https://doi.org/10.1243/09544054JEM1825
2. Li Y, Chen X, Liu Z, Sun J, Li F, Li J, Zhao G (2017) A review on the recent development of incremental sheet-forming process. Int J Adv Manuf Technol 92(5–8):2439–2462. https://doi.org/10.1007/s00170-017-0251-z
3. Xiao X, Kim J-J, Hong M-P, Yang S, Kim Y-S (2020) RSM and BPNN modeling in incremental sheet forming process for AA5052 sheet: multi-objective optimization using genetic algorithm. Metals 10(8):1003. https://doi.org/10.3390/met10081003
4. Gaddam DKR, Ansari MD, Vuppala S, Gunjan VK, Sati MM (2022) A performance comparison of optimization algorithms on a generated dataset. In: ICDSMLA 2020. Springer, Singapore, pp 1407–1415
5. Pham QT, Nguyen HQ, Tran VX, Xiao X, Kim JJ, Kim YS (2021) Process parameter optimization for incremental forming of aluminum alloy 5052-H32 sheets using back-propagation neural network. In: Research in intelligent and computing in engineering. Springer Singapore, pp 585–594
6. Narayana GS, Ansari MD, Gunjan VK (2022) Instantaneous approach for evaluating the initial centers in the agricultural databases using K-means clustering algorithm. J Mob Multimedia, pp 43–60
7. Szachogluchowicz I, Sniezek L, Slezak T, Kluczyński J, Grzelak K, Torzewski J, Fras T (2020) Mechanical properties analysis of the AA2519-AA1050-Ti6Al4V explosive welded laminate. Materials 13(19):4348. https://doi.org/10.3390/ma13194348
8. Rashid E, Prakash M, Ansari MD, Gunjan VK (2021) Formalizing open source software quality assurance model by identifying common features from open source software projects. In: ICCCE 2020. Springer, Singapore, pp 1375–1384
9. Kingma D, Ba J (2014) Adam: a method for stochastic optimization. In: International conference on learning representations

IoT-Based Environmental Parameter Monitoring Using Machine Learning Approach

M. M. Kashid, K. J. Karande, and A. O. Mulani

Abstract Today environmental protection is essential for people to make certain safe and rich existence. Tracking necessities vary widely, rely on location, extending to specialized applications that needed flexibility. The proposed system defines the implementation of an Internet of Things that can be evolved into a selection of programs and consists of the flexibility required to interchange and enhance without the necessity to systematization complicated infrastructure. The answer is primarily based on impartial Wi-Fi sensor nodes, small Wi-Fi receivers linked to the net, and a cloud structure that gives information storage and transport to faraway clients. The answer allows local administrators to not simplest screen the present day scenario via their cellular phones but additionally to reveal faraway Internet Web sites. All evaluation is saved at distinctive stages to ensure secure compliance and access to information saved within the event of a community breakdown or reach ability. The proposed device is beneficial for measuring temperature, humidity, etc. Predicting this parameter is using machine learning methods that include methods such as regression, editing, etc. Pre-data processing is required to clear the data using error rate, data verification, etc. Machine learning models are very powerful and accurate to work with data guesses.

Keywords IoT · Machine learning · NodeMCU ESP8266 · Temperature sensor · Humidity sensor · ThinkSpeak cloud

M. M. Kashid (✉) · A. O. Mulani
Department of Electronics Engineering, SKN Sinhgad College of Engineering, Pandharpur, Maharashtra 413304, India
e-mail: madhurikashid.ingole@gmail.com

K. J. Karande
SKN Sinhgad College of Engineering, Pandharpur, Maharashtra 413304, India

A. Kumar et al. (eds.), *Proceedings of the International Conference on Cognitive and Intelligent Computing*, Cognitive Science and Technology,
https://doi.org/10.1007/978-981-19-2350-0_5

1 Introduction

Conservation has grown to be one of the most important concerns of just about each country within the last few years. Despite the fact that the extent of industry should be growing uncontrolled in recent many years, the current scenario is absolutely converting closer to environmentally pleasant answers. Water and air standards are important for maintaining a healthy balance between a healthy environment and a good atmosphere. It is also vital to notice that by searching on the most well organized industrial product both pollution and use of herbal sources can be decreased.

Procedures, consisting of boiling, drying, binding, and so on, are practiced by means of almost all kinds of current industries. The ones approaches are liable for a huge quantity of gas extraction and waste water discharge. Even though maximum factories have their own sewage remedy plants, it is essential to degree the level of wastewater discharged into a sewage system. In truth, clean air is a fundamental necessity in normal life. Air pollutants affect human fitness and is taken into consideration a prime problem global, in particular in lands, wherein the gas and oil industries are discovered anywhere. In step with the US Environmental Safety Organization (USEPA), air satisfactory is determined with the aid of measuring sure gases that affect human health, namely carbon monoxide (CO), floor-degree ozone (O_3), and hydrogen sulfide (H_2S) [1, 2].

The primary cause of environmental tracking is not always just to collect data from many locations around; however, additionally to offer the data needed via scientists, planners, and policy makers, in order that individuals who make selections about dealing with and improving the environment can do so. Imparting useful information to cease customers, great efforts are being made to improve air pleasant in both areas internal and out. Natural surroundings and environmental tracking represent an essential category of sensory community packages. Currently advances in low-power Wi-Fi technology have generate technological conditions for the creation of many functional small sensory devices, which may be used to listen and see visual events [3, 4]. Wireless neighborhood networks (WSNs) are presently a lively studies site due to a huge range of packages consisting of military, medical, environmental tracking, security, and civilian. Many environmental tracking examples of WSNs have already been presented within the literature and designed for a multiplicity of purposes.

1.1 Problem Statement

The environmental monitoring system requires certain technical necessities including a high level of system integration, overall performance, reliability, productiveness, accuracy, sturdiness, flexibility, and many others, and WSN technology presents a reliable answer. As IoT technology brought reasonably-priced, low-power hardware, and information may be recorded and captured on a Web server.

1.2 Objectives

1. To establish communication between sensing unit and processor and processor and communication module.
2. To monitor environmental conditions like temperature and humidity.
3. To update the monitored current environmental condition along with sensor value on the ThinkSpeak cloud.

1.3 Scope

The IoT device gives immediate notification in actual time. Consequently, the temperature monitoring device permits organization to track the environmental parameter on comfy Web/cellular-based platform and gets rid of redundant task like taking manual readings, consequently saving time and raising short selection making [5, 6].

The system can be used in industries where short distance communication is required. Device can controlled more comfortably.

2 Methodology

2.1 Block Diagram

IoT is presently energetic studies vicinity because of its extensive variety applications inclusive of navy, medical, environmental tracking, and protection. Many environmental tracking examples of WSNs are previously existed within the literature and generated for one-of-a-kind purposes.

The temperature and humidity sensor that be in communication, the usage of wireless technology depends on the identical hardware. The distinction among the some of them consisting within the communication protocols that grow to be utilized which include basic data communication tools particularly user datagram protocol (UDP) or hypertext transfer protocol (HTTP) [7, 8]. For the systematic design of the system based on wireless technology is represented in Fig. 1. System includes a special kind of microprocessor, configurable chip/integrated circuit that holds many components of a computer which is system (PSoC 3) developed by Cypress Semiconductor corporations, both temperature sensor and humidity sensor driven using power supply.

The special kind of microprocessor eliminates the chances of connecting along with set of protocols used in a communications network. Consequently, a processor inside evolved devices is in price with acting all accomplishment for the right functioning of the tool, specifically, electricity control, accession of information from the temperature and humidity sensing unit, and communication [9, 10].

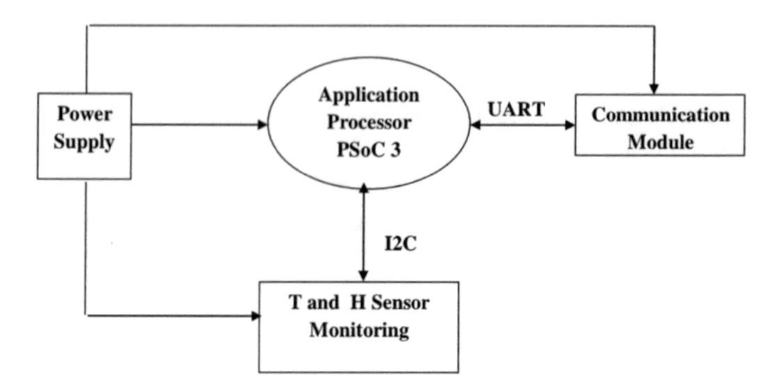

Fig. 1 Block diagram

The serial link for the transmission of information to base station among through the PSoC 3 device, communication unit, and application programming interface (API).

Selected Wi-Fi module is composed in its low-energy operation, providing 4μA at some stage in turn OFF and quick all over transference of 210 electrical pulses. For the improvement of future inexpensive Wi-Fi units primarily depends on IEEE 802.11 principals together with NodeMCU specific if can increase the variety of viable answers. Such as the ESP8266 from Expressif, will multiply the range of possible solutions. However, with a current larger than 20 μA in deep sleep mode, newer approaches for achieving power efficiency in the designs using it will have to be investigated [11].

2.2 Environmental Parameter Monitoring System Based on UDP

UDP-based system used for tracking the temperature and humidity inside the environment. This is the benefit of prevailing IEEE 802.11 framework, which sends measured temperature and humidity sensing data to a cloud-based server platform. It offers the chances of measuring the data from all system with a Web server [12].

In Fig. 2, the use of UDP permits the less-power consumption of wireless sensors, due to its wireless nature. User datagram protocol having smaller packet size, extended operation speed, and lower delay in comparison with TCP/IP protocol. A loss in transmission reliability, due to the fact there may be no renowned data received for the packets being dispatched [13, 14]. The cloud-based server is residing within ThinkSpeak cloud or an UDP listener strolling on a computer, which could explicate the acquired facts, put that in a data-base, and offer the chances of measuring and proceeding them constant with the person's wishes, through Internet server.

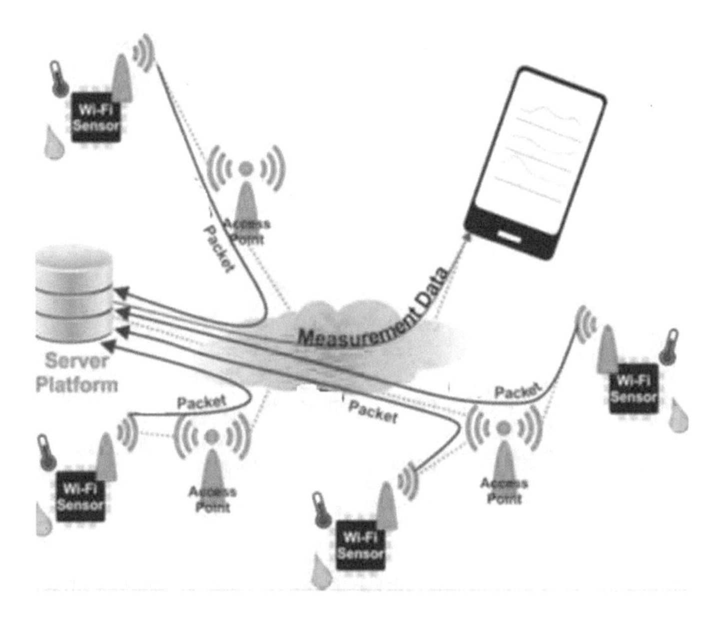

Fig. 2 Environment tracking system by using UDP communication

2.3 *Server Selection*

Internet of Things (IoT) is recent technology in which huge variety of embedded systems (things) are related to the Internet. In Fig. 3, those linked systems interact with humans and distinctive topics and usually offer measured static records to server platform and server computing assets in which the static records are prepared and examined to achieve significant perception. Low cost and effective device connectivity permits this trend [15, 16]. IoT applications inclusive of environmental tracking

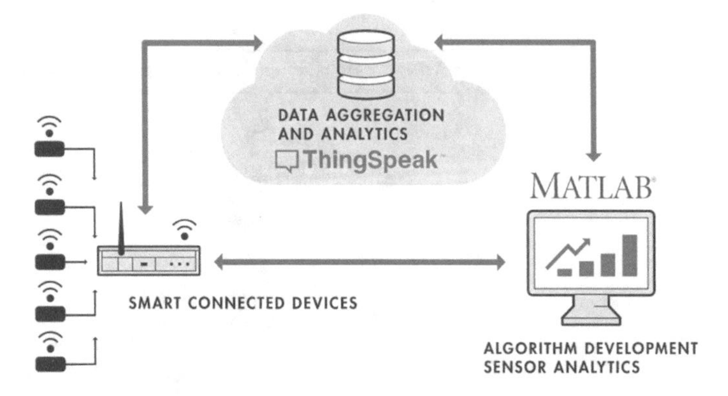

Fig. 3 ThinkSpeak server operation

and manage, fitness monitoring, automobile fleet tracking, business tracking and control, and domestic automation [17].

3 Result

3.1 Flowchart

See Fig. 4.

3.2 Supervised Machine Learning Models

Linear Regression

In linear regression, independent variable having linear relationship with the dependent variable.

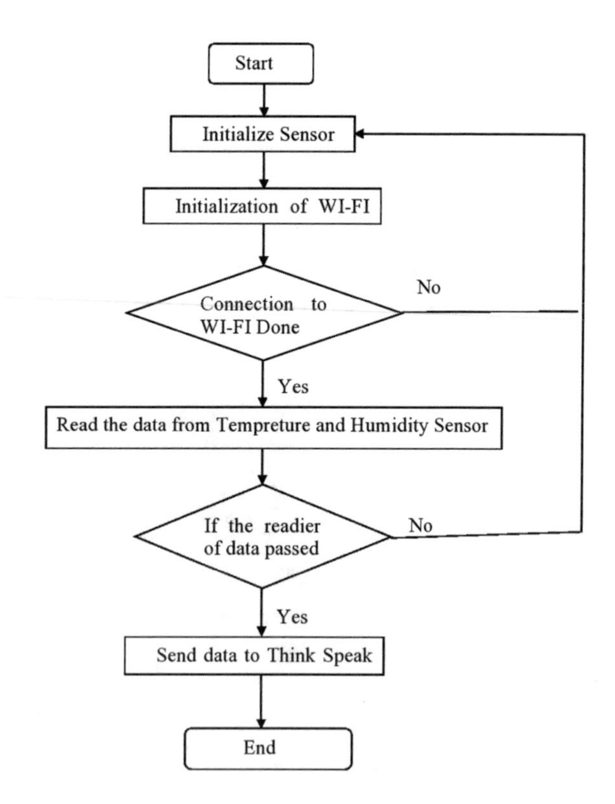

Fig.4 Flow chart of proposed system

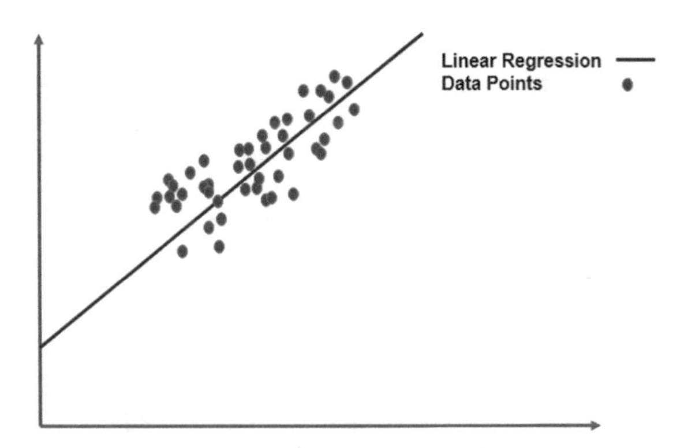

Fig. 5 Simple linear regressions technique

The straight line in Fig. 5 is the most appropriate line. The main goal of line deceleration assumes the set of measurements provided. Plan for most appropriate trend line the representation in all way.

Objective function

$$\mathbf{Y} = \mathbf{b0} + \mathbf{b1X} + \mathbf{e} \tag{1}$$

The most appropriate line depend on the linear Eq. (1) which is given above.

- Y represents dependent variable
- b0 represents a touchline for y-axis
- b1 represents slant of line.
- x represents the independent variables that determines Y.
- e represents error in the result.

The objective function gives good values for b0 and b1 to create the slant line for data points. This is done by transforming problem into a deprecation problem to gain the best values for b0 and b1. Error is deprecation in this problem between the actual value and the predicted value.

Simple Linear Regressions Model Implementation

The following are basics steps involved in the models implementations.

1. Loading Data
2. Data Checking
3. Cutting Data
4. Train and Divide Data
5. Generate Model
6. Check accuracy.

Evaluation of models with respect to temperature and humidity datasets is made and it is used to predict humidity for training a testing datasets with dependent variable is temperature. The model implementation is done in the python and tools used is numpy, pandas, sklearn, matplotlib library from it. Some sample datasets given to it to predict the humidity for some testing and training datasets. The model have 20% of the data set as testing and 80% is for training datasets. The best fit line is plotted on training data set gave the 84.44% accuracy for the prediction of the humidity.

4 Conclusion

Environmental monitoring is a complicated parameter as the environmental situations can without difficulty exchange from point to point even at small distances. The technique is used to monitor the environmental conditions at remote locations effectively because of reduction in power consumption, size, flexibility, IoT access, etc. The system can be extended to have observations at different locations with smallest distance between them.

5 Future Scope

We can extend this system to have monitoring of various other parameters like gas and air quality. It will be helpful to get the predictions of all these observations. We can also predict the weather condition from monitored parameter. We can able to implement this system such a way that, if we can program Web server to get the live predictions for all these parameters. It is also possible that we can use deep learning algorithms with machine learning models to get the more precise results.

References

1. Roopa ML et al (2021) IOT based real time weather predication system using NODEMCU 12-E ESP266 and lab view. Int J Eng Res Technol (IJERT) ISSN: 2278-0181 Published by, www.ijert.org NCCDS—2021
2. Verma G et al (2020) Real time weather prediction system using IOT and machine learning. 978-1-7281-5493-0/20/$31.00 ©2020 IEEE
3. Kompella S et al (2020) A novel environmental monitoring system for real time using Arduino and node JS. Int J Recent Technol Eng (IJRTE) 8(5). ISSN: 2277-3878
4. Kumar S, Ansari MD, Gunjan VK, Solanki VK (2020) On classification of BMD images using machine learning (ANN) algorithm. In: ICDSMLA 2019. Springer, Singapore, pp 1590–1599
5. Gunjan VK, Kumar S, Ansari MD, Vijayalata Y (2022) Prediction of agriculture yields using machine learning algorithms. In: Proceedings of the 2nd international conference on recent trends in machine learning, IoT, smart cities and applications. Springer, Singapore, pp 17–26
6. Abhyankar V et al (2019) Portable autonomous rain prediction model using machine learning algorithm. In: International conference on vision towards emerging trends in communication and networking (ViTECoN), IEEE

7. Rashid E, Ansari MD, Gunjan VK, Khan M (2020) Enhancement in teaching quality methodology by predicting attendance using machine learning technique. In: Modern approaches in machine learning and cognitive science: a walkthrough. Springer, Cham, pp 227–235
8. Sasikumar R et al (2018) Environmental monitoring system using IoT. Int J Curr Eng Sci Res (IJCESR) 5(4). ISSN (print): 2393-8374, (online): 2394-0697
9. Kumar S, Ansari MD, Naik MV, Solanki VK, Gunjan VK (2020) A comparative case study on machine learning based multi-biometric systems. In: Advances in cybernetics, cognition, and machine learning for communication technologies. Springer, Singapore, pp 353–365
10. Chauhan D et al (2018) Data mining technique for weather prediction: a review. Int J Recent Innov Trends Comput Commun
11. Mois G et al (2017) Analysis of three IoT based wireless sensors for environmental monitoring, published in IEEE Transactions on Instrumentation and Measurement
12. Swami SS, Mulani AO (2017) An efficient FPGA implementation of discrete wavelet transform for image compression. In: 2017 International conference on energy, communication, data analytics and soft computing (ICECDS 2017), pp 3385–3389
13. Mulani AO, Mane PB (2016) Area efficient high speed FPGA based invisible watermarking for image authentication. Indian J Sci Technol 9(39). https://doi.org/10.17485/ijst/2016/v9i39/101888
14. Mulani AO, Mane PB (2017) An efficient implementation of DWT for image compression on reconfigurable platform. Int J Control Theor Appl 10(15)
15. Shinde G, Mulani A (2019) A robust digital image watermarking using DWT-PCA. Int J Innov Eng Res Technol (IJIERT) 6(4)
16. Nagane UP, Mulani AO (2021) Moving object detection and tracking using Matlab. J Sci Technol 6(1). https://doi.org/10.46243/jst.2021.v6.i04.pp63-66
17. Mulani AO, Shinde GN (2021) An approach for robust digital image watermarking using DWT-PCA. J Sci Technol 6(1). https://doi.org/10.46243/jst.2021.v6.i04.pp59-62
18. Kunjumon C et al (2018) Survey on weather forecasting using data mining. IEEE
19. https://thingspeak.com/pages/learn_more

Comparative Analysis of Machine Learning Approaches for Crop and Yield Prediction: A Survey

A. Ashwitha, C. A. Latha, V. Sireesha, and S. Varshini

Abstract Agriculture is an integral part of the economy in most countries, and it provides the primary source of livelihood, income, food, and employment to most rural populations. The food and agriculture organization (FAO) reported that the agricultural population share in the total population is 67%. Agriculture in a country contributes to 39.4% of the GDP, and agricultural goods account for 43% of all the exports. Therefore, enhancing crop production is seen as an essential aspect of agriculture. Machine learning, data mining, and deep learning are the essential analytical technologies that support accurate decision-making in crop yield prediction, which includes some of the assisting conclusions on which crop to grow and the decisions regarding the crops in the growing season on the agricultural land. A mixture of machine learning, data mining, and deep learning algorithms are applied to support crop yield prediction research. The algorithms include several classifications, regression, and clustering techniques. Data in the agricultural field is enormous, considering various parameters and represented in structured/unstructured form. Hence, there is a need for an efficient technique to process these data and discover potential information. This paper mainly focuses on the algorithms that can be used to predict the most suitable crop and estimate the crop yield, which assists the farmers in selecting and growing the most profitable crop and thereby reducing the chances of loss and hence increasing the productivity and the value of his farming area.

Keywords Agriculture · Big data analytics · Convolutional neural network · Crop prediction · Deep learning · Plantain tree cultivation

A. Ashwitha (✉) · C. A. Latha · V. Sireesha · S. Varshini
Department of ISE, MSRIT, Bengaluru, India
e-mail: ashwitha.a@msrit.in

Department of CSE, AMCEC, Bangalore, India

Visvesvaraya Technological University, Belgavi, India

A. Kumar et al. (eds.), *Proceedings of the International Conference on Cognitive and Intelligent Computing*, Cognitive Science and Technology,
https://doi.org/10.1007/978-981-19-2350-0_6

1 Introduction

The agricultural sector in India is critical to the Indian economy. The vast majority of the Indian population is engaged in some form of farming activity. This sector feeds the ever-increasing Indian population. Agricultural practices must become efficient and sustainable enough to match the growing needs of the nation. However, there is a significant risk attached to farmers' decision-making process to choose the right crop for their land-based knowledge and experience. This decision is affected by various scientific and socio-political factors like temperature, rainfall, humidity, soil pH, soil type, nutrients, fertilizer requirements, capital investment, history of the area, and many more. Even after choosing a particular crop based on naked-eye observation of vast farmlands, there remains a possibility that the crop might fail. It can lead to enormous amounts of losses leading to severe issues like debt and farmer suicides.

The following are the motivation to perform the survey of various machine learning approaches for crop yield prediction.

- Comparative analysis of crop and yield prediction helps predict the optimal crop and the yield using different parameters like rainfall, temperature, fertilizers, pesticides, and other atmospheric conditions and parameters.
- The agricultural data is subject to several machine learning, data mining, and deep learning techniques, and the survey of these techniques are made to recommend the crops more accurately.
- This analysis will help the farmers to make the correct decision to sow the crops to increase the yield rate.

2 Related Work

A precise crop harvest forecast typical can assist agriculturalists to elect on pardon to cultivate and produce. There are different methods to crop harvest forecast. This appraisal artifact has examined the usage of machine learning cutting-edge crop harvest forecast in Table 1.

3 Comparison of Crop and Yield Prediction Algorithms

The comparative analysis of various algorithms of regressors, classifiers, clustering, and neural network are shown in Table 2. The table provides the name algorithm, description, benefits, and limitations of various algorithms.

Table 1 Survey of crop yield prediction and analysis

	Paper name	Year	Algorithms used	Application	Limitations
1	Crop Yield Prediction based on Indian Agriculture using Machine Learning [1]	2020	Laso, Kernel Ridge and ENet	Development of a Web App to provide the yield of the crop in the year of his choice	This work does not give the correlation between temperature and precipitation
T	Prediction and Analysis of Crop Yield using Machine Learning Techniques [2]	2020	Decision Tree, Multilinear Regression	Agricultural dataset from 1999 to 2014 used to predict all crops' yield in all states in India	This approach calculates yield for a shorter period without considering the climatic factors
3	Crop Yield Analysis using Combinatorial Multivariate Linear Regression [3]	2020	Multivariate Linear Regression	This paper analyzes the accuracy of yield prediction for all possible combinations	This analysis is carried out only for rice for a particular region
4	Crop Prediction Using Machine Learning [4]	2020	KNN, Decision Tree, Naive Bayes	This approach solves agricultural problems by monitoring soil properties	IoT devices are not used to get the real-time values of the soil
5	Analysis of crop yield prediction using data mining technique [5]	2020	K-Nearest Neighbor (KNN)	Creates a web application to predict the crops' yield	Multiple strategies are used to improve the efficiency of the proposed algorithm
6	Crop Prediction Using Machine Learning [6]	2020	Naive Bayes classifier	Crop prediction for district level with climatic parameters	Decision trees show poor performance with dataset has variations
7	Crop Yield Prediction Using Machine Learning Algorithms [7]	2019	RNN and LSTM	The work determine the temperature using a Sequential model like Simple RNN	The results are not presented in any form as a web application mobile application
8	Crop Analysis and Profit Prediction using Data Mining Techniques (Id:39) [8]	2019	A decision tree (ID3 algorithm)	The result shows which crop is the best to cultivate during a particular season	This paper uses ID3 only for analysis
9	Improvement of Crop Production Using Recommender System by Weather Forecasts [9]	2019	ANN(artificial neural networks)	This paper deals with developing a model which is used for the district-wise agricultural data analysis	The prediction accuracy of the model can be enhanced by using various other regression models

(continued)

Table 1 (continued)

	Paper name	Year	Algorithms used	Application	Limitations
10	Prediction of Crop Yield Using Data Mining [10]	2019	DBSCAN	A user-friendly web page was developed for predicting crop yield	Suitable strategies to improve the efficiency is needed

4 Analysis of Various Models

Performance analysis of various machine learning models and their sensitivity and specificity along with the percentage of test errors are shown in the following Figs. 1, 2, 3, and 4, respectively.

5 Conclusion

Agricultural practice is one of the most crucial application areas in most developing nations like India. The use of analytical features of various technologies in agriculture can impact the decision-making of many parameters, and it helps the farmers to produce better and improved yields. The papers considered dealt with multiple regression techniques, Advanced regression algorithms, RNN and LSTM, Decision Tree, Multilinear Regression, K-Means clustering, Rolling mean and Poly Regression, Big Data paradigm, Deep Convolutional Regression Network (DCRN), ANN (artificial neural networks), ID3, and Density-based clustering algorithm (DBSCAN) for agriculture crop yield prediction. The main motive of this work is to suggest a precise scheme to forecast the ideal yield and estimate the crop harvest forecast given numerous farming limits since the accretion of previous farming information. This work compares multiple information excavating, machine erudition, and deep learning algorithms that can be useful in choosing the most accurate algorithms which can be used to forecast the most excellent appropriate harvest and calculating the estimated crop yield considering various environmental parameters.

Table 2 Comparison of Different Algorithms Used in Crop and Yield Prediction

Algorithm	Description	Benefits	Limitations
Regressors			
Linear Regression	A simple linear model assuming linear relation among two variables	Overfitting can be decreased by regularization, and it has a considerably lower time complexity	Linear regression is used for the mean/average value
Lasso Regression	Lasso regression is an advanced form of linear regression does shrinkage to shift the values	It can produce better results when the number of rows in the dataset is lesser than the number of attributes of dataset	Regularization of the dataset reduces the number of less significant attributes
Kernel Ridge Regression	This algorithm combines ridge regression with the kernel	It is a more restricted model compared to the linear model	Regression lacks the fundamental concept of support vectors
Multiple Linear Regression(MLR)	This technique uses independent variable to determine dependent values	It is capable of determining the relative influence of one or more predictor variables on the criterion value	MLR depends solely on the dataset being used
Classifiers			
Decision Tree (ID3)	A decision tree acts like a decision support tool, where decision taken at each branch	It performs well on both linear and non-linear problems. Feature scaling is not required	The decision tree results are not significant for small datasets
Random Forest	This algorithm involves the creation of numerous decision trees	It helps in reducing overfitting and improves accuracy. It performs well with both continuous and categorical values	A large number of decision trees make it challenging to interpret the results
XGBoost	Decision tree algorithm which uses a gradient boosting framework	It is capable of handling missing values and prevents the model from overfitting	It is more likely to be overfitting than bagging. It needs a robust dataset with hyper parameters

(continued)

Table 2 (continued)

Algorithm	Description	Benefits	Limitations
Naive Bayes	This classifier works on the Naive assumption that the presence of a feature is independent of another set of features	It performs exceptionally well with the categorical input variables in comparison to numerical variables It can easily predict the class of a test dataset	Loss of categorical variables unobserved in the training dataset can seriously affect the results as it is naive enough to assign zero probability
Clustering algorithms			
k-Nearest Neighbors	The clustering algorithm is based on the k-nearest data points	It is effortless to implement the multi-class problem It is simple and quite intuitive	The quality of data finds the accuracy. Furthermore, with the increase in size, the calculations increase
k-Means Clustering	This algorithm classifies data points into clusters based on distance from a centroid	It can scale large datasets It easily adapts to new examples and is relatively simple to implement	k-Means work only on numerical data. It assumes that the clusters are spherical
DBSCAN	Solidity Centered Spatial Grouping of Submissions with Noise is recycled	It can handle clusters of various shapes and sizes It is resistant to noise	The DBSCAN algorithm is not suitable for data having varying spatial densities
Neural Networks			
Artificial Neural Network (ANN)	This algorithm is designed on the basic concept of neurons	It is fault-tolerant It can produce the output even with incomplete information	ANNs are both hardware and compute-intensive. There is no fixed structural layout to the network
Self-Organizing Map (SOM)	SOM is a type of ANN trained on unlabeled data to give a two or three-dimensional	It helps in interpreting the data mapping efficiently It is capable of organizing large and complex datasets	The main disadvantage of SOM is that the weights of the neuron must be precise

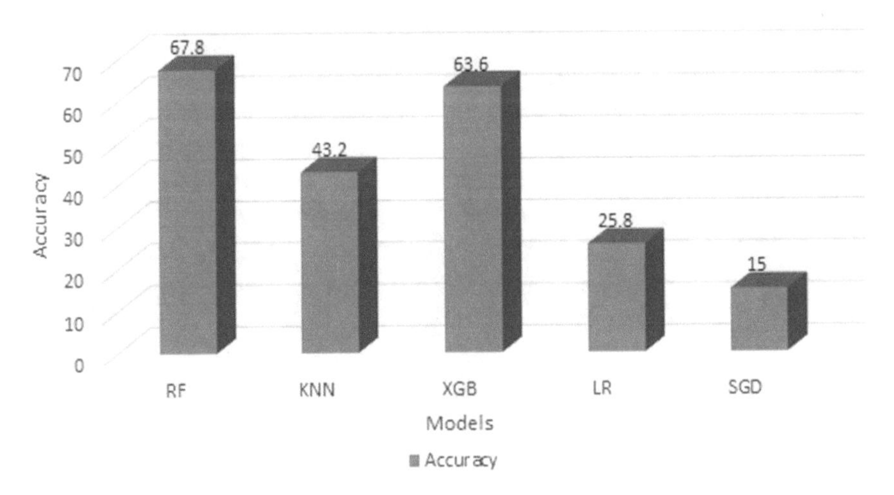

Fig. 1 Comparison of various model with accuracy

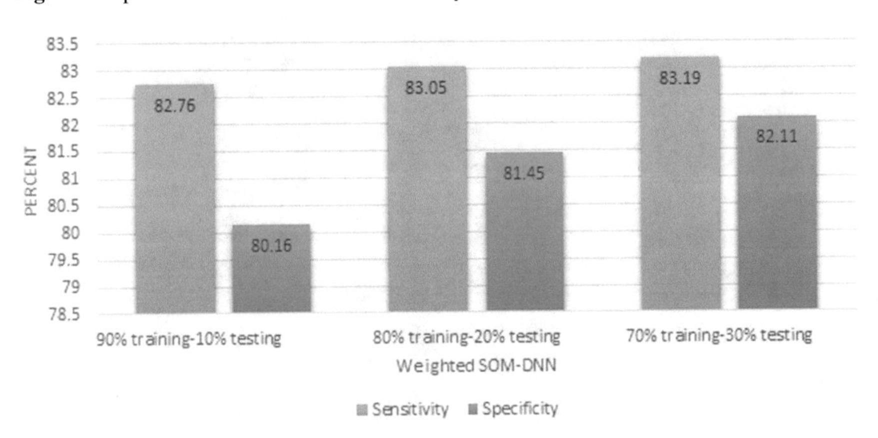

Fig. 2 Sensitivity and specificity of the models

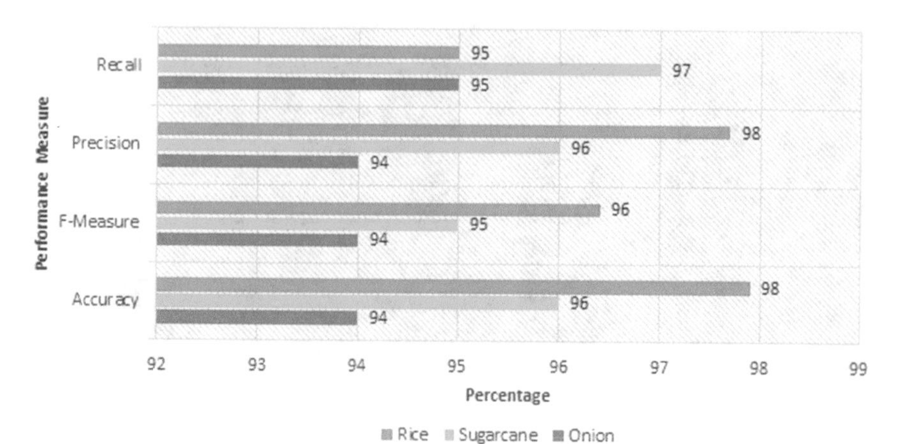

Fig. 3 Analysis of various models

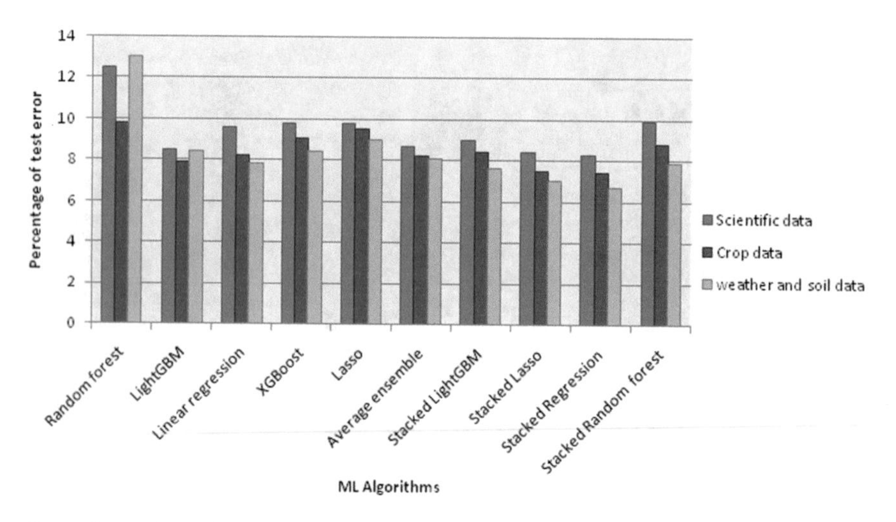

Fig. 4 Comparison of various models

References

1. Nishant PS, Sai Venkat P, Avinash BL, Jabber B (2020) Crop yield prediction based on Indian agriculture using machine learning. In: 2020 International conference for emerging technology (INCET), Belgaum, India, pp 1–4. https://doi.org/10.1109/INCET49848.2020.9154036
2. Manoj GS, Prajwal GS, Ashoka UR, Krishna P, Anitha P (2020) Prediction and analysis of crop yield using machine learning techniques. Int J Eng Res Technol (ijert) ncait 8(15)
3. Rani DE, Sathyanarayana N, Vardhan BV, Goud OSC, Crop yield analysis using combinatorial multivariate linear regression
4. Thomas KT, Varsha S, Saji MM, Varghese L, Thomas EJ (2020)Crop prediction using machine learning. Int J Fut Gener Commun Netw
5. Fathima, Sowmya K, Barker S, Kulkarni S (2020) Analysis of crop yield prediction using data mining technique. Int Res J Eng Technol (IRJET)
6. Patil P, Panpatil V, Kokate S (2020) Crop prediction using machine learning. Int Res J Eng Technol (IRJET)
7. Nigam A, Garg S, Agrawal A, Agrawal P (2019) Crop yield prediction using machine learning algorithms. In: 2019 Fifth international conference on image information processing (ICIIP). IEEE, pp 125–130
8. Akshatha A, Shailesh Shetty S, James AP, Saseendran AM, Poojary CM (2019) Crop analysis and profit prediction using data mining techniques (Id:39). Int J Eng Res Technol (ijert) rtesit 7(8)
9. Kamatchi SB, Parvathi R (2019) Improvement of crop production using recommender system by weather forecasts. Procedia Comput Sci 165:724–732
10. Kabeer N, Loganathan D, Cowsalya T (2019) Prediction of crop yield using data mining. Int J Comput Sci Netw (IJCSN)

Data Transfer Through Blockchain

Anasuya N. Jadagerimath, K. R. Nataraj, M. Mithali, Napa Pavani, and A. Kusuma

Abstract Blockchain is a sort of database that has become popular due to its decentralized nature. It consists of a collection of blocks in which if one block's hash value changes, all other blocks' hash values are updated immediately; this allows for a safe and secure data transfer in which a hostile user cannot gain access to the blockchain. The ledger becomes immutable once the transaction is updated to the blockchain. Decentralizing the blockchain eliminates the need for a third party in the transaction. Every node in a peer-to-peer network will have a copy of the blockchain and will follow a consensus procedure, ensuring that the blockchain is distributed and resistant to attacks. When data must be passed to another person, security is paramount. As a result, we may use blockchain to solve this problem and guarantee trust and reliability. This article discusses the concept of creating a Web-based application that uses blockchain technology to transfer data.

Keywords Blockchain · Immutable ledger · Consensus protocol · Timestamp · Hash · Decentralization

1 Introduction

The blockchain was created to serve as the public ledger for the cryptocurrency bitcoin's transactions. The blockchain-enabled bitcoin to solve the problem of double spending without the use of a trusted authority or central server. Other apps have been inspired by the bitcoin design. Blockchain applications have now surpassed bitcoin. Blockchain is becoming a game-changer in a variety of industries, including BFSI, health care, education, real estate, supply chain and logistics, and the Internet of Things, to mention a few.

A blockchain database is a special kind of database. The manner it saves information differs from that of a traditional database; blockchains store data in blocks

A. N. Jadagerimath (✉) · K. R. Nataraj · M. Mithali · N. Pavani · A. Kusuma
Department of Computer Science, Don Bosco Institute of Technology, Bangalore, Karnataka, India
e-mail: anasuyaprakash@dbit.co.in

A. Kumar et al. (eds.), *Proceedings of the International Conference on Cognitive and Intelligent Computing*, Cognitive Science and Technology,
https://doi.org/10.1007/978-981-19-2350-0_7

that are then connected together. As new information is received, it is entered into a new block. Once the block has been filled with data, it is chained onto the previous block, resulting in a chronological chain of data. A blockchain can hold a variety of data, but the most prevalent application so far has been as a transaction ledger.

Cryptography is one of the most important features of blockchain, as it ensures that transactions are secure, verified, and verifiable. Immutability is another critical aspect, as any approved records are irreversible and cannot be modified. Timestamped, trusted, anonymous, distributed, and programmable are some of the other major qualities.

The way that the data is structured differs significantly between a traditional database and a blockchain. A blockchain is a system that accumulates data in groups, often known as blocks that contain sets of data. The blocks have specific storage capabilities, and when they are filled, they are linked onto the previous block, establishing a data chain known as a "blockchain." All additional information added after that newly added block is compiled into a new block, which is then added to the chain after it is filled.

A database organizes the data into tables, but a blockchain organizes the data into linked pieces (blocks). As a result, all blockchains are databases, but not every database is a blockchain. When the blockchain is implemented in a decentralized manner, this method creates an irreversible data timeline. When a block is created and appended to the chain, it becomes permanent and part of the chronology. When a block is added to the blockchain, a precise timestamp is assigned to it.

Flask is nothing but a Python-based micro Web framework. When it comes to designing Web applications, Flask gives developers a lot of options. It includes tools, libraries, and mechanisms that allow you to build a Web application, but it does not impose any dependencies or tell you how the project should look.

The flask is classified as a micro or lightweight structure, as previously indicated. A micro-framework is typically a framework that has little or no dependencies on external libraries. In any event, there are benefits and drawbacks to using the flask framework to construct a Web application as a developer. Some of the advantages of adopting flask as your Web application framework include a low dependency on updates and the ability to spot security flaws. While one disadvantage of utilizing flask is that by adding extensions, you may end up performing additional work for yourself or increasing the list of dependents.

2 Related Work

As we all know, blockchain technology was created to facilitate bitcoin financial transactions, but it has now developed to include smart contracts and decentralized apps. It has also opened the door for it to be beneficial in a variety of other fields for improved security and maintainability, and research will continue to make the greatest use of it. In this section, we will discuss on the blockchain technology, decentralized applications, and cryptography or the security of blockchain.

2.1 Blockchain Technology

There are two parts to each block in the blockchain network: a header and a body as referred in [1]. By establishing a hash value, these blocks are linked together in a cryptographic chain. Figure 1 depicts the basic structure of a blockchain utilized in bitcoin and Ethereum. The block header contains the block number, a timestamp, a hash of the previous block (i.e., a cryptographic link that creates a chain and is tamperproof), a random nonce [i.e., used to solve the proof-of-work (PoW)], and the Merkle tree root (i.e., encodes transactions/data in the block in a single hash for rapid data verification; any data modification will change the hash value, making it easy to check data integrity). The transactions are safeguarded by cryptographic digital signatures (e.g., users' private keys) in the block body and include records of asset or data transfers, smart contracts, and broadcast messages. Only those with the cryptographic keys may validate the transaction's data, timing, and user (i.e., data privacy). As a result, integrity, privacy, and validity of data are all ensured using cryptographic methods.

The blockchain has following key characteristics as defined in [2]:

- *Decentralization:* Each transaction in traditional centralized transaction systems must be validated by a central trusted agency (e.g., the central bank), resulting in cost and performance bottlenecks at the central servers. In contrast to the centralized option, blockchain does not require the use of a third party. Consensus algorithms are employed in blockchain to keep data consistent across a distributed network.
- *Persistency*: Transactions can be validated fast, and honest miners would not accept invalid transactions. Once a transaction is incorporated in the blockchain, it is nearly hard to erase or rollback the transaction. Blocks containing incorrect transactions might be identified right away.

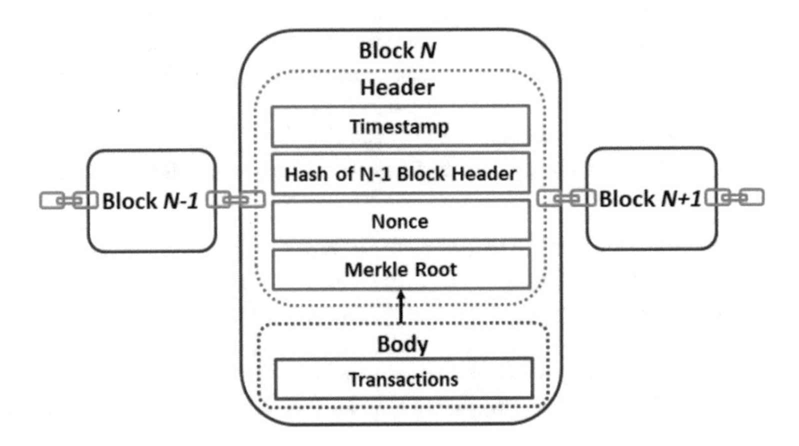

Fig. 1 General blockchain structure

- *Anonymity*: Each user interacts with the blockchain using a randomly generated address that hides the user's true identity. Due to the inherent restriction, blockchain cannot ensure full privacy preservation.
- *Auditability*: The unspent transaction output (UTXO) model is used to store data on user balances on the bitcoin blockchain: Any transaction must reference some previously unspent funds. The state of those referred unspent transactions changes from unspent to spend once the present transaction is recorded into the blockchain. As a result, transactions could be easily tracked and validated.

2.2 Decentralized Applications

DApps are digital applications or programs that reside and execute on a blockchain or peer-to-peer network of computers rather than a single computer and are not under the jurisdiction of a single authority. Unlike traditional applications, which run their backend code on centralized servers, DApps run their backend code on a decentralized peer-to-peer network.

The frontend code and user interfaces of a DApps can be developed in any language that can communicate with the backend. DApps are often open source, decentralized, rewarded through the provision of tokens to those who validate the DApp and adhere to a community-defined protocol. DApps can be built on distributed computing platforms like Ethereum or bitcoin. A blockchain system stores and executes decentralized applications.

Because of the single point of failure (SPOF) problem, the vulnerability of centralized systems has been challenged. Decentralized systems that are built in a distributed manner, on the other hand, have a data synchronization problem. To put this in another way, the participants present in the decentralized ledger system must reach some consensus, or agreement, on each message that is transmitted to them. On their decisions, the honest peers have a majority agreement. Nonetheless, intruders might use the Sybil attack to take control of a significant portion of the public P2P system by posing as many identities, in the blockchain-powered decentralized ledger, this could result in a significant "Double Spending" problem. To tackle the problem of double spending, proof-of-work (PoW) [3] is utilized.

DApps are characterized by four features, according to the definition of DApps in [4]:

- *Open Source*: Because of the trusted nature of blockchain, DApps must make their code open source in order for third party audits to be possible.
- *Internal Cryptocurrency Support*: Internal currency is the fuel that keeps a DApp's ecosystem running. A DApps can use tokens to quantify all credits and transactions between system users, including content providers and consumers.
- *Decentralized Consensus*: Transparency is built on the foundation of consensus among decentralized nodes.

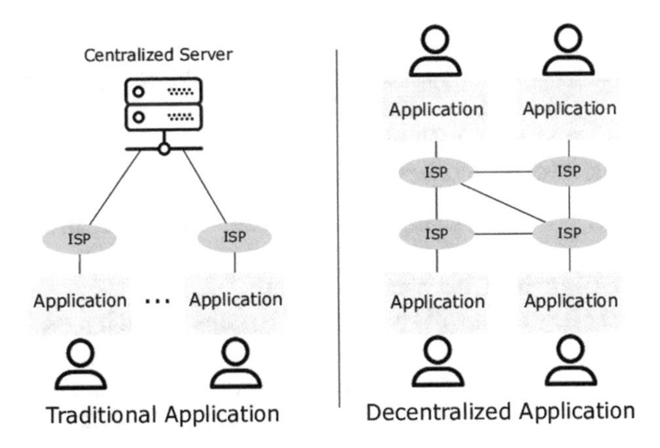

Fig. 2 Centralized application versus decentralized application

- *No Central Point of Failure*: Because all components of the apps will be hosted and operated on the blockchain, and a fully decentralized system should have no single point of failure.

The contrast between a centralized or traditional application and a decentralized application is seen in Fig. 2. With a centralized application, all logic for carrying out the required tasks is contained in a single cluster of servers. The cluster receives your request, processes it, and saves what it needs, and then responds correctly. All nodes in the network receive the logic you have implemented within your contract with a decentralized application (or DApp). Once the contract is mined, all nodes on the blockchain have the same contract-based logic preserved on their copies of the blockchain. Once you have interacted with this contract, you must process the same signal using the deployed logic, save the same data, and return the same signal. As a result, blockchain systems rely on encryption, peer-to-peer networking, and consensus mechanisms to provide decentralized application architecture.

2.3 Security of Blockchain

Blockchain technology creates a data structure with built-in security features. It is based on encryption, decentralization, and consensus principles, which maintain transaction trust. The data in most blockchains or distributed ledger technology (DLT) [5] which is organized into blocks, with each block containing one or more transactions. Each new block in a cryptographic chain connects to all the blocks before it in such a way that tampering is practically impossible. A consensus process validates and agrees on all transactions within the blocks, ensuring that each transaction is truthful and correct [6].

Asymmetric-key algorithms and hash functions are the two types of cryptographic algorithms used in blockchains. Hash functions are employed to give each participant with the capability of a single view of the blockchain. The SHA-256 hashing method [6] is commonly used as the hash function in blockchains.

Cryptographic hash functions provide the following benefits to the blockchain.

- *Avalanche effect*: A slight change in the data can result in a significantly different output.
- *Uniqueness*: Every input has a unique output.
- *Deterministic*: Any input will always have the same output if passed through the hash function.
- *Quickness*: The output can be generated in a very small amount of time.
- Reverse engineering is not possible, i.e., we cannot generate the input by having the output and the hash function.

Hash functions play an important role in connecting blocks and ensuring the integrity of the data stored within each block. Any change to the block data can cause inconsistency, causing the blockchain to become invalid. This requirement is achieved by the property of the hash functions called the 'avalanche effect'. In asymmetric-key cryptography, the private key is generated using a random number algorithm; whereas, the public key is created using an irreversible process. The advantage of the asymmetric encryption technique is that it contains separate public and private keys that may be sent through unprotected channels.

3 Proposed Design

The proposed design is to create a safe and secure platform to exchange data through blockchain. The first step is user authentication, if the user is a new customer, he will have to sign up to the portal. If the user is already an existing customer, he has to provide his credentials to login. The user can compose, check inbox, and see the sent data. He can log out of the portal whenever he wants to. After validation of the user, the user can send data to another user through blockchain and a hash value will be created for the transaction which will be used to check if the blockchain is valid. Following the completion of the transaction, the block is mined and added to the existing blockchain, where it is validated to reflect the transaction. The same blockchain is updated to all the other nodes in the P2P distributed network maintaining a consensus protocol.

Once the user is authenticated, he will be able to send data to another valid user through the Web-based application built using flask framework which implements blockchain to transfer the data for better security and reliability. If the blockchain is valid, the transaction will be created and if the blockchain is not valid, the transaction becomes unsuccessful as shown in Fig. 3, and the user will have to send the data again to make a successful transaction.

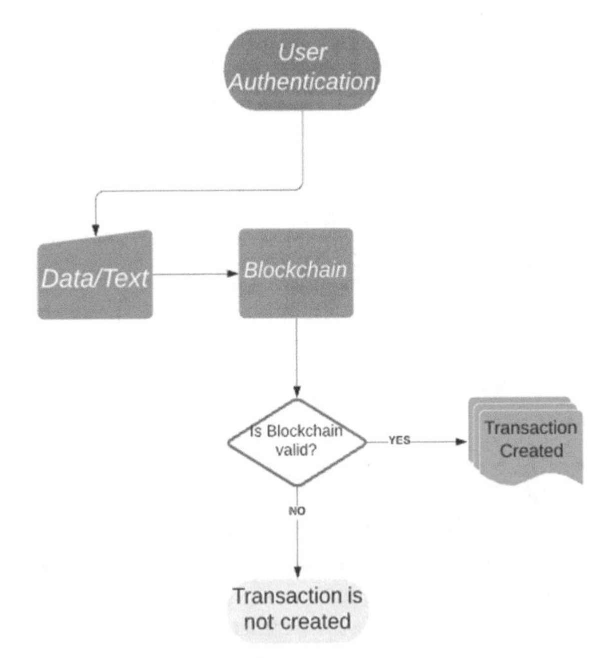

Fig. 3 Flowchart of data transfer through blockchain by creating a Web-based application

4 Implementation

In this system, we have tried to implement the blockchain which has been known for transactions, to securely transfer data. Each block in the blockchain will contain block number, timestamp, nonce, previous hash, and transaction.

Block number depicts the ID of the block created. Timestamp is the time when the block is created. Nonce depicts the proof that is, the previous hash of the current block links to the hash of the previous block, or its parent block, effectively securing the chain. SHA-256 algorithm is used for generating hash value for each block. Every node in the network will have a copy of the blockchain, which boosts the security of the system.

A consensus mechanism is a fault-tolerant mechanism that is used in computer and blockchain systems to achieve the necessary agreement on a single data value or a single state of the network among distributed processes or multi-agent systems. The transaction in Fig. 4 represents the sender address, receiver address, and the data to be transferred through the blockchain, and the process is as shown in Fig. 4 which is the block diagram of data transfer through blockchain.

The main functions of this system are:

- *User authentication*: If there is a new user then he must register first and then login. If an existing user logins, then his credentials are validated and then only allowed to transfer the data through blockchain to another valid user who exists in the database.

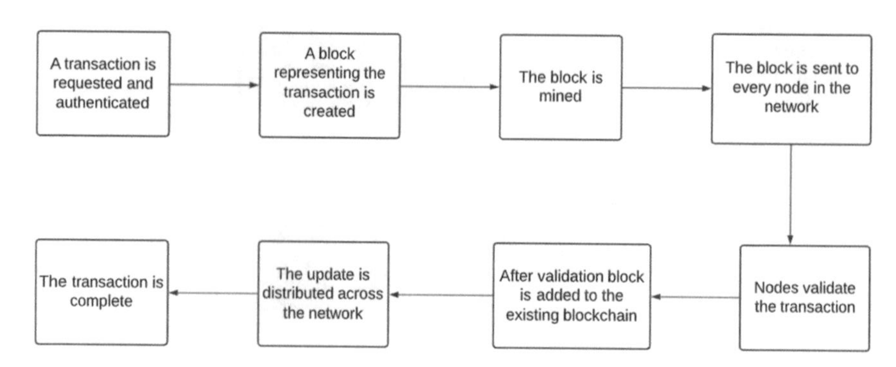

Fig. 4 Block diagram of data transfer through blockchain

- *Create block and add transaction*: After the user is authenticated, he can send the data along with the recipient address so as to who should receive it. The data, the sender address and the receiver address, is added to the transaction list, and then a block is created with block number, timestamp, nonce or proof, previous hash, and the transaction present.
- *Proof-of-work*: It is the number or piece of data that the miners have to find in order to mine a new block. Miners need to solve the problem and to solve they have to find specific number that will be exactly the proof-of-work (proof). It is used for validating the transactions and mining new blocks. It is hard to find but easy to verify. Due to proof-of-work, the transaction that represents the data in this system can be processed peer-to-peer in a secure manner without the need for a trusted third party. A given set of data can only generate one hash, so how do miners make sure they generate a hash below the target? They alter the input by adding an integer, called a nonce ("number used once"). Once a valid hash is found, it is broadcast to the network, and the block is added to the blockchain.
- *Chain validity*: The previous has value of the present block should be equal to the hash of the previous block except for the genesis block which is the first block of the blockchain which has no transaction. Each block in the blockchain should have correct proof-of-work.
- *Mining*: It is the process of mining a block. To mine a block, the miner needs to solve the proof-of-work problem by finding the proof based on the previous proof that is the last proof present in the last block. After getting the proof or nonce means that mining is done successfully and we need to get the previous hash value to create the block. So after mining, a new block will be created.
- *Consensus protocol*: It guarantees for the stable operation of blockchain systems. Nodes in the P2P distributed network agree on a certain value or transaction through the consensus protocol. Each node will have a copy of the blockchain making it decentralized. If the blockchain in one node adds a new block with the transaction, given the blockchain is valid then the other node's blockchain in the network is updated as well. If a hacker tries to change the data in a block the hash

value also changes and the blockchain becomes invalid because previous hash value should match the hash value of previous block.

5 Conclusion

Blockchain technology protects a network's integrity by codifying "truth" and, perhaps most importantly, allowing two parties to transact anything without "tokenization." This paper concludes that blockchain technology and data security can be very safe for protecting data and transferring data from one user to another. The fundamental principles of blockchain technologies are discussed, as well as an overview of a blockchain implementation that may be used to transmit data between peers utilizing blockchain by developing a Web-based application for user-friendly interaction and transfer.

References

1. Ochoa JJ, Bere G, Aenugu IR, Kim T (2020) Blockchain-as-a-Service (BaaS) for battery energy storage systems. IEEE J
2. Zheng Z, Xie S, Dai H, Chen X, Wang H (2017) An overview of blockchain technology: architecture, consensus, and future trends. In: 2017 IEEE 6th International congress on big data
3. Prashanth Joshi A, Han M, Wang Y (2018) A survey on security and privacy issues of blockchain technology. Math Found Comput 1(2):121–147
4. Cai W, Wang Z, Ernst JB, Hong Z, Feng C, Leung VC (2018) Decentralized applications: the Blockchain-empowered software system. IEEE Access, vol 6
5. Pop C, Cioara T, Anghel I, Antal M, Salomie I, Blockchain based decentralized applications: technology review and development guidelines
6. Li X, Jiang P, Chen T, Luo X, Wen Q (2017) A survey on the security of blockchain systems. Future Gener Comput Syst

Cancer Detection and Diagnosis Using Artificial Intelligence

Harnoor Kaur Khehra, Tanisha Saggar, Mansi Kashyap, and Bhupinder Singh Mavi

Abstract Healthcare sector in today's era is swapping drastically with the latest technologies like artificial intelligence, machine learning, and deep learning. These technologies have notable determined very aspects of the hospital. Many precise tasks such as image recognition for example classifying skin biopsy lesions, determining diabetic retinopathy severity, and detecting brain tumors can be done with the help of engineering science and present day technology called artificial intelligence. With such an amazing technology, there has been a great engrossment in developing AI diagnostics solutions for the detection and diagnosis of cancer. With the immediate requirement of AI systems, there will be great change in the healthcare sector. This advanced and extraordinary technology will play a great role in stopping the rapid growth of cancers, low awareness among the extended population, and lack of sufficient services and clinical expertise so that AI systems can assist the clinicians in this domain. Analysis of big datasets can be done very easily and with less cost and time by using AI technologies. Many deep learning algorithms, AI solutions are developed by researchers to detect and diagnose cancer precisely and to handle the skin datasets smartly. The advancement and endorsement of AI algorithms need large volumes of well-structured data, and these algorithms must work with changing levels of data quality. It is very important to understand how AI algorithms function to stop the severe conditions of cancers like breast cancer, gastric cancer, prostate cancer, and many more. The clinicians must understand this technology deeply and how it perfectly fits into everyday clinical practice and how patients are calmly addressed. This paper surveys how different types of cancers are detected and diagnosed using the latest AI technology as AI provides good support to doctors for the detection and diagnosis. It also plays a clear and supreme role in aiding patients in cancer detection in the future and will be thrilling to check the benefits that arise for patients and doctors from its use in everyday practices.

H. K. Khehra (✉) · T. Saggar · M. Kashyap · B. S. Mavi
Baba Banda Singh Bahadur Engineering College, Fatehgarh Sahib, India
e-mail: harnoorkhehra28@gmail.com

A. Kumar et al. (eds.), *Proceedings of the International Conference on Cognitive and Intelligent Computing*, Cognitive Science and Technology,
https://doi.org/10.1007/978-981-19-2350-0_8

Keywords Artificial intelligence · Machine learning · Deep learning cancer · Algorithms · Diagnosis · Breast cancer · Prostate cancer · Gastric cancer

1 Introduction

Cancer is a pugnacious disease with a low median survival rate. Cancer ranks as a leading cause of death and an important barrier to increasing life expectancy in every country of the world [1]. The data estimates that in 112 of 183 countries rank first and second and further 23 countries rank third and fourth [2]. Quick detection of cancer is the key toward accurate diagnosis and prediction of prognosis of the disease, which step up the patient's survival rate [3].With the advancement in the technology of AI, the problem can be solved easily and quickly. Using this extensive and extraordinary intelligence with machines will provide an edge for proper diagnosis and prognosis of the cancer [4] (Fig. 1).

It has been estimated by the global company for testing on cancer so as 1 within 5v humans spread cancer once in their whole life globally, and 1 in 8 men and 1 in 11 females pass away from having cancer. [2]

There are 16 million cancer victors in the USA according to estimates done by the National Cancer Institute (NCI). This figure is set to rise to 26 million by 2040. Unfortunately, 50% of the cancer sufferer will undergo drug-connected side effects which involved cardiovascular dermatologist gastrointestinal and neurological.

The prospective for AI as a strong positive radical in cancer pharmacotherapy is stupendous. In wide tasks, the appeals (s) are good to fall in the three main parts:

(1) Medication selection and toxicity prediction.

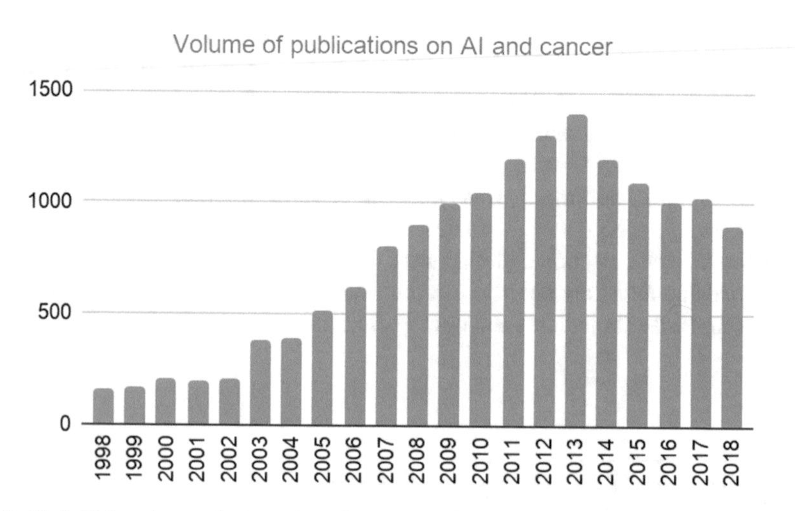

Fig. 1 Embellishes the steady extension of the use of AI technology in cancer analysis over the past two decades

(2) Medication pairing
(3) Medication repurposing.

Emphasizing the attractive potential of AI-based approaches in terms of drug selection determines whether they respond favorably to the treatment of a particular cancer [5].

Current advances in radiology, also computer systems, have follow in a quick rise in the prospective use of AI for numerous quantity of functions in the field of breast radiology. These AI appeals highly applied for detection and forecast surgery response and prognosis. Breast cancer presently is done using a radiology methods called CADe and CADx. Study shows a multiple work has been done in this field in the several last decades [6].

Today one third of all male cancers is contributed by lung cancer and prostate cancer. Cancer is detected often at the later stages when it has damaged or compromised one or more essential body organs. The key factor for cancer treatment is the early diagnosis of the cancer (Table 1).

Study on breast cancer by Sun et al. in China in 2018 in 1980 patients done using multimodel DNN. Study by Park et al. and Delen et al. in the USA in SEER in the years 2013 and 2005 using semi-supervised learning model and ANN and DT, respectively, having accuracy: DT (93.6%), ANN (91.2%).Another study on gastric cancer, Biglarian et al. in hospital in Iran in 2011 done using Cox Proportional Hazard, ANN having accuracy of 83.1% and by Zhu et al. in hospital in China in 2013 on 289 patients done using ANN technique having accuracy of 85.3%. Lynch et al. did study lung cancer on 10,442 patients in MOCDTBS in the USA in 2017 using the cross-breed dummy of Relief GA ANFIS whose validity was 93.81%. Another study done by Sepehri et al. in SEER in France in 2018 on 101 patients using GBM SVM technique, having accuracy RMSE (32,15.05) for GBM and SVM. Yu et al. did study lung cancer on 168 patients in a hospital in Italy using SVM with RFE and RF techniques and having accuracy of 71% and59%. Cancer is hard to detect at early stages.

2 Related Work

2.1 *Clinical Deployment of AI for Prostate Cancer Diagnosis*

Prostate cancer is detected by Health company name The Lancet Digital Health and by some persons Liron Pantanowitz and colleagues. They used artificial intelligence (AI) for determine it in entire slide picture of core needle biopsies. Their work provides insights into how these algorithms result and various pre-analytical difference is also done. Also, study shows the computer-aided design (CAD) tools can impact operation of pathology in the upcoming future. Deep learning models are trained by the CAD system in which cases are tested.

Table 1 Illustrates the data collected by authors on detecting different types of cancers

Type of cancer	Authors	Year	Country/region	Number of patients in study	Age (Years)	Study population	Methods	Results
Prostate cancer	Kuo et al.	2015	Taiwan	100	75	Hospital	Fuzzy neural network	
	Zhang et al.	2017	USA			TCGA	SVM model	Average accuracy (66%)
Breast cancer	Sun et al.	2018	China	1980	61		Multimodal DNN	
	Park et al.	2013	USA	162,500	N/A	SEER	Semi-supervised learning model	
	Delen et al.	2005	USA	433,272	60.61	SEER	ANN and DT	Validity: DT (93.6%), ANN (91.2%)
Gastric cancer	Biglarian et al.	2011	Iran	436	58.43 ± 13.02	Hospital	Cox proportional hazard, ANN	TP (83.1%),
	Zhu et al.	2013	China	289	63.20 ± 10.75	Hospital	ANN	TP: ANN (85.3%)
Lung cancer	Lynch et al.	2017	USA	10,442	N/A	MOCK TBS	Hybrid model of relief F-GA-ANFIS	Accuracy(93.81%),AUC (0.9)
	Sepehri et al.	2018	France	101	N/A	SEER	GBM, SVM	RMSE(32,15.05) for GBM, SVM
	Yu et al.	2016	Italy	168	N/A	Hospital	SVM with RFE and RF	Validity (71%, 59%)

A study on prostate cancer by Kuo et al. in a hospital in Taiwan in 2015 was done using a fuzzy neural network technique. Another study on prostate cancer by Zhang et al. in TCBA in the USA in 2017 using the SVM model shows 66% accuracy

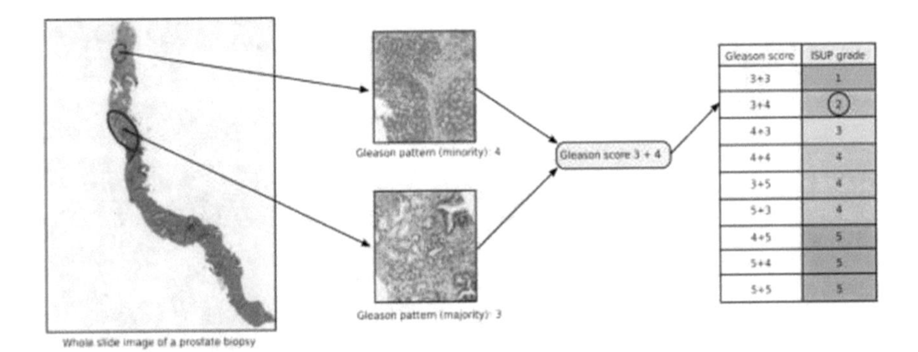

Fig. 2 Illustration of the Gleason grading process for an example biopsy containing prostate cancer [7]

In the survey and according to Fig. 2, the Gleason grading system helps to proceed the diagnosis of males with thyroid cancer using samples from a thyroid biopsy. Also with another parameters, it is incorporated into a strategy of thyroid cancer staging which predicts diagnosis and helps guide surgery. On the other side, tissue samples are needed from the thyroid and examined in the microscope. The process refers to how uncommon your thyroid cancer cells shows and how likely the cancer is too early and how swiftly it spreads. An inner Gleason grade means that the cancer is slower extended and not aggressive [8]. Thyroid cancer will be diagnosed in Australia and which have approximately the same number [9]. The introduction of AI brings a revolution in prostate cancer. Thyroid cancer proceed a large healthcare burden, thus being ideal for AI variation. Eventually, males could be naturally registered to undergo a machine-read MRI (Fig. 3).

2.2 Breast Cancer Screening

Breast cancer is the most common cause of cancer death in women. Over half a million women in the world die due to breast cancer. As in most forms of cancer, this type of cancer also requires early detection to be able to cure the disease. To reduce breast cancer-related impermanence, screening of breast cancer is done. This process is called mammography and has been introduced worldwide to detect breast cancer at the early stages. With screening and improvements in the treatment, the breast cancer rate mortality rate has reduced to 30%. Unfortunately, breast cancer still remains the number one cause of death of women due to cancer in 2017. The USA estimated 252,710 new cases and 40,610 deaths due to breast cancer. Physical examination of breasts is done to detect breast cancer in women; this method is known as mammography. Through this process, it is likely to notice a lump in the breasts and may detect other lesions in the breast. Another method is to take a sample of fluid for microscopic analysis called biopsy. With faster advancements in the technology

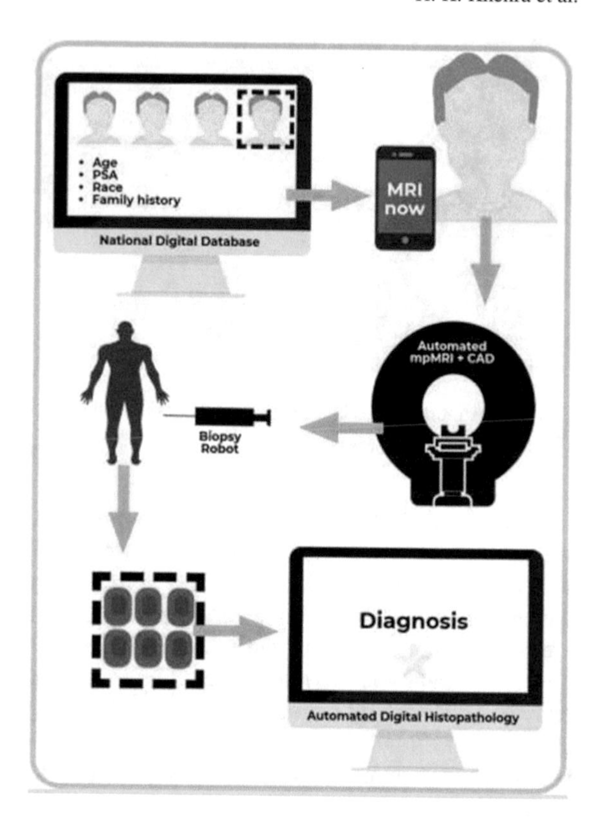

Fig. 3 Diagram illustrating a suggest future-affected intelligence-based pathway for thyroid cancer diagnosis

of artificial intelligence, breast cancer screening is being impacted at a higher rate. Digital mammography being the method used for screening of breast cancer. Various AI algorithms being developed for screening [10]. Studies are being conducted in development of AI algorithms in improving the diagnosis of breast cancer [11]. Algorithms are based on CNN are being used for mammography and classification of lesions based on DBT [12]. Google-trained AI being trained on 91,000 mammograms of women in the UK and USA performs better than a radiologist in detecting breast cancer. Also Google Health team trained AI on another 28,000 mammograms and resulted in lesser errors [13].

Many cancer detection algorithms using artificial intelligence have been made for mammography. Even though validation in a truly representative screening cohort is still lacking, many software algorithms are now performing at a level comparable to radiologists when assessing mammograms.

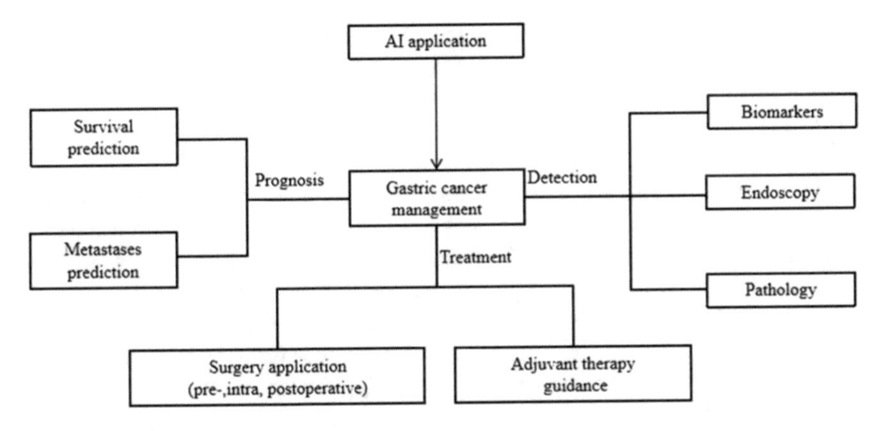

Fig. 4 Shows that artificial applications play a very important role in gastric cancer management

2.3 *Gastric Cancer*

Gastric cancer is also commonly known as stomach cancer. It is the type of cancer that develops in the lining of a stomach. It can occur in both males and females. A gastric atrophy examination is done for detection of abnormal tissue in the lining of the stomach. A biopsy of the abnormal tissue is taken and sent to a pathologist for histological examination of the tissue under the microscope. Biopsy with further histological analysis is used for confirming the presence of cancerous cells in the stomach. Due to its high-rise computational and efficiency skills, various AI algorithms are being developed to detect gastric cancer. Due to non-specific symptoms of gastric cancer, it cannot be detected at early stages.

This Fig. 4 shows us how AI applications help us to manage gastric cancer.

For treatment, we have surgery application (pre-intra and postoperative) and adjuvant therapy guidance. For detection, we have three phases: biomarkers, endoscopy, and pathology. For Prognosis, we have two predictions survival and metastases. In this way, AI applications have the gargantuan role to play in gastric cancer management.

3 Challenges and Future Scope

Artificial intelligence, AI, seems over be making a mark in using data of medical sciences and drawing significant extractions from it. AI is advancing at a very faster pace in every aspect of scientific research. As AI is developing, and its capabilities are increasingly high, the transition in AI into clinical routine practice is emerging. The use of ML and artificial neural networks in artificial intelligence has made many solutions to be more quick and precise. With the use of AI algorithms, computers are able to do thousands of biopsies in just a matter of seconds; whereas, doing a single biopsy can take 10 days by a pathologist. By this, we can say machines can do work at a fast pace and with greater accuracy level. From diagnosis to prognosis, AI is becoming more accurate by describing results and survival rates in a precise manner. Imaging analysis with AI needs substantial improvement in demonstrating comparable efficiencies. With human-controlled studies, these can supplement the clinical workflows. For widespread use in daily routines of medical practice, clinical utility in trials and education of clinicians, technologists and physicians are required. The detection of cancer at primary stages and that too more accurately helps in improving the predictions of various complicated cancers. Next generation vision and final output: As per the growth and claims mentioned above, the advancement of knowledge in the field of furious application of precise type of oncology (different type of cancers), this type of precision requires further advancement in AI and oncology to reduce the gap and also addressing the future references. Artificial intelligence will be a great help to overcome the challenges in the field of different types of cancers. AI can provide advanced preventive measures to overcome cancers. Machine learning provides a positive hand with AI to give a great platform for the treatment of different types of cancers. Artificial intelligence will be a great tool for the doctors who perform chemotherapy on cancer patients. With the help of AI, doctors can perform specific treatment easily rather than to go for unnecessary surgeries.

4 Conclusion

The healthcare system is the fundamental part of society guaranteeing that each and every individual gets fruitful diagnosis and treatment. Effective researches are conducted to ensure that individuals battle new diseases, viruses, and other disorders easily. Health is a significant and critical factor for deciding the capabilities of any individual as healthy life is a need for everyone. Today, the world has become automated and swiftly advancing in the field of technology so emerging new technology, machinery, robots, etc. in this field has become essential and inevitable. Diagnosis, treatment, medicine prescription, etc. have become more efficient and less time-consuming due to this advanced technology. It also reduces the need of a trained workforce and works swiftly and rapidly so that the problem is solved in less time. Thus, it is very important and compulsory to implement various technologies in health care for the advancement of medical sciences. So, in this paper, we discussed how AI is used in an outstanding and remarkable way for treating various cancers like prostate cancer, breast cancer, and gastric cancer. This noteworthy technology has provided a different impact and will regular start to cancer diagnosis among male in Australia 2020 is shown in this paper. There is also an infographic illustration of an advance future AI build pathway for thyroid cancer detects is shown AI which is used in the treatment and detection of breast cancer. Implementation of the technology of artificial intelligence in prognosis of gastric cancer is discussed. Survey paper also discusses the challenges and future scope of AI in cancer and a complete diagram for AI in precision oncology. So, in the end, the survey paper concluded that this dynamic, super-powerful technology can make significant improvements in detection of cancer at early stages, support advanced clinical decision-making, and lead the way to superior health results. There are high possibilities that artificial intelligence in health care is minimizing health problems and making life at ease.

References

1. Sung H, Ferlay J, Siegel RL, Laversanne M, Soerjomataram I, Jemal A, Bray F (2021) Global cancer statistics 2020: GLOBOCAN estimates of Incidence and Mortality international for thirty six Cancers in 185 Countries, CA Cancer J Clin, pp 1–41
2. https://www.uicc.org/news/globocan-2020-new-global-cancer-data
3. Huang S, Yang J, Fong S, Zhao Q (2020) Artificial intelligence in cancer diagnosis and prognosis: opportunities and challenges
4. https://www.kaggle.com/c/prostate-cancer-grade-assessment
5. Mirnezami R (2020) Cancer detections and treatment decisions using artificial intelligence. Artif Intell Healthc, pp 117–141
6. Patil S, Moafa IH, Alfaifi MM, Abdu AM, Jafer MA, Raju L, Raj AT, Sait SM (2020) Reviewing the role of artificial intelligence in cancer. Asian Pac J Cancer Biol 5(4)
7. Janowczyk A, Leo P, Rubin MA (2020) Clinical deployment of AI for prostate cancer diagnosis
8. Prostate cancer in Australia statistics. https://www.canceraustralia.gov.au/affected-cancer/cancer-types/prostate-cancer/prostate-cancer-australia-statistics
9. Dustler M (2020) Evaluating AI in breast cancer screening: a complex task. Lancet Digit Health 2(3):E106–E107
10. Kim HE, Kim HH, Han BK, Kim KH, Han K, Nam H, Lee EH, Kim EK (2020) Changes in cancer detection and false-positive recall in mammography using artificial intelligence: a retrospective, multireader study
11. Ono H, Yao K, Fujishiro M, Oda I, Nimura S, Yahagi N, Iishi H, Oka M, Ajioka Y, Ichinose M, Matsui T, Guidelines for endoscopic submucosal dissection and endoscopic mucosal resection for early gastric cancer
12. Antaa JA, Dinis-Ribeiro M (2020) Early gastric cancer and Artificial Intelligence: Is it time for population screening? Best Pract Res Clin Gastroenterol
13. Yoon HJ, Kim JH (2020) Lesion-based convolutional neural network in diagnosis of early gastric cancer. Clin Endos 53(2)

Soft Robotics-Fingered Hand Based on Working Principle of Asymmetric Soft Actuator

Hiep Xuan Trinh, Phung Van Binh, Le Duc Manh, Nguyen Van Manh, and Ngo Van Quang

Abstract This study presents a prototype of the soft-fingered hand based on the operating principle of an asymmetric soft actuator, which is designed as a tube with the different thickness and stiffness of two sides and is activated by the pneumatic actuation. Based on such design, four soft fingers were fabricated from the silicone rubber material by using the molding method and then were assembled to the connectors and base to complete the soft-fingered hand. The ability of the proposed soft-fingered hand is validated by conducting the simulations and gripping experiments. The simulation and experimental results show that under the pneumatic activation, the soft fingers have a good bending deformation, and the proposed hand can grip several objects with different shapes, sizes, weights.

Keywords Soft robotics · Soft gripper · Asymmetric soft actuator

1 Introduction

Soft robotics has been introduced as a novel frontier in robotic research. In general, soft robotics is fabricated by soft materials such as silicone elastomers, urethane, hydrogels, and so on [1]. The deformable properties of soft material enable soft robotics is much inherently compliant and withstand large strains in comparison with the traditional rigid robot. Thus, soft robotics has many advantages such as adaptable to the unpredicted surrounding environment, safe for human interactions, lightweight, and so on [2, 3]. The area research of soft robotics is a wide range from locomotion, manipulation, actuating, or sensing [4–6]. In which, the development of the soft-fingered robotics hand recently is an emerging area in soft robotics research and achieved significant results. The soft robotics hand has exposed many benefits on the dexterous manipulation, grasping delicate objects in unstructured environments [7, 8]. Many studies focused on proposing, developing soft-fingered

H. X. Trinh (✉) · P. V. Binh · L. D. Manh · N. V. Manh · N. V. Quang
Le Quy Don Technical University, 236 Hoang Quoc Viet, Cau Giay, Ha Noi, Viet Nam
e-mail: hieptx@mta.edu.vn

A. Kumar et al. (eds.), *Proceedings of the International Conference on Cognitive and Intelligent Computing*, Cognitive Science and Technology,
https://doi.org/10.1007/978-981-19-2350-0_9

hand for adapting various tasks. One of the first authors who proposed the gripping mechanism based on the softness design is Suzumori et al. In [9], they used four micro-actuators to construct a soft gripper for grasping various objects. Pneumatic and hydraulic actuating mechanics have recently widely utilized in the design of soft gripper, due to their advantages such as low cost, high gripping efficiency, simple fabrication, easy control, and fast responses [10, 11]. For instance, the authors in [12] presented the soft robotic gripper with the bellow soft pneumatic actuator for delicate manipulation and collecting the sample fragile species on the deep reef. Park et al. [13] proposed the soft gripper based on the electrohydraulic actuator, fabricated by polyethylene film, and silicone material. However, most pneumatic or hydraulic mechanics-based soft-fingered hand consists serial chambers with specially designed thin walls to generate the gripping shape under pressurization activation. The thin walls design causes easily the failures such as torn, wears, or bulge. Moreover, the construction with many chambers leads to the complication of mathematical models and the difficulty to fabricate and calculate exactly the curvature shape or contact gripping force. In [14], Ho et.al proposed the novel concept of soft-fingered hand with contact feedback based on the idea of morphological computation with the unbalanced deformation mechanism. In this design, the soft finger structure includes a softer layer much thicker than a stiffer layer and the gripping energy is generated from the elastic energy of the pre-stretched softer layer. Nonetheless, this design requires the external motor to pull or release the tendon string for controlling the opened state of soft fingers.

In this paper, based on the operating principle of asymmetric soft actuator, we present the soft-fingered hand that is easy to design, fabricate, and model. The robotic hand consists of four soft fingers; each finger is a simple asymmetric tube and is activated by air pressure. The remainder of this paper includes the principle, design, fabrication, simulation, and preliminary gripping test of the soft-fingered hand.

2 Principle and Design of Soft-Fingered Hand

2.1 Principle of Soft-Fingered Hand

In this study, the soft finger is operated based on the principle of the asymmetric soft actuator that has a cross-section of the asymmetric hole [15]. Due to the different thickness and stiffness between two sides of the asymmetric tube, under the actuation of internal pressure, the actuator is curved in the direction of the thicker side. The cross-section of the soft finger is circular, with different thicknesses t_1, and t_2 of the top and bottom side, where is smaller than (Fig. 1). The thicker side is partly flat to improve the gripping contact area. The bending effect of the soft actuator can be considered as the bending of the cantilever beam. In which, the actuator is closed at the free end; the air pressure is input into the soft actuator from the fixed end through a small tube. Under the pneumatic actuation, the different thicknesses lead

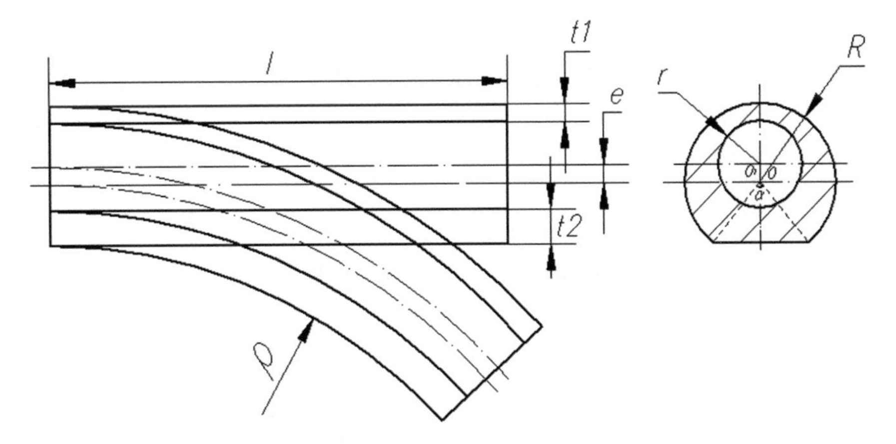

Fig. 1 The curved principle of the asymmetric soft actuator and its cross-section

to the different expansions between two sides, and the eccentricity causes the bending moment. The bending curvature of the soft actuator depends on its geometrical design and the pressure value.

2.2 *Design of Soft-Fingered Hand*

The robotics hand consists of four soft fingers, a rigid connector, and a rigid base (Fig. 2). The rigid connector is an assembly of two parts. It has a hole with a diameter of 20*mm* to fix the soft finger. Four rigid connectors are assembled to rigid based by using the screws. All parts of the rigid connectors and rigid based are designed by the SolidWork software, then fabricated by using a 3D printer. The soft finger is an eccentric tube; a free end is closed, and the fixed end has a small hole for directing the compressed air into the tube. The prototype soft finger has a dimension of designed parameters as $l = 100$, $R = 8$, $r = 5$, $e = 2$, $t_1 = 2$, $t_2 = 4mm$. The overall dimension of the soft finger is similar to the size of the human finger.

3 Fabrication

The soft fingers were made from silicone rubber of RTV 225 and fabricated with two 3D printing molds. The first mold is the assembly of three parts and is used to fabricate a main part of the soft finger (Fig. 3a, b). The remained mold (Fig. 3c) is used to seal the fixed end of the finger. The fabrication process of the soft finger by molding method is depicted as in Fig. 4 and can be summarized as follows. First, two components of the liquid silicone rubber were mixed with the ratio of 1:1; then,

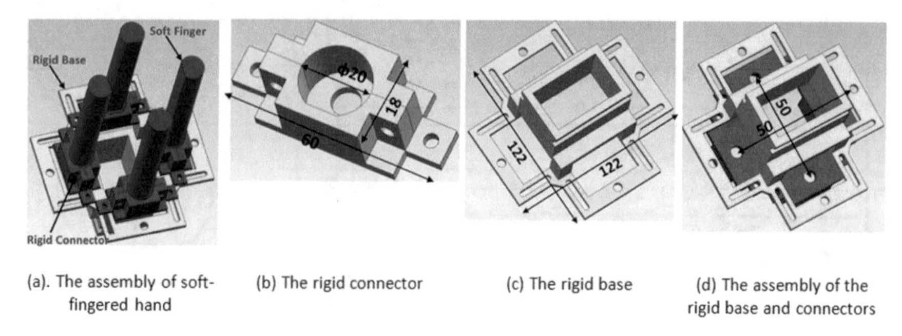

(a). The assembly of soft-fingered hand (b) The rigid connector (c) The rigid base (d) The assembly of the rigid base and connectors

Fig. 2 The design of soft-fingered hand

the mixed liquid silicone is poured into the assembly mold and is cured at room temperature after ten hours. After the curing process, by removing the assembly mold, we get the main part of the soft finger. At this stage, the soft finger still has an open end. This end was closed with the same mixed silicone by using the second mold and a similar curing method. Finally, after taking out that mold, we get the soft finger. The soft fingers were then fixed to the connector and the base to complete the soft-fingered hand.

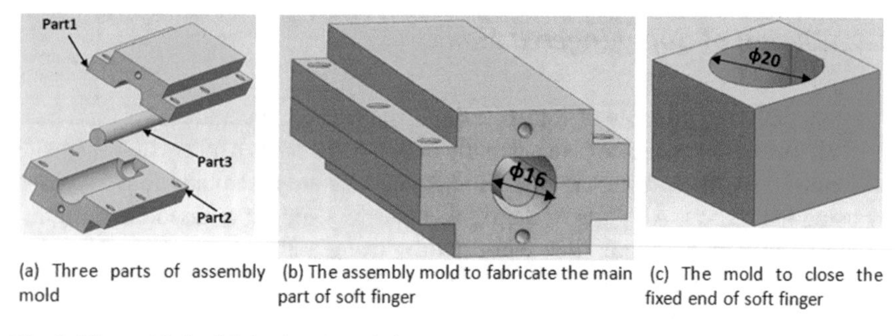

(a) Three parts of assembly mold (b) The assembly mold to fabricate the main part of soft finger (c) The mold to close the fixed end of soft finger

Fig. 3 The molds for fabricating the soft finger

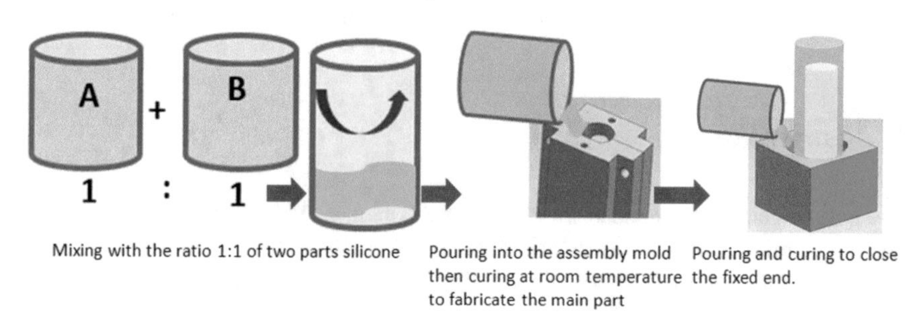

Mixing with the ratio 1:1 of two parts silicone Pouring into the assembly mold then curing at room temperature to fabricate the main part Pouring and curing to close the fixed end.

Fig. 4 The schematic of fabricating process

4 Simulation and Preliminary Gripping Test

To clarify the principle of the soft finger and the effect of geometrical parameters on its bending curvarute, we conducted the simulation with the commercial software Abaqus. The soft finger was modeled as a deformable solid with the hyper-elastic material. The Yeoh model with the parameters of $C_{10} = 0.1$, $C_{20} = 0.02$, $C_{30} = 0.0002$, $D_1 = D_2 = D_3 = 0$ was used in the simulations [16]. The FE model was meshed with the linear hexahedral elements of type C3D8R. The bending angle α of the soft finger is determined through the positions of two nodes at the fixed and free ends (Fig. 5). To investigate the effect of the designed structure on the bending curvature of the soft finger under the air pressure activation, we conducted the simulations with various cases of the finger's geometrical design as shown in Table 1. The bending angle α of these soft fingers with different pressure values are shown in Fig. 6.

From Fig. 6, we can see that when the pressure value increases, the bending angle of the soft finger does not increase linearly, and it grows up noticeably with high pressure. It is due to the physical nonlinearity of the silicone material and the geometrical nonlinearity of the large deformation. In the comparisons of the bending angle with the different designed parameters under the same pressure conditions, we see that the values of bending angle rise with the increase of the eccentricity e, thickness t_2, and the decrease of thickness t_1.

We conducted the experiments to test the preliminary gripping ability of the soft-fingered hand. The hand was mounted onto the SCORRA ER-14pro robotics arm. The soft fingers were activated by air pressure through the pump system. Several

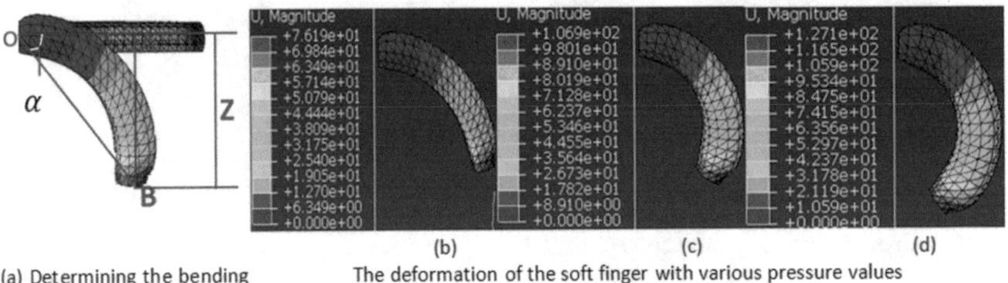

Fig. 5 The deformation of soft finger

Table 1 The various cases of finger's geometrical parameters.

Parameter	R (mm)	r (mm)	e (mm)	t_1 (mm)	t_2 (mm)
Case1	8	5	2	2	4
Case2	8	5	2.4	1.6	4.4
Case3	8	5	2.8	1.2	4.8
Case4	8	5	3	1	5

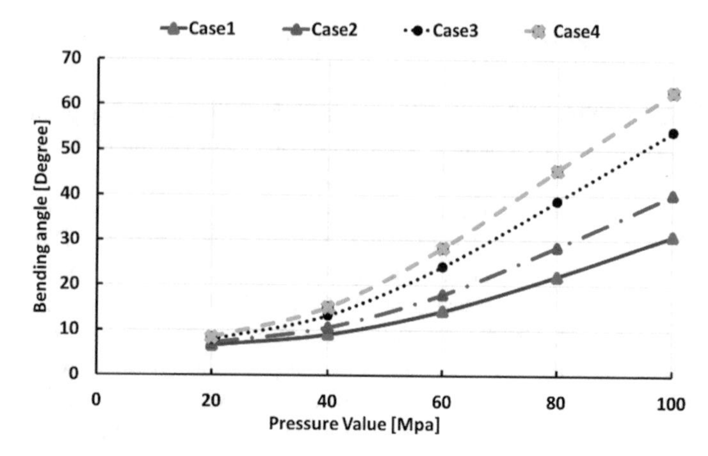

Fig. 6 The bending angle's values with different pressure of various case designs

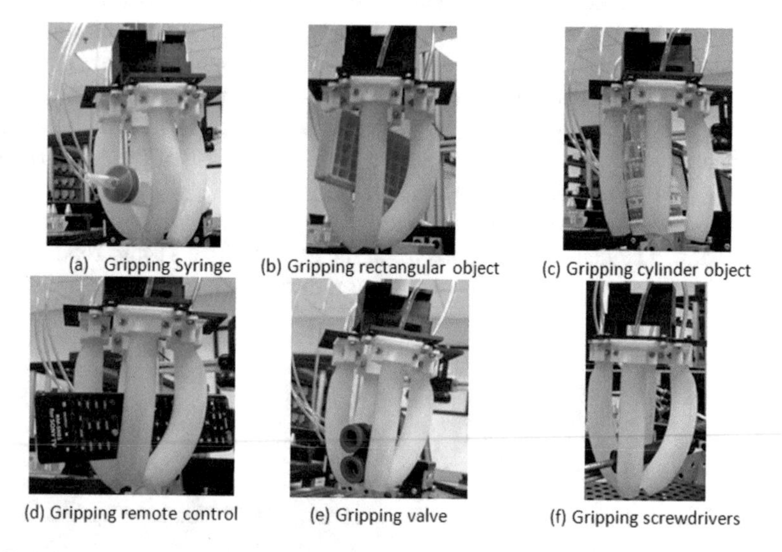

Fig. 7 Gripping tests of the soft-fingered hand

objects with different shapes, sizes, and weights were used for gripping tests, and the results were shown in Fig. 7. This figure indicates that the proposed hand has a good gripping ability with different objects.

5 Conclusion

The study presented in this paper is our first step toward a soft robotics hand that has a simple design, easy to fabricate, modeling and control. In the future, we will address some aspects to improve the performance of the soft-fingered hand. First, we will focus on proposing a novel mathematical model to estimate the bending deformation of the soft finger under the pneumatic actuation. This model would build based on the deformation laws of the nonlinear material and consider the geometrical nonlinearity of the large deformation. Then, based on the mathematical model, optimization algorithms would be proposed to improve the soft finger's design for obtaining desired bending shapes. That attempts to imitate the shape of human fingers when handling operations. Finally, a rigid nail will be embedded to the free end to limit the tip inflation, which will make better pick-and-place performance of the soft finger. The technical pre-stressed deformation will be used to extend the gripping area of the soft-fingered hand.

References

1. Rus D, Tolley M (2015) Design fabrication and control of soft robots. Nature 521:467–475
2. Chiwon L, Myungjoo K, Yoong K et al (2017) Soft robot review. Int J Control Autom Syst 15(1):1–13
3. Iida Fumiya, Laschi Cecilia (2011) Soft robotics: challenges and perspectives. Procedia Comput Sci 7(7):99–102
4. Lin HT, Leisk GG, Trimmer B (2011) GoQBot: a caterpillar-inspired soft-bodied rolling robot. Bioinspir Biomim 6:026007
5. Calisti M, Giorelli M, Levy G, Mazzolai B, Hochner B, Laschi C, Dario P (2011) An octopus-bioinspired solution to movement and manipulation for soft robots. Bioinspir Biomim 6:036002
6. Stefan M, Benjamin T, Randall M et al (2010) Highly sensitive flexible pressure sensors with microstructured rubber dielectric layers. Nature Mater 9:859–864, 036002
7. Amend J, Cheng N, Fakhouri S, Culley B (2016) Soft robotics commercialization: jamming grippers from research to product. Soft Robot 3(4):213–222, 036002
8. Deimel R, Brock O (2015) A novel type of compliant and underactuated robotic hand for dexterous grasping. Int J Robot Res
9. Suzumori K, Iikura S, Tanaka H (1991) Development of flexible microactuator and its applications to robotic mechanisms. In: Proceedings of IEEE International Conference on Robotics and Automation Sacramento, pp 1622–1627
10. Boyras P et al (2018) An overview of novel actuators for soft robotics. In: Actuators, vol 7, p 48
11. Robertson M et al (2016) Soft pneumatic actuator fascicles for high force and reliability. Soft Robot 4(1):23–32
12. Galloway KC, Becker KP, Phillips B, Kirby J, Licht S, Tchernov D, Wood RJ, Gruber DF (2016) Soft robotic grippers for biological sampling on deep reefs. Soft Robot 3(1):23–33, 036002
13. Park T et al (2020) Electrohydraulic actuator for a soft gripper. Soft Robot 7(1):68–75

14. Ho V, Hirai S (2017) Design and analysis of a soft-fingered hand with contact feedback. IEEE Robot Autom Lett 2(2):491–498, 036002
15. Ganesha U et al (2010) Robotic gripper driven by flexible micro actuator based on an innovative technique. In: IEEE workshop on advanced robotics and its social impacts, KIST, Seoul, Korea
16. Martins P et al (2006) A comparative study of several material models for prediction of hyperelastic properties: application to silicone-rubber and soft tissues. Strain 42:135–147, 036002

Indian Food Image Classification with Transfer Learning

B. E. Manjunathswamy, G. Shreyas, M. P. Tejaraj, and A. Fatima Shafaq

Abstract As image classification is the process of classifying and labeling groups in pixels or vector by image using certain rules. In today's world, the convolutional neural network (CNN) is the most widely utilized technology for image separation. The image for labeling and separation in this part comes from an Indian cuisine database that employs a number of transfer learning techniques. Because food is such a vital component of human life, providing us with essential nutrients for growth and energy, it is imperative that everyone maintain track of their eating habits, as they can be harmful to the body if ignored. A healthy lifestyle is exemplified by the food classification. In this paper, pre-trained models are employed instead of starting from scratch, which saves time, money, and calculations while also providing good results. Indian food database of the 12 classrooms with 710 pictures in each class used for training and certification.

Keywords Convolution neural network · Food classification · Dataset

1 Introduction

Computer vision and neural networks are a hot new IT machine learning technology. With the development of neural networks and the ability to read images such as pixel values, many companies rely on this method to get more details. Support vector machine (SVM) and decision tree (DT) are some of the methods that can be used to categorize an image; however, KNN separation is a global feature utilized in artificial neural networks, SVM, and random forest planning may fail to discriminate when there are big databases. Because the convolution neural network can readily manage a vast quantity of data and accurately identify it, it was given priority over the program. In order to train CNN for image classification, you can either start from scratch or use the notion of transfer learning. Transfer of reading by an in-depth learning process in which the model is trained to read and retrieve information and apply it to the

B. E. Manjunathswamy (✉) · G. Shreyas · M. P. Tejaraj · A. F. Shafaq
Don Bosco Institute of Technology, Bangalore, Karnataka, India
e-mail: manjube24@gmail.com

A. Kumar et al. (eds.), *Proceedings of the International Conference on Cognitive and Intelligent Computing*, Cognitive Science and Technology,
https://doi.org/10.1007/978-981-19-2350-0_10

following process. Separation is gaining less popularity due to their knowledge of health and nutrition among people.

According to the World Health Organization, 1.9 billion older persons over the age of 18 years are overweight. Since 1975, the global obesity rate has risen. By 2020, 39 million children under the age of five were overweight or obese. According to statistics, 95% of individuals do not follow any healthy eating plans because they are strong and restrict people from eating their regular food [1, 2]. Older people who want to take care of the food they eat, patients who want to look after their health through proper nutrition on a variety of dietary restrictions and especially young people who want to track calorie-rich foods to maintain body fat, the importance of food segregation has increased. Before that it was very difficult to count calories food, however, thanks to CNN has been easier. Through this study, efforts were made to divide the Indians food pictures in their lessons using transfer learning. Because of their efficiency in learning and categorizing complicated features, image separation using in-depth reading approaches such as the convolution neural network receives a lot of attention. Comparisons were made between models in terms of accuracy and loss of validation. Phase 2 represents related work; Phase 3 represents methodology, Phase 4 system design, and Phases 5, 6 conclusion.

2 Related Work

In-depth, reading has proven to be extremely useful when coping with enormous amounts of data. In the field of food image identification, there have been numerous developments in the literature in recent years. Many attempts have been made to recognize and identify local food photographs using a proposed novel configuration for an in-depth learning network. The study employed a local Malaysian food dataset that was gathered from publicly available Web sources such as regularly used search engines. In comparison with the previous ways, CNN's method appeared to gain points with far greater accuracy. The depth of the network, on the other hand, was clearly important for the model's higher performance [1]. This study's Indian food database was compiled from many sources, including this one. Identification, the only release, and segmentation of picture data, for example, are provided in a predictive model for the separation of Thai photos by fast food. The authors of [3] used Inception v3 with the Thai fast food dataset (TFF) to separate images with an accuracy of 88.33 percent. Several pre-trained TFF models separate the Inception v3 model from Google, the well-known Oxford Visual Geometry Group's VGG16 and VGG19, based on the residual network used here (VGG).

Because they perform effectively with a huge number of photographs and provide high accuracy results, these types are widely employed in image recognition, classification, and processing. Only 3×3 and 2×2 layers are employed throughout the network in VGG16 and VGG19. The VGG model demonstrates the importance of network depth, as deep networks learn more features and produce better outcomes. The Softmax activation separator layer is followed by two completely linked layers with 4096 layers each. The image input size for both the VGG16 and VGG19 networks is 224 by 224 pixels. Both models are noteworthy in their simplicity, with only three resolution layers connected to improve network depth. The number of weight layers in the network is represented by the digits "16" and "19." Due to performance and memory limits, the upper levels of the VGG16 and VGG19 formats have additional 3x3 layers that are more versatile than the lower layers. Lower layers have more localization and so control storage and computer difficulties, whereas upper layers with several specification levels slow things down and use more GPU memory. To address this, contemporary models such as Google Startup and ResNet limit local adjustment before adding any convolutional layers early in the process [2].

The architecture designed by Szegedy et al. [1] is referred to as GoogleNet. Start with a deep development of neural convolution created by code. The first module's objective is to operate as a multi-level compiler, integrating 1*1, 3*3, and 5*5 convolutions into a single network module and passing the output on to the next layer in the network. It tied for first position in the ImageNet Large Scale Visual Recognition Challenge 2014 (ILSVRC14) with acquisition and division. The imaginary effect of dense, easily accessible things is thought to be created through networks of viewing and covering [3, 4]. In comparison with other trustworthy networks, minor benefits were seen with a little forethought. In this study, partitions were created starting with V3 (the most recent version).

Food-11, Food-100, and even online archives are used to create the database. The results suggest that well-designed transfer learning models outperform creating a network from the ground up [5].

3 Proposed Methodology

Indian Food Database: The Indian food database was used in our research. It is divided into 12 main culinary categories, each with 700 sample images. Because there are images in which there is a lot of food in one image, the database naturally has a lot of noise. There are a lot of colors in the image samples, and some of them are labeled erroneously. Images of sample cuisine from the Indian food database are shown in the diagram below (Fig. 1).

The CNN model is supplied with image setup training, and test databases are used for validation. The suggested route is depicted in Fig. 2.

Training the CNN Classifier Using Pre-trained Models

Even with normal data augmentation processes, a large portion of computer vision's issues is that good accuracy is difficult to accomplish. Over the model size, create these networks with many parameters. As a result, rather than qualifying, transfer of learning is used. The initial layers will gain edges, the middle levels will gain shape, and the last layers will gain some high-quality data features as a result of the transfer. Many computer vision tasks like picture categorization can benefit from this

Fig. 1 Indian food dataset samples

form of transfer learning. The photos were utilized for training in 80% of the cases and verification in 20% of the cases. 30 epochs with a 32-bit batch size were used to run each model. On the other side, we employed the stochastic gradient descent optimizer to regenerate and tighten the model limits so that loss activity could be reduced.

Validation and Testing: The model is first trained on a train database (a sample of data used to model the model), then validated on a validation dataset (a sample of data used to provide impartial testing of model equity in the training data while adjusting hyperparameters), and finally tested on test data.

Calories Extraction from a Classified Image: Finally, we can use our classification to estimate the calorie amount of food obtained through the Internet. Web cutting may be done with a correct Python for any text to extract the nutritional information of a unique image from the Web and provide it to the user.

Convolutional Neural Network (CNN)

Convolutional neural framework is one of the principal categories for the photos affirmation and pictures portrayals [6]. Articles disclosures, affirmation faces, etc., are a bit of the regions where CNNs are commonly utilized. Figure 3 shows the neural network with various convolutional layers. In uncertainty, the possibility of significant learning CNN models *t* can be used for train and attempted, every data picture will be adhered to the course of action of convolution layers with procedures (Kernals), Pooling, totally related layers (FC) by applying Softmax work can arrange an article with probabilistic characteristics runs some place in the scope of 0 and 1. The underneath figure is a complete stream of CNN to process an information picture and requests the articles subject to values.

Yann Le Cun's creation of the CNN in 1994 resurrected artificial intelligence and deep learning. Since then, we have come a long way in this discipline. The first neural network, known as LeNet5, had a validation accuracy of only 42 percent.

Fig. 2 Proposed methodology

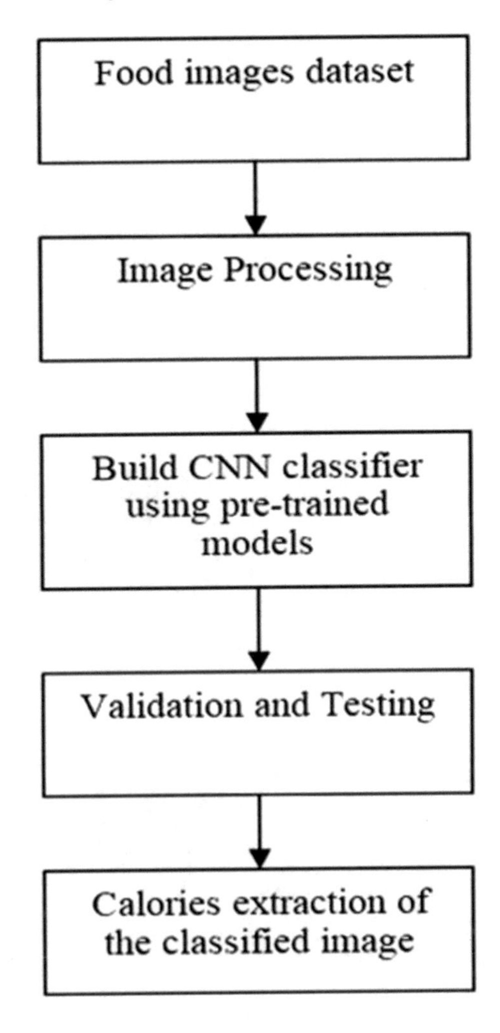

Almost all of the world's largest IT companies now rely on CNN for better results [7]. Before delving into the concept of CNN's "functionality and working," it is important to understand how the human brain recognizes an object despite its differences in features. A complex layer of neurons exists in our brain, each layer stores some information about the object, and the neurons extract all of the object's features and store them in our memory. The next time we view the same object, the brain compares the stored features to recognize it. However, it is easy to mistake it for a basic "IF-THEN" function, which it is to some extent, but it has an additional characteristic that gives it an advantage over other algorithms: self-learning. While it cannot replace a human brain, it can give it a run for its money. To detect the calories in food, the image is processed using the basic CNN. The data training in our CNN model must adhere to the following guidelines (Fig. 4).

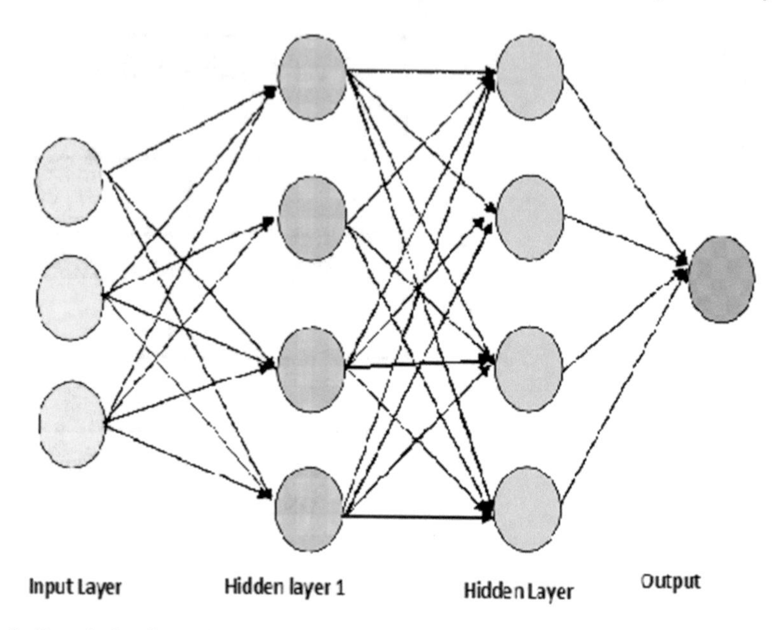

Fig. 3 Convolutional neural network layers

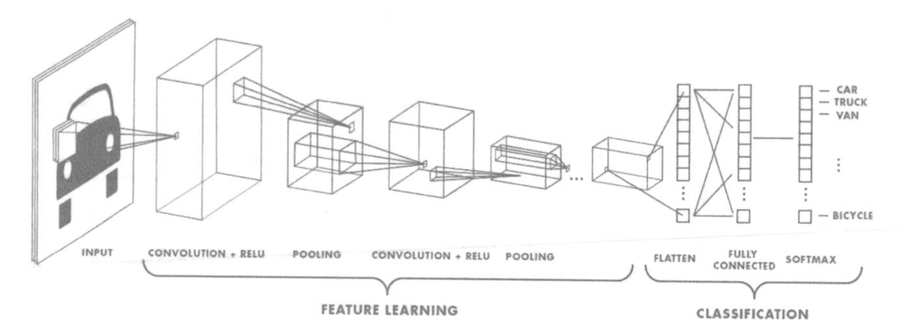

Fig. 4 CNN layers

Convolution Layer

This layer entails scanning the entire image for patterns and converting the results into a 3x3 matrix. Kernel is the name given to the image's convolved feature matrix. The weight vector is the name given to each value in the kernel.

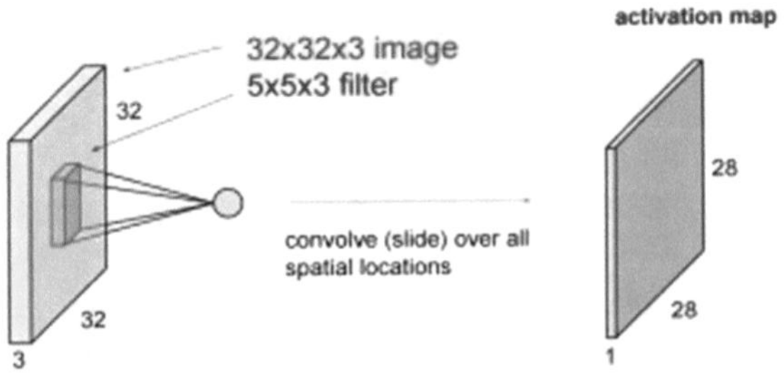

Fig. 5 Convolution layer

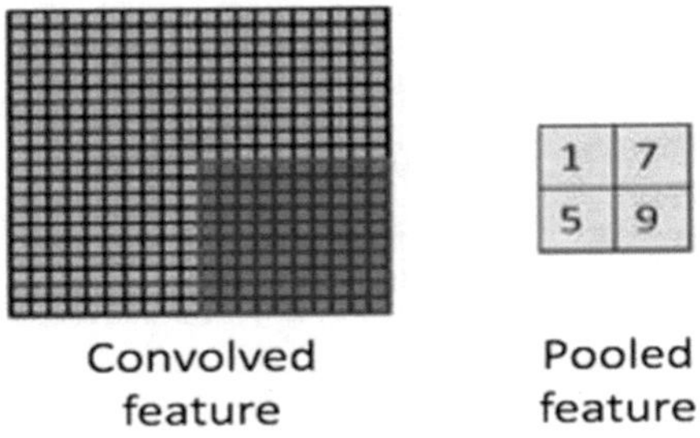

Fig. 6 Pooling layer

Pooling Layer

Following the convolution, the image matrix is divided into four non-overlapping sets of four rectangular segments. Pooling can be divided into two categories: maximum and average. The maximum value in the relative matrices region that is taken is determined by max pooling. In the relative matrix region, average pooling returns the average value. The pooling layer's key benefit is that it improves computer performance while reducing the chances of over-fitting (Figs. 5 and 6).

Activation Layer

It is the part of convolutional neural networks where the values are normalized, or fit into a specific range. The convolutional function employed is ReLU, which only accepts positive values and rejects negative ones. It is a result of the cheap computational cost (Fig. 7).

4 System Design

The process of establishing the architecture, components, modules, interfaces, and data for a system in order to meet specific criteria is known as systems design [5]. It is possible to think of systems design as the application of systems theory to product creation (Fig. 8).

The framework can be comprehensively sorted into following significant stages:

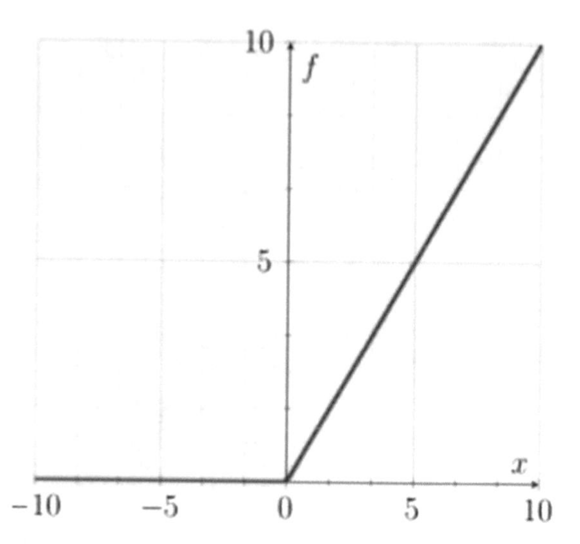

Fig. 7 ReLU function

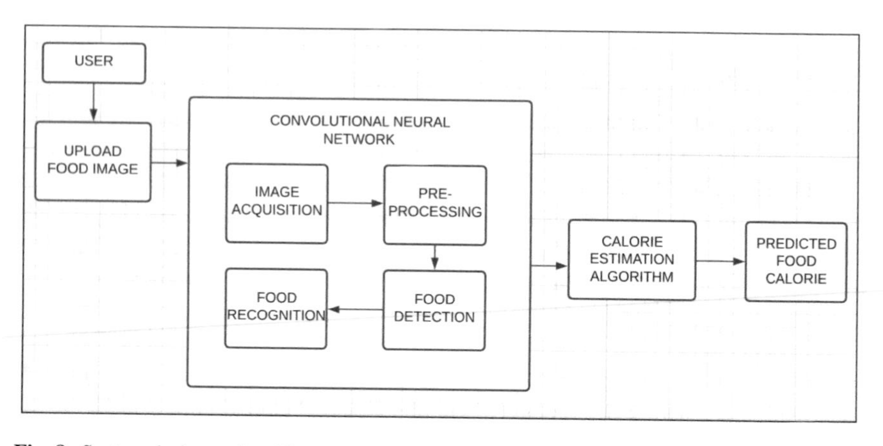

Fig. 8 System design and architecture

Acquisition of image

Images are obtained either by lens or by secretly deleting them from the contraction. Whatever the source may be, it is very important that the image of the data is transparent and cautious. An incredible picture is needed for this.

Pre-processing of Image

In this process, the photo is standardized by clearing the commotion, as it may confuse the evaluation. Similarly, the image given as the information may not be of standard size as required by the figure, so it is vital that the image size needed is obtained.

Data Storage Aspect to Preserve Information Images for Testing and training

If controlled learning will occur, as is the case here, it is important to prepare datasets. The sample database is the images collected during the photo procurement process.

Classifier to Classify the Food

The classifier used here is the last layer of the system which gives the true probability of each experience. The project involves two major parts such as image preparation unit and grouping unit. The object processing system enhances the image by removing the clatter and noisy bits. The food image will then be isolated into different segments to isolate the image from running the mill after the image features are evacuated to check whether or not the food is contaminated.

5 Experiments and Results

The tests were performed using Anaconda Navigator, a machine learning and research tool. Jupyter notes environment that does not need to be set to be used. A Google-provided free research tool that aids in the execution of programming that requires high GPU performance. The model is already developed on top of another database, such as ImageNet, and we need to add layers to it in order to meet our section's requirements. Specifically, the number of classes in the section.

Obtained result

All of the models' results are listed below. The verification accuracy parameter is utilized here (Fig. 9).

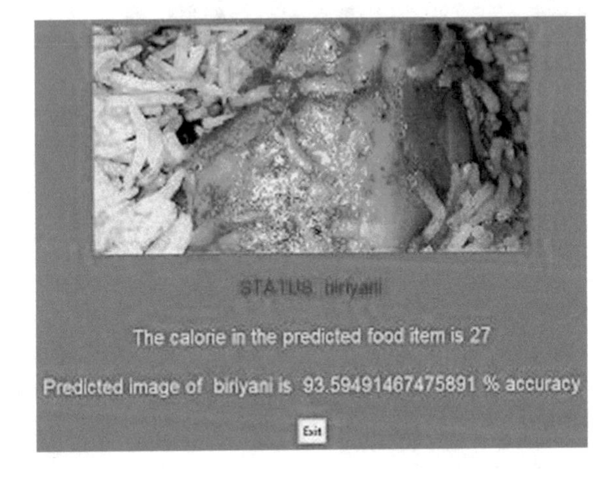

Fig. 9 Accuracy calorie of models is predicted

6 Conclusion

In this study, the convolutional neural network was used to segregate food photographs from their participants and forecast calorie image calories using transfer's learning process. In terms of future enhancements, removing audio from the database can improve the division function. The same study might be conducted on a big database with a high number of classrooms and photographs in each class, as large databases improve accuracy by learning more features and lowering the loss rate.

References

1. Szegedy C, Vanhoucke V, Ioffe S, Shlens J, Wojna Z (2016) Rethinking the inception architecture for computer vision. In: Proceedings of the IEEE conference on computer vision and pattern recognition, pp. 2818–2826, 27–30 June 2016. ISSN: 1063-6919
2. Attokaren DJ, Fernandes IG, Sriram A, Srinivasa Murthy YV, Koolagudi SG (2017) Food classification from images using convolutional neural networks. In: Proceedings of the 2017 IEEE region 10 conference (TENCON), Malaysia, 5–8 Nov 2017
3. Hnoohom N, Yuenyong S (2018) Thai "Fast Food Image Classification Using Deep Learning". In: 2018 International ECTI Northern section conference on electrical, electronics, computer and telecommunications engineering (ECTI-NCON). https://doi.org/10.1109/ECTI-NCON.2018.8378293
4. Yiğit GO, Özyildirim BM (2018) Comparison of convolutional neural network models for food image classification. J Inf Telecommun 2(3):347–357. https://doi.org/10.1080/24751839.2018.1446236, ISSN: 2475-1839
5. Subhi MA, Ali SM (2018) A deep convolutional neural network for food detection and recognition. In: IEEE-EMBS conference on biomedical engineering and sciences (IECBES). https://doi.org/10.1109/IECBES.2018.8626720
6. Bernstein et al (2014) Imagenet large scale visual recognition challenge
7. Berg JDA, Fei-Fei L (2018) Large scale visual recognition challenge 2010, net.org/download,2010, Online; accessed 29 Jan 2018

Performance Analysis of Rung Ladder-Structured Multilevel Inverter with PV Application

S. Muthubalaji, B. Divya Devi, and S. Sangeetha

Abstract Raising environmental concerns raised the usage of renewable energy sources. Inverter being main part of energy conversion system converts DC current produced from renewable source like solar PV cell to AC current used by load centers. Development of topologies that work efficiently when sourced with dynamically varying input is in high demand in the market. This paper deals with one of the newly developed reduced switch topology that provides flexibility of producing higher voltage levels while decreasing requirement of separate DC sources. Presently, though many of reduced switch topologies developed, least number of them ensures their flexibility and efficient working with real-time applications such as solar, wind, and industrial applications. Thus, in this paper, steady-state and transient analysis of newly developed rung ladder-structured multilevel inverter is studied with PV as source, and working of topology under real-time scenario is analyzed. A simple modulation technique is proposed for topology to trigger the switches. The analysis is explored in Simulink MATLAB 2018a, and results are discussed.

Keywords Rung ladder structure MLI · Steady-state analysis · Transient analysis · PV source

1 Introduction

These days PV installations are rapidly growing among remote areas as well as cities due to encouragement provided under green energy. Two-level converters can be used to fulfill the purpose, but we need to bare the system complexity and economic cost of large filters to obtain quality power from the system. Various multilevel inverters have been introduced to solve these problems. Because of the reduced voltage stress and harmonic components, such circuit topologies not only have compact filters but they also increase power-conversion efficiency and quality [1].

S. Muthubalaji · B. Divya Devi (✉) · S. Sangeetha
Department of EEE, CMR College of Engineering & Technology, Hyderabad, Telangana 501401, India
e-mail: bdivyadas03@gmail.com

A. Kumar et al. (eds.), *Proceedings of the International Conference on Cognitive and Intelligent Computing*, Cognitive Science and Technology,
https://doi.org/10.1007/978-981-19-2350-0_11

Multilevel inverters (MLIs) use a combination of power semiconductors and capacitor voltage sources to generate voltages with stepped waveforms at their output. The addition of the capacitor voltages, which reach high voltage at the output, is enabled by the commutation of the switches, whereas the power semiconductors must only endure decreased voltages. Diode-clamped (neutral-clamped), capacitor-clamped (flying capacitors), and cascaded multi-cell with independent DC sources are just the three different topologies proposed for multilevel inverters. In addition, for multilevel inverters, numerous modulation and control schemes have been developed or implemented, including multilevel sinusoidal pulse width modulation (PWM), multilevel selective harmonic elimination, and space-vector modulation (SVM) by various researchers in their respective publishing that are presented below [2].

2 Related Work

Roy et al. [3] proposed a single-phase MLI topology based on the concepts of the L-Z source inverter and the switched capacitor multilevel inverter with a single DC source that possesses all the advantages of both the inverters. As input voltage is boosted in two steps, it provides more flexibility to control boosting factor, size, cost and complexity of the inverter.

Hsieh et al. [4] presented a novel topology used in the seven-level inverter. An input voltage divider composed of three series capacitors C1, C2, and C3 is used as input to H-bridge. In this paper, several triangular carriers are distributed by phase disposition technique. The advantage of phase disposition technique is uncomplicated to realize and provide less total harmonic distortion.

A nine-level inverter module based on single-stage switched capacitor circuit is established for cascaded MLI is proposed by Lee [5]. The proposed S3 CM topology requires only single DC source with a voltage boosting gain of two. Therefore, it is capable of generating more levels and higher voltages up to twice the DC source by using switches with low voltage rating.

Sidorov et al. [6] proposed an improved symmetrical four-level sub-module as a basic cell for generating multiple DC voltage levels. The sub-module proposed reduces the number of conducting switches and gate driver requirements compared to the widely used half-bridge sub-module.

Panda et al. [7] investigated on an active power filter (APF) comprised of a transformer-less MLI for power conditioning in three-phase three-wire distribution network. The system generates compensation filter currents based on instantaneous active and reactive current component (id–iq) method and DC-link voltage regulation using a PI controller.

Bana et al. [8] reviewed on a number of recently developed MLIs used in various applications to assist with advanced current research in the field and help in the selection of suitable inverter for various applications; significant understanding on these topologies is clearly summarized based on the three categories, i.e., symmetrical, asymmetrical, and modified topologies.

Vanaja et al. [9] presented a novel design of an asymmetric multilevel inverter with a very few semiconductor switches for a single-phase grid-connected photovoltaic (PV) system. The proposed structure when connected in the cascaded form generates more voltage levels.

Thamizharasan et al. [10] developed a PWM strategy with a view to achieve explicit control of voltage and accrue a minimum total harmonic distortion in a MLI of any preferred level. The multi-objective digital formulation tailors to facilitate discrimination of pulses for various levels and equalize the switching losses in all the power switches in the MLI.

Kannan et al. [11] presented a new configuration of switches for H-bridge multi-level inverter (MLI) to provide five voltage levels as output and nine-level inverter in cascaded fashion. The proportional integral (PI) and fuzzy logic controller (FLC) are used to reduce the THD at the output.

Dhanamjayulu et al. [12] effectively compare a symmetric-hybridized cascaded multilevel inverter and an asymmetric multilevel inverter utilizing a switched capacitor unit for 17-level inverters. The output voltage is increased by using a bidirectional switch at the midpoint of a dual-input DC source.

Acharya et al. [13] explore the steady state and dynamic performance of a cascaded H-bridge multilevel inverter (CHBMLI) with the help of closed loop controllers for both linear load and nonlinear load. The steady state is controlled by voltage controller, while transient/dynamic period can be reduced with the help of current controller.

Tjokro et al. [14] constructed an asymmetrical inverter of 11-level to overcome complexities created by conventional topologies and limit THD for grid-tied PV systems. Lakshmi et al. [15] analyzed the performance of single-phase and three-phase symmetrical and asymmetrical multilevel inverters for distributed generating systems and discussed key to save cost, size of power switches, and methods to reduce conversion losses.

Suresh et al. [16] discussed various cascaded multilevel inverter structures for medium voltage that provide high-quality outputs and presented a novel topology that uses less number of sources to further improve the quality of voltage and current.

Rao et al. [17] presented an effective technique for eliminating voltage harmonics in the multilevel inverter using an adaptive neuro-fuzzy interference system. The voltage variations of MLI are obtained from real-load voltages, and voltage variations are applied to ANFIS.

Corzine et al. [18] developed a control methodology for cascaded multilevel inverter that switches at fundamental frequency by formulating in terms active and reactive power using P-Q theory. Sasikala et al. [19] work focuses on optimum controller for MLI in SMPS systems. The closed loop control system with PI and PR is developed, and performance comparisons are presented.

Jeevananthan et al. [20] presented a modulation technique that switches power devices equal number of times at all levels. The variable carrier frequency band approach creates as many clusters as the number of carrier frequencies that deteriorates harmonic spectrum.

In most of the developed topologies, constant DC sources are used as input for analyzing performance of the circuit. In real time, when renewable energy is used as input, circuit behavior shows much variation and place new challenges. To know behavior of circuit in real time, it necessitates study of circuit with renewable energy sources as input. MLIs are mostly used for energy conversion in renewable energy conversion systems such as solar. Contribution of a topology in real time is not discussed for many of efficient and newly developed topologies that constraint the applicability of research. Steady state and transient analysis of topology provide detailed performance of circuit for solar application.

3 Methodology

Figure 1 depicts the solar-fed modified rung ladder MLI. In this, the modified rung ladder multilevel inverter is supplied by the solar panel. The MPPT technique is implemented to harvest the maximum power from the PV source.

3.1 Modeling of Solar Panel

A single-diode model PV array consists of five parameters is used as DC input to the inverter. It consists of light-generated current source (IL), diode, series resistance (R_s), and shunt resistance (R_{sh}) that represents irradiance and temperature-dependent I-V characteristics of module (Fig. 2).

The diode I-V characteristics for a single module are defined by the equations:

$$I_d = I_0[\exp(V_d/V_T) - 1] \tag{1}$$

$$V_T = (kT/q) * nI * N_{cell} \tag{2}$$

where

I_d Diode current (A);
V_d Diode voltage (V);
I_0 Diode saturation current (A);

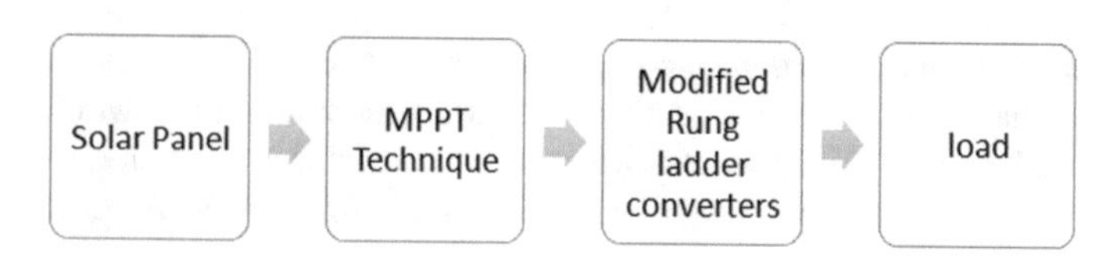

Fig. 1 Block diagram of solar-fed rung ladder MLI

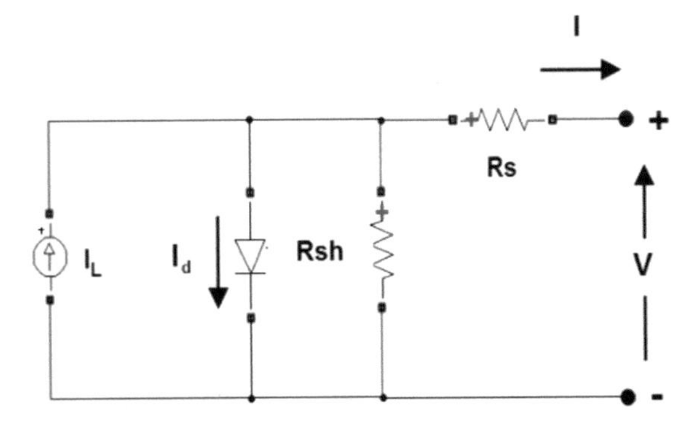

Fig. 2 Circuit diagram of PV array

T	Cell temperature (K);
k	Boltzmann constant $= 1.3806\text{e}^{-19}$ °C;
q	Electron charge $= 1.6022\text{e}^{-19}$ °C;
nI	Diode ideality factor, a number close to 1.0;
N_{cell}	Number of cells connected in series in a module.

3.2 P&O MPPT Technique

Perturbation an observation method is been used due to its ease of implementation and effective in maximum power tracking. It is simple technique to track maximum power of PV panel. Here, output voltage is increased or decreased to attain the maximum power point of PV panel. The control signal is provided by adding or subtracting step size to the duty cycle. The flowchart of P&O algorithm is shown in Fig. 3.

Here, the oscillation of MPP can be minimized by reducing the step size of D.

4 Rung Ladder Topology

The flexible rung ladder-structured multilevel inverter (FRLSMLI) is an expanded H-bridge construction that connects specific cells in place of the load to form a ladder shape. A source inclusion and bypass cell (SIBC) or a four-level creator cell can be included in the rung (FLCC).

The topology's generalized structure, shown in Fig. 4, consists of voltage modules/cells, SIBC, and FLCC, which are placed in rungs of the ladder structure. The ladder construction is created by adding supplemental rungs to a traditional H-bridge. The innovative cells are positioned between two vertical arms, where the load is checked. The FLCC has two SDCs (V_1 and V_2) that can produce four different DC levels (V_1, V_2, $(V_1 + V_2)$, and $(V_1 \sim V_2)$. Switches S_1 and S_2 are cascading switches,

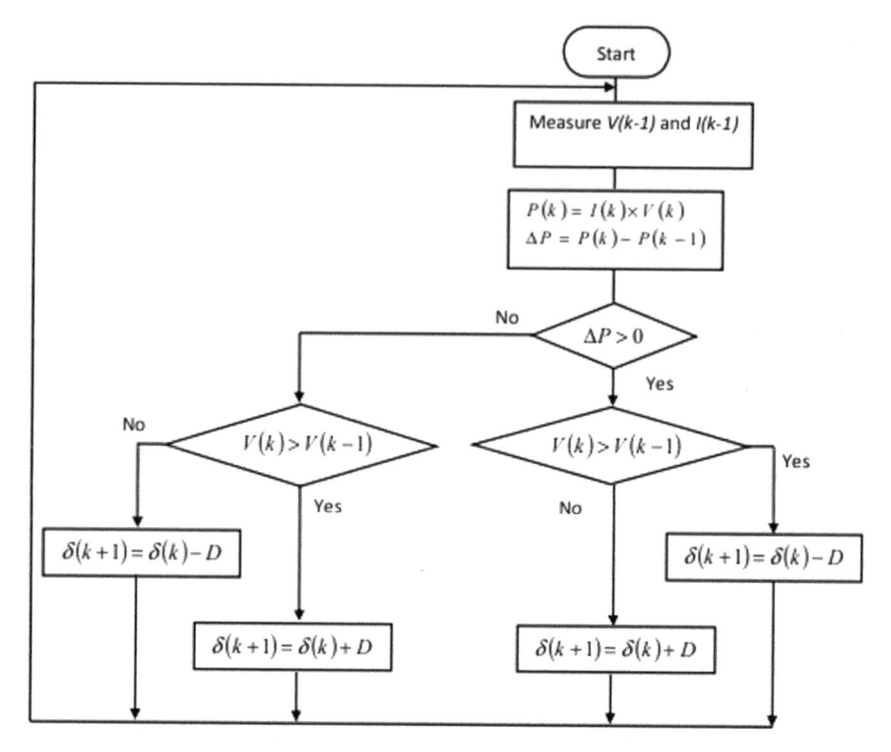

Fig. 3 Flowchart of P&O algorithm

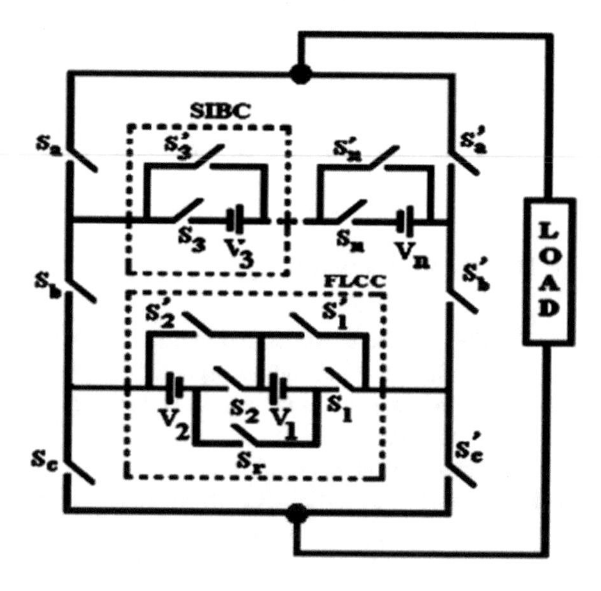

Fig. 4 Generalized structure of rung ladder MLI

but switches S_1' and S_2' are bypassing switches. S_r is the reverse connection switch that controls the last level (V_1 ~ V_2) and hence works as a subtraction facilitator.

In other words, the S_r is a bidirectional switch that subtracts the voltage source (V_2) from the voltage source (V_1). The SIBC is a single-source cell, with S_3 acting as a cascade switch and S_3' acting as a bypass switch. To achieve the appropriate number of voltage levels, a sufficient number of these cells can be subsumed (m).

5 Modified Rung Ladder Structure

See Fig. 5.

5.1 Switching States of Topology

The operation modes for extracting various values of the output voltage are shown in Figs. 6, 7, 8, 9, 10, 11, and 12. The asymmetrical combination of SDCs values can be written as V_1:V_2:V_3 = 1:3:3 = 100:300:300 V for ease of comprehension. The path to 100 V is depicted in Fig. 6. Similarly, Fig. 7 shows the switching combination for 200 V, which can be achieved in one of two ways, either $\pm$ ($V_2 - V_1$) or $\pm$ ($V_3 - V_1$). By referring to the appropriate figures, you can understand the mode diagrams and the many alternatives for obtaining a given level. With the three SDCs, a 15-level output can be achieved.

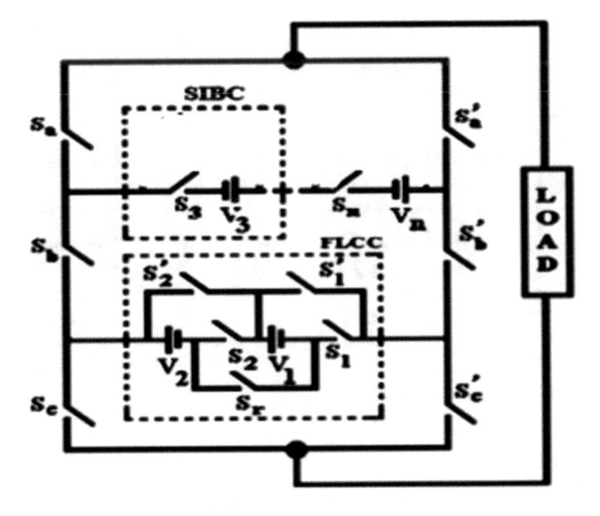

Fig. 5 Generalized structure of modified rung ladder MLI

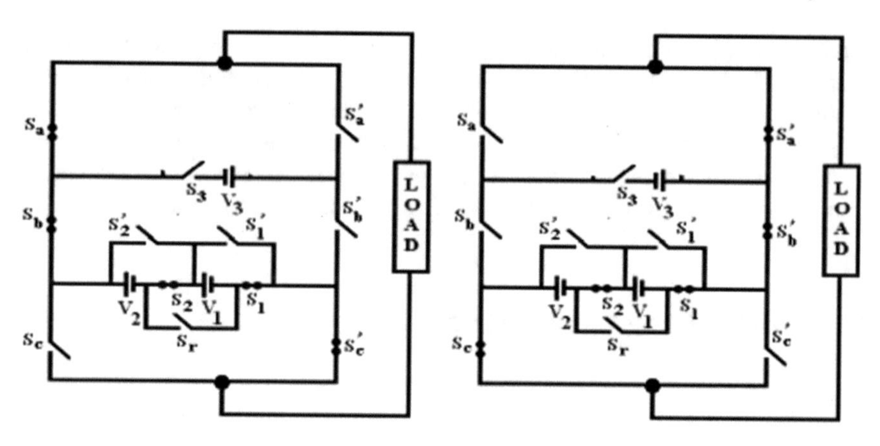

Fig. 6 Operating mode $+V_1/-V_1$

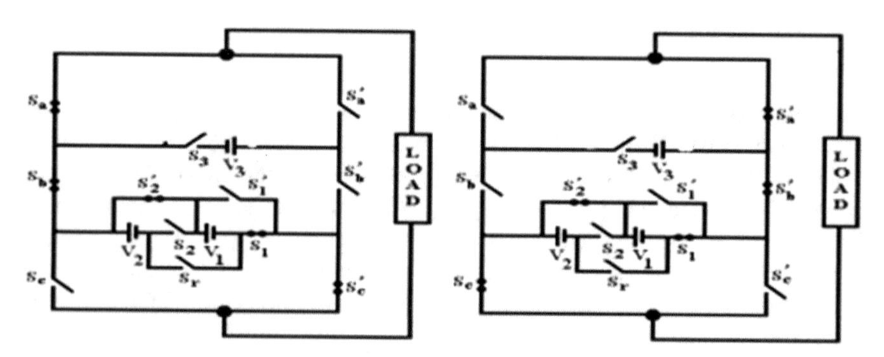

Fig. 7 Operating mode $+(V_2 - V1)/-(V_2 - V_1)$

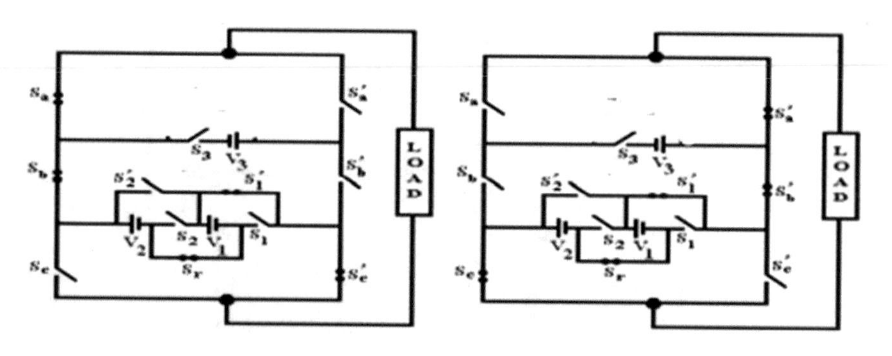

Fig. 8 Operating mode $+(V_2)/-(V_2)$

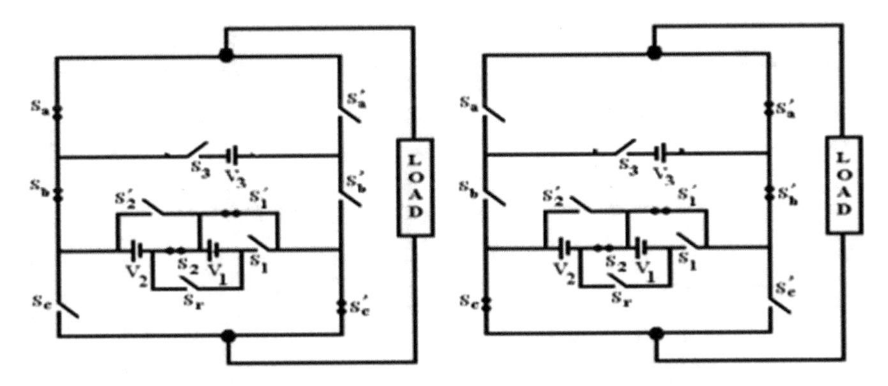

Fig. 9 Operating mode $+(V_1 + V_2)/-(V_1 + V_2)$

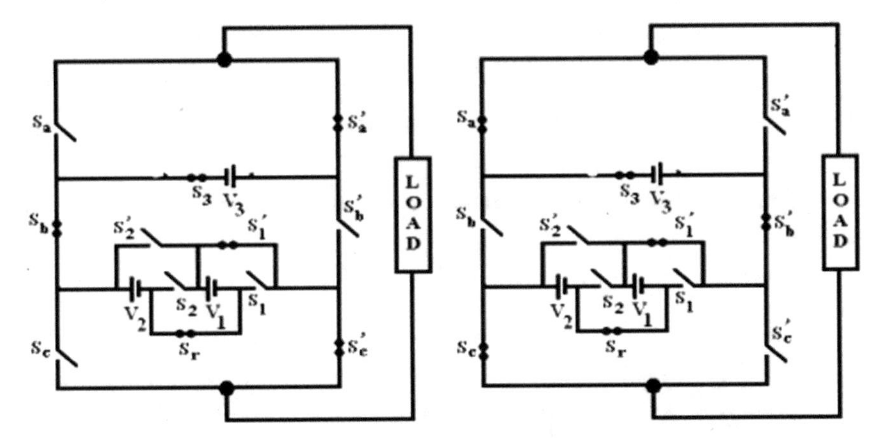

Fig. 10 Operating mode $+(V_2 + V_3 - V_1)/-(V_2 + V_3 - V_1)$

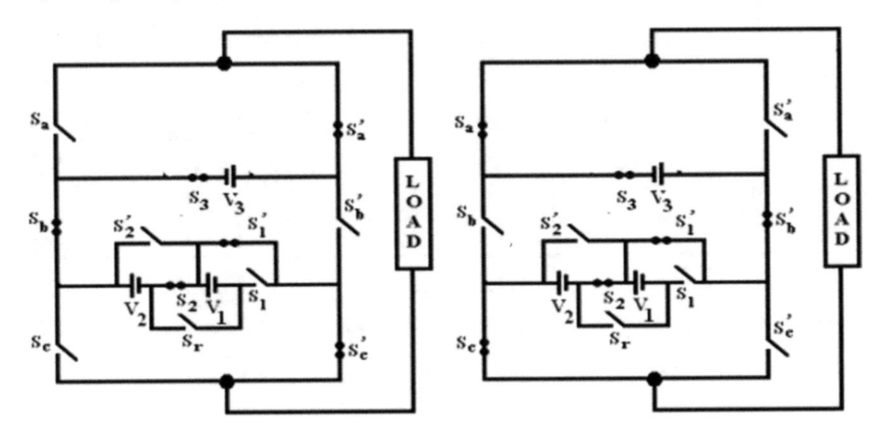

Fig. 11 Operating mode $+(V_2 + V_3)/-(V_2 + V_3)$

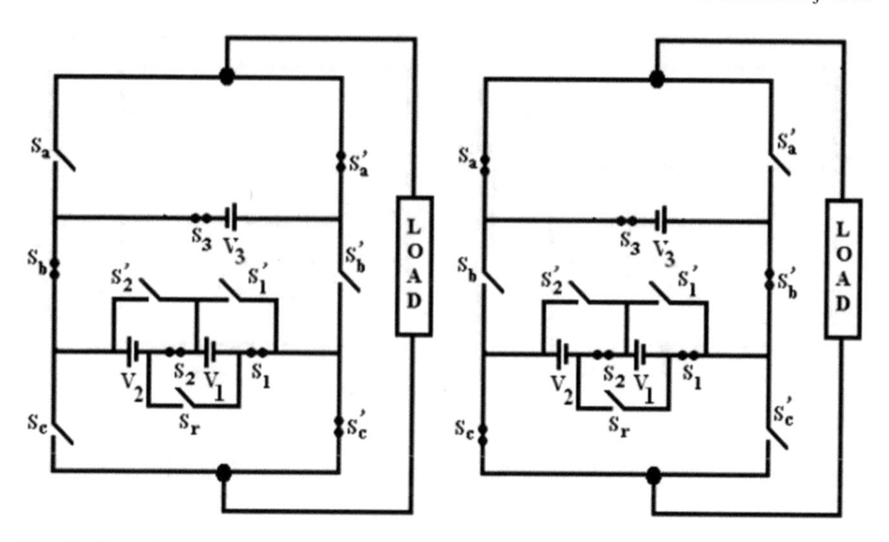

Fig. 12 Operating mode $+(V_1 + V_2 + V_3)/-(V_1 + V_2 + V_3)$

5.2 Switching Table and Switch Conduction Table

The conduction period of each switch is shown (Table 1).

Table 1 Conduction time for each switch in a circuit

Switches	In Positive half cycle							In Negative half cycle						
Sa &Sa'	■	■	■									■	■	■
Sb & Sb'								■	■	■	■	■	■	■
Sc & Sc'	■	■	■	■	■	■	■							
S1 & S1'		■	■		■	■			■	■		■	■	
S2		■	■		■	■				■	■		■	■
S2'	■							■						
S3				■	■	■						■	■	■
Sr		■			■				■			■		

1. Top four columns, darken squares for S_n' switches conduction timing and no color squares for S_n switches
2. For bottom four columns, darken squares indicate switch working time

5.3 *Modulation Technique*

A simple multicarrier level-shifted pulse width modulation technique is used to provide gate pulses to the inverter switches as shown in Fig. 13.

The modulation circuit is designed using logic gates and simple simulation blocks. All switches are divided in two groups, and separate logic is designed for each group. The switching logic table of switches is displayed in Table 2. The logic sequence is implemented using basic logic gates OR, NOT, XOR, and XNOR.

6 Results and Discussion

The modified topology is implemented with input PV array along with P&O MPPT technique. Fifteen-level output is obtained from modified rung ladder structure. To understand the functioning of inverter, it is operated in ternary scheme of unsymmetrical configuration with 1:3:9 ratios. The minimum input voltage of 34 V is considered to produce maximum voltage level 450 V. The resistance of 100 Ω is used as load for the system (Fig. 14).

To analyze the behavior of modified circuit in real-time scenario, varying irradiance is generated using signal builder block in Matlab Simulink. The output of inverter is stable for small varying irradiance. That can be investigated from voltage and current waveforms presented in Fig. 15.

To realize harmonic content in the system, FFT analysis is conducted, and result is presented in Fig. 16. The system shows harmonic content of 9.34% for fifteen-level output.

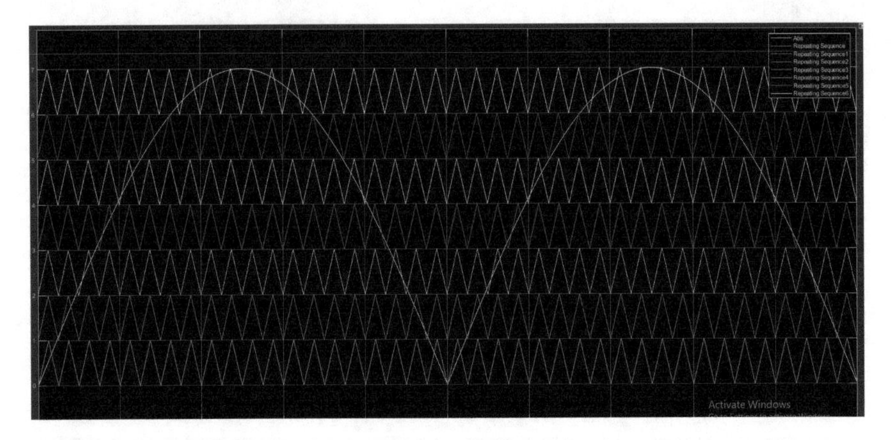

Fig. 13 Level-shifted seven carrier-based sinusoidal PWM

Table 2 Logic sequence of switches in circuit

Group A switches						Group B switches						
S_a	S_b	S_c	S'_a	S'_b	S'_c	S_1	S_2	S'_1	S'_2	S_3	S_r	V_o
0	1	0	1	0	1	1	1	0	0	1	0	$+(V_1 + V_2 + V_3)$
0	1	0	1	0	1	0	1	1	0	1	0	$+(V_2 + V_3)$
0	1	0	1	0	1	0	0	1	0	1	1	$+(V_1 + V_3 - V_2)$
1	1	0	0	0	1	1	1	0	0	0	0	$+(V_1 + V_2)$
1	1	0	0	0	1	0	1	1	0	0	0	$+(V_2)$
1	1	0	0	0	1	0	0	1	0	0	1	$+(V_2 - V_1)$
1	1	0	0	0	1	1	0	0	1	0	0	$+V_1$
1	1	0	0	0	1	0	0/1	1	1	0	0	0
0	0	1	1	1	0	1	0	0	1	0	0	$-V_1$
0	0	1	1	1	0	0	0	1	0	0	1	$-(V_2 - V_1)$
0	0	1	1	1	0	0	1	1	0	0	0	$-(V_2)$
0	0	1	1	1	0	1	1	0	0	0	0	$-(V_1 + V_2)$
1	0	1	0	1	0	0	0	1	0	1	1	$-(V_1 + V_3 - V_2)$
1	0	1	0	1	0	0	1	1	0	1	0	$-(V_2 + V_3)$
1	0	1	0	1	0	1	1	0	0	1	0	$-(V_1 + V_2 + V_3)$

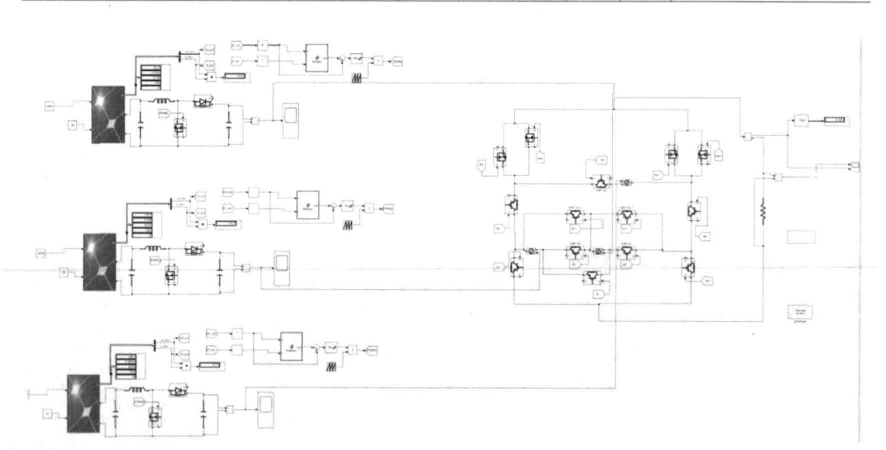

Fig. 14 Simulink model of proposed system

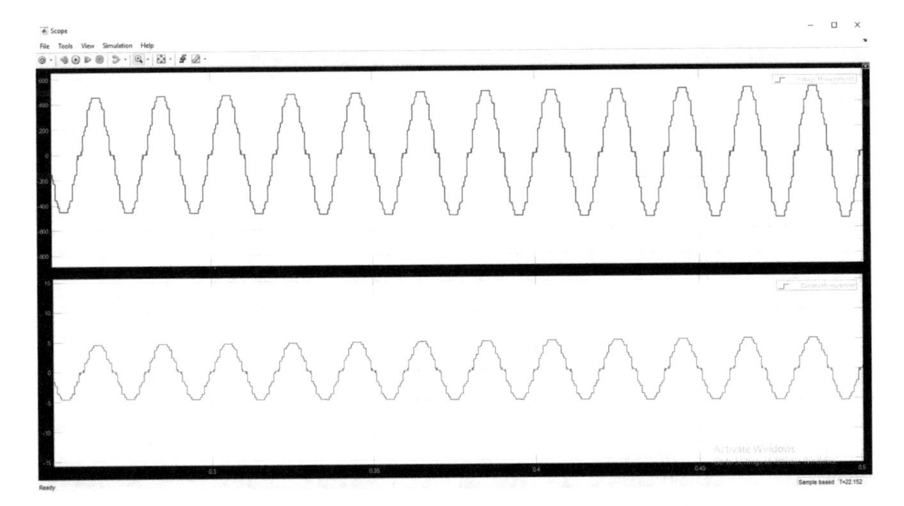

Fig. 15 Voltage and current waveforms of proposed system

6.1 FFT Analysis of Modified Rung Ladder MLI

Table 3 presents the comparison between existing system and proposed system.

The modified rung ladder structure MLI provides better THD performance and ease the need of modulation requirement on comparison to rung ladder structure. Modified structure is laid out with slight reduction in switch count on comparison to rung ladder structure.

7 Conclusion

A modified rung ladder structure is tested for solar application by providing PV array as input under steady-state and transient conditions. The P&O MPPT technique is implemented to draw maximum power from the PV array under variable irradiation condition. The modified topology is implemented using simple-level-shifted PWM technique and provides less harmonic content, i.e., 9.34% compared to rung ladder topology 13.99%. Though switch count is slightly decreased, it is working, and application with solar input is satisfactory. Future scope of study is to reduce the switch count and THD content further.

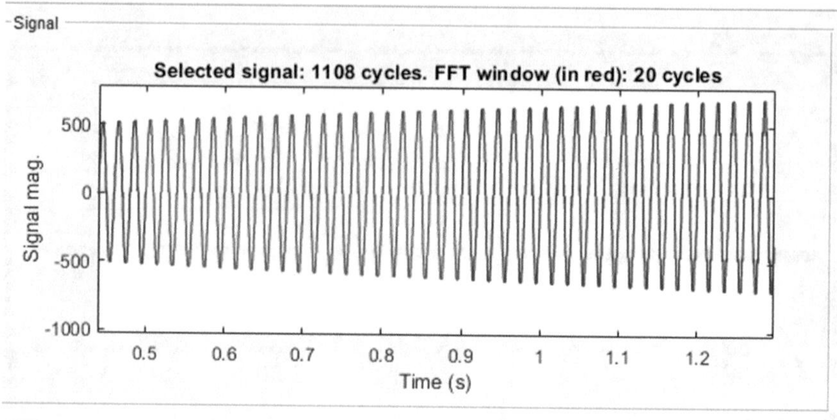

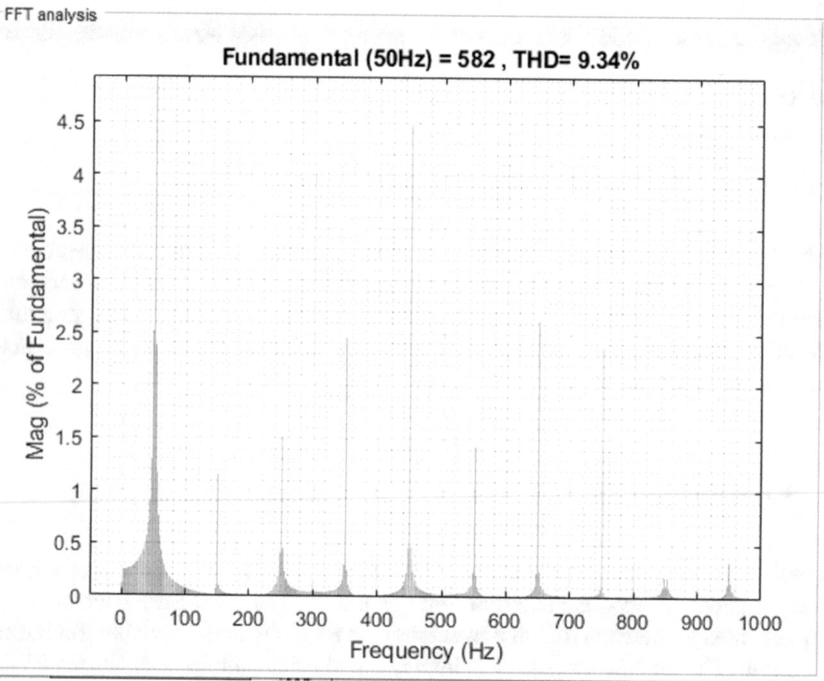

Fig. 16 FFT analysis of modified rung ladder MLI

Table 3 Comparison between existing and proposed systems

Parameter	Rung ladder system	Modified rung ladder system
THD	13.99%	9.34%
Modulation technique	Complex (switching circuit for each switch)	Simple (level-shifted PWM technique)
Switch count	13	12

References

1. Kim J, Kwon J, Kwon B (2018) High-efficiency two-stage three-level grid-connected photovoltaic inverter. IEEE Trans Ind Electron 65(3):2368–2377. https://doi.org/10.1109/tie.2017.2740835
2. Rodriguez J, Lai J-S, Peng FZ (2002) Multilevel inverters: a survey of topologies, controls, and applications. IEEE Trans Ind Electron 49(4):724–738. https://doi.org/10.1109/tie.2002.801052
3. Roy T, Sadhu PK, Dasgupta A (2017) A new single-phase multilevel inverter topology with two-step voltage boosting capability. https://doi.org/10.6113/JPE.2017.17.5.1173
4. Hsieh C-H, Liang T-J, Chen S-M, Tsai S-W (2016). IEEE
5. Lee SS (2018) Single-stage switched-capacitor module (S3 CM) topology for cascaded multilevel inverter. IEEE
6. Lee SS, Sidorov M, Lim CS, Idris NRN, Heng YE (2017) Hybrid cascaded multilevel inverter (HCMLI) with improved symmetrical 4-level submodule. IEEE
7. Panda AK, Patnaik SS (2014) Analysis of cascaded multilevel inverters for active harmonic filtering in distribution networks. Elsevier Ltd., pp 0142—0615
8. Bana PR, Panda KP, Naayagi RT, Siano P, Panda G (2019) Recently developed reduced switch multilevel inverter for renewable energy integration and drives application: topologies, comprehensive analysis and comparative evaluation. IEEE
9. Vanaja DS, Stonier AA (2019) A novel PV fed asymmetric multilevel inverter with reduced THD for a grid-connected system. https://doi.org/10.1002/2050-7038.12267
10. Thamizharasan S, Sudha LU, Baskaran J, Ramkumar S, Jeevananthan S (2015) Carrier-less pulse width modulation strategy for multilevel inveters
11. Kannan C, Mohanty NK, Selvarasu R (2017) A new topology for cascaded H-bridge multilevel inverter with PI and fuzzy control
12. Dhanamjayulu C, Meikandasivam S (2018) Implementation and comparison of symmetric and asymmetric multilevel inverters for dynamic loads
13. Acharya AK, Kumar K, Chowdary KVVSR, Sahu PK (2020) Enhancement of dynamic performance of a single-phase cascaded H-bridge multilevel inverter using closed loop controllers. IEEE
14. Tjokro C, Pratomo LH (2018) Design and simulation of an asymmetrical 11-level inverter for photovoltaic applications. In: Proceedings of 2018 5th international conference on information technology, computer, and electrical engineering (ICITACEE)
15. Lakshmi Ganesh K, Chandra Rao U, Rambabu C, Bhaskar A (2012) Performance analysis of single phase and three phase symmetrical and asymmetrical multilevel inverters
16. Suresh Y, Venkataramanaiah J, Panda A, Dhanamjayulu C, Venugopal P (2017) Investigation on cascade multilevel inverter with symmetric, asymmetric, hybrid and multi-cell configurations. Ain Shams Eng J 8(2):263–276. https://doi.org/10.1016/j.asej.2016.09.006
17. Nageswara Rao G, Chandra Sekhar K, Sangameswararaju P (2018) An effective technique for reducing total harmonics distortion of multilevel inverter. J Intell Syst 27(3):433–446. https://doi.org/10.1515/jisys-2016-0210
18. Lu S, Corzine K (2007) Advanced control and analysis of cascaded multilevel converters based on P-Q compensation. IEEE Trans Power Electron 22(4):1242–1252. https://doi.org/10.1109/tpel.2007.900471
19. Sasikala K, Krishnakumar R (2020) An improved response of multi level inverter based PR controlled SMPS. Front Energy Res 8. https://doi.org/10.3389/fenrg.2020.00001
20. Jeevananthan S, Paramasivam S, Thamizharasan S (2012) A novel invariable carrier frequency multi-level inverter PWM for balancing switching transitions and better distribution of harmonic power. Int J Power Electron 4(3):290. https://doi.org/10.1504/ijpelec.2012.046601

GDPR: A Bibliometric Analysis

Mugdha Kulkarni, Arnab Mondal, and Krishna Kumar Singh

Abstract EU adapted "General Data Protection Regulation" and serves as privacy law and of human rights law, currently the GDPR data security law is enforced in the EU as from 2018. As the name itself suggest, we all know it is a very complex law of 88 pages long including 173 recitals and 99 articles. To comply with the recent law, organizations have equipped required arsenal. In this article, a detailed bibliometric analysis is conducted, different quantitative metrics are pulled out such as most productive and influential authors, most cited work, work languages, top involved countries and popular keywords. This paper has summarized all the papers that have been published on this particular topic in last few years since till July, 2020 and gives an overall background of various work and application in this field.

Keywords GDPR · Bibliometric analysis · Survey · Scopus

1 Introduction

GDPR is the enforced data security law for EU people. Legislation is targeted at securing natural persons and storing their personal records. GDPR as a law serves as a direct statement by the EU that any data exchanged on social media will be covered. Individual records, consisting of names or home addresses, are such representation of data, privacy of which we wish to secure. Nevertheless, one should realize that many more obvious pieces of data will act as personal data and be used as an indicator to identify political or sexual identity in others. Without any effective protection or legal approach, personal data sharing is always a risk. Here GDPR and E-privacy bill of

M. Kulkarni (✉) · A. Mondal · K. K. Singh
Symbiosis Centre for Information Technology, Pune, India
e-mail: mugdha@scit.edu

A. Mondal
e-mail: arnab.mondal@associates.scit.edu

K. K. Singh
e-mail: krishnakumar@scit.edu

A. Kumar et al. (eds.), *Proceedings of the International Conference on Cognitive and Intelligent Computing*, Cognitive Science and Technology,
https://doi.org/10.1007/978-981-19-2350-0_12

United Kingdom give users the power to control their own data [1]. Under the GDPR, when it comes to collecting or processing data of any EU citizen, any organization regardless of their locations, have to maintain the transparency of their process, terms of services, privacy policy and consent withdraw because in this law, customer has more power in their hand to control their data and right to know the process [2]. In GDPR, the type of data that are covered is personal identifiable data including name, age, gender, ethnic background, address, birthdate, social and govt. number, online platform generated data, IP address, contact number, email ID, health-related data, biometric data and political data [3]. The companies whose base is in EU region or even though it is not based in EU but still processing the data of EU citizen have to comply with this GDPR law in which they must have to be a representative in EU. Bibliometric analysis is very popular in the field of library and information science, and now it is also trending in research publication also where we can conduct a quantitative analysis on the research done on any particular topic or field [4]. It is used to conduct quantitative analysis on academic literature. Using bibliometric methods, citation analysis is being done by graphical and network diagram representation. Bibliometric analysis also being done to understand the impact and research growth rate of any particular topic [5]. Bibliometric analysis is very useful for new researchers to understand the importance and impact of the field in which they are putting their first step, they can have a clear image which are the area of their field is vast and covered and which areas are left. They can also picturize strength and weakness of their field and can also gauge the effectiveness of their publication by identifying the trend and future growth of their current areas [5]. This paper is focused on GDPR, a detailed bibliometric analysis has been conducted in order to do that we have fetched the database from Scopus. How many papers are published in this topic, what is the yearly number of it and what is the trend has been shown. An overview is also given on GDPR in the beginning. Different bibliographic terminologies have been performed and discussed on the available database and also discussed and concluded on our findings. The paper has been structured with below discussed points, at first a brief overview is given on GDPR including its principles and consequences. In the next section, the procedure of this analysis is mentioned after that several bibliometric factors is discussed. Next, findings and their backgrounds are mentioned, and lastly, the article has been discussed and concluded.

2 Overview of GDPR

GDPR is born of complex cyber-attacks, advancements in technology and data misuse issues. It is not just one of the other mechanisms or standards for data security but there are several others. This legislation is the top legislative priority of 2018 for businesses, also among US companies, and is considered one of the most important laws on information security and privacy of current time [6]. The legislation grants data subject's rights over their personal data and creates responsibilities for every entity worldwide that handles data of an EU data subject, making the applicability of the legislation follow data rather than following a data subject or physical location [2]. GDPR demands all the entity that handles data subjects' personal data apply adequate protection and organizational measures to safeguard the confidentiality, fairness and quality of the information processed [3]. GDPR came into effect in 2016 and became enforceable on 25 May 2018. There are seven key principle which are clearly defined and actions must be taken by the organizations in order to comply with this law.

1. Personal data in relation to persons should be handled lawfully, fairly and in a transparent way.
2. The purpose of the data collection should be legal and consensual, and customer should be aware about the purpose; if the collected data is being used for other than the original purpose without any knowledge of the customer, then it should be considered as illegal or not compatible with the original purpose.
3. Only limited data can be collected which will only meet the main purpose.
4. Data which are being collected need to be updated, corrected any incorrect and inaccurate data should be erased properly.
5. Data can be stored for the limited period only whenever the main purpose is over, and organization cannot store customer's data without their knowledge, which should be considered illegal.
6. Data need to be processed in such a manner where integrity and appropriateness of the data can be maintained, proper method, process and protection should be taken by the organization to maintain the integrity.
7. Processor of the data always should be accountable for the points that are mentioned above.

GDPR let customer takes control of their data and added eight more privacy terms called privacy rights [2, 3]. Organizations that have grown accustomed to being hit with minor data breach fines or data abuse will be surprised to learn that GDPR sets two tiers of penalties, with the lower level paying up to Ten Million GBP or 2% of the annual income of an organization, whichever is higher [7]. For violations of controller or processor responsibilities the lower level tier is applied, and the upper level is applied for violations of rights and freedom of the data owner. The upper level tier fines up to twenty million GBP for breaches or 4% of the annual income of a company, whichever is greater [7]. Not only are the criteria and scope for GDPR incredibly broad, but the fines and penalties that organizations may face as a result of non-compliance are unlike any fines and penalties that a regulatory body has imposed before. Which is why it is so important to grasp this ground breaking data privacy regulation.

3 Research Method

To run the analysis, bibliometric data has been collected from the respiratory of Scopus, to collect all the details we have used the keyword "GDPR". Then all the

Table 1 Distribution of document types in Scopus

Document type	Total numbers	Contribution (%)
Conference paper	729	46.91
Conference review	68	4.37
Review	69	4.44
Article	575	37
Article in press	3	0.19
Book chapter	45	2.89
Book	3	0.19
Note	27	1.73
Editorial	18	1.15
Short survey	8	0.51
Letter	8	0.51
Erratum	1	0.06

title and abstracts of each papers have been thoroughly studied and where we did not find our respective keyword or not similar with our general data protection regulation (GDPR) topic, that has been eliminated. After this screening process, 1554 papers were selected which are related to our topic. Microsoft Excel and VOSviewer have been used as Bibliometric analysis tool. VOSviewer is one of the freely available software/websites. Using those tools, we have performed several indicating factors for bibliometric analysis on our extracted data such as most productive and influential author, most cited documents, popular sponsors, institution and organizations and most used keywords (Table 1).

In this part, we have put the picture of different results of different type of analysis [8]. We have discussed points like research growth, highly cited papers, countries, highly-used keywords, organizations in this particular topic of general data protection regulation (GDPR).

4 Research Growth

The general data protection regulation (GDPR) is gaining vast attention nowadays due to its implementation in European parliament. Though the law has been enforced in 2018, but it was tabled in 2016 in European parliament. As we have used GDPR as our Keyword so, from the way back, research papers are being published in this particular topic or the domains related to it like Internet of Things, data privacy, law, blockchain, information security, etc. From when the talk was started regarding personal data security in 2012, we can see as little as only one paper were published in this area, till 2016 there was not much growth, from 2017 actually, the ball started rolling in this field. In 2017, 2018, 2019 and 2020 it got some hike. In 2019, the greatest number of papers were published but 2020 is running so it can overtrump

Table 2 Yearly growth

Year	Count
2012	1
2013	2
2014	6
2015	1
2016	33
2017	100
2018	443
2019	654
2020	314

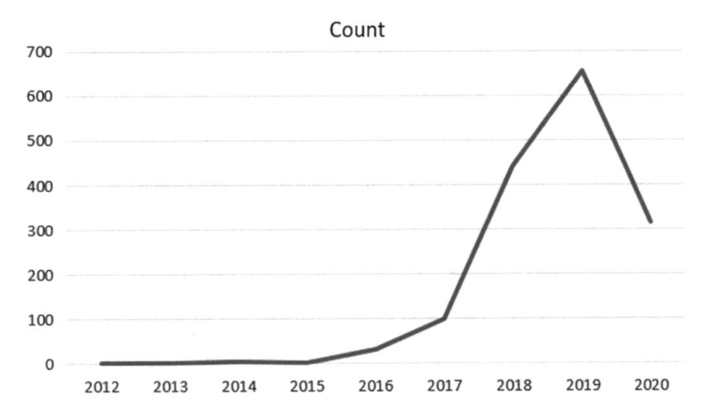

Fig. 1 Yearly growth

2019. In the beginning, there was not much work in this field but after the significant data scam like Cambridge Analytica when need arise several things implemented and people imparted about this domain. Here in Table 2 and Fig. 1, the growth has been shown.

5 Most Productive and Highly Cited Author

Authors who contributed extensively in this feild are considerd productive author. Bibliometric data has been processed in the mentioned tool and most productive authors have been extracted from Scopus database and arranged on the highest number of papers publication, where the minimum number publication was set to seven. We got thirteen authors who are meeting the criteria. So, Lewis D. is on the top of the list with 17 publications, Pandit H. J. is at second with 16 publications, Palmirani N. is at third with 11 publications, Bartolini C. is at number fourth with ten publications, Rossi A., Fabiano N., Kirrane S. are at fifth place with nine publications

Table 3 Most productive authors

Author	Publication count	Citation count
Bartolini C.	10	58
Fabiano N.	9	23
Ferreira A.	7	5
Kirrane S.	9	27
Lenzini G.	8	12
Lewis D.	17	56
Lopes I. M.	7	7
Malgieri G.	7	63
Oliveira P.	7	7
O'Sullivan D.	8	27
Palmirani M.	11	69
Pandit H. J.	16	48
Rossi A.	9	58

each, O Sullivan D. and Lenzini G. is at sixth with eight publications each and Lopez I. M., Oliveira P., Ferreira A., Malgieri G. with seven publications each. Here we have attached the Table 3, network visualization of Co-authorship is showing only five of them are connected rest of thirteen are not connected in this network in Fig. 2.

By saying highly cited author, it is meant who is the author is more influential in this work field, whose work is most frequently being followed by another research fellow. In order to do that, we have sorted author according to their total citation count, as we can see Lewis D, who was top on the list of the productivity where his name is in the ninth position and Pandit H. J. is at fifteenth position. Here we have not set any minimum criteria like what we did in productivity list, it has been observed that those authors who have a smaller number of works in that field, have been cited most by other. As citation counts are good for many authors here, we have kept top 25 author in the list in Table 4.

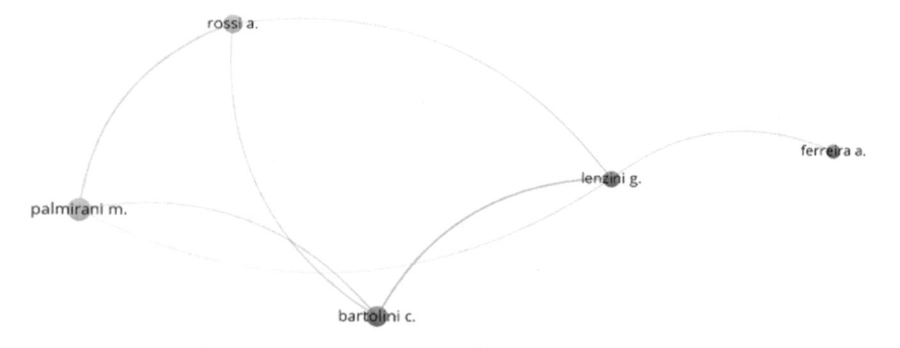

Fig. 2 Network visualization of most productive authors

Table 4 Most influential authors

Author	Publication count	Citation count
Chen T.	2	84
Liu Y.	2	84
Tong Y.	2	84
Yang Q.	2	84
Palmirani M.	11	69
Malgieri G.	7	63
Bartolini C.	10	58
Rossi A.	9	58
Lewis D.	17	56
Alepis E.	6	54
Patsakis C.	6	54
Politou E.	6	54
Urquhart L.	4	49
Pandit H. J.	16	48
De Hert P.	5	45
Martoni M.	5	45
Robaldo I.	4	45
Kieseberg P.	3	43
Mcauley D.	4	43
Papakonstantinou V.	3	43
Holzinger A.	2	42
Weippl E.	2	39
Veale M.	3	38
Fischer-Hübner S.	6	37

6 Discipline Wise Analysis

In the respiratory of Scopus, they have categorized ten top most subject area in their indexing list but in our case, the we have kept it only to general data protection regulation (GDPR) we have ignored some publications in "physics", "biochemistry, genetics and molecular biology" and "physics and astronomy" because it is not relevant to our topic which is given in Table 5. Most works are related to computer science because data protection means information protection which is nowadays is being done digitally, and the law is mainly focused on computer-based digital data, Apart from that there are also some research work is seen in health care as well, in social science, like the topic on right to privacy. There are many papers which can be clubbed in more than one discipline.

Table 5 Disciplines

Discipline	Count	Contribution (%)
Computer science	1016	36.44
Social science	463	16.60
Engineering	347	12.45
Mathematics	236	8.46
Business, management and accounting	195	6.99
Decision sciences	192	6.88
Medicine	179	6.42
Biochemistry, genetics and molecular biology	61	2.18
Economics, econometrics and finance	52	1.86
Physics and astronomy	47	1.69

7 Top Source Journal

We have identified top ten journals in which works have been published mostly on this mandatory law enforcement in information technology and digitalized business. The function of a research journal is to present new and particularly important research in the medical and other fields of science [8]. The journal of research in social sciences often serves as an opportunity to examine different research methods and approaches. In the journal name called "Lecture Notes in Computer Science Including Subseries Lecture Notes in Artificial Intelligence and Lecture Notes in Bioinformatics" has the greatest number of papers published followed by Ceur Workshop Proceedings, Computer Law and Security Review holding 2nd and 3rd place, respectively. All this paper caters in several domains like data privacy, social science, cyber law, health data protection, etc. It is given in Table 6, total publications in each of the journal.

Table 6 Source journal

Journal name	Publication count
Lecture notes in computer science including subseries lecture notes in artificial intelligence and lecture notes in bioinformatics	143
Ceur workshop proceedings	57
Computer law and security review	53
ACM international conference proceeding series	37
IFIP advances in information and communication technology	34
European data protection law review	32
Communications in computer and information science	30
Advances in intelligent systems and computing	27
International data privacy law	25
Computer fraud and security	21

8 Country-Wise Analysis

From the extracted data, we have prepared a list of those countries which has more interest in this field. In Table 7, it has given below. We have also showed network visualization between countries in Fig. 3. United Kingdom is on the top of the list with 213 publications and also most cited, Germany comes second in the list with 199 publications and 340 citations and USA is coming third in the list with 160 publications and 444 Citations. We have also mentioned some countries who have shown some interest on this topic. United Kingdom is sharing the network with almost top 25 countries.

Table 7 Productive countries

Country	Publication	Citation
United Kingdom	213	910
Germany	199	340
United States	160	444
Italy	138	385
Netherlands	119	325
Belgium	87	297
France	83	191
Spain	66	123
Greece	59	147
Ireland	57	129
Portugal	51	40
Austria	50	112
Switzerland	38	41
Finland	37	98
Australia	35	94
Sweden	34	90
Norway	33	74
Czech Republic	32	39
Luxembourg	32	82
China	28	109
Poland	23	25
Romania	22	71
India	21	18
Canada	19	40
South Korea	18	31

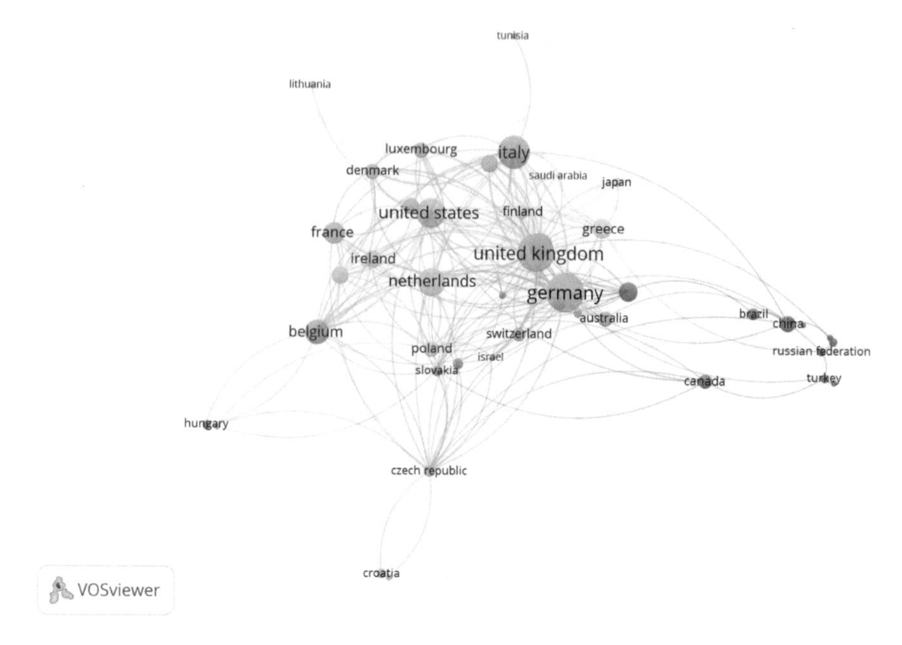

Fig. 3 Network visualization of countries

9 Institution and Funding Wise Analyse

On the basis of Scopus respiratory, here we have prepared a list of institution from where a good number of efforts has been observed. Top institution along with their total number of publications and top funders are given in Table 8. KU Leuven and University of Luxembourg both are having 27 publications. Ku Leuven is one of the old universities in Europe based in Belgium and comes under top 100 university in the world. University of Luxembourg is one of the renowned and costliest university. Also added language wise analysis in Table 10.

Highest co-author ship with total citation is also given in Table 9 and Fig. 4 (Table 10).

10 Top 20 Most Influential Publication

In this section, we have identified thirty most valuable publications, these publications are most choice by other people. In the Table 11, we have mentioned the author name, title and the total number of citations of them. "EU regulations on algorithmic decision-making and a "right to explanation"" is the most cited work with the number of 134, only ten papers have citation count more than 40 and approximately 19 [9] papers are having citation count of more than 20.

Table 8 Institutions and funding sponsors

Institution	Publication	Funding sponsor	Publication
KU Leuven	27	Horizon 2020 framework programme	70
University of Luxembourg	27	European commission	51
Trinity College Dublin	25	European regional development fund	38
Vrije Universiteit Brussel	22	Engineering and physical sciences research council	27
Alma Mater Studiorum Università di Bologna	21	Science foundation Ireland	24
Tilburg University	18	FundaÃ§Ã£o para a CiÃªncia e a Tecnologia	20
University of Oxford	18	National science foundation	19
Wirtschafts universität Wien	17	Horizon 2020	17
University of Piraeus	17	European research council	12
Universidade do Porto	16	Bundesministerium fÃ¼r Bildung und Forschung	11

Table 9 Organizations

Organization	Documents	Citations
Adapt Centre, Trinity College Dublin, Dublin and Ireland	14	45
Mandat International, Geneva and Switzerland	6	1
University of Luxembourg and Luxembourg	5	2
Vienna University of Economics and Business, Vienna and Austria	7	23

11 Keyword Analysis

Using the tools and bibliometric data, we have listed out the top most very frequently used keywords used by the authors. We have set the limitation to minimum of 70 times of usage in any publication and we got 20 keywords with the number of their occurrence is given in Table 12 and also shown the network visualization in Figs. 5 and 6,. GDPR, general data protection regulations, data privacy are top most used keywords found in publications, apart from that Internet of Things, law and information security are some popular keywords used by authors. There are four clusters are clearly visible in cluster 1 and 2, there are eight keywords and in Cluster 3 and 4 and there are two keywords each given in Table 13. And the connection strengths are also very prominent as shown in Fig. 5. How relatable are these keywords in the domain of GDPR and its related works have been discussed.

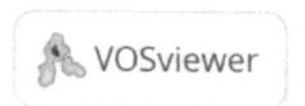

Fig. 4 Network visualization of organizations

Table 10 Languages

Language	Count
English	1490
German	24
Italian	11
Spanish	10
French	9
Portuguese	5
Czech	3
Dutch	3
Bosnia	2
Hungarian	2

Earlier, we have already given an overview on GDPR, in this section, we will try to discuss it difference aspects/ domain where is has more impact like data protection, data privacy, law, etc.

Table 11 Top 20 cited publications

Title	Authors	Cited by
EU regulations on algorithmic decision-making and a "right to explanation"	Goodman B., Flaxman S.	134
Federated machine learning: concept and applications	Yang Q., Liu Y., Chen T., Tong Y.	84
The impact of the EU general data protection regulation on scientific research	Chassang G.	64
What the GDPR means for businesses	Tankard C.	58
Meaningful information and the right to explanation	Selbst A.D., Powles J.	55
EU general data protection regulation: changes and implications for personal data collecting companies	Tikkinen-Piri C., Rohunen A., Markkula J.	53
The effect of the general data protection regulation on medical research	Rumbold J. M. M., Pierscionek B.	49
Forgetting personal data and revoking consent under the GDPR: challenges and proposed solutions	Politou E., Alepis E., Patsakis C.	43
Normative challenges of identification in the Internet of Things: privacy, profiling, discrimination and the GDPR	Wachter S.	41
Artificial intelligence as a medical device in radiology: ethical and regulatory issues in Europe and the United States	Pesapane F., Volonté C., Codari M., Sardanelli F.	40
The intuitive appeal of explainable machines	Selbst A. D., Barocas S.	38
Estimating the success of re-identifications in incomplete data sets using generative models	Rocher L., Hendrickx J. M., de Montjoye Y.-A.	36
Current advances, trends and challenges of machine learning and knowledge extraction: from machine learning to explainable AI	Holzinger A., Kieseberg P., Weippl E., Tjoa A.M.	36
A blockchain-based approach for data accountability and provenance tracking	Neisse R., Steri G., Nai-Fovino I.	33
The right to data portability in the GDPR: towards user-centric interoperability of digital services	De Hert P., Papakonstantinou V., Malgieri G., Beslay L., Sanchez I.	32
The law of everything. Broad concept of personal data and future of EU data protection law	Purtova N.	31

(continued)

Table 11 (continued)

Title	Authors	Cited by
PrOnto: privacy ontology for legal reasoning	Palmirani M., Martoni M., Rossi A., Bartolini C., Robaldo L.	29
A process for data protection impact assessment under the European general data protection regulation	Bieker F., Friedewald M., Hansen M., Obersteller H., Rost M.	28
Why a right to legibility of automated decision-making exists in the general data protection regulation	Malgieri G., Comandé G.	26
Privacy by design: informed consent and Internet of Things for smart health	O'Connor Y., Rowan W., Lynch L., Heavin C.	25

Table 12 Most used keywords

Keyword	Occurrences
Article	72
Artificial intelligence	87
Big data	97
Blockchain	87
Data handling	71
Data privacy	522
Data protection	241
EU	111
GDPR	628
General data protection regulation	108
General data protection regulations	500
Human	125
Information management	91
Information systems	74
Information use	73
Internet of Things	103
Laws and legislation	157
Personal data	88
Privacy	316
Security of data	135

12 Blockchain

Blockchain is in peer to peer-based ledger technology where data resided into a block, and each block is secured with a specific identical key, which is a distributed ledger [10]. It is hard to tamper the data which resides in every block [10]. Blockchain

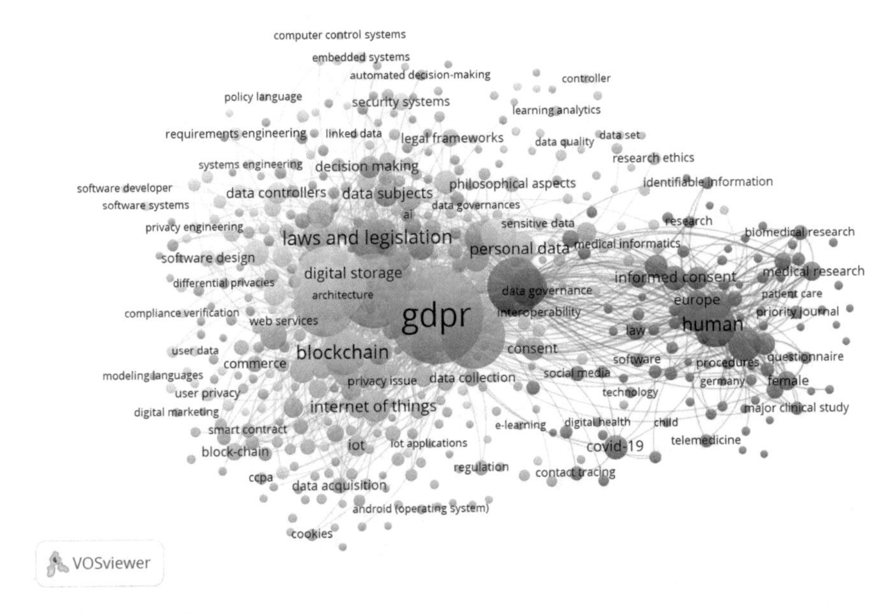

Fig. 5 Network visualisation of all the keywords

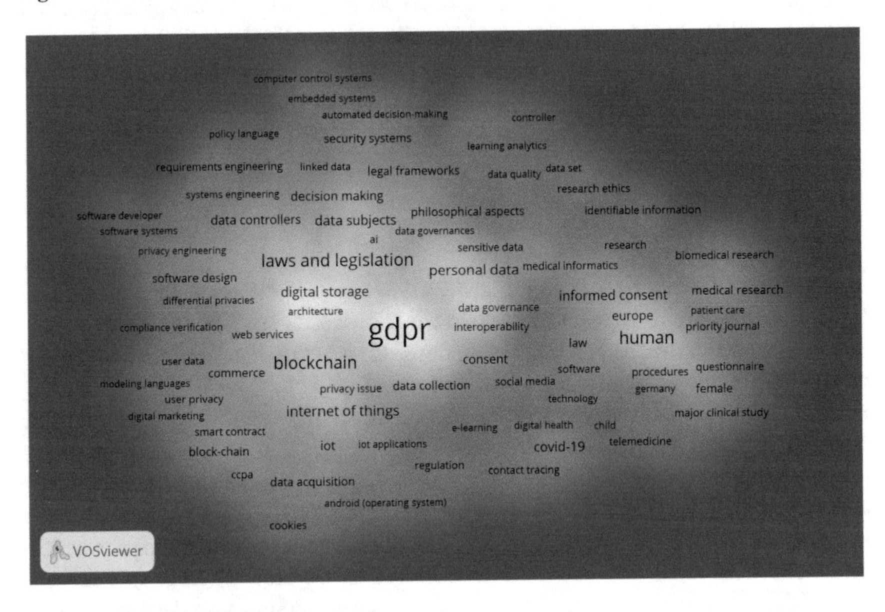

Fig. 6 Density visualization of keywords

Table 13 Clusters

Cluster 1	Cluster 2	Cluster 3	Cluster 4
Article, data protection, EU, GDPR, general data protection regulations, human, personal data and privacy	Data handling, data privacy, general data protection regulation, information management, information system, information use, law and legislation and security of data	Artificial intelligence and big data	Blockchain and Internet of Things

has a drawback when it comes to GDPR; it has an advantages and disadvantages as well, because in blockchain, data cannot be destroyed or tampered but according to GDPR's right to be forgotten, there is no chance to erase the customer's data when they withdraw the consent [11]. For example, social media site like Facebook collect data from customer in order to do that Facebook have to take consent from customers but study shows that Facebook's consent collection mechanism is not compliant with GDPR [11]. So here blockchain-based technology can be implemented for online social media where people will have more processing power on their data [11]. Not only in social media wherever personal identifiable data (PII) is being stored can be integrated with blockchain technology like banking and finance sector, supply chain and logistics, ID cards, smart device orders, etc. [3].

13 Internet of Things

The Internet of Things (IoT) is a hiked technology that allows data concentrated research across all IT farms, and also, IoT is focused on smooth and smart manufacturing and operations for big size industries, where technological complexity has increased due to manifold uses of smart devices [12]. The GDPR will be far-reaching effect on IOT devices [13]. The scientists now routinely fetch data for research purpose, behaviour studies, education, energy, transport, safety and safety through mobile devices and dedicated sensors. Companies use smart device's data and serviced devices in order to gain insight of the product [14]. They are keen to take these concerns seriously and work to find solutions on it for the confidence and trust of an IoT mechanism in global industrial context [14]. GDPR is one of the proactive measures for the risks associated with these data and any as per requirements mentioned in GDPR any company should assign data protection officers (DPO) to ensure the protection of privacy of customer's [15]. Technological flaws are increasing and cyber-attacker also finding their new way, so risks that are involved with IoT devices are to be addressed to comply with GDPR.

14 Big Data

Big data is an analytic technique which organization used for business analytics purpose, a large and complex data set, both organized and unorganized processed, to bring out business value out of it [16]. Using this technique, organization do measure their market performance and customer insight. The most importantly GDPR legislation restricted several ways of data collection as it asks organization to follow data minimization which means only consented data will be analysed which sets a boundary to gain more insight [17]. With the data collected immediately being updated, no long time is needed to maintain such data, which is one of the problems to be addressed by the GDPR. In addition to the secure storage of the large-scale data collected, customers who want to remove it and switch to another supplier need to be addressed. Businesses of all sizes must agree that customers gain more control over their own personal data [18].

15 Artificial Intelligence

The global economy and society are rapidly changing for artificial intelligence. Various countries are already seeing the benefits of AI from accelerating the development of pharmaceutical products to automating factories and farms [19]. Sadly, the EU's data processing regulations are increasingly becoming clear that the potential of AI in Europe will be reduced. Data provides the foundations for AI, and European companies will not be able to use technology to their full potential with serious restrictions on how they use it. GDPR limits how organizations can use data to make automated decisions [2]. Article 22 states that the data subject has the rights to have a human review of that decision wherever companies use AI to reach a substantial decision about individuals, such as whether to offer a loan [20]. This requirement is making the use of AI hard and ineffective in automating many processes, because it is required for every individual who opts out of the automated process to develop a redundant manual process [21]. Automated decisions are costly for people to review. One of the main reasons for AI is to reduce the amount of time people take to process large numbers of data, but that does indeed require people to continue to involve themselves [19].

16 Information System

Information system is that where organizations critical information assets reside and this information asset can contain consumer or customer's data which can be sensitive from company perspective too. Several key aspects are mentioned in GDPR article 32 [22]. Regarding the security posture requirement especially mandatory for data inventory, storages and processing of European personnel also under GDPR article 30 it is mandatory to all organization to keep record of all data processing or procedural activities [22, 23]. All the organization should map their data, they should make data map for their reference it is mandatory in GDPR which will help any organization to for data governance. And it should contain all the technical component and data asset. Several tools/applications are available in the market to capture data map them and manage the inventory [22]. If data are not properly mapped and managed the inventory, when it comes to right to be forgotten and individual will ask to revoke their consent then problem will arise if not properly followed. GDPR has actually change the approach towards the information system [2].

17 Data Protection

Data protection is actually the main objective of GDPR and establish such a uniform law framework which will be followed globally. In 2016, European parliament adapted GDPR replacing the old data protection directive because the old data protection directive was not fitting with the recent advance edge technology [24]. In this current law, the customers have been given more power to control their own PII data, they actually have right to ask the organization about the transparency of their data security posture [18]. At the beginning, most of the organizations were not familiar with the complexities and consequences of the GDPR except some big organizations and GDPR has forced them also to create some new roles like data protection officer (DPO), IT manager, project manager, etc. [24]. Lack of awareness of GDPR and also the use of emerging technologies have been found and in case of emerging technologies like AI, VR, Big data and cloud, it is pretty necessary to be protected from data breach where GDPR is providing the framework [2].

18 Privacy

Data rights and principles are already mention in earlier in this article [7]. The GDPR is just simply forward about data privacy law. The law in other way asks organizations to gain customers confidence and trust by giving them more power on controlling their own data like putting some more transparency to the processes and storages, where customer should provide their data only on need-to-know basis for a particular span of time [23]. GDPR also asks companies their privacy policy and terms and condition to keep clean and simple for layman [3].

19 Personal Data

In GDPR, the type of data that are covered are personal identifiable data including name, age, gender, ethnic background, address, birthdate, social and govt. number, online platform generated data, IP address, contact number, email ID, health-related data, biometric data and political data [3]. The companies whose base is in EU region or even though it is not based in EU but still processing the data of EU citizen, have to comply with this GDPR law there they must have to be a representative in EU [18, 24]. There are seven key principle which are clearly defined and actions must be taken by the organizations in order to comply with this law [2, 3].

20 Human

Significant changes have been seen in human resource domain as well company should adhere to GDPR when they are processing candidate data from screening and recruiting purpose [25] and the process should made clear to the candidate and there should be policy enforced. The previous Labour Law was not focused on this particular data protection which is main concern here [25]. Any company should always have a framework to collect and process personal data, here their own policy is also required. Candidate can or should check all the checks and balances it is their right under the GDPR [2].

21 Cybercrime

The rising usage of Internet and digitalisation all over the world, it has also go together with noteworthy increases in cybercrime. We can see increase in ways of cybercrime. The issues and challenges can be talked through framing a good cyber security policy ensuring the problem covers cloud computing, digitalisation, ecommerce social media, mobile phone applications and other inventions in digital technology [26]. Also, in case of information security, organization need to verify level of awareness of cybercrime. Required and necessary security measures can really be helpful in making the cyber environment protected, robust and reliable. [27]. In the environment digitalisation, cybercrime increasing with rapidity. Online banking, ecommerce and online shopping shares the sensitive information and which arises the threat of cybercrime [28].

22 Discussion

In this paper, a detailed bibliometric analysis has been conducted structures, and developments in this area have been detected with bibliometric analysis. Widely used bibliometric analysis repositories such as Scopus were used. Where 1554 publications in Scopus have been indexed. Research number of this field is increasing in each year as it is a vast law and must have to comply with for many businesses rather than that technological advancement of information technology actually widening the scope of research on this particular domain. So, it is expected that publications will increase in upcoming year as well. Lewis D. is the most productive author with 17 publications and 56 citations where Chen T., Liu Y., Tong Y., and Yang Q. are the most influential authors with having 84 citations each. In computer science discipline, most of the paper has been published with 36.44% contribution. Top journal is Lecture Notes in computer with 143 publication count. UK, Germany, USA, Italy and Netherlands has contributed most in this field. Almost all the papers are published in English Language. Institution wise KU Leuven and University of Luxembourg has contributed most with 27 publications each but when considered organization wise then Adapt centre, Trinity college Dublin, Dublin, Ireland is most productive with 14 documents and 45 citations. EU regulations on algorithmic decision-making and a "right to explanation" most cited work with 134 citations where the paper is talking about the business aspect of GDPR and its complexities. GDPR, general data protection regulations, data privacy are top most used keywords found in publications. The top papers were ranked according to the bibliometric analysis. The analysis is conducted on total citations. Then the most common visualization GDPR keywords are shown. The most popular words such as GDPR, general data protection, and data privacy regulations and they share vary strong connection in the network. This is only a bibliometric analysis. Numbers, however, are quantity does not mean quality but citations. In addition, we covered, however, Scopus is widely used for bibliometric research. Some other sources include open access journals.

23 Conclusion

This paper provides the structure inherent in the GDPR publications with bibliometric analyses. This is a very important study for emerging areas such as GDPR so that the publication hierarchy in this field can be assessed by the research community. After a detailed study on bibliometric data of GDPR field, it is been observed that interest of people who are associated with this field is growing through out every year. To comply with GDPR is in great demand now and one of thé main reasons is the customer, and people are becoming more aware about privacy and data security. This increasing demand of data owners and technical advancement is the main concern for any organizations, so the scope in this field is increasing as well because of this new advancement in people and technology. It is expected that percentage of research growth will increase in various domain like computer science, data engineering, health care, social engineering, cyber security, law and legal, etc. As seen in the analysis, after this, we will be looking for text analysis to understand the nature of the papers and how relatable is GDPR and in which manner which will give a qualitative concise of GDPR related work.

References

1. Gregory VW (2017) First the GDPR, now the proposed ePrivacy regulation. J Internet Law:3–11
2. Addis C, Kutar MS (2018) The general data protection regulation (GDPR), emerging technologies and UK organisations: awareness, implementation and readiness. In: UK academy for information systems conference proceedings, p 29
3. Menon M (2019) GDPR and data powered marketing: the beginning of a new paradigm. J Market Dev Competitiveness:73–84
4. Ellegaard O, Wallin J (2015) The bibliometric analysis of scholarly production: how great is the impact? Scientometrics:1809–1031
5. Patra SK, Bhattacharya P (2006) Bibliometric study of literature on bibliometrics. DESIDOC J Libr Inf Technol:1–6
6. Kotsios A, Magnani M, Vega D (2019) An analysis of the consequences of the general data protection regulation on social network research. ACM Trans Social Comput:1–22
7. Allen DW, Berg A, Berg C (2019) Some economic consequences of the GDPR. Econ Bull:785–797
8. Garfield E (2006) The history and meaning of the journal impact factor. JAMA:90–93
9. Goodman BAFS (2017) European union regulations on algorithmic decision-making and a "right to explanation". AI Mag:50–57
10. Zheng Z, Xie S, Dai H (2017) An overview of blockchain technology: Architecture, consensus, and future trends. In: 2017 IEEE international congress on big data (BigData congress), pp 557–564
11. Ahmed J, Yildirim S, Nowostaki M, Ramachandra R, Elezaj O, Abomohara M (2020) GDPR compliant consent driven data protection in online social networks: a blockchain-based approach. In: 3rd international conference on information and computer technologies (ICICT), pp 307–312
12. Riel A, Kreiner C, Macher G, Messnarz R (2017) Integrated design for tackling safety and security challenges of smart products and digital manufacturing. CIRP Ann:177–180
13. Allegue S, Rhahla M, Abdellatif T (2019) Toward GDPR compliance in IoT systems. In: International conference on service-oriented computing, pp 130–141
14. Bourgeois J, Kortuem G, Kawsar F (2018) Trusted and GDPR-compliant research with the internet of things. In: Proceedings of the 8th international conference on the internet of things, pp 1–8
15. Wachter S (2018) Normative challenges of identification in the internet of things: privacy, profiling, discrimination, and the GDPR. Comput Law Secur Rev:436–449
16. Gruschka N, Mavroeidis V, Vishi K, Jensen M (2018) Privacy issues and data protection in big data: a case study analysis under GDPR. In: 2018 IEEE international conference on big data (big data), pp 5027–5033
17. Greene T, Shmueli G, Ray S, Fell J (2019) Adjusting to the GDPR: the impact on data scientists and behavioral researchers. In: Big data, pp 140–162
18. Livraga G, Viviani M (2019) Data confidentiality and information credibility in on-line ecosystems. In: Proceedings of the 11th international conference on management of digital ecosystems, pp 191–198
19. Wachter S, Mittelstadt B, Floridi L. Transparent, explainable, and accountable AI for robotics

20. Torre D, Abualhaija S, Sabetzadeh M, Briand L, Baetens K, Goes P, Forastier S (2020) An AI-assisted approach for checking the completeness of privacy policies against GDPR. In: Proceedings of the 28th IEEE international requirements engineering conference (RE'20)
21. da Conceicao Freitas M, da Silva MM (2018) GDPR in SMEs. In: 2018 13th Iberian conference on information systems and technologies (CISTI), pp 1–6
22. Diamantopoulou V, Tsohou A, Karyda M (2020) From ISO/IEC27001: 2013 and ISO/IEC27002: 2013 to GDPR compliance controls. Inf Comput Secur:18
23. Boban M (2018) Cyber security foundations for compliance within GDPR for business information systems. In: Economic and social development: book of proceedings, pp 541–553
24. Carey P (2018) Data protection: a practical guide to UK and EU law. Oxford University Press, Inc.
25. Josimovski S, Kiselicki M, Pulevska-Ivanovska L (2019) Social media screening: impact of GDPR and Macedonian legal framework. J Contempor Econ Bus Issues:75–85
26. Broadhurst R, Chang LY (2013) Cybercrime in Asia: trends and challenges. In: Handbook of Asian criminology, pp 49–63
27. Gunjan VK, Kumar A, Avdhanam S (2013) A survey of cyber crime in India. In: 2013 15th international conference on advanced computing technologies (ICACT), pp 1–6
28. Gunjan VKKA, Rao AA (2014) Present & future paradigms of cyber crime & security majors-growth & rising trends. In: 2014 4th international conference on artificial intelligence with applications in engineering and technology, p 2014
29. Degerman I, Eckerbom J, Gu H (2019) How do B2B companies approach CRM and the management of customer data in today's era of social media and GDPR? In: A multiple case study, p 65
30. Tankard C (2016) What the GDPR means for businesses. Netw Secur 2016(6):5–8
31. Tikkinen-Piri C, Rohunen A, Markkula J (2018) EU general data protection regulation: changes and implications for personal data collecting companies. Comput Law Secur Rev:134–153
32. Purtova N (2018) The law of everything. Broad concept of personal data and future of EU data protection law. Law Innov Technol:40–81
33. Goddard M (2017) Viewpoint: the EU general data protection regulation (GDPR): European regulation that has a global impact. Int J Market Res:703–705
34. Pormeister K (2017) Genetic data and the research exemption: is the GDPR going too far? Int Data Privacy Law
35. Neisse R, Steri G, Nai-Fovino I (2017) A blockchain-based approach for data accountability & provenance tracking. In: Proceedings of the 12th international conference on availability, reliability and security, pp 1–10
36. O'Connor Y, Rowan W, Lynch L, Heavin C (2017) Privacy by design: informed consent and internet of things for smart health. Proc Comput Sci:653–658
37. Palmirani M, Martoni M, Rossi A, Bartolini C, Robaldo L (2018) PrOnto: privacy ontology for legal reasoning. In: International conference on electronic government and the information systems perspective, pp 139–152
38. Pandit HJ, Fatema K, O'Sullivan D, Lewis D (2018) GDPRtEXT—GDPR as a linked data resource. In: European semantic web conference, pp 481–495
39. Pesapane F, Volont'e C, Codari M, Sardanelli F (2018) Artificial intelligence as a medical device in radiology: ethical and regulatory issues in Europe and the United States. In: Insights into imaging, pp 745–753
40. Politou E, Alepis E, Patsakis C (2018) Forgetting personal data and revoking consent under the GDPR: challenges and proposed solutions. J Cybersecur:tyy001
41. Pritchard A et al (1969) Statistical bibliography or bibliometrics. J Document:348–349
42. Radulescu A et al (2018) Users' social trust of sharing data with companies: online privacy protection behavior, customer perceived value, and continuous usage intention. In: Contemporary readings in law and social justice, pp 137–143
43. Yang Q, Liu Y, Chen T, Tong Y (2019) Federated machine learning: concept and applications. ACM Trans Intell Syst Technol (TIST):1–19

44. Urquhart L, McAuley D (2018) Avoiding the internet of insecure industrial things. Comput Law Secur Rev:450–466
45. Tesfay WB, Hofmann P, Nakamura T, Kiyomoto S, Serna J (2018)Privacyguide: towards an implementation of the EU GDPR on internet privacy policy evaluation. In: Proceedings of the fourth ACM international workshop on security and privacy analytics, pp 15–21
46. Shabani M, Borry P (2018) Rules for processing genetic data for research purposes in view of the new EU general data protection regulation. Euro J Human Genet:149–156
47. Selbst AD, Barocas S (2018) The intuitive appeal of explainable machines. Fordham L Rev:1085
48. Selbst A, Powles J (2018) Meaningful information and the right to explanation. In: Conference on fairness, accountability and transparency, p 48
49. Rumbold JMM, Pierscionek B (2017) The effect of the general data protection regulation on medical research. J Med Internet Res:e47
50. Rocher L, Hendrickx JM, De Montjoye Y-A (2019) Estimating the success of re-identifications in incomplete datasets using generative models. Nat Commun:1–9
51. Malgieri G, Comande G (2017)Why a right to legibility of automated decision-making exists in the general data protection regulation. Int Data Privacy Law
52. De Hert P, Papakonstantinou V, Malgieri G, Beslay L, Sanchez I (2018) The right to data portability in the GDPR: towards user-centric interoperability of digital services. Comput Law Secur Rev:193–203
53. Crabtree A, Lodge T, Colley J, Greenhalgh C, Glover K, Haddadi H, Amar Y, Mortier R, Li Q, Moore J, Wang L (2018) Building accountability into the internet of things: the IoT databox model. J Reliab Intell Environ:39–55
54. Holzinger A, Kieseberg P, Weippl E, Tjoa AM (2018) Current advances, trends and challenges of machine learning and knowledge extraction: from machine learning to explainable AI. In: International cross-domain conference for machine learning and knowledge extraction, pp 1–8
55. Mannhardt F, Petersen SA, Oliveira MF (2018) Privacy Challenges for process mining in human-centered industrial environments. In: 2018 14th international conference on intelligent environments (IE), pp 64–71
56. Murmann P, Fischer S (2017) Tools for achieving usable ex post transparency: a survey. IEEE Access:22965–22991

E-Health Startups' Framework for Value Creation and Capture: Some Insights from Systematic Review

Imon Chakraborty, P. Vigneswara Ilavarasan, and Sisira Edirippulige

Abstract Information technology-driven health startups are an emerging phenomenon. They have a significant interest in the current care delivery systems and promise to add value to address inadequacies in the healthcare landscape. The present paper reviews the current literature on e-health startups to understand their business frameworks, offered services, and challenges. PRISMA guidelines were followed to find and review the relevant literature. The search identified in a final total of seventeen articles. The majority of existing studies offer inadequate insights on value creation and capture. The present study proposes a conceptual framework with success factors and classifies them based on value creation and capture. This study is the first to extensively examine the literature to appraise and analyze knowledge on e-health startups' frameworks in the setting of value creation in the health space. It also offers future research directions.

Keywords Digital health · Telehealth · Telemedicine · Ventures · Health-tech · Framework · Success factor · Service delivery

1 Introduction

Information technology (IT) has significantly changed our lifestyle in recent years. The so-called technological revolution and digitization have altered the work and organization culture. This revolution influences the way we work with others and has made significant impacts on the workplace. The healthcare system is no exception. Healthcare organizations, clinicians, and patients have observed several changes due

I. Chakraborty (✉)
UQ-IIT Delhi Academy of Research, Indian Institute of Technology Delhi, New Delhi, India
e-mail: imon.chakraborty@uqidar.iitd.ac.in

P. Vigneswara Ilavarasan
Department of Management Studies, Indian Institute of Technology Delhi, New Delhi, India

I. Chakraborty · S. Edirippulige
Centre for Online Health, The University of Queensland, Brisbane, Australia

A. Kumar et al. (eds.), *Proceedings of the International Conference on Cognitive and Intelligent Computing*, Cognitive Science and Technology,
https://doi.org/10.1007/978-981-19-2350-0_13

to the integration of new technology [1]. Technological innovation and integration might be a revolution in a healthcare ecosystem, especially in service delivery, but there are noticeable differences in various parts of the world. The government in emerging nations faces challenges in healthcare service delivery due to resource deficiency, for instance, limited primary health centers and inadequate facilities of hospitals for secondary and tertiary care [2].

E-healthcare has been used to address healthcare accessibility issues, especially service delivery needs in rural and remote communities in both developed and developing nations. Some of those solutions, for instance, 'Teladoc', 'CareClix' in the US, and 'Babylon Health' in the UK, succeed extensively in serving underserved communities. These technology-based innovative solutions have primarily emerged from health-tech ventures, which are also recognized as e-health startups. These are usually small companies working in an uncertain environment and dealing with various challenges to create services or products for filling the gap in the healthcare system. In today's world, numerous e-health startups are emerging, but only a few sustain because of the high failure rate in creating a value-based robust business model.

Encyclopedia Britannica uses Peter and Glasser's explanation as a definition of e-health, also called e-healthcare [3]—'Use of digital technologies and telecommunications, such as computers, the Internet, and mobile devices, to facilitate health improvement and health care services'. Here, e-healthcare is conceptualized as the healthcare service delivery using telecommunication technologies. Further, 'e-health startups' represent the information technology-driven health startup that uses the Internet and mobile devices to deliver healthcare services. Such startups continuously evolve to create and capture value in the e-health space.

The existing literature depicted various business frameworks but did not emphasize health-tech startups' value creation and capture aspects in their propositions, especially considering success factors [4]. Value creation and capture are critical concepts in the management literature, especially in the entrepreneurship arena related to creating sustainable ventures. Extensive literature discussed these concepts and highlighted that value depends on the level of analysis, such as individual, organizational, and social. Lepak et al. thoughtfully described value creation and capture as a multi-dimensional perspective in their study. Value creation represents the understanding of a service or product value by a consumer who shows the desire to exchange monetary amounts for receiving that intended value. For capturing the value, a startup's attributes should be unique and difficult enough to imitate by the competitors. It limits the value slippage and generates continuous revenue [5]. The present study utilizes the same concepts in the context of startups' products or services where the intended consumer shows the willingness to pay for the value offered by the startups. For a startup, the value creation activity comprises the proposition of intended value, co-creation with consumer feedback, and establishing effective communication channels to transfer the intended value to target consumers. The value capture primarily focuses on revenue generation to sustain the venture.

Success factors can drive the path of value creation and capture of a venture. These factors are the elements necessary for an organization's or project's success.

D. Ronald Daniel proposed the concept of success factors in 1961. This concept was redefined as critical success factors by Rockart [6]. The success factors denote the essential elements for a startup's value creation and capture. Performance measurement is an essential aspect of a business's or project's effective outcomes. It is generally a non-financial measurement of an organization's activities. The indicators assess the outcome regarding value creation and capture of an e-health startup.

Presently, we sought to investigate the evidence of e-health startups. Here, evidence refers to frameworks of the existing startups, their components, services, and the challenges faced in service delivery. To our knowledge, no current review represents the same in the value creation context. This study undertakes to address the gap and contribute to research knowledge by proposing a conceptual framework and success factors.

2 Methods

We searched Scopus, Web of Science, IEEE Xplore, and Embase databases without any initial time frame to March 2020. We performed several iterations to validate the search keywords—'health-tech' OR 'e-health' OR 'health ICT' OR 'health IT' OR 'health technology' OR 'med-tech' OR 'medical technology' OR 'digital health' AND startup OR startups OR startups OR firms OR firm OR businesses OR business OR incubator. The authors used those keywords in the title, abstract, and keywords fields of every mentioned database to preserve the intent of the query.

2.1 Screening Strategy and Selection Process

We sticked to the preferred reporting items for systematic reviews and meta-analyses guidelines for this review. We used Rayyan, a free Web-based tool [7], for the systematic review. To identify potential articles, we screened the title and abstract using Rayyan. Then, we assessed the full-text eligibility to finalize the articles for this study.

2.2 Data Extraction and Analysis

We extracted the data in an excel sheet that includes country, year of publication, study type, services offered, framework (if any), success factors (if any), and challenges (if any). IC (one author) assessed the full text, while SE and VI (other authors) checked the extracted datasheet. We discussed to resolve any disagreements.

Systematic reviews in business research commonly use explanatory synthesis. We followed a realistic approach [8] to perform the knowledge synthesis on existing e-health startups. This process gives flexibility in assumptions, as the same intervention may represent differently. We considered this method to understand the concepts that help to make a dynamic explanation rather than focusing on general findings of what does or does not serve. We endeavored to represent the existing business frameworks, success factors, and the challenges to sustain in the healthcare space. Then, we portrayed a conceptual framework for e-health startups.

3 Results

We performed the search in four databases and retrieved 721 records. We scanned 537 unique titles and abstracts after removing 184 duplicates. With a conflict discussion, we came up with 28 articles for full screening. The PRISMA flowchart describes the detailed search results in Fig. 1. We finally considered 17 studies in this systematic review. We have excluded 11 full-text articles because the studies represented duplicate content ($n = 1$), systematic reviews ($n = 1$), technology focused ($n = 4$), B-B startups ($n = 2$), and others ($n = 3$). We found a duplicate study at the advanced stage of PRISMA methodology because the title and abstract are different, but the findings were the same. We removed a systematic review article because the review represents the assumption of authors based on other studies for a particular context that might not be suitable for different settings. We eliminated the studies that discussed technological aspects in detail and B-B med-tech startups that are not involved in health service delivery direct to the consumer. We also rejected the studies with context mismatch, such as health-related commercial activity on the Internet as a part of business strategy, not focusing on e-health startups.

3.1 Characteristics of Included Studies

We considered 17 articles that were from 11 journals, five conference proceedings, and one book chapter. We found that the selected studies were between 2005 and 2019. The included papers were from eight developed countries and predominantly from the US ($n = 5$) and Switzerland ($n = 4$). We saw that studies had mainly used a qualitative analysis approach.

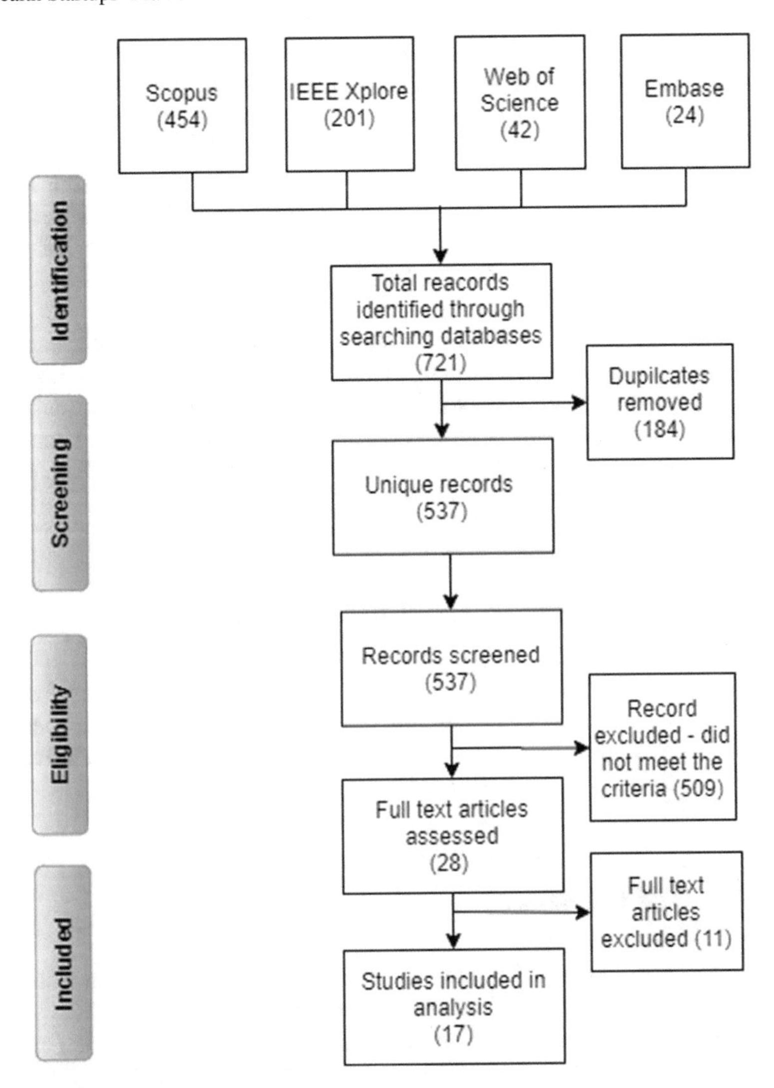

Fig. 1 PRISMA flowchart of the article selection process

3.2 *Details of the Existing Startups*

We found a few frameworks for sustainable startup development representing the components in detail, such as technology, profit formula, people, policy, and governance. Our analysis detected six articles with essential elements for e-health startups. A few common elements were understanding the actual need, a clear objective, revenue, quality, and efficiency [9]. Besides, we found that most of the included articles were descriptive. They did not engage in the validation of their proposed frameworks.

We got nine articles that represented e-health startups, such as 'My referrals', 'Health for me', 'O'care clouds', 'Patients like me', '23 and ME', 'Flatiron health', 'Thai public healthcare', 'MyoTel', 'CeHRes roadmap', 'TMM South Africa', 'Appollo telemedicine', 'Sky health', 'Myca nutrition', and 'Arvind ophthalmology'. The mentioned startups are mostly for-profit startups. Only 'TMM South Africa' was a non-profit venture for e-healthcare services. However, eight remaining articles represented information about startups and a few conceptual frameworks but did not discuss the propositions with existing ventures to direct them for sustainable value creation.

We found that e-health startups covered different services, i.e., chronic care, whiplashes, shoulder pain, nutrition, ophthalmology, eye, on-demand telemedicine, teleradiology, and teledermatology [10]. Nevertheless, based on the current literature, startups did not focus on various e-healthcare services in specialized treatment. Given that, startups have various opportunities for value creation by focusing on underexplored service areas using e-health.

3.3 Startups Challenges in E-Healthcare

We found that most of the health-tech startups were dealing with various obstacles in value creation. Several studies have discussed the challenges faced by them, for instance, IT investment, complex value networks, reimbursement, health records, data security, policy, and regulations [11].

3.4 Proposed Framework

The results from different studies showed that every e-health business has a different frame and components while trying to capture similar kinds of services. We came up with four concepts that emerged from all the components described in the existing frameworks. The concepts are stakeholder, service, technology, and revenue. These concepts lead to the startup's value generation, which depicts the process of creation to capture. We considered the complete aspect of e-health startups based on the literature and proposed a simplified conceptual framework in Fig. 2.

To provide clarity on the concepts in value generation, we have introduced success factors and performance indicators. Our framework, components, and corresponding success factors assist in creating and capturing the value of an e-health startup. Key performance indicators measured the effectiveness of proposed value for sustainable startups. We have represented the proposed success factors in a Venn diagram to indicate the factors essential for value creation, capture or both (Fig. 3). We have also highlighted the success factors associated with each concept using color code in the Venn diagram.

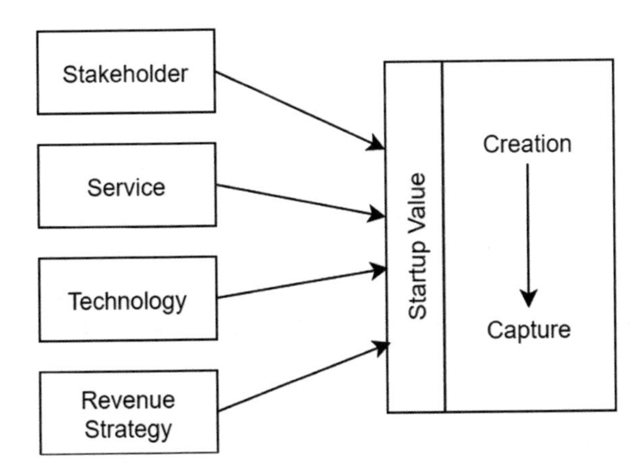

Fig. 2 Simplified value creation and capture framework for e-health startups

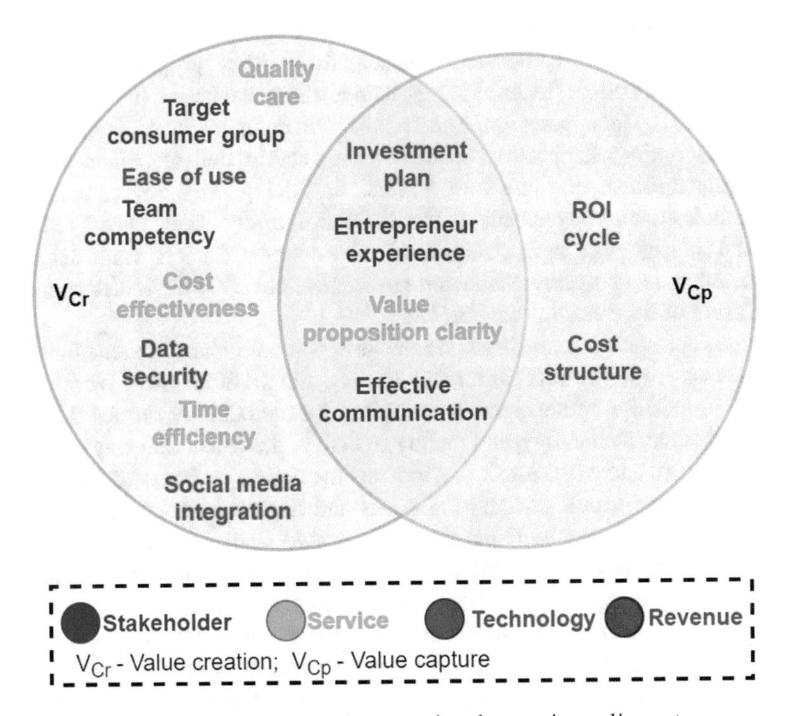

Fig. 3 Success factors associated with each concept in value creation and/or capture

Besides, we found that only creating and capturing value might not be enough to sustain in the healthcare industry unless it also measures the performance. To address this effectively, we proposed a few performance indicators, for instance, stakeholder satisfaction, service accuracy, system performance, team performance, and revenue index.

4 Discussion

The systematic review examines e-health startups for the aspect of value creation. In this section, at first, we elaborately explained the existing framework, its components, and possible factors in creating value in the health space. Next, we represented a few challenges faced by e-health startups. We found that merely proposing a framework is not good enough to create and capture value in healthcare. In this context, we explained our proposed framework with success factors and possible indicators for better direction in value creation and value capture of e-health startups.

4.1 Principal Findings

We observed various components of the existing e-health startups' frameworks. We found the simplest structure with four essential components: service, technology, organization, and finance. Service represented the value proposition and target consumer identification. Technology explained the technical architecture, service delivery process, and application. The organization components depicted the stakeholders' descriptions, their roles, and interaction in service delivery. Finance apprised the investment source, cost structure, and risk for pricing. We observed that another study includes two more elements in the STOF framework. They considered governance and societal value as an extension of value creation. Societal value defined the network value, and customer value and governance observed the issues concerning the diffusion of the e-health process [12].

A study presented a business framework for on-demand telemedicine that comprises value proposition, resources, process, and profit formula. We found that the value proposition offers creation credibility for a product or service. Resources define the people, technology, and facility to deliver the value. The way of working together helps to address the task that induces the delivery process of the proposed value. The profit formula covered the assets and fixed cost structure to reach the revenue of the services. We found that another two studies represented the profit formula as a cost structure and margin to create the velocity for covering the value [13]. One study composed the profit by formulating the cost structure of all the operations for a venture and revenue stream as an income generation source [14].

Besides, another study confirmed that three kinds of business patterns help to create consumer value. They showed e-health as a freemium, crowd-based, and multisided market. In freemium, basic services are free while they charge for premium. We found that the multisided market can create value based on the interaction among partners. In the one-sided market, consumers showed the willingness to pay based on the value of the platform. We noticed that the multisided market value of one partner influenced the value of other partners to measure the platform capability. Crowed-based e-health motivates individuals to produce value for the organization for free. This crowd-based pattern also included the review and comments of an online community [15] that helps consumers choose any services.

Most studies have focused on establishing a framework with few components to create value. But we found that few of the articles represented the essential elements that influence e-health startups' development. We captured a study that discussed five case studies by proposing real need and cost-effectiveness as critical considerations [14]. Another study reflected innovation in e-healthcare, outsourcing assistance, marketing strategy, and win–win partnership for the driving source of health-tech startups. We got a few studies that propose only a few essential business features without offering a framework, for instance, focus, growth management, service quality, market landscape, and network management [13].

We found a revenue model [16] that analyzed a case of 'health for me'. Actors, activities, resources, customer relationships, cost structure, and revenue stream were the business model components. The study explained each component and associated challenges in the two-sided market. The analysis suggested that the selection of the model should depend on the following: the innovativeness to attract new users; the dependence level between platform owner and service provider; the location, requirements, and customers' expectations regarding service delivery.

It was surprising to find that only one article talked about the concept of success factors and performance indicators associated with the business. The article further introduced a balanced scorecard to examine the value of the telehealth system in a healthcare organization. We found that business drivers should have a cost–benefit analysis of the project, a critical SWOT assessment, and benchmarking between the current system. SWOT assessment gives the strength and weaknesses of the internal organization along with external opportunities and threats [11].

Identification of the providers' needs and seekers' demands is imperative to forecast the level of activity required to achieve the return of investment within a specific time. But we found existing e-health startups are unable to fill these values due to several challenges. An analysis showed that myofeedback-based teletreatment was more costly than conventional treatment due to IT investments and operating charges. The study reported that labor cost savings could not compensate for IT costs associated with teletreatment [17].

We found that adequate business model engineering might reduce the risk of failure and help in the value deployment process. The management mechanism should target specific regulations to reach the goal as a part of the policy objective. The components of that mechanism could be financial regulations, licensure, and liability [4].

4.2 Analysis of the Proposed Framework

Most of the articles drew the framework with components. Few of them tried to explain with a case study. But the included studies were inconsistent in finding a robust framework with the success factor for value creation and capture of an e-health startup. Our proposed framework tried to strengthen by introducing the success factors for each component and indicators to measure the performance. We understand that identifying target consumers, entrepreneur's experience, team competency, and effective communication between providers and seekers have significant effects on startups' value creation [18]. Leveraging technological architecture for e-health startups should provide an ease-of-use platform, data security, and integration of social media platforms. E-health startups must have a clear vision of value proposition with quality measurement, cost-effectiveness, and service efficiency. This clear vision might help to reduce hospital visits and increase patient awareness, which influences the startups toward a sustainable goal. Revenue is also an essential consideration for the same. Before embarking on an e-health startup, revenue strategy must be transparent regarding investment plan, cost structure, and return on investment cycle [19].

Performance indicators in our proposition can effectively measure the need and satisfaction of the stakeholders in the e-healthcare space and can also capture intended value effectiveness. Continuous assessment of the activities will help to enhance the service quality and platform performance. Every aspect of establishing a value can sustain when measuring the revenue index for health-tech startups in the short and long-run goals [20].

4.3 Limitations and Recommendation for Future Research

Every systematic review captures a specific topic under a specified boundary. Our systematic review is no exception. Here, we only considered the papers written in English. So, the search results are bound to only one language that might not cover some significant publications in other languages. As the field of this study is interdisciplinary, we captured the articles from four bibliographic databases. But we did not consider other health space databases, gray literature, and magazines. This might be a drawback to obtaining a few existing e-health startups. This study talked about an emerging field that is changing rapidly. There is always a possibility of bias and information overlooked in qualitative research. Nevertheless, to downplay it, one author (IC) extracted the data from existing literature, and other authors (SE, VI) verified them. Several systematic reviews in the health space follow the quality assessment test. As this is an emerging field and the number of publications is relatively less, so we bound this study without a quality check.

The objective of a systematic review is to produce a complete, exhaustive summary of existing literature to address a specific problem that opens future opportunities. Research should include papers beyond the English language and other modes of publication, for instance, magazines, and newspaper articles, to gather more significant input. We found only a few studies on challenges. Therefore, it is also advisable to inquire more about the specific difficulties encountered by e-health startups. Our proposition is a conceptual framework that suggests further exploration, theoretical strengthening, and statistical validation for a good acceptance.

5 Conclusion

The present systematic review identified that most of the startups' frameworks did not consider success factors and performance indicators to evaluate the platform efficiency for sustainable value creation in the healthcare market. Drew on the synthesis of findings from existing literature, this study conceptually proposed a value creation and capture framework with success factors. In the broader context, research regarding e-health startups is emerging. This study provides a rich picture of the startup framework in e-healthcare. It might help healthcare entrepreneurs to mold their startups toward an appropriate path to a greater extent. From the researchers' perspective, this knowledge synthesis will further look forward to more theoretical grounding on the study proposition.

References

1. Haluza D, Jungwirth D (2015) ICT and the future of health care: aspects of health promotion. Int J Med Inform 84:48–57. https://doi.org/10.1016/j.ijmedinf.2014.09.005
2. Hossain N, Yokota F, Sultana N, Ahmed A (2019) Factors influencing rural end-users' acceptance of e-health in developing countries: a study on portable health clinic in Bangladesh. Telemed e-Health. 25:221–229. https://doi.org/10.1089/tmj.2018.0039
3. E-health| health care|Britannica. https://www.britannica.com/science/e-health
4. Kijl B, Nieuwenhuis LJM (2011) Deploying e-health service innovations—an early stage business model engineering and regulatory validation approach. Int J Healthc Technol Manage 12:23–44. https://doi.org/10.1504/IJHTM.2011.037219
5. Lepak DP, Smith KG, Taylor MS (2007) Value creation and value capture: a multilevel perspective. https://doi.org/10.5465/amr.2007.23464011
6. Rockart J (1979) Chief executives define their own data needs. Harvard Bus Rev 57(2):81–93
7. Ouzzani M, Hammady H, Fedorowicz Z, Elmagarmid A (2016) Rayyan—a web and mobile app for systematic reviews. Syst Rev 5:1–10. https://doi.org/10.1186/s13643-016-0384-4
8. Mary Q (2002) Evidence-based policy: in Search of a method. ISSN: 1356-3890
9. Sprenger M, Mettler T (2016) On the utility of e-health business model design patterns. In: 24th European conference on information system. ECIS 2016
10. Van Dyk L, Wentzel MJ, Van Limburg AHM, Van Gemert-Pijnen JEWC, Schutte CSL (2012) Business models for sustained e-health implementation: lessons from two continents. Proc Int Conf Comput Ind Eng CIE 2:889–904
11. Vannieuwenborg F, Van der Auwermeulen T, Van Ooteghem J, Jacobs A, Verbugge S, Colle D (2017) Bringing e-care platforms to the market. Inform Heal Soc Care 42:207–231. https://doi.org/10.1080/17538157.2016.1200052
12. Wass S, Vimarlund V (2016) Business models in public e-health. In: 24th European conference on information system. ECIS 2016
13. Muhos M, Saarela M, Foit D, Rasochova L (2019) Management priorities of digital health service start-ups in California. Int Entrep Manage J 15:43–62. https://doi.org/10.1007/s11365-018-0546-z

14. Cheng A, Chen S, Mehta K (2013) A review of telemedicine business models. Telemed e-Health 19:287–297. https://doi.org/10.1089/tmj.2012.0172
15. Alasmari A, Zhou L (2019) How multimorbid health information consumers interact in an online community Q&A platform. Int J Med Inform 131:103958. https://doi.org/10.1016/j.ijmedinf.2019.103958
16. Vimarlund V, Mettler T (2017) Business models in two-sided markets (analysis of potential payments and reimbursement models that can be used). Elsevier Inc. https://doi.org/10.1016/B978-0-12-805250-1.00009-5
17. Kijl B, Nieuwenhuis LJ, Huis in't Veld RM, Hermens HJ, Vollenbroek-Hutten MM (2010) Deployment of e-health services—a business model engineering strategy. J Telemed Telecare 16:344–353. https://doi.org/10.1258/jtt.2010.006009
18. Sehlstedt U, Bohlin N, de Maré F, Beetz R (2016) Embracing digital health in the pharmaceutical industry. Int J Healthc Manage 9:145–148. https://doi.org/10.1080/20479700.2016.1197513
19. Mas JP, Hsueh B (2017) An investor perspective on forming and funding your medical device start-up. Tech Vasc Interv Radiol 20:101–108. https://doi.org/10.1053/j.tvir.2017.04.003
20. Kapur PK, Nagpal S, Khatri SK, Yadavalli VSS (2014) Critical success factor utility based tool for ERP health assessment: a general framework. Int J Syst Assur Eng Manage:133–148. https://doi.org/10.1007/s13198-014-0223-8

A Review of Scene Text Detection and Recognition of South Indian Languages in Natural Scene Images

Vishnuvardhan Atmakuri and M. Dhanalakshmi

Abstract One of the key difficulties of computer vision is the localization and retrieval of words or phrases from natural scene images. It is a technique that is used to recognize and isolate the desired text from the images. There are researchers who explored this field and concluded with good results, they mainly concentrated on the English language, but there is a need to work on local/regional languages, Country like India is having a vast portion of the rural area so there is a need to highlight the different Indian languages and also addresses the official scripts and their Unicode ranges. In this paper, various detection or recognition approaches are highlighted especially for the scene images containing South Indian languages.

Keywords End-to-end text recognition · Deep learning · Text detection · Text recognition

1 Introduction

Text is an effective mode of communication and has a significant impact on our lives. It can be used to transmit information by incorporating it in documents or scenes. Texts in scene images provide a high-level semantic information that aids in the analysis and perception of the corresponding environment. Finding or locating the text in scene images is useful in different areas such as robotics, image search, instant translation, human computer interaction, and semantic natural scene understanding. Numerous challenges are raised during the process of detecting and recognizing the text from natural scene images such as scene complexity, text diversity,

V. Atmakuri (✉)
JNTUH, Hyderabad, India
e-mail: vishnuvardhan.a@vvit.net

Department of CSE, Vasireddy Venkatdri Institute of Technology, Nambur, Guntur, Andhra Pradesh, India

M. Dhanalakshmi
Department of Information Technology, JNTUH Jagityala, Hyderabad, India

A. Kumar et al. (eds.), *Proceedings of the International Conference on Cognitive and Intelligent Computing*, Cognitive Science and Technology,
https://doi.org/10.1007/978-981-19-2350-0_14

distortion factors, and multilingual environments. Keeping these challenges in mind, researchers across the globe striving to improve the accuracy of recognition.

Regarding the text exists in images there are various modalities such as scene text, printed text, handwritten text, and text in video frames, captions in video and images which are shown in Fig. 1.

Out of all these, text in a scene images gives important information such as contextual clues, assistance for the visual impaired people, and navigation in unknown areas, so significant research takes place on this area, i.e. detecting and recognizing the text from scene images. Different stages are involved in this entire process, text detection, text localization, text extraction and enhancement, and text recognition. Figure 2 shows how the input image is converted into the text in the form of a flow diagram.

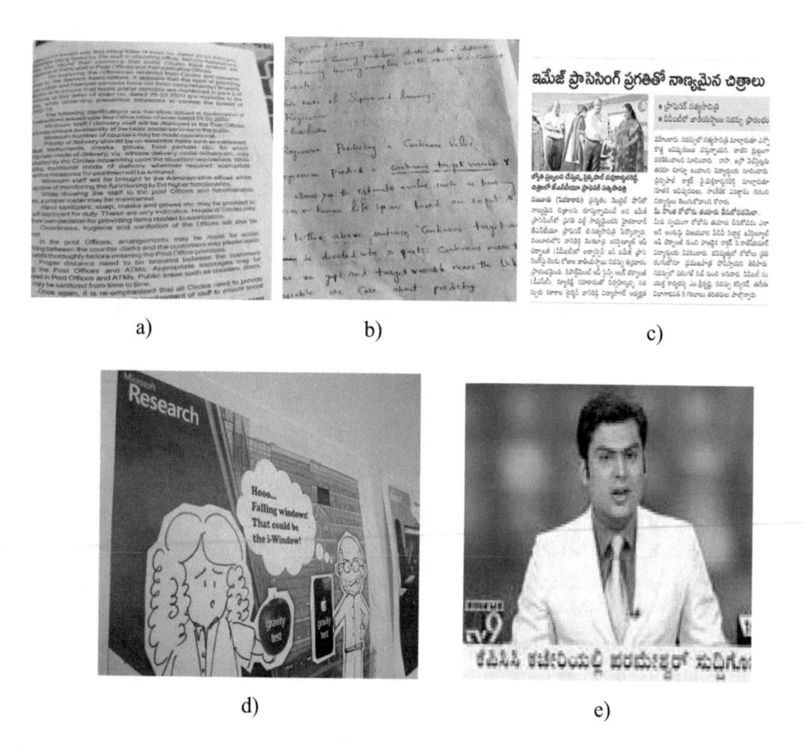

Fig. 1 Visual illustration of different modalities of text—**a** printed text, **b** handwritten text, **c** newspaper, **d** scene text, and **e** text in a frames of a video [1, 2]

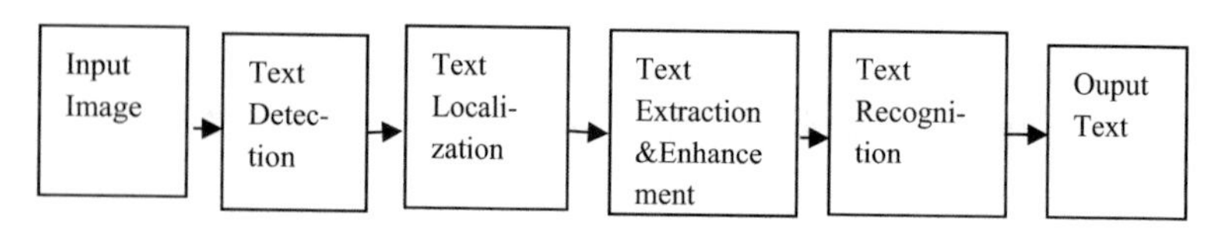

Fig. 2 Flow diagram of text information extraction process [3]

Fig. 3 Text is detected and finally recognized by the machine [4]

Fig. 4 Examples showing the images from IIIT-R text datasets. **a** Linear non-horizontal text image and **b** curved text image [5]

Initially, research is focussed mainly on the identifying and recognizing the English text, example is shown in Fig. 3, that too in horizontal orientation.

Later, the work is progressed towards non-horizontal but linear, curved, and so on (Fig. 4), as the successful results obtained, slowly the work is progressed to other languages like Chinese, Japanese, and Korean.

In general, scene text detection and recognition involve two sub-problems such as text detection and text recognition, specifically few researchers do both, i.e. text detection followed by text recognition known as end-to-end recognition. In [6], Prof. X. Bai mentioned scene text recognition and end-to-end recognition paid less attention compared to the text detection in the ICDAR 2017 Keynote. In order to encourage the people working on this area, competitions are conducted by ICDAR [7] and CBDAR [8] for every 2 years. Recent competitions mostly focussed on multilingual texts, as India is a country of that type so there is a need to detect and recognize the text from the scene images containing Indian languages specifically South Indian side. The main reason for shifting towards this trend is even though there are so many languages are widely used throughout the world, but the existing methods are mainly focussed on to recognize the English textual content in the natural scene images. In the present situation, where the people are navigating the entire world and come across with so many languages, in India also people from different states speak different languages so it is crucial and necessary requisite to develop systems that can tackle multilingual texts and assist the people.

2 Overview of Indian Languages and Scripts

There are more than 720 dialects derived from around 13 scripts that are used in India. Out of these 720 dialects, currently, there are 22 official languages of India can be read and write with the help of scripts. Indic scripts are represented as a 16 bit representation of a symbol, termed as UNICODE. Indic scripts are inherited from Indian Standard Code for Information Interchange (ISCII). Its derivative ISCII uses an 8 bit representation of a symbol. In some approaches [9], script identification takes place prior to the recognition of the text, so to get the better understanding the official 13 scripts, 22 languages and their Unicode ranges are shown in Table 1.

2.1 *Languages in South India*

Multiple languages are used in different parts of South India, for communication English language is the best choice but there are some limitations in understanding it as there is some kind of peculiarities in pronunciation. Telugu, Tamil, Kannada, and Malayalam are the languages used here in the southern part of India common point between them is they are all Dravidian languages, so they share structural aspects shown in Fig. 5.

Table 1 Unicode ranges for major Indian scripts [10, 11]

Script	Language(s)	Unicode begin	Unicode end
Ol-chiki	Santhali	0x2D80	0x2DAF
Arabic	Urdu, Kashmiri	0x0600	0x06FF
Devanagari	Hindi, Sanskrit, Marathi, Santhali Nepali, Konkani, Maithili, Sindhi, Dogri	0x0900	0x097F
Bengali	Bengali, Assamese, Manipuri	0x0980	0x09FF
Gurumukhi	Punjabi	0x0A00	0x0A7F
Gujarati	Gujarati	0x0A80	0x0AFF
Oriya	Oriya	0x0B00	0x0B7F
Tamil	Tamil	0x0B80	0x0BFF
Telugu	Telugu	0x0C00	0x0C7F
Kannada	Kannada	0x0C80	0x0CFF
Malayalam	Malayalam	0x0D00	0x0D7F
Meiti Mayak	Manipuri	0xABC0	0xABFF
Sharda	Dogri, Kashmiri	0x11180	0x111DF

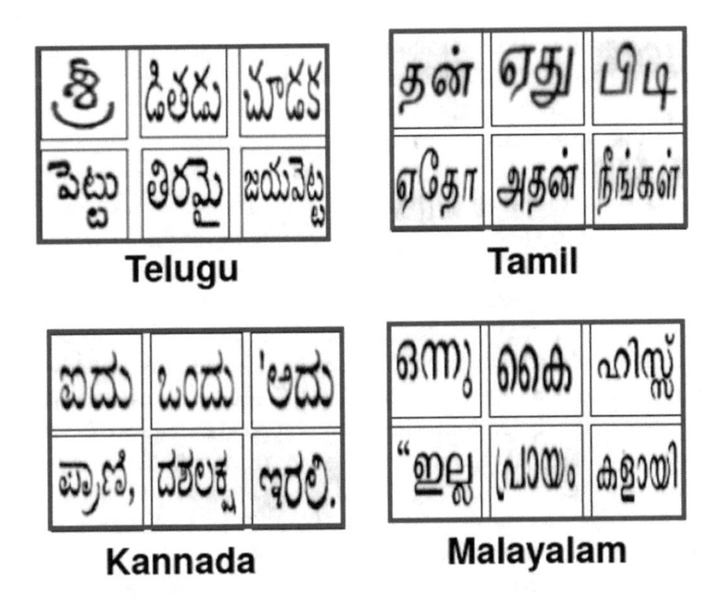

Fig. 5 Visual illustration of South Indian scripts [12]

3 Scene Text Detection and Recognition of Indic Scripts

To recognize text in a scene image containing Indian languages, there are mainly three approaches: (a) segmentation (b) hidden Markov models [13] (HMM), and (c) recurrent neural networks [14]. First approach faces difficulty in the cases where there is a connection between the characters, so later, recognition module depends on segmentation free approaches like 2 and 3. Compared to the HMM, RNN is the best choice.

There has been a little progress on the scene text detection and recognition in Indic scripts. Out of those, some of the prominent ones are highlighted here.

In [15] with the usage of Unicode fonts for Indic scripts from a huge vocabulary, artificial images are generated by varying parameters such as texture and colour, visually trying to make these synthetic images appear to be real scene images. A hybrid CNN-RNN architecture is employed, and it is trained completely on synthetically rendered word images. Later, after the training, the network is tested on IIIT-ILST dataset which is a new real scene text dataset.

In [16], regional languages (specifically Hindi, Telugu, and Kannada) are extracted from an image in an offline mode; initially, low-quality images are corrected by using existing deep learning method, and then, deep convolutional sequence is used to detect Indian languages followed by finding the meta information to identify GPS coordinates and finally proceeds to find the regional language. It is not robust to occlusion, blur, and small font.

In [17], Kannada text is detected in the scene images, and it is achieved in two steps: In the first step, for the sharpening of the edges and also to extract texture information, integration of Gabor filter and wavelet transform takes place. Wavelet entropy is applied to the final Gabor filtered images in second step to detect the text regions based on the average energy calculated.

In [18], texture information and sharp edges are extracted by the application of 2D wavelet transform and Gabor filters. To classify foreground, background, and true text pixels, K-means clustering is adopted with the k-value $= 3$. Locations of each and every connected component are attained by applying the linked list approach and morphological operations. Finally, for an input image, wavelet entropy is calculated to deduce the true text region of an image containing all South Indian languages.

In [19], MSER is applied to get the initial candidates. For the purpose of refinement of initial candidates, localized binarization technique is used. Stroke width transform is applied to filter out the non-text, SVM with RBF used for further classification of text from non-text, finally Grab cut algorithm is applied to improve the recall.

By introducing a new concept called iterative nearest neighbour symmetry, convex and concave deficiencies are investigated in [20]. To choose a candidate plane from eight planes to represent meaningful information. They suggested a new idea of mutual nearest neighbour pair (MNNP) components identification to identify the representatives of texts based on the outward gradient direction of components. Introduced a new notion for character recognition based on the angular relationship between fusion and high-frequency wavelet sub-bands to determine an automated window according to character size. With the use of an HMM model, proposed a combination of statistical-textures and spatial information-based features in the contourlet wavelet domain for recognition.

In [21], Gabor filter is applied on the image to extract the uncertainty features, and 2D wavelet transform is applied to find the locations of the text information. Finally, the non-text information is removed using the textual information based on the edge information.

In [11], Gaussian low-pass filter is used to reduce the noise and to smoothen the image, to get the texture features and edge information, a single level 2D discrete wavelet transform is used then the level set function executed continually to detect the true text regions. Finally, unsupervised K-means clustering algorithm and Gaussian mixture model are employed to obtain the segmented characters.

4 Datasets and Evaluation Protocol

Regarding the Indian languages, most of the researchers use their own custom dataset in order test and train their framework, and only few datasets are available publicly such as multiscript robust reading competition (MRRC) [9] contains 167 training and 167 testing images with all kinds of challenges and this dataset used by [11, 15, 21]. Uses IIIT-ILST dataset [22] contains thousands of images captured through smart phones or camera occurring in various contexts such as local markets, billboards, navigation and traffic signs, banners, and graffiti, and for the purpose of training, this dataset covers a wide variety of naturally occurring image-noises and distortions.

KAIST [23] used by [16] it provides the different perspectives of the world captured in coarse time slots (day and night), in addition to fine time slots (sunrise, morning, afternoon, sunset, night, and dawn).

Coming to the evaluation protocol, majority of the works uses the default evaluation measures such as precision, recall, and *f*-measure, however, some recent methodologies, like [15], employ two basic evaluation criteria to evaluate text recognition tasks: The Levenshtein distance [24] is indeed the least number of single-character edit operations (insertions, deletions, and substitutions) required to transform one word to another. Word error rate (WER) [25] is a word metric that, like the character metric, is based on Levenshtein distance but at the word level, i.e. a minimum number of single-word operations prescribed to modify one text to other. Different criteria are used in [26] for the purpose of evaluation where detection rate, false positive rate, and miss detection rate are calculated.

5 Conclusion

Scene text detection and recognition of native languages are really essential to the people located at the rural backgrounds. In the past scene, text recognition was done by the efficient algorithms in the computer vision, as after the inclusion of deep learning in this research, then overall task was simplified at every stage in the pipeline. After getting success of usage of the deep learning for recognizing English text, researchers were focussed on recognizing the text of Indian languages. But, as on date, only a small portion of progress was taken place in this dimension. Hindi, Devanagari, Telugu, Malayalam, and Kannada were the prominent Indic scripts were recognized, still there is need to explore more on this research area.

References

1. Huang R, Xu B (2019) Text attention and focal negative loss for scene text detection. Int Joint Conf Neural Netw (IJCNN) 2019:1–8. https://doi.org/10.1109/IJCNN.2019.8851959
2. Boaz TK, Prabhakar CJ (2013) A novel approach for detection and localization of caption in video based on pixel pairs. In: National conference on challenges in research & technology in the coming decades (CRT 2013), pp 1–6. https://doi.org/10.1049/cp.2013.2488
3. Soni R, Kumar B, Chand S (2019) Text detection and localization in natural scene images based on text awareness score. Appl Intell 49:1376–1405. https://doi.org/10.1007/s10489-018-1338-4
4. Zhu Y, Yao C, Bai X. Scene text detection and recognition: recent advances and future trends. https://doi.org/10.1007/s11704-015-4488-0
5. Sain A, Bhunia AK, Roy PP, Pal U. Multi-oriented text detection and verification in video frames and scene images. https://doi.org/10.1016/j.neucom.2017.09.089
6. http://u-pat.org/ICDAR2017/keynotes/ICDAR2017_Keynote_Prof_Bai.pdf
7. https://dblp.org/db/conf/icdar/index.html
8. http://www.wikicfp.com/cfp/program?id=375

9. Multi-script robust reading competition ICDAR 2013. http://mile.ee.iisc.ernet.in/mrrc/index.html
10. Ding K, Liu Z, Jin L, Zhu X (2007) A comparative study of GABOR feature and gradient feature for handwritten Chinese character recognition. In: International conference on wavelet analysis and pattern recognition, Beijing, China, 2–4 Nov 2007, pp 1182–1186
11. Basavaraju HT, Aradhya VNM, Pavithra MS et al (2020) Arbitrary oriented multilingual text detection and segmentation using level set and Gaussian mixture model. Evol Intel. https://doi.org/10.1007/s12065-020-00472-y
12. Tulsyan K, Srivastava N, Mondal A, Jawahar CV. A benchmark system for Indian language text recognition. https://doi.org/10.1007/978-3-030-57058-3_6
13. Natarajan P, MacRostie E, Decerbo M (2005) The BBN Byblos Hindi OCR system. DRR 2005
14. Mathew M, Singh AK, Jawahar CV (2016) Multilingual OCR for Indic scripts. DAS 2016
15. Mathew M, Jain M, Jawahar CV (2017) Benchmarking scene text recognition in Devanagari, Telugu and Malayalam. https://doi.org/10.1109/ICDAR.2017.364
16. Nag S, Ganguly PK, Roy S. Offline extraction of indic regional language from natural scene image using text segmentation and deep convolutional sequence. https://doi.org/10.1007/978-981-13-2345-4_5
17. Aradhya VNM, Pavithra MS, Naveena C (2012) A robust multilingual text detection approach based on transforms and wavelet entropy. Proc Technol 4:232–237
18. Pavithra MS, Aradhya VNM (2014) A comprehensive of transforms, Gabor filter and k-means clustering for text detection in images and video. Appl Comput Inform
19. Bosamiya JH, Agrawal P, Roy PP, Balasubramanian R (2015) Script independent scene text segmentation using fast stroke width transform and grab cut. In: 2015 3rd IAPR Asian conference on pattern recognition
20. Raghunandan KS, Shivakumara P, Roy S, Hemantha Kumar G, Pal U (2018) Multi-script-oriented text detection and recognition in video/scene/born digital images. IEEE Trans Circ Syst Video Technol. https://doi.org/10.1109/TCSVT.2018.2817642
21. Naveena C, Ajay BN, Manjunath Aradhya VN (2019) Transform-based text detection approach in images. In: Satapathy S, Bhateja V, Somanah R, Yang XS, Senkerik R (eds) Information systems design and intelligent applications. Advances in intelligent systems and computing, vol 863. Springer, Singapore. https://doi.org/10.1007/978-981-13-3338-5_40
22. https://cvit.iiit.ac.in/research/projects/cvit-projects/indic-hw-data
23. https://sites.google.com/site/pedestrianbenchmark/
24. Levenshtein V (1966) Binary codes capable of correcting deletions, insertions and reversals. Soviet Phys Doklady
25. Klakow D, Peters J (2002) Testing the correlation of word error rate and perplexity. Speech Commun
26. Manjunath Aradhya VN, Pavithra MS (2016) A comprehensive of transforms, Gabor filter and k-means clustering for text detection in images and video. Appl Comput Inform 12(2):109–116. ISSN 2210-8327. https://doi.org/10.1016/j.aci.2014.08.001

Prognasticating Stock Value Asset Using Machine Learning

B. E. Manjunathswamy, Riya Chauhan, A. Shireen Bano, and G. Shreevarshini

Abstract The paper aims to predict future stock values. The stock market check is an exceptionally fascinating errand that joins high substances of how the budgetary exchange limits, and what unconventionality can be prompted in market in different conditions to get better accuracy. Evaluating the stock value assets accurately will minimize the risk and bring a maximum amount of profit for all the stakeholders. In the paper, we are going to focus mainly on the three machine learning algorithms, which are more feasible and give high accuracy. When stockholders evaluate shares, they use a lot of quantitative and technical analysis, which is not always accurate. However, utilizing this prediction model and specific machine learning techniques, it can be improved. Algorithms for learning the model are based on linear regression, LSTM, ARIMA, and sentiment analysis. In this study, models for predicting stock prices in various markets employing assets daily.

Keywords Prognostication stock · ML · Linear regression · LSTM · ARIMA · NSE

1 Introduction

The stock market is a crucial aspect in a country's economic development. The right time to buy and sell the shares depends on predicting the trends in the stock market. To achieve high accuracy, prediction is made based on historical data (previous stock values), and the models developed using sentiment analysis and machine learning algorithms. The prediction is assumed to be robust, accurate, and more efficient. The model should try to consider real-life scenarios. It is also expected to focus on all the variables which might affect the stock values and accuracy.

Stock market price prediction might look like a random process; but it is not. The stock price varies from time to time and gives a linear curve. Usually, most people focus on those stocks whose prices are expected to rise in the further future and

B. E. Manjunathswamy (✉) · R. Chauhan · A. Shireen Bano · G. Shreevarshini
Don Bosco Institute of Technology, Bangalore, India
e-mail: manjube24@gmail.com

A. Kumar et al. (eds.), *Proceedings of the International Conference on Cognitive and Intelligent Computing*, Cognitive Science and Technology,
https://doi.org/10.1007/978-981-19-2350-0_15

buy only those stocks whose prices have gone high. The uncertainty and the risk in the stock market make people doubt and refrain from investing in stocks. Thus, the accuracy in predicting the future stock values is very important. The datasets consist of details like the close price, data, volume, open price, and various variables, which are required for prediction [1].

The paper mainly emphasizes, analyzing the historical data and tweets, based on which it will predict the future stock value. This paper creates a Web-based app using sentiment analysis of tweets and machine learning. The front end of the Web-based app is WordPress and flask. It forecasts the stock price value for the next seven days based on NSE as input by the user. Three algorithms are used for predictions, which are LSTM, ARIMA, and linear regression.

Finally, it will forecast whether the stock price will climb or fall in future. This research proposes a prediction model that will assist a trader in trading with reduced ambiguity and risk of losing money based on a gut feeling. The model relies on historical data, which is simply a collection of stock prices. The dataset is separated into a test set and a train set, which are then fed into the data's machine learning model. Our predictive model's main goal would be to maximize a user's investment earnings while also identifying the safe stock to investing [2].

2 Machine Learning in Stock Prediction

The prediction has moved up into the technological domain with the advancement in digitization, the intelligence, which powers the system to learn and enhance from past experiences without being programmed again and again. Lately, most of the researchers are using more machine learning algorithms. The predictions are used to predict future stock values [3].

In the industries, many machine learning models ranging from linear regression to sophisticated auto-LSTM and ARIMA have been applied for prediction. In this research, we employ ARIMA, the linear regression technique, and the LSTM to create a model that can help predict the average stock prices [4].

The most efficient way to apply machine learning algorithms for stock forecasting is to provide Most of the stock brokers while deciding to buy shares utilizes the fundamental, specialized, or time series analysis. Overall, these techniques did not give much accuracy and could not be trusted completely, so there emerges the need of using a certain technique that could give more accuracy to financial exchange predictions. For accurate results, the methodology used was implemented as machine learning.

Only by considering characteristics such as precision, speed, and accuracy will a model be successful. However, one of the most essential reasons to apply machine learning in the stock market is to improve prediction precision and accuracy so that stock brokers may be confident in their decisions to sell or buy shares. The model should be created in such a way that the stock broker's profit from trading his securities can be maximized. What could happen, what has happened, and what should we do? All of these question should be answered by the model [5, 6].

3 Tools and Technologies Used in Developing the Model

- **WordPress**

WordPress is a free and open-source Website-building.

- **Platform 2. Scikit-Learn**

It is machine learning Python library, which chiefly includes arrangement, relapse, and bunching calculations like help vector machines or SVM, inclination boosting, irregular woods, and k-nearest neighbor.

- **Flask**

Flask is popular framework, which is used for developing Web pages.

- **TensorFlow**

In machine learning, TensorFlow is free and open-source software. It can be used for various ranges of tasks but has a specific focus on training and conclusion of deep neural networks.

- **PHP, CSS, JavaScript**

Are used for designing Web pages.

- **Python IDLE**

It is a Python supervisor where in we will really execute the calculation to separate informational indexes, perform controls, and anticipate the normal out comes with exactness.

4 Architecture Designs

See Fig. 1.

Three parts contribute to the predictive model's architecture

(1) **Data Abstraction**

In this model, the method tries to fetch stock market data of the National Stock Exchange (NSE) through Yahoo finance API, which is being targeted and will be

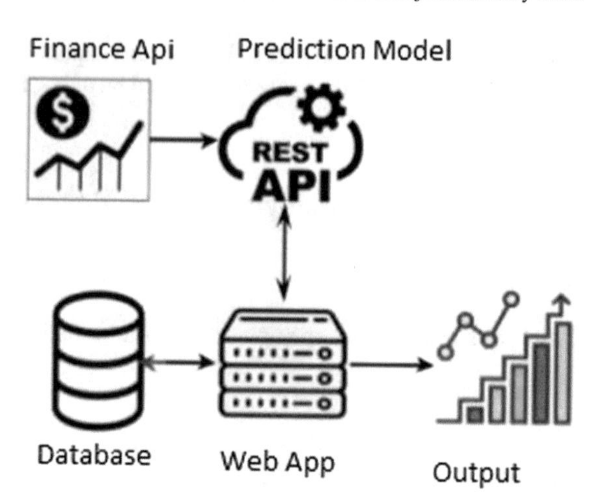

Fig. 1 System architecture

saved in the Python object. This object contains the raw data that must go through the data cleansing. Data cleansing is a process of distinguishing irrelevant, incorrect, and improper data format to required data format. This will remove all the corrupt data, which makes the data cleaner that leads to results that are more accurate. Once the data are clean, then it is fed to the specific algorithm model and utilized in testing and training of the model. And it will be rendered in the WordPress app through representational state transfer API (REST API) via GET method. Where all the HTML components are rendered to the Web app, and the output of the graphs will be displayed [7].

(2) **Prediction Model**

This module entails estimating the stock price using predetermined datasets obtained from the data abstraction module. Linear regression is a technique for establishing a link between a dependent variable and an independent variable. It has the formula $y = a + bx$. The model's purpose is to determine the best fit the actual estimated return of our stock to the predicted value [base paper] with the least sum of squared errors. The key advantage of utilizing this model to predict stock market movements is that the claims are based on a significant quantity of data and are generally based on the market's long-term history. As a result, LSTM controls error by assisting RNNs by retaining knowledge from earlier stages, allowing for prediction that is more accurate. The autoregressive integrated moving average (ARIMA) model forecasts future data using non-stationary data. Non-stationary data, such as stock market data, are used as an input in this model. In comparison with linear regression, ARIMA gives findings that are more accurate.

(3) **Web Page Rendering and Database**

The database used in this module is MySQL, which stores the computed results and is fetched by the front-end application. The database also maintains the user details for verification purposes. The front-end application is a WordPress app, which is deployed in the amp server.

The WordPress app is a Web-based application that gives all the basic functionality, which are required for a Website to be up and running. Once the WordPress application is up and running, whenever the user hits the search button for any stock, the WordPress application sends the get request to Python-flask application, which in turn communicates to Yahoo finance API, to get the data. After the arrival of data, the data are cleaned, and it is supplied to the machine learning models. Later, a static HTML page is rendered from flask app to the WordPress, which contains the predicted output for that company's stock.

5 Methodology and Implementation

The autoregressive integrated moving average (ARIMA) model uses non-stationary data to forecast future data. This model uses non-stationary data as an input, such as stock market data. When compared to linear regression, ARIMA yields more precise results. It included symbols such as open, close, low, high, and volume. Only one company's data were considered for demonstration purposes [8, 9]. All of the data were in JSON format, and it was cleansed to get only the essential fields in a certain format before being read and turned into a data-frame using Python's Pandas library. By segregating data depending on the symbol field provided by the user, the data for one company were extracted. Following that, the data were normalized using the Python sklearn. Metrics module and the data were split into training and testing sets. Although there are various algorithms for predicting stock prices in machine learning, we will focus on three algorithms [10, 11].

(1) **Linear Regression Mode**

We try to partition the dataset into a train set and a test set using the linear regression algorithm. Linear regression is a technique for establishing a link between two variables. A dependent variable, an independent variable, and a variable that is unaffected, it goes like this: $y = ax + b$. The model's purpose is to determine the line that best fits the actual average value of our stock to the predicted value [base paper] with the least sum of squared errors. Because the stock market is closed on Saturdays and Sundays, the model will only work five days a week after it is installed.

In general, the regression-based model is used to predict continuous values from a set of independent variables. For predicting continuous values, regression employs a given linear function:

$$v = a + bx + \text{error} \tag{1}$$

where V is a continuous variable, K denotes known independent values, and a and b denote coefficients.

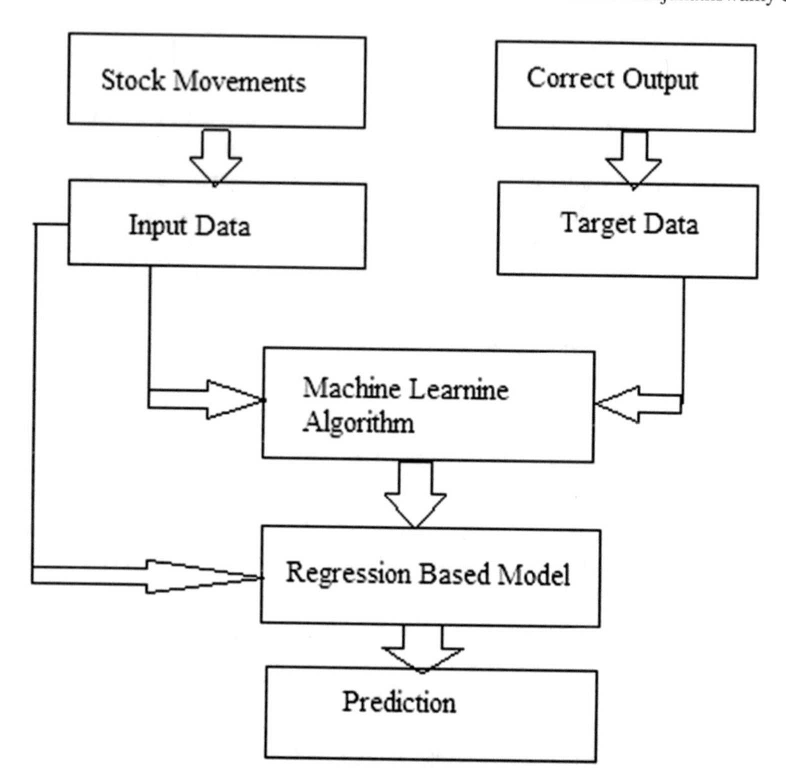

Fig. 2 Regression-based linear regression mode flowchart

Words as shown in Fig. 2, the paper uses the gradient descent linear regression approach to predict correct values by minimizing the error function.

On this data, linear regression is done as guided by the preceding Eq. 1, and the relevant predictions are created. Low, open, high, close, and volume were the variables examined in the regression.

(2) Long Short-term Memory (LSTM) Model

The evolved version of recurrent neural networks (RNNs) is LSTM, which keeps information from prior states. Recurrent networks do not just have one-way neural connections; neurons can also send input to the previous or same layer in a multidirectional manner. The key benefit of employing this model in stock market forecasting is that the predictions are based on enormous amounts of data and are generally based on the market's long-term history. As a result, LSTM aids RNNs by retaining knowledge from earlier stages, improving prediction accuracy, and reducing error [3] (Fig. 3).

(3) Autoregressive and Moving Average

A general modification of an autoregressive moving average (ARMA) model yields the autoregressive integrated moving average (ARIMA) model. ARIMA (p, d, q) is a

Fig. 3 Model of the long short-term memory (LSTM) network model

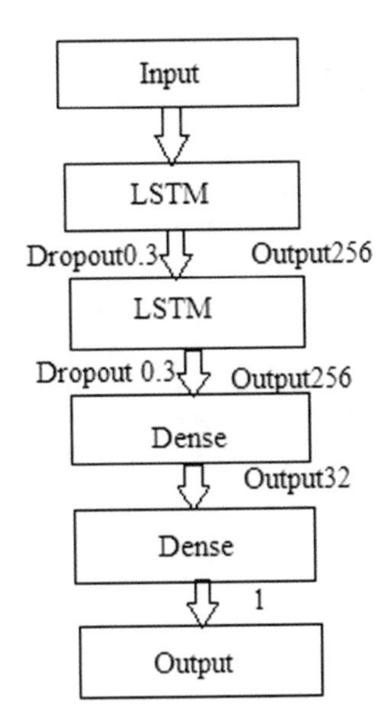

model type in which p signifies the autoregressive portions of the dataset, d denotes the integrated parts of the dataset, and q denotes the moving average parts of the dataset, and p, d, q are all non-negative integers [2].

6 Experimental Results

Figure 4 represents dash board, Fig. 5 is a stock market data representation, Fig. 6

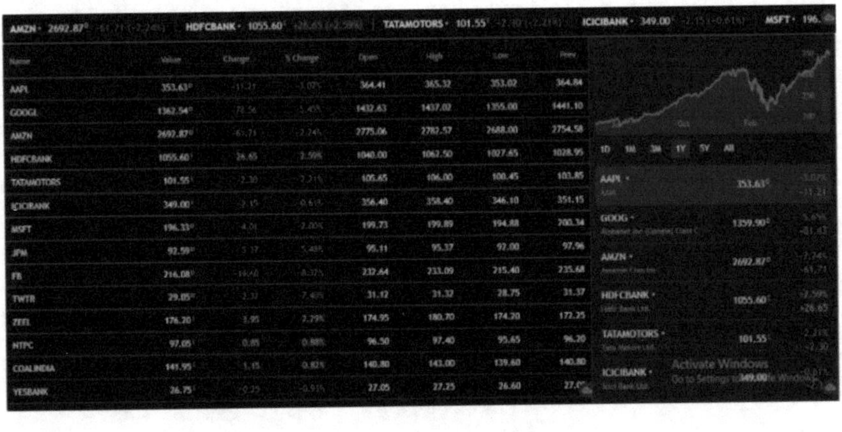

Fig. 4 Dashboard

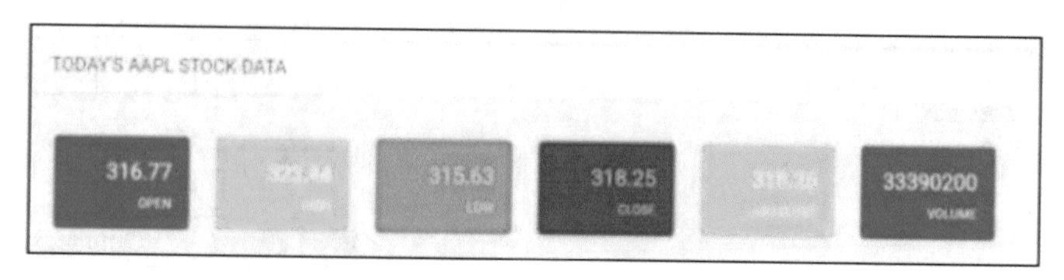

Fig. 5 Stock data

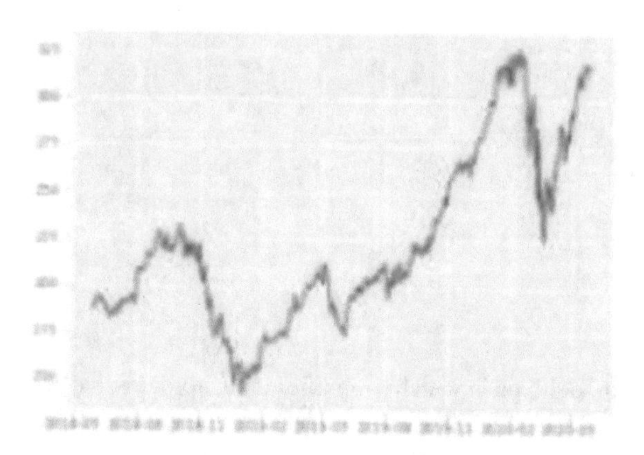

Fig. 6 Recent trend in APPL stock price

recent trend in APPL stock price, Fig. 7 ARIMA model accuracy, Fig. 8 LSTM model accuracy, Fig. 9 linear regression model accuracy results as it can be grasped in the figure given above, the model is trained using LSTM, linear regression, and ARIMA algorithms. The output gives separate individual graph by which customer can compare with and go with best choice along with the graphs the open price, closing price helps customer to invest their money in stock market.

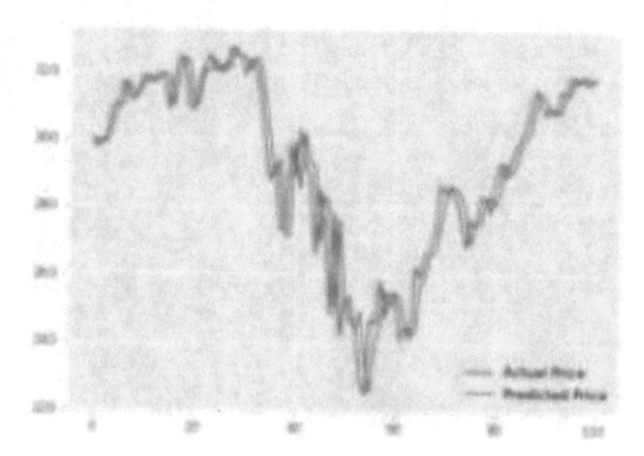

Fig. 7 ARIMA model accuracy

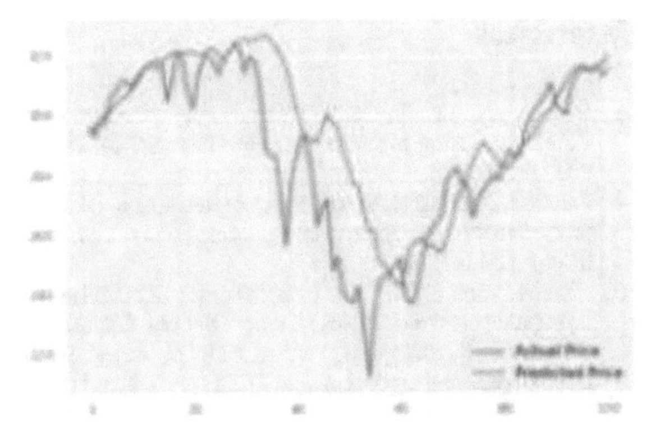

Fig. 8 LSTM model accuracy

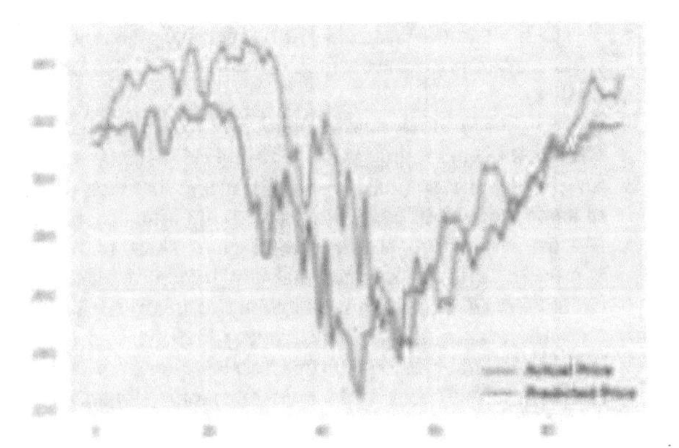

Fig. 9 Linear regression model accuracy

7 Conclusion

The model is very much efficient when compared to existing system. In existing system, only linear regression algorithm is used for training, but in proposed system, the model is trained with three different algorithms inclusive of sentiments taken by tweets. The model even predicts the next seven days' stock price, which was not present in existing system. The model may not be perfectly accurate but has a very close range of decisions, which can be accepted and implemented in real-time stock market. In coming days, model can be enhanced by storing datasets into cloud architecture like AWS or IBM cloud.

References

1. Soujanya R, Akshith Goud P, Bhandwalkar A, Anil Kumar G (2020) Evaluating future stock value asset using machine learning. 33(Part 7):4808–4813. https://doi.org/10.1016/j.matpr.2020.08.385
2. Parmar I, Agarwal N, Saxena S, Arora R, Gupta S, Dhiman H, Chouhan L (2010) Stock market prediction using machine learning. https://www.researchgate.net/profile/Lokesh-Chouhan/publication/331345883
3. Mondal P, Shit L, Goswami S (2014) Study of effectiveness of time series modeling (arima) in forecasting stock prices. Int J Comput Sci Eng Appl (IJCSEA) 4(2)
4. Hiba Sadia K, Sharma A, Paul A, Padhi S, Sanyal S (2019) Stock market prediction using machine learning algorithms. Int J Eng Adv Technol (IJEAT) 8(4). ISSN: 2249-8958
5. Goel S, Chatterjee P, Sable R (2019) Empirical study on stock market prediction using machine learning. In: International conference on advances in computing, communication and control (ICAC3), 20–21 Dec 2019
6. Yeh C-Y, Huang C-W, Lee S-J (2011) A multiple-kernel support vector regression approach for stock market price forecasting. Expert Syst:2177–2186. https://doi.org/10.1016/j.eswa.2010.08.004
7. Radu I (2015) Stock market prediction. In: 19th international conference on system theory, control, and computing. ISBN: 978-1-4799-8481-7
8. Singh J, Khushi M (2021) Feature learning for stock price prediction shows a significant role of analyst rating. https://www.mdpi.com/journal/asi, https://doi.org/10.3390/asi4010017
9. Sharma DB, Singh U (2017) Survey of stock market prediction using machine learning approach. In: 2017 international conference of electronics, communication and aerospace technology (ICECA), pp 506–509.https://doi.org/10.1109/ICECA.2017.8212715
10. Nikou M, Mansourfar G, Bagherzade J (2019) Stock price prediction using DEEP learning algorithm and itscomparison with machine learning algorithms, pp 1–11. https://www.researchgate.net/journal/Intelligent-Systems-in-Accounting-Finance-Management-1099-1174, https://doi.org/10.1002/isaf.1459
11. Bharathi S, Geetha A (2017) Sentiment analysis for effective stock market prediction. Int J Intell Eng Syst:146–154. https://doi.org/10.22266/ijies2017.0630.16

Public Policy on Science, Technology, and Innovation for Transformation and Its Relationship with Financial Education

Daniel Cardona Valencia, Sebastián Franco Castaño, and Lemy Bran-Piedrahita

Abstract The public policy of science, technology, and innovation (STI) has been an important engine for economic growth in several countries of the world, since it has allowed responding to the demands of the macroeconomic structure, as well as to the implementation of programs and strategies to promote attitudes of science, technological development, innovation, and competitiveness. Over the years, this approach has evolved, and today, we speak of a third innovation framework characterized by the concept of transformation and the inclusion of new actors, such as society. These proposals have recently become popular with the establishment of the sustainable development goals (SDGs) set by the UN. However, within the outreach strategies of STI policies, the role of financial education as a generator of development has not been clearly identified. The objective of this study is to validate the scope of this inclusion policy toward the SDGs and its relationship with financial education based on a bibliometric follow-up. As a result, it is possible to highlight the growing relationship between these two topics and the development of basically empirical experimentation and the inclusion of society as an actor participating in the transformative model.

Keywords STI · Innovation · Financial literacy

D. C. Valencia (✉)
Instituto Tecnológico Metropolitano, Universitat Politècnica de València, Medellín, Colombia
e-mail: danielcardona@itm.edu.co

S. F. Castaño
Instituto Tecnológico Metropolitano, Medellín, Colombia
e-mail: sebastianfranco@itm.edu.co

L. Bran-Piedrahita
Corporación Universitaria Americana, Medellín, Colombia
e-mail: lbpiedrahita@americana.edu.co

A. Kumar et al. (eds.), *Proceedings of the International Conference on Cognitive and Intelligent Computing*, Cognitive Science and Technology,
https://doi.org/10.1007/978-981-19-2350-0_16

1 Introduction

The public policy of science, technology, and innovation—STI has been an important engine for economic growth in various countries around the world [1, 2]. In industrialized countries, these policies have operated based on division of roles and interactions between different actors of the innovation system such as decision-maker entities (government), productive sectors (companies), and knowledge generators (universities, technological development centers, research centers, among others); without including society as an active entity to a large extent.

This behavior responds to the demands of the macroeconomic structure as well as to the implementation of programs and strategies to promote attitudes of entrepreneurship, innovation, and competitiveness, without considering, until recently, development, human welfare, social equity, and environmental protection [3–5]. This trend reflects the inherent complexity of STI processes, which has led to the emergence of more radical approaches over the last decade, which go beyond strategies that subscribe within an essentially economic nature, [6] and delve into an exclusionary model incapable of addressing the new social and environmental challenges of humanity [3, 7].

In this sense, some nations have reflected the need for a different innovation dynamic, capable of responding to the plurality of social and environmental demands [6–8]. They have raised the interest in strengthening, coordinating, and harmonizing current policies toward a more inclusive and sustainable environmental, economic, and social development of their territories, through a process of stakeholder involvement. The great challenge is to introduce a governance approach that generates systemic capacities in the design and implementation of public STI policies and that contemplates the intervention of different actors (HEIs, business, social, and public sector) in the territories, in order to ensure a more inclusive and sustainable development and to take into account family and personal finances as generators of responsible and informed capital flows [9–13].

According to the above, in recent years, there has been an increase in the field of innovation policy studies, which aim not only at entrepreneurship, competitiveness, industrialization, and economic growth of a region or nation; but also to social and environmental welfare, in which the actors of the system come to play a preponderant role and where its exercise rests on processes that are carried out through collaboration between different actors, highlighting the role to be played by different actors and economic intervention from the use of money by society [11, 14, 15]. In this sense, this study seeks, from a bibliometric monitoring of scientific articles, to establish trends and identify indicators of quality, quantity, or structure that provide relevant information on the thematic development.

2 Methodology

The main results of the study are presented below, considering the methodology used. A bibliometric follow-up is proposed based on a search equation in the Scopus database: ALL ("transformative innovation" OR "frame three" OR "frame innovation" OR "innovation policy" OR "inclusive policy" AND "financial literacy" OR "financial education" OR "financial inclusion") AND PUBYEAR > 2008. With this equation, we seek to prioritize terms related to the subject of study and to establish relationships between the key terms: transformative innovation, the third innovation framework, and inclusive policy with financial literacy in recent years.

Bibliometrics is proposed to analyze the data, defined as a measure of texts and information [16]. This technique includes statistical and mathematical methods that can be used to analyze and measure quantity and quality relationships in publications, with which trends can be defined and decisions can be made following patterns that may not be evident and that can determine scientific advances and developments [17]. The Lotka, Zipf, and Bradford's law described by Potter [18] is used to generate assertions.

3 Current Trends in Scientific Production

Figure 1 shows the number of scientific publications over the years on the topics of interest. According to this graph, between 2009 and 2020, about 166 documents have been published with an average of 13.8 publications per year. This behavior reflects the level of importance that these topics are having in the academic or scientific community and a growing interest expressed by the accumulation of publications on the subject.

Among the main countries with related scientific publications are United States who has the highest number of records to date with 32, followed by United Kingdom with 23, Italy 21, Germany, and China with 15 and 14 publications. Figure 2 shows

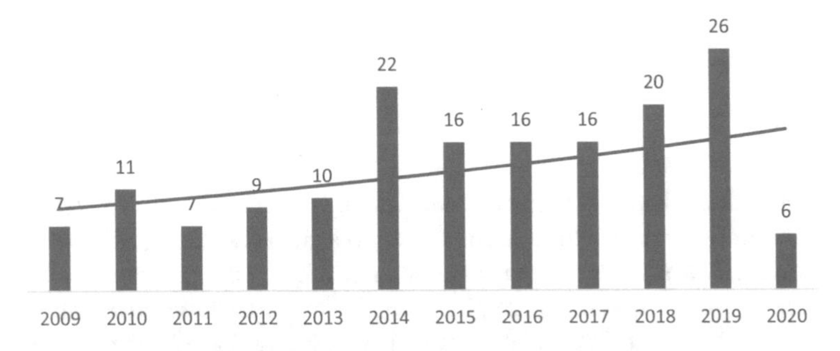

Fig. 1 Number of publications versus year of publication. *Source* Own elaboration based on Scopus®

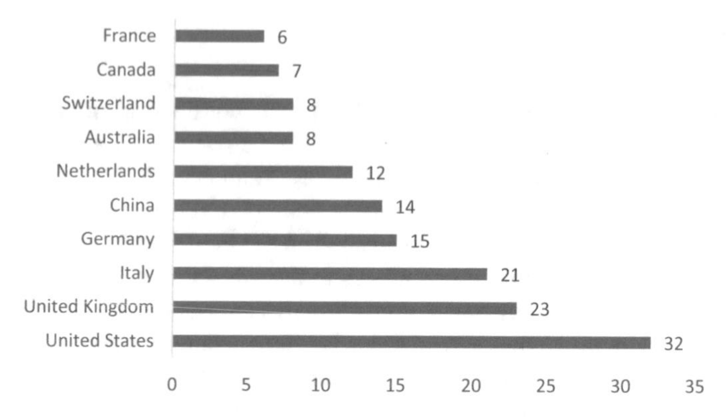

Fig. 2 Publications by country. *Source* Own elaboration based on Scopus®

the top 10 leading countries with scientific publications worldwide according to the topics of analysis.

Another of the instruments and indicators constructed in this study focused on the degree of collaboration, coordination, and influence of the main actors or agents in accordance with common objectives. In this line, several networks were elaborated which facilitated the visualization, representation, and analysis of the behavior of these actors and agents. This influence, as well as the collaboration among them, can be visualized by the size of their nodes (the larger the size, the greater the influence) and their colors (indicating the collaboration among the actors or agents and their relationships).

On the other hand, the study considered the main authors with scientific publications related to the topics under analysis. Kitagawa, F. together with Etzkowitz, H. and Gebhardt, C. are among the authors with the largest scientific records up to the date of analysis with 7, 6, and 5 published papers, respectively, focused on the development of governance policies, levels of innovation adaptation, design of education programs from the National Innovation System, evaluation of the dynamics of collaboration between social groups. Figure 3 shows the top 10 researchers by number of publications in the database.

4 Discussions

The use of studies related to finance as a determining factor in STI policies, framework three of innovation and development of SDGs has a great growth, especially in recent years, however, documents of high-research impact have not been generated; due to the particular use of the topic for empirical studies in communities in countries such as the United States, England, and Italy, and such studies have been of a reflexive, explanatory, or predictive nature. The subject from the vision of these communities and adapted to each country's own policies.

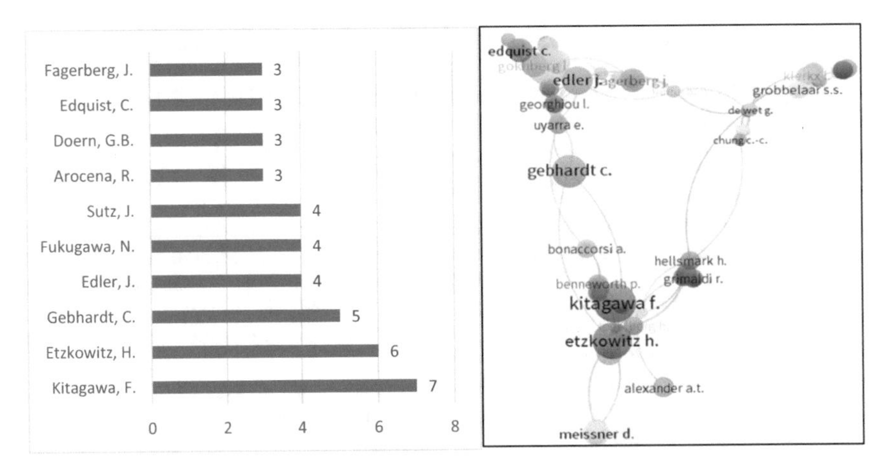

Fig. 3 Principal authors by number of publications. *Source* Own elaboration based on Scopus®

Despite being a global topic, academic production is focused on very few countries because of analysis in highly technical markets with high liquidity, especially in English and Mandarin. The topic is considered relevant and of current interest, as it is a topic inherent to times of crisis.

5 Conclusions and Recommendations

The academic production is dispersed in different institutions and journals, so it is not possible to define concentrations of knowledge. The most relevant and cited authors are Kitagawa and Etzkowitz who particularly contribute from very different aspects, the former with a more reflexive character and the latter from the qualitative and experimental.

It is also evident the publication in networks of few authors, with a tendency for publications of 2 and 3 authors, making the correlation between authors difficult, although on the other hand, important connections are found between the studies of social innovation and topics such as green economy and sustainability as tools that are increasingly related to the management of development and attention to stakeholders.

References

1. CONPES (2015) Política Nacional de Ciencia, Tecnología e innovación 2015–2025. Documento Borrador CONPES, pp 1–161
2. Kaplinsky R (2013) Innovation knowledge development early 21 st century in low and middle income economies

3. Schot J, Boni A, Ramirez M, Steward F (2018) Through transformative innovation policy, pp 1–8
4. Schot J, Steinmueller WE (2016) Framing innovation policy for transformative change: innovation policy 3.0
5. Schot J, Steinmueller WE (2018) Three frames for innovation policy: R&D, systems of innovation and transformative change.Res Policy 47:1554–1567. https://doi.org/10.1016/j.respol.2018.08.011
6. Andoni E, Hannot R, Andoni I (2017) Politicizing responsible innovation: responsibility as inclusive governance. Int J Innov Stud 1(1):20. https://doi.org/10.3724/SP.J.1440.101003
7. Hernández JLS, Pérez CD (2016) Innovación para el desarrollo inclusivo: Una propuesta para su análisis. Econ Inform 396:34–48. https://doi.org/10.1016/j.ecin.2016.01.002
8. Coenen L, Hansen T, Rekers JV (2015) Innovation policy for grand challenges. Econ Geogr Perspect 9:483–496
9. Malheiros TF, Marques RC (2016) Inclusive governance: new concept of water supply and sanitation services in social vulnerability areas. Utilities Policy 43:124–129. https://doi.org/10.1016/j.jup.2016.06.003
10. Ramírez M (2016) Inclusive innovation, capacity building and social. Palmas 37:151–158
11. Sengupta P (2016) How effective is inclusive innovation without participation? Geoforum 75:12–15. https://doi.org/10.1016/j.geoforum.2016.06.016
12. Zurbriggen C (2015) Hacia una nueva gobernanza: cocreación de iniciativas innovadoras para un desarrollo agrícola sostenible
13. Valencia DC, Calabuig C, Villa E, Betancur F (2021) Financial inclusion as a complementary strategy to address the SDGs for society. In: Nhamo G, Togo M, Dube K (eds) Sustainable development goals for society, vol 1. Sustainable development goals series. Springer, Cham. https://doi.org/10.1007/978-3-030-70948-8_6
14. Fal J, Oliveros A (2012) El pensamiento económico y social de la CEPAL: La fusión de los enfoques estructuralista y schumpeteriano. Notas para una construcción social alternativa, pp 19–76
15. Fressoli M (2015) Movimientos de base y desarrollo sustentable: la construcción de caminos alternativos. Cienc Invest 65:56–68
16. Norton MJ (2001) Introductory concepts in information science. In: ASIS monograph series
17. Van Raan AFJ (2005) Fatal attraction: conceptual and methodological problems in the ranking of universities by bibliometric methods. Scientometrics. https://doi.org/10.1007/s11192-005-0008-6
18. Potter WG (1981) Introduction to library trends 30 (1) summer 1981: Bibliometrics

Evaluation of Word Embeddings for Toxic Span Prediction

N. Kiran Babu and K. Hima Bindu

Abstract Toxic span prediction task is to predict the span of toxic content in social media posts, to identify objectionable parts of them. This task is dependent on word embeddings that represent the tokens as vectors and the model used for span prediction. This paper describes our experiment to evaluate word embeddings ranging from static and contextual to transformer-based embeddings for toxic span detection. Motivated by the top performers of SemEval 2021 Task 5, we chose to follow the sequence tagging approach for span prediction, hence, a stacked BiLSTM layer followed by CRF is used to predict the toxicity of each token. Our model with RoBERTa embeddings, fine-tuned on toxicity classification task has achieved the highest performance of 70.26 $F1$ score among all the word embeddings, which is on par with the top two performers of this task that are ensemble models. Our experiments compared word embeddings in terms of $F1$ score, training, and inference time of the model.

Keywords Toxic span · Contextual word embeddings · Transformer embeddings · BiLSTM-CRF · Sequence tagging

1 Introduction

Anonymous presence of users in social media platforms has resulted in toxic content [1–3]. As a result of this toxic communication, Internet users may have severe psychological impacts. Detecting toxic spans are critical for determining why a social media post is toxic, leading to interpretable models. Toxic span can be defined as a sequence of toxic words that make the entire text toxic. SemEval-2021 Task 5[1] is the first task aiming at identifying toxic span prediction. A character-level annotated dataset is

[1] https://sites.google.com/view/toxicspans.

N. Kiran Babu · K. Hima Bindu (✉)
National Institute of Technology, Tadepalligudem, Andhra Pradesh, India
e-mail: himabinduk@nitandhra.ac.in

N. Kiran Babu
e-mail: kiranbabu.sclr@nitandhra.ac.in

A. Kumar et al. (eds.), *Proceedings of the International Conference on Cognitive and Intelligent Computing*, Cognitive Science and Technology,
https://doi.org/10.1007/978-981-19-2350-0_17

Table 1 Samples from toxic span dataset[2]

Toxic span	Text
[8, 9, 10, 11, 12, 13, 14, 15, 16, 17, 18, 19, 20, 21, 22, 23, 24, 25, 26, 27, 28, 29, 30, 31, 32, 33, 34, 35, 36, 37, 38, 39]	*Another **violent and aggressive immigrant** killing a innocent and intelligent US Citizen.... Sarcasm*
[0, 1, 2, 3]	***Damn**, a whole family. Sad indeed*
[12, 13, 14, 15, 16, 17]	*Trump is an **animal***

released by the organizers of this task. The task is to identify the toxic span given the social media posts with zero-based indexing. Few samples from the training dataset are shown in Table 1.

Toxic span prediction can be modeled as a sequence tagging task and span boundary prediction task. The top performers of this competition are ensemble models that classify tokens as toxic or not and predict the span based on this classification. From the literature, it is observed that the BiLSTM-CRF model has achieved remarkable results for the sequence tagging task [4], synonymous to the token classification task. So, we choose to follow the sequence tagging approach.

In most of the NLP problems, word embeddings play a key role because these embeddings carry syntactic and semantic information about the input words. As pre-trained language models have become the state of the art for NLP tasks, and from the leader board of the SemEval-2021 Task5 challenge, it is observed that pre-trained transformers performed well upon fine-tuning. However, these pre-trained language models are complex and memory hungry. Hence, we perform an extrinsic evaluation to evaluate transformer embeddings against static and contextual embeddings for the downstream task in terms of $F1$ score, training time, and inference time.

The rest of the paper is organized as follows: In Sect. 2, we provided the background and related work on toxicity and toxic span prediction. Section 3 briefly describes the multiple word representations, and Sect. 4 describes the model architecture. Section 5 provides experiment details and result analysis of the proposed model with the selected word embeddings, and Sect. 6 concludes the paper.

2 Related Work

Toxic comment classification or hate speech detection has drawn a lot of attention from researchers in the past decade. Many machine learning and deep learning models exist in the literature [2, 3, 5–7] for this classification task. Recently, few researchers [8–12] have focused on recognizing the words or phrases that make a post abusive/toxic. SemEval task 5 has fueled the research in this direction.

Classification accuracy is not a complete measure of model performance, and explainable models are needed to gain trust on the model predictions. To test the

[2] https://github.com/ipavlopoulos/toxic_spans/tree/master/data.

explainability of models, Mathew et al. [12] provided the first benchmark dataset with annotated rationales. The author tested the interpretability of the models using attention score, LIME values. Instead of extracting rationales as a post-processing method, one can directly train the model to predict toxic spans.

Toxic span prediction problem can be modeled as sequence tagging task [4] and span boundary prediction (SBP) task [13–15]. The SBP [13] approach was used to extract span for the question-answering system, and it can extract a single span only. Li et al. [15] used two classifiers for each token to predict multiple spans.

The following methods are found in the literature which aim at token level classification. The RNNSL [14] model uses a BiLSTM network with random embeddings as input, and a classification head is used to predict the class label of each word. A word that is part of the toxic offset is assigned with a toxic label. Spacy[3] is an open-source library and is generally used for the named entity recognition and POS-tagging tasks. It can be used to predict custom NER tags. The least performing model in [11] at the leader board of SemEval task 5 predicts the random words as part of the toxic span. The next simple approach with little better performance is the lexicon-based approach [9] using an abusive/toxic dictionary. This model can predict explicit toxic spans only and is unable to identify implicit toxic spans, thereby not performing well.

In bag-of-words-based model [16], a toxic dictionary is built with their toxic frequency to get the toxicity ratio of each token. In the inference step, the toxic span is predicted using a threshold on the toxicity ratio. With fine-tuning of toxic frequency and toxicity ratio, its performance is similar to the spacy-based model. Karimi et al. [16] also employed a CharBERT [17] to handle out-of-vocabulary (OOV) words. The ensemble [16] of both the bag-of-words model and the CharBERT model improved the prediction performance.

Khan et al. [11] fine-tuned the BERT-base-cased model with a task-specific classification head to identify toxic tokens, and the cased model is used to detect offensive usage of proper nouns (e.g., *Muslim*). He also employed different data augmentation and multi-task learning approaches but none of them improved the performance. In post-processing of this method, if a token is identified as toxic the entire word is considered as toxic in toxic span offset construction. This may raise some false predictions. For example, arrogant is tokenized and *a, ##rog, #ant,* and similarly, toxicant is tokenized as *toxic, #ant*. When both of these are considered toxic in training, the model might learn that the token *#ant* is toxic. The model trained in this way might predict all the words ending with *#ant* as toxic. Ranasinghe et al. [9] fine-tuned the transformer models (e.g., RoBERTa, XLNet, XLM-R) with the language modeling task, and the fine-tuned model is trained on the token classification task. Apart from this model performance, the author did not provide the results regarding the importance of fine-tuning with the language modeling task.

Chhablani et al. [14] experimented with token classification, span prediction, and sequence tagging approaches using different transformer embeddings. Among them, token classification and sequence tagging approaches with SpanBERT and

[3] https://spacy.io.

BERT embeddings are the top performers. Xiang et al. [8] shown that multi-task learning models are performing well for text classification tasks, but span prediction approaches have much better performance compared to the models trained using multi-task learning approaches. The top two performers [18, 19] of the SemEval-2021 Task 5 used an ensemble of token classification models with transformer embeddings.

3 Word Representations

Word embeddings have changed the face of research in NLP. Word embeddings are the dense vector representation of words, and they have significantly improved the performance of a wide range of NLP applications. These embeddings are learned using huge amounts of unannotated textual data. These vector representations capture hidden linguistic information such as word analogies and semantics, inferring relationships like "*king:man* as *queen:woman*". Similar words will lie in close proximity in the embedding vector space. Distributed representation of words [20] reduced the dimensionality of one-hot representation and capture the semantic and syntactic relationship between words.

3.1 Static Embeddings

Static embeddings map each word to a fixed size vector. Word2vec [20] is the first successful word representation, and it uses a neural network to learn the embeddings either using CBOW (predicts current word given the context words) or skip-gram model (predicts the context words given the current word) methods. CBOW is faster and represents highly frequent words better, while skip-gram can represent rare words better even with a small amount of data. Word2vec solely uses linguistic information from the local context. GloVe [21] uses word co-occurrence to create embeddings and captures global information. FastText [22] embeddings were trained similar to word2vec on sub-words (n-grams) and can handle OOV words. BytePair embeddings [23] can embed any word by splitting words into sub-words and looking up their embeddings. Character embedding [4, 24] allows you to add character-level word embeddings during model training. ConceptNet Numberbatch [25] is conceptual embeddings that were built not only considering their surrounding words but also their meaning using ConceptNet. These embeddings benefit from ConceptNet's semi-structured common-sense information.

These static embeddings do not consider the context of a word while generating embeddings. Hence, the word "*bank*" has the same embeddings for both "*river bank*" and "*bank deposit*".

3.2 Contextual Embeddings

Contextual embeddings provide each word a representation dependent on its context, allowing for the capture of word usage in a variety of settings. ELMo [26] is the first proposed neural network architecture to create contextual embeddings and uses a CNN to extract initial embeddings and a stacked BiLSTM network to extract contextual embeddings. ELMo uses next word prediction as a language modeling task to train the word embeddings. Similar to ELMo, deep learning model Flair [27] uses a BiLSTM architecture and learns the contextual embeddings using the next character prediction task. It is a contextual language model operating at the character level, carrying information about word morphology.

Even though ELMo is using character convolution network to handle OOV words, it can generate representations at the word level only. Flair generates representation at the character level and considers the context in one direction only. To get contextual representation from both directions, concatenation of both flair forward and backward models is needed.

3.3 Transformer Embeddings

BERT [28], the first kind of transformer, uses bidirectional transformer architecture to learn interaction between words during pre-training using masked words prediction and next sentence prediction tasks. Transformer-based embeddings give syntactic and semantic information of the tokens with an embedding size of 768 which is much smaller compared to the Flair embedding. Even though ELMo and Flair are contextual embeddings, the training and inference time in downstream tasks are quite high because of the sequential nature of LSTM used in these models. Transformers overcome this problem by using positional encoding and a multi-head self-attention mechanism. To make balance over both word-level (ELMo) and character-level (Flair) embeddings, transformer models used subword embedding to handle OOV words.

Similar to BERT, multiple transformer models were released, namely SpanBERT [29], RoBERTa [30], and ConvBERT [31]. RoBERTa [30] is an extended version of BERT with hyper-parameter fine-tuning. It is pre-trained on 160 GB corpus with a dynamic masking approach, while the original BERT is trained on 16 GB corpus with the fixed masking approach and it eliminated the next sentence prediction task. SpanBERT [29] is trained on masked span prediction task, whereas original BERT is trained on masked word prediction and next sentence prediction. ConvBERT [31] is a modified model of the original BERT where self-attention blocks are replaced with mixed attention blocks. Dynamic span-based convolution heads are introduced in ConvBERT to learn local dependencies. Hence, it has a reduction in the pre-training time. ELECTRA [32] uses a generator-discriminator kind of approach to pre-train the model. In ELECTRA, the discriminator loss is calculated over all the input tokens instead of masked tokens in BERT.

The use of static embeddings on downstream tasks do not need the trained model, and it needs only the vectors. Whereas in contextual and transformer embeddings, the pre-trained model is needed to get word representations.

4 Model

We considered the sequence tagging approach for toxic span prediction. As the BiLSTM-CRF model has shown significant performance for sequence tagging, we designed our model by following [4]. Based on the recommendations of [4], we used two BiLSTM layer network followed by the CRF layer. The model is shown in Fig. 1.

Initially, the embedding layer is used to extract the vector representation of each token. After constructing embeddings for the input sequence, a two-layer stacked BiLSTM [33] network is used to get the interaction between tokens in a given input sequence. On top of the BiLSTM network, a conditional random field (CRF) [34]

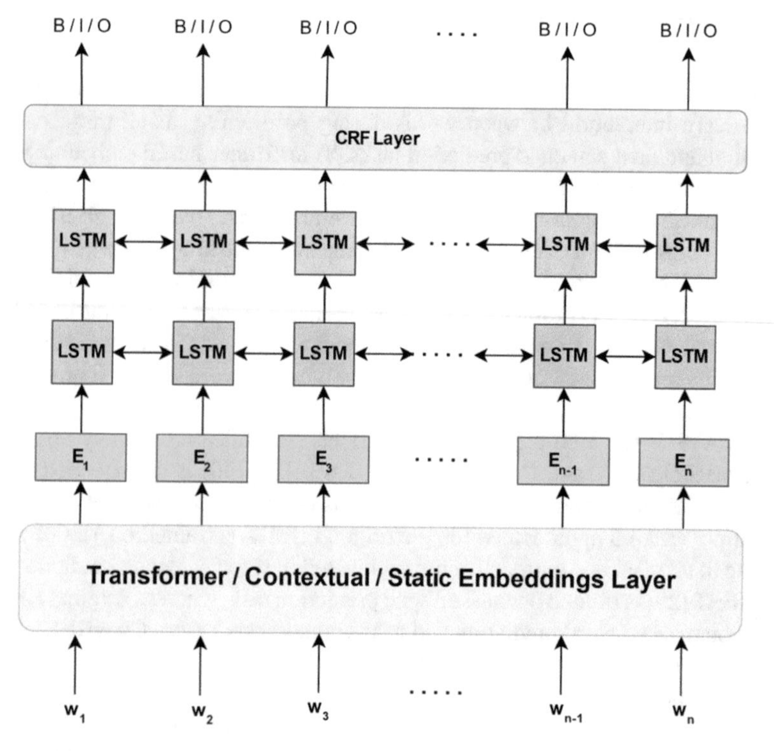

Fig. 1 Model architecture

layer is used to predict the toxicity tags for each token. CRF layer decodes the probable label sequence using the Viterbi decoding algorithm.

Given the input sequence $I = w_1, w_2, w_3 \ldots w_n$ of length n, the model predicts the label for each token as follows:

$$E_i = \text{embedding}(w_i) \tag{1}$$

$$h_i = \text{BiLSTM}(W_2 \cdot \text{BiLSTM}(W_1 \cdot E_i + b_1) + b_2) \tag{2}$$

$$\text{label}_i = \text{CRF}(h_i) \tag{3}$$

where w_i is the word in the input sequence and E_i is the output of the embedding layer. E_i is fed to the stacked BiLSTM layer to get the hidden representation of the input w_i. W_1 and b_1 are weights and bias of the first BiLSTM layer, and W_2 and b_2 are weights and bias of the second BiLSTM layer. $label_i$ is the label predicted by the CRF layer, using hidden representation h_i.

5 Experiments

Experiments were conducted to answer the following questions concerning the toxic span prediction task:

1. Can a single model achieve the $F1$ score of the top-performing ensemble models?
2. Which word embeddings can give the best $F1$ score?
3. What is the tradeoff between $F1$ score, training time, and inference time?
4. Is it possible to achieve a reasonable $F1$ score with a faster model?

5.1 Dataset

The dataset used in this paper was released by SemEval-21 Task5. It consists of 10 k posts annotated at span level by crowd annotators. The characteristics of the dataset, such as the distribution of the dataset for train, test, and validation tasks, average lengths of the posts and the toxic spans, and the average number of toxic spans per post are shown in Table 2.

Table 2 Dataset statistics

	#Posts	Average post length (in characters)	Average span length (in characters)	Average #toxic span
Train set	7939	204.56	10.77	1.39
Test set	2000	186.40	13.10	0.92
Trial set	690	199.48	7.90	1.43

5.2 Embeddings

Word embeddings are the state of the art for word representation in NLP. The word embeddings used in this study range from static word embeddings (GloVe [21], FastText [22], BytePair [23], ConceptNet Numberbatch [25]), contextual embeddings (Flair [27], ELMo [26]) to transformer-based embeddings (BERT [28], RoBERTa [30], SpanBERT [29], ConvBERT [31], ELECTRA [32]). The word embedding dimensions of each of them as used in this experiment are shown in Table 3.

Table 3 Results on TSD test dataset with different word embeddings and model training time and average inference time

Embeddings		Dimension	$F1$ score	Training time per epoch (s)	Average inference time on test set (s)
Static	GloVe	300	65.41	230	14.4
	ConceptNet	300	67.50	229	14.7
	FastText	300	67.80	233	30.3
	FastText + Char	350	68.70	421	32.5
	FastText + BytePair	400	68.79	261	30.7
Contextual	ELMo	1024	67.78	621	265
	Flair forward + backward	4096	68.66	983	378
	Flair forward + backward + FastText + BytePair	4496	69.09	1287	387
Transformer	BERT (layer-1)	768	66.62	420	72
	BERT	768	68.67	440	76
	SpanBERT	768	69.09	421	73
	RoBERTa	768	69.26	422	85
	ELECTRA	768	69.23	428	74
	ConvBERT	768	69.79	421	93
	ToxicRoBERTa	768	70.26	440	91

5.3 Pre-processing

The pre-processing step removes URLs, merges repeated strings into a single character (e.g., "??????" is converted to "?") and removes the punctuation at the beginning and end of the text which is uninformative for span prediction. The pre-processor removes the leading and trailing whitespaces in toxic offsets (e.g., in the text "you are an **idiot**", leading whitespace is also marked as part of toxic span) and also removes the singleton offsets (e.g., in a text "h**a**te speech" only "***a***" is marked as a toxic offset) by considering them as annotator inconsistency. After these steps, the text is tokenized using a custom tokenizer[4] to keep the words like "a$$" intact, instead of tokenizing it as "a", "$", "$". After tokenization, each token is assigned a begin toxic (B-T) if it is the start of toxic span and inside toxic (I-T) if it is inside the toxic span, otherwise, the outside (O) tag is assigned following the CoNLL[5] text format.

5.4 Evaluation Metric

The model performance is evaluated using a variant of the $F1$ score as proposed in [34] to measure the overlap between the ground truth and the predicted toxic span. For a document d, S_d is the set of toxic character offsets predicted by the system, and G_d is the set of ground truth annotations. Then, the $F_1^d(G)$ score of the system for the document d is defined as

$$F_1^d(G) = \frac{2 * P^d(G) * R^d(G)}{P^d(G) + R^d(G)}, \text{ where}$$
$$P^d(G) = \frac{|S_d \cap G_d|}{|S_d|} \text{ and } R^d(G) = \frac{|S_d \cap G_d|}{|G_d|} \tag{4}$$

When a document has no ground truth annotation ($G_d = \emptyset$), or the system outputs no character offset prediction ($S_d = \emptyset$), then Eq. 5 is used.

$$F_1^d(G) = \begin{cases} 1 \; G_d = S_d = \emptyset \\ 0 \; otherwise \end{cases} \tag{5}$$

Finally, the arithmetic mean of $F_1(G)$ over all the documents of an evaluation dataset is the $F1$ score of the system.

---

[4] https://www.nltk.org/_modules/nltk/tokenize/treebank.html.

[5] https://www.clips.uantwerpen.be/conll2003/ner/.

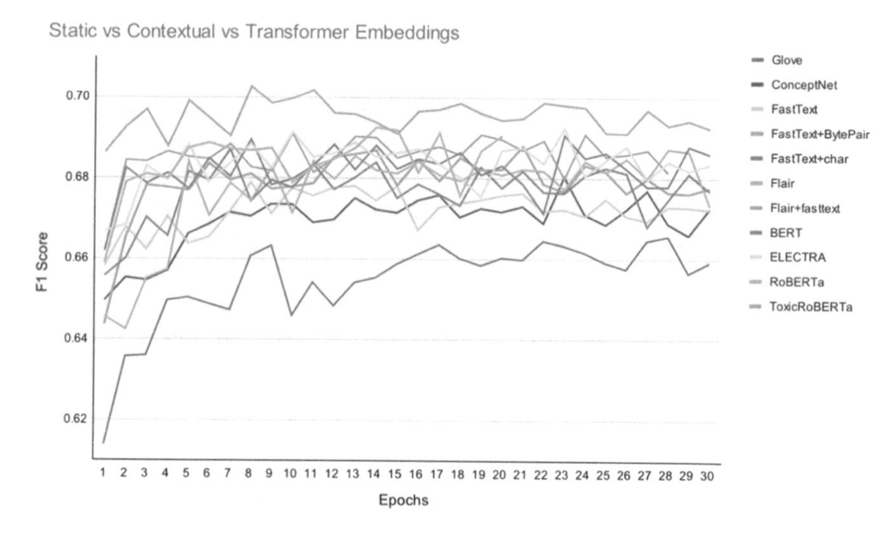

Fig. 2 Model performance with word embeddings

5.5 *Results*

All our experiments are conducted on an HP ProLiant server with Intel Xeon Processor and two Nvidia Quadro P2200 GPUs. All the models are implemented using PyTorch[6] and Flair NLP [35]. To train the models, the Flair model trainer is used with a learning rate of 0.0005 and AnnealOnPlateau scheduler. The number of epochs is tuned using the validation dataset. The maximum length of the sequence is set to 512 and made the embedding layer non-trainable except for character embeddings. The short length sequences are padded, and longer sequences are truncated.

As shown in Table 3 and Fig. 2, GloVe [21] is the least performer out of all the embeddings due to its static nature. From Fig. 2, it is observed that ConceptNet [25] embeddings clearly outperform Glove. The model using FastText (word2vec) [22] embeddings has achieved a maximum of 67.80 $F1$ score, but it required 30 epochs. Stacking of FastText and BytePair or Char embeddings showed a better performance with 5–6 epochs only. Stacking of FastText and character embeddings gave 3.2% improved $F1$ score at a cost of 45.3% and 55.6% increase in train and inference time, respectively. Similarly, stacking of FastText and BytePair embeddings gave 3.38% improved $F1$ score over GloVe at a cost of 11.8% and 53% increase in train and inference time, respectively.

The transformer model gives different levels of context embeddings at different layers of the transformer [36]. It is observed that taking embeddings of three layers (top, middle, bottom) and concatenating them improved the model performance. As shown in Table 3, compared to using the BERT layer-1 embeddings, extracting three layers of BERT [28] embeddings improved the $F1$ score by 2% at a cost of

[6] https://pytorch.org/.

Table 4 Model performance

Model	$F1$ score
HITSZ-HLT [18]	70.83
S-NLP [19]	70.77
ToxicRoBERTa + BiLSTM + CRF (ours)	70.26

4.5% increase in training time and 5.2% increase in inference time. Hence, we used embeddings from three layers for all the remaining transformer-based experiments.

All the variants of BERT have consistently shown better performance due to pre-training and fine-tuning. ToxicRoBERTa embedding is the top performer in our experiments because it is fine-tuned on the dataset which is a parent dataset of this toxic span dataset. As shown in Table 4, HITSZ-HLT [18] is the top performer in the SemEval Task5 and used an ensemble of BERT + CRF, BERT + LSTM + CRF, and BERT + Span. Similarly, the second-best performer [19] used an ensemble of three transformer models which makes the system complex for both training and inference. With a single transformer model and BiLSTM-CRF network as classification head, we got 70.26 $F1$ score.

One critical issue with neural networks is overfitting. When the model is trained using more than the required number of epochs, it learns patterns specific to the train set. As a result, the model is unable to perform well on new datasets. Figure 2 shows the learning curve of the model over the increased number of epochs. After a certain number of epochs, the model starts to overfit or no improvement in performance. We used the validation dataset to tune the number of epochs required.

The training and inference times increase with the complexity associated with the embeddings model, and it is observed that the selection of word embeddings plays a vital role in the toxic span prediction. From Figs. 2 and 3, it is observed that the

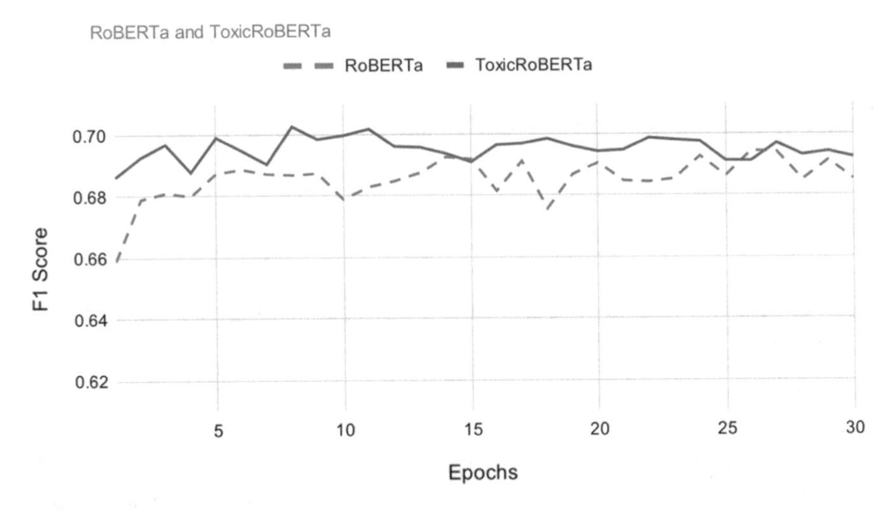

Fig. 3 RoBERTa versus ToxicRoBERTa

ToxicRoBERTa embeddings give a sufficient margin over the RoBERTa and static embeddings and shown the importance of fine-tuning the transformer on the task-specific dataset. By compromising 1.53% $F1$ score over ToxicRoBERTa embeddings, we got 4.5% and 64% improvement in training and inference time, respectively, using a stack of FastText and character embeddings. Similarly, by compromising 1.47% $F1$ score over ToxicRoBERTa embeddings, we got 40% and 66% improvement in training and inference time, respectively, using a stack of FastText and BytePair embeddings. When compared with static embeddings (Glove), Flair embeddings and their stacked variants have shown 3.6% of improvement in the $F1$ score at the cost of 82% and 96% increase in the training and inference times, respectively. However, when compared to transformer-based embeddings, the $F1$ score is similar to BERT variants, with the exception of ToxicRoBERTa. But both the times as reported in Table 3 are quite high for Flair embeddings when compared with transformer variants, with raise in at least 67% for training time and a 79% raise for inference times. Hence, to build a faster model, we recommend FastText + BytePair embeddings and ToxicRoBERTa for the best performance.

6 Conclusion

We experimented with static, contextual, and transformer-based word embeddings to predict toxic span using the BiLSTM-CRF model. Compared to the Flair contextual embedding, transformer-based models perform equally well with less training and inference time. It is observed that the fine-tuning of transformers on task-specific datasets improves the toxic span prediction performance, as demonstrated by the embeddings of ToxicRoBERTa. Our simple and single transformer-based model led to a comparable state-of-the-art $F1$ score that only ensemble models were able to achieve for the toxic span prediction task. It is also shown that by compromising over 1.47% of $F1$ score over ToxicRoBERTa, we got a huge improvement in training and inference time using stacking of FastText and BytePair embeddings. We recommend using FastText + BytePair or task-specific fine-tuned transformer embeddings.

References

1. Waseem Z, Hovy D (2016) Hateful symbols or hateful people? Predictive features for hate speech detection on Twitter. In: Proceedings of the NAACL student research workshop. Association for Computational Linguistics, Stroudsburg, PA, USA, pp 88–93
2. Founta A-M, Chatzakou D, Kourtellis N et al (2019) A unified deep learning architecture for abuse detection. In: Proceedings of the 10th ACM conference on web science. In: Proceedings of the 10th ACM conference on web science (WebSci'19). Association for Computing Machinery, New York, NY, USA, pp 105–114
3. Chakrabarty T, Gupta K, Muresan S (2019) Pay attention to your context when classifying abusive language. In: Proceedings of the third workshop on abusive language online.

Association for Computational Linguistics, Florence, Italy, pp 70–79
4. Lample G, Ballesteros M, Subramanian S et al (2016) Neural architectures for named entity recognition. In: Proceedings of the 2016 conference of the North American chapter of the association for computational linguistics: human language technologies, pp 260–270
5. Waseem Z (2016) Are you a racist or am i seeing things? Annotator influence on hate speech detection on twitter. In: Proceedings of the first workshop on NLP and computational social science. Association for Computational Linguistics, Stroudsburg, PA, USA, pp 138–142
6. Davidson T, Warmsley D, Macy M, Weber I (2017) Automated hate speech detection and the problem of offensive language. In: Proceedings of 11th international conference web and social media, ICWSM 2017, pp 512–515
7. Zhang Z, Robinson D, Tepper J (2018) Detecting hate speech on Twitter using a convolution-GRU based deep neural network. In: Lecture notes in computer science (including subseries lecture notes in artificial intelligence and lecture notes in bioinformatics). Springer Verlag, pp 745–760
8. Xiang T, Macavaney S, Yang E, Goharian N (2021) ToxCCIn: toxic content classification with interpretability. In: Proceedings of the 11th workshop on computational approaches to subjectivity, sentiment and social media analysis, pp 1–12
9. Ranasinghe T, Zampieri M (2021) MUDES: multilingual detection of offensive spans. In: Proceedings of the 2021 conference of the North American chapter of the association for computational linguistics: human language technologies: demonstrations, pp 144–152
10. Gia Hoang P, Thanh Nguyen L, Nguyen K (2021) UIT-E10dot3 at SemEval-2021 Task 5: toxic spans detection with named entity recognition and question-answering approaches. In: Proceedings of the 15th international workshop on semantic evaluation (SemEval-2021). Association for Computational Linguistics, online, pp 919–926
11. Khan Y, Ma W, Vosoughi S (2021) Lone pine at SemEval-2021 task 5: fine-grained detection of hate speech using BERToxic. In: Proceedings of the 15th international workshop on semantic evaluation (SemEval-2021). Association for Computational Linguistics, online, pp 967–973
12. Mathew B, Saha P, Yimam SM et al (2020) HateXplain: a benchmark dataset for explainable hate speech detection. arXiv:2012.10289
13. Seo M, Kembhavi A, Farhadi A, Hajishirzi H (2017) Bidirectional attention flow for machine comprehension. In: Proceedings of ICLR 2017
14. Chhablani G, Sharma A, Pandey H et al (2021) NLRG at SemEval-2021 task 5: toxic spans detection leveraging BERT-based token classification and span prediction techniques. In: Proceedings of the 15th international workshop on semantic evaluation (SemEval-2021). Association for Computational Linguistics, online, pp 233–242
15. Li X, Feng J, Meng Y, et al (2020) A unified MRC framework for named entity recognition. In: Proceedings of the 58th annual meeting of the association for computational linguistics. Association for Computational Linguistics, online, pp 5849–5859
16. Karimi A, Rossi L, Prati A (2021) UniParma at SemEval-2021 task 5: toxic spans detection using character BERT and bag-of-words model. In: Proceedings of the 15th international workshop on semantic evaluation (SemEval-2021). Association for Computational Linguistics, online, pp 220–224
17. Ma W, Cui Y, Si C et al (2020) CharBERT: character-aware pre-trained language model. In: Proceedings of the 28th international conference on computational linguistics, pp 39–50
18. Wang C, Liu T, Zhao T (2021) HITMI&T at SemEval-2021 task 5: integrating transformer and CRF for toxic spans detection. In: Proceedings of the 15th international workshop on semantic evaluation (SemEval-2021). Association for Computational Linguistics, online, pp 870–874
19. Nguyen VA, Nguyen TM, Quang Dao H, Huu Pham Q (2021) S-NLP at SemEval-2021 task 5: an analysis of dual networks for sequence tagging. In: Proceedings of the 15th international workshop on semantic evaluation (SemEval-2021). Association for Computational Linguistics, online, pp 888–897
20. Mikolov T, Sutskever I, Chen K et al (2013) Distributed representations of words and phrases and their compositionality. In: Advances in neural information processing systems

21. Pennington J, Socher R, Manning CD (2014) GloVe: global vectors for word representation. In: EMNLP 2014—2014 conference on empirical methods in natural language processing, proceedings of the conference. Association for Computational Linguistics (ACL), pp 1532–1543
22. Joulin A, Grave E, Bojanowski P, Mikolov T (2017) Bag of tricks for efficient text classification. In: Proceedings of the 15th conference of the european chapter of the association for computational linguistics: volume 2, short papers. Association for Computational Linguistics, Valencia, Spain, pp 427–431
23. Heinzerling B, Strube M (2018) BPEmb: tokenization-free pre-trained subword embeddings in 275 languages. In: Proceedings of the eleventh international conference on language resources and evaluation (LREC 2018)
24. dos Santos C, Guimarães V (2015) Boosting named entity recognition with neural character embeddings. In: Proceedings of the fifth named entity workshop. Association for Computational Linguistics, Beijing, China, pp 25–33
25. Speer R, Chin J (2016) An ensemble method to produce high-quality word embeddings. Computing research repository. arXiv:1604.01692
26. Peters ME, Neumann M, Iyyer M et al (2018) Deep contextualized word representations. In: Proceedings of the 2018 conference of the North American chapter of the association for computational linguistics: human language technologies, vol 1 (long papers), pp 2227–2237
27. Akbik A, Blythe D, Vollgraf R (2018) Contextual string embeddings for sequence labeling. In: Proceedings of the 27th international conference on computational linguistics, pp 1638–1649
28. Devlin J, Chang M-W, Lee K, Toutanova K (2019) BERT: pre-training of deep bidirectional transformers for language understanding. In: Proceedings of the 2019 conference of the North American chapter of the association for computational linguistics: human language technologies, vol 1 (long and short papers). Association for Computational Linguistics, Minneapolis, Minnesota, pp 4171–4186
29. Joshi M, Chen D, Liu Y et al (2019) SpanBERT: improving pre-training by representing and predicting spans. Trans Assoc Comput Linguist 2020:64–77
30. Liu Y, Ott M, Goyal N et al (2019) RoBERTa: a robustly optimized BERT pretraining approach. arXiv:1907.11692
31. Jiang Z, Yu W, Zhou D et al (2020) ConvBERT: improving BERT with span-based dynamic convolution. In: Proceedings of the 34th conference on neural information processing systems (NeurIPS 2020)
32. Clark K, Luong M-T, Le QV, Manning CD (2020) ELECTRA: pre-training text encoders as discriminators rather than generators. In: Proceedings of the eighth international conference on learning representations (ICLR 2020)
33. Schuster M, Paliwal KK (1997) Bidirectional recurrent neural networks. IEEE Trans Signal Process 45(11):2673–2681
34. Lafferty J, McCallum A, Pereira FCN (2001) Conditional random fields: Probabilistic models for segmenting and labeling sequence data. In: Proceedings of the eighteenth international conference on machine learning (ICML'01), pp 282–289
35. Akbik A, Bergmann T, Blythe D et al (2019) FLAIR: an easy-to-use framework for state-of-the-art NLP. In: NAACL, 2019 annual conference of the North American chapter of the association for computational linguistics (demonstrations), pp 54–59
36. Ethayarajh K (2019) How contextual are contextualized word representations? comparing the geometry of BERT, ELMo, and GPT-2 embeddings. In: Proceedings of the 2019 conference on empirical methods in natural language processing and the 9th international joint conference on natural language processing (EMNLP-IJCNLP). Association for Computational Linguistics, Hong Kong, China, pp 55–65

Error Notification Management for Synchronization of ES10b Interface in RPM

R. Swathi, Bibhuti Ranjan Mishra, and K. A. Nethravathi

Abstract Remote profile management is a feature of eSIM for consumer solution that allows mobile network operators to perform profile management functions on their profiles remotely. RPM commands are carried out through the ES10 interface, between the RSP server and the eSIM. Currently, a report, referred to as notification, is generated by the eSIM to the RSP server indicating that the profile management operation was successfully performed in the eSIM. The main objective of this paper is to address the failure scenarios of RPM commands. Definition and characterization of these notifications allow the eSIM to generate notifications in error scenarios. These error notifications help maintain synchronization between the RSP server and the actual state of the eSIM. This eventually aides the RSP server to reinitiate the command to complete the operation.

Keywords eSIM · Remote profile management · ES10b interface

1 Introduction

An eSIM in a GSMA-approved environment allows consumers to set up a contract with their chosen mobile network operator to download and enable profiles [1]. Each profile comprises of the operator and subscriber data that belong to network operators. Once the profile is installed and activated, the device is now able to connect to that operator's network.

R. Swathi (✉) · K. A. Nethravathi
R. V. College of Engineering, Bangalore, India
e-mail: agalyaswathi@gmail.com

K. A. Nethravathi
e-mail: nethravathika@rvce.edu.in

B. R. Mishra
N. X. P. Semiconductors, Bangalore, India
e-mail: bibhutiranjanmishra@nxp.com

A. Kumar et al. (eds.), *Proceedings of the International Conference on Cognitive and Intelligent Computing*, Cognitive Science and Technology,
https://doi.org/10.1007/978-981-19-2350-0_18

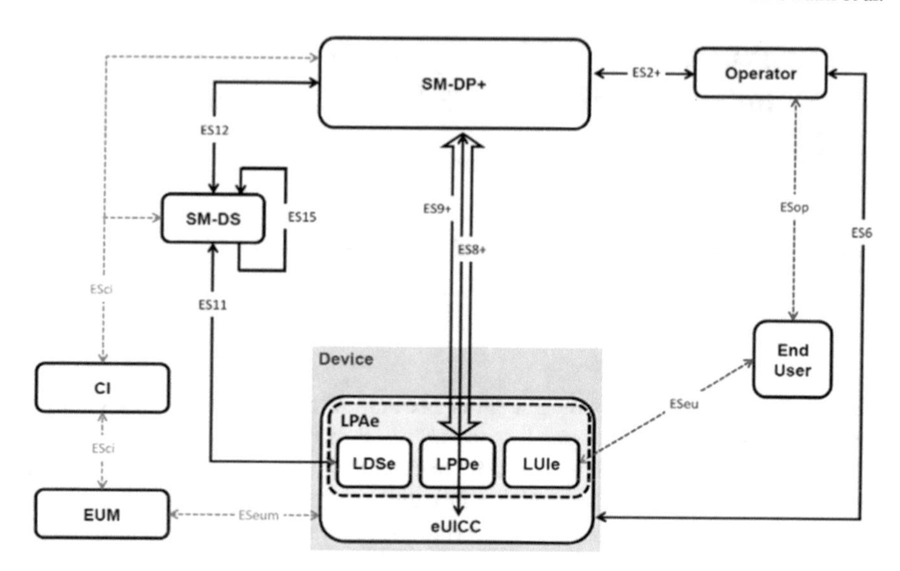

Fig. 1 Remote SIM provisioning system [3]

In local profile management [2], the end user requests for profile management operation to be performed through a user interface called local user interface (LUI). Similarly, to allow MNOs access to their profile, remote profile management is introduced, where the RSP server initiates the request for a profile management operation to be performed on behalf of the MNO. The ES2+ interface is used by the MNO to indicate to the RSP server to initiate the RPM procedures as seen in Fig. 1. The managing RSP server prepares the RPM package. Depending on the configuration of the profile by the MNO, there may be several managing RSP servers.

An RPM package contains one or more RPM commands. Each RPM command requires end-user confirmation which is enforced by the local profile assistant (LPA).

User confirmation on multiple RPM commands for consecutive operations is combined to simplify the user experience and avoid repeated input steps for the end user. In the case of combined verifications, it shall be clear to the end user what RPM commands will be executed, and a blanket confirmation for multiple RPM command will be obtained from the end user.

RPM initiation procedure is used by the MNO to issue an RPM command(s) or to audit the eUICC. Once RPM initiation is completed, the LPA downloads the RPM package using the ES9+ interface. This procedure is referred to as RPM download. Additionally, in this procedure, end-user consent is obtained. This is followed by RPM execution, where the LPA transfers the RPM package to the eSIM if end user accepts the execution of the RPM package. The eSIM verifies the authorization of the RSP server, and only, then sequentially executes the RPM command(s) contained in the RPM package. These commands are Enable Profile, Disable Profile, Delete Profile, List Profile Info, Update Metadata, and Contact PCMP.

Enable Profile RPM command is used to remotely enable a profile already downloaded and installed on an eSIM, and Disable Profile command is used to remotely disable an enabled profile already downloaded and installed on an eSIM. Delete Profile command is used to remotely delete a profile already downloaded and installed on an eSIM. List Profile Info is used to list the profiles and their current states pre-installed or previously downloaded and installed on an eSIM. Update Metadata is used to remotely update the Profile Metadata of a profile already downloaded and installed on an eSIM. Contact Profile Content Management Platform (PCMP) is used to fetch the PCMP address configured in the enabled profile. PCMP is a platform owned by the profile owner, used to manage the content of an enabled profile.

The RPM commands in the RPM package are executed sequentially until the end is reached or an error is encountered for a command. Each executed command produces an RPM command result.

The eSIM executes the RPM command(s) contained in the RPM package in the received order and generates the load RPM package result. The operator receives the load RPM package result. Based upon the outcome, the operator can choose to initiate additional RPM operations.

A failure during execution of RPM commands occurs due to the following factors. The modem is unable to process the command at that instance of time. A software fault could cause a crash of the LPA, host device, and the baseband processor. The end user could switch off the power, and in case of a removable eSIM card, the end user may remove the battery.

To combat these failures, notification for these error scenarios is generated.

2 Methodology

2.1 Maintaining the Integrity of the Specification

Remote profile management and remote eSIM management procedures will follow the specification [4]. The functionalities and behavior of the eSIM will remain the same as expected in the specifications [5, 6].

2.2 Generation of Failure Notifications

When the eSIM encounters a failure while processing a RPM command for reasons not addressed in the specification, the eSIM follows the sequence, Fig. 2, and generates a notification with the corresponding error code, where a field is added to specifically indicate the RPM error reason. The LPA queries the eSIM for the pending notifications list. The eSIM provides the LPA with the pending notifications list. The LPAd establishes a transport layer security (TLS) secure channel with the relevant RSP

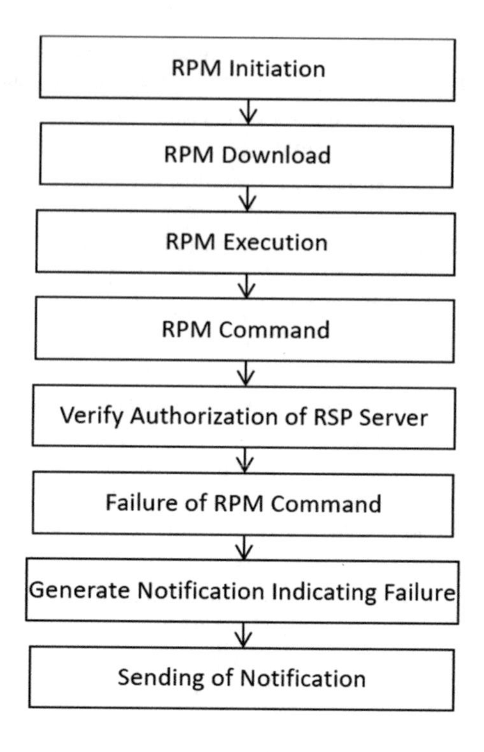

Fig. 2 Block diagram implementation of error notification management in RPM

server. The LPA sends each notification to the RSP server. The RSP server acknowledges notification reception. The LPA calls to remove notification that has been acknowledged. The eSIM removes the notification from the pending notifications list.

3 Implementation

All possible use case scenarios are identified. The characterization of error notification is defined by the error codes and reflected in the ASN.1 coding.

3.1 Error Codes

The error codes defined are as follows:

- Modem busy
- User consent
- LPA issue
- Undefined error.

Modem busy is likely to occur when the baseband processor is unable to process the command. The terminal response is indicated with a fatal error or warning status.

When user confirmation is not combined and the user has to consent to each command in the RPM package, the user may cancel the operation.

In case of software crash, the LPA services are temporarily unavailable. The error code generated must be lpaissue.

Whenever these scenarios are encountered by the eSIM during the exception of RPM commands, notifications are generated with their respective error code.

3.2 ASN.1 Coding

The ASN.1 coding has been updated to accommodate notifications with the proposed error codes for remote profile management

```
ListNotificationResponse ::= [40] CHOICE { -- Tag 'BF28'
notificationMetadataList SEQUENCE OF NotificationMetadata,
listNotificationsResultError INTEGER {
undefinedError(127)
}
}
NotificationMetadata ::= [47] SEQUENCE { -- Tag 'BF2F'
seqNumber [0] INTEGER,
profileManagementOperation [1] NotificationEvent,
notificationAddress UTF8String, -- FQDN to forward the notification
iccid Iccid OPTIONAL
errorReason ErrorReasonRPM
}
NotificationEvent ::= BIT STRING {
notificationInstall (0),
notificationEnable(1),
notificationDisable(2),
notificationDelete(3)
}
ErrorReasonRPM ::= INTEGER {
modembusy(1),
userconsent(2),
lpaissue(3),
installFailedDuetoUnknownError(127)
}
```

4 Testing and Validation

Test case scenario is written to validate the card functionality and behavior.

1. An external tool is used to simulate the RSP server.

2. A virtual interface is used to establish a connection between RSP server and eSIM [7].
3. The RSP server issues a RPM command, mentioned above, to the eSIM.
4. As the eSIM is processing the command, a reset command is triggered, to mimic the power loss and loss of connection.
5. The eSIM generates a failure notification, which is sent to the SMP-DP+.
6. The notification received at the RSP server is verified.

5 Conclusion

Remote profile management is a new feature of remote SIM provisioning. Initially, when only local profile management was supported by eSIM, in case of a failure due to reason mentioned in section ASN.1 coding, the user can issue a retry with ease. However, in remote profile management, the RSP server is not aware of a failure in RPM command at eSIM. This leads to a discord between the eSIM and the RSP server. The methodology proposed in this paper will help maintain synchronization between the RSP server and eSIM by sending a report called notification to the RSP server. The RSP server is now aware of the state of execution of the RPM command and the state of the eSIM.

Acknowledgements We would like to thank Jashin S. G., Karthik Dharmaraj, and Harsha Selvaraj for their wholehearted support, suggestion, and invaluable advice. We would also like to thank Anish Damodaran for this thorough review and feedback.

We take this opportunity to express our gratitude to the Department of Electronics and Communication, R. V. College of Engineering.

References

1. GSMA (2018) eSIM whitepaper, version 4.11. GSMA Association, Mar 2018
2. GSMA (2020) SGP.22 remote SIM provisioning (RSP) technical specification, version 2.2.2. GSMA Association, June 2020
3. GSMA (2017) SGP.02 remote provisioning architecture for embedded UICC technical specification, version 3.2; 27 June 2017
4. GSMA (2018) SGP.22 remote SIM provisioning (RSP) technical specification, version 2.2.1. GSMA Association, Dec 2018
5. eUICC consumer SGP.23 compliance test suite, version 1.6. GSMA Association, 12 Oct 2019
6. SGP.26 RSP test certificates definitions, version 1.4 GSMA Association, 31 July 2020
7. ETSI TS 102 v13.1.0 smart cards; UICC-terminal interface

Use of AI/ML to Telecom Service Delivery Model to Change the Customer Perception About Service Performance

Saumya Choudhury

Abstract With the new technologies like 5G coming to the market, the complexity of the mobile network seems to increase inescapably the launch of new services which earlier were not into being with the older technologies. Technologies like software defined wide area networking (SD-WAN) and network functions virtualization (NFV) got introduced to meet the ever-rising customer expectation, whereas artificial intelligence/machine learning (AI/ML) got introduced to network functions to increase the service provider's need to increase the intelligence of their network design and planning, optimization, day to day operations, etc. Continuous research is going on AI/ML techniques to make the communication networks and systems more scalable, controllable, and manageable to enable more autonomous operations. With the use of AI/ML into the network automation platforms, like open network automation platform (ONAP) can deliver efficient, time sensitive, and reliable service operations to enhance the service performance and thus resulting to customer satisfaction. However, while choosing the scenario analysis service providers need to be careful to not fall into trap of classical data analytics techniques like big data tools with different statistical analysis and a dose of marketing. Intelligent network is much more than that where intelligent network is able to predict the network outage and can automate *the* healing, can predict vulnerability threat to the network and can take the smart preventive action based on the type of the threats, etc. In this article, I am going to talk about how AI/ML can be scaled to change the telecom operation scenario in future to reduce the operational cost, increasing revenue, and customer experience.

Keywords AI/ML · Telecommunication · Automation strategy · Automated business operations · Business functions

S. Choudhury (✉)
Lovely Professional University, Jalandhar-Delhi, Grand Trunk Rd, Phagwara, Punjab 144001, India
e-mail: e08saumya@iima.ac.in

A. Kumar et al. (eds.), *Proceedings of the International Conference on Cognitive and Intelligent Computing*, Cognitive Science and Technology,
https://doi.org/10.1007/978-981-19-2350-0_19

1 Introduction

AI/ML is not new to the telecom world but the key factors that have led to the acceleration of investment in this field are the following:

1. Breakthroughs in the neural networking theory [1].
2. Public cloud (AWS, Azure, and GCP) availability to make computing ability highly available, scalable, cost efficient, and pay-per-use [2].
3. Massive data sets produced due to IOT devices [2].
4. Need to run the operations 24/7 without any interruptions [3].
5. Newer use cases like manufacturing automation (example: car industries), driverless car, remote home surveillance, remote operations, etc. [3].

All the above shows telecom in diversified in many ways, and with several data types, the interest in AI/ML is renewed, due to this much complexity arises in telecom network. The workload is growing, and we are experiencing a paradigm shift to have a predictive, proactive, self-learned, and self-adaptive system to handle all these changes.

Almost all the major telecom service providers use AI/ML to improve the customer service and experience through the virtual assistants and chatbots. Customer care department in telecom gets number of support requests for set up, installation, connectivity, incidents, maintenance, etc. Virtual assistants can help the journey of the customer to self-service with the automation and scaling to handle these support requests, reduces the operational expenditure and improves customer experience. In one example, after introducing its chatbot TOBi, Vodafone saw a 68% improvement in customer experience in positive direction [4].

AI/ML can make the speech and voice services ability to the chatbots which can handle 80–90% issues independently without any human intervention [4]. Not only are these technological advancements used in service or marketing, but it can also handle the internal company operations like supply chain, sales and marketing, procurement, and other enterprise functions of telecom service providers, thus gives the agility required by the service provider to have a new go-to-market strategy to take on their competitors.

2 Challenges for the Use of AI/ML in Service Operations

Based on the points discussed above the telecom service providers who cannot take the full advantages of AI/ML will lose the ground rapidly to their competitors who can—as we already have seen these in many industry verticals like auto manufacturing like Volvo, Mercedes or in telecom gear manufacturers, and/or service providers like AT&T, Verizon, Reliance, Bharti-Airtel, etc. [5]. With the outbreak

Table 1 Factors determining the too hyperopic or too myopic objectives

No#	Too hyperopic	Too myopic
1	The work package is too long to complete where the end product can only matter and interims can draw only a very limited benefit	Some important and/or niche challenge gets resolved by the automation project how the root cause analysis is not done to get to the root of the business situation and put a solution to it
2	Many heads are there whose inputs are valuable and many of them has different objectives and no clear accountability. This can jeopardize the outcome of the project	The business head for the target group does not feel connected as the project does not bring value to him/her, and the actual group from across a specific value chain has very little to say about it
3	All the business functions and technical architectures are completely redesigned and have big bang approach with total digital transformation	A silo solution has been created which has limited capability to integrate with upstream and downstream systems, thus business value generated is not percolated to many directions but confined to minimum number of group(s)

of COVID-19, the service providers are now racing to build the automation infrastructure because of the sudden disruption of the business model which force them to come up with the new normal condition. Now, the challenges are how to put those skills into being—to draw the larger benefits in line with the modified organization vision.

Now, how to put forward the automation projects undertaken by the organization—is it too myopic or too hyperopic so that results are affecting the organization or a proper balance strike in between to get the maximum benefit out of it. Following are the factors which need to be considered while taking up the automation projects to make the organization more agile [5, 6] (Table 1).

3 Methodology to Make the Improvements in Service Operations Through AI/ML

Following steps need to be undertaken to make the improvements happen in service operation:

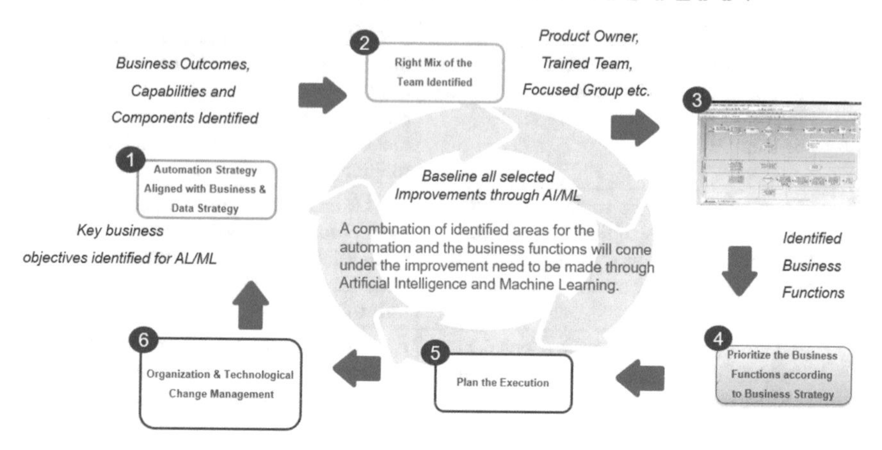

1. **Automation strategy**

The first the AI/ML strategy across the organization needs to be in place, which can only be done by proper discovery of the areas of the organization to be automated. This step is very challenging as to identify the areas and getting the buy-ins from the CXOs is the most important thing to have a successful run of the project. Once this part is over, getting the project sponsor and creating the team to execute will not be a very big issue.

The domain area chosen should be big enough to show case the benefits, and the scope should be broken down into multiple pieces with value backed approach. This will help to show the benefits in a smaller time along with working for a larger time and larger benefits collectively coming from each of the broken-down pieces. One example: a telecom service provider redesigned its' business processes to raise the customer experience (which spans across all the touchpoints through which the company interacts with its' customers), using AI/ML to understand and address the customer's unique needs which leads to the new go-to-market strategy and reduced the time to execute marketing campaigns by 75%. They became the market leader with their new capability and gain more share of wallet of their customer as early mover to the market. This also helps the company to lower customer churn by three percentage points. The company expects those improvements to add $70 million to its bottom line by the end of 2021 [5].

The strategy targeting three main factors:

- Business Outcomes—the business benefits that the customers can expect as a result of the different activities.
- Capabilities—the proficiency across the organization due to the AI/ML transformation.
- Components—the business and technical assets are required and how much of the existing assets (in terms of technology and data) can be reused.

2. **Right mix of the team (internal and external)**

The right mix of the team starts with finding out the right sponsor and with the help of the sponsor identifying the right people to execute the project. Following are the areas to be covered during this period [7]:

- A business champion who knows how to define the value in telecom business vertical operations.
- A dedicated product owner.
- A team of trained AI/ML practitioners such as data science and automation experts.
- A focused group of frontline developers and testers for intelligent automation.

The team responsible for AI/ML initiatives should have all the above-mentioned resources in the team. In many scenarios, the internal team needs to be trained with certain technical and/or business skills which are mandatory for the project. There are two ways to handle this situation. Either to hire new resource with the required skill set or hire external consultants to train the internal resources for a period to help them gaining the proper knowledge.

3. **Identify the business functions to be automated and how (vanilla automation, predictive, or proactive)**

As I mentioned earlier, as a part of the strategy creation we need to identify the business outcome, whereas here the idea is to identify the exact business functions of the service operation which can come under AI/ML scope and how they are related to the set of specific business outcomes. Following table describes how the business functions in telecom operations are related to business outcomes [7–9] (Table 2).

With the identification of business functions, it is imperative to define whether intelligent automation (AI/ML) will be required or vanilla automation (robotic process automation or RPA) where the only need is to mimic the human interactions. This is important as cost (development and licensing) is a prime factor over here.

Table 2 Co-relation between business functions and business outcomes

Business functions though AI/ML	Business outcomes
1. Smart network design 2. Robust and dynamic deployment of IT and network 3. Fully data-driven operations 4. Intelligent optimization 5. Seamless security 6. Agile management of the applications 7. Innovation through collaborative data collection	1. Improved customer experience 2. New technological transformation 3. Cost optimization 4. Revenue growth 5. Accelerated go-to-market (GTM) and time-to-market (TTM) 6. Agile service creation 7. Trusted and secure business

Table 3 Maturity of the business functions

Basic level	Intermediate level	Advanced level
• Limited capabilities. Focus on automation/ML internal use cases that are less complex to implement and where the benefits/return on investment (ROI) are easier to identify • Foundational skills on data analytics and automation • Processes defined for key deliveries with some degree of automation and ML • Limited data availability with some focal points and point solutions	• End to end capabilities. Well defined approach to deploy technology and use cases (on-premises, cloud, or hybrid) is applied • Basic cognitive skills in place • Processes based on predictive and other data science capabilities • There is an organization-wide data lake to improve data gathering, management, and availability	• End to end capabilities. AI/ML and automation capabilities infused across or influencing services, resources, and operations • Advanced cognitive skills in place to make critical business decisions using AI/ML • Fully data-driven and predictive processes enabling the discovery and operationalization of new data insights and use cases • Advanced data management combining multiple sources with a standardized and well-governed approach to assure quality, format, and ownership of data types

4. **Prioritize the business functions**

Prioritizing the business functions is identifying the functions which can be taken from the basic to intermediate to advanced level and how they are interrelated with other business functions. Following is an example of state transition of the business functions while maturity changes [7, 8, 10] (Table 3).

5. **Plan the execution**

Based on the prioritization of business functions and reimagining the business as usual (BAU) of domains, the execution plan will be chalked out. There are many interconnectivities and interdependencies amongst the different domains, so the plan has to be drawn in such a way that the critical path gives least down time and lesser costs. Typically, in telecom, we find that the service providers are best served by applying first principles or design-thinking techniques and working backward from a key business objective/function to stich the plan of actions.

6. **Change management**

To support the new AI/ML-based business processes and the corresponding developed models, a new mindset like agile ways of working and interdisciplinary collaboration will be required which can bring a significant organization change. My thorough research on this field shows that the companies expecting the highest returns on AI need to enact efficient change management practices. This can only happen when a top-down model like the changes facilitated by CEOs and top executives can help employees to follow the model desired behaviours, and that such efforts work best then [5, 7, 11].

This change management cannot be fructified unless there is a complete buy-in from CXOs and get their equal participation. They need to be heavily invested into the progress of the transformations in terms of investment, team's performance, producing the expected result, etc. The more the involvement the more is the chance of the success, and the more the organization is ready to change. All these transformation and organization change management which is percolated down from top can create the desired domino effect.

Most of the telecom service providers these days building cloud-based platform for storing the federated and non-federated data from the existing transactions and customer service systems so in future, it could easily be used by data engineers and data scientists than data from the age-old legacy warehouse system. New analytics workbench along with trained data scientists can deploy new models faster, and tools that streamlined data collection, analysis, and model building are very much in the picture almost for all big service providers. The AI-driven customer-value-management system helps to have the paradigm shift when it comes to process the unstructured data. It can the capability and necessary business functions to apply more complex approaches and work more efficiently. While prioritizing the requirement of technology investments, the ecosystem should always be there in the mind and a clear picture what will require and when, how the value realization can happen.

In order to get the domino effect, the change in doing the regular business and the real transformation of the organization is absolutely mandatory. It is no longer only digital transformation, but the idea is to have an agile organization where transformation to automation and system intelligence is imbibed into the culture.

4 Conclusion

Based on the above-mentioned point and the situation analysis it can be concluded that many factors are there to finalize the scope of the automation (AI/ML) solution. Following is the depiction of that (Table 4).

Table 4 The whole ecosystem of the AI/ML transformation model

Organization	Business process	Infrastructure	Capability	Business outcome
• Delivery model—the service delivery organization structure/layers based on business/technical capabilities • Business and technical skills like data science, project management, automation, customer experience, etc.	• All the service operation processes need to be streamlined to have a clear idea what is required and what steps need to be followed. Some of the examples are incident management, change management, etc.	• The infrastructure, tools, and applications required to implement the requirements and actualize the business value. Some examples are workflow application, automation and AI/ML platform, cloud native infrastructure, etc.	• The main capabilities identified as the business functions like smart network design, fully data driven operations etc.	• The expected business outcomes which are the very beginning of assuming the automation and AI/ML projects like revenue growth, cost optimization, etc.

References

1. Crawshaw J (2018) AI in telecom operations: opportunities & obstacles. https://www.guavus.com/wp-content/uploads/2018/10/AI-in-Telecom-Operations_Opportunities_Obstacles.pdf
2. Taylor T (2021) AI in telecom: top strategies and use cases, Feb 2021. https://www.bizops.com/blog/ai-in-telecom-top-strategies-and-use-cases
3. Artificial intelligence & data science in telecom: how can telcos gain? https://www.sytoss.com/blog/ai-in-telecom-how-can-telcos-gain/
4. Merr B (2019) The amazing ways telecom companies use artificial intelligence and machine learning, Sept 2, 2019. https://www.forbes.com/sites/bernardmarr/2019/09/02/the-amazing-ways-telecom-companies-use-artificial-intelligence-and-machine-learning/?sh=13e9861b4cf6
5. Fountaine T, McCarthy B, Saleh T (2021) Getting AI to scale, don't try to change everything at once, but do begin with something important, May–June 2021. https://hbr.org/2021/05/getting-ai-to-scale
6. Iansiti M, Lakhani KR (2020) Competing in the age of AI, how machine intelligence changes the rules of business, Jan–Feb 2020. https://hbr.org/2020/01/competing-in-the-age-of-ai
7. Medrala M (2020) How AI can improve your business: successful AI use cases in telecoms in the area of assurance, June 18th 2020. https://www.comarch.com/telecommunications/blog/successful-ai-use-cases-in-telecoms-in-the-area-of-assurance/
8. Artificial intelligence in telecom—from hype to reality. https://www.ericsson.com/en/blog/2019/6/ai-in-telecom
9. Verma A. Powering telecom with AI. AI practice and thought leadership. Wipro. https://www.wipro.com/holmes/powering-telecom-with-ai/
10. 4 areas where AI is transforming the telecom industry By Liad Churchill AI in customer service posted on Dec 20, 2020. https://techsee.me/blog/artificial-intelligence-in-telecommunications-industry/
11. Banerjee S, Candelon F, Lorenzo R, Cacouros A, Harguil S (2020) Transforming telcos with artificial intelligence, June 25, 2020. https://www.bcg.com/en-in/publications/2020/transforming-telecommunications-companies-with-artificial-intelligence

Feature Extraction for Image Processing and Computer Vision—A Comparative Approach

Radha Seelaboyina and Rajeev Vishwkarma

Abstract Digital image processing is a technique to process image digitally. This means, image is represented using pixels and all the algorithms are implemented using that value. A digital image consists of a finite variety of parts remarkably named as image parts, pixels. Pixel value determines the type of image, i.e. whether it is grey scale or RGB. Computer vision is a part of deep learning in which processing is done on images. The application of image processing includes robotics, object detection, weather forecasting, etc. In this paper, the main goal is to focus on different feature extraction techniques applied by computer vision and digital image processing. Image features are important input for any image processing tasks. Features include blobs, corner, edges, etc.

Keywords SIFT · Image processing · Artificial intelligence · Pixel

1 Introduction

Computer vision includes another branch of experience called artificial intelligence(AI) whose objective is to emulate human intelligence. The world of image analysis or image understanding is in between image process and computer vision. There is no general agreement among analysis community relating to the borderline between image process, image analysis and computer vision. Generally, a distinction is formed by process image process as a discipline during which each the input and output of the method area unit pictures. It is a method that involves primitive operations [1] like reduction of noise, distinction sweetening, image sharpening, etc. Image analysis could be a method that is characterised by the very fact that its inputs area unit typically pictures, however, its outputs area unit attributes extracted from those pictures (e.g. edges, contours, and therefore, the identity of individual objects) and at last, computer vision could also be outlined as a method that involves "making

R. Seelaboyina (✉) · R. Vishwkarma
Dr. A.P.J. Abdul Kalam University Indore, Indore, India
e-mail: radha.seelaboyina@gmail.com

A. Kumar et al. (eds.), *Proceedings of the International Conference on Cognitive and Intelligent Computing*, Cognitive Science and Technology,
https://doi.org/10.1007/978-981-19-2350-0_20

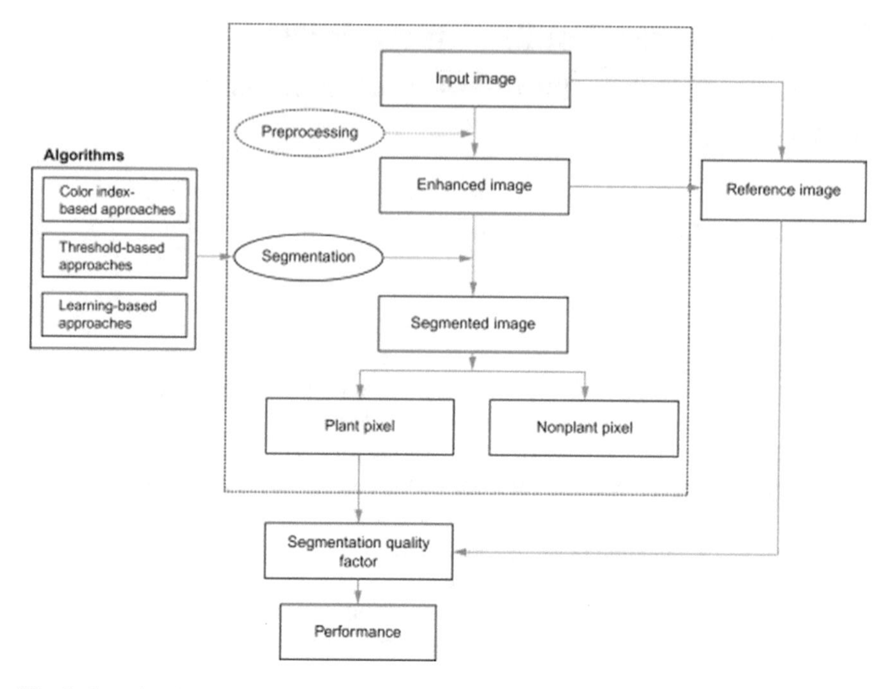

Fig. 1 Steps involved in image processing (*source resource* Sciencedirect.com)

sense" of associate in nursing ensemble of engravers to the sector of process digital pictures with the resistance of a data processor. A digital image consists of a finite variety of parts remarkably named as image parts, image parts, pels or pixels. Pixel is the term most generally accustomed denote the essential part of a digital image [3]. Like human imaging system, digital imaging devices will acquire digital pictures and store them for process. However, in contrast to human, WHO will capture pictures just for visible band of spectrum, and imaging devices will capture pictures covering whole vary of spectrum. Thus, digital image process encompasses a large sort of application areas within the domains of life science, remote sensing, traffic observance, document analysis and retrieval, etc. [2]. On the opposite hand, computer vision is another domain of analysis space, wherever the final word objective is to use system to emulate human vision by learning from the setting and having the ability to form inferences on specific things and take necessary actions on the premise of the inferences. This space includes another branch of experience called artificial intelligence (AI) whose objective is to emulate human intelligence the world of image analysis [6] (Fig. 1).

2 Parallel Architectures for Image Processing

Image understanding is in between image process and computer vision. There is no general agreement among analysis community relating to the borderline between image process, image analysis and computer vision. Generally, a distinction is formed by process image process as a discipline during which each the input and output of the method area unit pictures. It is a method that involves primitive operations like reduction of noise, distinction sweetening, image sharpening, etc. [7]. Image analysis could be a method that is characterised by the very fact that its inputs area unit typically pictures, however, its outputs area unit attributes extracted from those pictures (e.g. edges, contours, and therefore, the identity of individual objects) and at last, computer vision could also be outlined as a method that involves "making sense" of associate in nursing ensemble of recognized objects, as in image analysis and activity the psychological feature functions call related to vision. Based on the preceding discussions, it is seen that a logical place of overlap between image process and image analysis is that the space of recognition of individual regions or objects inside a picture [9]. Thus, during a broader sense, digital image process encompasses processes whose inputs and outputs area unit pictures and also the processes that extract attributes from pictures, up to and together with the popularity of individual objects. To clarify the thought any, allow us to take into account the matter of machine-driven recognition of text inside a general scene image [10]. The features of a picture containing the text, pre-processing that image, segmenting the individual characters, describing the characters within the sort of feature values appropriate for computer process and recognizing those individual characters will somewhat be thought- about inside the scope of digital image process. Creating sense of the content of the image could also be viewed as being within the domain of image analysis and even computer vision, reckoning on the extent of complexity of the matter reminiscent of the extent of expected answer implicit by the statement "making sense". Applications of digital image process [4, 5]. There are a unit various applications of digital image process staring from early 20th century until currently. One in every of the oldest applications of digital pictures, as according in literature, was causing photos by submarine cable between London and the big apple. In 1920, introduction of the Bartlane cable image gear improved the time needed to move an image across the ocean specialised printing instrumentation coded the photographs for cable transmission so the photographs were reconstructed at the receiving finish [11]. Visual quality of those early digital photos was improved towards the tip of 1921 by introducing a way supported photographic copy. The first Bartlane systems were capable of secret writing pictures in S distinct grey levels, that was increased to fifteen grey levels in 1929 [8]. Though the examples simply cited involve digital pictures, the evolution of digital image process is intimately tied to the event of the data processor. Owing to its inherent nature, digital pictures need most cupboard space and their process needs most process power that progress within the field of digital image process had been extremely obsessed on the event of digital computers [14]. Though the thought of a computer was planned long back, what is known as a contemporary data processor came solely in Forties with the introduction of 2 key ideas by John von Neumann: (1) a memory to carry a keep programmed knowledge, and (2) conditional branching. These two concepts area unit the muse of a central process unit (CPU), that is, that the heart of computers nowadays. Beginning with the John von Neumann design, there have been a series of key advances from the invention of the semiconductor by Bell Laboratories in 1948 to this use of radical massive scale integration (ULSI) in nineteen eighties that semiconductor diode to computers powerful enough to be used for digital image process. Initially, the primary computers are powerful enough to hold out meaningful image process tasks appeared within the early Sixties, and therefore, the first potential digital image process victimisation is those computers for rising pictures from an area probe began at the reaction propulsion laboratory [7, 8]. Golden State in 1964 once the photographs of the moon, transmitted by Ranger seven, were processed by a computer to correct varied varieties of image distortions. The invention of computed tomography (CT) within the early seventies is one in every of the foremost necessary events within the application of image process in life science. Another rising space of application is image registration, that is, that the method of orienting two or additional pictures (the reference and perceived images) of a similar scene taken at completely different time, from completely different viewpoints, and/or by completely different sensors [5]. This has monumental applications in life science field from the sixties till this days, the sector of image process has fully grown quickly additionally to applications in medication and within the house programs, digital image process techniques, area unit currently utilised in a broad vary of applications. Study of the aerial and satellite imagery is another space of the applying of image process [14]. In physics and allied fields, image process techniques area unit accustomed enhance ensuing pictures of the experiments in areas like high-energy plasma's and microscopy [12]. Archaeological use of image process includes restoration of blurred pictures that were the sole obtainable records of rare artefacts lost or broken once being photographed. Alternative productive applications of image process techniques may be found within the field of bioinformatics, astronomy, chemistry, medical specialty, enforcement, defence, industrial applications, etc. The same examples illustrate the cases wherever the results of process area unit supposed for human interpretation [11].

3 Applications of Digital Image Processing

Another major space of application of digital image process techniques is in determination issues, managing machine perception which can have very little similitude to visual options that humans use in deciphering the content of a picture [4]. Bio-metric recognition is a vital and wide application space wherever digital image process plays a vital role. Automatic face recognition, fingerprint recognition, palm recognition and tissue layer recognition have been the well-researched areas during this domain. Content based mostly image retrieval is another domain of analysis during which pictures area unit categorised and indexed according their content specified the specified class of pictures may be retrieved as and once required in future [13]. Sign language and angle analysis are the area unit as wherever import of arm/body movement and countenance area unit analysed and high level linguistics are inferred. Alternative typical area unit as of machine perception that utilises image process techniques are optical character recognition(OCR), robotics, military signal process, meteorology and environmental assessment. Human motion tracking with additional advancement in image process techniques, evolution in electronic devices and development of extremely economical sensors and imaging devices, automatic police work has become a growing demand from completely different body authorities over the world. Human motion tracking is one in every of the requirements during this domain. Several survey reports have been printed on computer vision based mostly human motion trailing. Real-time trailing of human motion has been according in[2]. A 3D model based mostly human motion trailing has been printed in associate in nursing analysis and synthesis of human movement is according in[15]. Police work in computer game is additionally according in numerous literature's trailing human motion during a scene and tally variety of interactive individuals has been according in alternative connected applications like gesture recognition, prediction of motion trajectories and artificial intelligence are necessary analysis areas of digital image process [3]. Pedestrian behaviour analysis is another application space-associated with human motion trailing is pedestrian behaviour analysis and collision rejection. A report on pedestrian collision rejection systems.

4 Conclusion

This paper gives information about different image processing and computer vision algorithm. Nowadays, feature extraction is very important and crucial tasks for implementation of any algorithm that is based in images. Day by day number of feature extraction algorithm is developing. This paper the main goal is to focus on different feature extraction techniques applied by computer vision and digital image processing. Image features are important input for any image processing tasks. Features include blobs, Corner Some are modification of traditional digital image processing algorithm. Computer vision has a wide future and it mostly works on images. Computer vision uses SIFT, GLOH, SURF, PCA SIFT for extracting feature space at different scale space.

References

1. Allusse Y, Horain P, Agarwal A, Saipriyadarshan C (2008) GPUCV: a GPU-accelerated framework for image processing and computer vision. In: International symposium on visual computing. Springer, pp 430–439
2. Ayache N (1995) Medical computer vision, virtual reality and robotics. Image Vis Comput 13(4):295–313

3. Bhandarkar SM, Arabnia HR, Smith JW (1995) A reconfigurable architecture for image processing and computer vision. In: VLSI and parallel computing for pattern recognition and artificial intelligence. World Scientific, pp 29–57
4. Cootes TF, Taylor CJ (2001) Statistical models of appearance for medical image analysis and computer vision. In: Medical imaging 2001: image processing, vol 4322. International Society for Optics and Photonics, pp 236–248
5. Dickmanns ED, Zapp A (1987) Autonomous high speed road vehicle guidance by computer vision. IFAC Procs 20(5):221–226
6. Fung J, Mann S (2005) Openvidia: parallel GPU computer vision. In: Proceedings of the 13th annual ACM international conference on multimedia, pp 849–852
7. Fung J, Mann S (2008) Using graphics devices in reverse: GPU-based image processing and computer vision. In: 2008 IEEE international conference on multimedia and expo. IEEE, pp 9–12
8. Gunjan VK, Janga Reddy M, Shaik F, Hymavathi V (2018) An effective user interface image processing model for classification of brain mri to provide prolific healthcare. Helix J 8(3):2129–2132
9. Jiang X (2009) Feature extraction for image recognition and computer vision. In: 2009 2nd IEEE international conference on computer science and information technology. IEEE, pp 1–15
10. Koutis I, Miller GL, Tolliver D (2011) Combinatorial preconditioners and multilevel solvers for problems in computer vision and image processing. Comput Vis Image Understand 115(12):1638–1646
11. Mairal J (2010) Sparse coding for machine learning, image processing and computer vision. Ph.D. thesis, Cachan, Ecole normale supérieure
12. Prasad PS, Pathak R, Gunjan VK, Ramana Rao HV (2020) Deep learning based representation for face recognition. In: ICCCE 2019. Springer, pp 419–424
13. Remagnino P, Jones GA, Paragios N, Regazzoni CS (2002) Video-based surveillance systems: computer vision and distributed processing
14. Weickert J (2000) Applications of nonlinear diffusion in image processing and computer vision
15. Wu B, Xu C, Dai X, Wan A, Zhang P, Yan Z, Tomizuka M, Gonzalez J, Keutzer K, Vajda P (2020) Visual transformers: token-based image representation and processing for computer vision. arXiv:2006.03677

Methods for Lung Cancer Detection, Segmentation, and Classification

Supikha Jain, Sanjeev Indora, and Dinesh Kumar Atal

Abstract Detection of lung cancer has been a trending research area today as lung cancer detection automation is beneficial if detected in the early stage. If the malignancy in the lung nodules is identified at the beginning phase, then survival chances may increase. Lung cancer starts within the cells covering the bronchi and lung parts like bronchioles or alveoli. Machines that have generally been used to diagnose cancer within the body are X-Ray, CT, MRI, PET, and SPECT. Radiologists have proposed different lung cancer prediction models like feature fusion mechanism, reference-model, 3D-CNN, optical flow methods, and cloud-based 3DDCNN CAD system. Picture handling procedures give a helpful quality device for improving automatic as well as manual analysis. When human beings breathe, airflow goes in through the mouth and nose and down airways. Deep down in the interior lungs, the airways end in clusters of little sacs holding the air within the lungs—these little sacs contain the air within the lungs, encircling small blood vessels. Here, oxygen from the lungs proceeds into the bloodstream to infuriate tissues all over the body. This paper overviews the various lung cancer detection, segmentation, and classification methods used earlier. This study also enables us to measure the challenges, benefits, and limitations of existing techniques.

Keywords 3D Deep convolutional neural network (3DDCNN) · Machine learning · Decision making systems

S. Jain · S. Indora
Department of Computer Science and Engineering, Deenbandhu Chhotu Ram University of Science and Technology, Murthal, Sonipat, Haryana, India

D. K. Atal (✉)
Department of Biomedical Engineering, Deenbandhu Chhotu Ram University of Science and Technology, Murthal, Sonipat, Haryana, India
e-mail: dineshatal.bme@dcrustm.org

A. Kumar et al. (eds.), *Proceedings of the International Conference on Cognitive and Intelligent Computing*, Cognitive Science and Technology,
https://doi.org/10.1007/978-981-19-2350-0_21

1 Introduction

The human body consumes a certain measure of water and oxygen every day to endure. The human body works effectively if there is a proper and adequate supply of oxygen. When taking in, air enters through the nose and mouth and goes into the lungs through the windpipe. Alveoli assimilate oxygen into the blood from the breathed-in air and eliminate carbon dioxide through blood when inhale exposed. Take oxygen and free of carbon dioxide is the fundamental capacity of the lung. The lung has two sponges, i.e., the right lung possesses three lobes, and the left lung possesses two sections called lobes. Each lobe has the same function bring oxygen and remove carbon dioxide. The segment of projection or whole lobe can be taken out as a treatment for lung cancer. Around the globe, people are dying due to many diseases, and cancer is one of the most hazardous diseases. In addition, the second most important cause of death because of cancer is lung cancer. It is hard to distinguish cancer in the lungs from its start since its indications show up at the high-level stage, where the possibility of endurance is exceptionally low. Modern three-dimensional medical image processing offers the potential for significant advances in science and technology as higher constancy pictures are created. Smoking is the reason for about 85% of cancer in the lungs, yet there are exact confirmations like arsenic inside water and beta-carotene appurtenance, increasing inclination toward illness [1]. Mainspring for lung cancer is smoking–active or passive, vigorous family record of lung cancer, chronic lung infections, and other respiratory diseases, subjection to harmful gases and chemicals like radon gas, asbestos, silicon, and environmental pollution. Lung cancer is based on CT scans. They take a different form which is DICOM format, addressed as ".dcm." DICOM contrasts from other picture designs in that it bunches data into informational indexes. DICOM record comprises a header and picture informational collections stuffed into a solitary document. DICOM format has some impediments: document sizes are huge, and extraordinary programming is needed to survey them closely on the computer [2]. The tests to diagnose lung cancer may encompass: CT scans, MRI, sputum cytology, fiber optic bronchoscopy, and transthoracic fine-needle goal, while lab tests are: blood test, biopsy, endoscopy, mediastinoscopy, and video assisted thoracoscopic surgery (VATS). Stages of image mining for lung cancer diagnosis are data preprocessing, segmentation, feature extraction, and classification were the steps to distinguish either the lung nodules is benign (non-cancerous) or malignant (cancerous) [1].

2 Image Mining Stages for Diagnosis of Lung Cancer

2.1 Preprocessing Stage of Lung Cancer

For automatic diagnostic of lung cancer, image preprocessing is the first step to take. It simplifies the identification and classification and also improves the overall

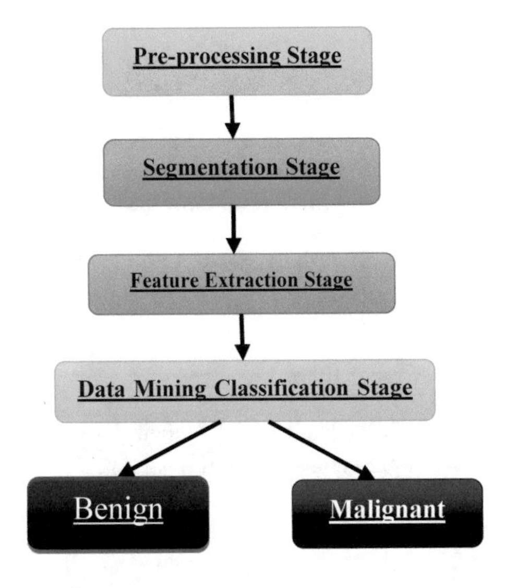

Fig. 1 Image mining stages for diagnosing of lung cancer

precision of diagnosis [2]. Preprocessing is a method of enhancing image data by removing undesired distortions or enhancing certain image features in preparation for further processing. It is also used to enhance image features like lines, borders, and textures. It can easily divide image content into two categories: wanted and unwanted content [3].

Steps which can be used for preprocessing the image

A. Median Filter

A median filter is used to minimize the noise from images and signals, which is a nonlinear digital filtering method. One of the essential preprocessing techniques for improving outcomes is noise reduction, which is done by median filter.

$$\widehat{f}(x, y) = \underset{(s,t)\in S_{xy}}{\text{median}}\{g(s, t)\} \tag{1}$$

B. Average Filter

The average filter is also commonly known as the mean filter. Average (or mean) filtering is a technique for "smoothing" images by decreasing intensity variations between adjacent pixels. As it passes pixel by pixel through the image, the average filter replaces each value with the average value of nearby pixels, including itself.

C. Gabor Filter

A Gabor filter combines a Gaussian filter with a sinusoidal term. It is used to examine texture patterns in terms of pixel values. Therefore, it is beneficial for looking at the lung cancer texture within the image [4].

Gabor filter in two dimensions:

$$X' = x\cos(\theta) + y\sin(\theta);$$
$$Y' = -x\sin(\theta) + y\cos(\theta);$$

Gaussian provides weights and sinusoidal provides directionality.

$$g(x, y; \lambda, \theta, \psi, \sigma, \gamma) = \exp\left(-\frac{x'^2 + \gamma^2 y'^2}{2\sigma^2}\right)\cos\left(2\pi\frac{x'}{\lambda} + \Psi\right) \quad (2)$$

D. Weiner Filter

Wiener filter is a special domain linear filter. As a result, we get an image that is uncorrupted and with minimal error.

$$w[\mathrm{n}] \rightarrow \boxed{G(\mathrm{z}) = \sum_{i=0}^{\mathrm{N}} a_i \mathrm{z}^{-\mathrm{i}}} \xrightarrow{\mathrm{x}[n]} \overset{-}{\rightarrow} \mathrm{e}[\mathrm{n}] \quad (3)$$

E. Fast Fourier Transform

The fast Fourier transform (FFT) technique divides signals into the frequency in a specified time or location. FFT is a frequency domain function that can be used to generate images from the response of signals such as ultrasound and magnetic resonance (MR) images [5].

$$x[k] = \sum_{n=0}^{N-1} [n] e^{\frac{-j2\pi kn}{N}} \quad (4)$$

2.2 Segmentation

Lung segmentation is usually the initial stage in interpreting a lung CT image and is essential in identifying lung disease. The segmentation system is made up of several steps that eventually lead to segmenting the lung tumor. The technique of dividing an image into two sections, one of wanted and another is unwanted. Because of the different texture properties of lung tissue, automatically detecting the tumor is a critical task for the machine [6]. Image segmentation are of two types:

(a) Thresholding Approach
(b) Marker-Controlled Watershed Segmentation Approach.

A. Threshold segmentation

A typical segmentation technique for separating an object from its surroundings is thresholding. Thresholding is a technique for splitting an image into smaller segments, or junks, by establishing their boundaries with at least one color or greyscale value.

B. Marker-Controlled Watershed Segmentation Approach

This method for segmenting objects with closed contours and ridges as boundaries is robust and flexible. There are two types of markers utilized in this: internal markers and external markers. A smoothing filter is used to filter an image as part of the preprocessing scheme. This step can reduce the significant number of irrelevant details, which is the cause of over-segmentation.

2.3 *Feature Extraction*

A dimensionality reduction procedure reduces the large set of raw data into manageable groups for processing. A large number of variables in data sets demand a lot of computational resources to process. The dimensional measures themselves are described as follows:

A. Perimeter

$$\mathbf{P} = \mathbf{S_n S_1}| + \sum_{i=1}^{n-1} |\mathbf{S_i S_i} + 1| \tag{5}$$

where $s = \{s_1 \ldots s_n\}$ is a set of the boundary points.

B. Eccentricity

Aspect ratio is measured by using eccentricity. It is the length of major-axis to the length of minor-axis. The principal axes method or the smallest bounding rectangle approach can be used to calculate it.

C. Entropy

$$\mathbf{Entropy} = -\sum\sum \mathbf{p(i,j) logp(i,j)} \tag{6}$$

where p is the number of gray-level co-occurrence matrices in GLCM.

D. Contrast

$$\mathbf{Contrast} = \sum\sum \mathbf{(i-j)^2 p(i,j)} \tag{7}$$

where $p(i, j)$ = pixel through location (i, j).

E. Correlation

$$\mathbf{Correlation} = \sum_{\mathbf{i,j=0}}^{\mathbf{n}} \mathbf{P_{ij}} \frac{(\mathrm{i}-\mu)(\mathrm{j}-\mu)}{\sigma^2} \tag{8}$$

F. Energy

$$\mathbf{Energy} = \sum\sum (\mathbf{p(i,j)})^2 \tag{9}$$

G. Homogeneity

$$\mathbf{Homogeneity} = \sum_{\mathbf{i,j}} \frac{\mathbf{P(i,j)}}{1 + |\mathbf{i} - \mathbf{j}|} \tag{10}$$

2.4 *Feature Selection*

It selects those attributes that contribute most to the prediction variable and delivers the desired outcome, either automatically or manually. It is characterized as the proportion of the number of essential highlights effectively chosen to the all-out number of highlights. It is estimated regarding rate (%). The numerical recipe for include determination rate is estimated as follows.

$$\mathbf{FSR} = \frac{\text{Number of correctly selected features}}{\text{Total number of features}} \times 100 \tag{11}$$

2.5 *Classification*

The task of categorizing any object within a particular category called class is characterized as classification. Majorly lung cancer is classified into four parts: squamous cell lung cancer, adenocarcinoma, large cell lung cancer, and small cell lung cancer. Males are more likely to get squamous cell lung cancer, which has a central origin and appears early. Female adenocarcinoma is the most frequent form, with a peripheral origin and manifests late. The least common type of lung cancer is large cell lung cancer, which has a peripheral origin.

3 Performance Analysis

3.1 Mean Square Error

Mean square error appears when anyone is trying to predict numeric and continuous data. Mean square error is a loss function it measures the amount of error. Therefore, it is needed to minimize the MSE. It is opposite to the accuracy because there is a wish to maximize the accuracy. While calculating MSE: firstly, take actual and predicted values. Then, find out the error by subtracting the actual and predicted value. One advantage of using MSE is even negative error becomes positive either predicted value is less than or more than the actual. The error does not cancel out; that is the one reason for taking square. Then, find out square error and sum up them divide by a number of the element, then the value of MSE is obtained.

$$\sum_{i=1}^{n} \frac{(w^{T}\, x(i) - y(i))^{2}}{n} \tag{12}$$

3.2 Peak Signal-To-Noise Ratio

The peak signal-to-noise ratio (PSNR) is the ratio of a signal's maximum possible power to the power of corrupting noise.

$$\mathrm{PSNR} = 10\log_{10\left(\frac{\mathrm{MAX_f}}{\sqrt{\mathrm{MSE}}}\right)} \tag{13}$$

3.3 Structural Similarity Index (SSIM)

It is a perceptual metric that measures the degradation of image quality due to processing such as data compression or transmission. It is a complete reference metric that requires two images—a reference image and a processed image.

$$\mathrm{SSIM}(x, y) = \frac{\left(2x\mu_x\mu_y + c_1\right)\left(2\sigma_{xy} + C_2\right)}{\left(\mu_x^2 + \mu_y^2 + c_1\right)\left(\sigma_x^2 + \sigma_y^2 + c_2\right)} \tag{14}$$

3.4 Mean Absolute Error

It is a measure of errors between paired observations that describe the same phenomena. In this, many minor errors are equal to one large error.

$$\text{MAE} = \frac{1}{n}\sum_{i=1}^{n}|x_i - x| \tag{15}$$

4 Performance Evaluation Parameters

The performance analysis of the systems is characterized utilizing four estimates true positive (TP), true negative (TN), false positive (FP), and false negative (FN).

True positive: Arrhythmia identification harmonizes with the choice of doctor.

True negative: Both classifier and doctor recommended nonappearance of arrhythmia.

False positive: The system names a solid case as an arrhythmia one.

False negative: Framework names an arrhythmia as normal.

4.1 Accuracy (Ac)

The numerical value of accuracy represents the proportion of true positive results (both true positive and true negative) in the selected population.

$$\text{Accuracy} = \frac{\text{TP} + \text{TN}}{\text{TP} + \text{TN} + \text{FP} + \text{FN}} \tag{16}$$

4.2 Sensitivity (Sn)

It (true positive rate) measures the proportion of positives that are correctly identified (i.e., the proportion of those who have some condition (affected) who are correctly identified as having the condition).

$$\text{Sensitivity} = \frac{\text{TP}}{\text{TP} + \text{FN}} \tag{17}$$

Table 1 Comparing accuracy, sensitivity, and specificity

S. no	Title	Accuracy	Sensitivity (%)	Specificity
1	[7]	98.51%	98.4	92%
2	[8]	93.71%	92.96	98.52%
3	[9]	84.15%	83.96	84.32%
4	[10]	95.5%	99.9	*
5	[11]	*	86	91%
6	[12]	*	98.33	97.11%
7	[13]	93.25%	93.12	91.37%
8	[14]	*	71.4	91.9%

asterisk (*) stands for data not available or calculated for the particular study

4.3 Specificity (Sp)

It (True Negative rate) measures the proportion of negatives that are correctly identified (i.e., the proportion of those who do not have the condition (unaffected) who are correctly identified as not having the condition).

$$\text{Specificity} = \frac{\textbf{TN}}{\textbf{TN} + \textbf{FP}} \tag{18}$$

4.4 Positive Predictivity (+P):

The probability of having the disease based on a positive screening test is known as positive predictive value

$$\text{Positive predictivity} = \frac{\textbf{TP}}{\textbf{TP} + \textbf{FP}} \tag{19}$$

5 Lung Cancer Dataset Description

The lung dataset is an extensive collection of data from the PLCO study used for lung cancer screening, incidence, and death analysis. The collection includes CT and PET-CT DICOM images of lung cancer patients and XML comment documents indicating tumor location. The images are taken from the patients who have a low risk of lung cancer and go through norm of care lung biopsy and PER/CT. There are

Table 2 Different tests with their advantages and disadvantages

S. no	Test	Description	Advantages	Disadvantages
1	Bone scan	It is a nuclear imaging treatment that involves injecting a small amount of radioactive material into a vein to highlight specific locations [15]	One test for the entire body, low radiation exposure, and sensitive evaluation	Radiopharmaceuticals and a gamma camera are required, which are not generally available
2	Computed tomography	To obtain cross-sectional images of the body, computer and rotating X-ray machines are used [16]	Very good in evaluating solid organ	Risk to the patient because of the high radiation dose, very expensive
3	Positron emission tomography	It is a nuclear imaging technique that involves injecting radioactive tracers into the body, which are then absorbed by the tissue	It depicts how organs and tissues works	The radioactive components utilized in the scans is harmful and only few times PET imaging can be performed
4	Chemotherapy	It is a treatment that utilizes strong chemicals to kill your body's fast-growing cells	It can help live longer by shrinking cancer or slowing its growth	It has varied effects on various people, and it may or may not work for some
5	Immunotherapy	It is a kind of cancer treatment in which the immune system is used to detect and fight cancer cells	Create powerful immune responses that target cancers all across the body, including those that are difficult to reach	Muscle aches, shortness of breath, leg swelling, sinus congestion, migraines, and weight gain due to fluid retention
6	Radiation therapy	Cancerous cells are targeted by high-energy waves, which may disrupt the cell's internal functioning	Allows the delivery of a high radiation dose to a specific area	Treated volume is small

many online platforms from where we can easily collect the dataset, like LIDC-IDRI, Automatic Nodule Detection 2009, LUNA16, The Cancer Imaging Archive, etc.

6 Discussion and Conclusion

This study represents various methods that have been used for lung cancer detection. Different steps for lung cancer detections are described for the detailed description of the same. Out of all imaging techniques, CT scans have been found more suitable and reliable for detecting even very small nodules in the lungs; LDCT of the chest is especially effective for diagnosing lung cancer at its earliest, most treatable stage. CT is fast, which is essential for patients who have trouble holding their breath. CT scanning is painless and noninvasive.

References

1. Nasr M, Atif A, El-Mageed A (2016) Using image mining techniques for optimizing the treatment methods of lung cancer. J Multidiscip Eng Sci Technol 3(1):3159–3199 [Online]. Available: www.jmest.org
2. Research article pre-processing and segmentation techniques for lung cancer on CT images (2016)
3. Bari M, Ahmed A, Sabir M, Naveed S (2019) Lung cancer detection using digital image processing techniques: A review 38(2):351–360. https://doi.org/10.22581/muet1982.1902.10
4. Zhang J, Li D, Zhao Y, Chen Z, Yuan Y (2015) Representation of image content based on RoI-BoW. J Vis Commun Image Represent 26:37–49. https://doi.org/10.1016/j.jvcir.2014.10.007
5. Avinash S, Manjunath K, Senthilkumar S (2017) Analysis and comparison of image enhancement techniques for the prediction of lung cancer. In: RTEICT 2017—2nd IEEE International Conference Recent Trends Electron. Inf Commun Technol Proc 2018, pp 1535–1539. https://doi.org/10.1109/RTEICT.2017.8256855
6. Muthazhagan B, Ravi T, Rajinigirinath D (2020) An enhanced computer-assisted lung cancer detection method using content based image retrieval and data mining techniques. J Ambient Intell Humaniz Comput. https://doi.org/10.1007/s12652-020-02123-7
7. Masood A et al (2020) Cloud-based automated clinical decision support system for detection and diagnosis of lung cancer in chest CT. IEEE J Transl Eng Heal Med 8(October):1–13. https://doi.org/10.1109/JTEHM.2019.2955458
8. de Pinho Pinheiro CA, Nedjah N, de Macedo Mourelle L (2020) Detection and classification of pulmonary nodules using deep learning and swarm intelligence. Multimed. Tools Appl 79(21–22):15437–15465. https://doi.org/10.1007/s11042-019-7473-z
9. Song QZ, Zhao L, Luo XK, Dou XC (2017) Using deep learning for classification of lung nodules on computed tomography images. J Healthc Eng 2017. https://doi.org/10.1155/2017/8314740
10. Baby YR, Ramayyan Sumathy VK (2020) Kernel-based Bayesian clustering of computed tomography images for lung nodule segmentation. IET Image Process 14(5):890–900. https://doi.org/10.1049/iet-ipr.2018.5748
11. Magalhães Barros Netto S, Corrêa Silva A, Acatauassú Nunes R, Gattass M (2012) Automatic segmentation of lung nodules with growing neural gas and support vector machine. Comput Biol Med 42(11):1110–1121. https://doi.org/10.1016/j.compbiomed.2012.09.003

12. Lee SLA, Kouzani AZ, Hu EJ (2010) Random forest based lung nodule classification aided by clustering. Comput Med Imaging Graph 34(7):535–542. https://doi.org/10.1016/j.compmedimag.2010.03.006
13. Xiao X, Qiang Z, Zhao J, Qiang Y, Wang P, Han P (2019) A feature extraction method for lung nodules based on a multichannel principal component analysis network (PCANet). Multimed Tools Appl 78(13):17317–17335. https://doi.org/10.1007/s11042-018-7041-y
14. Machado RF et al (2005) Detection of lung cancer by sensor array analyses of exhaled breath. Am J Respir Crit Care Med 171(11):1286–1291. https://doi.org/10.1164/rccm.200409-1184OC
15. Van den Wyngaert T et al (2016) The EANM practice guidelines for bone scintigraphy. Eur J Nucl Med Mol Imaging 43(9):1723–1738. https://doi.org/10.1007/s00259-016-3415-4
16. Fullerton D, Ph D, Ezekiel J (1982) Innovations and notes of computed tomography scans for treatment

A MapReduce Clustering Approach for Sentiment Analysis Using Big Data

Mudassir Khan, Mahtab Alam, Shakila Basheer, Mohd Dilshad Ansari, and Neeraj Kumar

Abstract The modern era of organizations are generating huge amount of data by digitalizing their way of promoting services and products. The companies trying to know what customers are saying in terms of products through reviews in social media analytics constitutes a prime factor to enhance the success of big data era. However, the social media data analytics is a very complex discipline due to subjectivity in the textual review and the productivity in their complexity. Different approach of framework to tackle this problem is proposed: The first stage discussed, in this paper, the sentiment analysis of social media through the different approaches of machine learning. The second stage is used to discuss the challenges faced for processing the sentiment analysis of social media. Then, an overview of the case study presented for the two stages sentiment analysis in different approaches with the help of customer reviews. The machine learning approaches will be followed to analyze the sentiment analysis of social media in processing the big data.

Keywords MapReduce · MapReduce clustering · Social media analytics · Deep learning · Sentiment analysis

---

M. Khan (✉)
Department of Computer Science, College of Science & Arts Tanumah, King Khalid University, Abha, Saudi Arabia
e-mail: mudassirkhan12@gmail.com; mkmiyob@kku.edu.sa

M. Alam
Department of Computer Science, Mewar University, Masaka, Nigeria

S. Basheer
Department of Information Systems, College of Computer and Information Science, Princess Nourah Bint AbdulRahman University, Riyadh, Saudi Arabia

M. D. Ansari
Department of Computer Science & Engineering, CMR College of Engineering & Technology, Hyderabad, India

N. Kumar
Department of IT, Dr. B.R Ambedkar National Institute of Technology, Jalandhar, Punjab, India

A. Kumar et al. (eds.), *Proceedings of the International Conference on Cognitive and Intelligent Computing*, Cognitive Science and Technology,
https://doi.org/10.1007/978-981-19-2350-0_22

1 Introduction

The fast-growing social media analytics and its emergence are quiet trending and the fast-growing field of scientific research. The popularity of Internet communities is getting huge response from the millions of users and business owners to receive or start and speared public information at feasible convenient value. The limitless magnification of social media popularity among Internet and social media users is producing a massive volume of information; regardless of it has been outlined as social media and implemented using big data analytics techniques. The advance and modern social media frameworks show much executable of generated information configurations (e.g., tweets, text, audio, video, image, geo-locations, and animations/graphics). The enormous volume of data generated from social media platforms into various forms of unstructured and structured data. The huge and uncontrolled number of social media data popularity focused to keep the data. The success of social media utilizes, establish facilitate and newest comfort for analyzing different particulars and ornaments in communications. The sentiment of users among social media has become the challenge to analyze it. As per the different data sources, data are being processed by different sources to analyze the sentiment of processed to data. Even many organizations are using social media technologies to get the information about their product sentiments from users' point of view. As per the sources, social media has become main tool and technology to put their views.

2 Related Work

The Impact of schoolroom Behaviors and Student Attention on Written Expression [1], this examination looked for after to take a gander at schoolroom practices and a spotlight as indicators of composing execution among third-grade understudies getting a tier one execution input intercession. Data concerning the schoolroom direct of eighty third-grade understudies (39 folks, forty one females) was assembled before mediation began through usage of 2 educator report measures: the enlightening performance rating scale and as needs be the central mental methodology issue on the strengths and weaknesses of ADHD symptoms and standard behavior rating scale. Those understudies UN association were settled through this screening to be ineligible for participation during this assessment completed unmistakable instructional exercise activities consigned by their instructors every single through datum set.

R elatedly, the study additionally got some information about the quantity of long periods of showing background in math. One conceivable clarification for my discoveries might be that school grade-year scores are getting the nature of showing groups, which additionally is identified with student achievements. Simultaneously, these distinctions are not enormous. Further, standard mistakes are bigger in Model C than in Model B, as I would anticipate given increasingly constrained variations in my primary indicator factors. At long last, I find that assessment in Model D, which supplant school fixed impacts with school-by-year fixed impacts, are comparable in greatness to those in Model C. This shows year impacts do not drive results.

Khan, M (2019), introduced the concepts of MapReduce clustering in advanced incremental big data [2] processing. In this incremental method, author has given a clustering technique for processing big data. The author has given a detailed view how to analyze and process big data through MapReduce clustering technique [3, 4]. The author has worked on the various techniques of data processing through the MapReduce technique by using big data and social media analytics [5, 6].

3 Problem Description

The vast unique online informational collective develops slowly following several period as smart components focused gently included and persistence passages with some opinion erased otherwise will be altered. Utilizing the methods instrumentality, [7] the available platforms for secure mass data handling of available online information through the various techniques, for example, social network datasets, Google's percolator, can achieve efficient updates. This capability, however, occurs at the cost of dropping compatibility with the most simple and easy programming models rendered by non-incremental [8] systems, e.g., MapReduce, clustering and further importantly, essentials the programmer to enhance the application-specific dynamic/incremental step-by-step procedures, finally the enhanced increasing algorithm and code complexity [9].

The task-level coarse-grain incremental advanced processing system, Incoop, is not openly convenient. Therefore, we are unable to contrast i2MapReduce with Incoop [2]. Instead, we can think to contrast i2MapReduce with existing MapReduce model on Hadoop.

4 Proposed System

The proposed i2MapReduce, an extension to the different MapReduce clustering approaches that reinforces the fine-grain incremental advanced processing for existing available one-step and iterative computation. As we have discovered the result to the available preceding outcomes, i2MapReduce integrate the following three novel characteristics.

4.1 Cluster Configuration

The available system programs were accomplished on a cluster that contained of nearly [10–12] 1800 nodes. Each node had two 3 GHz Intel Xeon processors with hyperthreading installed, 6 GB of memory [7], two 250 GB IDE disks, and a high-speed Ethernet link. The nodes were lay out in a two-level tree-shaped switched network topological manner with approximately 150–200 Gbps of total bandwidth available at the root node. All of the nodes were in the similar hosting provision, and therefore, the round-trip time between any batches of nodes was less than a millisecond.

Out of the 6 GB of memory, approximately 1–1.5 GB was [2] restrained by other focused tasks executing on the cluster. The programs were running on a weekend [2] afternoon, when the CPUs, disks, and network were mostly idle as per the convenience.

4.2 Fine-Grain Incremental Advanced Processing Using MRBG-Store

Far from Incoop, i2MapReduce focused kv-pair level fine-grain incremental processing to keep down the volume of recompilation approximately feasible. The replica that depends on the kv-pair level information proceed and background subordination in a MapReduce figuring out as a bipartite graph, and carried as MRBGraph.

4.3 Unrestricted Iterative Computation with Acceptable Augmentation to MapReduce API

Our ongoing draft provides unrestricted support, counting not only one-to-one, in the same way one-to-many, many-to-one, and many-to-many uniformity. Escalate the Map API to permit users to undoubtably point out the loop-invariant formation of data and come up with a [7] Project API function to exhibit [2] the uniformity from Reduce to Map. While some users required to marginally change their step-by-step procedures in order to get complete benefit of i2MapReduce.

4.4 Gradual Processing for Iterative Computation

Gradual iterative processing is considerably more stimulating than gradual one-step toward processing because even a slightly updates may generate to influence a huge section of transitional states following several iterations. To overcome this study, this paper suggests discussing the converged state from the earlier solutions of discussed problem and take on an alternate propagation control (CPC) mechanism. In addition to intensify the MRBG-Store [7] to greater brace the approach design in gradual iterative processing [9].

4.5 MRBG-Store

A MRBG-Store is planned to maintain the compact positions in the MRBGraph and hold up well organized questions to fetch compact positions for incremental processing.

The MRBG-Store hold up the sustaining and recapture of compact MRBGraph positions for gradual processing. Inside the MRBG-store user perceives two main necessities on the MRBG-Store. First, the MRBG-Store necessity of incrementally repository the progressing the MRBGraph [13–15]. To study a flow of works that gradually restores the outcomes of a big data mining methods. Simultaneously input information progress, the transitional portion in the MRBGraph will also spread. It is observed that to store the entire MRBGraph of each following job is wasteful. First, the number of users would like to acquire and place only the updated section of the MRBGraph. Second, the MRGB-Store must produce well organized recoupment of keep sections of specified reduce occurrences. For gradual reduce outcomes, i2MapReduce re-evaluates the reduce instance engaged with each substitute MRBGraph edge. For a substitute edge [7], it inquiries the MRGB-Store to acquire the updated sections of the in-edges of the related K2, and unite the updated sections with the recently enumerated edge changes as shown in Fig. 1.

5 Advantages of the Proposed System

- To our understanding, i2MapReduce is the *first MapReduce-based outcomes that accurately carry incremental iterative solutions.*
- A MRBG-Store is planned to safeguard the compact sections in the MRBGraph and protect accurate queries to recover compact sections for gradual processing [16].
- Disparate Incoop, i2MapReduce reinforces kv-pair level compact gradual processing to reduce the quantity of re-production of outcomes as much as possible.
- The existing approach contributes general basis support, as per the counting not only one-to-one, but also one-to-many, many-to-one, and many-to-many compatability [16].

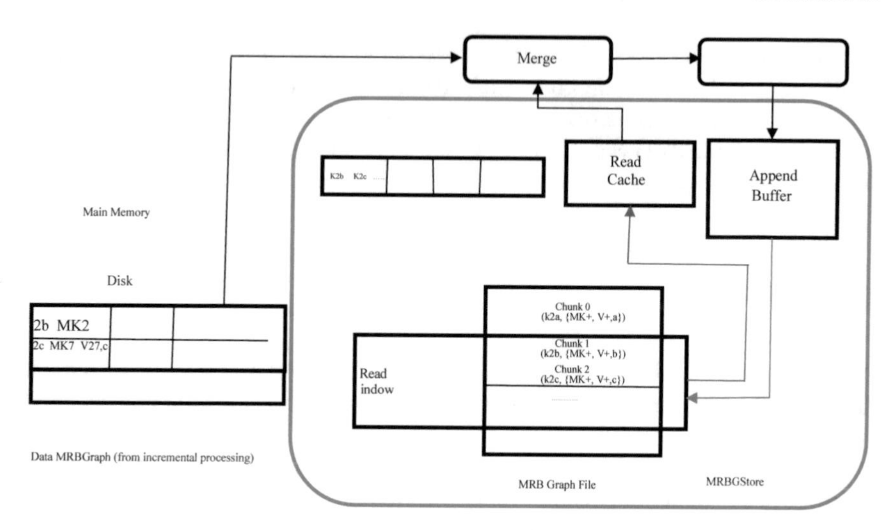

Fig. 1 MRBG-Store

- Users upgrade the Map API to authorize users to comfortably convey loop-invariant structure information, and user suggests a [2] project API responsibility to communicate the comfortably from Reduce to Map.

6 Conclusion

We have worked on I2MapReduce Sustainable Ounce Incremental exemption based absolutely on MapReduce contour. The proposed system of k/v couple degree exceptional-ounce incremental dispensation to reduce the volume of collection and MRBG-Stock to construct upgrade quires for appealing the best-grain positions for gradual grasping and to make it available for the first class-ounce orders in MRBG. This configuration syndicates an excellent-modicum advance device, a UpToDate-motive should be iterative perfect, and it is based on authentic systems trained at best-ounce incremental iterative outcomes. The available physical devices probes to exhibit that I2MapReduce repository significantly minimize the track duration to reinvigorate big data shifting results outcome as associated to re-calculation on jointly simple MapReduce. We are scanning cost-conscious implementation of data optimization that rationally builds utilize of the MRBGraph country and sets the leading implementation approach build on online cost analysis. The MapReduce technique is widely used to classify the sentiment of the use behavior. This clustering approach has an added advantage for the analysis of social media information based on user reviews.

References

1. Hamsho N (2017) The impact of classroom behaviors and student attention on written expression. Theses - ALL. 145
2. Khan M, et al (2019) Map reduce clustering in incremental Big Data processing. Int J Innovative Tech Exploring Engineering (IJITEE) ISSN: 2278–3075 9(2):4205–4211
3. Khan M, Malviya A, Yadav S, (2020) Big Data approach of sentiment analysis of Twitter data using k-means clustering approach. Int J Mech Produc Eng Res Develop (IJMPERD) 10(3):6127–6134
4. Khan M, Basheer S (2020) Using web log files the comparative study of Big Data with map reduce technique. In: 2020 International Conference on Intelligent Engineering and Management (ICIEM) 978-1-7281-4097-1/20/$31.00 ©2020 IEEE
5. Khan M, et al (2020) Big Data and social media analytics: A challenging approach in processing of Big Data. Springer, Singapore. https://doi.org/10.1007/978-981-15-7961-5_59
6. Khan M, et al (2020) Challenges and uses of Big Data analytics for social media. Springer Nature Singapore Pte Ltd. https://doi.org/10.1007/978-981-15-1420-3_118
7. Raconteur Media (Ed.) (2013) Big data. The Times [supplemental material], September 4
8. Mislove A, et al (2007) Proc 7th ACM SIGCOMM Conf Internet Measurement, ACM, , pp 29–42
9. Kalil T (2012) Big data is a big deal. Retrieved from http://www.whitehouse.gov/blog/2012/03/29/big-data-big-deal
10. Khan M, Ansari MD (2019) Security and privacy issue of Big Data over the cloud computing: A comprehensive analysis. Int J Recent Tech Engineering (TM) 7(6s):413–417
11. Khan M, and Kalra DB (2018) An inspection on Big Data data computing. Int J Eng Sci Res (Special Issue/Article No-52), 326–329
12. Khan M, Malviya A (2020). Big data approach for sentiment analysis of twitter data using Hadoop framework and deep learning. In: 2020 International Conference on Emerging Trends in Information Technology and Engineering (ic-ETITE) 978–1–7281–4142–8/20/$31.00 ©2020 IEEE https://doi.org/10.1109/ic-ETITE47903.2020.201
13. Gold MK (2012) Debates in the digital humanities. University of Minessota Press
14. Deters R, Lomotey RK (2014) Int J Business Process Integration and Management 6:298–311
15. Mayer-Schönberger V, Cukier K (2013) Eamondolan/Houghton Mifflin Harcourt, New York
16. Lohr S (2012) The age of big data. The New York Times, February 11. Retrieved from http://www.nytimes.com/2012/02/12/sunday-review/big-datas-impact-in-the world.html?pagewanted=all&_r=0

Language Translation for Impaired People

A. Ashwitha, Akash S. Tonse, M. Bharat Kumar, and Akash K. Hegde

Abstract Gesture-based communication is a visual language utilized by both discourse disabled and hearing-impeded individuals as their first language. Individuals experiencing hearing and communicating in handicaps utilize communication via gestures as a straightforward method of correspondence among one another and furthermore with others, yet tragically few out of every odd one of the ordinary individuals can comprehend communication via gestures consequently. It brings about an absence of correspondence and seclusion. To the extent both discourse disabled and hearing-weakened individuals are concerned, approaching gesture-based communication is extremely fundamental for their social, passionate, etymological, and social development. Objective of this proposed work to overcome any barrier between both discourse disabled and hearing-weakened individuals with the coming of new innovations. Proposed application will change the voice of the client over communication with the assistance of normal language preparing semantics.

Keywords Impaired people · Sign language · Vocal communication · Semantics

1 Introduction

Communication via gestures is a sort of language that utilizes hand developments, outward appearances, and non-verbal communication to convey. It is utilized overwhelmingly by the hard of hearing and individuals who can hear yet cannot talk. However, it is additionally utilized by some conference individuals, regularly families and family members of the hard of hearing, and translators who empower the hard of hearing and more extensive networks to speak with one another [1].

Correspondence through signaling is a sort of correspondence used by people with upset hearing and talk. People use motion-based correspondence movements as a technique for non-verbal correspondence to convey their insights and sentiments [2].

A. Ashwitha (✉) · A. S. Tonse · M. B. Kumar · A. K. Hegde
Ramaiah Institute of Technology, Bangalore, India
e-mail: ashwithaprashanth7@gmail.com

A. Kumar et al. (eds.), *Proceedings of the International Conference on Cognitive and Intelligent Computing*, Cognitive Science and Technology,
https://doi.org/10.1007/978-981-19-2350-0_23

However, non-endorsers find it difficult to see, in this manner arranged correspondence through signals interpreters are needed during clinical and authentic game plans, informational, and instructional gatherings. Over the span of ongoing years, there has been a growing interest in translating organizations [3]. Various techniques, for instance, video inaccessible human unraveling using quick Internet affiliations, have been introduced. They will in this manner give an easy-to-use motion-based correspondence translating organization, which can be used, yet has huge requirements [4].

Existing system: A good deal of study has been done on the topic of Indian Sign Language, to help create systems that can help in the betterment of lives of people with speech and hearing impairment. There are multiple projects and systems that have been developed and are being developed to recognize the Indian Sign Language, they make use of different kinds of technologies that will help in achieving it [5]. They have made use of different machine learning techniques, for example, they classify single and double handed ISL using algorithms of machine learning, and they have used k-nearest neighbor classification, ANN-based and convolution neural networks-based ISL recognition techniques have been used. There are a very few systems that have been developed that convert audio to Indian Sign Language, for example, a system that converts audio to ISL gloss using WordNet. There are other language texts that are converted to ISL. But not many systems are created that converts voice/speech/audio to Indian Sign Language [6] Proposed system: The application relies upon changing over the sound signs and has the opportunity to message using talk to message API (Python modules or google API) and thereafter using the semantics of natural language processing to break down the substance into more unassuming sensible pieces. Informational collections of predefined gesture-based communication are utilized which the application can utilize to show the changed over sound into the communication via gestures. It is a word reference-based machine translation.

The motivation of this project is to decrease the barrier of communication between the impaired people. Deaf people will be able to communicate just like normal people through the help of this project. They can understand others' messages once the audio input is changed to video format. The output format is very easy to understand, and there would be no need for complex devices to solve an easy problem.

Designing a software which will be used to convert audio spoken by normal people into sign languages used by vocally and hearing impaired people. The sign language generated will be in the form of video which will be displayed in the software.

Objective.

- The main requirement is to translate audio to sign language.
- To develop a parser that converts English text to its processed form.
- To translate a complete speech audio file to its corresponding Indian Sign Language video.
- To take comparatively less time for translation of audio to Indian Sign Language.

2 Literature Survey

This chapter records the survey of many journals and articles presented in the Table 1

3 Modeling and Implementation

This chapter deals with the system architecture as shown in Fig. 1, data flow diagram in Fig. 2, taking input Fig. 3, output video 1 Fig. 4, output video 2 Fig. 5, output video 2 and Fig. 6 listing useless words to remove them, respectively.

4 Testing, Results, and Discussion

Testing is the principal part of any venture improvement cycle. A task is inadequate without effective testing and execution. Various test cases for our application are shown in the following Tables 2, 3, and 4, respectively and application front view is shown in the Fig. 7.

5 Results

See Figs. 3, 4, 5, 6, and 7.

6 Conclusion

We have been successful in implementing the sign language interpreter. After successful user testing, it has been found that the new system has overcome most of the limitations of the existing systems especially for ISL. As the ISL is new and very little advancement has been done in this subject, numerous new recordings for various words can be added to the word reference to augment its degree and assist with imparting better utilization of this language.

The current system operates on a basic set of words and in order to extend the system, many new words can be included in the dictionary in future, and specialized terms from different fields can be incorporated too. This project can be made as a mobile application, so that user can install the application into their mobile phones or laptops and can access it easily.

Table 1 Literature Survey

	Paper name	Author name	Algorithm used	Application	Limitation
1	Motionlets matching with adaptive kernels for 3D IndianSign language recognition	Kishore, D.A, Sastry,A C. S, & E. K, P.V. Kumar	They create atwo-phase algorithmfor device translations thatmaintain many regions of three-dimensional signlanguage motion information	Anapplication that recognizes Indiansign language indications. It is generated3D motion captured data, which is thenusedto recognizesign language	Themodel translates sign languages doesnot convert text to sign languages
2	Awearable systemfor recognizing American sign language using IMU and surface EMG sensors	Wu, Jian, Lu, Sunand Roozbeh Jafari	The best subset of highlights from countless different highlightsis selected, and 4 common different algorithms are researched for device designs	Handgestures areusedto detectsignals performedby bothspeech- impaired and hearing impaired people into speech	Using hand- held sensors and talking would not have the same level of precision
3	Avatar- based sign language interpretation for weather forecastand otherTV programs	oh J, Kim B, Kim M, Kang S, Kwon H, Kim I, Song Y	They studied the previous 3 years' worth of weather forecasting documentsfrom the variety of sources to determine the frequency of each word	For both speech-impairedand hearing-impaired people toseethe weather forecast withasign language	This system works only with a weather forecasting system

(continued)

Table 1 (continued)

	Paper name	Author name	Algorithm used	Application	Limitation
4	Glove-based continuous ArabicSign language recognition in user-dependent mode	Tubaiz N, Shanableh T, Assaleh K	Modified K-Nearest neighbors classifier	Continuous Arabicsign language recognition (ArSL)	Sensor readings cannot be visually checked for manual labeling,
5	Intelligent mobile assistantfor hearing impairersto interact with the society in Sinhala language,	Yasintha Perera, Nelunika Jayalath, Shenali Tissera, oshani Bandara, Samantha Thelijjago da	Instant messaging, mobile application, voice recognition, natural language processing, graphic interchange format introduction	The significance is that it allows hearing- impaired individualsto communicate when they are longdistance apart. This app will close the divide between hearing impaired people	The file formatis not compatible
6	Sign language recognition system using deep neural network, advanced computing, and communicationSystems (ICACCS)	Surejya Suresh, Haridas T.P Mithun, M.H Supriya	CNN structure and a summary of the planned construction. The planned construction was studiedusing 2 plans, the 1st of whichuse the stochastic gradient optimizer and the second of which usedthe Adam	The convolutional neural network wasusedto create a basic versionof the sign recognition plan, which was successfully tested	It is not user friendly

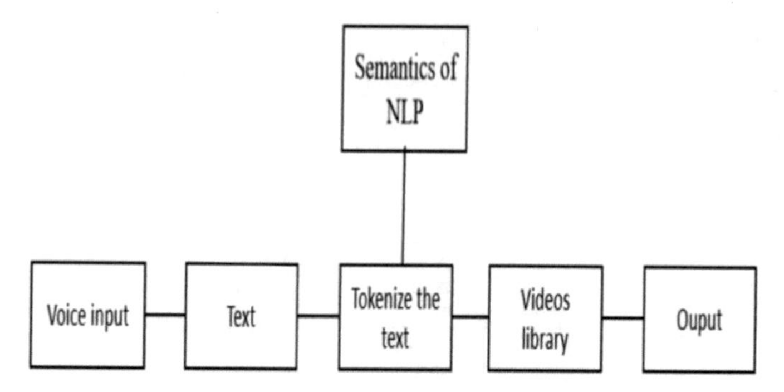

Fig. 1 System architecture

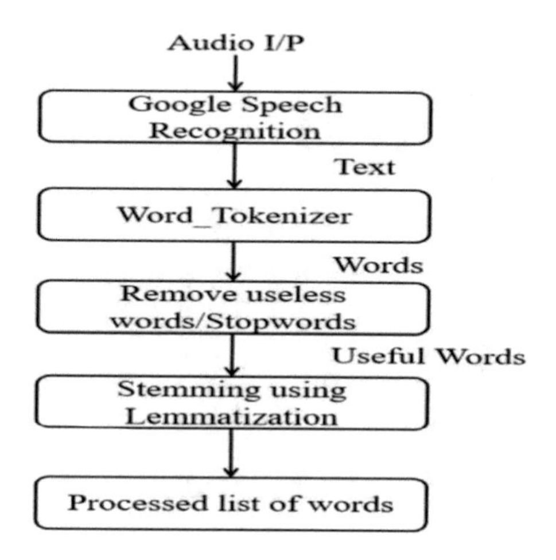

Fig. 2 Audio to processed text dataflow

```
akash@akash-VirtualBox: ~/Downloads/new_project
File Edit View Search Terminal Help
akash@akash-VirtualBox:~$ cd Downloads/
akash@akash-VirtualBox:~/Downloads$ cd new_project/
akash@akash-VirtualBox:~/Downloads/new_project$ python3 main.py
ALSA lib pcm_dmix.c:1052:(snd_pcm_dmix_open) unable to open slave
ALSA lib pcm.c:2495:(snd_pcm_open_noupdate) Unknown PCM cards.pcm.rear
ALSA lib pcm.c:2495:(snd_pcm_open_noupdate) Unknown PCM cards.pcm.center_lfe
ALSA lib pcm.c:2495:(snd_pcm_open_noupdate) Unknown PCM cards.pcm.side
ALSA lib pcm.c:2495:(snd_pcm_open_noupdate) Unknown PCM cards.pcm.surround71
ALSA lib setup.c:547:(add_elem) Cannot obtain info for CTL elem (MIXER,'IEC958 P
layback Default',0,0,0): No such file or directory
ALSA lib pcm.c:2495:(snd_pcm_open_noupdate) Unknown PCM cards.pcm.hdmi
ALSA lib pcm.c:2495:(snd_pcm_open_noupdate) Unknown PCM cards.pcm.hdmi
ALSA lib pcm.c:2495:(snd_pcm_open_noupdate) Unknown PCM cards.pcm.modem
ALSA lib pcm.c:2495:(snd_pcm_open_noupdate) Unknown PCM cards.pcm.modem
ALSA lib pcm.c:2495:(snd_pcm_open_noupdate) Unknown PCM cards.pcm.phoneline
ALSA lib pcm.c:2495:(snd_pcm_open_noupdate) Unknown PCM cards.pcm.phoneline
ALSA lib pcm_dmix.c:1052:(snd_pcm_dmix_open) unable to open slave
Say something
Recognizing...
['I', 'want', 'some', 'water']
```

Fig. 3 Taking input

Fig. 4 Output video 1

Fig. 5 Output video 2

Fig. 6 Listing useless words to remove them

Table 2 Live voice

Test case #	1
Test case name	Live voice
Description	Users can use a microphone to give the voice input. This input gets converted to text, and then, the sign language video will be displayed on the screen otherwise it says could not hear properly
Expected output	Sign language video is displayed
Actual output	Sign language video is displayed
Remarks	Pass

Table 3 Recorded voice

Test case #	2
Test case name	Recorded voice
Description	A pre-recorded video file can be selected by the user and it gets converted to text, and then, the sign language video will be displayed on the screen otherwise it says could not hear properly

Table 4 Display of sign language video

Expected output	Sign language video is displayed
Actual output	Sign language video is displayed
Remarks	Pass

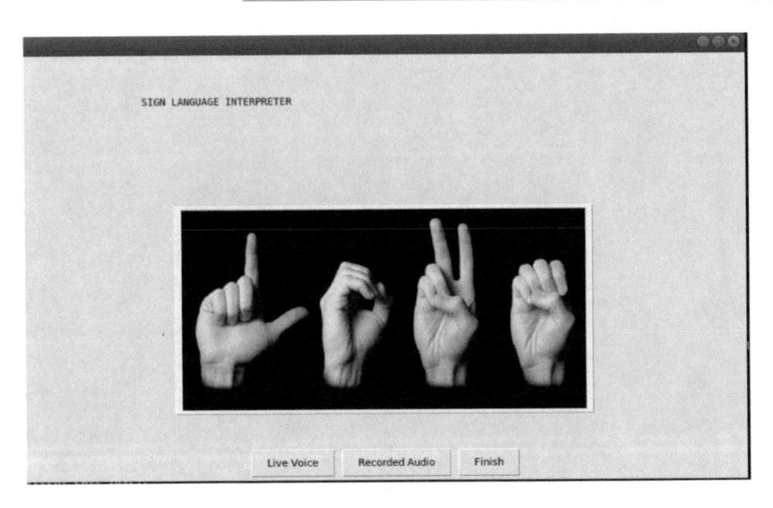

Fig. 7 Application interface

References

1. Kishore PVV, Anil Kumar D (2018) Motionlets matching with adaptive kernels for 3D Indian Sign Language recognition. IEEE Sensors J 2. https://doi.org/10.1109/JSEN.2018.2810449
2. Wu J, Sun L, Jafari R (2016) A wearable system for recognizing American sign language in real time using IMU and Surface EMG Sensors. IEEE J Biomedical Health Informatics 20(5), Sept
3. Jeon S, Kim B, Kim M (2015) Avatar-based sign language interpretation for weather forecast and other TV programs. SMPTE Motion Imaging J 126(1), Jan–Feb
4. Tubaiz N, Shanableh T, Assaleh K (2015) Glove-based continuous Arabic sign language recognition in user-dependent mode. IEEE Trans Human-Machine Syst 45(4), Aug
5. Shinde A, Kagalkar RM (2015) 'Sign language to text and vice versa recognition using computer vision in Marathi'. Int J Comp Appl 118(13):1–7
6. Poddar N, Somavanshi V (2015) Study of sign language translation using gesture recognition. IJARCCE, February. https://doi.org/10.17148/IJARCCE.2015.4258

A Novel Approach for Protecting Against Neural Fake News in Social Media Network

Sudesh Rao, Krishnaraj Rao, S. Santhosh, and Krishna Kaushik

Abstract Fake news is a major problem of the modern society. Hence, it is an increasingly important concern on social media. The paper uses CNN to detect the complex format of the context with the help of convolutional neural networks (CNNs). Long short-term memory (LSTM) is used in sequence predictions. It allows the reading of a sequence from both front-to-rear and rear-to-front with the bi-directional LSTM. An LSTM-recurrent neural network is used in this paper as a model for detecting fake news.

Keywords Natural language processing · Machine learning · Bi-directional LSTM · RNN · Text analysis

1 Introduction

Fake news refers to purposefully created false article that is circulated through the social media network and thereby consumed and shared by people on daily basis. Priorly, social media platforms like Facebook, Twitter, and Instagram authorize all kinds of news to reach people without proper supervision [1]. Social media allows people to communicate and share the information online. These technologies changed the way in which people interact and communicate. In olden days if we had to pass any information, we had to meet them personally and share the information. Due to improvement in technologies with low cost, we can deliver the information very quickly. As popularity of social media increased, people started misusing it. Facebook and Google use new tools to combat misinformation on the social media networks. At present, Facebook uses techniques of artificial intelligence which aids in shielding the consumer from false information. Even Google uses AI methods for detecting the fake news. It is hard to identify track and control the fake news. In such situation, a good solution would be making use of advanced digital technologies. It is difficult to create awareness about the fake news problem as it takes time and

S. Rao (✉) · K. Rao · S. Santhosh · K. Kaushik
NMAMIT, SJEC, NMAMIT, SUCET, Karkala, Karnataka, India
e-mail: sudesh.rao@nitte.edu.in

A. Kumar et al. (eds.), *Proceedings of the International Conference on Cognitive and Intelligent Computing*, Cognitive Science and Technology,
https://doi.org/10.1007/978-981-19-2350-0_24

uses large amount of resources. Hence, we can educate people to think critically before using and posting any digital content [2]. A number of approaches are used to identify the complication of fake news on social media like analyzing the origins of the news content, examining patterns of distribution and graphs are used to identify correlations among various aspects of the news. It also uses the machine learning algorithm for classifying and analyzing the accounts and subjects that shares such content or news.

2 Literature Survey

Information about events is disseminated to the public through the media. Through social networks or Websites, rapid dissemination of information is made possible by the rapid development of the Internet. Unverified or false information is diffused across social networks without any regard for the credibility of the information. Thus, society has been challenged by dissemination of fake news across the internet.

Text characterization is a primary assignment in numerous NLP applications. Customary content classifiers regularly depend on numerous human-planned highlights, like word references, information bases, and exceptional tree bits. As opposed to conventional strategies, we present an intermittent convolutional neural organization for text characterization without human-planned highlights. In our model, we apply an intermittent design to catch relevant data quite far when learning word portrayals, which may acquaint impressively less commotion contrasted and conventional window-based neural organizations. We likewise utilize a maximum pooling layer that consequently decides which words assume key parts in text arrangement to catch the critical segments in messages. We lead probes four regularly utilized datasets [3].

William Kai Shu et al. published a paper related to detect fake news on social media by using data mining techniques [4]. In this paper, they inspected the fake news problem by analyzing the existing literature in two phases: characterization and detection. They used a different data mining techniques to detect fake news. The immeasurable expansion of fake news can create a significant negative impact on individuals and society. First, fake news can disturb the authenticity balance of the news ecosystem. Second, they create chaos among the public which may initiate quarrels, riots, violence, protests, delinquency, crime, etc., in the society or between individuals.

Bourgonje et al. proposed a method to discover the fake news based on detecting stance of headlines to articles [5]. They present a system for stance detection of headlines with respect to their corresponding article bodies. In this paper, their goal is to battle against fake news and they called it as stance detection. Their challenge is to find paragraph with the regard to matching content from another content.

Z Khanam et al. proposed method to detect fake news using different machine learning techniques [6]. In this paper, they have mentioned different machine learning methods to identify fake news. They have used supervised machine learning algorithms to discover the fake news by using different tools like Python scikit-learn, NLP, etc.

3 Problem Statement

Fake news can be of several forms, such as accidental errors pledged by news creators, absolute deceitful stories, or the stories that are cultivated to mislead and persuade reader's opinion whereas faux news could have varied sorts of adverse consequences that will create turmoil between the folks, government, and organizations since it differs from the facts. Cyber technology's wide reach and quick unfold contributes to its hazard. False messages through faux news on cyber have been accepted by states and establishments in addition to people for varied motives in miscellaneous forms. Usually, historic data are formed and unfold through social media to realize planned finish. On the opposite hand, it should additionally involve chronicle of a real truth but being consciously exaggerated. This might additionally involve titling the Webpages with fabricating title or tag-lines so as to know interest of readers. Such distortion could lead in committing crimes, money scams, political gain, financial gain related to click-bait, etc. All these will disturb the importance of fourth estate. Any mistreatment of electronic media will pose a threat of supplying fake news, thereby carrying forward unfolding of such news. Hence, it is very necessary to observe the credibleness of the news and online content but a significant downside is to differentiate between the people and bots engaged in diffusion of faux news.

4 Objective

Fake news is typically created to deceive and attract readers. It is usually generated for commercial and political purposes. This project's primary goal is to reduce the dispersal of fake news from the Web and social media. The current research interest is based on automatic detection the fake news and analysis of news articles as the dissemination of fake news is one of the major concerns of the society. In this project, we use bi-directional LSTM to detect the fake news and prevent the expansion of the fake news in society.

5 Methodology

(a) ***RNN***

Recurrent neural networks will find the advanced patterns within the matter information. It is a conjointly a feed-forward artificial neural network, wherein the output from the previous steps is taken as input to the present step.

RNN has a frenzied—memory|| that stores all the data. The RNN has a crucial feature referred to as hidden state that remembers the data of next few sequences. Here, the hidden layer activation at a particular time depends on the previous time [7] (Fig. 1).

Here, $y1$, $y2$, $y3 \ldots yt$ represent the input words. $x1$, $x2$, and $x3$ represent the anticipated next word. $h0$, $h1$, $h2 \ldots ht$ hold the knowledge of the previous inputs words. The equations required for training.

$$\mathrm{Ht} = f(w(hh)\mathrm{ht1} + \mathrm{w}(hy)yt) \tag{1}$$

$$\mathrm{Yt} = \mathrm{softmax}(w(s)\mathrm{ht}) \tag{2}$$

where the Eq. (1) stores the knowledge about the previous words within the sequence and Eq. (2) is going to calculate the expected word vector at a given time step t. The bottom layers of a RNN network suffer from vanishing gradient problem. Hence, there are some solutions to beat this. For example, by making use of rectified liner unit (ReLU) activation function alternatively, we are able use LSTM.

(b) ***LSTM***

It is quite similar to RNN. However, its highly economical than RNN since it will store the word for more amount of time and provides the correct results. There are three gates of LSTM particularly such as forget gate, input gate, and output gate (Fig. 2).

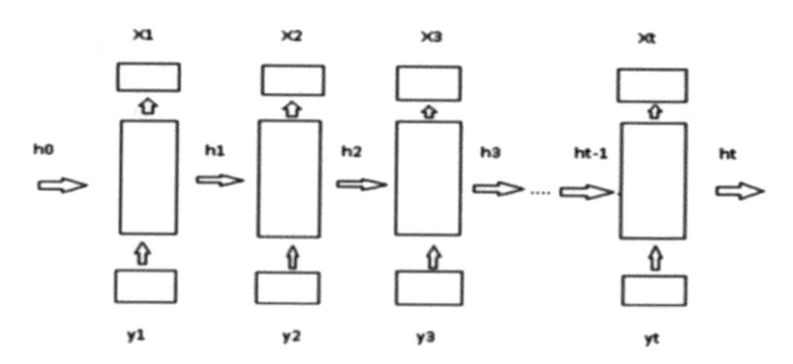

Fig. 1 RNN system design

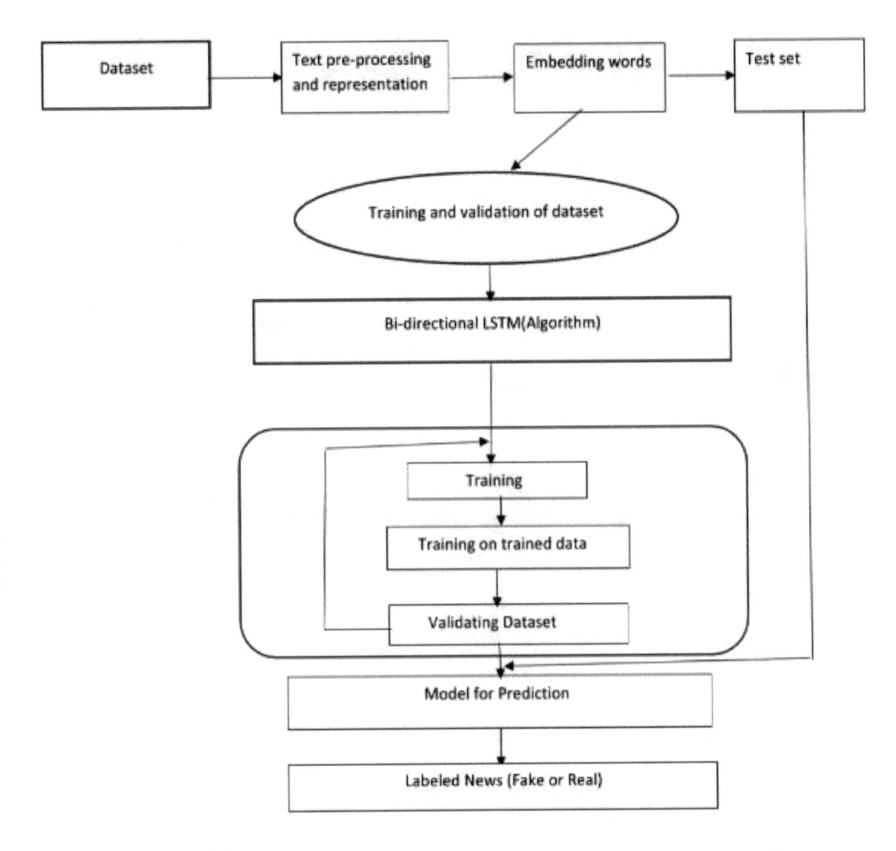

Fig. 2 Architectural diagram

6 Architectural Design

At first, we will take dataset. Then, it is sent to text pre-processing in which all the special characters and unwanted characters are removed. Then, it goes to tokenization in which it finds the vector format and considers a specific number to each and every word in the dataset. If words are repeated, the same numbers which was allotted before will be used again and again. It is a one dimensional array. From there it moves to word embedding after that it goes to bi-directional LSTM which will do the training to the given data. The training process comprises of three steps: training, validation, and testing. This will make the accuracy rate high. First, we will train the model, and then, we will be testing the model on the trained data, then it will construct hyper parameter from the data. This is nothing but model tuning parameter, in the sense, we will get to know number of neurons in which the above 3 steps are executed in iterative mode, until we get the accuracy as expected. After that the data goes to prediction, and it gives predicted label to the news article and model accuracy.

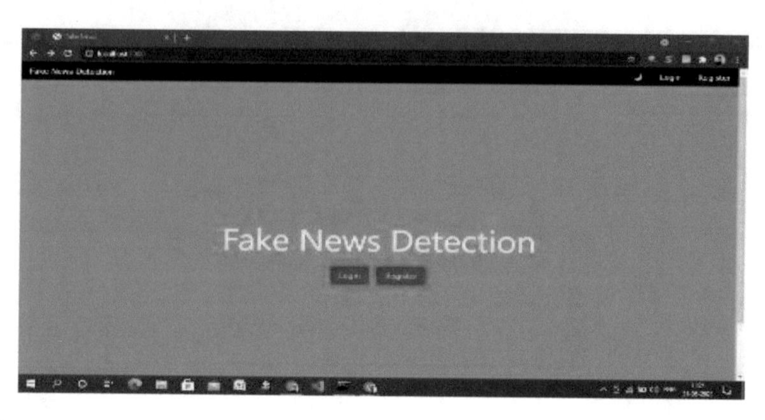

Fig. 3 Login phase

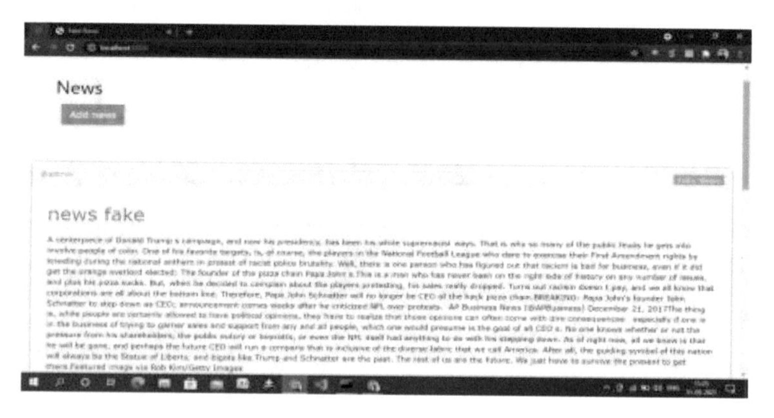

Fig. 4 Fake news detection phase

7 Results

User can login to this module by using their credentials as shown in the above Fig. 3. If we post any fake news, it shows that news is fake as shown in Fig. 4. If news is fake, admin can delete the news as shown in Fig. 5.

8 Conclusions

Fake news goes viral rapidly on the Internet and its found everywhere on the Web. Therefore, the goal of this project is to find the inaccurate and fake news. We have

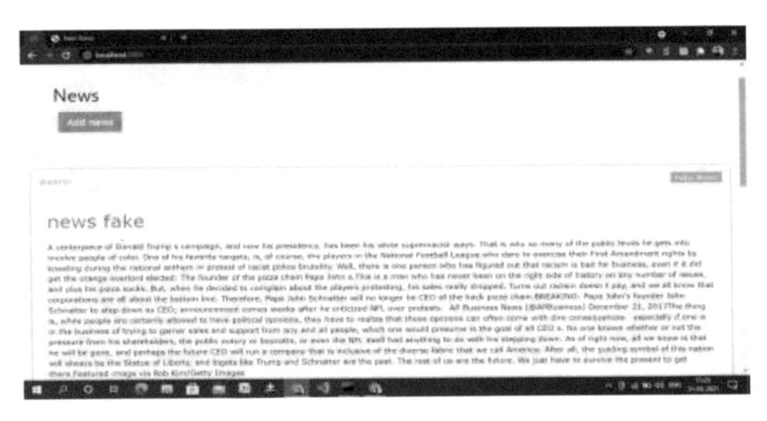

Fig. 5 Fake news detection result

used two algorithms, namely RNN and Bi- LSTM for training the models for prediction. Bi-LSTM has the very best accuracy compared to RNN. Hence, bi-LSTM is employed to predict the fake news.

References

1. Allcott H, Gentzkow M (2011) Social media and fake news in the 2016 election. J Economic Persp 31(2):211–236
2. CNN Business. https://edition.cnn.com/2019/04/21/tech/sri-lanka-blocks-socialmedia/index.html
3. Aldwairi M, Alwahedi A (2018) Detecting fake news in social media networks, EUSPN 2018, vol 141. Elsevier. https://doi.org/10.1016/j.procs.2018.10.171
4. Shu WK, Sliva A, Wang S, Tang J, Liu H (2017) Fake news detection on social media: a data mining perspective. Research Gate, August. https://doi.org/10.1145/3137597.3137600
5. Bourgonje P, Moreno Schneider J, Rehm G (2017) From Clickbait to fake news detection: An approach supported detecting the stance of headlines to articles. In: Proceedings of the 2017 EMNLP Workshop, pp 84–89
6. Khanam Z, Alwasel BN, Sirafi H, Rashid M (2020) Fake news detection using machine learning approach. ASCI-2020. https://doi.org/10.1088/1757-899X/1099/1/012040
7. Yin W, Kann K, Yu M, Schütze H (2017) Comparative study of CNN and RNN for language processing, research gate, February

Robust Brain Diseases Classification Using CNN and Soft Computing Techniques

Deipali Vikram Gore, Ashish Kumar Sinha, and Vivek Deshpande

Abstract The speedy growth and technology in the brain imaging domain technologies are regularly advanced for a crucial task in investigating and concentrating the distinct aspects of the brain anatomy and roles. The image processing methods using computer vision technology have received significant interest in the domain of medical science to present effective computer-aided diagnosis (CAD) for early disease detection. Brain tumor detection and classification importantly become essential for early detection and treatment to reduce the mortality rate. We propose a novel CAD system for automatic brain diseases detection and classification using the automated deep learning mechanism from the input MR images. The raw MR images are first pre-processed to improve the quality using the median filtering. The brain tumour segmentation is then performed using the adaptive thresholding algorithm to determine the region of interest (ROI). We design the convolutional neural network (CNN) layers for automatically learning the features using the ResNet50 model. As the dimensions of automatically learning features are higher, we applied the principal component analysis (PCA) for its reduction. For the classification, we applied different soft computing techniques such as artificial neural network (ANN), support vector machine (SVM), and K-nearest neighbours (KNN). The simulation results show the proposed model received a higher accuracy than the literature.

Keywords Brain diseases · Brain tumor · Convolutional neural network · Features extraction · Segmentation · Soft computing

D. V. Gore (✉) · A. K. Sinha
Kalinga University, Raipur, India
e-mail: deipali.gore@gmail.com

V. Deshpande
Vishwakarma Institute of Information Technology, Pune, India

A. Kumar et al. (eds.), *Proceedings of the International Conference on Cognitive and Intelligent Computing*, Cognitive Science and Technology,
https://doi.org/10.1007/978-981-19-2350-0_25

1 Introduction

Detecting brain tumors diseases is a very challenging process, from the complex structure of human brain. The brain tumor has been detected by a radiologist through a comprehensive analysis of MR images, which takes substantially a longer time. The manual practices do not give reliable solution to predict brain diseases as they may take time as well as the possibility of inaccurate results which may lead to severe problems with the patients [1, 2]. Thus, it is important to have an early prediction of such brain diseases using MR images. Since from last two decades, several studies reported designing computer-aided diagnosis (CAD) systems to acquire MR images and predict brain diseases automatically. In some systems, if the MR image is of a diseased brain, then its severity is also detected for the appropriate treatment purpose. The main goal of such systems is to assist the radiologist to have a second opinion regarding the presence or absence of tumor and initiate the specific treatment [3].

A computer-aided diagnosis (CAD) framework is created for programmed identification of brain tumors through MRI. The CAD framework can give a superior component to recognize brain tumors than the customary techniques. The framework comprises of two phases. The primary stage has brain image obtaining, pre-processing, and upgrade [4, 5]. The subsequent stage comprises of division, feature extraction, classification, and execution investigation. Pre-processing and upgrade methods are utilized to work on the identification of dubious districts in MRI. The pre-processing and improvement techniques eliminate film artefacts and high-recurrence parts from the MR image [6]. Division portrays the partition of the dubious area from the foundation MR image utilizing different advance methods. Textural features are removed and chosen from the dubious area and afterwards delegated typical or harmful [7].

Picture division plays an extremely crucial part in brain tumor identification. The mistaken image division might prompt some unacceptable expectations. There are three kinds of division strategies like manual, self-loader, and completely programmed. Manual division requires the radiologist to utilize the multi-methodology data introduced by the MR Images alongside physical and physiological information acquired through preparation and experience. Not withstanding, this is more error inclined as necessities the full aptitude and tedious cycle, thus it is not utilized any longer. The self-loader techniques require the connection of the patient for three purposes; statement, mediation, or input reaction, and assessment to fragment the picture [8]. The self-loader brain tumour division techniques are less tedious than manual strategies and can acquire productive outcomes, anyway, still inclined to intra-picture and between picture/client inconstancy challenges. Consequently, momentum brain tumour division research has principally centred around completely programmed strategies that are relied upon to work paying little heed to pictures changeability [9–13].

The recent progress of the Internet of Things (IoT) [14, 15] across the different applications including the smart healthcare systems has received significant attention as in such IoT-empowered systems the medical images are transferred remotely to the

CAD systems [16–18]. The IoT and deep learning integrations for automatic CAD systems provided popular nowadays. The deep learning techniques were recently applied for automatic brain tumour segmentation; however, such techniques only focused on automatic tumour ROI extraction. This approach may not be reliable for a longer duration as the technology grows. In this paper, we focus on automatic features extraction from the already segmented brain ROI rather than focusing on automatic ROI extraction from the raw MR images. The raw MR images are first pre-processed using effective and lightweight median filtering, and then, we applied a thresholding-based ROI extraction algorithm on the pre-processed MR image. The CNN has been applied to extract the automatic features from the ROI image using the ResNet50 deep learning model. The extracted features were further reduced using the PCA. For the classifications, we applied the SVM, KNN, and ANN techniques for their performance analysis. Section II presents the study of the related works. Section III presents the methodology of the proposed model. Section IV presents the simulation results and section V presents the conclusion.

2 Related Work

Recently, several deep learning-based methods have been proposed for automatic brain tumor segmentation and disease detection. This section presents a review of some recent works on the same.

In [19], authors have proposed a fully automatic system for brain tumor segmentation and detection of diseased tissue from the input MRI and fluid-attenuated inversion recovery (FLAIR) images.

In [20], authors have proposed another automatic brain tumor detection framework that used the novel threshold-based image segmentation algorithm. They designed the segmentation algorithm according to the collaboration of learning automata and beta mixture systems.

In [21], authors have focused on conducting the review on various deep learning-based image segmentation techniques using brain MRI. They investigated the performances of deep learning systems for the segmentation of brain lesions and brain structures.

In [22], the completely automated CAD system for brain disease detection using deep learning was proposed. They designed a DeepSeg system that consists of encoding and decoding parts. The CNN had used in the encoding part for the automatic extraction of spatial information. The outcome of the encoder was fed to the decoder part to produce the probability map. They used the models such as ResNet and DenseNet in CNN.

In [23], authors have proposed an automated deep learning system to classify brain tumors. They have used the DenseNet201 model for fine-tuning and training using imbalanced data learning deep transfer. They have extracted the features from the average pool layer.

In [24], authors have introduced another recent deep learning-based for the automatic brain tumor classification problem. They have designed deep dense inception residual model using ResNet V2. The outcomes proved the proposed model has improved the classification accuracy.

In [25], authors have proposed the pre-processing mechanism to focus on small lesions of the MR image than a complete image. Then, in each small part, they focused on smaller brain image parts using the Cascade CNN (C-CNN) deep learning model.

In [26], authors have proposed a multi-level features extraction technique followed by features concatenation for early brain tumor detection. They have used two deep learning systems such as DensNet201 and Inception-v3 for automatic features extraction.

In [27], authors have proposed automatic brain tumor detection and classification using deep learning features and different classifiers. They designed the transfer learning and utilized the various pre-trained deep CNN models for features extraction from the MR images. For the classification, different machine learning classifiers were applied.

The proposed model is closely related to the approach presented in [27] and partially to other kinds of literature. We mainly focused on MR image pre-processing first, which is not addressed in [19–27], then designed the robust CNN layers using optimal ResNet50 pre-trained model rather than using various other models. We newly introduced, features optimization to handle the high-dimensional CNN features effectively for accuracy improvement. Then, the machine learning algorithms have been applied.

3 Proposed Methodology

Figure 1 shows the functionality of the proposed model that mainly depends on functions such a pre-processing, adaptive segmentation, automatic features extraction, and classification. The proposed CAD system has been specially designed for MRI brain images. As there is the possibility that input MR images may have noises and artefacts while scanning and storage processing, therefore, we first aimed at suppressing such noises and artefacts from that image. In a pre-processing step, we performed the image quality improvement using appropriate filters. In the adaptive ROI extraction phase, we aimed to accurately locate and extract the tumor-specific regions using the dynamic thresholding mechanism. The automatic CNN architecture proposed using the pre-trained ResNet50 model for features learning. The features are reduced using the PCA. The multi-class machine learning algorithms were then applied for the classification of brain tumors.

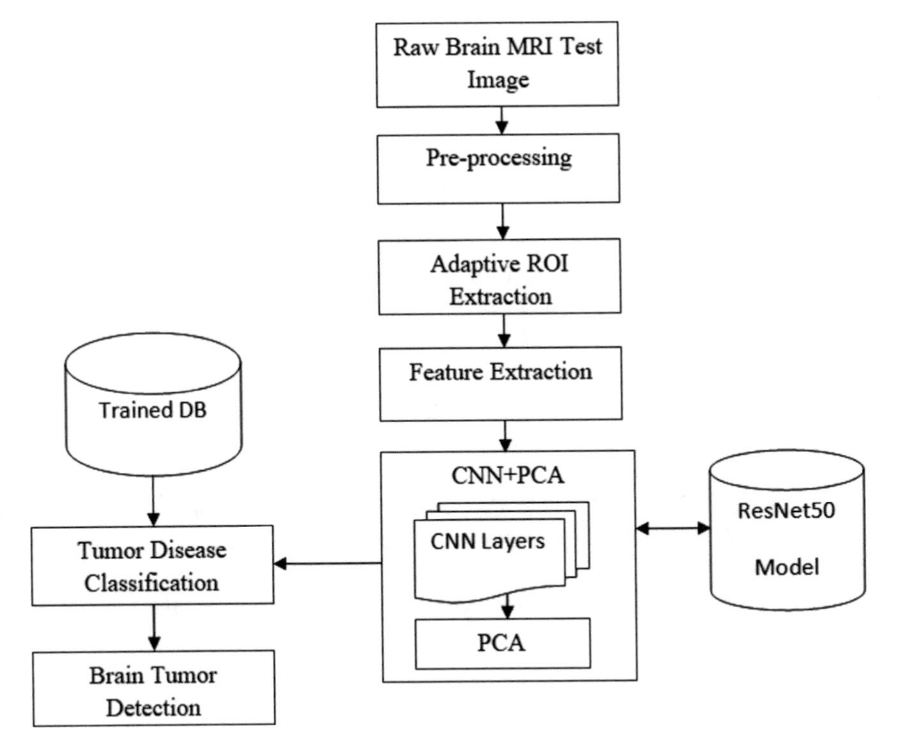

Fig. 1 Proposed automatic brain tumour classification system

3.1 Image Pre-Processing

Let image I be the test raw MR image as the input to the proposed system, we are first required to enhance the quality of the image. As the input MR images are generally represented in low-contrast form, we first applied the contrast limited adaptive histogram equalization (CLAHE). The CLAHE improves the contrast according to its dynamic nature of histogram equalization as:

$$I1 \leftarrow \text{CLAHE}(I) \tag{1}$$

However, improve the contrast of raw MR images may further introduce the artefacts components while balancing the image contrasts. Therefore, to suppress such artefacts, we applied the 2D median filtering technique using the optimal window size w (3×3) as:

$$I2 \leftarrow \text{MEDFILT}(I1, w) \tag{2}$$

3.2 Image Segmentation

For accurate tumor segmentation, we have exploited the adaptive threshold-based image segmentation algorithm in this paper. Extracting the ROI or tumour regions accurately from the MR images is a challenging problem. Recently, several deep learning methods proposed, however, as such a technique takes a long time for the segmentation of tumors, the disease classification step is left intact. In this paper, we divide the image segmentation and features extraction steps. For image segmentation, we have re-designed Otsu's method, where the image threshold value is dynamically generated first. The adaptive threshold for each image is computed as:

$$T \leftarrow \frac{(\max(I2) + \mathrm{mean}(I2))}{\gamma} \tag{3}$$

where γ is the scaling factor that is used to estimate the threshold value in the range of 0 to 255. We set value of this factor to 2.5 according to the accuracy obtained by fine-tuning.

After discovering the threshold value for input image, we must segment the image into ROI and non-ROI parts as:

$$I3 \leftarrow \begin{cases} I2(i, j), I2(i, j) \geq T \\ 0, \text{Otherwise} \end{cases} \tag{4}$$

where $I3$ represents the ROI image that shows the actual tumour regions and rest all parts represented as black. This is very simple and effective mechanism to segment any MR image.

3.3 CNN Features Extraction

This is an important part of the proposed model. We aim to automatically extract the features of the ROI image with the minimum computational requirement. We designed the robust CNN layers to load the pre-processed ROI image and automatically learn its features using the pre-trained ResNet50 model. The CNN layers consist of layers such as the input layer, convolutional layer (Conv), batch normalization layer (BNL), rectified linear unit (ReLU) layer, and max pooling layer (MPL). The layers of a total of 4 produced the features for the input image. We start with the 8 filters of size 15×15 and end up with 62 filters of size 15×15. This setting ensures the extraction of detailed deep features from each smallest part of brain tissues in less computational requirements. Each layer in the proposed CNN consists of sub-layers such as the:

- Input image layer that loads the image of size 224×224

- $Conv(.)$ Layer: with 8 to 64 filters of size 15 × 15.
- Batch normalization layer (BNL),
- Rectified linear unit (ReLU) layer, and
- Max pooling layer (MPL).

The systematic design of layers significantly reduces computational efforts. To address the parameter explosion problem, we have applied batch normalization and ReLU after the convolutional filter. The third max pooling layer generates the automated features vector for further proceedings. The proposed CNN takes input I3 and performs the consolidated one squashing function as per design of each layer as:

$$F_j^l = \tanh\left(\text{pool}_{\max}\left(\text{ReLU}\left(\sum_i x_j^{l-1}(I3) * k_{ij}\right) + b_j^l\right)\right) \tag{5}$$

where F_j^l is outcome of CNN features extraction model as features set using convolutional layer l of jth input, x_j^{l-1} represents the previous convolutional layer features maps of $I3$, k_{ij} represents ith trained convolutional kernels, and b_j^l represents the additive bias. The function tanh(.) represents the activation function, $\text{pool}_{\max}(.)$ represents the operation of max pooling for features extraction, and ReLU(.) represents the operation of ReLU layer. The BNL is integrated with convolutional layer. As the extracted features are high-dimensional and redundant, we applied the PCA to get the reduced CNN features as:

$$F^{\text{PCA}} \leftarrow \text{PCA}(F) \tag{6}$$

3.4 *Classification*

For the classification, we applied different soft computing techniques called ANN, SVM, and KNN on both test and trained CNN features. The aim of these machine learning techniques must detect the type of brain tumor in the input MR image. For performance analysis, the dataset is divided into 70% (training) and 30% (testing and validation) for each machine learning classifier. For the performance evaluations of the proposed model using each classifier, we computed the metrics such as precision, recall, accuracy, specificity, and F1-score.

$$\text{Precision} = \frac{\text{tp}}{\text{tp} + \text{fp}} \tag{7}$$

$$\text{Recall} = \frac{\text{tp}}{\text{tp} + \text{fn}} \tag{8}$$

$$\text{Accuracy} = \frac{\text{tp} + \text{tn}}{\text{tp} + \text{tn} + \text{fp} + \text{fn}} \tag{9}$$

$$\text{Specificity} = \frac{\text{tn}}{\text{tn} + \text{fp}} \tag{10}$$

$$F1\text{score} = 2 \times \frac{\text{Precision} \times \text{Recall}}{\text{Precision} + \text{Recall}} \tag{11}$$

where tp stands for true positive, fp stands for false positive, fn stands for false negative, and fp false positive.

4 Simulation Results

We have implemented the proposed model using the MATALB tool under the I5 processor, 4 GB RAM, and Windows 10 OS. For performance analysis, we have used the MRI brain tumor dataset mentioned in [28]. This dataset is scalable and having total 3064 samples from 233 subjects. It consists of three types of brain tumors such as glioma (1426), meningioma (708), and pituitary tumor (930). The proposed model consists of pre-processing, segmentation, and CNN + PCA which have been investigated using ANN, SVM and KNN classifiers in this section.

Figures 2, 3, 4, 5, and 6 show the performance analysis in terms of accuracy, precision, recall, specificity, and F1-score parameters, respectively. From these results,

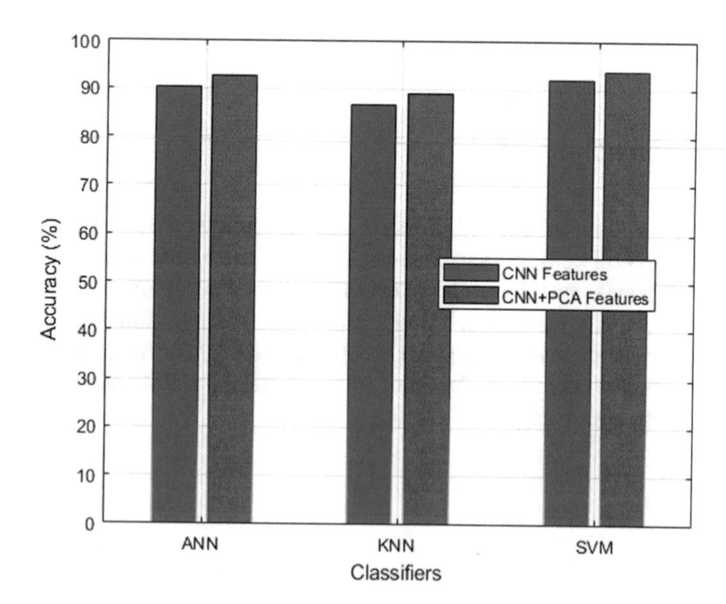

Fig. 2 Accuracy performance analysis

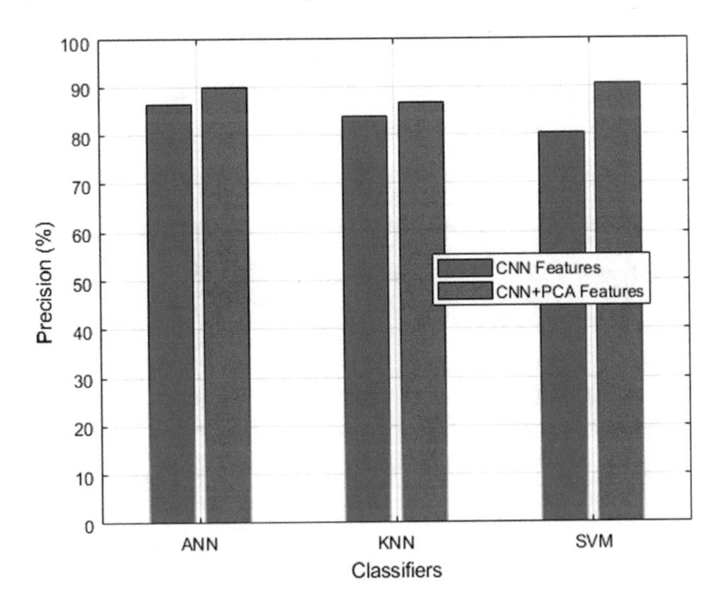

Fig. 3 Precision performance analysis

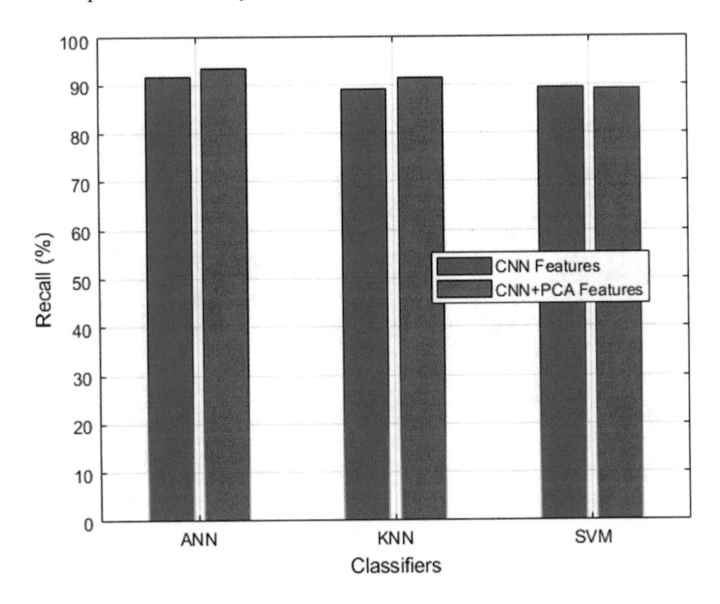

Fig. 4 Recall performance analysis

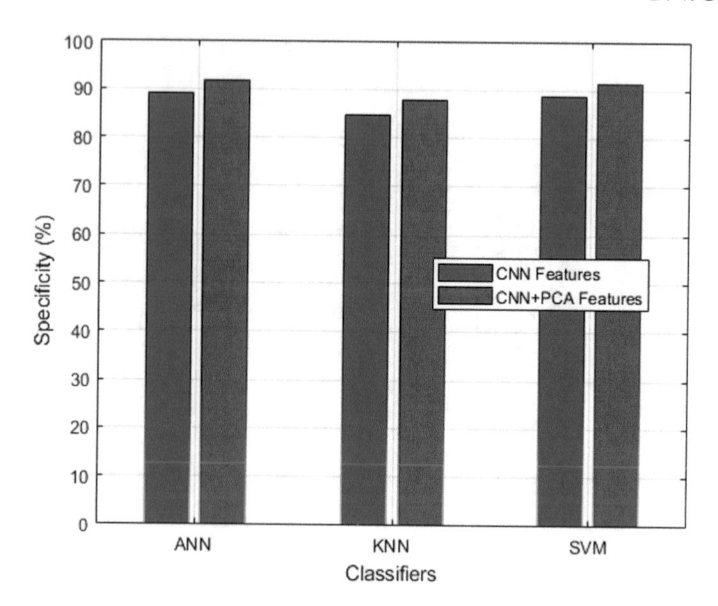

Fig. 5 Specificity performance analysis

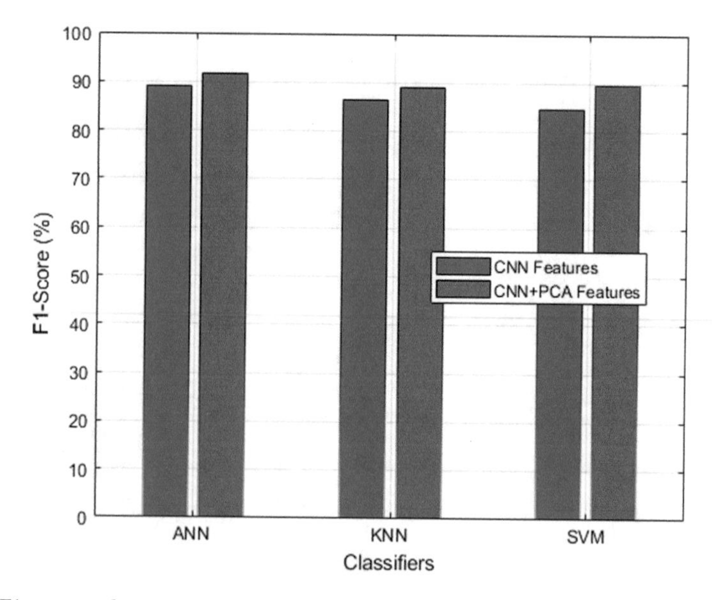

Fig. 6 F1-score performance analysis

we noticed that the proposed model CNN + PCA shows a higher performance than just using the high-dimensional CNN feature using each classifier. The proposed CNN model produced the optimal and higher accuracy for each classifier. The use of pre-processing and segmentation methods affects unique features extraction using CNN. Applying PCA further improved the accuracy of brain tumor detection.

Among the different classifiers, the performance of SVM shows a better accuracy and precision rate as compared to other classifiers. The ANN shows better recall rate, specificity, and F1-score performances. However, the KNN classifier has delivered the worst performance among all the classifiers. In general, the ANN classifier produced a higher performance than SVM and KNN classifiers.

5 Conclusion and Future Work

This paper proposed the novel framework for automatic brain tumor detection and classification using the combined approach of deep learning and machine learning techniques. Before applying the deep learning and machine learning methods, we first performed the MR image pre-processing and adaptive segmentation. The outcome of adaptive segmentation was the segmented ROI of the tumor image. The ROI image has fed as input to the proposed CNN model for automatic features extraction followed by the PCA for features reduction. For the classification, we applied three machine learning classifiers. The simulation results show this preliminary model delivered promising results for future extensions. In the future, we plan to extend this model by applying the deep learning classifier. We additionally suggest introducing the handcrafted features as well to estimate the tumor-specific features for accurate tumor grading.

References

1. Menze B et al (2015) The Multimodal brain tumour image segmentation benchmark (brats). IEEE Trans Med Imaging 34(10):1993–2024
2. Gordillo N, Montseny E, Sobrevilla P (2013) State of the art survey on MRI brain tumour segmentation. Magn Reson Imaging 31(8):1426–1438
3. White D, Houston A, Sampson W, Wilkins G (1999) Intra and interoperator variations in region-of-interest drawing and their effect on the measurement of glomerular filtration rates 24:177–181
4. Foo JL (2006) A survey of user interaction and automation in medical image segmentation methods. Tech rep ISUHCI20062, Human Computer Interaction Department, Iowa State University
5. Hamamci A et al (2012) Tumor-Cut: segmentation of brain tumours on contrast enhanced MR images for radiosurgery applications. IEEE Trans Med Imaging 31(3):790–804
6. Havaei M, Larochelle H, Poulin P, Jadoin PM (2016) Within-brain classification for brain tumour segmentation. Int J Cars 11:777–788
7. Prastawa M, Bullitt E, Gerig G (2009) Simulation of brain tumours in mr images for evaluation of segmentation efficacy. Med Image Anal 13(2):297–311

8. Bauer S, Wiest R, Nolte L, Reyes M (2013) A survey of MRI-based medical image analysis for brain tumour studies. Phys Med Biol 58:97–129
9. Liu J, Wang J, Wu F, Liu T, Pan Y (2014) A survey of MRI-based brain tumour segmentation methods. Tsinghua Science and Technology 19(6):578–595
10. Angelini ED, Clatz O, Mandonnet E, Konukoglu E, Capelle L, Duffau H (2007) Glioma dynamics and computational models: a review of segmentation, registration, and in silico growth algorithms and their clinical applications. Curr. Med. Imaging 3:262–276
11. Kwon D, et al (2014) Combining generative models for multifocal glioma segmentation and registration. In: Medical Image Computing and Computer-Assisted Intervention–MICCAI 2014. Springer, 763–770
12. Nowak RD (1999) Wavelet-based rician noise removal for magnetic resonance imaging. IEEE Trans Image Processing 8(10):1408–1419
13. Zhuang AH, Valentino DJ, Toga AW (2006) Skull stripping magnetic resonance brain images using a model based level set. Neuroimage 32(1):79–92
14. Mahajan HB, Badarla A, Junnarkar AA (2021) CL-IoT: cross-layer Internet of Things protocol for intelligent manufacturing of smart farming. J Ambient Intell Human Comput 12:7777–7791. https://doi.org/10.1007/s12652-020-02502-0
15. Mahajan HB, Badarla A (2018) Application of Internet of Things for smart precision farming: Solutions and challenges. Int J Adv Sci Tech, 37–45
16. Alhayani B, Abbas ST, Mohammed HJ, Mahajan HB (2021) Intelligent secured two-way image transmission using corvus corone module over WSN. Wireless Pers Commun. https://doi.org/10.1007/s11277-021-08484-2
17. Mahajan HB, Badarla A (2021) Cross-layer protocol for WSN-assisted IoT smart farming applications using nature inspired algorithm. Wireless Pers Commun. https://doi.org/10.1007/s11277-021-08866-6
18. Uke N, Pise P, Mahajan HB et al (2021) Healthcare 4.0 enabled lightweight security provisions for medical data processing. Turkish J Comp Mathematics 12(11)
19. Soltaninejad M, Yang G, Lambrou T et al (2017) Automated brain tumour detection and segmentation using superpixel-based extremely randomized trees in FLAIR MRI. Int J CARS 12:183–203. https://doi.org/10.1007/s11548-016-1483-3
20. Edalati-rad A, Mosleh M (2019) Improving brain tumor diagnosis using MRI segmentation based on collaboration of beta mixture model and learning automata. Arab J Sci Eng 44:2945–2957. https://doi.org/10.1007/s13369-018-3320-1
21. Akkus Z, Galimzianova A, Hoogi A et al (2017) Deep learning for brain MRI segmentation: state of the art and future directions. J Digit Imaging 30:449–459. https://doi.org/10.1007/s10278-017-9983-4
22. Zeineldin RA, Karar ME, Coburger J et al (2020) DeepSeg: deep neural network framework for automatic brain tumour segmentation using magnetic resonance FLAIR images. Int J CARS 15:909–920. https://doi.org/10.1007/s11548-020-02186-z
23. Sharif MI, Khan MA, Alhussein M et al (2021) A decision support system for multimodal brain tumour classification using deep learning. Complex Intell Syst. https://doi.org/10.1007/s40747-021-00321-0
24. Kokkalla S, Kakarla J, Venkateswarlu IB et al (2021) Three-class brain tumour classification using deep dense inception residual network. Soft Comput 25:8721–8729. https://doi.org/10.1007/s00500-021-05748-8
25. Ranjbarzadeh R, Bagherian Kasgari A, Jafarzadeh Ghoushchi S et al (2021) Brain tumour segmentation based on deep learning and an attention mechanism using MRI multi-modalities brain images. Sci Rep 11:10930. https://doi.org/10.1038/s41598-021-90428-8
26. Noreen N, Palaniappan S, Qayyum A, Ahmad I, Imran M, Shoaib M (2020) A deep learning model based on concatenation approach for the diagnosis of brain tumor. IEEE Access, 1–1 https://doi.org/10.1109/access.2020.2978629

27. Kang J, Ullah Z, Gwak J (2021) MRI-based brain tumor classification using ensemble of deep features and machine learning classifiers. Sensors 21(6):2222. https://doi.org/10.3390/s21062222
28. Cheng J, Huang W, Cao S, Yang R, Yang W, Yun Z, Feng Q (2015) Enhanced performance of brain tumor classification via tumor region augmentation and partition. PLoS ONE 10(10):e0140381. https://doi.org/10.1371/journal.pone.0140381

Social Media—A Key Pathway to Marketing Analytics

C. Immaculate Priya and Geeta Kesavraj

Abstract Social media is a dashboard of an individual, group, or society that offers all the information with or without the knowledge of the users, and it is one of the most prominent online activity. The nature of social media data is determined on real-time experience, and it enhances the marketers to pick this for their social media marketing. Hence, this research focuses on how social media information and its usage influence marketing analytics that strongly enhancing market analysis for the product or services improvement. The data from social media are gathered, analysed, tested, and the results of the analysis are used for marketing planning, designing, and decision-making. There are various third party tools offered in the market, but, however, there are analytical tools offered by social networking sites which would assist a lay person decision analysis. This research is performed to understand which segmentation focuses on social media data, and how the freely available data influence even the unfocused group to do a marketing analysis. It helps the social media users to understand the audience insights, emotions of the other social media users, helps to compare brands and represents brand positioning and mainly focus the trends that is currently adopted amongst the society. The various indispensable features of social media pave way to emergence of various marketing analytics tools, and this also impacts social media marketing since the social networking sites are one of the powerful medium which is common amongst demographic dimensions that carry advertising and awareness of various of brands, products, information, and services. The research results that social media is one of the core feature stimuli for marketing analytics which enables effective social media marketing.

Keywords Social networking · Customer engagement · Analytics tools · Advertising · Perception and recognition

---

C. I. Priya (✉) · G. Kesavraj
School of Management, Vel Tech Rangarajan Dr. Sagunthala R&D Institute of Science and Technology, Avadi, Chennai 600054, India
e-mail: immaculaterex@gmail.com

A. Kumar et al. (eds.), *Proceedings of the International Conference on Cognitive and Intelligent Computing*, Cognitive Science and Technology,
https://doi.org/10.1007/978-981-19-2350-0_26

1 Introduction

Social media is an indispensable organ of any human being. The culture of surfing through social media sites has become practice across countries, age, sex, race, etc. People have become habitual of using social media network sites in such a way that they do it as their daily activity and sometimes for few people, it has become breathe of their life. The usage of social media has reached the large population of world only through the telecommunication industry thanks to their attracting offers and discounts on the Internet services.

In this quick snack world, the way social media is used has taken a different diversion. It was initially introduced with the purpose of networking but fortunately more than networking, it is used as a tool or key factor that influences the various aspects such as

1. A researcher tool
2. To understand the consumer behaviour
3. To know a person completely just by the Facebook page not only their personal traits but also their professional front
4. It gives an overview or a graph to understand the wants and needs of the individual
5. A career builder
6. Marketing tool
7. Branding and positioning
8. Decision-making strategy.

The dimension of visualising the social media has completely changed in the upcoming generations, and social media has been taken up as a vital tool that the assist the researchers in their marketing analytics. To analyse various data from various sources such as Web, social media, and search engines assist the researcher in offering the insights through data analysis to make effective and frame marketing strategy. There are importantly three pillars that marketers use social media for their marketing and marketing analytics, such as

- Recognise the data
- Examine and extract the required information for Analysis
- Deliver the results interpreting the data.

Social media is especially important for research into computational social science that investigates questions the source of data is very much essential for data analytics. Corporates take their handy marketing tool to project their product in their market, one amongst the best online marketing tool is social media, through social media marketing business communicates/markets their products/services through images and videos to invoke customer attention and trigger an interest. Through blogging, posting, bookmarking, commenting the data and information are transferred from one person to many networks by which it will give raise to data creating a history of information of various customers.

The important attribute of social media is, that they are sentiment drivers, social media explicitly shows attributes, emotions, and behaviour of an individual or a group and this sentiment drivers are utilised by the marketing research analysts to predict the future and to get a clear image of customer perception towards the product/services. Depending on which topic modelling is performed which is a branch of machine learning technique that enables to identify a pattern or helps identify a dominant variable in the collected data and which is followed by visual analytics which interprets the data in a graphical or a pictorial representation which helps the management in decision-making.

1.1 Primary Objectives

1. The social media attributes influences marketing analytics.
2. The social media attributes also impact the social media marketing.

1.2 Secondary Objectives

1. To understand and analyse various trending marketing analytic techniques that utilise the data from social media.
2. To rationalise that the social media data impact social media marketing.

1.3 Scope of the Study

1. Broadly used tools and techniques that influence the usage of social media amongst the society.
2. Key factors that influence use of social media for their marketing analytics.
3. Designing and developing the innovative techniques by utilising social media information.

1.4 Significance of the Study

1. Due to the complexity of marketing and the reach out to people, it is essential to analyse their needs.
2. To deal with competitive market, marketers have to assist the customer in their decision-making viably.

1.5 Hypothesis of the Study

H0: Social media data are the stimuli for marketing analytics.

H1: Social media data are not the stimuli marketing analytics.

H0: Customer perception, engagement networking, advertisements, and brand recognition impact social media data.

H1: Customer perception, engagement networking, advertisements, and brand recognition does not impact social media data.

H0: Social media analytics impacts social media marketing.

H1: Social media analytics does not influence social media marketing.

1.6 Research Methodology and Data Collection

- Design of the study: Descriptive research design was adopted. [Survey Method].
- Study area: In India, questionnaire is collected through Google Form.
- Sampling Method: Stratified Proportionate Sampling—For study, 486 data were collected from 3 sets of population; illiterates, fresh graduates, and working professionals. From the 3 sets of population, researcher considered 100 samples of illiterates, fresh graduates, and working professionals summing up to 300 samples, thus forming a strata.
- Data Collection Method: Google Form.
- Study period: 1 year.

2 Review of Literature

Kangb conducted a study titled: "Exploring Big Data and Privacy in Strategic Communication Campaigns: A Cross-Cultural Study of Mobile Social Media Users' Daily Experiences" with the objective to understand individuals perception towards privacy issues and also to analyse whether the social media users usage behaviour on the culture related to big data. The study variables considered are social media experience, customer perception, and privacy issues. The author has chosen descriptive study, and data were analysed through experience sampling method from US and Taiwan as the rate of social media users and privacy issues are relatively high in these places and exploratory factor analysis were performed from 113 college students. The result indicates that individual perception got positively correlated with their privacy issues, also the participants were aware that individual data are collected as a big data to make strategic decision-making, hence this gave the mindfulness to individual to be cautious on the page information given and the collected consumer related big data that is used for strategic communication is offensive across the cultural differences and cultural specific differences will impact on social media users usage, and it is a threat towards privacy matters for social media users.

Hossein Jelvehgaran Esfahania conducted a study titled on "big data and social media: A scientometrics analysis" with the purpose of investigating the position and the growth of scientific studies for the effect of social networks on big data and how big data are used for modelling the social networks. Research methodology adopted was Scopus database considering frequently cited articles over the period 2012–2019 which was statistically analysed and characterised in diversified buckets upon applying mapping technic with Bibliometrix R-package and constructing data media for co-citation, coupling, scientific collaboration analysis, and coword analysis on topic of use of big data in social media that provided variables such as social media, social networking, big data, big data analytics scientometric, bibliometrix R-package, and the results suggested that researches have grown consistently since 2014 and decision support systems are the keynote identified from heuristics methods and thematic analysis suggested that big data analytics and Twitter can also be merged to add essence to the research and surveys showcased that United States has its major role in the ground of big data in social media, and Boyd et al. have received the uppermost citations, and many big data in social media studies have dispensed with combinatorial optimisation methods and conclude that meta-heuristics methods are widely used amongst the research to reach optimal solutions.

Abdul Jabbara conducted a study titled on "real-time big data processing for instantaneous marketing decisions: This research contains" with an objective to identify the suitable big data processing approaches for B2B industrial marketing for real-time display advertising, to analyse the impact of structured and unstructured data in big data processing techniques. The researchers have adopted a research methodology demonstrating a combination between big data, programmatic marketing and real-time processing, and how decision-making is performed for B2B industrial marketing organisations that relies on big data-driven marketing or big data-savvy managers and to explore successively suitable big data sources and effective batch and real-time processing linked with structured and unstructured datasets that influence qualified processing techniques. The handled research methodology was three main databases in this area (IEEE, Scopus, Science direct) published in the last 10 years acquisition tools and techniques; storage facilities; analytical tools and techniques to deal with problematisation and gap spotting approach which enhanced the variables such as real-time processing, batch processing, social media, programmatic marketing, decision-making, and big data. The first objective was evidenced by applying batch and real-time processing, in addition to real-time display advertising, the researcher explored the latency challenges and issues around data storage. The results suggested that both batch and real-time processing were suitable for the storage and interrogation of big data. However real-time processing within an apache storm configuration (to evidence the customer attention at the point of inspiration requires an Apache Storm configuration to provide appropriate adverts within the current user session.) is really suitable for B2B online display advertising within a programmatic marketing environment and batch processing with an optimised Hadoop and MapReduce solution is appropriate for the storage and manipulation of data for forecasting and long-term strategy purposes, and not really suitable for analysis of decisions that are required in real-time. The second objective states that there are availability of noises in the unstructured and the data analysis complexity is high, analysts describes that the challenge does not lies only with collection and storage of data whereas cleansing the data that meets the requirements and helps in decision-making reducing the latency issues is more puzzling.

Michela Arnaboldi conducted a study titled: "Accounting, accountability, social media, and big data: Revolution or hype?" with the objective to analyse the relationship between social media and big data and their accounting function and also to identify new performance indicators, governance of social media and big data information resources, and variation of information and decision-making processes. The research was conducted through paper surveys the prevailing popular literature identifying the variables such as social media, big data, accounting, and management control. The results suggest that from the objectified view social media and big data are considered as data and information upon interpreting with accounting function it appears that social media and big data are, however, enabling the alteration of, accounting practice and focal object engaging with accounting as a practice. New performance indicators are decision-making process, user engagement, predicting device, communication of information, visualisation, and governance of social media, and big data are very much necessary as that would enable towards strategic decision-making. Researcher also suggests that there is effect on alteration of information during its collection and analysis process and also researcher provokes that social media is being used by half of the population and hence it cannot be taken as measurement methodology for decision-making process.

3 Conceptual Framework

See Fig. 1.

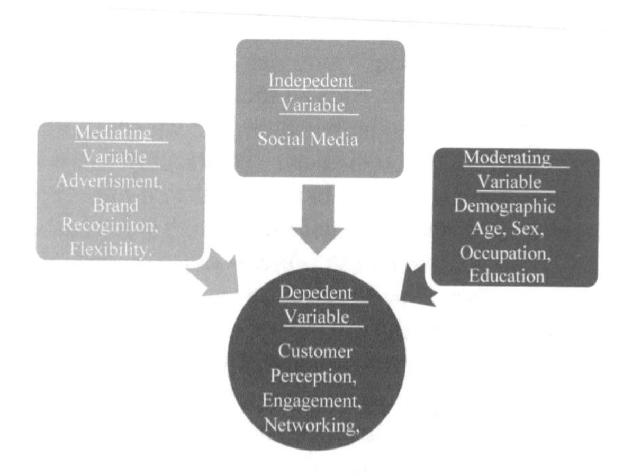

Fig. 1 Conceptual framework

4 Study Concepts

Data: It is a collection of information, so to choose the data from social media, it is very much important to understand the type of data. There are structured and unstructured data, social media data: (Bogdan Batrinca & Philip C. Treleaven). Social media is one of the richest and flexible source that assist the marketers to reach the base of the customer behaviour. Thus, social media data are one of the important tool that stimulates the researchers to analyse the behavioural science of a group, individuals, and society. Marketers have to collect the data from social media understanding the objective of the analysis. Social media data types are social network sites, Wiis, blogs, etc., scraping is an important terminology that is used when retrieving information from social media. Scraping is defined as process of collecting data from social media sites which will be usually a unstructured data which will be analysed through effective data analytics tools such as natural language processing, artificial intelligence, etc., there are also paid sources (Thomson Reuters and Bloomberg) that offer the researchers to access social media information for their research purposes.

Marketing Analytics: To put into simple understanding, it is the process of collecting marketing related data such as performance of the products, perception of any products/services from various marketing channels followed analysis and interpretation of data to frame marketing strategy or decision-making in order to improve the efficiency and high profits. Social media is one of the effective marketing tools through which marketers will be able to perform their analyses using various tools to analyse the obtained data. One of the best tools which is trendy with respect to marketing industry is the digital marketing analytics environment.

Marketing analytics has a bright future in the marketing world due to these three important core concepts:

1. Analysing the past data
2. Interpreting the present data
3. Prediction/stimulation of future

Social Media Analytics: This is one of the marketing analytics that directly scrutinises the data from social media. The core factor of social media analytics is their sentiment analysis. Sentiment analysis helps marketing research analyst to capture the attributes, expectation, motivation, perception, level of satisfaction upon which various technical tools such as Google analytics, SAS are applied to accurately measure their emotions graph and its all captured and analysed without the knowledge of the individual.

Big Data and Big Data Analytics: It is one of the emerging concepts and it is the future of one of the marketing analytics tools. It is identified as the business brain that has an effective impact on the marketing analysis. Since social media marketing a real-time marketing this particular tools comes up with accurate results with real-time analysis. It helps to identify the target audience the sentiment of the society, the taste of the customers and to make an efficient analysis which would pay way to state the art technological marketing strategy. As the name suggests big data, this tool can manage and analyse any big data analysis.

5 Analysis and Interpretation

Herein the independent data being categorical with three options and dependent data being metric having 5 pt. Likert scale, ANOVA test was performed to identify the significant difference in opinion amongst the respondents.

H0: Social media data are the stimuli for marketing analytics.

H1: Social media data are not the stimuli marketing analytics (Table 1).

The estimated significance value is greater than 0.05, meaning the null hypothesis is accepted, and the alternate hypothesis is rejected. Therefore, there is no significant difference in opinion amongst, illiterate, fresh graduates, and professionals for the question; social media data are the stimuli for marketing analytics.

Through the mean score, it can be well interpreted that illiterates, fresh graduates, and professionals all three admit that the social media data are the stimuli for marketing analytics.

Herein the independent data and dependent data being metric whose data are been obtained from a set of questions. Having found there is the relationship between both the variable a regression estimates was done.

H0: Social media analytics impacts social media marketing.

H1: Social media analytics does not influence social media marketing (Table 2).

The estimated R-square value is 0.867096, meaning the model has 86.7096% forecasting power. Further, the calculated ANOVA significance value is less than 0.005, indicating that the model is fit. Also, the coefficient significance value is less 0.05, meaning there social media analytics impacts social media marketing.

The regression equation is given by

Social Media Marketing = -0.12414 + (0.858514 x Social Media Analytics)

Having a set of demographic categorical data and independent metric variables, multivariable test was carried to identify whether there is a significant difference in opinion as well to identify the intercept.

H0: Customer perception, engagement networking, advertisements, and brand recognition impacts social media data.

H1: Customer perception, engagement networking, advertisements, and brand recognition does not impact social media data (Table 3).

From the estimated significance, value is greater than 0.05, meaning the null is accepted. That is, there is no significant difference in opinion amongst the respondents [age, gender, occupation, and education] for the question; do customer perception, engagement networking, advertisements, and brand recognition impacts social media data. Also from the intercept value obtained through multivariate test indicates that customer perception, engagement networking, advertisements, and brand recognition impacts social media data.

Table 1 ANOVA table on social media data acts as stimuli for marketing analytics

ANOVA

		Sum of squares	df	Mean Square	F	Sig	Illiterate –Mean	Fresh Graduates–Mean	Professionals-Mean
Social media Data are the stimuli for marketing analytics	Between groups	10.642	4	2.661	1.663	0.158	4.210	4.794	4.737
	Within groups	663.956	295	1.600					
	Total	674.598	299						

Source Primary data

Table 2 Level of impact of social media analytics on social media marketing

Summary output					
Regression statistics					
Multiple *R*				0.93118	
R^2				0.867096	
Adjusted R^2				0.86665	
Standard error				0.453037	
Observations				300	
ANOVA					
	Df	SS	MS	F	Significance *F*
Regression	1	399.0345	399.0345	1944.213	1.3E-132
Residual	298	61.16218	0.205242		
Total	299	460.1967			
	Coefficients		Standard error	*t* Stat	*P*-value
Intercept	– 0.12414		0.070516	– 1.76039	0.079367
Social media analytics	0.858514		0.01947	44.09323	1.3E-132

Source Primary data

6 Findings

- Social media data clearly state the researcher the audiences and their expectations and this enables the marketer to reach the consumer with appropriate marketing techniques.
- Demographic profile such as age, sex, occupation, and education has an impact over social media data that helps the marketing researcher in planning and designing the marketing strategies.
- From the analysis, it is clear that there is no significant difference in opinion amongst, illiterate, fresh graduates, and professionals for the question; social media data are the stimuli for marketing analytics.
- The social media data have an impact and clearly define the consumer behaviour which reflects their perception and level of engagement.
- Though the consumer becomes a member of the social networking sites unintentionally, but later, they become habitual of using social network sites and it becomes part of their life where they day never ends without using social media sites.
- The analysis interpretation states that customer perception, engagement networking, advertisements, and brand recognition impact social media data.
- Though it was initially accessed for pleasure, but it influences even a layperson to do market analysis without the purpose.

Table 3 Level of variables impact on social media data

Social media data		Type III sum of squares	df	Mean square	*F*	Sig
Intercept	Customer perception	1535.251	1	1535.251	931.104	0.000
	Engagement networking	1535.164	1	1535.164	1004.356	0.000
	Advertisements	1461.083	1	1461.083	992.537	0.000
	Brand recognition	1400.748	1	1400.748	771.664	0.000
Age	Customer perception	2.342	1	2.342	1.420	0.235
	Engagement networking	0.310	1	0.310	0.203	0.653
	Advertisements	1.622	1	1.622	1.102	0.296
	Brand recognition	0.094	1	0.094	0.052	0.820
Gender	Customer perception	11.879	4	2.970	1.801	0.132
	Engagement networking	5.779	4	1.445	0.945	0.440
	Advertisements	10.571	4	2.643	1.795	0.133
	Brand recognition	12.048	4	3.012	1.659	0.163
Occupation	Customer perception	9.098	3	3.033	1.839	0.143
	Engagement networking	3.530	3	1.177	0.770	0.513
	Advertisements	2.382	3	0.794	0.539	0.656
	Brand recognition	4.115	3	1.372	0.756	0.521
Education	Customer perception	4.745	3	1.582	0.959	0.414
	Engagement networking	6.141	3	2.047	1.339	0.264
	Advertisements	5.927	3	1.976	1.342	0.263
	Brand recognition	3.357	3	1.119	0.616	0.605

Source Primary data

- Social media data also provide lot of awareness, and it makes the people occupied that makes the consumers to consider the social media sites as their best companion.
- Hypothesis testing suggests that social media data are one of the best stimuli for marketing analytics.
- It clearly defines and measures the pulse and taste of the audience accurately.
- The best marketing analytics roots through the effective social media marketing.
- Since the usage of social media sites is expected only to elevate in the future the best medium of marketing would be social media marketing.

7 Suggestions and Conclusion

- Marketing analytics will assist the marketing research analysts to understand the customer insights which will help the business to tailor size their products/services.
- Performing marketing analytics through data obtained from social media will enhance data consistency, and also social media gives an opportunity analyse the SWOT of competitors.
- Marketing analytics will boost supply and demand optimisation and help the management to effectively plan the market strategy.
- Data are analysed mostly on opinion pooling which clearly defines the insights of customer needs and wants, consumer perception, consumer interest, the targeted audience, their interest, and brand positioning.
- Data insights additionally give a root design on customer engagement areas and the various attributes that draw then towards that particular product or services.
- Though there are various tools available in the market, it is very essential for the researcher to make a choice over the marketing analytics tools depending on their objective in order to plan, design, monitor, and execute the precise marketing strategy to portray in social media marketing.
- Due to the growing usage of social media users, it is essential to test the protocol using large data hence the usage of big data marketing tool can be used as it is very popular amongst marketing analytics world.
- Due to the large data management researcher should be also equipped with technology knowledge along with research analytics as the technology and marketing industry are correlated to result in the expected output.
- The young generation is looking for advanced and fast technology hence in order to cope up with the targeted audience, it is very essential to understand the audience across all ages and various demographic features only then the determined objective could be achieved satisfying the needs amongst the diversified society.

References

1. Bashar A, Ahmad I, Wasiq M (2012) Effectiveness of social media as a marketing tool: an empirical study. Int J Marketing, Financial Services & Management Res 1(11), November. ISSN 2277 3622.Faridabad
2. Baym NK (2013) Data not seen: The uses and shortcomings of social media metrics. First Monday 18(10)
3. Bruns A (2013) Faster than the speed of print: Reconciling 'big data' social media analysis and academic scholarship. First Monday 18(10)
4. Cao G, Wang S, Hwang M, Padmanabhan A, Zhang Z, Soltani K (2015) A scalable framework for spatiotemporal analysis of location-based social media data. Comput Environ Urban Syst 51:76–82
5. Ciulla F, Mocanu D, Baronchelli A, Gonçalves B, Perra N, Vespignani A (2012) Beating the news using social media: the case study of American Idol. EPJ Data Science 1(1):8
6. Evan Prokop (2014) 4 essential trends in social media marketing in 2014, Online Marketing, Social Media. Social media marketing world, 30 March. Retrieved from http://www.topranblog.com/2014/03/social-media-marketing-tremds/
7. Pate SS and Adams M (2013) The influence of social networking sites on buying behaviors of millennials. Atlantic Mark J 2(1):7
8. Jiang B, Miao Y (2015) The evolution of natural cities from the perspective of location-based social media. Prof Geogr 67(2):295–306
9. Levin N, Kark S, Crandall D (2015) Where have all the people gone? Enhancing global conservation using night lights and social media. Ecol Appl 25(8):2153–2167
10. Michael stelzner (n.d). The energy was sky high at Social Media Marketing World this year
11. Ordenes FV, Ludwig S, De Ruyter K, Grewal D, Wetzels M (2017) Unveiling what is written in the stars: Analyzing explicit, implicit, and discourse patterns of sentiment in social media. J Consumer Research 43(6):875–894
12. Sema P, Sivula M (2013) Does social media affect consumer decision-making? RSCH 5500: Business information and decision making, July 30
13. Ribarsky W, Wang DX, Dou W (2014) Social media analytics for competitive advantage. Comput Graph 38:328–331
14. Stieglitz S, Mirbabaie M, Ross B, Neuberger C (2018) Social media analytics-Challenges in topic discovery, data collection, and data preparation. Int J Inf Manage 39:156–168
15. Uldam J (2016) Corporate management of visibility and the fantasy of the post-political: Social media and surveillance. New Media Soc 18(2):201–219
16. Williams ML, Burnap P (2015) Cyberhate on social media in the aftermath of Woolwich: A case study in computational criminology and big data. Br J Criminol 56(2):211–238
17. Wood S, Guerry A, Silver J, Lacayo M (2013) Using social media to quantify nature-based tourism and recreation. Sci Rep 3:2976
18. Xie C, Bagozzi R, Meland K (2015) The impact of reputation and identity congruence on employer brand attractiveness. Marketing Intelligence and Planning, 124–146
19. Yang M, Kiang M, Shang W (2015) Filtering big data from social media-Building an early warning system for adverse drug reactions. J Biomed Inform 54:230–240
20. Zide J, Elman B, Shahani-Dennig C (2014) LinkedIn and recruitment: How profiles differ across occupations. Employee Relations, 583–604

A Survey of Satellite Images in Fast Learning Method Using CNN Classification Techniques

R. Ganesh Babu, D. Hemanand, K. Kavin Kumar, N. Kanniyappan, and V. Vinotha

Abstract It focused on identifying sickness from the picture on the various changes used to group the satellite picture. Improvement of CNN classification techniques implementation centres on the proposed strategy to concentrate on the convolution organization and its examination for the exhibition in different procedures and our technique utilizing CNN. For the most part, it consolidates four phases: picture division, affirmation of packed regions, iterative breaking down, and circulatory development. Among them, the inspiration driving affirmation of packed regions is to see what region is gathered. The iterative crumbling was to do the deterioration action reliably for a similar picture until each picture region in a bundled district was disengaged from each other. Each seed from the area associated with each image is collected region retrieved when needed, and each picture in the packed district was viewed. More experiences concerning this estimation are as follows. Regardless, picture division was recognized using the fixed edge picture division strategy reliant upon a normalized concealing differentiation. The image division can be recognized as the fast learning method relying upon the concealing differentiation. Also, to diminish the effect of encompassing light assortment, the concealing difference was normalized.

R. G. Babu (✉) · V. Vinotha
Department of Electronics and Communication Engineering, SRM TRP Engineering College, Tiruchirappalli, TN, India
e-mail: ganeshbaburajendran@gmail.com

D. Hemanand
Department of Computer Science and Engineering, Sriram Engineering College, Chennai, TN, India

K. K. Kumar
Department of Electronics and Communication Engineering, Kongu Engineering College, Erode, TN, India

N. Kanniyappan
Department of Electronics and Communication Engineering, Jerusalem College of Engineering, Chennai, TN, India

A. Kumar et al. (eds.), *Proceedings of the International Conference on Cognitive and Intelligent Computing*, Cognitive Science and Technology,
https://doi.org/10.1007/978-981-19-2350-0_27

Keywords Satellite images · CNN · Image classification · Fast learning method · Edge picture shading contrast

1 Introduction

The key to focusing is in the selection and synchronization of highlight emphases. Since the concentrates on accessible pictures on the collected left and also suitable images are a couple of excellent match point settings, coordinating must be performed by the following calculations for necessity, individuality, and different inclinations:

1. Removed the clustering structure of its chosen image from the conservative and liberal picture assortments. It is looked for a match between the centroid purposes of the left and right pictures and the requirement circumstances.
2. The complexity requirement circumstances separate the matched centroid directions of the left and right images;
3. To determine whether the match focuses on meeting the uniqueness slope imperatives;
4. It is obtained and completed the process of coordinating the picture's purposes, after which the programme was completed.

The test results revealed that when picking pictures that are not or insignificantly obstructed by offshoots and images, coordinating error could be regulated within 5 to 8 pixels using the technique shown in Fig. 1.

Such studies have attempted to develop tacit skills for inexperienced farmers relying primarily on environmental and photographic data [1–3]. Water-strain farming, which is done to reduce the amount of irrigation, is a complex cultivation technique based on the knowledge and intuition of professional growers [4]. Drought-related global water pressure has been determined as the most dangerous to vegetation. Occasionally green are exceptionally well during the evaluation of the vegetables that have been grown extensively.

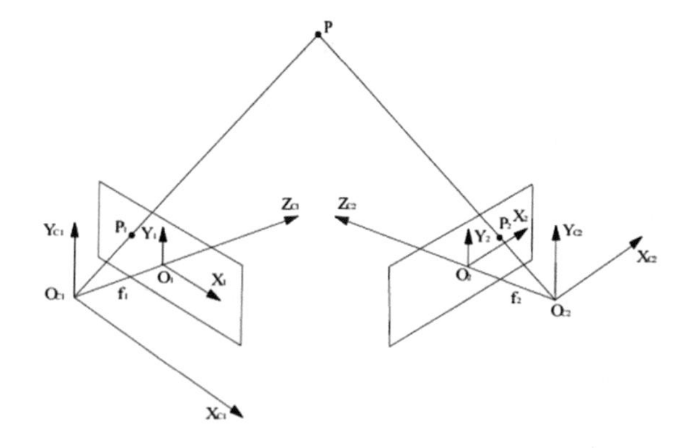

Fig. 1 Three-dimensional reconstruction for Curvelet

2 Implementation of Proposed Work

The root is located in the visual focus on picture organize framework $O_{k1}X_{k1}Y_{K1}Z_{K1}(O_{W1}X_{W1}Y_{W1}Z_{W1})$. It is located picture level of exterior as the clear central length f_1 picture aircraft focus arranges $O_1X_1Y_1$. The appropriate camera enables framework seems to be (u_{01}, v_{01}) its picture organizes $O_{k2}X_{k2}Y_{k2}Z_{k2}$. It appears to be good length f_2 picture aircraft focus part $O_{k2}X_{k2}Y_{k2}Z_{k2}$ of the learning experience. The pin-opening model is visualization model $(X_1, Y_1) and (X_2, Y_2)$; P position organizes by1 geometric connection in Eqs. (1), and (2) can be used to create the following formula as

$$Z1_{k1}\begin{bmatrix} X1_1 \\ Y1_1 \\ 1 \end{bmatrix} = \begin{bmatrix} f_1 & 0 & 0 \\ 0 & f_1 & 0 \\ 0 & 0 & 1 \end{bmatrix}\begin{bmatrix} X1_{k1} \\ Y1_{k1} \\ Z1_{k1} \end{bmatrix},$$

$$Z1_{k1} \neq 0, X1_1 = (u_{01} - u_1)fx, Y1_1 = (v_{01} - v_1)fy \tag{1}$$

$$Z1_{k2}\begin{bmatrix} X1_2 \\ Y1_2 \\ 1 \end{bmatrix} = \begin{bmatrix} f_2 & 0 & 0 \\ 0 & f_2 & 0 \\ 0 & 0 & 1 \end{bmatrix}\begin{bmatrix} X1_{k2} \\ Y1_{k2} \\ Z1_{k2} \end{bmatrix}, Z1_{k2} \neq 0,$$

$$X1_2 = (u_{02} - u_2)fx, Y1_2 = (v_{02} - v_2)fy \tag{2}$$

Throughout this situation, u_1, v_1, u_2, v_2 the left and right picture $u_{01}, v_{01}, u_{02}, v_{02}$ pixel orientations are individually arranged, the frame of the image of the right and left camera f_x, f_y are differently arranged, and the actual range of each pixel has planned the x, y axis.

The highlights are shown in Fig. 2 after completing the increase of a comparison network. The CNN provides the output by partitioning it by the exact number of pixels, then creating a checklist and placing the channel model there [5]. The element is then moved to all other points in the image, the framework's creation is obtained, and the means are rehashes for numerous screens. The channel is moved to every conceivable spot on the image with this layer. So, because framework for left image arranges is consistent with the universal enabling framework, the connection here between left and right picture arrange situations could be conveyed as follows:

$$\begin{bmatrix} X1_{k2} \\ Y1_{k2} \\ Z1_{k2} \end{bmatrix} = R1 \times \begin{bmatrix} X1_{k1} \\ y1_{k1} \\ Z1_{k1} \end{bmatrix} + s \tag{3}$$

If $R1$, s is references to the revolutionary framework the interpreting vector orientations within connections independently, homogenization can be conveyed as Eq. (3) if the condition Eq. (3) is met (4)

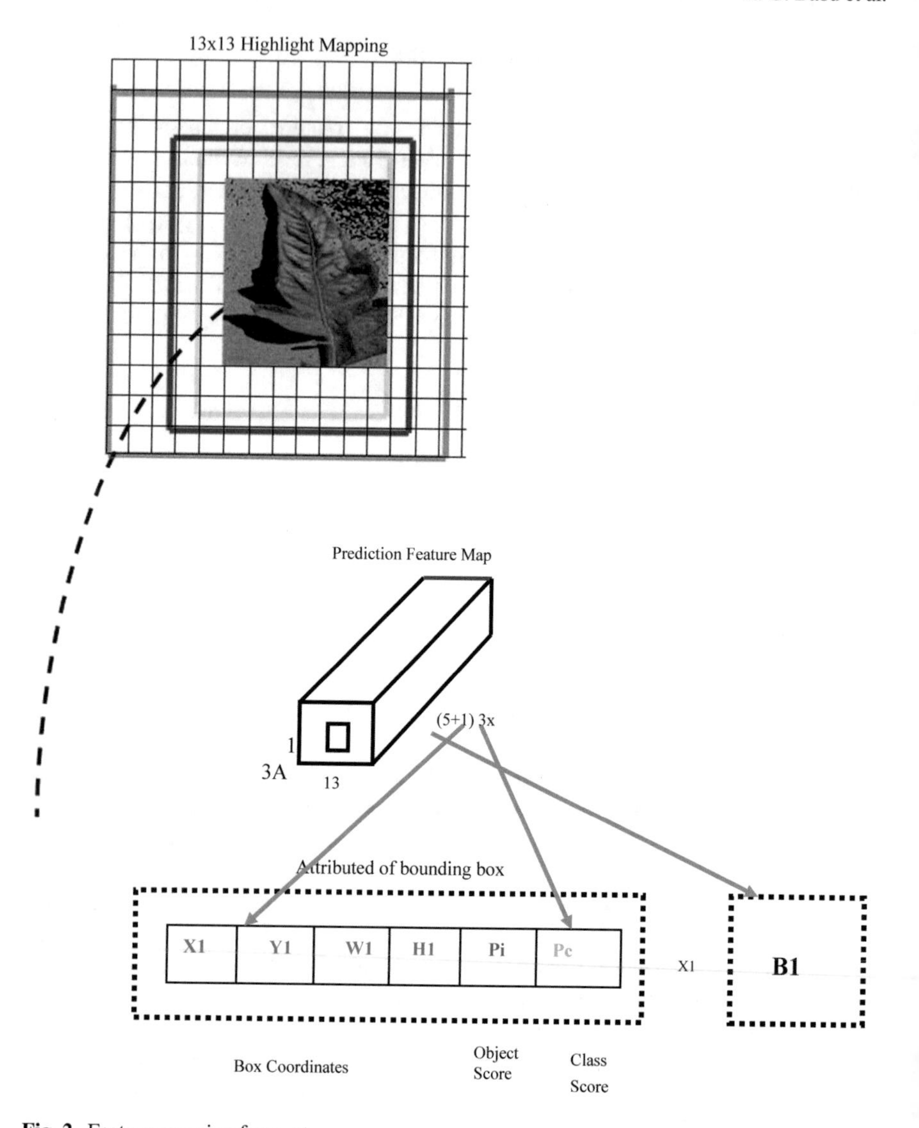

Fig. 2 Feature mapping forecast

$$\begin{bmatrix} X1_{k2} \\ Y1_{k2} \\ Z1_{k2} \\ 1 \end{bmatrix} = \begin{bmatrix} r_1 & r_2 & r_3 & s_x \\ r_4 & r_5 & r_6 & s_y \\ r_7 & r_8 & r_9 & s_z \\ 0 & 0 & 0 & 1 \end{bmatrix} \begin{bmatrix} X1_{k1} \\ Y1_{k1} \\ Z1_{k1} \\ 1 \end{bmatrix} \tag{4}$$

$$Z1_{k2}\begin{bmatrix} X1_2 \\ Y1_2 \\ 1 \end{bmatrix} = \begin{bmatrix} f_2 & 0 & 0 & 0 \\ 0 & f_2 & 0 & 0 \\ 0 & 0 & 1 & 0 \end{bmatrix}\begin{bmatrix} r_1 & r_2 & r_3 & s_x \\ r_4 & r_5 & r_6 & s_y \\ r_7 & r_8 & r_9 & s_z \\ 0 & 0 & 0 & 1 \end{bmatrix}\begin{bmatrix} X1_{k1} \\ Y1_{k1} \\ Z1_{k1} \\ 1 \end{bmatrix}$$

$$= \begin{bmatrix} f_2r_1 & f_2r_2 & f_2r_3 & f_2s_x \\ f_2r_4 & f_2r_5 & f_2r_6 & f_2s_y \\ r_7 & r_8 & r_9 & s_z \end{bmatrix}\begin{bmatrix} \frac{Z1_{k1}X1_1}{f_1} \\ \frac{Z1_{k1}Y1_1}{f_1} \\ Z1_{k1} \\ 1 \end{bmatrix} \tag{5}$$

Using equal system image models in conjunction with the actual focus situation [6–9], the visual model may be constructed as a condition function in Eqs. (5) and (6).

$$\begin{cases} X1_{k1} = \frac{Z1_{k1}X1_1}{f_1} \\ Y1_{k1} = \frac{Z1_{k1}Y1_1}{f_1} \\ Z1_{k1} = \frac{f_1f_2b1}{f_1X1_2 - f_2X1_1} \end{cases} \tag{6}$$

The picture points are represented by b_1 and b_2.
As the attribute match point, we will use $P_1(x_1, y_1) and P_2(x_2, y_2)$, $(u_1, v_1) and (u_2, v_2)$ as just a pixel groups together related image feature.

$$\begin{cases} x1_w = \frac{b1(u_1 - u_0)}{u_2 - u_1} \\ y1_w = \frac{b1(v_1 - v_0)}{(u_2 - u_1)} \\ z1_w = \frac{b1f1}{u_2 - u_1} \end{cases} \tag{7}$$

$b1$ is the length of standard images;
$f1$ is the picture's central length;
u_1, u_2 is the counting key focuses on three-dimensional orientations and is the pixels differentiation of coordinated component focuses.

3 Quick Enhanced Learning Technique Algorithm

Input: A dataset consists (both input $N1_{kts}$ and target $L1_{kts}$ data), two signal faults (b_i and s_i) to govern convergence, as well as a phase margin.

Output: The strengths of the levels as well as the cumulative sum errors sensitivities for data sets.

Step 1: Initialize value of in λ.

$$L1(\{b_i\}, \{s_i\}) = \frac{1}{N1_{kts}}\sum_i L1_{kts}(b_i, b_i^*) + \lambda\frac{1}{N1_{reg}}\sum_i b_i^* L1_{\text{reg}}(s_i, s_i^*) \tag{8}$$

In Eq. (8) λ is located to 12, $N1_{kts}$ is 256, with $N1_{\text{reg}}$ is 2400. The two different of an R.P.N. pounding may be objective at this position $L1_{\text{reg}} = L1_{\text{loc}}$.

$$L1_{\text{loc}}(s^u, v) = \sum_{i \in \{x,1y1,w1,h\}1} \text{smoth}_{L_1}(s_i^u - v_i) \tag{9}$$

$$smoth_{L1}(x1) = \begin{Bmatrix} 0.6x^2\text{ifx1} < 1 \\ \text{x - 0.6otherwise} \end{Bmatrix} \tag{10}$$

Step 2: In Eqs. (9) and (10), the shortfall of speed CNN is divided into two large squares (10). The central square represents the difficulty in generating R.P.N., whereas the following square represents the difficulty in preparing rapid CNN by the classifier. The overall capacity for the misfortune of faster CNN is depicted as follows:

$$L1_{\text{finel}} = L1(\{b_i\}, \{s_i\}) + L1\big(b, u, s^u, v\big) \tag{11}$$

The red chunk in Eq. (11) shows the difference between blind CNN with faster CNN. The overall organization structure is comparable, as can be seen. After Align in Masks CNN, there is indeed a "head" section.

Step 3: The main task is to increase Align's yield assessment, which would be more precise when anticipating Mask. Instead of using FPN-style SoftMax unhappiness in the informational meeting of masked branches, K Mask expectations diagrams are produced, and the commonly used paired cross-entropy sorrow preparing is used in creating Mask branch. Only one of the component maps corresponding to the regression coefficients classification adds towards Mask suffering when the K component maps surrender. Mask CNN's ability for preparing tragedy can be described as follows Eq. (12).

$$L1_{\text{finel}} = L1(\{b_i\}, \{s_i\}) + (L1_{cls} + L1_{\text{box}} + L1_{\text{mesk}}) \tag{12}$$

Step 4: The information picture vanishes during a slew of convolution layers that contribute to the construction. The RGB median respect registered to a training stage is removed from a fixed-size 224 224 RGB image. To preserve the spatial purpose underlying convolution, the channel has a 3× 3 open area. The convolution step is limited to 1 pixel and five max-pooling layers totalling 2 × 2 pixels spatial frequencies pooling. Three fully connected layers follow a stack of convolution layers, the first two of which include 4096 streams, each containing 1000 channels. Nonlinearity is present in all veiled layers provided by rectification. This strategy separates component descriptions by yielding the penultimate entirely connected layer. It describes a set of images that are used to train a CNN classifier using their comparing markers.

Step 5: The lowest value of the optimization process is used to determine the number of features. Whenever the error function achieves a different effects, iteration is halted, as demonstrated in Fig. 3.

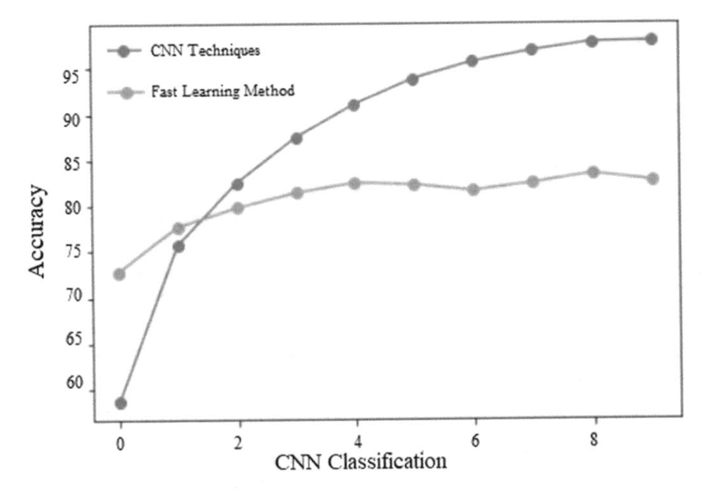

Fig. 3 Accuracy classification using CNN techniques and fast learning method

Step 6: Learning inaccuracy quantified by all layers.

Step 7: Once a fault value has been determined, and these operations are repeated multiple times Eq. (13).

$$\frac{1}{n1}\sum_{i=1}^{n} \max 1(0.2 - y_i(\omega^* x_i + p)) + \lambda \left\| \omega^2 \right\| \tag{13}$$

where x_i is all the ith location in the training dataset.

4 Conclusion

A CNN model recognizes images as data, performs the convolution process, and effectively disengages it to the most significant degree feasible, reducing the dimensionality of the information provided. Those who have started taking highlights directly affect the accuracy of the image they are working with it. The CNN algorithm has many layers, including convolution, REL, pooling, connection, reduction, and regularization. CNN would dissect the pictures piecemeal. Each component is denoted by the term "ports" or "channels." CNN uses weight association to isolate particular highlights from the data picture without losing spatial output information.

References

1. Karthika P, Vidhya Saraswathi P (2017) A survey of content based video copy detection using Big Data. Int J Scientific Res Sci Tech 3(5):114–118
2. Vijayalakshmi D, Nath M, Acharya OP (2020) Analysis of artificial intelligence based image classification techniques. J Inn Image Process 2(1):44–54
3. Choudhary R, Gawade SA (2016) Comprehensive survey on image enhancement method and a GUI. Sensing Imag 21(3):21–25
4. Karthika P, Saraswathi VP (2019) Image security performance analysis for SVM and ANN classification techniques. International J Recent Tech Eng 8(4S2):436–442
5. Yuan T, Zheng X, Hu X, Zhou W, Wang W (2014) A method for the evaluation of image quality according to the recognition effectiveness of objects in the optical remote sensing image using machine learning algorithm. PLoS ONE 9(1):1–7
6. Wiley V, Lucas T (2018) Computer vision and image processing: A paper review. Int J Artificial Intelligence Res 2(1):28–36
7. Ganesh Babu R, Karthika P, Elangovan K (2019) Performance analysis for image security using SVM and ANN classification techniques. Third IEEE International Conference on Electronics. Communication and Aerospace Technology. IEEE Press, Coimbatore, India, pp 460–465
8. Singh G, Mittal A (2014) Various image enhancement techniques—A critical review. Int J Innov Scientific Res 10(2):267–274
9. Rajput S, Suralkar SR (2020) Comparative study of image enhancement techniques. Int J Comput Sci Mob Comput 2(1):11–21

Artificial Intelligence Method Using Fast Enhanced Image Quality Evaluation

R. Ganesh Babu, L. Saravanan, N. Kanniyappan, G. Manikandan, and N. Poornisha

Abstract Advances and unique techniques for resolving actual difficulties are becoming increasingly rare in today's world. The essential requirement for food is growing in lockstep with the development of field workers. However, due to a loss of appropriate refined information and awareness of the estimated upheaval, those adjusting harvests and grains are occasionally wrecked wholly or partially. This paper aims to integrate a bit of horticulture using science with time to reduce the loss caused by bug attacks. With authorize thoughts, the "satellite picture" is selected as the central image, wherein sickness wasn't out of the norm and is distinguished using artificially intelligent computations, CNN, and creative computer skills. The typical reason is to alleviate the recognition difficulties that "satellite picture" growers encounter suitable for cultivation ground nowadays, particularly in India. Farmers have the alternative of entering the manifestations within the state of valuable images with people expecting the diseases. It performed a correlation analysis of various sequences of filter/methods employed in various processes, and the device confirmed the method's accuracy at 99.5%.

Keywords RBF kernel · Fast enhanced learning method · CNN · K-means clustering · NN classifier · SVM

R. G. Babu (✉)
Department of Electronics and Communication Engineering, SRM TRP Engineering College, Tiruchirappalli, TN, India
e-mail: ganeshbaburajendran@gmail.com

L. Saravanan · N. Poornisha
Department of Electronics and Communication Engineering, Rajalakshmi Institute of Technology, Chennai, TN, India
e-mail: saravanan.l@ritchennai.edu.in

N. Kanniyappan
Department of Electronics and Communication Engineering, Jerusalem College of Engineering, Chennai, TN, India

G. Manikandan
Department of Electronics and Communication Engineering, Dr.M.G.R Educational and Research Institute, Chennai, TN, India

A. Kumar et al. (eds.), *Proceedings of the International Conference on Cognitive and Intelligent Computing*, Cognitive Science and Technology,
https://doi.org/10.1007/978-981-19-2350-0_28

1 Introduction

In Japan, innovative farming practices for increasing plant output are becoming more popular. Nevertheless, farmer transition costs rise as farmers grow, making it more likely that expert farmers' ideas may be lost before younger farmers acquire them [1]. As a result, research on the propagation of better farming strategies has been carried out. This type of research aimed to develop the tacit understanding of professional producers focuses on ecological and photographic data [2]. River farming, which is carried out to reduce the amount of irrigation, is a complex agriculture technique that relies on the knowledge and sense of experienced farmers [3]. Drought-related worldwide water pressure has been determined as the most dangerous to vegetation. Uncommon greens are particularly well-known in terms of the vegetables that have been produced extensively.

2 Summary of Proposed Work

The goal of this study was to detect disease from a photograph [4]. The different modifications sometimes used to classify a satellite view are discussed in this section techniques for CNN classification improvement. The suggested methodology to describe the convolution network and its analysis for success in specific ways and our method employing CNN [5] focus on implementation. It mainly consists of four stages: image division, recognition of clustered areas, repetitive disintegration, and circular expansion [6–8]. The objective for recognizing grouped districts is to understand what constitutes a combined area. Gradual dissolution entailed repeating the disintegrating process for each concurrent picture till every image area in a grouped locale was separated from others.

As a consequence of recovering each seeding from district linked with each picture included within the grouping region, each image contained within the grouping region was also recognized. The following are some additional details about this analysis. To begin, the continuous edge picture split method was utilized to recognize image picture partitions using a standardized shade difference. The shading distinction, which itself is R-G, can be used to identify the picture division. The shading contrast was also standardized to reduce the impact of ambient light diversity. The uniform shading intensity was created with the help of Eq. (1).

$$D_c = \frac{255(2R + B)}{2(R + G + B)} \tag{1}$$

where D_c is the standardized shading contrast, the three-shade segments are R, G, and B. The picture divide was recognized using the constant limitation picture divisions method. The edge and s are adjusted to 160 depending on the tests behind the standardized shade difference of each pixel.

Second, the numerical morphological activity was used to prepare bunched locations to boost acknowledgment output. As a result, clustered locations should be considered then first, primarily. Bunched places might be recognized using the length and width of a foundation incased squares shapes of captured images since areas of grouping localities were larger than single image regions. The recognition rule for grouped image areas is as follows: in Eq. (2).

$$R_z = \begin{cases} R_p|\max(H_M, W_M) < T \\ R_q|\max(H_M, W_M) >= T \end{cases} \quad (2)$$

The image R_z implies everywhere it R_p appears. The physical stature and width of base incased square forms of this local independently indicate that this area is just a single image district H_M, W_M. T is indeed the edge of a larger one between H_M with W_M, inside the test shown in Fig. 1, it was located at 30 pixels.

Finally, the repeated disintegrating process for grouped areas was completed, as shown in Eq. (3).

$$R_{ki} = \begin{cases} R_s i = 1 \\ (R_{k(i-1)} \Theta Y) i = 1, 2, 3, \ldots n \end{cases} \quad (3)$$

where R_{ki} the image is produced for i times, R_s a digital picture before dissolution by the disintegrating activity. Θ denotes a process. The recurrent disintegration's conclusion state is D. The repeated disintegrating activity ends after n hours of dissolution activity. Y stands for a team that works. According to the status of the satellite picture, it is an oblate 44 building unit.

Fourth, seed districts obtained by iterative disintegrating were increased for an n period to hours of disintegrating activity is depicted in circulation expansions in Eq. (4).

$$R_u = R_{sn} \oplus nY \quad (4)$$

where R_u is all the picture locale for the seed district rebuilt via circulatory wide. Tomatoes are a clustered area that is regarded individually behind such a development. Finally, the base encloses rectangular shape was constructed, with the center of each observed image inside the grouped region. The centered image's arrangement should be registered using Eq. (5).

$$\begin{aligned} x_s &= \frac{1}{ns} \sum_{i=1}^{n_s} x_{si} \\ y_s &= \frac{1}{n_s} \sum_{i=1}^{n_s} y_{si} s = 1, 2, \ldots, m \end{aligned} \quad (5)$$

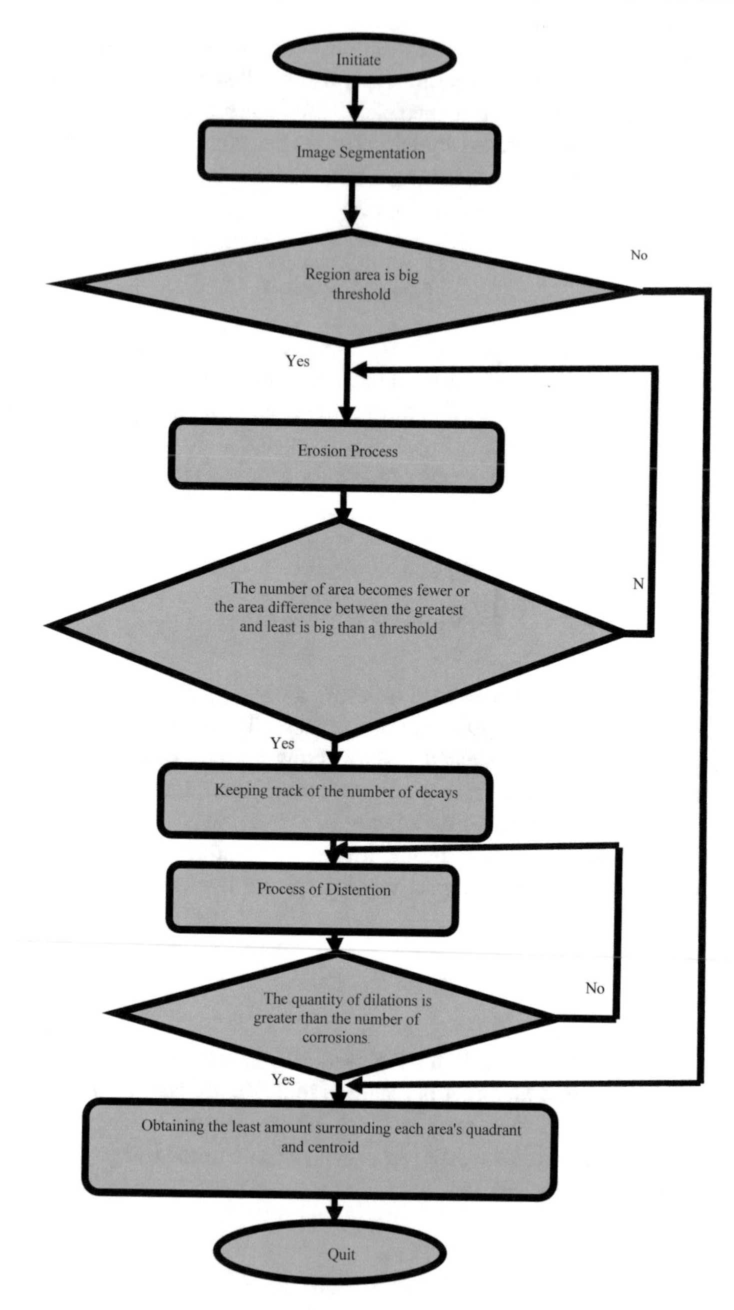

Fig. 1 Image recognition using a clustering algorithm

wherever x_s, y_s, the rth district's center is convenient. N_s is the number of pixels in the rth region. A similar picture is made possible by the ith pixel in the rth location, where m denotes the number of locations involved in the circulatory expansion.

There are apparent shading contrasts between the red of the image and the green of the branches. It is used to recognize picture division for images that have been registered as in Eq. (6).

$$D = R - G \tag{6}$$

where C denotes chromatic aberration, R denotes RGB color freedom, and G denotes RGB color freedom in the RGB color space. It accurately recognizes picture split in a variety of illumination circumstances, as shown in Fig. 1. Picture split calculation based on OTSU is used [9]. Before that, the color distortion requirement is rectified, as in Eq. (7), to keep chromatic fluctuation values within the range of 0–255.

$$D_e = (R - G + 255)/2 \tag{7}$$

As in Eq. 1, the standardized chromatic variable esteem is calculated (8). Picture split recognized after it has been recorded, as in Eq. (9).

$$S = 255(B + 2R)/(2G + 2R + 2B) \tag{8}$$

where S-regularize chromatic deviation, B-B constituent in RGB color freedom.

$$T_{(x,y)} = \begin{cases} 0 A_{(x,y)} < T \\ 1 A_{(x,y)} >= T \end{cases} \tag{9}$$

where $T_{(x,y)}$-Chordal abnormality value standardized of pixel to whom the picture organize is (x, y); $A_{(x,y)}$-Gray frequency of pixel to whom the picture organize is (x, y); (x, y).

There are almost equal proportion linkages between R sections and G segments of the similar picture [10, 11]. For one sort of coloring in varying illumination, the spectrum percent is considered to be similar. The pixels in photo district are all red. As in Eqs. (10) and (11), the band percentage is reported

$$k_{\mathrm{RG}} = R/G \tag{10}$$

$$k_{\mathrm{GB}} = G/B \tag{11}$$

k_{GB}-b and ratio.

Whenever the light is brighter, it is not easy to see pixels with R parts of 255 while using a picture segment technique based on a personal frequency percentage among R and G segments. However, it is possible to use a picture segment approach based on group percentages between G and B segments. As a result, two distinct types of population proportion regard are used to independently feel the image area of pixels, as represented by R esteems. Band percentages between R segments using G segments and band percentages between G segments using B segments. This type of picture splitting technique can be used to extend the self-versatile reaching to enlightened variety.

$$T(x, y) = \begin{cases} 1F < 255\text{with}k_{RG} > T_{RG} \\ 1F < 255\text{with}k_{GB} > T_{GB} \\ 0\text{others} \end{cases} \tag{12}$$

3 Algorithm for Fast Enhanced Learning Method

A CNN model takes photographs as input, performs convolution on them, and isolates the maximum extent possible, reducing the information's dimensionality. Extracted highlights have a direct impact on the precision with which images are handled. Convolution, ReLU layer, accumulating, fully connected, reduce, and normalization are some of the CNN model layers. The photographs would be examined piece by piece on CNN. Each component is referred to it as an element and channel. CNN uses the weighted network to focus the specific highlights from the data image without sacrificing the information as to its spatial strategy.

Step 1: User interface to load the image gathering and preprocessing the database: Obtaining both image and standard image. Seedlings villages dataset was used to create this dataset.

Step 2: The median filter was used to preprocess and remove noise from images of satellite image.

Step2: The median filter has been used to annotate and eliminate noise from satellite picture images.

Step 3: Analyze data noisy utilizing convolution filtering and the initial layer to accept an incoming signal.

Step 4: Using Kernel and TensorFlow, create a model for classifying healthy or unhealthy by subsampling and extracting features. Filter/methods for matrix classification were utilized in many processes, such as the device confirming our strategy's accuracy.

Step 5: Corrective action of plant noise formulation and construction improvements for classifying picture images as usual or infected, with 7 types of image illness included.

Step 6: Connect everything and evaluate how different classifiers work.

Step 7: Loss during the training as well as testing.

The highlights after finishing the comparison lattice increase are shown in Fig. 2. The CNN adds and parcels its result by measuring the number of pixels, creating

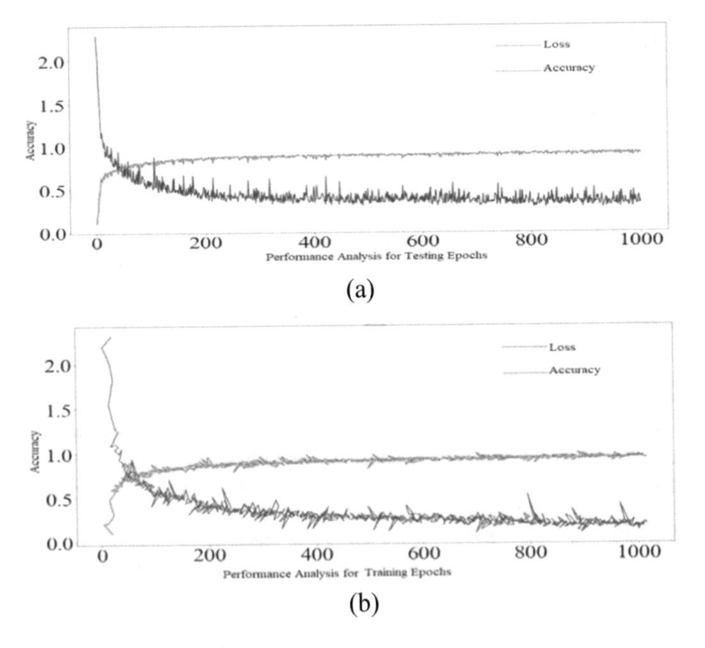

Fig. 2 Performance analysis for testing and training epochs

aid, and placing the channel's evaluation there. Starting from the beginning, it moves the portion to every remaining area of the picture, obtains the building's yield, and continues the process for different channels. This layer shifts the channel to every potential situation on the image.

4 Conclusion

This research focuses on the programmed positioning of picture irritations and illnesses concerning the environment. The recognition models created utilizing learning technology to identify images with bugs achieve an average characterization precision of 99.5%. Nonetheless, the overall elite is reliant on relatively high-quality test images. Future research focuses on confound estimates to separate image insects from illnesses linked to poor image resolution. A CNN-based model is used to detect the infection inside the picture. There are three convolution and pooling layers inside the suggested CNN-based technology, each with various outlets. The picture data from the plants village collection are used in the analysis. There are many 16 infection classes in the collection, and each class has a solid image. Furthermore, attempts to enhance a comparative framework on the same dataset to assess exactness are being examined.

References

1. Yuan T, Zheng X, Hu X, Zhou W, Wang W (2014) A method for the evaluation of image quality according to the recognition effectiveness of objects in the optical remote sensing image using machine learning algorithm. PLoS ONE 9(1):1–7
2. Shakya S (2020) Analysis of artificial intelligence based image classification techniques. J Innovative Image Process 2(1):44–54
3. Wiley V, Lucas T (2018) Computer vision image processing: A paper review. Int J Artificial Intelligence Res 2(1):28–36
4. Cui M, Zhang DY (2021) Artificial intelligence and computational pathology. Lab Invest 101:412–422
5. Akkus Z, Cai J, Boonrod A, Zeinoddini A, Weston AD, Philbrick KA (2020) Artificial intelligence in musculoskeletal imaging: current status and future directions. J Am Coll Radiol 16(9):1318–1328
6. Robertson S, Azizpour H, Smith K, Hartman J (2018) Digital image analysis in breast pathology—from image processing techniques to artificial intelligence. Transl Res 194:19–35
7. Choudhary R, Gawade S (2016) Survey on image contrast enhancement techniques. Int J Inno Studies Sciences Engineering Tech 2(3):21–25
8. Karthika P, Vidhya Saraswathi P (2020) IoT using machine learning security enhancement in video steganography allocation for Raspberry Pi. J Ambient Intelligence Humanized Computing 11(11):1–15
9. Karthika P, Vidhya Saraswathi P (2019) Image security performance analysis for SVM and ANN classification techniques. Int J Recent Technology Engineering 8(4S2):436–442
10. Ganesh Babu R, Karthika P, Aravinda Rajan V (2020) Secure IoT systems using raspberry Pi machine learning artificial intelligence. In: Smys S., Senjyu T., Lafata P. (eds) Proceedings of Second International Conference on Computer Networks and Inventive Communication Technologies. Lecture Notes on Data Engineering and Communications Technologies, Springer, Singapore, vol 44, pp 797–805
11. Ganesh Babu R, Karthika P, Elangovan K (2019) Performance analysis for image security using SVM and ANN classification techniques. In: Third IEEE International Conference on Electronics, Communication and Aerospace Technology, pp 460–465. IEEE Press, Coimbatore, India

A Survey: Recent Advances in Hierarchical Clustering Routing Protocols

Zoya Akhtar, Himani Garg, and Sanjay K. Singh

Abstract Wireless sensor networks have been an area of interest for researchers. It has exceptional adaptive capacity and provides a wide range of applications. Albeit, apart from its many advantages, one of the main quandary with WSNs is that that its sensor nodes are battery driven. In order to make the network energy efficient, routing algorithm is utilized. Hierarchical clustering makes use of the sensor nodes by grouping them into different clusters and appointing a cluster head from every group and this cluster head accumulates the data from its member nodes and deliver it to the base station. Since, over a decade, various improvements have been made in order to enhance the overall lifetime of the network. In this paper, we will discuss different advances being made overtime in hierarchical clustering routing protocols to amplify its lifespan.

Keywords WSN · Routing protocols · Energy efficient · Hierarchical clustering

1 Introduction

A wireless sensor network consists of a number of sensing nodes that collects the information and delivers it to the base station. These sensors nodes are capable of a variety of function, i.e. sensing, communicating and processing. The source of energy for these sensor nodes is their batteries and they communicate in a wireless environment. Due to the limitation of energy, energy efficient algorithm is a vital part for a wireless sensor network, i.e. the more efficient a routing protocol is better will be the lifetime of the network. Since, the major drawback in wireless sensor network is due to the battery life of the sensor node which is non-rechargeable. Therefore, in order to increase the lifetime of the network and decreasing the delay in the transmission of data, sensor nodes are combined together to form small groups which is basically known as clusters. This method of combining the sensor nodes is known as clustering and a sensor node is selected from every group and is appointed

Z. Akhtar (✉) · H. Garg · S. K. Singh
Department of Electronics and Communication, ABES Engineering College, Uttar Pradesh, Ghaziabad 201016, India

A. Kumar et al. (eds.), *Proceedings of the International Conference on Cognitive and Intelligent Computing*, Cognitive Science and Technology,
https://doi.org/10.1007/978-981-19-2350-0_29

as the cluster head or CH. The function of the cluster head is to deliver or transmit the data which it gathers from the other sensor nodes present in the cluster to the sink node or the base station. The cluster head has high energy and capability with respect to the other sensor nodes [1–5]. Some features must be kept in mind when selecting the cluster head, i.e. it should have greater residual energy than the rest of the sensor nodes and cluster head should be rotated periodically. In this paper, we will study the recent advances made in the hierarchical clustering routing protocol.

2 Literature Review

In wireless sensor networks, routing plays a very important role especially in environmental monitoring. The routing protocols are designed in a fashion that it can withstand several challenges like energy consumption, security issues, delays in the network, etc. A lot of research is ongoing in order to increase the efficiency of the wireless network. As the resources in the wireless sensor network are limited, so making a choice for selecting the best routing protocol is very crucial. Delivering of the packets containing data to the destination node from its source node should be done perfectly, i.e. without any interference in the route. The list of all the available paths towards the destination node must be contained in the routing tables. It is very important to update these routing tables periodically so that the stability must be maintained in the network. So, while designing these routing algorithm some factors should be taken in consideration (Fig. 1).

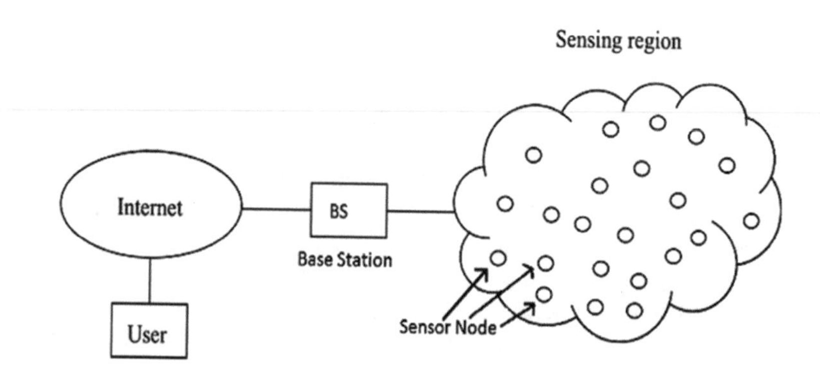

Fig. 1 WSN

2.1 Design Challenges in Routing Protocols

Connectivity in the network: As we know, that in a wireless sensor network, sensor nodes are densely deployed and some of these sensor nodes are isolated from one another so the connection between these sensor nodes should remain established at any cost even if a node goes into failure.

Energy efficiency: This is perhaps the most important factor in the wireless sensor network as it determines the lifetime of the network. We are all aware that the sensor nodes are battery-driven making its energy limited. So, the routing algorithms should be designed in such a way that it makes the wireless sensor network energy efficient which can be done by reducing the energy consumption when the data is being transmitted as the transmission of data consumes much more energy than its processing.

Cost: As in wireless sensor network, hundreds of thousands of sensor nodes deployed which can increase the cost of the network. So, the cost should be kept in mind while selecting the nodes to be deployed as it can increase the overall cost of the wireless sensor network.

Scalability: A wireless sensor network can contain a few sensor nodes to thousands of sensor nodes. The routing protocol selected should be able to maintain a large network, i.e. should be able to transmit and receive the data successfully.

Fault tolerance: In the wireless sensor network, the sensor nodes are thrown casually which can lead to damaging the sensor nodes and some sensor nodes battery can be exhausted after a while or any other factor that can affect the sensor nodes. So, the routing algorithm used should be able to find the another path for the data to be transmitted in case of node failure. So that it can reach the base station effectively. Data redundancy can be useful in some cases.

Aggregation of data: A lot of needless traffic can be caused by redundant data which is generated by the multiple sensor nodes. In order to avoid that, the similar data should be aggregated or perhaps combined together which can lead to the reduction in the number of the transmitted data. The function of data aggregators is to collect the data from different sensor nodes by suppressing them, i.e. by eliminating any similar data or by using the minimum, maximum or average. The function of data aggregator is assign to the nodes that are powerful.

Mobility of the nodes: Although, sensor nodes are mostly stable, and in some cases of the wireless sensor network, the mobility of the sensor nodes becomes a must. So, it can be challenging and the routing protocol used should be able to implement the function properly.

Sensor node position: The sensor nodes deployed can be randomly scattered in the network, i.e. there is no predetermined path for the sensor nodes or the sensor nodes can also be manually placed in the network and has a routing path that is predetermined.

Quality of service (QoS): The routing protocol should give best quality of service, i.e. delivery of the data, reliability, the lifetime of the network, consumption of the energy, loss of the packet, etc. The selected routing protocol should be able to deal with all the demands required for the application.

Hardware of the sensor nodes: The size of the sensor nodes varies, i.e. they are either be very small or can be of medium to large size. Typically, a sensor node contains four parts: a unit for sensing, a transceiver, a processor and the most vital unit, i.e. power unit. According, to the required some additional parts can be implemented in the sensor node, i.e. a GPS unit or mobiliser, etc. The storage part of the sensor node is linked to its computing unit (processor), and the communication of the sensor nodes is managed to this unit. Sometimes, the solar cell is equipped for the battery in the power unit.

Environmental surroundings: Sensor nodes should be able to withstand all the conditions as these nodes can be deployed in the remotest of the areas, forests, oceans, battlefield, can be placed on animals, etc.

Transmission media: The sensor nodes are connected to one another through wireless medium. This connection is formed by radio, infrared or the optical form.

Security: As these wireless sensor network can be used in sensitive areas, so, it is important for the routing protocol to provide the required security for the wireless network, so that, it can combat any external attacks or any unwanted access to the network.

These are some of the factors that should be kept in mind before selecting the routing protocol for the WSN. There are various classification of the routing protocol [3, 4].

3 Material and Methods

3.1 Classification of Routing Protocols

The classification of the routing protocols is given below:

Node centric routing protocol: In this routing protocol, some numerical identifiers are specified to the destination sensor node. This type of transmission is not expected in wireless sensor network.

Data centric routing protocol: In almost every wireless sensor network, the data which is sensed by the sensor nodes is far superior than the sensor node. So, in this type of routing protocol, the main focus in on the data transmission inspite on the collection of the data from particular sensor nodes.

Destination initiated routing protocols: When the origin of the route path is from the destination sensor node, then it is called as destination initiated routing protocol.

Source initiated routing protocol: In this, the source originates the route path and when it has the data, it advertises it.

3.2 Categories of Routing Protocols

There are mainly two categories of the routing protocols in the wireless sensor network i.e. network organisation or protocol operation.

On the basis of network organisation and they are classified as:

Flat routing protocols: It is mostly effective for small-scale networks and the function is divided equally among the sensor nodes. It is mostly implemented in networks that are flat in nature, where all the sensor nodes collect the data and then distributes it. The flat routing protocols mainly consist of three schemes: data centric, forwarding and flooding. In flat-based topology, as we all know that the sensor nodes have roles divided equally and all of them gathers the information. Particular identification to every sensor node is not possible in this type of wireless network because of the huge number of the nodes. Due to this, the base station sends the query to particular group of sensor nodes and then will wait for its response.

Location-based routing protocols: These are used when the location of the sensor nodes is required and it can be achieved by implementing a GPS chip in the sensor node these are also called as location aware routing protocols. The location required in this routing protocol is obtained by using some localisation mechanism. As in some of the wireless sensor network, the information of the position of the sensor node is a must so all of these sensor nodes are equipped with localisation devices. By using infrared, radio or acoustic signals, there exist various technique for location sensing mainly based on proximity. Depending upon the cost, complexity of deployment and range these localisation techniques differ. In small WSNs, flooding is used for spreading the base station location to every sensor node in the network. While in large WSNs, this is undesirable especially when there is multitude of sinks and sources. For delivering data from source node to its destination node, these routing protocol uses greedy forwarding approach (Fig. 2).

Hierarchical routing protocol: Sensors nodes are clustered together into different groups. It is by far the most energy efficient routing protocols. The information

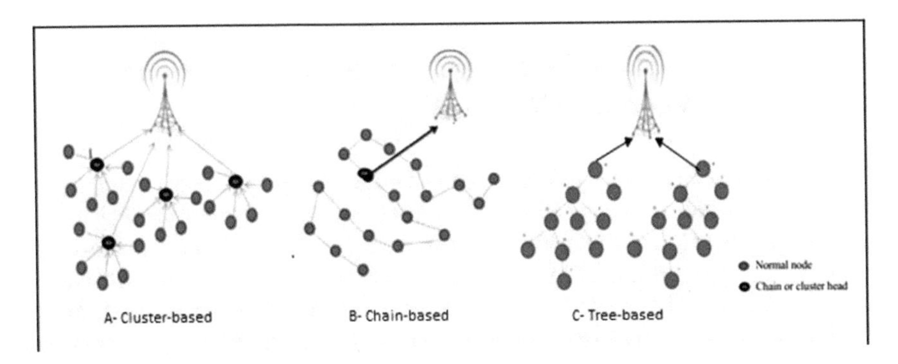

Fig. 2 Types of hierarchical clustering

received from the sensor node is aggregated reducing its energy consumption. Hierarchical routing protocols are further divided into three categories:

Cluster-based hierarchical protocol: In this type of routing algorithm, the sensor nodes are separated into different clusters, and every cluster has a cluster head (CH) and it communicates with the sink node and base station and the rest of the sensor nodes are termed as member nodes.

Chain-based hierarchical protocol: In a chain routing algorithm, the sensor nodes are fashioned in chained topology and among the sensor node one node will behave as a cluster head which will contact with the sink node or the base station.

Tree-based hierarchical protocols: In a tree-based routing protocol, all the sensor node are arranged in a tree like topology and all the sensor nodes send its data to its parent.

Cluster-based routing protocol is widely used because of its routing protocol are more energy efficient and is more scalable [2]. Here, are some of the recent advances in clustering protocols of wireless sensor network:

3.3 An Energy Efficient Clustering Algorithm Based on Residual Energy for Wireless Sensor Network (IECA) [5]

Based on spectral classification, a new method for clustering is introduced in this paper for improving the energy consumption rate in the wireless sensor network. Here, the k ways method is used, and cluster head is selected by a new proposed technique. In this routing algorithm, they specify cluster head after determining the clusters and each sensor node is given a unique identifier. Now, sink node or base station is placed far from the network, and every node has a GPS attached to it. The cluster head is elected on the basis of their ID and the residual energy of the sensor nodes.

3.4 Wireless Routing Clustering Protocol Based on Improved LEACH Algorithm (N LEACH) [6]

In this improved LEACH algorithm, the cluster head is appointed depending upon the remaining residual energy and the position of the sensor nodes and the probability of the advanced sensor nodes and remaining motes for cluster head election is done through the SEP algorithm. In this way, the cluster head is more uniformly distributed. As for the transmission phase, a hybrid technique is being implemented for the data communication that is the distance calculated among different sensor nodes or motes from its cluster head and from the sink node or base station. When it is found that the distance of the sensor node is closer to sink node, then it directly makes contact with the base station but if it is found that the distance from sensor node is quite far from sink node or base station then its cluster head will be in contact with base station. In this way, the clustering number are reduced which means less consumption of energy and increment in the lifespan of our entire network, i.e. the wireless sensor network. The result of the simulation shows that the dead nodes are more discretely distributed.

3.5 Energy Efficient and Cluster-Based Routing Protocol for WSN (Improved LEACH) [7]

This is another improved version of the Leach routing protocol, and is split into two parts: selection of the CH and the transmission phase. For the election of cluster head, a timer is used which is based upon the residual energy, its neighbouring sensor nodes, position of the sensor nodes and to the base station or sink node, and as for the data transmission phase, it used single-hop and multi-hop routings for the cluster head to establish communication with the sink node. When compared with the Leach routing protocol, this improved protocol shows that the lifespan of the network is extended by 15%.

3.6 HHCS: Hybrid Hierarchical Cluster-Based Secure Routing Protocol for Wireless Sensor Networks [8]

In this hybrid protocol for secure model based on hierarchical clustering, the nodes are segregated into separate groups known as clusters, and every group includes a cluster head (CH) and the remaining of the sensor nodes behaves as the members (CM) at level one. The simulation shows that this proposed HHCS method is more efficient and secure than the leach protocol.

3.7 Q Leach: An Energy Efficient Cluster-Based Routing Protocol for Wireless Sensor Networks [9]

This proposed Q leach routing protocol is a mixture of location-based and hierarchical routing algorithm. In this algorithm, restricted flooding concept is used to its full extent in which the sensor nodes located near the destination are used to broadcast the data or packet. Before deciding their advancement toward the destination node, each sensor nodes calculates its distance and forwarding zone information. These sensor nodes will display the packets they are containing and the process will be repeated at each midpoint sensor node until finally arriving at its destination. This

Q leach uses clustering algorithm for blending and data combination execution, so that, it can achieve the energy efficiency by minimising the energy consumption of the sensor nodes hat belongs in the cluster.

3.8 An Energy Aware and Distance-Based Hierarchical Routing Protocol in WSNs (EDPR) [10]

In this routing protocol, the cluster head (CH) should be rotated between all the sensor nodes. Hence, it is split into rounds, and in each round, it starts with the formation of cluster head by keeping clusters organised and the second phase is routing tree phase in which the routing tree is formed and the data which is aggregated is sent to the base station. In energy aware and distance-based hierarchical routing protocol, all the nodes are required to maintain a neighbouring table for data storing of its neighbour which includes the position to the base station and the residual energy.

3.9 Fuzzy C Means-Based Hierarchical Routing Protocol in WSN with Ant Colony Optimisation (FCM-ACO) [11]

This is another hierarchical routing protocol which uses fuzzy c means and ant colony optimisation. By using fuzzy c means, the sensor nodes are partitioned into fixed clusters by the sink node. The selection of the cluster head (CH) is due to its residual energy and the location from the base station or sink node. Data is aggregated by the cluster head after receiving it from its member nodes. By using the ant colony optimiser, an upper chain including cluster heads is established by the sink node and a super leader (SL) is also elected. The flow of data is through the upper chain which is merged at every sensor node except at the terminal ones to the super leader which transmits to the sink node.

3.10 Multilevel Hierarchical Clustering Protocol for Wireless Sensor Networks (MLHC) [12]

This is the heterogeneous hierarchical routing protocol, which is used to reduce the energy consumption. In multilevel hierarchical clustering, clustering of the wireless sensor nodes is employed more than one time at each round. Following the first level of clustering, the clusters developed high-level clustering with one another. At the second level, these second level cluster heads communicate with the sink node. The simulation result shows that this heterogeneous clustering increases the lifespan and scalability of the wireless sensor network (WSN).

3.11 *Energy Efficient Routing in WSN: A Centralised Cluster-Based Approach via Grey Wolf Optimiser [13]*

In this routing algorithm, the cluster heads are selected by the grey wolf optimiser. Grey wolf optimiser is a swarm intelligence algorithm which is based on the behaviour of the grey wolves. for the election of the cluster heads, the present residual energy of the sensor nodes and on the predicted energy rate of consumption. In order to make the network energy efficient, it uses the same clustering in several rounds. Hence, saving the extra energy which is required for the reformation of the clusters. It also uses dual hop for the cluster heads which are located far from the sink node and uses the single hop for the rest of the sensor nodes. It results in saving the energy rate of consumption of the wireless sensor network.

3.12 *ECH: An Enhanced Clustering Hierarchy Approach to Maximise Lifetime of Wireless Sensor Networks [14]*

This is an enhanced clustering approach, in which sleeping waking approach is used for overlapping data and neighbouring sensor nodes by which the redundancy in data is minimised which results in increasing the network lifetime. As the typical hierarchical routing algorithms, uses all sensor nodes for the data collection and transmission, it only uses the waking sensor nodes for this. This routing protocol is implemented in both homogeneous and heterogeneous network.

3.13 *A Heterogeneous Nodes-Based Low-Energy Adaptive Clustering Hierarchy in Cognitive Radio Sensor Network [15]*

In this routing algorithm, the base station is held responsible for updating the information globally which includes the ideal number of the clusters required and its average radius and then the sink node broadcasts it. After receiving this broadcasted message, each and cognitive node will calculate its competition radius, and after that, the competition for cluster head starts based upon the proposed rules. To optimize the distribution of the cluster heads across the wireless sensor network, the elected cluster heads are examined targeting the ideal number of clusters. The results show that the HLEACH is not only energy efficient but also balances the distribution of cognitive nodes (CN) between clusters which leads to prolong the wireless sensor network lifetime.

4 Results and Discussion

Here, the results are summarised in a tabular form (Table 1).

One thing to be notice here is that the criteria for the selection of the cluster head mostly based upon the residual energy or the distance from the sink sensor node and

Table 1 Recent advances in the routing protocols

Clustering algorithm	Nodes Capability	Cluster head selection	Standard routing	Other characteristics	Advantges
IECA	Homogenous	Residual energy, unique identifier	Not mention	Spectral clustering K ways approach	Better lifetime and energy efficient
N LEACH	Heterogeneous	Initial value of energy for sensor nodes	Leach, SEP	Not mention	More energy efficient than LEACH
IMPROVED LEACH	Homogenous	Residual energy, distance from base station	Leach	Not mention	Network lifetime extends about 15%
HHCS	Homogeneous	Residual energy	Not mention	Not mention	Consume less energy than Leach and provides security
Q LEACH	Homogeneous	Initial energy	Q-DIR, Leach	Not mention	Q LEACH consumes 57% less energy for sending packets than leach
EDRP	Homogeneous	Residual energy, distance from other nodes and sink node	EAP	Not mention	Better energy consumption than leach and EAP
FCM-ACO	Homogeneous	Residual energy, distance from B.S	Leach, PEGASIS	FCM, ACO	FCM-ACO outperforms LEACH, LEACH-C, PEGASIS and DFCP
MLHC	Homogeneous	Depend upon the no. of optimal clusters	LEACH	None	Better performance than Leach

(continued)

Table 1 (continued)

Clustering algorithm	Nodes Capability	Cluster head selection	Standard routing	Other characteristics	Advantges
Centralised cluster-based with GWO	Homogeneous	Residual energy, predicted energy consumption	Not mention	Grey wolf optimiser	Improved first node dead in all scenarios by 18% to 66%
ECH	Implemented for both heterogeneous and homogeneous	Depend upon the no. of optimal clusters as func. of the scalability, following the invariable distribution of waking sensor nodes in the area of wireless sensor network	Not mention	Sleeping waking mechanism for overlapping and neighbouring nodes	More energy efficient than other routing protocol
HLEACH	Heterogeneous	Mostly, CNs are selected for CH as they have more initial energy	Leach	Cognitive radio sensor network	In 1280th round, the no. of clusters in the network has been fallen to 6 and the dead CNs and SNs are evenly distributed in the HCRSN

there should be new advances in the election of the cluster head as most of the new routing protocols also uses the same old method for the selection of the cluster head.

5 Conclusion

In this paper, we have briefly studied different advances in the hierarchical clustering routing algorithm. We have studied different attributes of the clustering algorithm. Different routing protocols have been implemented by the researchers but for the selection of cluster head almost all of them consider the residual energy and the distance from the base station. There should be more study on the election of the cluster head in the network as it is one of the major parameter of any wireless sensor network and the comparison on the position of the base station or sink node outside the sensing area and at the centre of the sensing area.

References

1. Akyildiz IF, Su W, Sankarasubramaniam Y et al (2002) A survey on sensor networks. IEEE Comm Magazine 40(8):102–114
2. Matin MA, Islam MM (2012) Overview of wireless sensor network, wireless sensor networks—Technology and protocols. Mohammad A. Matin, IntechOpen, September 6. https://doi.org/10.5772/49376. Available from: https://www.intechopen.com/chapters/38793
3. Heinzelman WR, Chandrakasan A, Balakrishnan H (2000) Energy-efficient communication protocol for wireless microsensor networks. In: Proceedings of the 33rd Annual Hawaii International Conference on System Sciences, vol 1, p 10
4. Ameer Ahmed Abbasi MY (2007) A survey on clustering algorithms for wireless sensor networks. Computer Comm, 2826–2841
5. Jorio A, Elbhiri B (2018) An energy-efficient clustering algorithm based on residual energy for wireless sensor network. In: 2018 Renewable Energies, Power Systems &Green Inclusive Economy (REPS-GIE), pp 1–6
6. Yang T, Guo Y, Dong J, Xia M (2018) Wireless routing clustering protocol based on improved LEACH algorithm. In: 2018 IEEE International Conference on RFID Technology & Application (RFID-TA), pp 1–6
7. Zhao H, Zhou W, Gao Y (2012) Energy efficient and cluster based routing protocol for WSN. Eighth International Conference on Computational Intelligence and Security 2012:107–111
8. Deepa C, Latha B (2014) HHCS: hybrid hierarchical cluster based secure routing protocol for Wireless Sensor Networks. In: International Conference on Information Communication and Embedded Systems (ICICES2014), pp 1–6
9. Gnanambigai J, Rengarajan N, Anbukkarasi K (2013) Q-Leach: An energy efficient cluster based routing protocol for Wireless Sensor Networks. In: 2013 7th International Conference on Intelligent Systems and Control (ISCO), pp 359–362
10. Mazinani SM, Homayounfar B, Mazaheri MR (2012) An energy aware and distance based hierarchical routing protocol in WSNs. In: 2012 20th Telecommunications Forum (TELFOR), pp 1767–1771
11. Ghosh S, Mondal S, Biswas U (2016) Fuzzy C means based hierarchical routing protocol in WSN with ant colony optimization. In: 2016 2nd International Conference on Applied and Theoretical Computing and Communication Technology (iCATccT)
12. Ateya AA, Sayed MS, Abdalla MI (2014) Multilevel hierarchical clustering protocol for wireless sensor networks. International Conference on Engineering and Technology (ICET) 2014:1–6
13. Daneshvar SMMH, Alikhah Ahari Mohajer P, Mazinani SM (2019) Energy-efficient routing in WSN: A centralized cluster based approach via grey wolf optimizer. IEEE Access 7:170019–170031
14. El Alami H, Najid A (2019) ECH: An enhanced clustering hierarchy approach to maximize lifetime of wireless sensor networks. IEEE Access 7:107142–107153
15. Pei E, Pei J, Liu S, Cheng W, Li Y, Zhang Z (2019) A heterogeneous nodes-based low energy adaptive clustering hierarchy in cognitive radio sensor network. IEEE Access 7:132010–132026

COVID-19 Warning System

N. J. Anasuya, B. N. Yashasvi, Shubam S. Uttarkar, and T. R. Supreeth

Abstract The current pandemic has compelled everyone, at all levels, to make a quick and difficult reversal. Our lives are significantly influenced and altered, particularly in terms of how we behave and connect in various facets of our professional and personal life. For example, whilst the peak of the pandemic's first wave in Europe, people delayed seeking medical help out of fear of becoming infected, a fear that has yet to be completely dispelled. We are still in the midst of the pandemic, unsure when or how it will end. The paradigm is also horribly, extremely attention-grabbing when it comes to research projects within the framework of this scientific journal. The creation of research project processes can take a lengthy period in some cases. For example, it is usual for a project's look to take two months (or more) to develop, then six months (or more) to review, and finally two or three years to complete once authorised. The direct benefit to society is not always apparent. Sometimes, the outcomes are "just" intermediate stages (little steps) that will be adopted or modified (complemented) into something directly beneficial to society in the future.

Keywords Early warning and rapid response system (EWRRS) · Coronavirus disease · 2019 (COVID-19)

1 Introduction

Since the beginning of the Gregorian calendar month 2019, a completely new coronavirus, COVID-19 [1], has been registered in a cluster of infectious cases primarily characterised by respiratory illness [2]. The explosion has been sweeping the globe. By the conclusion of the Gregorian calendar month, China had eighty-five thousand confirmed coronavirus cases; rumoured verified cases of the coronavirus in the rest of the world had reached nine.8 million, with about forty-nine million deaths. According to recent rumours, COVID-19 caused a five-hitter in terms of morbidity. Severe instances can have a much greater morbidity [3], ranging from twenty seconds

N. J. Anasuya (✉) · B. N. Yashasvi · S. S. Uttarkar · T. R. Supreeth
Don Bosco Institute of Technology, Bangalore, India
e-mail: anasuyaprakash@dbit.co.in

A. Kumar et al. (eds.), *Proceedings of the International Conference on Cognitive and Intelligent Computing*, Cognitive Science and Technology,
https://doi.org/10.1007/978-981-19-2350-0_30

to sixty two. Fever, dry cough, exhaustion, shortness of breath, dyspnoea, palpitation, chest pain or pressure, and myalgia are all symptoms of myalgia., diarrhoea, and other signs and symptoms are common in COVID-19 patients. After ten days, severe symptoms such as shortness of breath and a slowed natural process may appear. Acute metastatic distress syndrome, which necessitates intubation and mechanical breathing, can develop quickly in more severely compressed patients. Any deterioration could lead to septic shock and organ failure. As a result, people with COVID-19 must keep a close eye on their symptoms and progress.

2 Objectives

The COVID-19 framework intends to evaluate perform the task and provide data to facilitate analysis of ongoing process against the SPRP1 and, as a result, the strategy update. The most important goal is to compile and maintain a database of indicators at the global and national levels to enhance strategic planning, operation tracking, and decision-making based on time and evidence, as well as to ensure support and transparency amongst donors, international organisation agencies, and response partners. This will enable United Nations agencies, other international organisation agencies, and partners to track go towards targets, if necessary, correct approach and action. The following are the precise goals:

- Measures important input, output, and result metrics at the COVID-19 response initiatives are being tracked at the global and country levels [4];
- Assess and analyse response efforts in a methodical manner;
- Compare activity results to the pandemic's epidemiological evolution;
- Assist in prioritising response actions and informing decision-making amongst all partners;
- Support and accelerate transparency and information sharing.

3 Structure

3.1 Geographical Scope

The pandemic has an influence on all countries4 around the world, requiring transportation through countries with a wide range of medical specialties, as well as resource convenience and knowledge systems. These countries also have a wide range of political and socioeconomic situations, as well as poor capabilities, conflict, and humanitarian situations. As a result, indicators have been regrouped as follows:

• International level: A set of essential indicators for all countries to monitor international and cross-cutting issues; • Countries: All countries affected by the COVID-19 pandemic; • Priority countries: Countries affected by the COVID-19 pandemic as defined by the GHRP.

3.2 Planning and Observance Desires

Indicators are used to inform decision-making in each of the design and monitoring functions, and they can be categorised as follows:

- Preparedness: The ability to respond in advance;
- Response: a short-term emergency component that focuses on activities;
- Situation: Indicators that are less reactive and provide a snapshot of the situation at a specific point in time, such as country needs that are assessed on a yearly or time period indicators with delayed coverage, as well as indications on a regular basis (e.g. data from national health systems). The influence of the COVID-19 pandemic on routine public health initiatives and scenario analysis is the main focus.

4 Methodology

4.1 Studying Methods and Objectives

This is a stepped-wedge mixed methods study examining the practicable of EWRRS, a multimodal screening programme that is interactive to aid in the early identification of patients who may move from a moderate to a severe or critical illness. The following are the objectives of the in-depth research:

1. To improve the detection of early indicators of coronavirus patients' clinical worsening and provide real-time data to the entire clinical team, giving in a greater awareness of patient deterioration across the team.
2. To use the rapid response system to provide time-to-time and cost-effective treatment recommendations for coronavirus patients, promoting active management of degradation, early detection of beneficial patient situations, and preventing the progression of degradation to a more serious level.
3. Using interactive, multimodular procedures, reduce medical personnel's employment, and risk of activity exposure throughout the pandemic.
4. To use huge knowledge analysis to detect COVID-19 advancement patterns and characteristics, thereby offering an EWRRS example for outbreaks of public

health problems in the future. Furthermore, such a system will almost certainly be expanded to be used as a grading EWRRS to respond to patients who are deteriorating outside of a social unit context.

4.2 *Investigate the Population and Environment*

Patients who have been diagnosed with COVID-19 according to the National Health Commission of China's New Coronavirus Respiratory Illness Prevention and Management Programme are eligible for this study (7th edition). [5] The following are the requirements for inclusion: adult patients (18 and up) with a clear medical history, fever, and/or metabolic symptoms traditional or recurrent white somatic cell count and/or WBC count; high suspicion of novel coronavirus pneumonia imaging characteristics; positive SARS-CoV-2 supermolecule detection and serum-specific antibodies; capacity to produce a signed consent type patients were excluded if their expected lifespan was less than 48 h, if they were pregnant or new, if they had end-stage diseases, or if they had immunological problems.

5 Result

The EWRRS is critical in a number of ways: It serves as an example of future public health outbreaks: an early detection, warning, and response system. It is frequently reborn as an everyday. An early warning system for hospital patients who are encountering emergency situations outside of the Intensive care unit (ICU) [6]. Through extensive data analysis, it aids in the detection of disease (COVID-19) progression trends and characteristics (Figs. 1, 2, 3, 4, and 5).

6 Conclusion

The new virus has led the human being life to challenging stage. As it is widely spreadable decease, prevention is the best medicine. The people need to improve their immunity and stay fit by doing regular physical exercise and maintaining proper food habits (avoiding junk food).

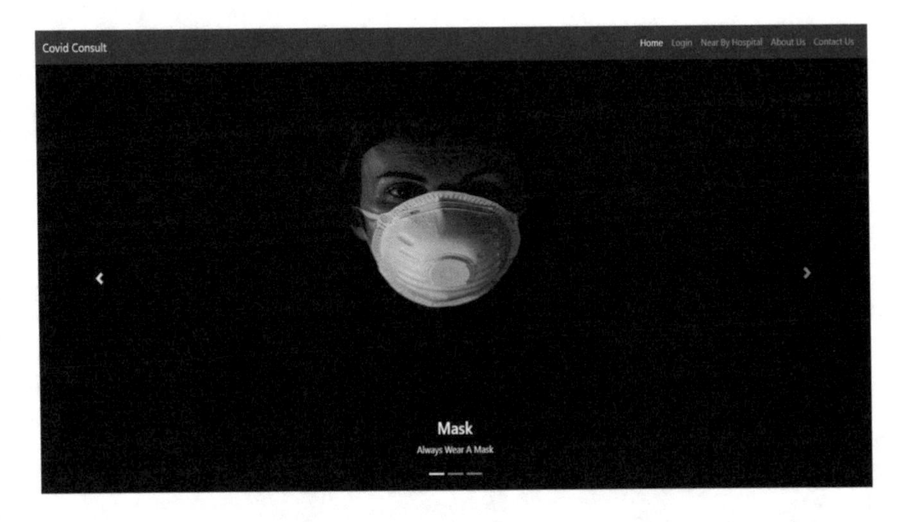

USER INFORMATION

UserName

Email

Mobile Number

Password:

Submit

Fig. 1 Representation of user information page

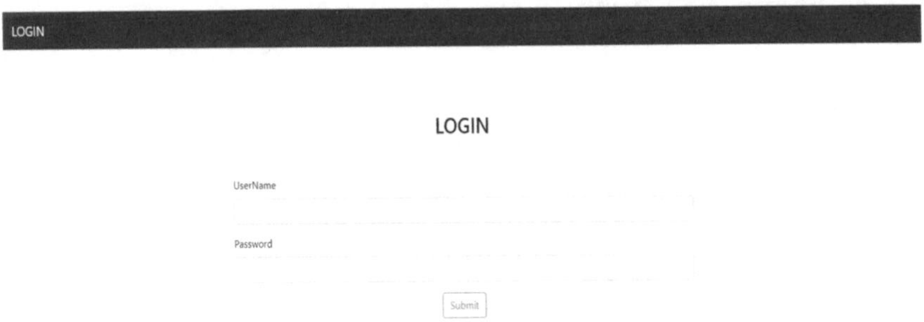

LOGIN

LOGIN

UserName

Password

Submit

Fig. 2 Login page

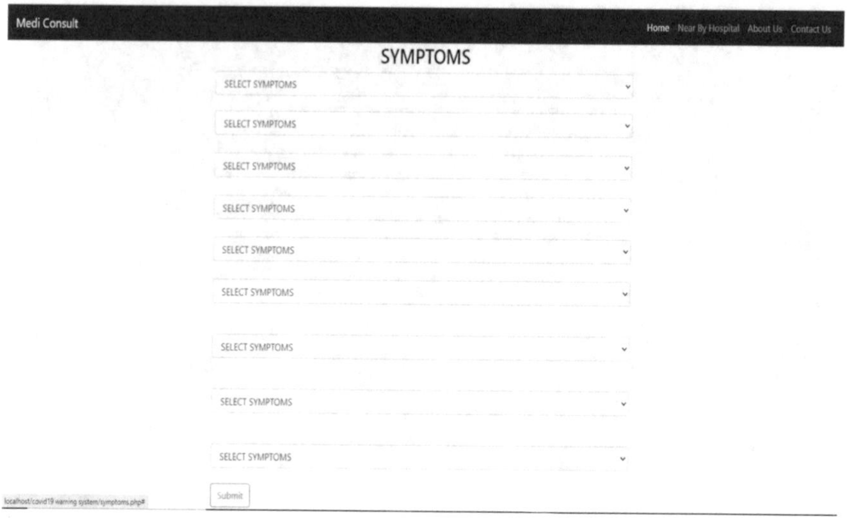

Medi Consult Home Near By Hospital About Us Contact Us

SYMPTOMS

SELECT SYMPTOMS
SELECT SYMPTOMS
SELECT SYMPTOMS
SELECT SYMPTOMS
SELECT SYMPTOMS
SELECT SYMPTOMS
SELECT SYMPTOMS
SELECT SYMPTOMS
SELECT SYMPTOMS

Submit

localhost/covid19 warning system/symptoms.php#

Fig. 3 Symptoms checker page

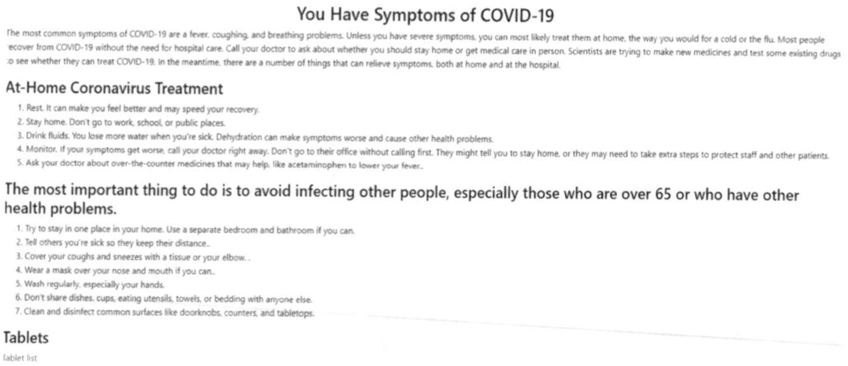

You Have Symptoms of COVID-19

The most common symptoms of COVID-19 are a fever, coughing, and breathing problems. Unless you have severe symptoms, you can most likely treat them at home, the way you would for a cold or the flu. Most people recover from COVID-19 without the need for hospital care. Call your doctor to ask about whether you should stay home or get medical care in person. Scientists are trying to make new medicines and test some existing drugs to see whether they can treat COVID-19. In the meantime, there are a number of things that can relieve symptoms, both at home and at the hospital.

At-Home Coronavirus Treatment

1. Rest. It can make you feel better and may speed your recovery.
2. Stay home. Don't go to work, school, or public places.
3. Drink fluids. You lose more water when you're sick. Dehydration can make symptoms worse and cause other health problems.
4. Monitor. If your symptoms get worse, call your doctor right away. Don't go to their office without calling first. They might tell you to stay home, or they may need to take extra steps to protect staff and other patients.
5. Ask your doctor about over-the-counter medicines that may help, like acetaminophen to lower your fever..

The most important thing to do is to avoid infecting other people, especially those who are over 65 or who have other health problems.

1. Try to stay in one place in your home. Use a separate bedroom and bathroom if you can.
2. Tell others you're sick so they keep their distance..
3. Cover your coughs and sneezes with a tissue or your elbow. .
4. Wear a mask over your nose and mouth if you can..
5. Wash regularly, especially your hands.
6. Don't share dishes, cups, eating utensils, towels, or bedding with anyone else.
7. Clean and disinfect common surfaces like doorknobs, counters, and tabletops.

Tablets

Tablet list

Fig. 4 Treatment page

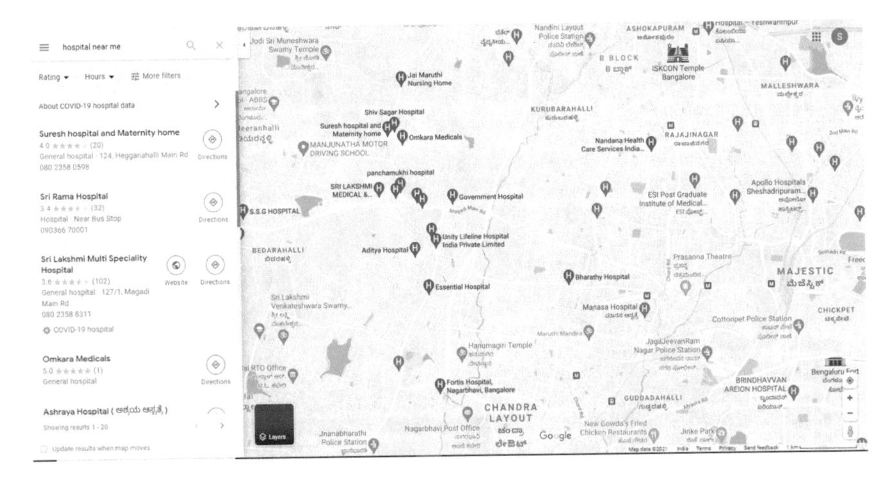

Fig. 5 Nearby hospital page

Acknowledgements The development to the proposed system was extremely supported by Don Bosco Institute of Technology. We present gratitude to our guide Dr. Anasuya N J, Professor, Department of Information Science & Engineering, DBIT. We thank the department and management for providing the resources to complete this work.

References

1. Yang X, Yu Y, Xu J et al (2020) Clinical course and outcomes of critically ill patients with SARS-CoV-2 pneumonia in Wuhan China. China Lancet Respir Med S2213–2600:30079–30085
2. Murthy S, Gomersall CD, Fowler RA et al (2020) Care for critically ill patients with COVID-19. JAMA 323:1499–1500
3. Amarasingham R, Moore BJ, Tabak YP et al (2010) An automated model to identify heart failure patients at risk for readmission or death using electronic medical record data. Med Care 48:981–988
4. Chan A-W, Tetzlaff JM, Altman DG et al (2015) SPIRIT 2013 Statement: defining standard protocol items for clinical trials. Rev Panam Salud Publica 38:506–514
5. Richardson S, Hirsch JS, Narasimhan M, et al (2020) Presenting characteristics, comorbidities, and outcomes among 5700 patients hospitalized with COVID-19 in the New York City Area. JAMA 2020 [Epub ahead of print]
6. Wang D, Hu B, Hu C et al (2020) Clinical characteristics of 138 hospitalized patients with 2019 novel coronavirus-infected pneumonia in Wuhan. China. JAMA 323:1061–1069
7. Fan Z, Chen L, Li J, et al (2020) Clinical features of COVID-19-related liver damage. Clin Gastroenterol Hepatol 2020 [Epub ahead of print]
8. Bangalore S, Sharma A, Slotwiner A et al (2020) ST-segment elevation in patients with Covid-19—a case series. N Engl J Med 382:2478–2480

Literature Survey on Depression Detection Using Machine Learning

Tushtee Varshney, Sonam Gupta, and Lipika Goel

Abstract Depression is a mental illness of the human body that continuously affects human activities such as thinking capacity and physical appearance of the body. The emotional feeling of feeling low and dull toward the situation breaks down the career growth. Psychologists face a major problem in detecting depression at early stages. Patients find them difficult to interact with and share every thought regarding feeling they have. Major depressive disorder is characterized by sadness, worthlessness, disturbed sleeping patterns and eating habits, and lethargy in activities that were once enjoyed. Social sites such as Facebook, Reddit, Twitter, Snapchat, etc., turn out a helpful way to express ideas and negative thoughts to feel free. Many researches have been completed on the dataset to detect depression. It turns out the part of the sentimental analysis by applying the machine algorithm such decision tree, random forest, naïve Bayes, Ensemble model, KNN, maximum entropy, etc. In this paper, the author studied various research to enhance and conclude the best algorithm and high accuracy, precision, and recall of depression detection.

Keywords Depression · Machine learning · Social sites

1 Introduction

Depression accounts for continuous negative feelings of worthlessness, low mood, and loneliness causing a change in mood and behavior. People with depression have long-term lasting unceremonious issues in mood changes like anger, anxieties, apathy, and loss of pleasure of doing work behavior changes such as irritability, being restless, and less social interaction. Sleep issues like an awakening, restless sleep, excess sleep and insomnia. Body effects like fatigue loss of appétit and cognitive

---

T. Varshney · S. Gupta (✉)
Ajay Kumar Garg Engineering College, Ghaziabad, India
e-mail: guptasonam6@gmail.com

L. Goel
Gokaraju Rangraju Institute of Engineering and Technology, Hyderabad, India

A. Kumar et al. (eds.), *Proceedings of the International Conference on Cognitive and Intelligent Computing*, Cognitive Science and Technology,
https://doi.org/10.1007/978-981-19-2350-0_31

issues like slowness in activity thoughts of suicide and repeatedly expanding time over thoughts.

Depression affects the person with sadness, worthless and disturbed sleeping patterns, change in appetite, and distraction from activities which creates an unproductive environment that results in a negative environment. Depression not only affects the adults but also affects the youngster which is working in their career growth. Factors playing a major role in depression are.

- Biochemical changes and neurotransmitters in the brain
- Genetics or family history
- Personality traits as people low confidence are at higher risk
- Environmental conditions like the continuous negative environment of violence and abuse.

Also, the COVID-19 pandemic landed up a large number of people in depression due to indoor lifestyles leading to less bodily interaction and less outdoor activities and more use of social media platforms and virtual mode with increased substances abuse like drugs, alcohol, smoking, etc. People reported anger issues due to built-up frustration, low mood, anxieties, and anger.

Studies done by WHO in depression result that due to pandemic of COVID-19 has given rise to psychotic disorder in approx. 5% of the adult population. Engagement in the drug activities like smoking is the result of lockdown. The loss of jobs and family issues resulted in a person with the loss of daily productive life. Detection of depression at minor stages protects the individual from many severe issues like scars, breathing issues, self-burning, and suicide. The less human interaction to the real world highly interacts them to virtual social media to share their feeling. People find themselves more comfortable and reliable to interact on the smartphone and this causes the use of social media like Reddit, Twitter, and Facebook. This raise in the use of social media attracted psychologists to interact with their patients and to deeply study their patient's mental conditions. The psychotic disorder needs the deep study of the patient thoughts and emotions.

The technology of knowing text emotionally is called the sentimental analysis of machine learning. Industry uses sentimental analysis in the receiving the users view over product, marketing of the product, and in medical, this help in knowing the polarity of positiveness and negativeness of the thoughts in the emotional expression. The text consists of words and is formed as sentences and collecting useful information from those big data using the libraries, and APIs are part of artificial intelligence. Using natural language processing, analysis of text and statistics is a major part of sentimental analysis. Text mining helps in predicting the human mind. Emotion detection sentiment analysis is used to segregate emotion behind the text. The advantage behind using the emotion-based analysis helps to know why the person feels this kind of thoughts toward the particular situation.

This survey paper is organized as follows: Sect. 2 contains the terminology words that help in identifying the domain, API, and field of machine learning; Sect. 3 gives a brief idea of the process to collect paper for the survey; Sect. 4 consists of the dataset used for different researches. Section 5 consists of the algorithm used by researchers in their studies. Section 6 consists of the result concluded by the researcher in their researches and comparison based on the mathematical terms accuracy, precision, and recall. Sections 7 and 8 consist of the discussion and conclusion from the survey done, respectively.

2 Terminology

Key helped the researcher to study depression detection using machine learning.

2.1 **Domain**: Social sites is the vast area that consists of emoticons, images, paragraphs, and memes to relate to emotional feeling. Review studies done signify that text is the major area covered the social user to express thoughts.

2.2 **Sentimental analysis**: Machine learning techniques to study the hidden emotion behind the text. Techniques used to identify the polarity of text from positive to negative. It includes both techniques machine learning and natural language processing.

2.3 **API**: It stands for application programming interface. It is the function that creates the application for interacting with different applications to process together. In researches, various APIs are used to extract useful information from social media and sites.

3 Systematic Literature Review

For the studies done before the literature survey on depression detection, the author concluded to some questions.

R Q 1. Which algorithm resulted in better accuracy of machine learning?

The algorithm applied to the text to detect depression results to accuracy.

R Q 2. Which social site is mostly used by people to express themselves?

As many social sites are used in this era, from this study, author tried to find out the most used social sites.

3.1 **Fact-Finding Process**: The author surveyed for this paper from search depression detection on Google Scholar engine.

3.2 **Sources**: Used IEEE, Springer, Google Scholar science direct for the collection of paper for study

3.3 **Study Method Criteria**: For this paper, the collection of data is done from the duration of 2019–2021 which involves the latest research in this field to get the knowledge of researches in this field.

3.4 **Research focus**: Research is done on the most used machine learning algorithm to detect depression and to know the various answer to question to enhance the knowledge of detecting depression from social sites. Below, Table 1 contains the paper collection of paper.

Table 1 Paper collection

Key	Title	Author
P1	Machine learning based on approach for detection of depression using social media using sentiment Analysis [14]	Geeta Tiwari, Gaurav Das
P2	Depression detection using machine learning [11]	Mr. P.Y. Kumbhar, Dr. Rajendra Dube, Mr. Sudhakar Barbade, Ms. Gayatri Kulkarni, Ms. Nikita Konda, Ms. Meghana Konkati
P3	A textual-based featuring approach for depression detection using machine learning classifiers and social media texts [6]	Raymond Chiong, Gregorius Satia Budhia, Sandeep Dhakal, Fabian Chiong
P4	Depression detection using machine learning techniques on Twitter data [9]	Govindasamy K, Palanichamy N
P5	Combining sentiment lexicons and content-based features for depression detection [5]	Raymond Chiong, Gregorious Satia Budhi, Sandeep Dhakal
P6	Real-time acoustic-based depression detection using machine learning techniques [15]	Bhanusree Yalamanchili; Nikhil Sai Kota; Maruthi Saketh Abbaraju; Venkata Sai Sathwik Nadella; Sandeep Varma Alluri
P7	Analysis of machine learning algorithms for predicting depression [10]	Purude Vaishali Narayanrao; P. Lalitha Surya Kumari
P8	Machine learning-based approach for depression detection in Twitter using content and activity features [3]	Hatoon AlSagri, Mourad Ykhlef
P9	Depression detection using machine learning [13]	K. Sudha, S. Sreemathi, B. Nathiya, D. RahiniPriya
P10	Using Twitter social media for depression detection in the Canadian population [12]	Ruba Skaik, Diana Inkpen
P11	Prediction of postpartum depression using machine learning techniques from social media text [7]	Iram Fatima Burhan Ud Din Abbasi, Sharifullah Khan, Majed Al-Saeed, Hafiz Farooq, Ahmad Rafia Mumtaz
P12	Mining Twitter data for depression detection [4]	Priyanka Arora, Parul Arora
P13	Depression detection and prevention system by analyzing tweets [8]	Jayesh Chavan, Kunal Indore, Mrunal Gaikar, Rajashree Shedge
P14	Detecting Arabic depressed users from Twitter data [2]	Salma Almouzini, Maher khemakhema, Asem Alagee

4 Dataset Taken for Studies

The different social sites are used by the researchers to collect the datasets and study them to detect depression. Table 2 is the summary of the dataset studied in the survey.

The survey concluded that the Twitter dataset is the most studied social site used for detecting depression.

Table 2 Dataset table

Keys	Datasets
P1, P2, P4, P5, P7, P8, P10, P12, P13, P14, P15	Twitter
P3	Facebook, Reddit, and Electronic diary
P6	Audio samples and video sample
P9	Questionnaire
P11	Reddit

5 Methods Taken for Studies

According to the survey, many algorithms of machine learning are used as classifier methods. The collection of methods from various papers contains the different algorithms as given in Table 3.

5.1 **Logistic Regression** is a statistical analysis of data. Machine learning approach where input data is classified based on the existing data. The relationship between the one or more independent data predicts the dependent data variable.

Table 3 List of classifier methods

Keys	Classifier methods
P1	Decision tree, K-nearest neighbor, ensemble model, and support vector machine
P2	Naïve Bayes, support vector machine, decision tree, random forest, and K-nearest neighbor
P3	Logistic regression, support vector machine, multilayer perceptron, decision tree, ensemble model, random forest, adaBoost, bagging predictors, and gradient boosting
P4	Naïve Bayes and Naïve Bayes tree
P5	Support vector machine, logistic regression, multilayer perceptron, decision tree ensemble model, bagging predictors, Random forest, adaBoost, and gradient boosting
P6	Logistic regression, support vector machine, random forest
P7	Support vector machine, K-nearest neighbor, ensemble and tree ensemble
P8	Support vector machine, decision tree, and Naïve Bayes
P9	Decision tree, Naïve Bayes, K-nearest neighbor, random forest
P10	Support vector machine, logistic regression, random forest, gradient boosting algorithm, gradient boosted decision tree, and eXtreme gradient boosting
P11	Support vector machine, multilayer perceptron, and logistic regression
P12, P13, P15	Naïve Bayes and support vector machine
P14	Random forest, Naïve Bayes and adaBoostM1, and LIBLINEAR

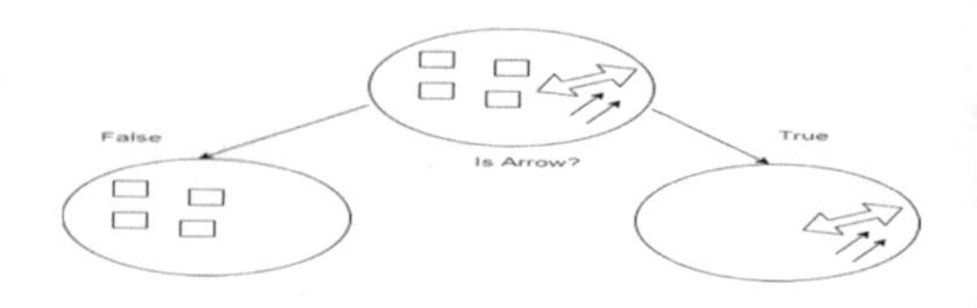

Fig. 1 Example of decision tree

5.2 **Support Vector Machine** is introduced by Vladimir Vapnik at AT and T Bell Laboratories. Used for classification and regression of data. Here the dataset is divided into two classes and based on the hyperplane design, the result is concluded.

5.3 **Naive Bayes** is the probabilistic model where the Bayes theorem is used to classify the dataset based on the dependent and independent variables.

5.4 **A decision tree** is a tree-based study done over the dataset where the dataset is separated based on the outcomes of the test and node containing the dataset (Fig. 1).

E.g. Is the figure arrow?

5.5 **Random Forest** is the most popular algorithm of machine learning where the complex problem is solved by creating a bunch of decision trees used for both purposes as classification of data and regression of data.

5.6 **MultiLayer Perceptrons** is a deep learning approach where data is passed through the neural network model which is made of the three-layer input layer, hidden layer, and output layer. The hidden layer is the mid-layer of the model that comprises many layers to give the best accurate result.

5.7 **K-Nearest Neighbor** algorithm is introduced by Evelyn Fix. Here datasets are classified based on the votes after applying some classification methods. Distance calculated mathematician formulas such as Euclidean, Manhattan, Minkowski, and Hamming distances are applied.

5.8 **Gradient Boosting** is used in both regression and classification. In this, present base learner is more effective than the previous learner. In this, misclassifies weights are not increased but the optimization of the loss function is done to overcome the error from the previous learners for computing prediction.

5.9 **Gradient Boosted Decision Tree** is a technique of ensembling the weak decision tree for giving more results.

5.10 **eXtreme Gradient Boosting** algorithm is used for increasing the accuracy, efficiency, and performance of the model.

5.11 **Ensemble Model** is a combination of various learning algorithms to obtain better predictive as a comparison to another constituent algorithm.

5.12 **Adaptive Boosting** is the sequential-based classification of data where the wrong classified data is again passed through the model to identify whether the result is misclassified.

5.13 **Bagging Predictors** is the predicting-based model where various versions are created based on multiple bootstrap replicates of the learning algorithm.

5.14 **Tree Ensemble Model** where data is classified based on the tree structure where various trees are created and combined to get better accuracy.

5.15 **LIBLINEAR** is the large linear classification SVM algorithm where minimizes a multivariant function by solving univariate problems.

6 Comparison of Baseline Method Proposed in Studies

In this section, we study the outcome of various classifier algorithms that are applied to the dataset taken by the author in their researches. Table 4 contains all the results of the review collection.

As the comparison is done by involving the literature survey of this paper, it can be visually concluded that the SVM has taken the maximum accuracy in the result on different types of datasets followed by other machine learning algorithms (Fig. 2).

Table 4 Result of method

Key	Result				
	Classifier	Accuracy	Precision	Recall	F1
P1	Decision tree	73.00%			
P2	Naïve Bayes	98.55			
P3	Logistic regression	92.61%	93.32%	72.21%	
P4	Naive Bayes and Naïve Bayes tree	Both accuracy 90–95% on a dataset basis			
P5	Ensemble, gradient boosting	98%		94%	
P6	Support vector machine	90%	95%	93%	94%
P7	Support vector machine, K-nearest neighbor, Ensemble, and tree Ensemble	~ 90%			
P8	Support vector machine	82.5%	73.913%	85%	79.06%
P9	Decision tree	98.24%			
P10	XGBoost	96.40%	0.956%	96.00%	95.80%
P11	Support vector machine	80.70%	82.85%	80%	
P12	Support vector machine	78.80%			
P13	Support vector machine and Naïve Bayes hybrid	85%			
P14	LIBLINEAR	87.50%		87.50%	

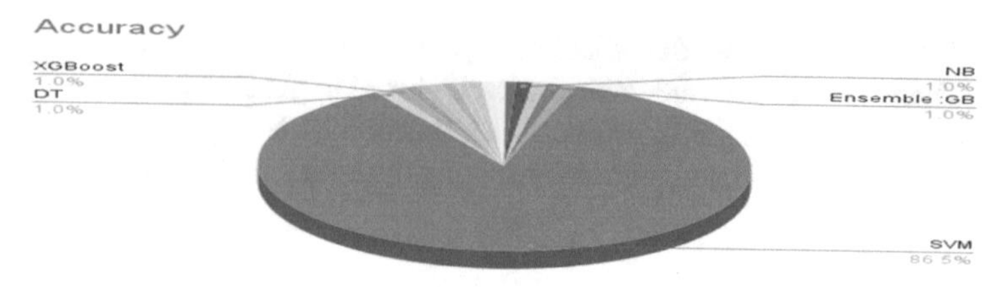

Fig. 2 Accuracy of algorithm

7 Discussion

A machine learning technique called sentimental analysis is used for studying the textual-based emotion of the person. Applying various machine learning algorithms such as support vector machine, Naïve Bayes, decision tree, random forest multilayer perceptron, etc., are used by the various researchers in their studies. Applying algorithms on various datasets collected from different social sites such as Twitter, Reddit, Facebook, and LiveJournals and other Web sites help in the emotional study. According to the survey done, many authors used small sample sizes for their research and variation in mathematical terms such as accuracy, sensitivity, recall, and precision. According to the framework studied, the identification of the techniques for preprocessing of dataset, classifier algorithm can be done so that the better accuracy of the model can be achieved and detection of depression at early stages can be done more accurately and precisely.

8 Conclusion

After the survey was done, it has been concluded that the people willingly, without restraint, and second thought share their feeling to feel free. Social sites such as Twitter, Reddit, and LiveJournals are the most popular way and various research has been done using machine learning algorithms and better models with good accuracy can be created.

References

1. Al Asad N, Pranto MAM, Afreen S, Islam MM (2019, November). Depression detection by analyzing social media posts of user. In: 2019 IEEE international conference on signal processing, information, communication and systems (SPICSCON), pp 13–17. IEEE
2. Almouzini S, Alageel A (2019) Detecting Arabic depressed users from Twitter data. Procedia Comput Sci 163:257–265
3. AlSagri HS, Ykhlef M (2020) Machine learning-based approach for depression detection in Twitter using content and activity features. IEICE Trans Inf Syst 103(8):1825–1832

4. Arora P, Arora P (2019 March) Mining Twitter data for depression detection. In: 2019 international conference on signal processing and communication (ICSC), pp 186–189. IEEE
5. Chiong R, Budhi GS, Dhakal S (2021) Combining sentiment lexicons and content-based features for depression detection. IEEE Intell Syst 36(6)
6. Chiong R, Budhi GS, Dhakal S, Chiong F (2021) A textual-based featuring approach for depression detection using machine learning classifiers and social media texts. Comput Biol Med 104499
7. Fatima I, Abbasi BUD, Khan S, Al-Saeed M, Ahmad HF, Mumtaz R (2019) Prediction of postpartum depression using machine learning techniques from social media text. Expert Syst 36(4):e12409
8. Gaikar M, Chavan J, Indore K, Shedge R (2019, March) Depression detection and prevention system by analysing tweets. In: Proceedings 2019: conference on technologies for future cities (CTFC).
9. Govindasamy KA, Palanichamy N (2021, May) Depression detection using machine learning techniques on twitter data. In: 2021 5th international conference on intelligent computing and control systems (ICICCS), pp 960–966. IEEE
10. Narayanrao PV, Kumari PLS (2020, March) Analysis of machine learning algorithms for predicting depression. In: 2020 international conference on computer science, engineering and applications (ICCSEA), pp 1–4. IEEE
11. PY K, Dube R, Barbade S, Kulkarni G, Konda N, Konkati M (2021) Depression detection using machine learning. Available at SSRN 3851975
12. Skaik R, Inkpen D (2020, December) Using twitter social media for depression detection in the Canadian population. In: 2020 3rd artificial intelligence and cloud computing conference, pp 109–114
13. Sudha K, Sreemathi S, Nathiya B, RahiniPriya D Depression detection using machine learning
14. Tiwari G, Das G Machine learning based on approach for detection of depression using social media using sentiment analysis. Depression 9(10):16
15. Yalamanchili B, Kota NS, Abbaraju MS, Nadella VSS, Alluri SV (2020, February) Real-time acoustic based depression detection using machine learning techniques. In: 2020 international conference on emerging trends in information technology and engineering (ic-ETITE), pp 1–6. IEEE

RESTful API Integrations in Telecom Billing System Management

K. K. Deepika, R. S. Ravi Sankar, and A. V. Satyanarayana

Abstract Connectivity is the capability to connect people to every corner of the world using various devices. It gives access to purchase, pin, pick anything from anywhere. But how does the data get from here to there, it is because of application programming interface (API). Over nineteen thousand APIs have been added to the online API directory programmable Web since 2005, and more are added every month. Using RESTful APIs as a major communication, core integrations are performed between telecom application services (customers, credits, payments, invoices) and an accounting application to maintain end-to-end customer billing transactions by creating two-way journal entries. This thesis work is facilitates future scope of RESTful API integrations and their key role in performance factors like customer satisfaction, customer retention, and conversion. A captious literature survey has been conducted on non-academic articles to prove the facts that API development and integrations provide major contribution for improving sales in small-scale as well as large-scale industries.

Keywords API · Python · QuickBooks · Hypertext transfer protocol · Metadata

1 Introduction

ERP is abbreviated as enterprise resource planning which constitutes all the activities of an undertaking or an organization. It is also referred as ERP coordinates inventory and it can be described as system software that efficiently allows the organization to manage and control various activities. It gives the gateway to effectively use the company's resources by connecting all the functional areas of the organization. The key aspect of the system is that all its respective users share a single database in all the departments for implementation of various applications. So everyone has the same inventory-related information. The main motive of enterprise resource planning is to support planning and execution systems, and the decisions they drive

K. K. Deepika (✉) · R. S. R. Sankar · A. V. Satyanarayana
Vignan's Institute of Information Technology, Visakhapatnam, Andhra Pradesh 530027, India
e-mail: kkdeepika@vignaniit.edu.in

A. Kumar et al. (eds.), *Proceedings of the International Conference on Cognitive and Intelligent Computing*, Cognitive Science and Technology,
https://doi.org/10.1007/978-981-19-2350-0_32

throughout your company. For example, a customer order is tracked as it moves through the process, and transactions are automatically updated. This gives people in each part of operations management the information they need to make appropriate decisions. And the system helps to keep track of the inventory needed for that customer order. As software and modules have been added to coordinate logistics, transportation, suppliers, and customers, ERP systems have moved beyond the organizational boundaries allowing planning and control for supply chain management. In fact, over the years, the E in ERP has come to represent, not just the enterprise, but the enterprises supply chain. This allows companies to analyze, monitor, and manage customer inventory throughout the entire supply chain adjusting as problems and issues come up. ERP systems can be used to effectively control inventory in global operations, but it is much harder. As with any successful project, the right input from the right resource must be available at the right time.

As the scope of ERP systems is demandingly increased, it need to be integrated with other CRM applications to provide cross functional solutions. Each CRM software is developed with respective API structure to expose data and applications to other ERP system. The API-led connectivity enables businesses to easily exchange data, main customer relationships, and reduce production costs as well. APIs provide an endpoint or a unequivocal URL where the user wanted to be expose the data or functions. Most API calls need authentication, and the API will have a confidential pin that recognize a particular user or application request. The server can then control or regulate if user is free to contact the data or not. In this project thesis, the CRM tools under cloud platforms are set to group their billing management and accounting principles with the international standards. Telecom companies have changed their operations toward customer management and retention, with enlarging the usage of technology and also various services based on technological characteristics contributing substitute channels such as automation in customer registration, payment, billing entries, wireless networking, and Internet banking.

2 Literature Survey

In a Nordic APIs blog post from 2014 [1], Bruno Pedro lists five benefits APIs will have for a business's revenue. Among these are increasing customer retention and creating "upsell" opportunities. The term "upsell" in this case refers to the act of persuading a user to purchase the required access to one or many features, i.e., customer conversion. In a similar article from 2016 [2], increasing retention and conversion was the number one benefit of adding an API to a service.

The keys to successful API strategies are discussed in an article by Mark Boyd from 2015 [3]. These include, among other things, developer support, developer community size, API discoverability, and tools to make the API easy to use and learn.

In June of 2009, only fifty-five percent of Twitter users directly used the actual service, while the remaining 45% of users used some third-party application connected to the Twitter API [4]. The most used third-party application was Tweet Deck with around 20% of the user base. In 2011, Twitter bought Tweet Deck for around forty million dollars [5].

This literature survey comprises mostly non-academic articles, as the academic literature relating to API development and key performance indicators is limited. It was also conducted while prioritizing customer retention and customer conversion, as these are the key performance indicators of interest. However, works relating to other key performance indicators are included as well.

A systematic literature review was performed on the opted ER system, CRM tools, and cloud computing units. From the obtained results, we can conclude that API research is of a non-theoretical basics and to a great extent focuses on the design and utilization of the technological dimensions. But it neglects many of the social principles like application of API in both business and managerial aspects which are considered to be equitably important.

3 Background

Usage of technological advancements to improve sales and product marketing, coax customers with highly upgraded features proved to be the best strategy.

Critical factors that correlate business success are eye-catching products to customer, unique features, detailed understanding of target projects, reduced costs, product specifications and requirements, etc.

Product marketing will have extended future if the key performance factors—customer satisfaction, customer retention, and customer conversion are regularly measured and applied to aimed project.

Telecom application seamless and configurable integration with QuickBooks accounting software give clients confidence of complete data mapping from users billing system to users accounting system. The integration supports both real-time and scheduled synchronization for.

New created customers to synchronized from telecom application to QuickBooks. All the invoice charges and taxes applied to created customer accounts from telecom application to QuickBooks. Telecom application payments and refunds via same customer account to QuickBooks.

3.1 API Architecture

Specifying the architecture of an API, as they requires an understanding of the purpose and capabilities of the system. While the RESTful API architecture is the most common and is at times powerful, the functionality of this API was too simple to correctly utilize the HTTP verbs as required. In addition to this, the nature of the QuickBooks integration causes the URLs used when calling the API to be hidden inside the QuickBooks. RESTful API's attributes to software architecture design constraints in form of group that bring about productive, steady, extensible, and modular systems. So REST is not a specific technology, but rather a data architecture and design methodology that produces predictable and consistent outputs and behaviors by receiving a set of standard methods called verbs and returning standardized structured data, typically JSON or XML, called the resource. Postman tool is used for testing an API.

3.2 *Task Automation Services*

In addition to the development of the API, an application for the online service iPaaS tool was built. The purpose of this application was to serve as the interface between the user and the API. IPaaS tool connect to any API and build workflows with drag-and-drop ease. This application was responsible for authenticating and authorizing the client. This was done as specified by the impasse tool, redirecting the authenticating client with the necessary URLs. This tool helps in integration between telecom application and QuickBooks application.

4 Requirements

The required functionality of the API described and developed in this work was automatization of survey distribution, which in turn contains the following functional requirements:

- Ability to authenticate the user.
- Ability to fetch surveys from the database.
- Ability to fetch users from the database.
- Ability to verify the user's access to a given survey.
- Ability to send emails given correct arguments.
- Ability to authorize the Zapier integration.
- Ability to authenticate the user via the API.
- Ability to request inputs from the user.
- Ability to call the API using the required data.

5 Design

- Use case 1 customer sync: Synchronize newly created customers from telecom application to QuickBooks application using iPaaS tool. Whenever a new customer is logged in telecom application, same customer should be synchronized to QuickBooks as a new customer account. This will act as prerequisite for loading invoice billing to same customer in QuickBooks.

- Use case 2 invoice sync: Synchronize invoice charges from telecom application to a single QuickBooks GL account using iPaaS tool. Need to synchronize invoice charges from telecom application to GL code mapped QuickBooks products in a scheduled batch process aligned with the accounting close period. All products are onboarded from telecom application to QuickBooks with common GL codes in both applications. After product mapping two-way journal entries should be created for each bill charged against specific product
- Use case 3 payment sync: Synchronize daily payments from telecom application to a single QuickBooks GL account using iPaaS tool. Whenever a new payment is logged in telecom application, same payment should be transferred as a two-way journal ledger entry to QuickBooks application and mapped to respective accounts based on payment type (cash, credit card, manual check). These payments are applied to invoices created for same customer.
- Use case 4 credit sync: Synchronize daily credits from telecom application to a single QuickBooks GL account using iPaaS tool. Whenever a new credit is logged in telecom application, same credit charges should be transferred as a two-way journal ledger entry to QuickBooks application. These credits are to be recorded after invoice creation in telecom application (Figs. 1 and 2).

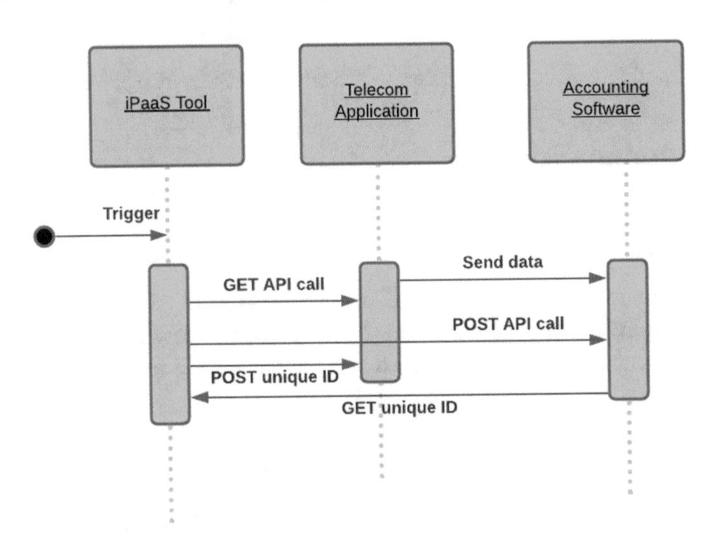

Fig. 1 Sequence diagram depicting the flow of the system

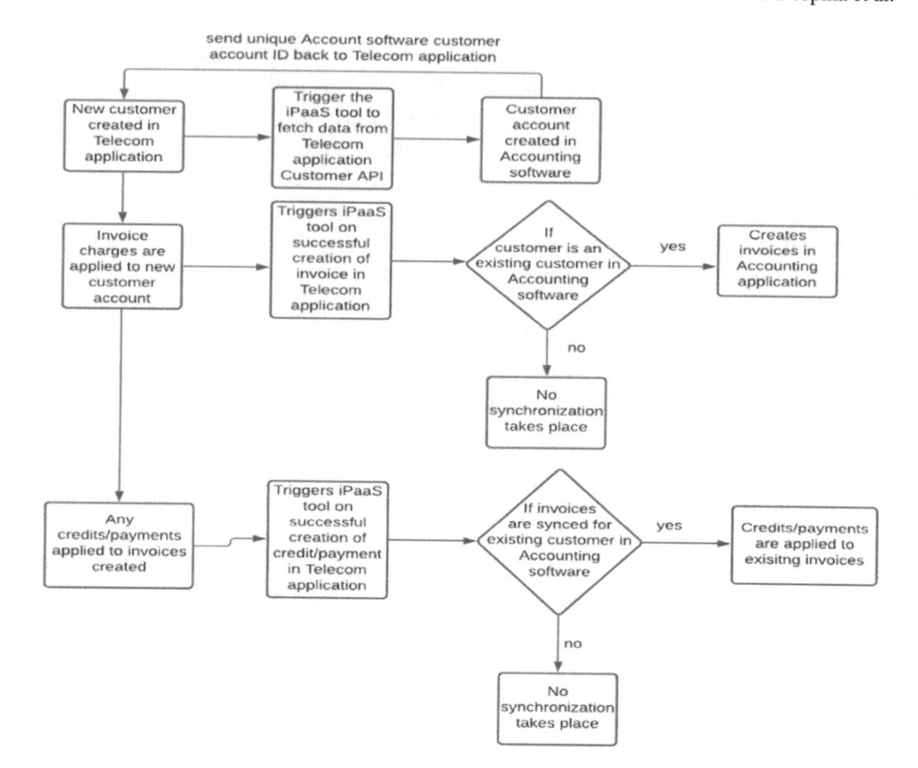

Fig. 2 User workflow process from telecom application to QuickBooks application

6 Results and Discussion

Customer Sync:

- The customer details (name, email-id, billing address, shipping address) are mapped from telecom application to QuickBooks application
- QuickBooks customer id is generated as a unique identification.

Invoice Sync:

- The total invoice amount of three different charges is mapped from telecom application to QuickBooks application.
- A two-way invoice entry is created in QuickBooks application.

Payment Sync:

- A manual cheque payment made by a customer is mapped from telecom application to QuickBooks.

Credit Sync:

- The credit amount is applied to current billing month of customer is mapped from telecom application to QuickBooks application.
- A two-way credit entry is created in QuickBooks application.

7 Conclusions

This paper has focused on the design, development, and evaluation of an API. It is intended to act as a case study for evaluating the effect API development has on a business from a sales and marketing perspective. The API, which is based on an RPC architecture, included one procedure, namely the ability to send surveys via email. In addition, a QuickBooks application was built to be used as user interface for the API is done. API calls for both telecom application and QuickBooks using iPaaS tool and Python are successfully implemented.

OAuth authentications, data mapping, and migration concepts are successfully implemented. In addition to the development of the API, a literature survey has been conducted on works relating to similar technologies and marketing aspects as this one, to highlight the effects API development can have. Works relating to the key performance indicators customer conversion and customer retention have been prioritized, but other works have been added.

Extraction of JSON metadata and data transfer among cloud services using various HTTPS methods and Python packages like urllib, json, etc., are developed using Python code. Automation of B2B, B2C transactions efficiently via cloud services and API integrations are successfully implemented.

References

1. Kuan H-H, Bock G-W, Vathanophas V (2008) Comparing the effectsof website quality on customer initial purchase and continued purchase at e-commerce websites. Behav Inf Technol 27(1):3–16
2. Product Benchmark Report (2017) Mixpanel
3. Santos W (2017, Nov.) Which API types and architectural styles are most used? (Accessed 8 May 2018).Online. Available https://www.programmableweb.com/news/which-api-types-and-architectural-styles-are-most-used/research/2017/11/26
4. Fielding RT (2000) Architectural styles and the design of network-based software architectures. Ph.D. dissertation, University of California, Irvine
5. Architectural styles in web services. IBM. (Accessed 8 May 2018). Online. Available https://www.ibm.com/support/knowledgecenter/en/SSMQ79_9.5.1/com.ibm.egl.pg.doc/topics/pegl_serv_access_overview.html
6. Burnett JJ (2003) Core concepts of marketing. John Wiley and Sons Ltd

Automated Image Captioning Using Machine Learning

P. Puneeth Kumar, Anoopkumar Kulkarni, and Y. V. Sanjay

Abstract The study of machine learning algorithms and processes is suited for each image and language process. The use of existing packages in implementing machine learning algorithms. Implementation of an Associate in Nursing algorithmic programme takes a photograph and explains it in full phrases. A list of domain experience and prospective models was developed after the demand analysis. Following that, a study of the available technologies that may aid with the creation of the appliance was conducted, and the look, the solution's implementation and validation began. A deep convolutional neural network is used to extract features, and transfer learning is combined with a repeating neural network for description generation in this study. The implementation is done with Keras with a TensorFlow backend. It was possible to obtain photo exploitation language is described by a trained model.

Keywords Image captioning · Machine learning · Model

1 Introduction

Over the last few decades, our daily lives have become increasingly reliant on technology. From providing quick answers via search engines to facilitating global, low-cost communication via Internet-based messaging programmes to tackling difficult engineering challenges, the Internet has made it possible.

Artificial intelligence [1, 2] is a field that has been rapidly increasing as a result of the ever-growing knowledge and data, as well as new technology. More and more businesses are attempting to include others try to benefit from the knowledge that may be acquired using these approaches by incorporating algorithms for data science and machine learning into their product development. [1].

P. P. Kumar (✉) · A. Kulkarni · Y. V. Sanjay
Don Bosco Institute of Technology, Bengaluru, India
e-mail: parthapuneeth@gmail.com

A. Kumar et al. (eds.), *Proceedings of the International Conference on Cognitive and Intelligent Computing*, Cognitive Science and Technology,
https://doi.org/10.1007/978-981-19-2350-0_33

In artificial intelligence, the algorithm type that is used, sometimes referred to as deep learning or machine learning, the algorithm employed determines this. Applications range from gaining insights into a company's from assisting in the construction of better detecting items from a satellite image using guiding systems (GPS applications and maps), we have got you covered.

The list could go on and on, ranging from translation to text-to-speech, there is something for everyone software, we have got you covered [2].

Computer science, mathematics, and statistics are the three disciplines that make up computer science most prominent subjects involved in the development algorithms for machine learning. The method distinguishes deep learning algorithms and machine learning algorithms are two different types of algorithms. Deep learning employs neural networks, which have enormous computational capacity but also necessitate a large the quantity of data and powerful technology to produce beneficial outcomes. Machine learning is a subclass of it, as well. Its basis is the creation of self-learning neural networks. Machine learning algorithms, on the other hand, are regarded traditional data-driven algorithms that attempt to learn from it so that they can use it in new ways. In the event where an algorithm for machine learning produces a poor forecast, the designer must alter retrain the model by changing the parameters if there is a deep learning algorithm that can tell if a forecast is excellent or bad. [3]. Hyperparameter tweaking is the term for this step.

It is necessary to feed good data to an algorithm in order for it to be accurate. The datasets clean toys are generally available for toy crafts and require little or no adjustment, whereas in most cases, real-life data are not available, dirty and necessitates extensive cleaning and normalisation. This stage is known as data preparation. It might consume up to 80% of a machine learning engineer's time work to complete. When it comes to unstructured information, the possibilities are endless like photos and text, even publicly available datasets must be changed to meet the neural network's inputs or standards. [4].

Data can be extracted from nearly anything in their environment. When it hears a noise, for example, it has the ability to express itself in natural language. Machines, on the other hand, are unable to accomplish this. Every sense that humans have can be put to use in this way. Consider what would happen if one of those senses was absent. When it comes to printed information, deaf persons can benefit from machine learning programmes that use to "read" to them, a text-to-speech technology will be used.

Images can also be utilised to extract data. When a person is shown a floral image, she or he will be able to recognise the thing since she or he has seen a flower previously.

However, if that individual is blind, they will be unable to perceive that image at all. What if they could acquire a portion without having to look at the image or have it described to them? Of the information without having to look at the image or be told what it means? In this case, machine learning has the ability to be useful [5].

The purpose of this research is to extract useful textual data in the form of short descriptions from a variety of sources, photographs. To ensure complete sustainability, the A text-to-speech engine is a programme that converts text into speech can be used to read the results. This manner, people with this impairment could have a fully independent experience. These people may feel left out as a result of the development of social media interactions, especially when their relatives and close friends post photos on the internet.

The rest of the paper is formatted as follows: A field is displayed in Sect. 2 research as well as various current systems, Sect. 3 presents the proposed notion, and Sect. 4 presents the conclusions.

2 Work on the Subject

A top-down and bottom-up attention device that work together are provided in this study, allowing to pay attention to estimated at the level of objects and other types of data prominent visual regions, which serves as a natural base for attracting attention consideration. [6].

Picture: The process of creating captions is called process of developing a text that describes an image. Furthermore, defining the material of a picture automatically is a key AI challenge that natural language processing (NLP) with computer vision is linked. We present a solution in this work quick overview of certain technical elements and methodologies for picture description generation. [7]. The study [8] discusses a new difficulty in picture captioning: how to successfully infuse feelings into generated captions whilst maintaining semantic correspondence between the visual and descriptive materials that are produced.

Khaing and Yu [9] present a comparison study of a model for deep learning picture caption creation that is based on attentiveness. In picture for producing sentence descriptions, the attention-based encoder-decoder networks are particularly valuable. In addition, the mechanism of attention pays to alterations in the encoder network's output and a more prominent portion of the image. For the decoder network's input, feature vectors are created by converting feature maps to feature vectors. Important processes for writing image descriptions are also included in the document as well as datasets that are commonly utilised and assessment measures for calculating performance.

This research focuses on picture to text translation, or to put it another way, making linguistic make sense of an image's contents investigating the various algorithms and the optimum combinations of them would be the initial stage in the model architecture process. Because there are so many options to choose from when it comes to most common problems, this section may be extensive.

Following the selection of the suitable collection of methods and models, the following step is to choose a relevant dataset. Many photo datasets are utilised for supervised learning on a wide scale, however, specifically for different the majority of them are sorts of applications that have been collected and labelled. An image dataset containing in an ideal world, in the form of written statements, labels that summarise the image's contents would be chosen. To gain knowledge, deep learning algorithms necessitate a large amount of data produce correct predictions, hence the dataset must include a large number of entries.

Following the selection of the architecture, models, and dataset, the dataset must be preprocessed to ensure that it has all of the algorithms' required inputs for a neural network that works with images. This normally requires scaling the images, but it also entails textual data tokenization and vectorization in the case of a language processing neural network.

The models must be implemented after the data have been preprocessed. Throughout the training part, they must keep an object-oriented mindset and make yourself visible and easy to utilise.

This section involves running fresh data points through the saved results and analysing them using certain metrics. If the results are satisfactory, the algorithm might be considered a success. Older machines can be used to train the algorithm, but if the dataset is too huge, it will fail may experience a "out of memory" problem. Although it is advised that you practise on a GPU, newer CPUs can also be used. The problem is that if the training is not done properly, it will not be effective. Technique is sophisticated, convergence will take a long time, ranging from days to weeks.

The findings' various milestones must also be stored during the training process. The learned weights are saved at these checkpoints and can be used to infer the method with new data later. During the training process, certain checkpoints of the findings must also be saved. The learned weights are saved at these checkpoints and can be used to infer the algorithm with new data later. Furthermore, in the event of a difficulty during training, the most recent successful iteration would be kept. The next natural step after training and maintaining, the algorithm's output is to put it to the test. This section involves running fresh data points through the saved results and analysing them using certain metrics. If the results are satisfactory, the algorithm could be regarded as a triumph.

3 Proposed Idea

One of the goals is to improve the quality of life for those who live in rural areas look into the broad subject of machine learning in order to figure out. In this instance, the ideal methods to adopt are its easy to settle for a good but not excellent algorithm or architecture when there are so many to pick from. Considerable study is required when dealing with a topic that requires deep learning and machine learning are two different types of learning technologies to tackle. A multitude of publications and blog postings are available, the bulk of which focus on the offered solutions' theoretical aspects.

Despite the fact that the programmable component of the application is not of the highest calibre for many data scientists. The code's quality becomes a priority when developing an application that will be used in production. Another objective is to develop software that can run in real time; no one wants to be kept waiting for more than a few seconds. This entails investigating and evaluating various platforms and packages for creating deep learning-based applications. As the industry grew, a multitude of tools were available, many of which optimised the compilation and computational components so that engineers could concentrate on enhancing and speeding up the algorithms. At the same time, integration must be considered in addition to performance. Given the fact that this app will be incorporated, into a number of larger projects, it should be built utilising an object-oriented approach to make it easy to integrate. As a result, the next subchapters detail the set of data, data preprocessing and architecture encoder-decoders, training phase, and outcomes The dataset, data preprocessing, network architecture encoder-decoders, training phase, and outcomes are all improved as a consequence provided in the following subchapters.

A. *The dataset*

MS-COCO is an acronym that stands for "Microsoft-Common Objects in Context," was utilised to train and test the model. It was first released in 2014, and the most recent upgrade was in 2015.

This dataset was generated expressly for this type of situation. It is a "large-scale item identification, segmentation, and captioning dataset," according to their Website. It is one of the most popular image-related datasets applications because it comprises over 80,000 tagged photos it also uses a "separate." The labelled dataset is used to train and test the decoder (captions).

Only 40,000 captions and their related photos were used in this study, which were divided an 80:20 split ratio was used to divide training and validation (testing) samples. This is in reaction to a performance issue that arose throughout the training procedure.

B. *The data preprocessing*

In every machine learning method, data preprocessing is a critical step. If you skip this step, the model will raise because it will not get the input it expects, it will generate an error.

Deep neural network encoder, the neural network encoder convolutional neural network encoder-decoder are deep convolutional neural networks (DCN) which is a type of neural network that encoders and deep recurrent neural network decoders are required for data preprocessing.

Because the Inception-v3 model is used in the deep convolutional neural network, and encoder is a type of neural network that uses deep convolutional neural networks to encode data photos must be expanded to fit the format that is necessary, i.e. (299, 299), as well as bringing the pixels into the $[-1, 1]$ range. For image processing, TensorFlow provides the "image" module, which allows images read into memory, scaled, and encoded as jpeg Keras' high-level API Inception-v3 has a method called "preprocess input" normalises the pixels in the specified range, as indicated previously.

The language generator decoder, which is a textual data, i.e. the captions, must be preprocessed before using the help of a deep recurrent neural network. In this section, the Keras module "preprocessing" and associated methods are used. The following are the steps taken:

```
In [58]:  1 caption_vector[0]

Out[58]: array([ 2, 354, 672,  2, 275,   3,   2,  81, 340,   0,   0,   0,   0,
                 0,   0,   0,   0,   0,   0,   0,   0,   0,   0,   0,   0,   0,
                 0,   0,   0,   0,   0,   0,   0,   0,   0,   0,   0,   0,   0,
                 0,   0,   0,   0,   0,   0,   0,   0])
```

Fig. 1 After preprocessing the caption data, an array was created

- Tokenization of captions, which involves dividing captions by white spaces and retaining just the words that are unique.
- There is a limit to the number of words that can be used in a vocabulary; the vocabulary in order to preserve memory is limited to the top 5000 words.
- Transforming text into a numerical sequence.
- Word-to-word translation.
- Increasing the length of all sequences to that of the most extensive.

The end result is a vector containing an integer sequence that has been padded to fit the size of the input dataset's longest caption. The following picture depicts the outcome (Fig. 1).

C. *The Architecture of Encoders and Decoders*

The encoder-decoder design, which employs two recurrent neural networks, is commonly utilised machine translation, for example, is a good example of this. This architecture's main purpose is to reduce the size of the second network's vector of input uses requires.

The first neural network in this situation is a convolutional network, and the second neural network is a recurrent neural network. The second neural network's vector output is identical to the first neural network's vector output i.e. a person's input neural network. The following is a schematic of the architecture in general:

According to the paper, the attention mechanism serves as a link between the encoder and the decoder. This is required since the data provided a single vector representation would otherwise be fed into the decoder. The attention approach allows the decoder to focus on the useful bits by providing information from all of the encoder's hidden states (Fig. 2).

(1) The encoder

This step produces a vector with the shape (8, 8, 2048), however, the application requires a vector with the shape (64, 2048). A model class comes in handy here called "CNN encoder" (convolutional neural networks) is a term used to describe a type of neural network. RNNs are a type of neural network that consisting of a single layer that is totally connected and activation of the ReLU gene (rectified linear units). In the feature extraction stage, the extracted features are read into memory and pass across this entirely interconnected layer.

Fig. 2 High-level encoder-decoder architecture

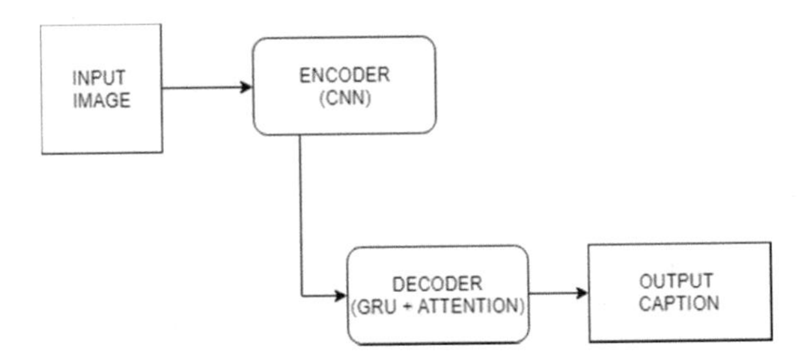

Putting it another way, the encoder is made up of features extracted. Using the Inception-v3 model for transfer learning, which is then transferred over a fully linked layer that additionally handles resizing.

(2) The decoder

The encoded features are taken from the encoder by the decoder for deep recurrent neural networks. The recurrent the gated recurrent unit (GRU) is a form of neural network. In order to guess the following word, the decoder evaluates the image.

The shape of the encoder's vector output, per layer, the number of neurons (units), as well as the size of the vocabulary are all factors to consider, are all inputs to the decoder during the data preparation phase. When calling, it is necessary to pass the extracted characteristics as well as the reset state.

The context vector and attention weight are received by the attention layer the encoder for deep convolutional neural networks, and it is considered as though it were a different model. The model's embedding is a concatenation of the retrieved context vector and transmitted across the gated recurrent unit network's layers.

Two completely connected layers and a gated recurrent unit layer make up the recurrent neural network architecture. The Keras API's module "layers" can be used to add all of these layers. The GRU layer is a gated recurrent unit layer that takes the units (number of neurons), it should have as an input when it is created. Type "dense" refers to fully connected layers (also known as "dense" layers). The units are fed into the first completely connected layer, whilst the vocabulary size is fed into the fully connected second layer.

D. *The preparatory stage*

The training period is self-explanatory. The algorithms in this technique learn to map the function parameters. This is the most difficult stage, both in terms of computing and programming. This entails running the data through the algorithm numerous times, which necessitates the setting of some parameters and the selection of some functions. The following are the logical stages that occur:

- Using the CNN encoder to extract features;
- Sending the encoder output to the decoder, together with the decoder input is set to 0 (zero) and the hidden state is set to 0 (zero) (the "start" token).

- The decoder's hidden state is fed back into the model in a loop, and the loss is calculated using the model's predictions.

A total of 40,000 entries were used for training, with a split of 80:20: training accounts for 80% of the budget, 20% for testing, and 20% for assessing.

The entire dataset can be submitted to the algorithm in one batch if the dataset is small enough. The dataset in question, however, is rather huge, and batching is required. Batching entails dividing into the data set many equal portions and giving each one to the computer algorithm one at a time. Because the batches must it is necessary to preserve proportionality between the number of batches and the number of occurrences in the dataset. The quantity of the batch employed in this study was 64.

Cross-entropy with logits was chosen as the loss function. Before passing each batch, it is set to zero to determine each batch's specific error values, the overall loss is set to zero for each period. We can determine if the model converged appropriately by visualising these data once the training procedure is completed.

Deep neural networks are a type of artificial intelligence [1] that uses neural "learn" how to from inputs to outputs, map a function (outputs). This is determined by putting the data through its paces various models multiple times in order to minimise the by error (loss) adjusting the number of epochs is a measure of how long a period of time has passed. The more epochs employed, the longer the algorithm takes to train, but the possibility of a superior the function of mapping grows. The algorithm has been tuned across two and twenty epochs, with the function of loss evolving in both cases, as shown in the diagram below (Fig. 3).

When looking at the plots, the number of times the algorithm has been applied on the data has increased can be seen as a substantial improvement.

During the training phase, the model's weights must be saved for later usage. The training procedure takes a long time, and all that matters are the weights in the end, or those resulting from the most cost-effective calculation. After every 100 batches, these are kept to maintain consistency a comprehensive backup. The model can be stored in a number of different formats, with the most popular being of which being

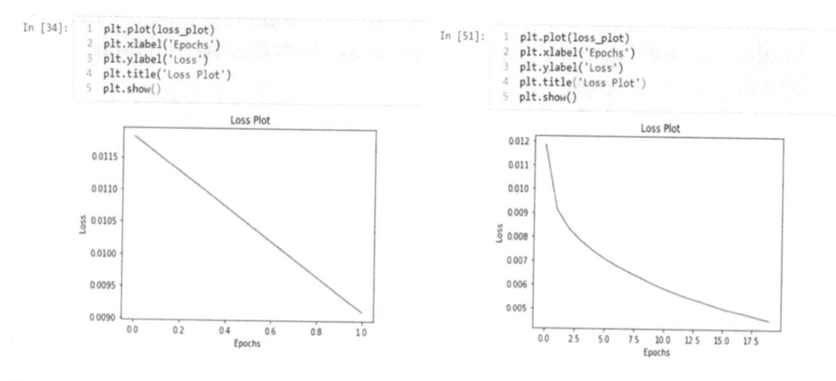

Fig. 3 There have been two epochs of loss evolution and twenty epochs of loss evolution

of which is a "pickle." The "reference point" attribute of type index is used to save the checkpoints in this project.

It is tempting to argue that because the model grew smoothly, it should produce good outcomes and on a logarithmic scale rate after looking at the loss graphs, but this is not this is constantly the case. The simplest method to thoroughly test the outcome is to feed it more data, previously unknown in this situation, the data points are photographs.

On a hexa-core CPU, it took about 26 hours to train the models on 32,000 instances for 20 epochs.

E. *Results*

Unfortunately, due to the restricted processing capability of the device on which the models are trained, it takes a long time to train for a limited number of epochs. To accomplish this, the size of the input dataset must be reduced. In this scenario, the number of photos input is reduced to 80, and the training and testing datasets are split in an 80:20% ratio. As a result, 64 photos are used for training. By selecting this proportionality, additional factors such as batch size can be preserved.

In other words, this is the bare minimum of information that may be used without reducing other values. It is vital to take it "one step at a time" and rule out each parameter as being incorrect.

There are still 20 captions to print, both real and anticipated, following the reduction of the number of training data. This is merely a quick technique to see if the algorithm is still working without having to wait a long time. Two photos are printed as a sanity check to see how captions grow over time for different types of images. The first image shows a woman is shown tennis court in the first photograph, Boats moored near a river are depicted in the second. The first is straight a forward. The second, on the other hand, is a little more difficult to grasp due to the large number of pieces it comprises. The outcomes are as follows:

After 20 epochs, the results of images 1 and 2 are shown. The outcomes are abysmal, but the method is effective. Following that, the total number of epochs has been increased to 50. As can be seen in the graph below, the results should improve slightly but not dramatically (Fig. 4).

The first image's prediction improved, correctly recognising the ball, whereas the forecast in the second image is erroneous. This indicates that the models have not been adequately trained over a sufficient number of iterations (Fig. 5).

The next step is to see if the models can ever converge, even if the quantity of input photos is reduced. The number of epochs is extended to 100 in this case, with the following results (Fig. 6).

Image 1 has a better outlook than predicted after 100 epochs and can be deemed a good prediction. If you just look at that image, you will see what I mean, you may believe that the prototypes have the converging that the epochs count required is adequate, However, if you look at image number two, you will notice that, it is clear that the prediction remains unfavourable, despite the fact that the white boat was

```
Image 1: C:\Users\test_ml\imgs\COCO_train2014_000000001375.jpg
Actual caption: a tennis player swings her tennis racquet on the court
Predicted caption:  a <end>

Image 2: C:\Users\test_ml\imgs\COCO_train2014_000000001453.jpg
Actual caption: a boat of women are pushing off of the shore
Predicted caption:  a <end>
```

Fig. 4 After 20 epochs, the results of images 1 and 2 are shown

```
Image 1: C:\Users\test_ml\imgs\COCO_train2014_000000001375.jpg
Actual caption: a tennis player swings her tennis racquet on the court
Predicted caption:  a ball on a ball on a ball on a ball on a ball on
                    a ball on a ball on a ball on a ball on a ball on
                    a ball on a ball on a ball on a

Image 2: C:\Users\test_ml\imgs\COCO_train2014_000000001453.jpg
Actual caption: a boat of women are pushing off of the shore
Predicted caption:  a tv in a tv in a tv in a tv in a tv in a
                    tv in a tv in a tv in a tv in a tv in a tv
                    in a tv in a tv in a
```

Fig. 5 After 50 epochs, the results of images 1 and 2 are shown

```
Image 1: C:\Users\test_ml\imgs\COCO_train2014_000000001375.jpg
Actual caption: a tennis player swings her tennis racquet on the court
Predicted caption:  a person hitting a ball on a tennis court <end>

Image 2: C:\Users\test_ml\imgs\COCO_train2014_000000001453.jpg
Actual caption: a boat of women are pushing off of the shore
Predicted caption:  a big white in a single in the middle of a bottle <end>
```

Fig. 6 Findings of pictures 1 and 2 are exhibited 100 epochs later

correctly recognised ("a big white"—6th diagram). The number of epochs can be extended to 200 because this is a dummy dataset with insufficient images to develop a generalised model to examine how the model reacts to complicated images. The expected outcome is that the predicted caption for the first image stays the same, whilst the second one does improves. The following are the outcomes (Fig. 7).

```
Image 1: C:\Users\test_ml\imgs\COCO_train2014_000000001375.jpg
Actual caption: a tennis player swings her tennis racquet on the court
Predicted caption:  a person hitting a ball on a tennis court <end>

Image 2: C:\Users\test_ml\imgs\COCO_train2014_000000001453.jpg
Actual caption: a boat of women are pushing off of the shore
Predicted caption:  a red many one in the middle of a surfboard <end>
```

Fig. 7 After 200 epochs, the results of images 1 and 2 are shown

The first image's projected caption did not alter, as expected, but the second image's focus switched out of the white yacht to the boats that are teeny-tiny, there is something for everyone, everyone will find something to their liking. Because the paddleboards are used in this area photograph something in the centre that resembles a surfboard, it improved in some ways (people). Even if it has improved, the caption is not quite right. Given that no algorithm can ever correctly detect everything, one might conclude that this is an outlier. The bad result can be attributed to the input image's poor quality.

After fine-tuning the settings, it is clear that the initial training iteration count the algorithm's convergence was never achieved since the threshold was too low. Increasing the number of epochs is, however, a difficult task computationally costly decision, but it is critical in this circumstance. After a specific number of epochs, the model is finished, the epochs' value can be calculated. In our case, it should be between 50 and 100, or a decent outcome.

4 Conclusion

The goal of the study was to create a deep learning-based application for automated image description whilst also conducting extensive research in the topic. This prototype will to achieve its purpose of assisting visually impaired people in making sense of information, and it will integrate with a text-to-speech engine is required in the future.

The research component of the study was successful, enabling readers to grasp a variety of machine learning and deep learning techniques. Potential loss function, feed-forward, error propagation through a neural network's backpropagation perceptron, at the same time, activation functions and other artificial neural networks are made up of basic parts investigated, the design of a convolutional neural network, this results in sense of elements of a picture by using matrix computations and dimensionality reduction. A time-series artificial neural network is a type of artificial neural network that analyses data over time is the recurrent neural network, allowing it to make judgments on its own at each stage based on the value of the previous step, allowing it to "memorise" values whilst computing.

Following extensive investigation into various models and architecture, a convolutional neural network-based composite model was chosen as the encoder. Transfer learning is used to reduce a recurrent neural network, as well as training time and computing complexity is used as a decoder, which is the cutting-edge technology for dealing with textual input, notably forecasting the word that will come next in a sentence (sequential prediction). Because the traditional because architecture cannot remember long words, as a link between the two models, the attention mechanism was introduced, with its output indicating the concatenation of the attention weights and the output of the attention mechanism encoder's features.

Despite the loss's good evolution, the overall model did not behave as expected after 20 epochs of training on 32,000 photos, implying that the tuning parameters needed to be adjusted. Because the technology available at the time did not allow for considerably more complicated and time-consuming tasks to be completed processes, to find the problem, the training set of 64 pictures was used to tune the parameters. The issue was that the algorithm had not yet reached a point of convergence and had not been trained for a sufficient number of epochs. Despite the short dataset, the model was able to converge after 100 training epochs. The finding is that it behaves typically, with some photos predicting captions well and others not so well.

General development and training can be done even if the hardware is not up to scratch, as stated in the "parameter adjustment" section, at least for prototype.

This application, as a prototype, can be enhanced in a number a variety of methods. The first step should be to get a more powerful computer machine for the prototypes to be trained on. This would have an impact how long it takes the model to learn, making it faster and allowing the programmer to do more improve performance of the model (in terms of terminology of forecasting).

To take a step forward farther, rather than captioning photos, the model may be adjusted to caption movies with numerous frames per second. This could be the case used to provide additional unstructured textual data, such as movie descriptions, that can be used to summarise videos.

References

1. Anderson J, Rainie L (2018) Artificial intelligence and the future of humans. Pew Research Center. Available online https://www.pewresearch.org/internet/2018/12/10/artificial-intelligence-and-the-future-of-humans/
2. Apoorva O, Sainath YM, Rao GM (2017) Deep learning for intelligent exploration of image details. Int J Comput Appl Technol Res 6(7): 333–337. ISSN: 2319–8656
3. Chollet F (2017) The limitations of deep learning. Deep learning with python, Section 2, Chapter 9, Essays
4. Adnan K, Akbar R (2019) An analytical study of information extraction from unstructured and multidimensional big data. J Big Data 6(1). https://doi.org/10.1186/s40537-019-0254-8
5. Vinyals O, Toshev A, Bengio S et al (2015) Show and tell: lessons learned from the 2015 MSCOCO image captioning challenge. IEEE Trans Pattern Anal Mach Intell 39(4). https://doi.org/10.1109/TPAMI.2016.2587640
6. Anderson P, He X, Buehler C, Teney D, Johnson M, Gould S, Zhang L (2018) Bottom-up and top-down attention for image captioning and visual question answering. In: 2018 IEEE conference on computer vision and pattern recognition (CVPR), pp 6077–6086
7. Shabir S, Arafat SY et al (2018) An image conveys a message: a brief survey on image description generation. In: 1st international conference on power, energy, and smart grid (ICPESG). https://doi.org/10.1109/ICPESG.2018.8384519
8. You Q, Jin H, Luo J (2018) Image captioning at will: a versatile scheme for effectively injecting sentiments into image descriptions. Computer science. Available online https://arxiv.org/abs/1801.10121
9. Khaing PP, Yu MT (2020) Attention-based deep learning model for image captioning: a comparative study. Int J Image Graph Signal Process 6:1–8

Traffic Sign Detection and Recognition Using Deep Learning

P. Puneeth Kumar, R. C. Kishen, and M. Ravikumar

Abstract The system of detecting and recognizing the traffic signs is an important component of the system of smart transportation. Having the ability to detect the signs related to traffic precisely and correctly will improve safety of driving. This particular paper provides a deep learning-based detecting and recognizing of signs related to traffic methods with the primary purpose of recognizing and classifying circular signs. To begin, a picture is pre-processed to information which is crucial. Second, Hough rework is utilized in police investigations and area mapping. Finally, deep learning is employed to map the road traffic signs that have been detected. This book proposes an image-based traffic sign detection and identification technology that is subsequently merged with a convolutional neural network to type traffic signals. Because of its major identification rate, CNN can be utilized to recognize a wide range of computer vision tasks.

Keywords Traffic sign recognition · Traffic sign detection · Deep learning · Convolutional neural network

1 Introduction

Recognizing the signs related to traffic might be a major component of Advanced Driver Assistance Systems and a prominent research direction in computer vision. This can be broken down into two types: traffic sign detecting and recognizing. The precision with which detecting step is performed will have a direct impact on the final identification results. Important messages about vehicle safety are conveyed by traffic signs, which include current conditions of traffic, prohibiting and allowing driving routes and their behaviors, road rights, and cueing messages which are cautioned, among other things. They will also help drivers determine the road condition and, as a result, the optimal routes of driving.

P. P. Kumar (✉) · R. C. Kishen · M. Ravikumar
Don Bosco Institute of Technology, Bengaluru, India
e-mail: parthapuneeth@gmail.com

A. Kumar et al. (eds.), *Proceedings of the International Conference on Cognitive and Intelligent Computing*, Cognitive Science and Technology,
https://doi.org/10.1007/978-981-19-2350-0_34

Traffic-related signs contain some stable qualities that can be used for detecting and categorizing, and also shape and color are crucial traits that will help the drivers acquire data of road. The colors of traffic-related signs in every country are essentially identical, with simple colors (red, blue, yellow, etc.) also pasted forms (circles, triangles, rectangles, etc.) External factors such as the weather tend to taint the image of traffic signs. As a result, recognizing traffic-related signs could be a difficult and worthwhile research topic in traffic engineering. Different traffic sign identification systems are also in the works. This paper proposes a CNN-supported transfer of learning approach [1]. A huge data set is used to train Deep CNN, and then a typical traffic training scenario is used to achieve successful regional convolutional neural network (RCNN) identification [2].

A multi-resolutive characteristics amalgamation fabric network capable of analyzing several useful features from small objects is designed to gain various data and improve detecting quality. The framework for detecting traffic signs is further molded into specified sequence of classification and reversion tasks [3], in order to achieve concurrent CNN detecting combined with traffic sign recognizing. In this project, the Hough Transform is utilized to identify also pretreat signs prior they are recognized, majorly improving accuracy and timeliness [4].

The three components of this text that implement traffic-related signs detecting and identifying are refining, detecting, and classifying and Fig. 1 displays detecting and recognizing method. In the refining step, the stable color picture is highlightened, and then color location is changed [5]. In stage of detecting, signs in roads are divided based on image look and color data, and signs which are circular are recognized using the Hough Transform. An image of the region of interest is created at this point, and location of traffic-related signs is decided. In recognizing and classifying step, retrieved and implemented traffic-related sign region is utilized as input, and discovered information is identified and classed using a convolutional neural network in deep learning [2].

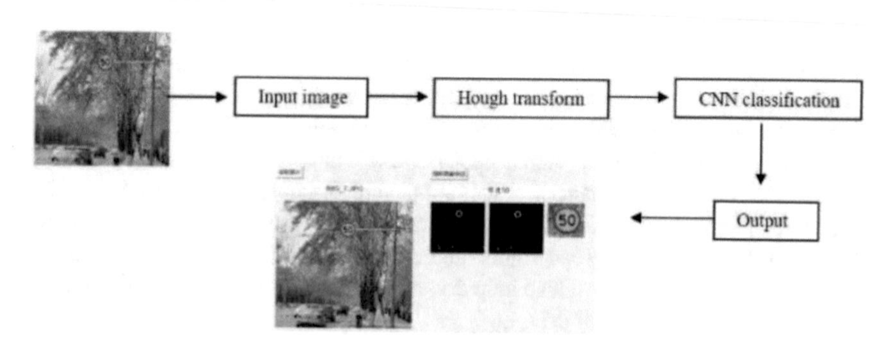

Fig. 1 The process of traffic signs recognition system

2 Stage of Pre-processing

A. *Image augmentation*

Pre-processing is used to remove any extraneous data from photographs and return only the important information. As a result of this operation, the exactness of the detecting and classifying phases of indications will be increased, resulting in improved real-time performance.

Image intensification is the most simple sort of picture pre-processing. Image intensification technique aims to make the image's jumbled information clear so that the important signals can be recovered [1]. The most common approaches for reducing noise from images in image improvement technologies are mean and median filtering. Mean filtering is an algorithm of filtering linearly that replaces original picture element value with the layout's average and provides the aimed picture element a layout that includes the picture element surrounding it. The mean filtering formula is:

The RGB color room of the image is transformed to HSV, i.e., more suitable for the upcoming phase. The RGB and HSV color spaces differ significantly, as shown in Fig. 2.

$$\mathrm{g(x, y)} = 1/2 \sum f(x, y) \tag{1}$$

where g(x, y) is current gray merit of refined picture, f(x, y) is the present picture element to refined, and m is the total of every picture elements in the template that contain the current pixel.

The accompanying formula demonstrates how simple it is to calculate mean filtering; however, mean filtering eliminates the image's details, i.e., the fundamental snag of mean filtering. Median filtering is nonlinear filtering procedure. In some cases, median filtering may be able to counteract the blurring of visual clarity induced by linear filtering such as mean filtering. Median filtering alternates the merit of the

Fig. 2 Color spaces RGB and HSV

Fig. 3 Comparison of median filtering

window's center space with the median value of each space in sliding room with an uneven number of points. Median filtering, in contrast to mean filtering, that removes the image's information, keeps the picture's details. Figure 2 depicts the median filter disparity figure (Fig. 3).

B. *Transformation of color spaces*

The RGB color scheme stands for red, green, and blue. It is a rather frequent color scheme. The red, green, and blue elements of the image represent the merit of the colors in the picture.

3 Recognition of Traffic Signage

The two basic components of sign recognition are detecting and classifying of signs, with detecting serving as foundation. The accuracy of the detection is directly proportional to the excellence in eventually classifying. In respective research, the Hough Transform is utilized in detecting, whereas CNN is utilized in identifying also classifying.

A. *Detecting of Signs*

The period of detecting main purpose is to retrieve the image's delight scope and prepare the picture for classifying. The color and structure data of traffic-related signs are two crucial pieces of data; individual traffic-related sign has its own color and structure; hence, respective article will focus on traffic sign detection utilizing these two pieces of data [2]. To construct a zone of interest based on color information, the H and S components of a picture that has been translated into HSV color space are extracted [5]. Hue is significant in segmentation because it displays greater irregularity in changing lighting situations and color congestion and resemblances in the background. A segmentation figure related on HSV room is shown in Fig. 4.

Fig. 4 Color space detection using RGB and HSV

$$F = r[R] + g[G] + b[B] \tag{2}$$

This is demonstrated by the formula above. The RGB color room is characterized by three foremost colors, with the three elements (R, G, and B) [6] being inextricably linked. Each change in component results in a change in the image's pixel color value, which is inconvenient for spotting traffic signals. Because traffic-related signs are revealed to the elements each year and may fade or be destroyed as a result of the weather's influence, processing the image in RGB will not aid upcoming behavior [5]. So, in the respective experiment, HSV color room is used, which consists of three components: Hue, Saturation, and Value, and each of these three components represents the picture's brilliance. HSV is closer to the picture viewed by homo eyes and has a broader color gamut than RGB. It is also fewer affected by changes in environmental light than RGB. As a result, the HSV color space outperforms the RGB color space in terms of traffic sign detection [6]. As a result, when it comes to pre-processing.

After severance, there is going to be some disturbance in the picture. To remove superfluous interference information, this work transforms the image after segmentation using the morphological operation in mathematics. Picture morphology is a technique for reducing picture data, preserving the image's basic shape, and removing extraneous structure. The image after segmentation applies the open operation since there are some tiny interference sites in image after color space segmentation based on HSV.

Open functioning, as previously explained, is effective in removing these little things. So, following the image segmentation, execute corrosion, and then expand processing [7] as shown in Fig. 5. After opening function in patterning, diffuse involvement data in space can be successfully erased, allowing for a more precise detection of the characteristic recurrence. Using morphological techniques, the duplicate involving data in space may be successfully deleted, allowing the distinctive region to be discovered more precisely [8] (Fig. 6).

- At last, the Hough Transform is frequently utilized to find circular indications [9]. The Hough Transform's mapped on the concept of joining fringe of picture

Fig. 5 Segmentation of the HSV color space

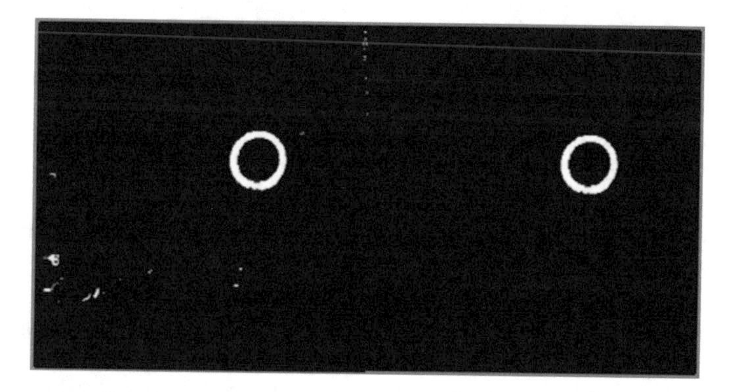

Fig. 6 Image after diffuse involved data removed

element to construct room cornered limits utilizing world picture properties. A picture is converted into a parameter space using the Hough transform [10, 11]. Using Hough, the difficult-to-solve global detection problem can be reduced to a simple local peak detection problem, making the changed result easy to spot. It has a modest influence on noise and curve discontinuities, which is a plus.

- Convert picture room into guideline room and brief guidelines room to fulfill the aim of recognizing picture corners. Figure 7 displays the Hough Transform detecting outputs for rounded traffic-related signs, which show good detection accuracy and precision. The radius of Hough round is lowered to create detected rounded room more visible, illustrating that the Hough circle's detection effect on circular traffic lights is good.

B. Classifying Signs

The detected jam indicators must be picturized and categorized. The traffic signs that have been discovered must be picturized and categorized. In respective study, CNN is mostly utilized categorization. In this, GTSRB dataset is utilized to train and test CNN, and it is the crucial aspect of CNN training and testing. CNN is a multi-layered

Fig. 7 Detection of Hough transform

network that mimics the human brain's functions. Individual layer CNN is made up of several neurons [8]. To accomplish a job, each neuron is given an input, and some actions and outputs are passed on to the upcoming neuron as input. The convolutional layer in a convolutional neural network is mostly used for feature selection. When it comes to categorization, CNN surpasses traditional machine learning.

In this study, the observed indications are classified using CNN, and delicate CNN divisor is created. Delicate CNN is made up of two convolutional layouts, pooling layouts, and complete connection layouts [12]. Kernel size of convolution layout is 5×5, number of convolution kernels equals 32, and the step size is 1. There are 16 concealed layouts points in initial convolution layout and 32 in the second convolution layout. Expanse of the character graphs is 32×32 pixels and 16×16 pixels. Kernel expanse of pooling layout is 2×2, concealed points of whole networked layout are 512 and 128, and hidden nodes of ending resulting layout are 43. Learning rate in text is initially set to 0.0001. Dropout (regularization) processing is a concealed layout of full network that prevents the network from becoming over-fit. Some node data is randomly removed throughout the training process to minimize over-fitting. Dropout makes node information to 0 to eliminate few eigenvalues. The character retrieval and classifying by CNN is depicted in Fig. 8.

The model presented in this paper is built on Keras, a Tensor Flow-compatible neural network toolkit, which provides an open source ultra-modern string pile. All of the suggested CNN layouts use ReLU as trigger task to intake difficult characters [13, 14]. The most common trigger tasks in NN are ReLU, can accelerate network model acquiring knowledge, and make it more suitable for real-time learning. A dropout layout is inserted to CNN simultaneously to avoid over-fitting. Dropout can

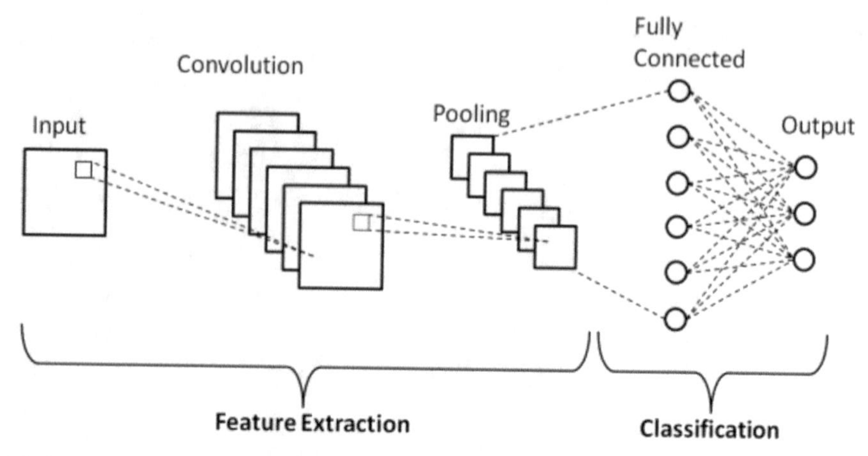

Fig. 8 Framework of CNN feature extraction

uplift the unpredictability of connected layout. GTSRB and own created signs related to traffic datasets utilized for training and testing across layouts. Complete dataset is categorized into two sections: training and test sets. 90% of the dataset is made up of training sets, whereas 10% is made up of test sets. When training model, model's precision improves with time, and the loss rate decreases as training time increases. The preciseness and training loss rate curve is depicted in Fig. 9.

The model's accuracy in police investigations and recognizing traffic signs is 98.2%, according to the trained CNN's examination data. Figure 10 shows the testing accuracy. The outputs of trials show network structure is capable of accurately identifying and recognizing traffic signs.

Traffic signs in the Dalian region are collected using a mobile and integrated into layered structure for detecting, recognizing, and categorizing in order to better measure network efficacy. As a result, as shown in Fig. 11, the last identifying and classifying outputs are displayed in a visible window. As you may observe in figure,

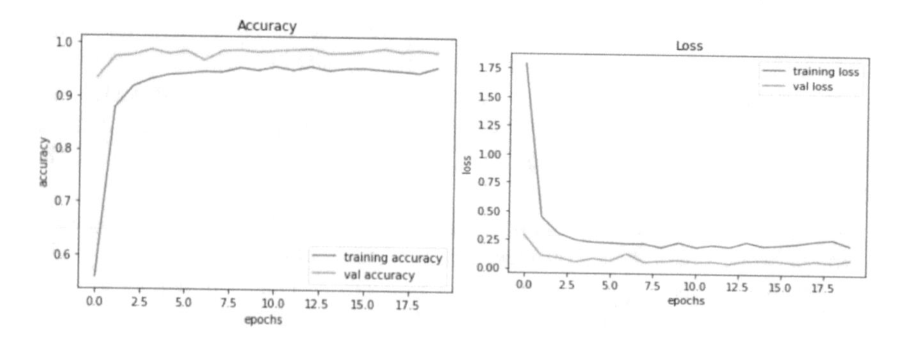

Fig. 9 Rate of accuracy and rate of loss

```
test accuracy : 0.9820000

true label : 2
computed_label : 2
```

Fig. 10 Testing accuracy

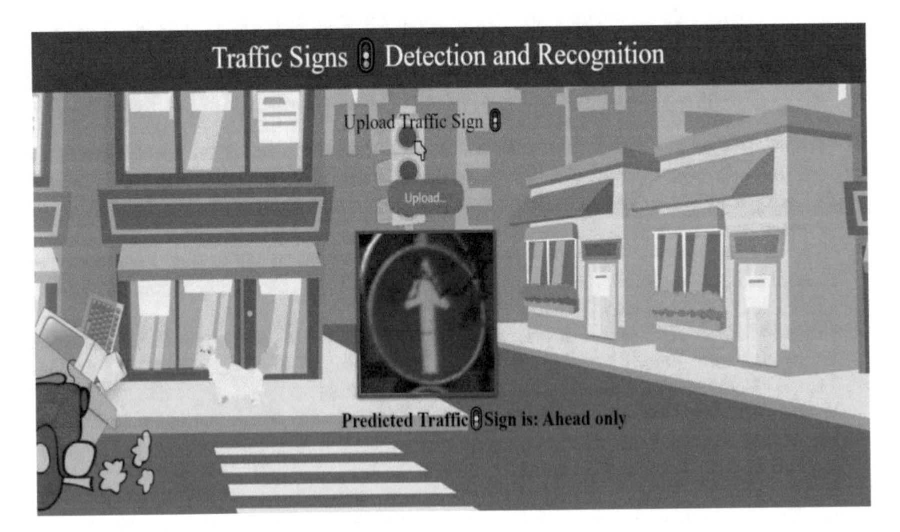

Fig. 11 Final detection of sign board

the layered structure could consistently and properly determine signs and categorize them accordingly.

4 Conclusion

The goal of this research is to offer a deep learning-based traffic sign recognition technique that focuses on rounded signs. Utilizing picture initialization, sign related to traffic detecting, recognizing, and classifying, the technique properly finds and determines signs related to traffic. According to the findings, this approach has 98.2% accuracy.

References

1. Yelmanov S, Romanyshyn Y (2020) A new technique of image enhancement by intensity transformation. In: IEEE 15th international conference on advanced trends in radioelectronics, telecommunications and computer engineering (TCSET). https://doi.org/10.1109/TCSET49122.2020.235440, pp 281–286
2. Wang CY (2018) Research and application of traffic sign detection and recognition based on deep learning. Int Conf Robots Intell Syst (ICRIS). https://doi.org/10.1109/ICRIS.2018.00047,pp.150-152
3. Zhu Z, Liang D, Zhang S, et al (2016) Traffic-sign detection and classification in the wild. In: IEEE conference on computer vision and pattern recognition (CVPR), pp 2110–2118
4. Chandrasekar L, Durga (2014) G Implementation of Hough transform for image processing applications. In: International conference on communication and signal processing. https://doi.org/10.1109/ICCSP.2014.6949962, pp 843–847
5. Ghimire D, Lee J (2010) Color image enhancement in HSV space using nonlinear transfer function and neighborhood dependent approach with preserving details. In: 2010 Fourth pacific-rim symposium on image and video technology. https://doi.org/10.1109/PSIVT.2010.77, pp 422–426
6. Xing M, Chunyang M, Yan W et al (2016) Traffic sign detection and recognition using color standardization and Zernike moments. Chinese Control Decis Conf. https://doi.org/10.1109/CCDC.2016.7531926,pp.5195-5198
7. Gomez-Moreno H, Maldonado-Bascon S, Gil-Jimenez P et al (2010) Goal evaluation of segmentation algorithms for traffic sign recognition. IEEE Trans Intell Transp Syst 11(4):917–930. https://doi.org/10.1109/TITS.2010.2054084
8. Kumar MA, Goud NS, Sreeram R, Prasuna RG (2019) Image processing based on adaptive morphological techniques. In: 2019 international conference on emerging trends in science and engineering (ICESE). https://doi.org/10.1109/ICESE46178.2019.9194641, pp 1–4
9. Ye H, Shang G, Wang L, Zheng M (2015) A new method based on Hough transform for quick line and circle detection. In: 8th international conference on biomedical engineering and informatics (BMEI). https://doi.org/10.1109/BMEI.2015.7401472, pp 52–56
10. Rong F, Du-wu C, Bo H (2009) A novel hough transform algorithm for multi-objective detection. In: Third international symposium on intelligent information technology application. https://doi.org/10.1109/IITA.2009.387, pp 705–708
11. Chen WW, Wu W (2019) Linear and circular extraction method based on Hough transformation. Electron Mass 383(02):25–27
12. Zhu Y, Zhang C, Zhou D et al (2016) Traffic sign detection and recognition using fully convolutional network guided proposals. Neurocomputing 214:19. https://doi.org/10.1016/j.neucom.2016.07.009,Pages758-766
13. Mazumdar A, Rawat AS (2019) Learning and Recovery in the ReLU Model. In: 2019 57th annual allerton conference on communication, control, and computing (Allerton). https://doi.org/10.1109/ALLERTON.2019.8919900, pp 108–115
14. Ide H, Kurita T (2017) Improvement of learning for CNN with ReLU activation by sparse regularization. In: International joint conference on neural networks (IJCNN). https://doi.org/10.1109/IJCNN.2017.7966185, pp 2684–2691

Identification of Road Traffic from Multiple Sources Using Modern Gaussian Approach

R. Jegadeesan, Ayush Choudhary, Abhinav Pundir, and Baha Ur Rehaan

Abstract Prediction of road traffic from multiple sources using Gaussian approach is most import in intelligent transport systems. Existing works are only focused on non-intrusive sensors that are very expensive. Sensors are detecting traffic conditions and image recognition, etc. The maintenance of these sensors is very difficult, and to address the issue, this paper aims to improve road traffic speed prediction by using tweet sensors and social media. This includes many challenges, including location uncertainty of low-resolution data, language ambiguity of traffic description in text, etc. To response these challenges, we provide a uniform modeling probabilistic framework called Topic Enhanced Gaussian Aggregation model (TEGPAM). It consists of three components location disaggregation model, traffic topic model, and traffic speed Gaussian model. The module is designed with the features of a typical social web base, with functions that are related to the proposed model, and different driving directions are referred to as different road links.

Keywords Gaussian process · Multiple sources

R. Jegadeesan (✉)
Jyothishmathi Institute of Technology and Science, Karimnagar Telangana, India
e-mail: ramjaganjagan@gmail.com

A. Choudhary
Tata Consultancy Services, Haryana, India

A. Pundir
Infosys, Chandigarh, India

B. U. Rehaan
TCS, Siruseri, Chengalpattu, India

A. Kumar et al. (eds.), *Proceedings of the International Conference on Cognitive and Intelligent Computing*, Cognitive Science and Technology,
https://doi.org/10.1007/978-981-19-2350-0_35

1 Introduction

Transportation system is the most important system nowadays. Existing works are mainly focused on very expensive, and they are using cameras and image recognition tools. Existing techniques are not fit for the current road conditions. To address this issue in this paper, we proposed social media and Car trajectory techniques.

Social media:

This is related to websites, e.g., Facebook and Twitter. People are exchanging the information by using these social media. Messages are sent about the traffic conditions. Such as struck in traffic road no.22 are posted by the driver, passengers can be viewed by the sensors. Meanwhile, it is a traffic authority registered on public accounts and post messages to inform public about traffic status.

Car trajectory data:

Car trajectory data is getting the location in which the application is installed in the driver's cellphone. It is used to map the location. For example, we can take Uber, Ola cabs, and google maps which is used to navigate the location. This makes the traveling easier and comfort. If the origin destination (OD) is passed on a map, then the route will be mapped from source to the destination with the time (min, sec). According to origin destination, trajectory is a sequence of links in which the segments of a road is divided. The traveling time of a road link is so-called trajectory travel time. If the road link is congested, then it may take longer trajectory travel time with longer traffic speed.

From the above example in Fig. 1, the question marks indicate that they are not covered by traditional speed sensors. But the traffic conditions are described in tweets.

(1) Speed sensor collects the speed observations.
(2) Trajectory sensor indicates the trajectory speed which is to be predicted or observed.
(3) Tweet sensor which covers multiple road links that is to be noted.

Challenges :

When we combine traditional traffic speed data (e.g., sensor data) with data of a new type (e.g., Twitter data and track data) to predict the speed of road traffic, technical challenges arise because of the characteristics of each data source:

Uncertainty location of low-precision data: Log and track data are called low-precision data because we cannot directly identify it in specific road links. Most tweets do not contain site tags, so the local language is the main directory, but it is ambiguous. For example, the term "Stuck in traffic on E 32nd St. Stay away!" The entire street is without precise road sites. At the same time, the travel time of the route is a complex measure based on the speed of the multiple links, which may vary greatly. A strategy is therefore needed to separate the data into specific road links.

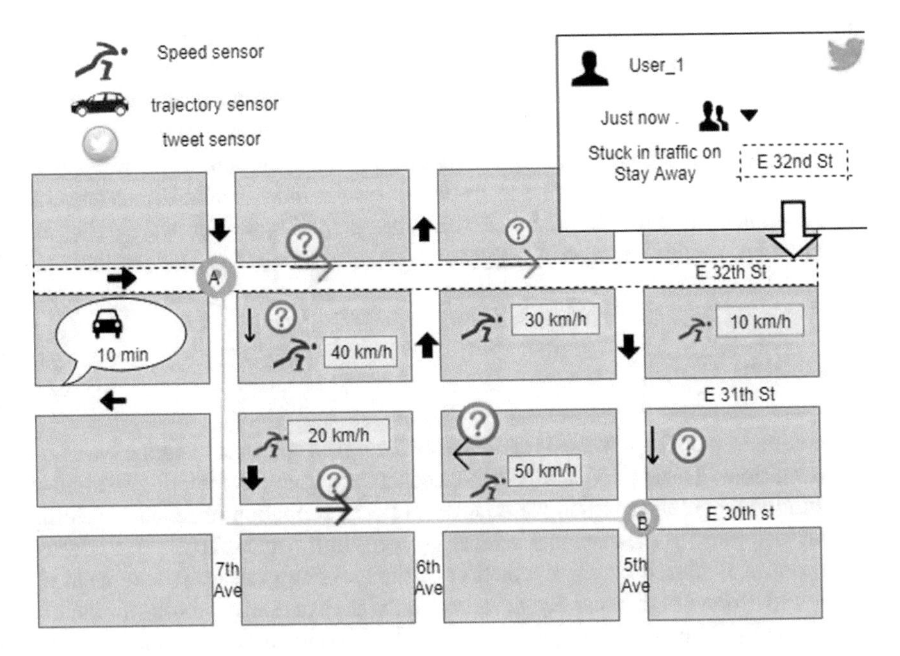

Fig. 1 Trajectory data with sensor process

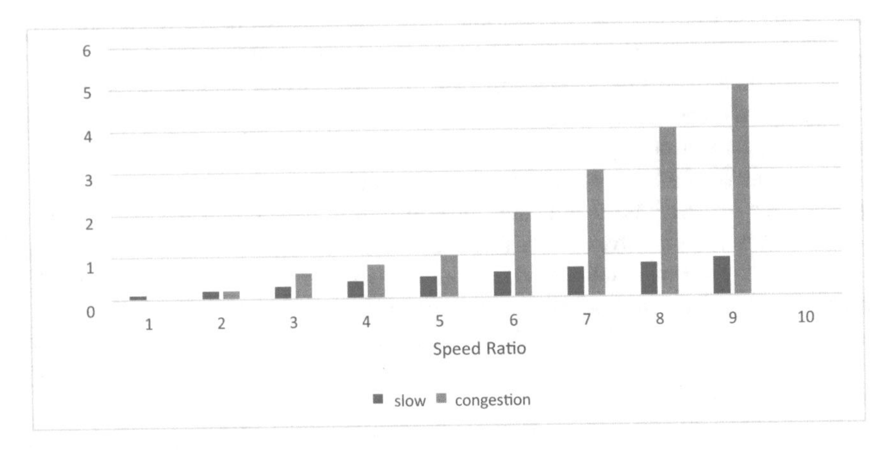

Fig. 2 Word frequencies

The linguistic ambiguity to describe traffic in Twitter: Expressions that illustrate traffic conditions are varied and may indicate different speed values. Figure 2 shows an example of frequency distribution of the word on the degree of congestion when people use busy words. At the same time, some words that are not directly related to traffic may have a strong impact on speeding, such as words that complain of bad weather. Therefore, a language model is required to capture patterns between separate descriptive words and continuous velocity values.

Multi-source data heterogeneity: Data sources across domains include various characteristics and have latent relationships with the speed of traffic on the road. For example, tweets contain latent, speed-based themes, and there was a negative correlation between traffic time and traffic speed for links. Therefore, a uniform framework is needed for the model of these characteristics and for aggregating the inherent relationships between heterogeneous data for industrial speed prediction.

2 Related Work

This non-intrusive sensor is used for detecting the traffic speed and volume based on road conditions. These fixed and portable sensors are used for detecting the image recognition technologies which were texted [1]. The formation of reliable welding links in electronic assemblies is a critical problem in the manufacture of surfaces. Strict control is placed on the deposition of the welding paste to reduce welding defects and achieve high assembly yield. Modeling the time series process for quality properties of welding paste using neural networks (NN) is a promising approach that complements conventional control planning schemes deployed over the Internet. We are studying the construction of a multilayered neural network to monitor the deposition performance of the welding paste. Neural network modeling not only provides insights into process dynamics, but also predicts the future behavior of the process. The data measurements collected are used on Ball network array (BGA) packages and Quadrilateral Flat Packet (QFP) packages to illustrate NN technology, and the expected accuracy of the models is summarized [2].

Real-time road traffic prediction is to address the smart transportation technologies. As the real-time route guidance describes from the point of view of network operators, travelers involves the first step, i.e., short-term traffic prediction [3], as to overcome the existed technique which has only for some time in short-term traffic prediction. Small transportations technologies can be used which is fastest and scalable to the urban networks. Hence, this provides the predictions of speed over minimal time interval [4]. Several models are proposed for short-term traffic flow. They are univariate and multivariate.

This univariate historical average and ARIMA (Auto Regression Integrated Moving Average) and in multivariate VARIMA (Vector Auto Regressive Moving Integrated Average) and STARIMA (Space-Time Auto Regressive Integrated Moving Average) [5]. Single-point short-term traffic flow forecasting is a major key role need to the operational network models. Seasonal ARIMA (Auto Regressive Integrated Moving Average) is parametric approach to time series. Non-parametric approach is also well suited for the single-point short-term traffic flow. Past researches have shown ARIMA (Auto Regressive Integrated Moving Average) to deliver the results that are statistically [6]. It is application of STARIMA (Space-Time Auto

Regressive Integrated Moving Average) for representing the traffic flow pattern. The traffic flow data are in the form of spatial time series and collected at the specific locations and different time intervals. Main characteristics of spatial time are incorporated. The STARIMA (Space-Time Auto Regressive Integrated Moving Average) use of weighted matrices estimates basis of distances among the different locations [7].

It is a novel model. GSTARIMA (Generalized Space-Time Auto Regressive Integrated Moving Average) aim is short-term traffic flow of forecasting in urban networks. Compared to traditional methods, GSTARIMA is more flexible model where parameters are designed to different per location. Proposed model forecasting experiment is based on actual traffic data in urban networks like China, etc. [8]. Existing models that are used in short-term forecasting are univariate in nature. Extension of univariate time series model, a multivariate is involved and huge computational complexities. A different type time series model is the structural time series model that is in multivariate nature. In this, it introduces a resource and computational simple and short-term algorithm is used [9]. The transportation literature is very rich for the prediction of travel time. It degrades the prediction performance of neural networks due to uncertainty in the operation of transportation systems and also represents the uncertainty associated with predictions. For traveling, in the bus and freeway times, it applies the delta techniques for construction of prediction intervals. It depends on hyper parameter. For the selection and adjustment of hyper parameter, genetic algorithm-based method is developed [10]. In modern Intelligent Transportation Systems research, long range decisions of traffic parameters that is flow and occupancy are essential elements. Different methods are used to predict, literature is one of the best suggested neural networks for it. Due to limited knowledge, networks' optimal structure has a specific dataset, where researchers have to rely on time consuming [11]. This paper presents an illustrative introduction to the use of contrast methods for inference and learning in graphical models (Bayesian networks and random domains). We provide a number of examples of graphic schemes, including the QMR-DT database, the Sini Belief network, the Boltzmann device, and many Hidden Markov models that are not useful for running fine-grained algorithms. Then, we introduce methods of variation and we are half a general framework for making different transformations based on specific duality. Finally, we return to the examples and explain how the algorithms of variation can be formulated in each case [12].

3 TEGPAM Model Design

In this paper, we introduce models that are Location disaggregation model, Traffic topic model, System model, and Traffic-Related Tweets.

Location disaggregation model :

To address the challenge of uncertainty in locating new gender data, this section provides a classification strategy for assigning low-precision data, which are tweets and paths, to specific road links. Because only 1% of tweets have geographic coordinates, most location information is extracted from a tweet text by specifying road names or alias.

System Model :

In this module, we are developing a system with the disassembly model for site uncertainty in Twitter and track data, the traffic theme model to obscure Twitter language and GP model to capture the spatial correlation of speed sensor data. In this module, we first develop the system creation powers required for the proposed model. The system provides the new user with registration and permission to log on. Authorized users can post their Tweets. Users are also given the option to post comments. The module is designed with the features of a typical social web base, with functions that are related to the proposed model.

Traffic topic data :

To meet the challenge of linguistic ambiguity and capture traffic description in tweets, it is proposed to model traffic theme. Through road logs that contain geographic coordinates, names, and aliases, we geo-tag to tweets via road links by matching their geographic content and text content with the front end of those links, which correspond to the direction of the outside and symbolize the header. Different driving directions are referred to as different road links.

Traffic-Related Tweets :

Our goal is to quickly predict the traffic of certain links in a particular time stamp using previous and current observations from multiple sources of data, including traffic sensor data, tweets, and paths. Tweets are aggregated in the same time period and cities via the Twitter REST search API. Traffic-related tweets are initially extracted by matching at least one of the predefined vocabularies developed by domain experts, which includes terms such as "traffic," "accident," "stuck," "crash," and then classify and filter them.

Estimation Maximization Algorithm 1:

This algorithm consists of both model parameters and variation parameters. This algorithm consists of two steps.

1. E-step: we maximize the lower bound with respect to variation parameters
2. M-step: we maximize the bound with respect to the model parameters

Input:

link speed observations, traffic-related corpus, travel times, link lengths

Output:

Model parameters.

1. Estimate of GP priori.
2. Initialize the model parameters.
3. Initialize the variation parameters.
4. Calculate the initial lower bound.
5. Repeat.
6. Repeat.
7. E-step: Fix model parameters and update the variation parameters.
8. until convergence;
9. Repeat.
10. M-step: Fix variation parameters and update the model parameters.
11. until convergence;
12. until lower bound.
13. Return model parameters.

TEGPAM Test Algorithm 2:

Input:

Learned model parameters, link length, link speed observations at time t*, traffic-related time t*, travel time at t*.

Output: posterior Gaussian .

1. Repeat.
2. Update variation parameters.
3. until convergence:
4. Return variation parameters.

4 Conclusion and Future Scope

This paper proposes a new potential framework for predicting road traffic speed with multiple cross-cutting data. The existing work is based mainly on speed sensor data, which suffer from data scarcity and low coverage. In our work, we have to deal with the challenges resulting from the explosive multi-source data, including location uncertainty and the ambiguity of language data homogeneity by using the site rating model traffic on a topic model for traffic speed Gaussian Process Model. Experience with real data shows the effectiveness and efficiency of our model. For future work, we plan to implement the kernel-based distribution public, so the traffic prediction framework can be applied in real-time large of traffic on the network.

References

1. Lin L, Li J, Chen F, Ye J, Huai J (2018) Road traffic speed prediction: a probabilistic model fusing multi-source data. IEEE Trans Knowl Data Eng 30(7)
2. Ho S, Xie M, Tang L, Xu K, Goh T (2001) Neural network modeling with confidence bounds: a case study on the solder paste deposition process. IEEE Trans Electron Packag Manuf 24(4):323–332
3. Yu X, Prevedouros PD (2013) Performance and challenges in utilizing non-intrusive sensors for traffic data collection. Adv Remote Sens 2:45–50
4. Min W, Wynter L (2011) Real-time road traffic prediction with spatial-temporal correlations. Transp Res 19:606–616
5. Kamarianakis M, Prastacos P (2004) Forecasting traffic flow conditions in an Urban network: comparison of multivariate and univariate approaches. Transp Res Rec 1857:74–84
6. Jegadeesan R, Ram S, Kumar MN (2013) Less cost any routing with energy cost optimization. Int J Adv Res Comput Networking Wireless Mobile Commun 1(1)
7. Jegadeesan R, Ram S, Janakiraman R (2013) A recent approach to organise structured data in mobile environment R.Jegadeesan et al/(IJCSIT) Int J Comput Sci Inf Technol 4(6):848–852 ISSN 0975–9646 Impact Factor 2.93
8. Jegadeesan R, Ram S (2013) Enrouting technics using dynamic wireless networks. Int J Asia Pacific J Res Ph.D. Res Scholar 1 Supervisor 2. 3, pp Print-ISSN-2320–5504 impact factor 0.433
9. Jegadeesan R, Ram S, Tharani MS (2013) Enhancing file security by integrating steganography technique in linux kernel. Global J Eng Design Technol G.J. E.D.T. 2(5):9–14, ISSN 2319–7293
10. Ramesh R, Kumar VR, Jegadeesan R (2014) NTH third party auditing for data integrity in cloud. Asia Pacific J Res vol I, issue XIII, ISSN: 2320–5504, E-ISSN-2347–4793 vol I, issue XIII, pp Impact Factor 0.433
11. Vijayalakshmi BJC, Jegadeesan R (2014) SUODY-preserving privacy in sharing data with multi-vendor for dynamic groups. Global J Eng Design Technol G.J. E.D.T. 3(1):43–47. ISSN 2319–7293
12. Jegadeesan R, Ram S, Karpagam T (2014) Defending wireless network using randomized routing process. Int J Emerg Res Manage Technol

Personality Prediction from Text of Social Networking Sites by Combining Myers–Briggs and Big Five Models

Gaurav Katare, Ankur Maurya, and Divya Kumar

Abstract An eager desire to predict the personality of a person is not as new after the invention of social media. It is useful for understanding the psychology and behavior of users on social networking sites. The researchers are exploring the usefulness of personality prediction for different purposes such as in organization development, marketing, health care, dating suggestions, and personalized recommendations. A lot of research is done in this field, but to achieve higher accuracy is still challenging. In this paper, we have combined Myers–Briggs and Big Five model to extract the features of a person for personality prediction. We analyzed and compared our method with many existing machine learning algorithms. The accuracy from our methodology is up to 80.39%.

Keywords Big Five model · Personality prediction · Machine learning

1 Introduction

A person's personality describes all the aspects of life. It depicts the thought pattern, feelings, properties predicted by them describe a person's behavior and also influence activities of daily life like emotions, motivations, preferences, health, and many others. The consistency along with the situation, individualism (all individuals have different behaviors), the stability based on time are three criteria that are important to signify a personality. As technology advances, the use of social sites such as Facebook, Instagram, and YouTube is also increased. This type of platform is used to express the sentiments, feelings, emotions, and expectations. Alike in doing so, users also share their personal information such as likes and dislikes and professions. All this information can be extracted from social networking sites easily. This extracted

G. Katare · A. Maurya (✉) · D. Kumar
Department of Computer Science and Engineering, MNNIT Allahabad, Prayagraj, Uttar Pradesh 211004, India
e-mail: ankur.maurya.pbh@gmail.com

D. Kumar
e-mail: divyak@mnnit.ac.in

A. Kumar et al. (eds.), *Proceedings of the International Conference on Cognitive and Intelligent Computing*, Cognitive Science and Technology,
https://doi.org/10.1007/978-981-19-2350-0_36

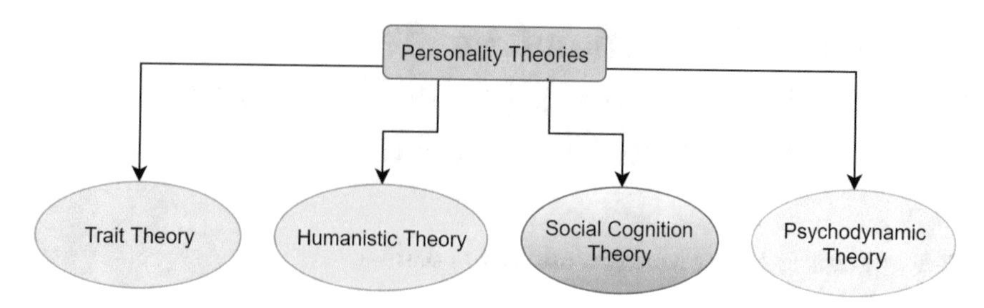

Fig. 1 Personality theories

data can give businesses an opportunity. It helps in connecting with their customers and is also used in understanding their needs. Thus, it can improve the quality of product or service which we want to get accordingly. Based on their posts, researchers have predicted health problems like mental or stress levels.

To understand the personalities, personality theory is made. The different types of personality theories [1] are given in Fig. 1.

- **Trait Theory:** It says that the personality of human personality is made of characteristics or traits which influence or make a person act in some specific way. Every traits are the blueprint of a person behavior such as a person can be introvert, extrovert, aggressive, social, ambitious, and many others.
- **Humanistic Theory:** It suggests that the full potential of a person can be achieved by shifting their basic needs to self-realization (Maslow's humanistic theory).
- **Psychodynamic Theory:** Psychodynamic theory describes personality in accordance of unconscious psychological processes (for example, desires and fears which are not completely known to us), and facts that a child experiences in childhood is important in shaping and building adult personality.
- **Social Cognition Theory:** The social interaction is the main aspect of cognition theory. It says that the behavior of a person is highly affected by the environment where he lives.

In this paper, Sect. 2 discusses the literature review. The proposed approach is given in Sect. 3. Results are given in Sect. 4. The conclusion is discussed in Sect. 5.

2 Literature Review

The prediction is done in research to get an early idea about anything whether a machine, technology, or human being [2, 3] for its improvement before being exploited or failure. The prediction of personality can be done using the speech of person [4, 5] or text. The personality prediction is done using the questionnaire method, semantic similarity, machine learning, and deep learning (Fig. 2).

Howlader et al. [6] proposed the topic modeling approach by using the term frequency-inverse document frequency (TF-IDF) and linear Dirichlet allocation

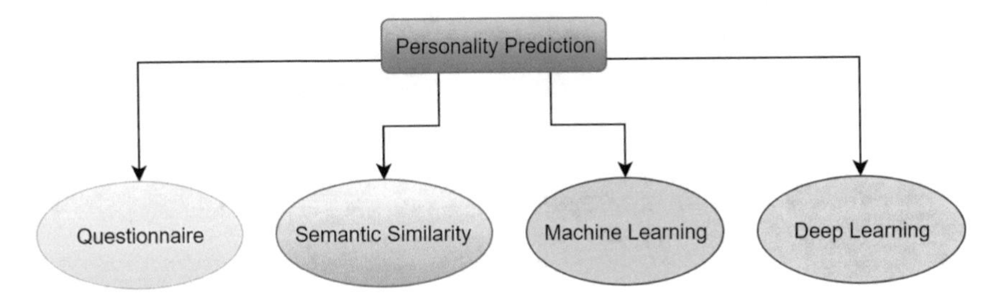

Fig. 2 Personality predictions

(LDA) as feature selection from Facebook data to predict personality. Tandera et al. [7] introduced deep learning methods. A review of personality prediction using online social media site is done by Kaushal and Patwardhan [8].

Hall et al. [9] examine the hypothesis that a prediction of personality from digital foot print data can make better by prediction of nuances and its score. Lynn et al. [10] studied the importance of message-level attention for five-factor personality to calculate the relative weight of users on social networking sites.

Cho et al. [6] proposed an algorithm for the vehicle before crash occur. This pre-crash algorithm has all the information about crash scenarios before a crash happened. The pre-crash algorithm decides if a crash has occurred or not based on the knowledge. After this if necessary, it sends signals to post-crash algorithm to deploy the airbag. Harish et al. [7] and Famiglietti et al. [8] performed rescue function to passenger as well as to the vehicle.

3 Proposed Approach

The steps of proposed approach are shown in Fig. 3.

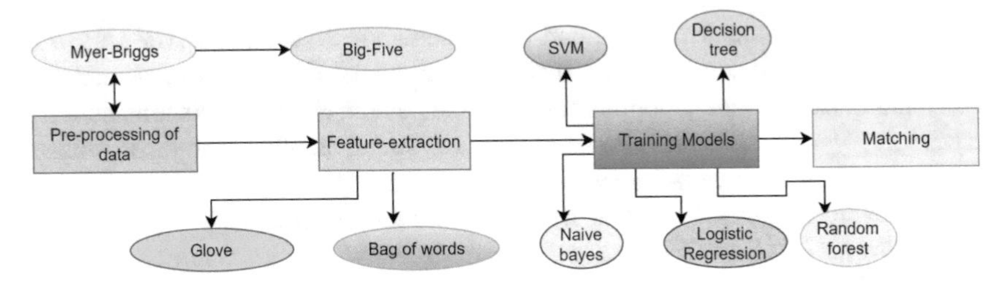

Fig. 3 Steps of proposed approach

	Subjective	Objective
Deductive	Intuition/Sensing	Introversion/Extraversion
Inductive	Feeling/Thinking	Perception/Judging

Fig. 4 MBTI

3.1 *Preprocessing of Data*

The preprocessing of data is done using Myers–Briggs and Big Five models.

3.1.1 Myers–Briggs

How people make a decision by perceiving the world is psychological preferences indicated from self-report questionnaire from the Myers–Briggs type indicator (MBTI). It was designed to identify the best-suited occupation for a person and lead healthier, happier lives.

From Fig. 4, the combinations as four pairs of preferences lead to 16 possible combinations.

3.1.2 The Big Five Personality Model

It is a kind of taxonomy for personality traits. This model is also known as OCEAN model. The Big Five personality traits are a suggested taxonomy, or grouping, for personality traits. It is also known as the five-factor model (FFM) and the OCEAN model. On applying actor analysis on personality surveys (used to facilitate understanding of human behavior), we can reveal semantic associations. The semantic association refers to some words which are used to tell the aspects of personality which are most commonly applied for the same person. Suppose someone narrates as conscientious is more likely to be careful and attentive rather than disorganized. The theory identifies five factors which are shown in Fig. 5.

The research [11] shows that the five-factor model "*captures some real attributes of the individual that are expressed as patterns of perception in the context of people's daily lives.*"

Five factors
Openness to experience (inventive/curious vs. consistent/cautious)
Conscientiousness (efficient/organized vs. easy-going/careless)
Extraversion (outgoing/energetic vs. solitary/reserved)
Agreeableness (friendly / compassionate vs. challenging / detached)
Neuroticism (sensitive / nervous vs. secure / confident)

Fig. 5 Five factors

Table 1 Co-occurrence matrix

	the	dog	sat	on	hat
the	0	1	0	1	1
dog	1	0	1	0	0
sat	0	1	0	1	0
on	1	0	1	0	0
hat	1	0	0	0	0

3.2 *Feature Extraction*

3.2.1 GloVes

Vector representation of words is obtained in GloVes feature extraction. SpaCy model uses GloVes vector to build semantic words. The ratio of word–word co-occurrence probability has a lot of potential to encode some form of semantics is the main idea behind this model. The semantic relationship can be formed using co-occurrence matrix which can be calculated using conditional probability of words. Example. Let's say we have a phrase "*the dog sat on hat.*" Based on its co-occurrence matrix (Table 1), we can form a semantic relationship.

3.2.2 Bag of Words

The number of times a word appears in a text is calculated by converting arbitrary text into vectors of fixed length. This is called the vectorization of words. A bag of words help us to extract features from the text for use in models.

For example, we want to vectorize the text (document) (given in Table 2).

"the dog sat in the cap"

Table 2 Vectorized data

Document	the	dog	sat	in	cap	with
the dog sat	1	1	1	0	0	0
the dog sat in the cap	2	1	1	1	1	0
the dog with the cap	2	1	0	0	1	1

"the dog sat"
"the dog with the cap"

- **Determine the Vocabulary**

The set of words in document set are listed. In above text, we will get six words "with," "the," "dog," "sat," "cap," and "in."

- **Count the frequency**

To vectorize our document set, we have to count the frequency of each word appearing.

Vectorized form of each documents is

- *the dog sat:* [1, 1, 1, 0, 0, 0].
- *the dog sat in the cap:* [2, 1, 1, 1, 1, 0].
- *the dog with the cap:* [2, 1, 0, 0, 1, 1].

3.3 *Classifier*

In this step, we train the different models for all five personality traits on preprocessed data and split the data into 80:20 ratio. We are using following classification models:

- SVM
- Decision tree
- Naive Bayes
- Logistic regression
- Random forest.

4 Proposed Approach

For personality prediction, we need to evaluate all five traits of a person with the help of text. All the traits except neuroticism are calculated using models trained on Big Five + Myers–Briggs; the text in Fig. 6 is given as input for personality prediction in our approach.

The result of personality prediction from above text is given in Fig. 7.

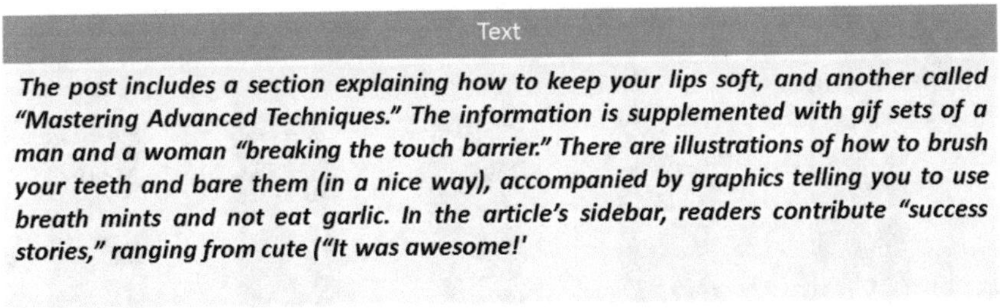

Text

The post includes a section explaining how to keep your lips soft, and another called "Mastering Advanced Techniques." The information is supplemented with gif sets of a man and a woman "breaking the touch barrier." There are illustrations of how to brush your teeth and bare them (in a nice way), accompanied by graphics telling you to use breath mints and not eat garlic. In the article's sidebar, readers contribute "success stories," ranging from cute ("It was awesome!'

Fig. 6 Input text for personality prediction

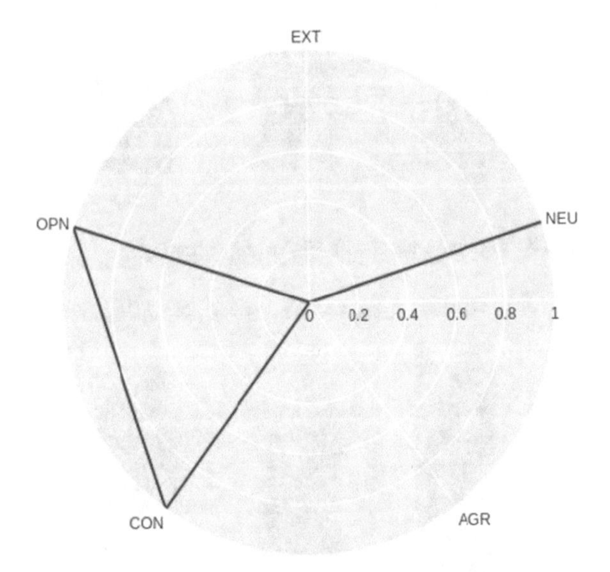

Fig. 7 Result of personality prediction

The blue line shows the personality of person who is open, neurotic, and conscientious. The accuracy graph of Big Five dataset with bag of words preprocessing and Big Five dataset with GloVe preprocessing is given in Figs. 8 and 9. In the graph, cEXT, cNEU, cAGR, cCON, and cOPN are five parameters of Big Five model.

The accuracy of our approach using Big Five + MBTI dataset and bag of words preprocessing is given in Fig. 10.

The accuracy using Big Five + MBTI dataset and GloVe preprocessing is given in Fig. 11. The highest accuracies from these model are obtained as: EXT: 77.18%, NEU: 61.74%, AGR: 75.51%, CON 70.34 %, OPN 80.39%.

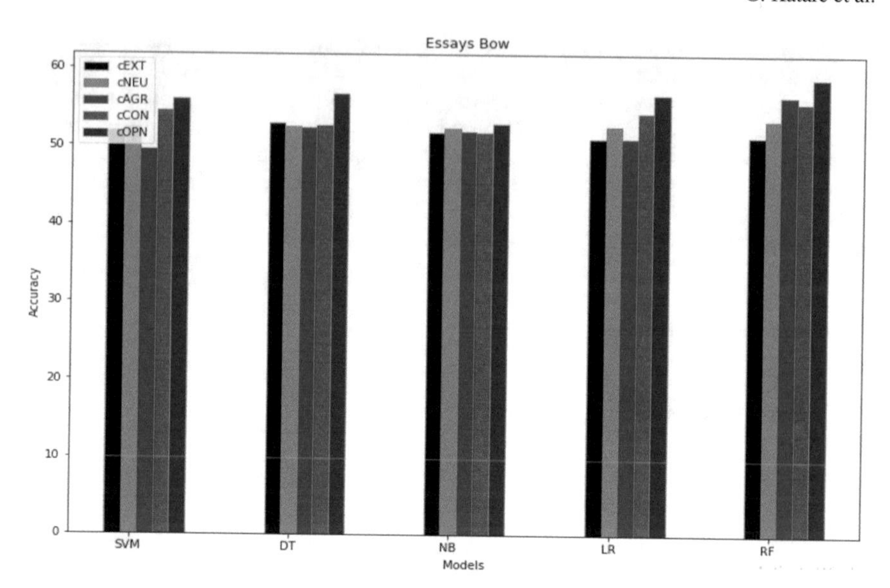

Fig. 8 Big Five dataset with GloVe preprocessing

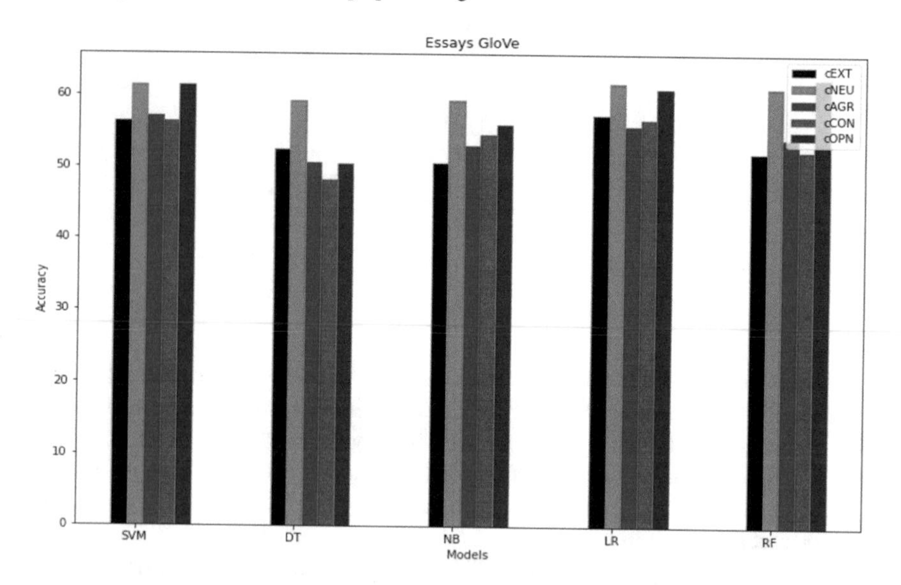

Fig. 9 Big Five dataset and bag of words preprocessing

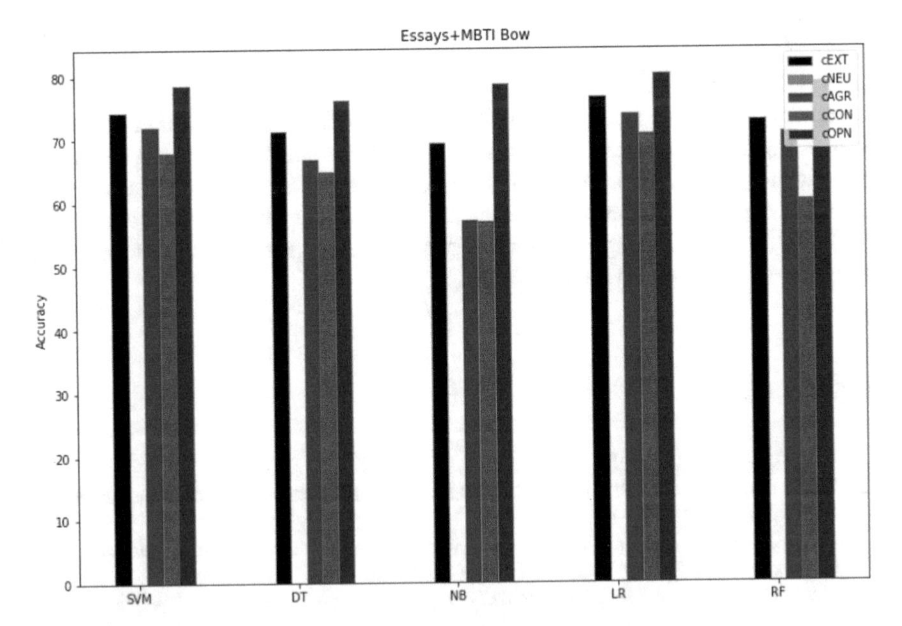

Fig. 10 Big Five + MBTI dataset and bag of words preprocessing

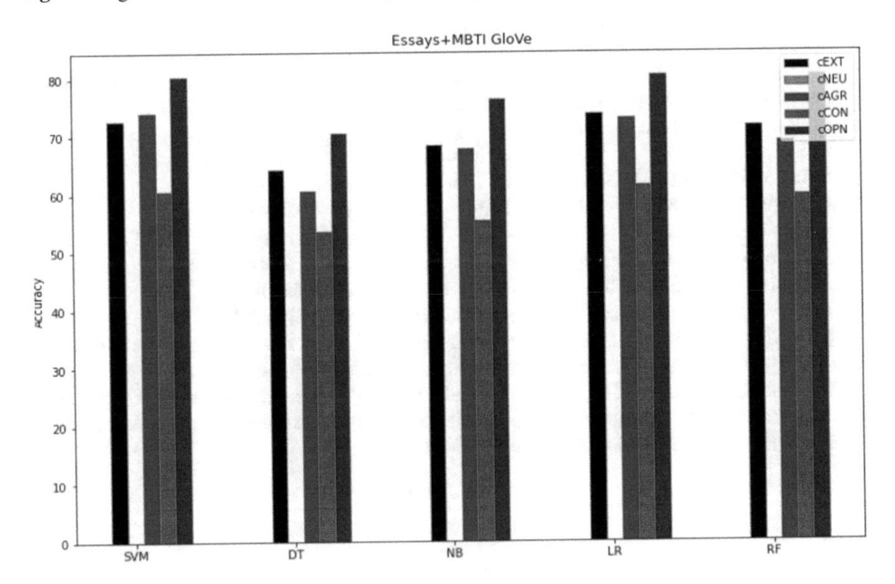

Fig. 11 Big Five + MBTI dataset and GloVe preprocessing

5 Conclusion

Based on our results, we concluded that by combining Big Five and MBTI dataset we are getting a more accurate model, and GloVe preprocessing is performing better than bag of word (BOW). Also, random forest is performing better than other classification models. Moreover, for unlabeled data, we need a more diverse dataset for better prediction accuracy. Using the positives of Myers–Briggs and Big Five models, the accuracy can be achieved up to 80.39%.

References

1. Peck D, Whitlow D (2019) Approaches to personality theory. Routledge
2. Maurya A, Kumar D (2020) Reliability of safety-critical systems: a state-of-the-art review. Qual Reliab Eng Int 36(7):2547–2568
3. Maurya A, Kumar D (2021) Translation of SysML diagram into mathematical petri net model for quantitative reliability analysis of airbag system. Int J Veh Des (In Press)
4. Maurya A, Aggarwal RK (2016) Speaker recognition for noisy speech in telephonic channel. In: 2016 2nd International conference on applied and theoretical computing and communi cation technology (iCATccT). IEEE, New York, pp 451–456
5. Maurya A, Kumar D, Agarwal RK (2018) Speaker recognition for Hindi speech signal using MFCC-GMM approach. Proc Comput Sci 125:880–887
6. Howlader P, Pal KK, Cuzzocrea A, Madhu Kumar SD (2018) Predicting Facebook-users' personality based on status and linguistic features via flexible regression analysis techniques. In: Proceedings of the 33rd annual ACM symposium on applied computing, pp 339–345
7. Tandera T, Suhartono D, Wongso R, Prasetio YL et al (2017) Personality prediction system from Facebook users. Proc Comput Sci 116:604–611
8. Kaushal V, Patwardhan M (2018) Emerging trends in personality identification using online social networks—a literature survey. ACM Trans Knowl Discovery Data (TKDD) 12(2):1–30
9. Hall AN, Matz SC (2020) Targeting item–level nuances leads to small but robust improvements in personality prediction from digital footprints. Euro J Personal 34(5):873–884
10. Lynn V, Balasubramanian N, Andrew Schwartz H (2020) Hierarchical modeling for user personality prediction: the role of message-level attention. In: Proceedings of the 58th annual meeting of the association for computational linguistics, pp 5306–5316
11. Rauthmann JF, Sherman RA, Nave CS, Funder DC (2015) Personality- driven situation experience, contact, and construal: how people's personality traits predict characteristics of their situations in daily life. J Res Personal 55:98–111

Reliability Analysis of Safety–Critical Airbag System with Bumper Guard During a Collision

Ankur Maurya and Divya Kumar

Abstract The failure of safety–critical systems can lead us to huge damage and catastrophic consequence. Airbags are such safety–critical systems whose failure can create a lot of trouble for human life or sometimes even death. These systems should be highly reliable and designed with great precision. In this paper, the reliability of airbag is computed using failure rate of sensors by modeling it into petri net. These are tested on three vehicle models SUV, pickup truck, and van. The bumper guard absorbs the jerk in case of vehicle collision which prevents the sensors to receive the impact of the collision, resulting in airbag non-inflation. When the bumper guard is used the reliability of deploying the airbag system is decreased by 9.91% to 50%, depending upon the velocities of the collision.

Keywords Safety–critical systems · Airbag systems · Reliability analysis · Bumper guard

1 Introduction

Safety–critical systems are designed to meet the high reliability requirements [1]. The failure of safety–critical systems is calamitous. Vehicle airbags are examples of such a system. The examples of other safety–critical systems are shown Fig. 1. Vehicle airbags require in-depth knowledge of parameters to build the system. The reliability of system is high if sensors have low-failure rate. The bumper guard also affects the reliability of airbag deployment. The bumper guard absorbs the jerk in case of vehicle collision which prevents the sensors to receive the impact of the collision, resulting in airbag non-inflation. The unemployment of airbags depends on the change in the velocity of the vehicle during a collision.

A. Maurya (✉) · D. Kumar
Department of Computer Science and Engineering, MNNIT Allahabad, Prayagraj, Uttar Pradesh 211004, India
e-mail: ankur.maurya.pbh@gmail.com

D. Kumar
e-mail: divyak@mnnit.ac.in

A. Kumar et al. (eds.), *Proceedings of the International Conference on Cognitive and Intelligent Computing*, Cognitive Science and Technology,
https://doi.org/10.1007/978-981-19-2350-0_37

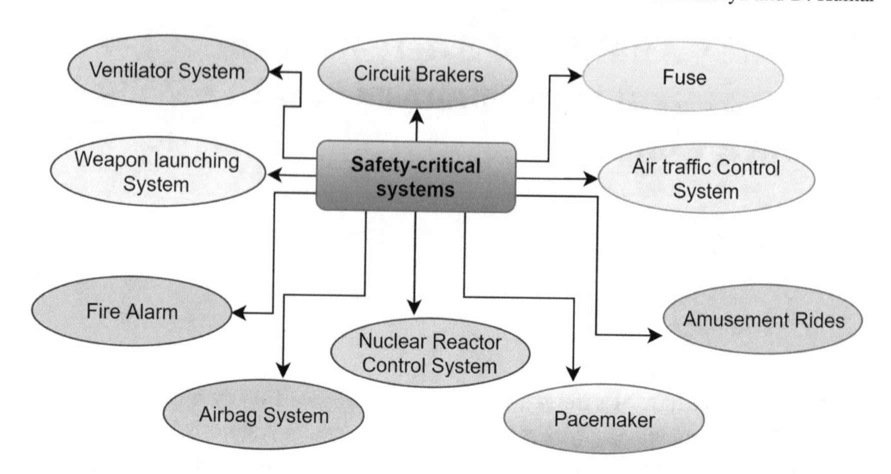

Fig. 1 Examples of safety–critical systems

The reliability of airbag is highly dependent on the working of its sensors. If sensors work properly and with very less failure rate, the reliability of the system will be very high.

The reliability is the probability of a system working in a given time without being failed. Airbag is expected to deploy in case of collision of vehicle occurs. A higher delta-v (change in velocity) is needed for airbag inflation during collisions. A bonnet in a car is designed to resist the shock made by accident. In the presence of a bumper guard, the jerk is distributed among all the components of the vehicle, made it very risky. There are various field where reliability can be used like nuclear power plant, ventilator, amusement rides, and IoT speech processing [2]. The literature review is given in Sect. 2. The important terms of this paper are studied in preliminaries section. The working and results are discussed in Sect. 4. Finally, the conclusion is given in Sect. 5.

2 Literature Review

Safety of passengers is very important during the collision of vehicles. Researchers carried a lot of work to improve the safety algorithm of vehicle, so that the airbag opens on time. In case of collision, an airbag is the last hope of passenger to save their life. Komari et al. [3] studied the method to calculate the threshold for acceleration and proposed an adjustment method for airbag module algorithm. They also analyzed different database on crashes and models of vehicles. The different crash report is studied on NASS database [4–6].

Wood et al. [6] worked on NASS database and analyzed airbag inflation on time-based methodology for different vehicle models. Over 6000 crash reports from the year 2000–2011, deployment threshold is quantified for Ford and general motor vehicles.

Cho et al. [7] proposed an algorithm for the vehicle before crash occur. This pre-crash algorithm has all the information about crash scenarios before a crash happened. The pre-crash algorithm decides if a crash has occurred or not based on the knowledge. After this if necessary, it sends signals to post-crash algorithm to deploy the airbag. Haris et al. [8] and Famiglietti et al. [9] performed rescue function to passenger as well as to the vehicle.

3 Preliminaries

The structure of airbag system is shown in Fig. 2. The important terms are discussed below.

- **Accelerometer**: The acceleration of vehicle is measured through accelerometer.
- **Crash sensor**: This sensor detects the crash of the vehicle when acceleration is differed by the threshold range.
- **Airbag controller**: The airbag is opened when it gets signal from airbag controller.
- **Current/Voltage sensor**: When the acceleration of vehicle go well with greater than 3 g ($g = 9.8$ m/s^2), the voltage sensor generates the current pulse.
- **Ignition unit**: It converts analog signal to digital signal after getting the command from airbag controller.
- **Airbag**: After getting the commands from ignition unit, the airbag deploys.

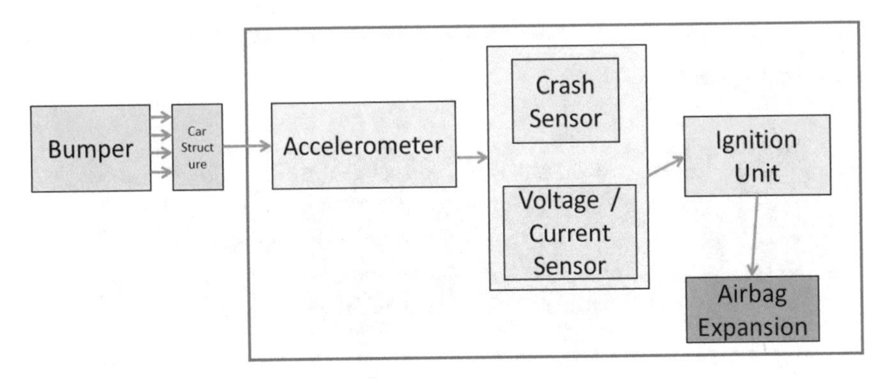

Fig. 2 Structure of airbag system

4 Working and Results

4.1 Reliability Calculation of Airbag System Using Petri Net Model

The reliability of airbag is calculated using petri net model by modeling failure of sensors in the airbag. The steps are shown in Fig. 3 Petri net contains five tuples (Places, transition, input, output, and marking). The meaning of places used in our analysis is shown in Fig. 3.

The transition is shown by T1, T2, T3, T4, T5, T6, and T7. Fe' is showing the failure events occur at any sensor. Example, if airbag controller is giving the signal to ignition unit there may be the chance that IU_unit fail to work (Fe' at T7). Similarly, using all the failure possibilities of sensors, the petri net model is made (Fig. 4), and reliability is calculated using time net tool [10] by modeling the scenarios of airbag system (see Fig. 5). The reliability of the airbag system after computation is 0.99138769 [11]. The token is initially at P_start.

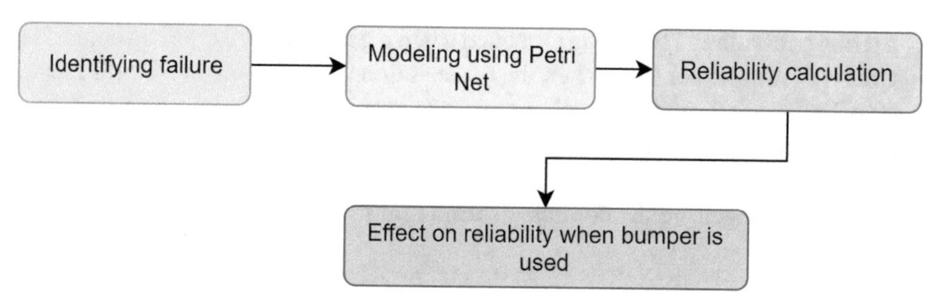

Fig. 3 Working steps

Places	Meaning
P_start	Staring point
Acc_START	Accelerometer is in working state
Airbag _controller	Airbag controller is working
IU_WORK	Ignition unit gets signal from air bag controller
VS_ON	Voltage sensor gets signal from airbag controller
CS_START	Crash sensor gets signal from controller when crash is detected
Air_OPEN	Airbag open after getting signals from airbag controller and ignition unit

Fig. 4 Meaning of places used in time petri net

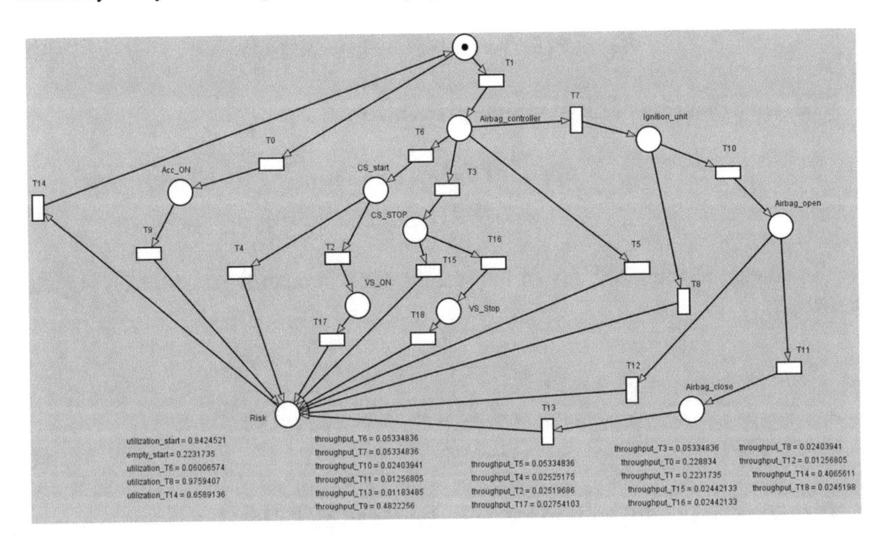

Fig. 5 Reliability using time petri net

4.1.1 Effect on Reliability if Bumper Guard is Used

The probability that airbag is not deployed during a collision for various vehicle model in presence of bumper is given in Fig. 6 [12]. If the velocity is high and the difference of velocities (Δv) is 18 or 19 for the sedan model, then the probability that airbag is not deployed is 0.10. Here, $N = 5436$ shows sample size, i.e., the number of times the collision occurs for a vehicle.

Reliability is defined using Eq. 1. The overall reliability of the system in presence of bumper (R_S), when the vehicle speed is higher and bumper is present (P(airbag failure)$= 0.10$ from Fig. 6) is calculated below (**collision = high**)

$$\text{Reliability} = 1 - \text{P(Failure)} \tag{1}$$

If R_a is the reliability of airbag system ($R_a = 0.99138769$). The new reliability of the system in presence of bumper can be defined using Eq. 1.2.

Severity	Sedan N=5,436	SUV (N=1,493)	Pickup Truck (N=990)	Probability of failure (with bumper)
High	[-18, -19]	[-21, -22]	[-21, -22]	0.10
Average	[-13, -14]	[-16, -17]	[-17, -18]	0.25
low	[-7, -8]	[-9 -10]	[-11 -12]	0.50

Fig. 6 Δv for vehicles

$$R_S = P(Ra) * (1 - P(\text{airbag failure})) \tag{2}$$

The new reliability of the system is given below

$$\begin{aligned} R_S &= (0.9913)(1 - 0.10) \\ &= 0.89217 \end{aligned} \tag{3}$$

The change in the reliability of the system when bumper is present (R_d) is given below

$$\begin{aligned} R_d &= 0.9913 - 0.89217 \\ &= 0.09913 \\ &= 9.91\% \end{aligned} \tag{4}$$

The overall reliability of the system is decreased by 9.91%, if bumper is present and collision occurs with high velocity. Similarly, we can calculate the change in the reliability if (**collision = average**).

$$\begin{aligned} R_S &= ((0.9913)(1 - 0.25)) \\ &= 0.743475 \end{aligned} \tag{5}$$

The change in the reliability of the system when bumper is present (R_d) is given below

$$\begin{aligned} R_d &= 0.9913 - 0.743475 \\ &= 0.247825 \\ &= 24.78\% \end{aligned} \tag{6}$$

The overall reliability of the system is decreased by 24.78% if bumper is present and collision occurs with high velocity. Similarly, we can calculate the velocity when collision is very low, which is exactly half of the overall reliability of the system. Because if the velocity of vehicle is low and there (Δv) is very less, then the chances of airbag deployment are lesser. The graphs of reliability in presence of bumper and without bumper are shown in Figs. 7 and 8. The reliability of individual vehicle at different velocities are shown in Fig. 9. The $(-4, -5)S$ represents sedan model where velocity change is from 4 to 5 when any collision occurs. Here, $S =$ sedan, $V=$ SUV, and $T =$ pickup truck.

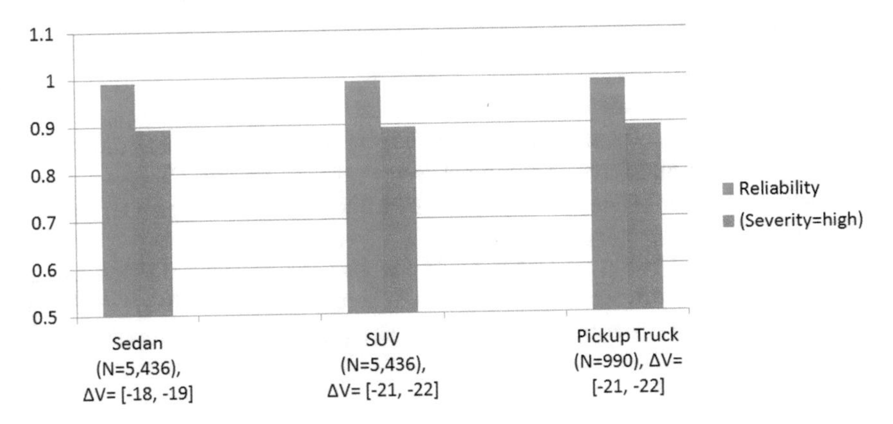

Fig. 7 Reliabilities when collision is high

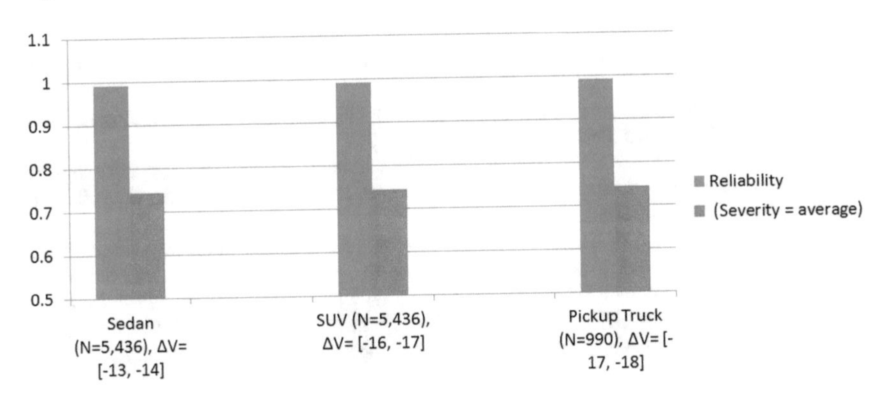

Fig. 8 Reliabilities when collision is average

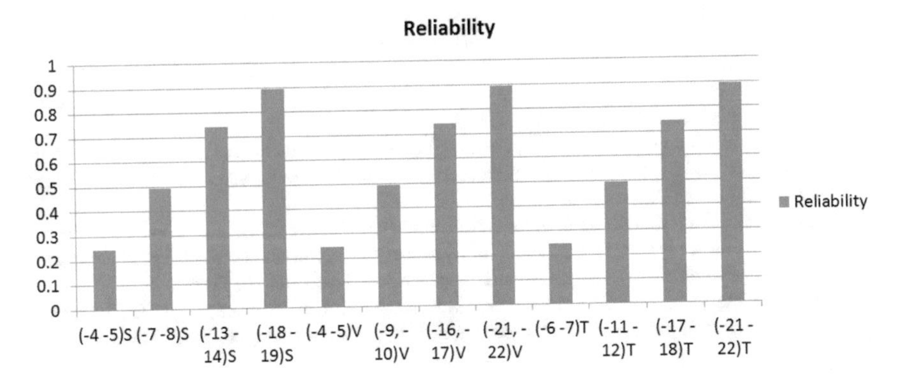

Fig. 9 Individual reliability of vehicle models

5 Conclusion

The high reliability of safety-critical systems is necessary to save the life and disastrous effect of failure. These systems should be properly designed. Since, the failure cannot be stopped completely but we good design can minimize the effect of calamitous failure. In this paper, we have computed the effect of the bumper on the deployment of airbag during collision.

In our study, we found that the reliability of airbag system 0.99138769 and is decreased with the presence of bumper guard. If bumper is present, the reliability of airbag system decreases from 9.91% to 50% based on the velocities of collision.

References

1. Maurya A, Kumar D (2020) Reliability of safety-critical systems: a state-of-the-art review. Qual Reliab Eng Int 36(7):2547–2568
2. Maurya A, Kumar D, Agarwal RK (2018) Speaker recognition for Hindi speech signal using MFCC-GMM approach. Proc Comput Sci 125:880–887
3. Komari O, Javidi M, Merrill Z, Toosi K (2020) Side airbags deployment range from analysis of event data recorder database of real-world incidents. In: SAE Int J Passenger Cars-Electron Electr Syst 12(07-12-02-0012):163–171
4. Kusano KD, Kusano SM, Gabler HC (2011) Automated crash notification algorithms: evaluation of in-vehicle principal direction of force (PDOF) estimation algorithms. In: Proceedings of the Third TRB international conference on road safety and simulation, Indianapolis, IN
5. Hampton CE, Gabler HC (2009) Influence of the missing vehicle and CDC only delta-v reconstruction algorithms for prediction of occupant injury risk. Biomed Sci Instrum 45:238–243
6. Wood M, Earnhart N, Kennett K (2014) Airbag deployment thresholds from analysis of the NASS EDR database. SAE Int J Passenger Cars-Electron Electr Syst 7(2014-01-0496):230–245
7. Cho K, Choi SB, Lee H (2011) Design of an airbag deployment algorithm based on precrash information. IEEE Trans Veh Technol 60(4):1438–1452
8. Haris O, Wicaksono S, Kurniawan B, Edytia G, Darmawan A (2020) Design & analysis of external airbag system at the Toyota Venza vehicle. In: 2020 6th international conference on computing engineering and design (ICCED). IEEE, New York, pp 1–5
9. Famiglietti N, Hoang R, Fatzinger E, Landerville J (2020) Evaluation of general motors event data recorder performance in semi-trailer rear underride collisions. SAE Int J Adv Curr Pract Mob 2(2020-01-1328):3496– 3512
10. Zimmermann A (2017) Modelling and performance evaluation with timenet 4(4):300–303
11. Maurya A, Kumar D (2021) Translation of SysML diagram into mathematical petri net model for quantitative reliability analysis of airbag system. Int J Veh Des (In Press)
12. Lee F, McCleery CH, Ngo C, Limousis-Gayda M, Hashish R (2019) Probability of frontal airbag deployment in bumper-bumper and underride collisions. Technical report, SAE Technical Paper

Prediction and Solution for Crop Disorder Using Machine Learning and Image Processing Techniques for Agricultural Domain

N. J. Anasuya, P. K. Shravani, and C. Lavanya

Abstract Crop disorders are the major quality constraints for the farmers of broad acre of crops in India. Microorganisms can be fungal, bacterial, viral, or nematodes and that can harm the crops or the plants above or below the ground. Identifying and knowing how and when to control the disorders are the challenge faced by the farmers. Silkworm is the backbone of sericulture. Commercial silk is obtained by cultivating various species of silkworm, of which Bombyx mori, originally from Asia, is the most commonly and commercially used. Mulberry is significant sole nourishment for mulberry silkworm (Bombyx mori). These mulberry plants include a high danger of yield disappointment and are over the highest expensive for production, so should be addressed well indeed. Initially, the images undergo acquisition, preprocessing, feature extraction, and at the final step, the classification result alongside appropriate pesticide or fertilizer and its proportion to be used, which will be displayed on the user interface. Here in this paper, we evaluate various features based on image processing: EHD (Edge Histogram Descriptor), HOG, and GLCM for automatic detection of mulberry leaf disorders using machine learning. These features are evaluated on various classification algorithms: SVM, KNN, and decision tree (DT). Our experimental evaluations indicate that EHD and KNN obtain the best accuracy of 97.5%. We find that KNN used with Euclidean distance formula gives the accurate result.

Keywords Mulberry leaves · EHD · HOG · GLCM · SVM · KNN and DT

1 Introduction

The Primary Industry Department and Regional Development has a strong research and has an focus to assist industries to reduce the impact of crop disorders on all crops. India is second largest producer of silk. Sericulture has following highlights: high work potential, gives energy to economic status of the town, minimized gestation

N. J. Anasuya (✉) · P. K. Shravani · C. Lavanya
Don Bosco Institute of Technology, Bangalore, Karnataka, India
e-mail: anasuyaprakash@dbit.co.in

A. Kumar et al. (eds.), *Proceedings of the International Conference on Cognitive and Intelligent Computing*, Cognitive Science and Technology,
https://doi.org/10.1007/978-981-19-2350-0_38

period and maximum returns, female friendly occupation, and eco-accommodating activity. India is the only country where all the five recognized commercial silks are made, mainly mulberry, tropical tasar, oak tasar, eri, and muga. Mulberry silk is manufactured more frequently in India from Karnataka, Andhra Pradesh, Tamil Nadu, Jammu and Kashmir, and West Bengal, whereas non-mulberry silks are manufactured in the states of Jharkhand, Chattisgarh, Orissa, and Northeastern. Silkworm provides mulberry silk. Bombyx mori L exclusively benefits from the leaves of mulberry. Silkworms are reared and domesticated indoors. Other uses of mulberry leaves are seen in the fields of health and skin care. These mulberry plants include a high pace of yield disappointment and are over the top expensive for creation, so should be dealt quite well.

The customary methodology for identification of diseased leaves is through naked eye, this method is not precise and brings about use of wrong pesticides or synthetic compounds, and hence, pulverizing the yield and farmers may watch just when the plant is completely contaminated which may cause an infection flare-up in the homestead. There is an option for consulting experts, but farmers living in remote villages may not be able to afford traveling expenses because the laboratory test results could take a few days, farmers have to travel many times, and government experts visit the farm of a few villages, but not regularly. Our goal is to overcome these problems using a farmer-friendly system where the result involves cure of the disease and the fertilizer or pesticide proportion to be used are displayed on the user interface.

2 Literature Survey

It is essential to perceive the past research done concerning this field to have the option to effectively progress the correct way. Prof. Sanjay B. has explained crop or plant leaf disorder detection and also explained the preprocessing step includes RGB to HSI color space conversion [1–3]. Kapil Prashar has implemented an expert system for recognizing crop diseases. The pooling and overlapping have been used with MLP with adaptable layering component. The proposed model localizes the disorder region with accuracy 96% [4]. K. Prashar has explained the double layered approach for the overlapping layer of classification to minimize the errors using the K-nearest neighbor (KNN) and support vector machine (SVM). In feature extraction, methodology used is color co-occurrence developed through GLCM [5]. Z. N. Reza, F. Nuzhat, N. A. Mahsa and M. H. Ali and Y. Dandawate and R. Kokare have discussed about DISORDERS IN LEAVES using image processing techniques. Srinivas and Hanumaji [6] have discussed the SIFT, SURF, and HOG [7, 8]. Namrata R. Bhimte proposed disease detection of cotton leaf by using GLCM, SVM [9]. The functions of classifiers with sickness proof give accuracy of 98.46%. Ahmad Nor Ikhwan Masazhar presented the oil palm leaf disease symptom detector using digital image processing and classification techniques. 13 sorts of features from leaf images are extracted using K-means clustering techniques. The result shows that accuracy of

97% is achieved from SVM for Chimera and 95% for Anthracnose [10]. Priyadarshini Patil here has compared the SVM performance, random forest, and ANN on potato leaves on the sam test dataset. ANN scores highest 92% accuracy along with 84% accuracy of SVM and RF of 79 [11].

3 Mulberry Diseases

Mulberry leaf belongs to the morus family that is majorly grown in south India where most of the sericulture industries are concentrated. The main food source being mulberry for silkworm *Bombyx mori* L. is a perennial crop. Since mulberry can be grown throughout the year, diseases and pests are susceptible for the plant. Different pathogens causing the diseases in mulberry leaves are fungi, bacteria, viruses, and nematode.

1. **Leaf Spot**

Moricola Cooke is a cause of the leaf spot disease. The fungus belongs to class Deuteromycetes Moniliales. In rainy season (June–December), the disease is extremely common and prevails until January–February. Depending on the season and variety, it decreases the leaf yield by around 10–20%.

Symptoms: At initial stage, formation of small brownish and irregular spots on the leaves is observed. At later stages, when disease gets severe, these spots enlarge and get combined with each other and lead to shot holes. Severely affected leaves turn yellow and eventually fall [12] (Fig. 1).

2. **Powdery Mildew**

Phyllactinia corylea Karst causes powdery mildew. It is the most common and widely spread disease. The disease causing fungus belongs to the class Ascomycota. Powdery mildew can be seen during the rainy and winter season (July–March). These mildew infected leaves should not be fed to silkworm as this affects the development

Fig. 1 Leaf spot

Fig. 2 Powdery mildew (white patches)

and growth of silkworm which gives poor cocoon yield and hence the depreciation of silk quality.

Symptoms: White powdery patches on the lower surface of leaves are appeared that make disease easily identifiable. When disease gets severe, the whole surface of leaf turns to blackish color and patches are spread on whole leaf [12] (Fig. 2).

3. **Leaf Rust**

Cerotelium fici causes leaf rust disease. Under the class *Irnperfect fungi*, the pathogen belongs to the order Uredinales and the family Uredinaceae. It is a common disease which occurs during the winter season (November–February). Diseases attack the matured leaves majorly. Rapid, premature defoliation of the leaves can occur in the presence of rust, resulting in leaf shortages during late age (adult silkworm) rearing.

Symptoms: The leaf rust disease is identified by the presence of numerous circular to oval head sized, blackish to brown spots on the surface of leaves. The leaves affected by the disease turn yellowish, and when disease becomes severe, the leaves wither away pre-maturely (Fig. 3).

4. **Leaf Curl**

The presence of Taeniothrips causes the curling of mulberry leaves. They damage the epidermal tissue, affected leaves look dull due to loss of moisture content, and they mature early. This disease is commonly observed in India and Sri Lanka. Though the presence of pests is seen throughout the year, outbreaks occur in summer season.

Symptoms: Disfigurement of the apical shoots in the affected leaves, decreased growth rate, wrinkling, and curling which later turns dark green in color. In extreme infestation, the leaves are light yellow. Parts affected become frangible. Symptoms are usually referred as Tukra disease (Bushy top) (Fig. 4).

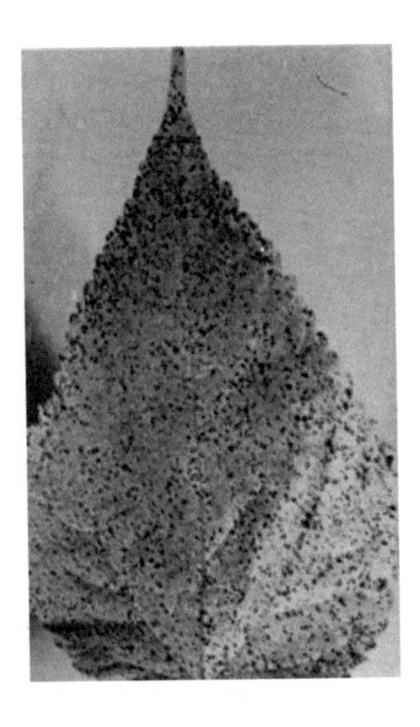

Fig. 3 Leaf rust

Fig. 4 Leaf curl

5. **Leaf Thrips**

Pseudodendrothrips Niwa mori cause thrips. In South India, this species of thrips is widely recorded to cause damage to mulberry.

Symptoms: Infected leaves show white streaks or blotches which on maturity become yellowish brown (Fig. 5).

6. **Nitrogen Deficiency**

Symptoms: Plant growth is slow and weak with less branching. Young tender leaves loose the pigment and become chlorotic (Fig. 6).

Fig. 5 Leaf thrips

Fig. 6 Nitrogen deficiency

4 Proposed System

The proposed scheme for the development of this system includes two stages:

1. Disease identification.
2. Disease management (Fig. 7).

1. **Disease identification is done using the steps as followed**:

A. Acquisition of image
B. Preprocessing of image
C. Extraction of feature
D. Classification of image

A. **Image Acquisition**: The images obtained from digital camera are subjected for further preprocessing.
B. ***Image Preprocessing***: The images obtained from the camera are subjected to preprocessing for increasing the quality of the images. The preprocessing steps include the following:

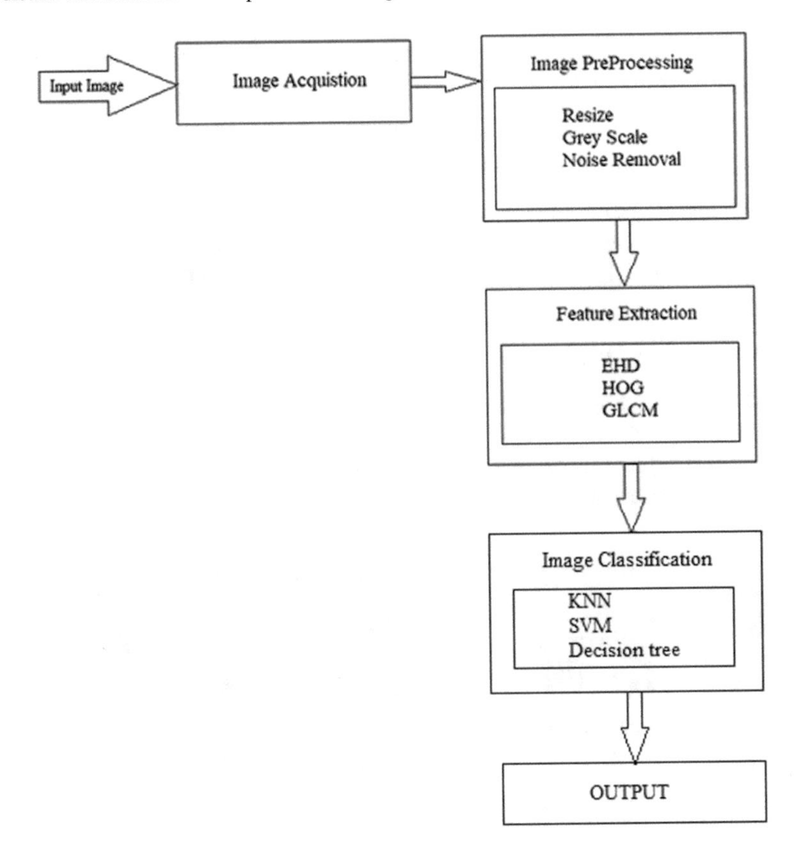

Fig. 7 Model architecture

Resize: The acquired images are fed to AI algorithm which fluctuate in their size; therefore, a base size should be established for all images fed into AI algorithm.

Grayscale: We need 24 bits to store an RGB color image, but gray scale image only 8 bit. Less memory with 33% is required to store gray scale image than RGB image.

Noise Removal: The visible grains and random variation of image intensity are a noise. Here, we use median filtering to remove the noise. Median filtration is a nonlinear process which helps to reduce impulsive noise. The median filter is broadly utilized in computerized image processing just because it preserves edge properties unlike Gaussian blur technique.

C. ***Feature Extraction***: It is a method that selects or combines variables into features, by reducing the amount of data that has to be processed, but still precisely describing the original dataset ***EHD***: This algorithm is broadly used for detection of shapes. It describes relative frequency of occurrence of each local area of edges of 5 types called image block. The sub-image is characterized as non-overlapping blocks dividing the image space into 4 * 4. The features of image block are defined by generating histogram for edge distribution every image block (Fig. 8).

Table 1 sums the total semantics of EHD with 80 histogram bins. Each histogram bin should be quantified as normalized.

HOG: Histogram of oriented gradients is a feature descriptor which is mainly used for extracting features from the image. This is usually used in computer vision tasks for object detection. HOG is a descriptor that concentrates on the structure or shape of an object.

Steps to Calculate HOG

- Preprocessing: Image will be resized depending on the size of the pixel.
- Gradient images are calculated: To find a descriptor of HOG, firstly we need to find vertical and horizontal gradient. Magnitude and direction are determined using these equations-

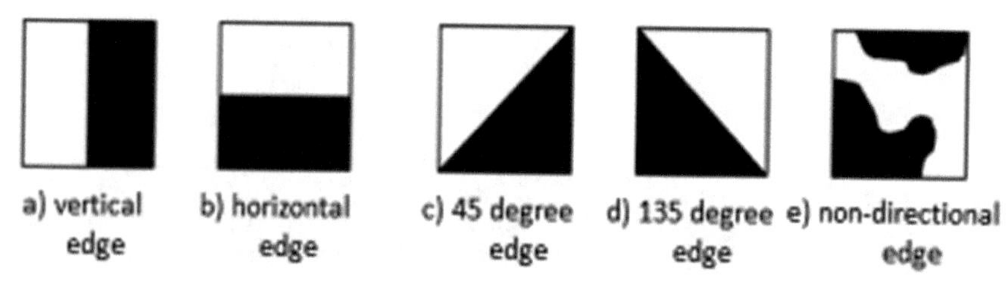

Fig. 8 Five types of edges in EHD

Table 1 Semantics of local edge bins

Histogram bins	Semantics
Bin Counts[0]	Vertical edge of sub-images at (0,0)
Bin Counts[1]	Horizontal edge of sub-images at (0,0)
Bin Counts[2]	45-degree edge of sub-images at (0,0)
Bin Count[3]	135-degree edge of sub-images at (0,0)
Bin Counts[4]	Non-directional edge of sub-images at (0,0)
Bin Counts[5]	Vertical edge of sub-images at (0,1)
:	:
Bin Counts[74]	Non-directional edge of sub-images at (3,2)
Bin Counts[75]	Vertical edge of sub-images at (3,3)
Bin Counts[76]	Horizontal edge of sub-images at (3,3)
Bin Counts[77]	45-degree edge of sub-images at (3,3)
Bin Counts[78]	135-degree edge of sub-images at (3,3)
Bin Counts[79]	Non-directional edge of sub-images at (3,3)

$$g = \sqrt{g_x^2 + g_y^2}$$
$$\theta = \arctan\frac{g_y}{g_x} \quad (1)$$

where magnitude is g and direction is θ.

- **Calculate HOG in 3 * 3 cells**: Here, image is divided into 3 * 3 cells and HOG is calculated for each of these cells. The gradient of image contains 2 values per cell which sums up to $3 \times 3 \times 2 = 18$. The histogram consists of 9 bins which corresponds to angle from 0, 20, 40, …, 180.
- **Block Normalization**: Normalization is need because gradients of normalized image are sensitive to overall lighting.
- **Calculation of Feature vector**: Final feature vector is calculated by combining complete image, vectors to make one vector.

GLCM: Gray level co-occurrence matrix is a method for extracting statistical texture features. This is created using "graycomatrix," and we can derive several statistics from GLCM using "graycoprops." The statistics mentioned below provide information about the texture of an image. The statistics are as follows:

- **Energy**: It is a measure of information when an operation is formulated under a probability framework like maximum a priori estimation with a Markov random fields

$$f7 = \sum\sum p(i, j)^2 ij \quad (2)$$

- **Mean:** Formula is used to calculate mean where $p(i, j)$ is value of pixel at a point (i, j) with size of an image $m * n$.

$$M = 1/MN \quad \sum i = 1^M \sum i = 1^M (P(i, j) - \mu)^2 \quad (3)$$

- **Standard Deviation**: SD is the estimation of a gray pixel value's mean square deviation from its mean value.

$$\sigma = 1/MN \quad \sum i = 1^M \sum i = 1^M (P(i, j) - \mu)^2 \quad (4)$$

- **Smoothness**: Relative smoothness measures gray level contrast which is utilized to find relatively smooth descriptors.

$$R = I - I/1 + \sigma^2 \quad (5)$$

- **Root Mean Square:** It gives value of RMS for either all column or row of input, with the vectors for a specified dimension of the input/for whole input.

$$\sqrt{\sum} i = 1^M |\mu\, ij|^2/M \quad (6)$$

- **Inverse Difference Moment:** It measures image texture. It ranges from 0.0 for highly textured image and up to 1.0 for un-textured image.

$$H = \Sigma i, jp(i, j)/1 + |i - j|. \tag{7}$$

- **Entropy:** It measures uncertainty associated with random variable

$$fx = -\sum \sum p(i, j) \log(pi, j) \tag{8}$$

- **Contrast**: It is the difference between dark and light areas of an image.

$$f_2 = \sum_{n=0}^{Ng-1} n^2 \left\{ \sum_{i=1}^{Ng} \sum_{i=1}^{Ng} p(i, j)\, |i - j| = n \right\} \tag{9}$$

- **Homogeneity:** It transfers the value to the GLCM diagonal which calculates the tightness of distribution of the elements in the GLCM.

$$f_9 = \sum_i \sum_j I/I + (i - j)^2 p(i, j) \tag{10}$$

- **Variance:** By using square window around a center pixel, variance map of an image is calculated.

$$f_{11} = \sum_i \sum_j (i - j)^2 p(i, j) \tag{11}$$

D. ***Image Classification:*** It refers to labeling the images into a number of pre-defined classes.

SVM: Multiclass SVM is implemented against all approach in which there is a binary SVM for every separate members of the class from the classes with other members. K SVM is constructed where K is the number of classes.

$$\begin{aligned} \min_{w^i, b^i, \xi^i} \quad & \tfrac{1}{2}\left(w^i\right)^T w^i + C \textstyle\sum_{j=1}^{l} \xi_j \left(w^i\right)^T \\ & \left(w^i\right)^T \phi\left(x_j\right) + b^i \geq 1 - \xi_j^i, \quad \text{if } \; y_j = i \\ & \left(w^i\right)^T \phi\left(x_j\right) + b^i \leq -1 + \xi_j^i, \quad \text{if } \; y_j \neq i \\ & \xi_j^i \geq 0, j = 1, \ldots, l \end{aligned} \tag{12}$$

The parameter of penalty is where the xii training data is mapped to higher dimensional space and C.

After solving Eq. (12), functions of K-decision are there.

$$\left(w^1\right)^T \phi(x) + b^1, \left(w^k\right)^T \phi(x) + b^k. \tag{13}$$

The largest value of decision function is in the class which is x.

$$\text{class of } \; x \equiv \arg\max_{i=1}\left(\left(w^i\right)^T \phi(x) + b^i\right) \tag{14}$$

KNN: The KNN algorithm follows non-parametric approach which is used for regression and classification. KNN classification shows the last output. The classified object is according to its neighbor's majority votes. The object which is near the K value is considered, where positive integer is K. We have implemented Euclidean, Minkowski, Chebyshev, Cityblock, and Cosine distance formulas (Fig. 9).

The KNN classifier execution is specially dictated by K and the separation matric connected. Gauge is influenced by determination of the affectability of the measure K, dedicated by the separation of Kth closest neighbor, and diverse K yields restrictive class probability. KNN classification is the straight forward and the foremost of all machine learning calculations.

Decision Tree: Supervised learning is used by decision tree algorithm. Decision tree is used to solve regression as well as classification problems. Representation of tree is used to solve the problem. One class represents the leaf node. Initially, we take

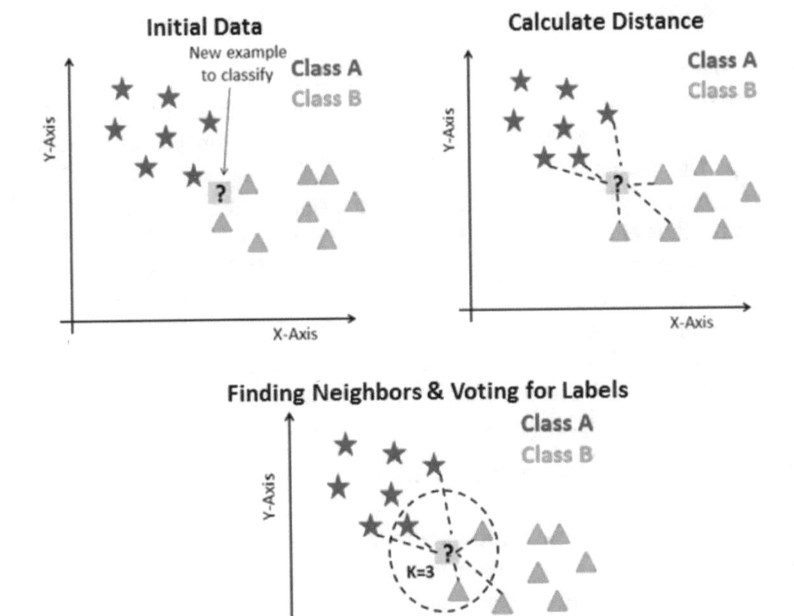

Fig. 9 Working of KNN

the full training data to be the root. Recursive distributions are made on the basis of attribute values. Information gain is used in each level for the identification of the root node attribute. To accurately define gaining information, we need to define a measure commonly used in entropy, the information theory.

The classification of test data is done through traversing the tree. Test is applied at the root node to decide which child node the record will come up next. This process is repeated until the record reaches leaf node. Each leaf node has a unique path from root to root. The path is a rule used to sort the record.

2. **Disease Management**

After the image undergoes acquisition, preprocessing, feature extraction, and classification, the result will be obtained in the user interface. This contains cure of the disease and the fertilizer or pesticide proportion to be used which helps the farmer to manage the disease.

5 Results

The implementation of our project is done in MATLAB. We have basically concentrated on 6 types of mulberry leaf diseases. All these datasets were captured from mobile camera. As the first step an input image is selected from the folder which undergoes image acquisition and preprocessing where it undergoes rescaling, converting into gray scale and removal of noise. The trained images are fed to feature extraction. The features of texture are extracted and further used for classification. Output is obtained at last which indicates the disease name along with its control measures for the better management of the crop (Figs. 10, 11, 12, 13, 14, and 15).

The algorithm proposed is applied on a database of 600 mulberry leaf images. During the training or experiment, the database was divided into two sets "the training set" containing 420 images and "the testing set" containing 180 images. For the feature extraction, we have used EHD, HOG, and GLCM. For classification, we have used SVM, KNN, and decision tree, where we have applied cross-validation for SVM. After applying these algorithms, we have obtained the following results:

Accuracy: Accuracy indicates the correctly predicted data points from all the data points.

Specificity: Specificity is a ratio between how much is correctly classified as negative and how much is the actual negative. Specificity = TN/(TN + FP).

Precision: Precision describes what proportion of positive identification was actually right.

Recall: Recall provides us what proportion of actual positives was right.

F1 score: F1 score is defined as the recall and precision harmonic mean. F1 Score = 2*(Recall * Precision)/(Recall + Precision).

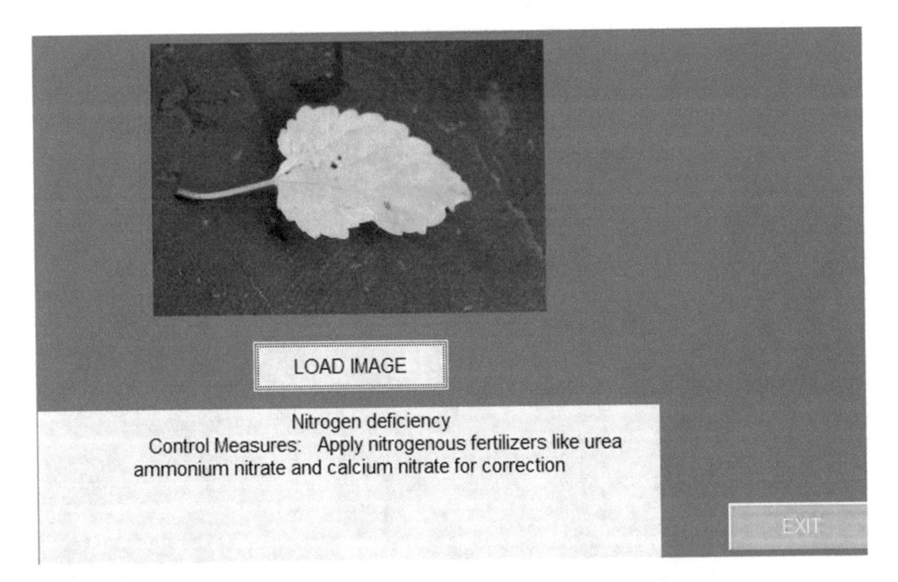

Fig. 10 Nitrogen deficiency

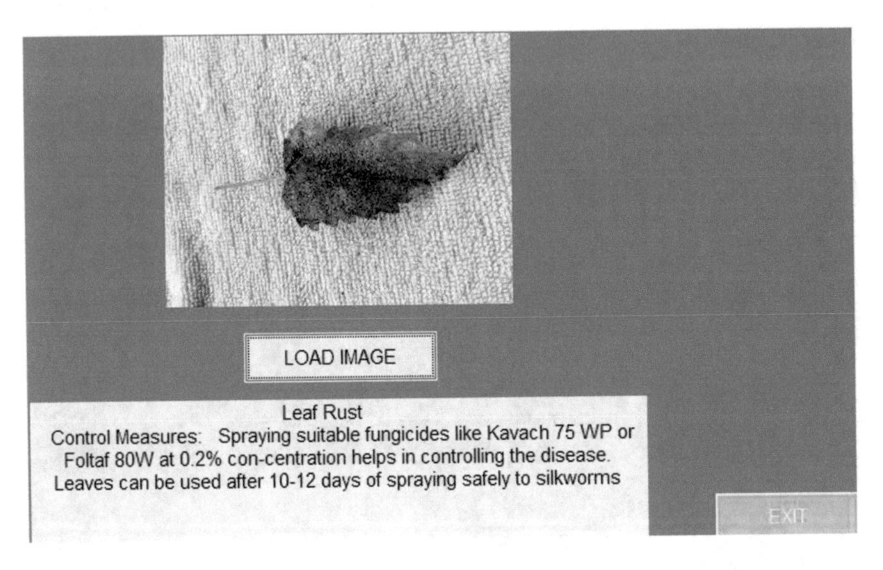

Fig. 11 Leaf rust

Here, we have developed the system which is useful for the farmers to prevent and identify the crop disorder in an easy way. The proposed paper evaluates various image processing-based features: EHD, GLCM, and HOG for detecting diseases automatically with the use of machine learning. These features are evaluated on various machine learning algorithms: SVM, KNN, and DT. Our experimental results

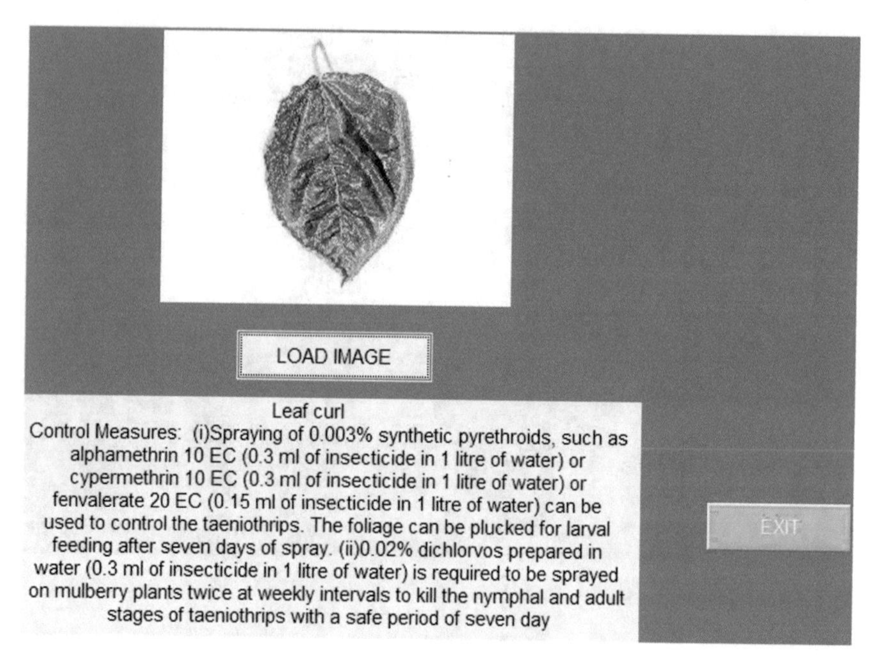

LOAD IMAGE
Leaf curl
Control Measures: (i)Spraying of 0.003% synthetic pyrethroids, such as alphamethrin 10 EC (0.3 ml of insecticide in 1 litre of water) or cypermethrin 10 EC (0.3 ml of insecticide in 1 litre of water) or fenvalerate 20 EC (0.15 ml of insecticide in 1 litre of water) can be used to control the taeniothrips. The foliage can be plucked for larval feeding after seven days of spray. (ii)0.02% dichlorvos prepared in water (0.3 ml of insecticide in 1 litre of water) is required to be sprayed on mulberry plants twice at weekly intervals to kill the nymphal and adult stages of taeniothrips with a safe period of seven day
EXIT

Fig. 12 Leaf curl

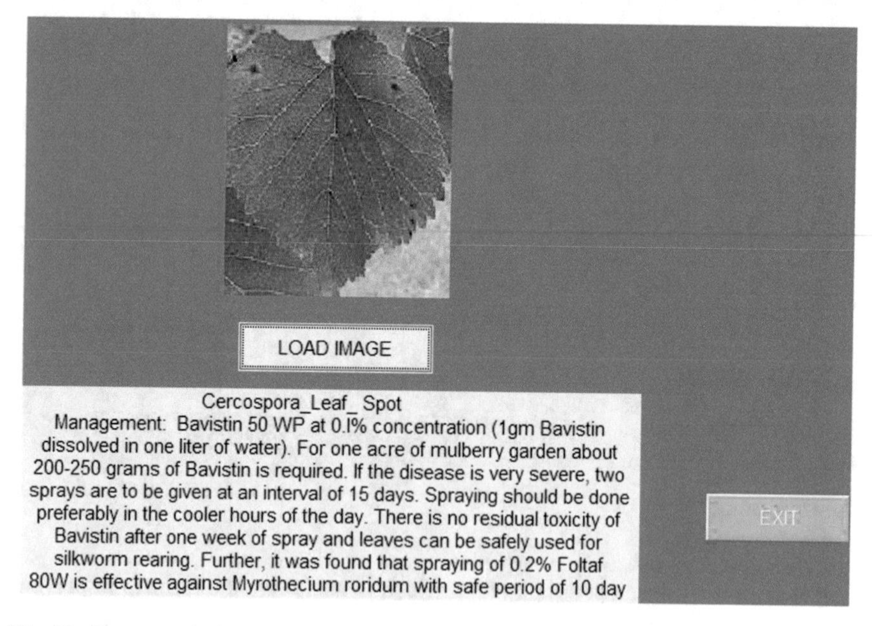

LOAD IMAGE
Cercospora_Leaf_ Spot
Management: Bavistin 50 WP at 0.l% concentration (1gm Bavistin dissolved in one liter of water). For one acre of mulberry garden about 200-250 grams of Bavistin is required. If the disease is very severe, two sprays are to be given at an interval of 15 days. Spraying should be done preferably in the cooler hours of the day. There is no residual toxicity of Bavistin after one week of spray and leaves can be safely used for silkworm rearing. Further, it was found that spraying of 0.2% Foltaf 80W is effective against Myrothecium roridum with safe period of 10 day
EXIT

Fig. 13 Cercospora leaf spot

Fig. 14 Leaf thrips

LOAD IMAGE
Powdery Mildew
Control Measures: Spraying of KarathaneEC (0.2%) or Bavistin (0.2%) is effective for the control of disease. If disease is severe, two sprays are required with an interval of 15 days. While spraying the fungicides, the lower surface of the leaves should thoroughly be drenched. The leaves can be fed to silkworm after 10 days of spraying
EXIT

Fig. 15 Powdery mildew (white patches)

indicate that EHD with KNN gives the best result with 97.50% of accuracy, 0.9778 of specificity, 0.9801 of precision, 0.9690 of recall, and 0.9746 of F1 score. GLCM with KNN gives less result with 73.75% of accuracy, 0.9169 of specificity, 0.7406 of precision, 0.7418 of recall, and 0.7412 of F1 score. The combination of other feature extraction algorithms with different classifier algorithms along with their performance measures is mentioned in Table 2, and the graph is represented in Fig. 16.

From the above result, it is observed that EHD with KNN gives 97.50% accuracy, 0.9778 specificity, 0.9801 precision, 0.9690 recall, and 0.9746 F1 score. It is observed that for this combination of algorithm we have obtained better precision which indicates that the ratio of the correctly positively predicted data by our program to all positive data is accurate.

HOG with DT gives 85.00% accuracy, 0.9806 specificity, 0.8221 precision, 0.8433 recall, and 0.8326 F1 score. It is observed that for this combination of algorithm we have obtained better specificity which indicates the true negative rate which means the proportion of actual negatives, which got predicted as the negative.

GLCM with DT gives 83.25% accuracy, 0.9723 specificity, 0.8123 precision, 0.8265 recall, and 0.8193 F1 score. It is observed that for this combination of algorithm we have obtained better specificity which indicates the true negative rate which means the proportion of actual negatives, which got predicted as the negative.

Table 2 Result comparison

Algorithm	Accuracy (%)	Specificity	Precision	Recall	F1 score
EHD + KNN	97.50	0.9778	0.9801	0.9690	0.9746
HOG + DT	85.00	0.9806	0.8221	0.8433	0.8326
GLCM + DT	83.25	0.9723	0.8123	0.8265	0.8193
GLCM + SVM	82.71	0.8333	0.9684	0.8241	0.8905
EHD + DT	78.50	0.9640	0.7645	0.7798	0.7721
EHD + SVM	75.56	0.5769	0.9192	0.7904	0.8499
HOG + KNN	74.75	0.9391	0.7762	0.7785	0.7774
GLCM + KNN	73.75	0.9169	0.7406	0.7418	0.7412

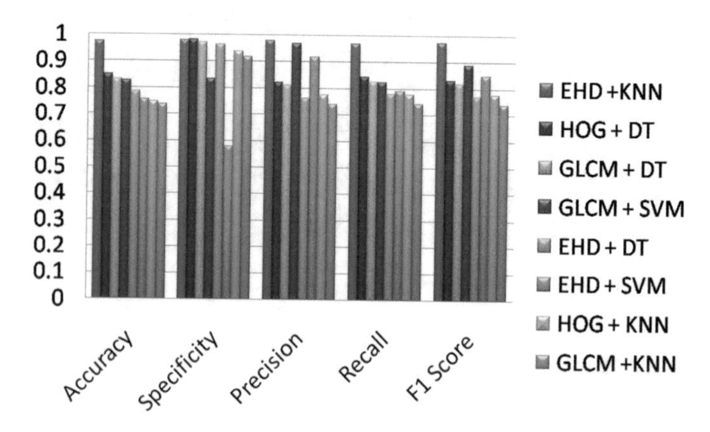

Fig. 16 Result comparison

GLCM with SVM gives 82.71% accuracy, 0.8333 specificity, 0.9684 precision, 0.8241 recall, and 0.8905 F1 score. It is observed that for this combination of algorithm we have obtained better precision which indicates that the ratio of the correctly positively predicted data by our program to all positive data is accurate.

EHD with DT gives 78.50% accuracy, 0.9640 specificity, 0.7645 precision, 0.7798 recall, and 0.7721 F1 score. It is observed that for this combination of algorithm we have obtained better specificity which indicates the true negative rate which means the proportion of actual negatives, which got predicted as the negative.

EHD with SVM gives 75.56% accuracy, 0.5769 specificity, 0.9192 precision, 0.7904 recall, and 0.8499 F1 score. It is observed that for this combination of algorithm we have obtained better precision which indicates that the ratio of the correctly positively predicted data by our program to all positive data is accurate.

HOG with KNN gives 74.75% accuracy, 0.9391 specificity, 0.7762 precision, 0.7785 recall, and 0.7774 F1 score. It is observed that for this combination of algorithm we have obtained better specificity which denotes the TN rate which means the proportion of actual negatives, which got predicted as the negative.

GLCM with KNN gives 73.75% accuracy, 0.9169 specificity, 0.7406 precision, 0.7418 recall, and 0.7412 F1 score. It is observed that for this combination of algorithm we have obtained better specificity which indicates the true negative rate which means the proportion of actual negatives, which got predicted as the negative.

6 Conclusions

In this paper, we have explained the details of the work which is useful for farmer to identify and prevent the leaf or crop diseases in an easy way. The proposed paper evaluates various image processing-based features: EHD, GLCM, and HOG for automatic detection of mulberry leaf diseases using machine learning. The used dataset combination of EHD and KNN with Euclidean distance formula gives the best result. This can be enhanced to implement in android application also. With better dataset, we can obtain better accuracy, specificity, and precision.

References

1. Dhaygude SB, Kumbhar NP (2013) Agricultural plant leaf disease detection using image processing. Int J Adv Res Electr Electron Instrum Eng 2(1):599–602 ISSN 2278-8875
2. Ananthi SVV (2012, February) Detection and classification of plant leaf diseases. Int J Res Eng Appl Sci 2(2) ISSN: 2249-3905
3. Kajale RR (2015) Detection & Recognization of plant leaf diseases using image processing and android O.S.s. Int J Eng Res General Sci 3(2), Part 2, 6–9
4. Prashar K, Talwar R, Kant C (2019) CNN based on overlapping pooling method and multi-layered learning with SVM & KNN for American cotton leaf disease recognition. In: International conference on automation, computational and technology management (ICACTM),

London, United Kingdom, pp 330–333. https://doi.org/10.1109/ICACTM.2019.8776730
5. Puspha Annabel LS, Annapoorani T, Deepalakshmi P (2019) Machine learning for plant leaf disease detection and classification—a review. In: International conference on communication and signal processing (ICCSP), Chennai, India, pp 0538–0542. https://doi.org/10.1109/ICCSP.2019.8698004
6. Srinivas D, Hanumaji K (2019) Analysis of various image feature extraction methods against noisy image: SIFT, SURF and HOG. J Eng Sci 10(2):32–36 ISSN: 0377-9254
7. Reza ZN, Nuzhat F, Mahsa NA, Ali MH (2016) Detecting jute plant disease using image processing and machine learning. In: 3rd international conference on electrical engineering and information communication technology (ICEEICT), Dhaka, pp 1–6. https://doi.org/10.1109/CEEICT.2016.7873147
8. Dandawate Y, Kokare R (2015) An automated approach for classification of plant diseases towards development of futuristic decision support system in Indian perspective. In: International conference on advances in computing, communications and informatics (ICACCI), Kochi, pp 794–799. https://doi.org/10.1109/ICACCI.2015.7275707
9. Bhimte NR, Thool VR (2018) Diseases detection of cotton leaf spot using image processing and SVM classifier. In: 2018 Second international conference on intelligent computing and control systems (ICICCS), pp 340–344. https://doi.org/10.1109/ICCONS.2018.8662906
10. Md Kamal M, Masazhar ANI, Rahman FA (2018) Classification of leaf disease from image processing technique. Indonesian J Electr Eng Comput Sci 10(1):191–200 ISSN: 2502-4752. https://doi.org/10.11591/ijeecs.v10.i1
11. Patil P, Yaligar N, Meena MS (2017) Comparision of performance of classifiers—SVM, RF and ANN in potato blight disease detection using leaf images. In: IEEE international conference on computational intelligence and computing research (ICCIC), 1–5
12. Masazhar ANI, Kamal MM (2017) Digital image processing technique for palm oil leaf disease detection using multiclass SVM classifier. In: IEEE 4th international conference on smart instrumentation, measurement and application (ICSIMA), pp. 1–6. https://doi.org/10.1109/ICSIMA.2017.8311978,2017

An Intelligent Air Pollution Vehicle Tracker System Using Smoke Sensor and GPS

K. H. Asha, K. Abhijna, Saniya Tabassum, and Shafiya Shaur

Abstract In today's modern world, automobiles have become an integral element of everyone's life. In contemporary modernized urban life, situations and circumstances have a significant impact in automobile utilization. When considering both sides of the coin, this has its own set of consequences, one of which being air pollution. Every vehicle produces emissions; however, the primary issue arises when the emissions exceed the defined limits set by India's pollution control board. The main cause of this infraction of emission standards is caused by inadequate vehicle maintenance and culminates in inadequate combustion of fuels supplied to the engine. Although vehicle emissions cannot be totally prevented, they can be significantly reduced. Using the advancement of semiconducting sensors used to monitor a myriad of gases, this project seeks to use those semiconductor sensors at vehicle emission outlets to detect the level of pollutants and display that level with a meter. When the pollution or emission level exceeds the previously specified threshold level, the car will emit a buzz to indicate that the limit has been exceeded, and the vehicle will come to a halt after a predetermined period of time, allowing the driver to park his or her vehicle. When the timer runs out, the gasoline supply to the engine is shut off, a fee is incurred, and the vehicle must be towed to a repairman or a service facility center nearby. A Raspberry Pi monitors and controls the entire process's synchronization and execution. When this project is enhanced as a tangible endeavor project, it will perk society and aid in the reduction of air pollution.

Keywords Sensor · GPS · GSM · Raspberry pi

1 Introduction

One of the primary environmental concerns is air contamination, which is crucial challenge that humanity is dealing with. Air pollutants from taxies, cars, and buses result in the damage of ground level ozone and other respiratory problems like asthma

K. H. Asha (✉) · K. Abhijna · S. Tabassum · S. Shaur
Don Bosco Institute of Technology, Bengaluru, India
e-mail: asha.kh06@gmail.com

A. Kumar et al. (eds.), *Proceedings of the International Conference on Cognitive and Intelligent Computing*, Cognitive Science and Technology,
https://doi.org/10.1007/978-981-19-2350-0_39

attacks. And also upsets off another different weather cycle. Vehicles, apart from industries, were some of the most prominent sources of pollutants. Automobiles are the most popular mode of transport. Throughout many urban environments, there seems to be a reservoir of subsurface o3. Defilements induce chronic cough and breathlessness are mostly manifestations of asthma. Additionally, pollutants from automobiles have been shown that lung tissue is weakened, and respiratory disorders are exacerbated like asthma. As a result, detecting all such defilements also keeping a continuous measurement of defilements generated by automobiles is becoming increasingly important in order to reduce this problem. The discharge of pollutants like carbon dioxide and nitrogen oxides can be monitored using semiconductor gas sensors. Discovering and regulating these gases is a vital task. As a result, a suggestion is made in this proposed system that would aid in reducing car pollution. So, by closely examining the operation of gasoline engines in automobiles and combining it with practical comprehension of microcontrollers and their outlyings, a methodology may be presented so that it help us track and minimize pollution. As a result, if pollution from a vehicle exceeds a predetermined threshold, the system will take the appropriate actions.

2 Literature Survey

Over the years, the government has enacted a number of measures to limit excretion, the majority of which have proven to be ineffective. The Central Pollution Control Board is responsible for monitoring pollution of the environment, which is part of the Ministry of Environment and Forests [1], sets the standards and schedule for implementation. This report is the most comprehensive and systematic review to date of the scientific literature on emissions, exposure, and health effects from traffic-related air pollution. These were followed by the requirement of a catalyzed converter which is connected for gasoline automobiles and the introduction of unleaded gasoline to the market. Car manufacturers were unprepared for this transformation, and as a result, the Euro II deadline was not enforced. The standards were initially introduced, and they were based on European laws. Since then, all around the world, increasingly harsh requirements have been implemented. The sequence of two-stroke engines for multiple wheels, the suspension of development of various previous model cars, and the implementation of e-controls are all regulations overseeing transportation emissions. Many evaluation sports have happened in the last decade in order to develop semiconductor gas sensors. This model focuses on examining modules: a smoke detector, a microcontroller, and gasoline. The pollutants level is determined by the smoke detector. The microcontroller compares the amount of pollution to the government's allowable limits. This air pollution may have a considerable impacts on the world environment if it is not regulated properly, and this process is typically intended to control air pollution. When the pollution or emission level surpasses the formerly predetermined limit, the vehicle emits a buzz to signal that the limit has been exceeded, and this information is sent to the traffic control room, together with the vehicle number, owner details, and the vehicle's location via GPS. The author described an embedded technique for monitoring and alerting vehicle harmful gases [2]. GSM and GPS are then used to develop a process for monitoring dangerous gases. For cars, an automated control system for air pollution detection has been refined. The GSM and GPS systems are employed to send and receive data as well as pinpoint the adjacent workstation. An author chronicles numerous automobile sensors such as safety sensors, distance sensors, night vision sensors, and so on. The government created laws and regulations for vehicles in the last few decades. Under the Ministry of Environment, the Central Pollution Control Board establishes emission standards, which were first implemented in India. Increasing knowledge of these health effects will prove valuable as well as according to a recent report from the Health Effects Institute on traffic-related air pollution. It has also frequently been claimed that cycling in heavy traffic is unhealthy, more so than driving a car. The purpose of GPS-GSM BASETRACKING SYSTEM by the authors Abid Khan and Ravi Mishra is to create composition of a new methodology that is consolidated with GPS-GSM to come up with the ensuing features such as SMS tracking and geolocation updates, real-time tracking and instantaneous communication. In most car markets, tailpipe and evaporative emissions regulations are legislated to some extent. Diesel engine tailpipe emissions are regulated as well, with a concentration on particulate and NOx emissions. The emission of fuel vapor from the vehicle during operation and while parked is taken into account by evaporative emission limits [3].

3 System Design

3.1 Existing System

The GSM modem delivers a text message to the designated sim card informing them of the incident. This permits it to stay informed of incident scenarios and notify traffic police or service centers as quickly as an event occurs. Atmel's AT89S52 microcontroller serves as the basis in this project [4].

All of the SMS alert system's functionality is delivered by this microcontroller. Filtering of signals at the inputs is also handled by it. This project is unique as in that doing so it uses its buzzer to alert neighbors, and it also sends a notification SMS to registered cell phone number.

3.2 Proposed System

The elements of the proposed system include the following: microcontroller, motor driver, global navigation satellite (GPS), and the universal mobile telecommunications system (GSMC) seem to be two distinct types of global positioning systems

(GSM), DC, smoke detector motor, and Raspberry Pi was used to detect the noise and the level of air pollution in the environment [5]. IoT based on metamorphic changes with a sensing device. When the smoke level exceeds the lopped mark, the smoke sensor will detect this and transfer the values to the Raspberry Pi (1.5 ppm). A buzzer will be generated, along with an alarming message, to the user's registered phone number, and the same will be displayed on the LCD screen, signaling that the vehicle's emission is high. If the sensed data goes above the threshold, a warning message is given to the owner using GSM .When the smoke level exceeds the threshold level, the smoke sensor will detect it again after a fixed period of time and transmit the values to the Raspberry Pi (1.5 ppm). The vehicle will stop after 1 km, as per a buzzer and an alarming message sent to the user's registered phone number. The same will be displayed on the LCD display. After the vehicle has come to a stoppage, the GPS will track the GPS values and transmit them to the RTO (cloud). RTO can track the vehicle's location and seize them using these GPS values. To restart the vehicle, the user should make payment, restart the vehicle, and then visit to the nearest service center to get the vehicle repaired.

4 Requirements

4.1 Hardware Requirements

Raspberry Pi, LCD, GSM, GPS, RFID, DC motor, power supply, buzzer, and smoke sensor.

4.2 Software Requirements

Software requirements also provide brief description of software features that are required for the software to work successfully and with minimal defects. Software requirements for the project are Dropbox, Arduino compiler, along with flash magic are employed in the implementation.

5 Block Diagram

See Fig. 1.

Entire vehicle tracker is made up of the following components:

- Raspberry Pi
- Power Supply
- GSM

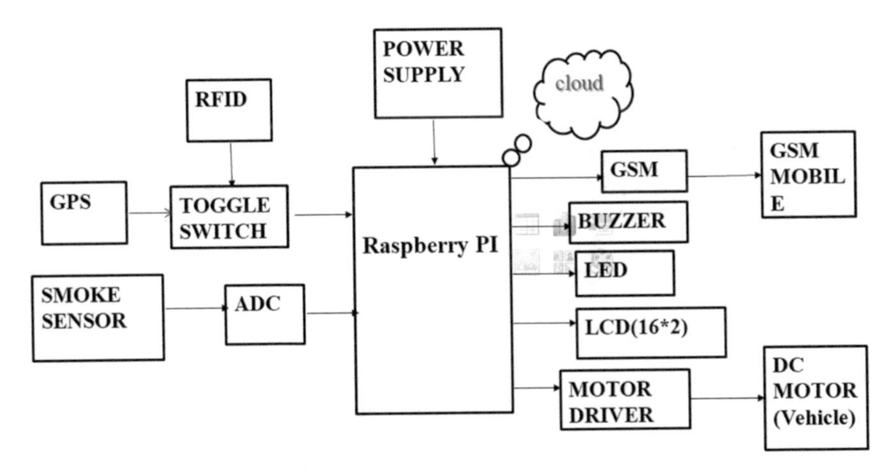

Fig. 1 Intelligent vehicle tracker methodology

- GPS
- Smoke sensor
- Buzzer
- DC motor
- RFID
- LCD.

Raspberry Pi

A Raspberry Pi is an economical device which functions as an entire desktop. It is an excellent little device to learn to write code in languages like Scratch and Python. It does everything a desktop computer can. The Raspberry Pi foundation has devised a small, minimum, and programmable computer [6, 7] (Fig. 2).

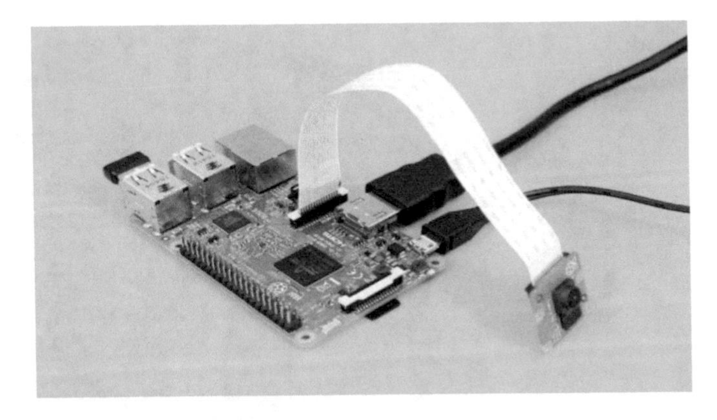

Fig. 2 Raspberry pi

Power Supply

A power fount is an electrical stratagem that provides current to connected appliances. A power supply's primary job is to metamorphose one type of electronic vigor into the other. As an outcome, electric power granting tools are occasionally interrupted. Electric add on adapters are just what they are termed. Others are independent. Power source found in automobiles are just an example of the latter. Desktop computers and consumer electronics devices were also different sorts of consumer electronics devices. Their power source is 5 V [2].

GSM

It is a GSM/GPRS wireless modem designed for reporting between a PC and a network. A GSM modem accepts a subscriber identity module (SIM) card and basically acts like a mobile phone. Receiving, sending, and deleting SMS messages in a SIM are all actions that a GSM MODEM may accomplish. SIM phonebook entries can be read, added, and searched. Make a voice call, accept it, or reject it (Fig. 3).

GPS

Device that employs a satellite-based navigation system to determine the latitude and longitude position values of an object on the ground. A GPS receiver is a navigation device, which could be able to receive information from GPS satellite (Fig. 4).

Smoke Sensor

MQ5 sensor senses the defilements emitted from the combustion of fuel from vehicles. The sensor can sense NH3, NOx, alcohol, Benzene, smoke, CO2, and some other gases, so it is perfect gas sensor [8] (Fig. 5).

Buzzer

Buzzers do have broad array of applications and can also be placed on a vast array of equipment. Typical uses of buzzer and beeper or alarm device confirm the inputs and

Fig. 3 System for networked communication

Fig. 4 Global positioning system

Fig. 5 MQ5 Sensor

device outputs by making sound. It may be mechanical, piezoelectric, and electromechanical component. Buzzers are regularly utilized to have an acoustic indication of a mechanical device's stage to a user or operator (Fig. 6).

DC Motor

It is the device which runs the rotator motor in a specified manner. It is used as a vehicle motor for project implementation and affirmation (Fig. 7).

Fig. 6 Buzzer

Fig. 7 10 DC Motor

RFID

Radio frequency identification gadget used to illustrate the remittance exercise using optical scanner [3] (Fig. 8).

LCD

It displays the information on crystal screen. It is an electronic display module and finds a wide range of applications (Fig. 9).

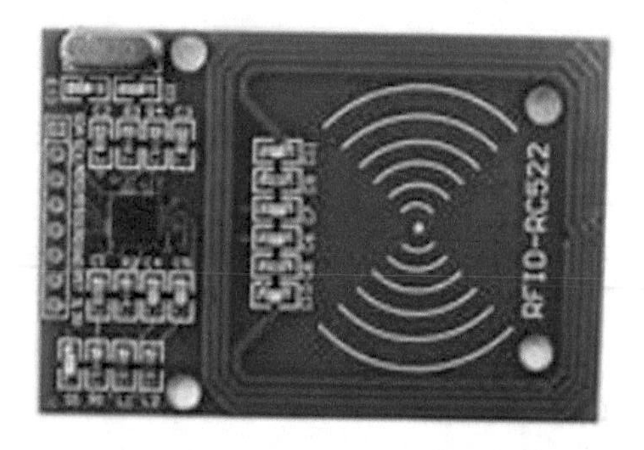

Fig. 8 RFID

Fig. 9 LCD

6 Results

Initially when the system is given the power supply LCD will provide VEHICLE EMISSION TEST as display (Figs. 10 and 11).

This is the entire view of the system (Fig. 12).

MQ135 sensing values within threshold level (Fig. 13).

Alert message displayed on LCD when emission level is beyond the threshold value (Fig. 14).

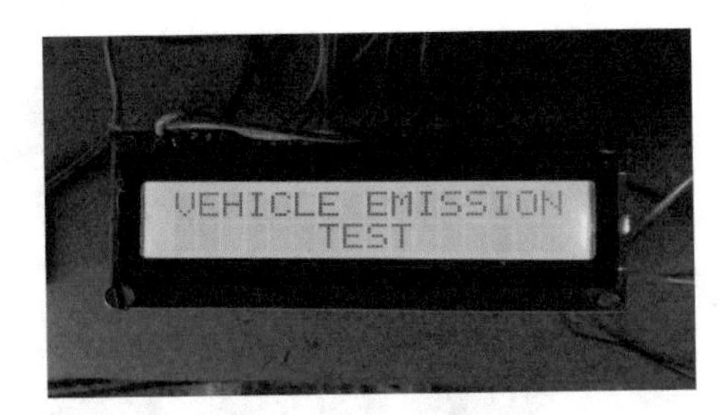

Fig. 10 Initial LCD reading

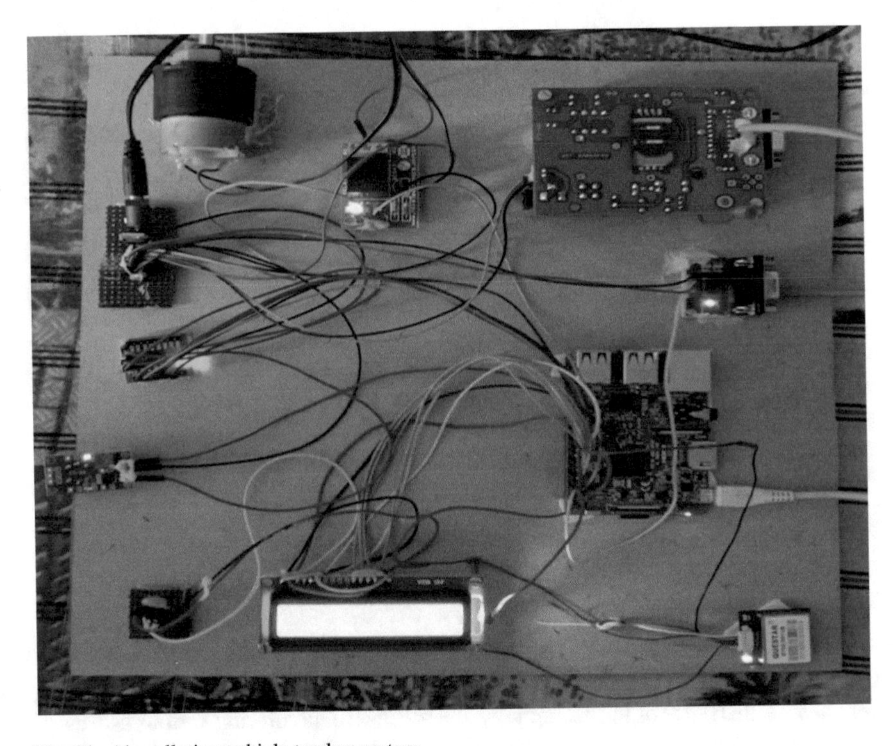

Fig. 11 Air pollution vehicle tracker system

VEHICLE EMISSION

Fig. 12 Smoke sensor readings on LCD

VEHICLE EMISSION
IS HIGH

Fig. 13 Warning message on LCD screen

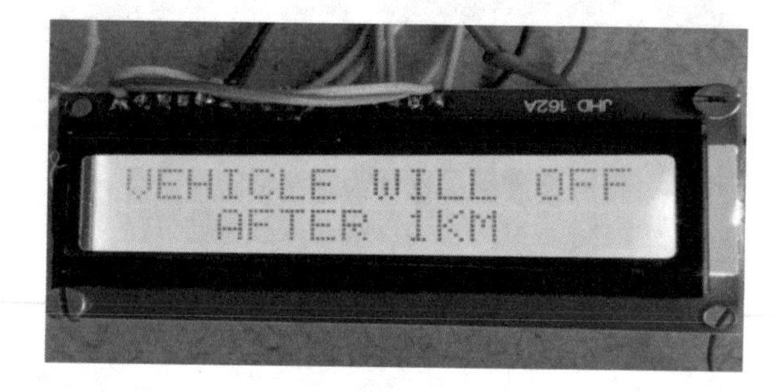
VEHICLE WILL OFF
AFTER 1KM

Fig. 14 Final warning on LCD screen

Warning message that vehicle will off after one km (Fig. 15).
Warning messages sent to vehicle owner (Fig. 16).
Owner vehicle latitude and longitude position uploaded to cloud (Fig. 17).
Owner details sent to RTO through messages.

7 Conclusion

One of the main reasons for the increase in pollution is the widespread use of polluting vehicles. With the use of IoT technology, the process of monitoring various aspects of environment such as air quality monitoring issue has been enhanced. The proposed

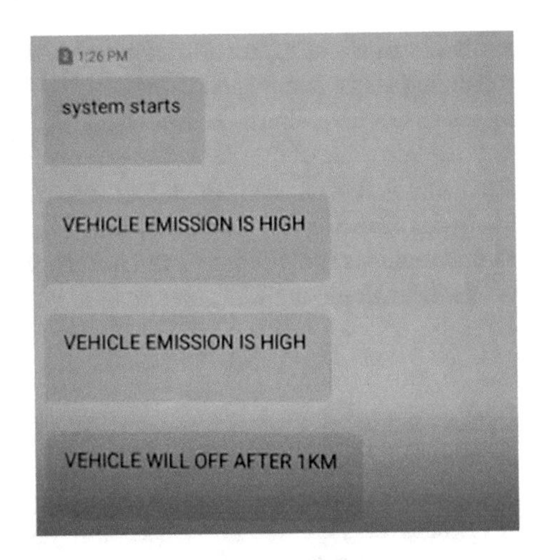
1:26 PM
system starts
VEHICLE EMISSION IS HIGH
VEHICLE EMISSION IS HIGH
VEHICLE WILL OFF AFTER 1KM

Fig. 15 Warning messages sent to user

```
2021-06-26 16:30:25,1303.1093,07730.3193
2021-06-26 16:30:32,1303.1093,07730.3193
2021-06-26 16:34:31,1303.1103,07730.3189
2021-06-26 16:34:39,1303.1103,07730.3189
2021-06-28 12:11:22,1303.1070,07730.3165
2021-06-28 12:11:32,1303.1070,07730.3165
2021-06-28 12:23:20,,,,,06532,,V,N*48
```

Fig. 16 GPS values sent to cloud

RTO
Edit
VEHICLE NO:KA04 C 1234 FINE AMOUNT:500
VEHICLE AMOUNT TO BE PAID
PAYMENT SUCCESSFUL
Send

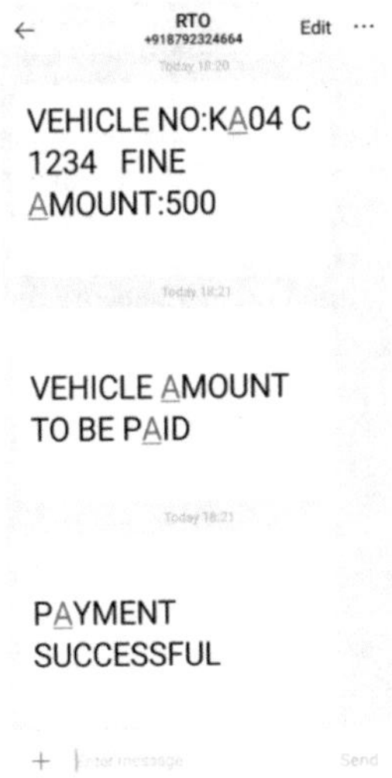

Fig. 17 Messages to RTO

method will assist us in addressing this issue, allowing us to maintain an environmentally pleasant atmosphere. In order to save our environment, the pollution caused by gases discharge must be decreased. The proposed approach is the most effective way to keep track of the gases released by the vehicle. In the existing systems, microcontroller had been used which requires external Wi-Fi connection and no time was given for reducing the emission level. The proposed methodology uses Raspberry Pi that includes in-built Wi-Fi and also time will be given for reducing the emission levels through paying fine amount.

References

1. Chandrasekaran SS, Muthukumar S, Rajendran S (2017) Automated control system for pollution detection in vehicles. In: 2013 4th International conference of electronics, communication and aerospace technology (ICECA)
2. Jadhav N, Nichal S, Patil M, Patil A (2017) Smart vehicle tracking using GPS. Int Res J Eng Technol (IRJET), pp 66–69
3. Rathod M, Gite R, Pawar A, Singh S, Kelkar P (2017) An air pollutant vehicle tracker system using gas sensor and GPS. In: 2017 International conference of electronics, communication and aerospace technology (ICECA)
4. Kumar S, Satya Prabha G, Vara Prasad Rao PV (2015) Automated control system for air pollution detection in vehicles. Int J Maga Eng Technol Manage Res, pp 1163–1169
5. Gadade US, Banale VV, Kadu PS (2020) Automated vehicle accident reporting and air pollution detection with advance tracking system. J Telecommun Switch Syst Networks
6. Mattos EERY, Terrones Vasquez BA, Vilchez Orcon WR, Puente Camasca RL (2020) IoT-based virtual platform for recording information under single execution environment in a Circuits and Devices Lab. In: 2020 IEEE world conference on engineering education (EDUNINE)
7. Akash SA, Menon A, Gupta A, Wakeel MW, Praveen MN, Meena P (2014) A novel strategy for controlling the movement of smart wheelchair using internet of things. In: 2014 IEEE global humanitarian technology conference—South Asia Satellite (GHTC-SAS)
8. Kavitha D, Chinnasamy A, Sangeerani Devi A, Shali A (2021) Safety monitoring system in mining environment using IOT. J Phys: Conf Ser

Securing Sensor Data on Internet of Things (IoT) Devices

R. Aparna, Abhinav Kumar Mallick, and Utkarsh Sahay

Abstract The amount of personal devices has been increasing at an unprecedented rate for the past few decades. Also, all these devices are becoming smarter (containing microprocessor, AI enabled), and all are having Internet connectivity. This has made the life much simpler and smarter, but it has also increased the privacy risks. One such sensitive data is health data. There has been multiple cases of breach into health data storage systems. Personal IoT devices such as fitness bands are very close to the user and keep track of many sensitive data such as heart rate, number of steps walked, and the places visited. Efficient methods to secure Internet of Things in medical services are a need of the hour. The innovation and research in the field of Internet of Things for health care promise to meet that demand. It has brought rapid developments in medical treatment, services, and health care, but the improvements in this field are facing challenges related to security and authentication. In this paper, we propose a novel method of securing sensor data on IoT devices. We make use of RSA encryption technique; the public key of the user is transferred to the IoT device at the time of connection, and this key is used to encrypt the sensor data. Only, this encrypted data are synchronized with the user's account.

Keywords Internet of Things · Internet of Medical Things · RSA · Public-key cryptography

1 Introduction

The *Internet of Things (IoT)* is a collection of interrelated devices, comprises of mechanical and digital systems, the user, and the ability to transfer data over a network without requiring any human interaction.

Personal IoT refers to the overall system (devices and their communication with the sensors) and related results, usance and services that relate to direct use by a

R. Aparna (✉) · A. K. Mallick · U. Sahay
Department of Information Science and Engineering, Siddaganga Institute of Technology, Tumakuru, Karnataka, India
e-mail: raparna@sit.ac.in

A. Kumar et al. (eds.), *Proceedings of the International Conference on Cognitive and Intelligent Computing*, Cognitive Science and Technology,
https://doi.org/10.1007/978-981-19-2350-0_40

person and/or for the direct benefit of humans. Difference between IoT and personal IoT can be something like a person may not be involved directly. A device that is used by a consumer for various purposes including entertainment, education, information, and lifestyle enhancement is referred as connected consumer device. Personal IoT data may be used for a variety of purposes including marketing for the respective industries involved as well as leveraged by various third parties.

Across the globe, demand for secure and faster IoT medical services is on the rise. The IoT for medical services promise innovation to meet that demand. It is dramatically redefining medical services, and patient outcomes. The *Internet of Medical Things (IoMT)* is a humongous revolution in the healthcare industry, and with IoT, it has quickly established itself as an important and critical part of modern healthcare [1–3].

In currently available personal IoT devices, the sensor data are not encrypted; instead, it is directly transferred to the mobile app using Bluetooth. This kind of implementation raises concern about the security of personal data. This paper describes a novel mechanism to secure data from different sensors of an IoT device so that even if there is a hack, the data remain secure. The sensor data are encrypted using RSA encryption and then stored on the device [4]. Only, this encrypted data are synced with any other device or the cloud [5]. The mobile app allows the user to register and access the sensitive data. The main contribution of our research is an encryption mechanism for data transmitted from IoMT such as fitness bands. Another contribution is the android application to display the decrypted data only on the devices logged in by the user. For security purpose, the transmitted data must be encrypted at the IoT device itself, and the decryption process will be done at the user end, i.e., the android application [6, 7]. Encryption transforms the original data (Plaintext) into an 'unreadable' data (Ciphertext, which is scrambled data), whereas decryption performs the reverse operation, i.e., transforms the ciphertext into plaintext. Encryption and decryption processes require random cryptographic keys shared between sender and receiver, and these keys can be generated using a key generation function.

The paper is organized as follows: Literature survey is discussed in Sect. 2, Problem definition and proposed solution to the problem are discussed in Sects. 3 and 4 give implementation details, and Sect. 5 concludes the paper.

2 Literature Survey

To provide privacy and security for the data, different types of cryptographic techniques are used, among which encryption and decryption techniques are most common. Encryption techniques are divided into asymmetric and symmetric encryption. The asymmetric encryption technique uses two keys, a public key (for encryption) and a private key (for decryption). The symmetric encryption technique uses only one key which is the private key which is used for both encryption and decryption. Symmetric encryption [8] is extremely secured, simpler, and faster as it uses only one key for both encryption and decryption.

There are other symmetric encryption techniques, e.g., data encryption standard (DES), double DES, and triple DES. DES makes use of 64-bits block size plaintext at a time and a 56-bits key to generate a 64-bits ciphertext. Analysis on DES algorithm shows that it is quite weak encryption algorithm, and it can easily be cracked using brute-force technique [9, 10]. Thus, one cannot consider it as a secure technique for encrypting sensitive medical data. The triple data encryption standard (3DES) technique is similar to DES except the key size can be 112 or 168 bits, but it considers the same 64-bits block size as input and produces 64-bit ciphertext. It is stronger than DES, but still, it can be easily cracked.

This paper utilizes RSA encryption system [11], which is one of the awry encryption procedures. It utilizes two keys, open key, also called as public key and private key dependent on two enormous prime numbers. Any user can make use of the open key to encrypt the message; however, just somebody with the information of the private key can decrypt the message. Breaking RSA encryption is known as the RSA problem. The intricacy of the RSA encryption depends on the calculating issue, i.e., to locate the prime elements of a number. There are as of now no distributed strategies to overcome the framework if a huge enough key is utilized. The usage of RSA is less regularly used to straightforwardly encode client information.

The IoT application depends on a system including a progression of heterogeneous sensors and gadgets, which can continually watch the encompassing condition and gather the information. This heterogeneity is reflected in the crude information gathered by that sort of framework. Accordingly, IoT's significant level application assignments to decipher the information and identify true occasions are increasingly perplexing. What's more, information heterogeneity prompts an absence of interoperability between IoT applications. Semantic Web (SW) innovation has been generally embraced for displaying and incorporating information from different sources on the Web. Be that as it may, this sort of procedure requires a lot of processing assets, particularly in situations including countless sensors. To address such a test, Al-Osta et al. [12] proposed a lightweight semantics comment approach that can be actualized on asset-compelled IoT passage associating with constrained sensors. To assess the methodology, a progression of tests utilizing middleware model has been finished.

3 Problem Definition

In this section, first, existing system is discussed, and then, the system we are proposing will be elaborated.

3.1 Existing System

The Internet of Things is the augmentation of Internet network into physical gadgets and regular articles. It is installed with several physical gadgets that connects to the

Internet and different types of equipment (for example, sensors). These gadgets can impart and associate with others over the Internet, and one can operate the gadgets from remote locations, verify their status, and can be controlled remotely [8].

The meaning of the Internet of Things has developed because of the intermingling of different innovations, continuous examination, AI, commodity sensors, and embedded systems. Conventional fields of embedded systems, remote sensor systems, control frameworks, computerization (counting home and home automation), and others all add to empowering the Internet of Things. In the consumer market, IoT innovation is most synonymous with items relating to the idea of the "smart home," covering gadgets and apparatuses, (for example, lighting installations, indoor regulators, home security systems and cameras, and other home appliances) that help at least one normal biological systems, and can be controlled by means of gadgets related with that environment, for example, cell phones and smart devices [8].

Providing security and privacy to these devices (gadgets) is one of the important task, and the concept of IoT has faced prominent criticism, in this regard [8].

Security concerns have been raised about the rapid development and growth in usage of IoT devices without ample focus on the complex security challenges involved and the dearth of regulatory changes required to handle this phenomenon. The technical security concerns have commonalities with those of network servers, workstations, and smartphones. But, with the increasing usage and popularity of IoT devices, challenges distinctive to these have surfaced. The concerns can arise due to various reasons; the prominent causes are insecure Web interface, insufficient authentication/authorization, insecure network services, lack of transport encryption, insufficient security configurability.

3.2 Proposed System

In the proposed system, an android app has been developed that allows the users to share their health-related information with the doctor in a secure manner and uses cryptographic approach to provide security for the users. In the implementation, when the user registers on the app, a key is generated for the login id. Only, this key can be used to decipher the data. The data received from the sensors are encrypted and stored on the device. Only, this encrypted data are synced with the server. It can also be decided by the user to share the information with a doctor or with any person of trust. Once the data are shared, the key for that data is also attached with the ID of the person with whom it is shared with. This is achieved by ensuring that all the sensor data flow through the user-authorized app to any other place (app or user) on the OS level.

In the proposed implementation, a cryptographic approach is used to introduce security, authentication/authorization to improve the non-existent data security in IoT devices. We also aim to improve data confidentiality by providing users with the option to encrypt data being collected by the devices and provide permissions for a third party to access the data. The data can be stored on cloud-based platform to improve its portability and hence making it accessible on other devices (which are authorized by the user).

The implementation uses Rivest–Shamir–Adleman (RSA) algorithm to provide encryption of data on target devices. In RSA algorithm, the encryption key is public and is different from the decryption key which is kept private. In the proposed system, as soon as the user registers on the platform, encryption and decryption keys would be generated for that user. The system is then ready to perform cryptographic operations on user's data. The data would then be encrypted with the public key or the encryption key thus generated. This encrypted data would then be synchronized with the data storage servers or database. Users will be able to view their data on demand as it will be decrypted with their personal private keys on their devices.

We also have an option to share data with third parties, based on user authorization of the same. The user would be able to authorize several other users or applications to use his/her data, pertaining to which the private key or the decryption key for that data will be attached with the ID of the concerned party. This will allow other users to view and utilize the shared data. Inculcating this keeps end-user data private, improves his/her data privacy and simultaneously allowing free usage of the concerned data by trusted parties to perform analysis, or however, they wish to manipulate it.

4 Implementation

An android application has been implemented for the problem defined above [6]. The implementation of the android application mainly consists of the following major parts.

The android app has a login/sign-up screen that allows users to register or to sign up. Whenever a new user registers or signs up on the android application, a pair of keys is generated for that particular user. The pair of keys thus generated is unique and only with the use of this key can the encrypted data be decrypted. The app then monitors data being collected by the specified IoT devices. This data, being collected, is then encrypted with the help of the keys assigned to the user at the time of registration. RSA algorithm has been used for performing cryptographic operations.

The encrypted data are then stored on the local system and are then synchronized with the cloud [13]. This makes the application portable and user data more secure and easily accessible on other devices of their choice. Here, a pair of keys are generated which act as a public key and a private key. The public key is used for encrypting the data and the private key for decrypting it. The private key is stored in such a manner that it can be accessed by the app only that too for the purpose of decrypting the data.

The android application also provides an option for sharing of data with third parties. The user would be able to authorize several other users or applications to use that data. After which, the private key or the decryption key for that data will be attached with the ID of the person or the application with which the data are shared. This will allow other users to view and utilize the shared data.

The graphs in Figs. 1 and 2 represent the relation between various parameters of original file and the encrypted file. There are four fields representing original file size, encrypted file size, percentage change in size, and time taken to encrypt the file, respectively. The original file size is the size of file before any cryptographical operations are performed on it. The encrypted file size column shows the size of

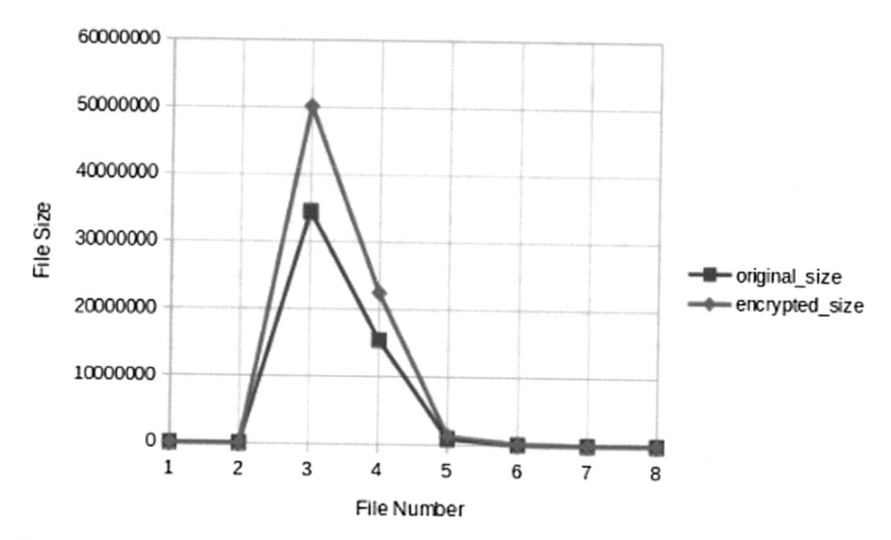

Fig. 1 Graph showing change in file sizes for image files before and after encryption

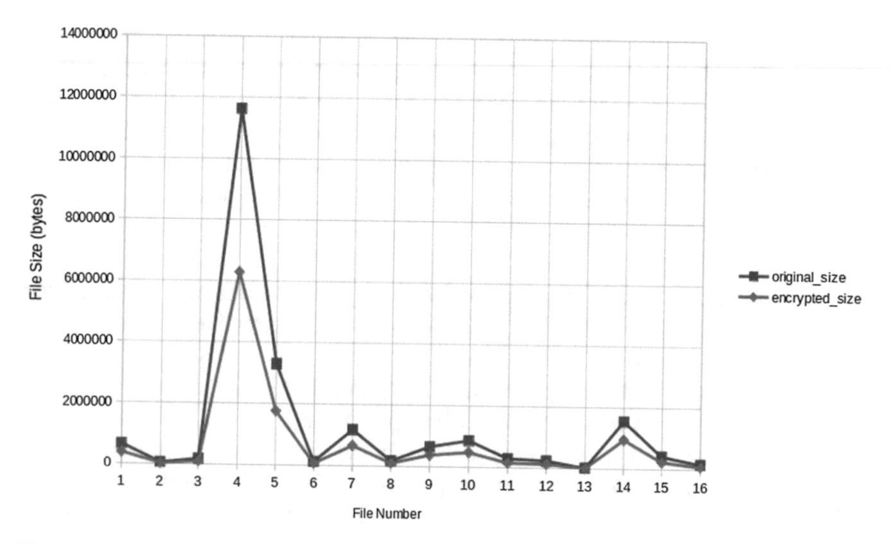

Fig. 2 Graph showing change in file sizes for and text files before and after encryption

the same file after encryption. The percentage change is calculated by calculating the difference between original file size and encrypted file size. The time taken to encrypt column shows the time required by the system to encrypt the file. In case of text files, the encrypted file size reduces by anywhere between 38 and 49%. This implies that after encryption, there is a reduction in the size of the file. The time taken to encrypt the files also varies proportionally with the original file size. The larger the file, more time it requires to encrypt. For very large text files, around 10 Mb, the time taken is around 20s (Tables 1 and 2).

Table 1 Details regarding encryption of image files

Original file size (bytes)	Encrypted file size (bytes)	Percentage change in size	Time taken to encrypt (seconds)
11,165	15,020	32.5275	0.0523400306702
4919	7512	52.7139	0.0245800018311
34,508,414	50,077,016	45.1153	157.187961817
15,478,214	22,462,464	45.1231	83.7759029865
860,513	1,239,724	44.0680	4.57197403908
71,706	104,448	45.6614	0.373322963715
3783	5464	44.4356	0.023008108139
23,744	32,768	38.0053	0.11828494072

Table 2 Details regarding encryption of text files

Original file size (bytes)	Encrypted file size (bytes)	Percentage change in size	Time taken to encrypt (seconds)
657,287	380,928	−42.04540	1.23898601532
52,139	29,356	−43.69665	0.112913131714
177,817	102,400	−42.41270	0.341831922531
11,626,439	6,278,488	−45.99818	20.0791721344
3,293,491	1,762,648	−46.48086	5.61558794975
123,783	66,904	−44.95057	0.21481900024
1,175,713	658,092	−44.02613	2.24571204185
177,817	102,400	−42.41270	0.409587144852
657,288	380,928	−42.04549	1.359126091
867,184	484,696	−44.10690	1.8191678524
300,788	160,428	−46.66409	0.596852064133
244,262	135,852	−44.38267	0.490957021713
25,267	12,972	−48.66030	0.075508117675
1,553,205	955,052	−38.51088	3.44151306152
421,335	240,300	−42.96699	0.833518028259
169,856	88,064	−48.15373	0.32085609436

For images, the size of the encrypted files increases. The increase in size is directly proportional to the size of the original file size. The percentage change in the image size after encryption lies in between 32 and 52%. The percentage change does not depend on the image size. There is no direct correlation between percentage change in size and the original size of the image. The time required to encrypt an image also directly depends on the size of the image. The larger the image, more time it takes to encrypt. The maximum time elapsed for image encryption is 157 s for a file of 32 Mb.

5 Conclusion and Future Enhancements

In this paper, we have proposed an efficient approach toward securing the sensitive data on IoMT devices. This will secure the data present on these personal devices by encrypting them; it will restrict any unauthorized third-party access to personal sensitive data. It also provides platform independence by backing up user data on a cloud platform and also synchronizes the data so that it is easily accessible through other devices with permission to access the data. This will also remove the vulnerability of the data being stolen from the cloud services as all the files are encrypted using every user's private key and secure the data overall. This will also introduce a level of opacity between the user and the IoMT manufacturer and also mitigate the possibilities of breaches into medical data storage systems.

The proposed work can be extended to a lot of other personal IoT devices and can be used for securing data on them.

References

1. Wu F, Wu T, Yuce MR (2019) Design and implementation of a wearable sensor network system for IoT-connected safety and health applications. In: 2019 IEEE 5th world forum on internet of things (WF-IoT), Limerick, Ireland, 2019, pp 87–90
2. Garg N, Wazid M, Das AK, Singh DP, Rodrigues JJPC, Park Y (2020) BAKMP-IoMT: design of blockchain enabled authenticated key management protocol for internet of medical things deployment. IEEE Access 8:95956–95977
3. Merabet F, Cherif A, Belkadi M, Blazy O, Conchon E, Sauveron D (2020) New efficient M2C and M2M mutual authentication protocols for IoT-based healthcare applications. Peer-to-Peer Netw Appl 13(2):439–474
4. Ferencz K, Domokos J (2018) IoT sensor data acquisition and storage system using Raspberry Pi and Apache Cassandra. In: 2018 International IEEE conference and workshop in Óbuda on electrical and power engineering (CANDO-EPE), Budapest, 2018, pp 000143–000146
5. Huang S, Chang H, Pan J (2014) Sensor dispatching methods for gathering data in rechargeable wireless mobile sensor networks. In: 2014 IEEE world forum on internet of things (WF-IoT), Seoul, 2014, pp 479–484
6. Thiyagarajan M, Raveendra C (2015) Integration in the physical world in IoT using android mobile application. In: 2015 International conference on green computing and internet of things (ICGCIoT), Noida, 2015, pp 790–795

7. Sarkar S, Gayen S, Bilgaiyan S (2018) Android based home security systems using internet of things (IoT) and firebase. In: 2018 International conference on inventive research in computing applications (ICIRCA), Coimbatore, 2018, pp 102–105
8. Elminaam DSA, Kader HMA, Hadhoud MM (2008) Performance evaluation of symmetric encryption algorithms. IJCSNS Int J Comput Sci Network Security 8(12):280–286
9. Coppersmith D, Johnson DB, Matyas SM (1996) A proposed mode for triple-DES encryption. IBM J Res Dev 40(2):253–262
10. Stallings W (2011) Cryptography and network security-principles and practices, 7th edn. Prentice Hall of India
11. Rivest RL, Shamir A, Adleman LM (1983) cryptographic communications system and method. US Patents, Patent No. US4405829
12. Ajay DM, Umamaheswari E (2016) An initiation for testing the security of a cloud service provider. In: Smart innovation, systems, technologies. Springer, Switzerland, pp 35–41
13. Nikolov N, Nakov O (2019) Research of communication between IoT cloud structure, android application and IoT device using TCP sockets. In: 2019 X National conference with international participation (ELECTRONICA), Sofia, Bulgaria, 2019, pp 1–4

COVID-19 Vaccination Twitter Data Sentiment Analysis Using Natural Language Processing

A. V. Mohan Kumar, Jnanavi, and Harshita S. Bhat

Abstract Statistics from Twitter, due to the emergence of social structures and social media, sentiment analysis has become relevant to a growing range of applications. In India, the vaccine stress for prevention in competition with COVID-19 began on January 16, 2021. Tweets about vaccines such as Covaxin, Sputnik, Pfizer and BioNTech are included in the statistics. The paper reflects everyone's views, which were used to provide feedback on the COVID-19 vaccination in January 2021. With COVID-19 as the keyword, we employed the Naive Bayes classifier to do sentiment analysis. Also, utilizing the API technique, we manually crawl the data for statistics using the retrieved entries taken from Twitter. We can extract the subjective facts of a file using device learning techniques and natural language processing and attempt to classify it based on its polarity, which includes first-rate, neutral and terrible. It is a useful evaluation to consider whether we might also, additionally, if most people expect excellent things from it, its stock markets will rise, and so on. Sentiment analysis is definitely a long way from being solved due to the language's complexity (objectivity/subjectivity, negation, lexicon, syntax and so on), but it is also why it is a great experience to work on. In this project, you will use an approach of developing a model based entirely on possibilities to try to classify messages from Twitter into feelings. In micro-blogging platform, people can express themselves quickly and spontaneously using any means of sending tweets.

Keywords Twitter · COVID-19 vaccine · NLP · Tweetpy

1 Introduction

COVID-19 rolled out all the exhilarating new sensations of culture and relief to be followed throughout the course of a 365-day period of sickness and misery. Previous methods for estimating or identifying the ones had a number of drawbacks, including

---

A. V. Mohan Kumar (✉) · Jnanavi · H. S. Bhat
Department of Information Science and Engineering, Don Bosco Institute of Technology
Kumbalgodu, Bengaluru 560074, India
e-mail: avmohan@dbit.co.in

A. Kumar et al. (eds.), *Proceedings of the International Conference on Cognitive and Intelligent Computing*, Cognitive Science and Technology,
https://doi.org/10.1007/978-981-19-2350-0_41

a limited number of phrases and a stuck sentiment orientation. Even with a large set of statistics, they no longer perform well. Reliability is still an issue. Sentiment assessment is a tool for learning about a project in which we want to figure out how others feel about a particular file.

We can extract the subjective facts of a file using device learning techniques and natural language processing and attempt to classify it based on its polarity, which includes first-rate, neutral and terrible. Few of the previous solutions were intended to deal with the severe traumatic circumstances that might arise from a wide range of situations. Furthermore, in order to improve class capabilities, the device obtaining knowledge of methods used through these systems no longer simply takes advantage of the enormous number of statistics sets and statistics available through the Internet. In this study, we attempt to classify polarity as first-rate, awful or impartial using a Naive Bayes set of rules. Fortunately, we can show positivity rate feedback using natural language techniques (NLP), which includes approaches like sentiment evaluation and word cloud visualization. The method is demonstrated below by someone tweeting statistics referred to as All COVID-19 Vaccine Tweets. The data is gathered using tweepy, a Python programme that gives access to the Twitter API in the event that we need to establish a Twitter application, and, as a result, a developer account is created and the credentials are accessed. Due to the fact that the statistics are all pre-compiled in this situation, it is far on hand for anyone who wishes to receive their immunization. Tweets about vaccines are included in the statistics, as are vaccines like Covaxin, Moderna, Sputnik V and others. Every vaccine, every historical period, is utilized to call into question current tweets. The primary goal of the explanatory records is to assess a section of this assignment, familiarize yourself with some of the body records' columns and begin to formulate research questions. To begin, load the Pandas usage records and some number one statistics and then collect data from the user's Twitter account so that we can get the following details for each tweet: clean text and complexity, with 1 indicating first-rate and 0 indicating unbiased and indicating awful.

Figure 1 depicts the path in four stages. To begin with, tweets are included in the data collection process. Then, there is statistics annotation, followed by statistics pre-processing, which saves your time by removing terms and filtering tokens. The

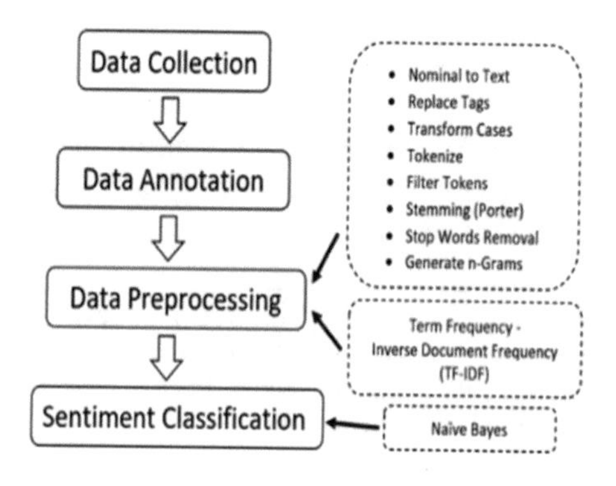

Fig. 1 Shows the workflow of the categorization process

sentiment class will next pick a region in which first-rate, awful and unbiased will be examined in a bar plot using a Naive Bayes classifier. This is the final certification, and it will show an ordinary overall performance evaluation.

2 Literature Review

Applying sentiment assessment for Twitter application is a way of developing the style and recognizing some trials and their potential applications. There are many ups and downs while using this sentiment assessment as regards processes.

Twitter is one of the most widely used and popular social media sites. Almost every enthusiastic or social person tends to communicate his or her opinions in the form of remarks, and this is viewed as the primary source of sentiments. These remarks not only communicate people's feelings, but also provide insight into their moods. Because the text on these media is unstructured, we must first pre-process it. Six pre-processing approaches are employed, and then, features are retrieved from the pre-processed data. Bag of Words, TF-IDF, word embedding and NLP (natural language processing)-based features like word count, noun count and so on are only a few examples of feature extraction approaches. We looked at the influence of two features in this paper. On the SS-Tweet dataset of sentiment analysis, TF-IDF word level and N-gram were used. They discovered that using TF-IDF word level (term frequency-inverse document frequency) features improve sentiment analysis performance by 3–4% over using N-gram features. Analysis is performed using six classification algorithms (decision tree, support vector machine, K-nearest neighbour, random forest, logistic regression, Naive Bayes) and F-score, accuracy, precision and recall [1].

This survey primarily focuses on sentiment analysis of Twitter data, which is beneficial for analysing the information contained in tweets, where opinions are highly unstructured and heterogeneous and can be either positive or negative, or even neutral in some cases. They conducted a survey and comparative analysis of existing opinion mining techniques, such as machine learning and lexicon-based approaches, as well as evaluation metrics. They have conducted research on Twitter data streams using a variety of machine learning algorithms, including Naive Bayes, max entropy and support vector machine. This paper also discusses the general challenges and applications of sentiment analysis on Twitter [2].

Work on sentiment evaluation which observed the usage of a proper technique is the paintings via way of means of L. Barbosa, J. Feng. The paper offers a technique to hit upon sentiment of newspaper headlines, in reality partly the usage of the equal grammar formalism that later could be offered and used on this paintings, but without the combinatorial common sense technique [3], additionally gift greater well-known techniques, inclusive of a technique for constructing a tremendously masking map from phrases to polarities primarily based totally on a small set of nice and terrible seed phrases. This approach has been followed via way of means of this, because it solves the venture of polarity values at the lexical stage pretty elegantly and could be very loosely coupled to the domain.

Ajay Bandi and Aziz Fellah, "Sentiment Analysis on Twitter within the 365 days of 2017" used each Corpus-primarily based totally approach and another dictionary approach based strategies which are used to extract features [4]. They are used at every level of language gaining knowledge of ends up a very huge advantage but predominant disadvantage ends up it emerges which does now no longer attempt to provide interpretation, a critique. So with the inference, the given tweets are of a below score that are categorized as horrible, and taking of another set of tweets as a first-rate price are categorized as first-rate [5].

Bifet and Frank hypothesized that customers suspect that many reviewers are authors or different biased parties. They observed marginal (terrible) effect of 1 super megacelebrity critiques is more than the (nice) effect of five super megacelebrity critiques [6]. The consequences propose that new varieties of consumer verbal exchange at the Internet have an essential effect on consumer conduct. For every product, reviews are very important for customers, similarly to producers. Previous approaches had consistent finite statistics devices and are now no longer capable of taking massive statistics devices, which led to low accuracy and the lexicon method ending up the predominant drawback.

Nanda Kumar, Mohan Kumar in "Twitter Sentiment Analysis the usage of Aspect-primarily based totally Bidirectional Gated Recurrent Unit" with another name called was self-attention part of mechanism where International Related Journal of Intelligent and Systems by the publisher of Intelligent Networks [7] and Systems Society with volume 13 and issue 5 with pages of 97–110.

Nanda Kumar, Mohan Kumar "A Survey on Challenges and Research Opportunities in Opinion Mining" on film critiques-primarily based totally packages via way of means which in details explains about all the aspects of sentimental analysis and taking different journals and discussed about the advantages and disadvantages and other aspects in which the sentimental analysis can be further refined and work can be carried out [8].

Mohan Kumar, Nandakumar, "Sentiment Analysis Using Robust Hierarchical Clustering Algorithm for Opinion Mining On movie Reviews-based Applications" in used Natural Language (NLP) primarily based totally technique to beautify the sentiment category via way of means of including semantics in function vectors [9] the usage of ensemble techniques for category wherein the person does now no longer want to learn in the way to use the interface become a benefit and reliability stays trouble.

Brandon Joyce, Jing Deng, Exploring Sentiment Analysis on Twitter Data in used small and half way approach of version where category of sentiment which explored tweet features [10], in which the SVM accuracy has advanced within the range of 1.5–3.5 and J48 might also additionally need to provide an accuracy improvement was advantage and does not perform well when having large dataset was a major drawback.

Well-known researcher Hermanto et al. used data mining approach for Twitter sentiment analysis in used examining the emotions of clients utilizing statistics mining classifiers in which advantage end up directed closer to writing surveys [11] and the predominant drawback is paying little head to their category.

3 Implementation

Figure 2 depicts the shape diagram containing modules. For developing the machine, wonderful methodologies were used. They are as follows:

Module 1: Twitter API

First, Twitter software is created and tweets are accumulated from the Twitter database. The statistics end up accumulated using a Python application known as *Tweepy* and TextBlob, which allows someone to get entry to the Twitter API on the occasion that they have got efficaciously created a Twitter Developer account and purchased get entry to credentials which includes getting entry to a token, get in this case, due to the fact the live character tweets are taken [12].

Module 2: Data Pre-processing

Collected tweets are stored as statistics set and are pre-processed and parsed through a manner of a method of having rid of now no longer unusual place, unwanted

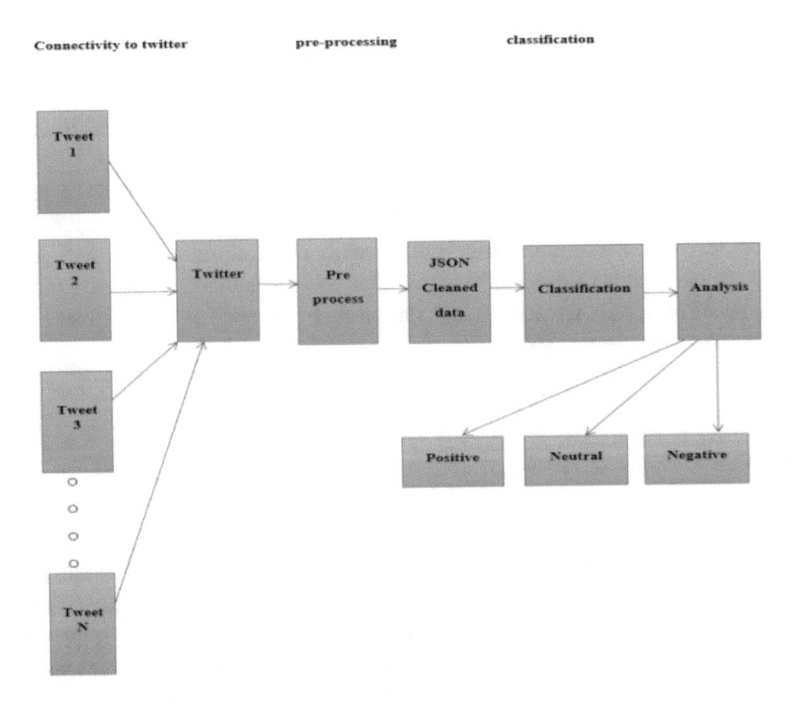

Fig. 2 Architecture diagram

terms, symbols, characters and numbers and convert them. After pre-processing, the emotions will be analysed through a manner of a method using a natural language processing tool. Each sentence is provided with a sentiment price. Based absolutely on this sentiment price, the statistics are catalogued as first-rate or horrible [13].

Module 3: Naive Bayes Approach

The approach used is Naive Bayes method which is a category approach especially as textual content mining that can be applied in sentiment assessment. Naive Bayes is typically applied in class techniques, specifically Twitter, using some strategies which incorporate some methods like types in Naïve Bayes like maximum entropy method [14].

The predominant function of this algorithm category is to acquire a robust speculation of any circumstance or event. Calculating the opportunity agencies of Naive Bayes makes use of the Bayes set of rule technique via way of means of the usage of equations.

$$p(x|y)\,p(Y)/P(X) = P(Y|X) \tag{1}$$

Equation (1) suggests that Y can be specific class of X which is records of some undefined class and even $P(Y|X)$ being chance of any instance speculation, even they are preceding chances of some category, and $P(X)$ can be opportunity of class Y [15].

Module 4: Sentiment Classification

Sentiment evaluation is the manner of figuring out the feedback of analysing sentiment and where approach of differentiating the polarity of the texting report and word in order that divides may be defined as nice, terrible or neutral analysis can be done [16].

The sentiment category is then given in a bar plot for the COVID-19 vaccine classifying as positive, negative and neutral which depicts highest positive rate tweets for vaccine. So that the application comes with text mining and product used is vaccine tweets with various vaccine types. The sentiment analysis can be depicted in any charts like bar plot, pie chart, etc. Comparing in terms of accuracy gives highest rating for positive tweets [16].

4 Technologies Applied

Wordcloud Visualization: Word clouds are a high-quality starting point to tap into customer comment statistics. But they offer more precious insights even as you combine them with a specific text assessment gadget, which incorporates sentiment assessment. Word clouds have emerged as an honest and visually appealing visualization method for text. They are applied in several contexts as a technique to provide a defined manner of a method of distilling text proper all of the manner right down to those terms that appear with most frequency.

Typically, this is finished in a static way as herbal text summarization. We expect, however, that there is a larger cap potential to this smooth however powerful visualization paradigm in text analytics. In this artwork, we find out the usefulness of word clouds for preferred text assessment duties.

Text Blob: Coming to the sentiment analyzation a part of this artwork, we can take all the tweets which are informative and few of the tweets are being opinion related where partitioning to greater part where it can be separated into devices of the entire similar many tweet with good features. And these can be extracted using package of Python and bundle refer to as TextBlob, which employs API credentials by using NLP technique which includes techniques like parts of speech, class, translation and more. TextBlob sentiment assessment is involved with some processing which employed two key major thing metrics: polarity and subjectivity. The score for a polarity is go along with some glide within the values [−1.0, 1.0], and it presents the emotion content of an assertion. For subjectivity, it is a go along with the glide inside the values of [0.0, 1.0], in which 0.0 will be very purpose and 1.0 will be very subjective. It is a Python library with a smooth interface to perform some NLP duties.

5 Tools Used

Anaconda Navigator: Anaconda is loose and open-supply combination of the Python and R programming languages for medical computing (statistics science, device gaining knowledge of applications, massive-scale statistics processing, predictive analytics, etc.), that is used to simplify package deal manipulate and use the software.

Jupyter Notebook IDE: Jupyter Notebook is free and open sourced software based totally utility which lets in you to create and percentage files containing stay code, equations, visualizations and narrative textual content. The IDE additionally consists of record cleansing and transformation, numerical simulation, statistical modelling, record visualization and plenty of others.

Data: For this venture, we used the COVID-19 vaccine tweet statistics from character tweets; coming to the data collection frequency, initial statistics ended up by merging all from tweets about the vaccines. This ended up with inside the primary day's instances a day, until I recognized approximately the modern tweet quota and then collection (for all vaccines) stabilized properly nowadays, all through the morning hours (GMT). The statistics include tweets referring to all vaccines mentioned. For each and every vaccine, there is a relevant searching for term ends up used to impeach the present-day tweet together with smooth text and sophistication (Fig. 3).

Fig. 3 COVID-19 vaccine tweets

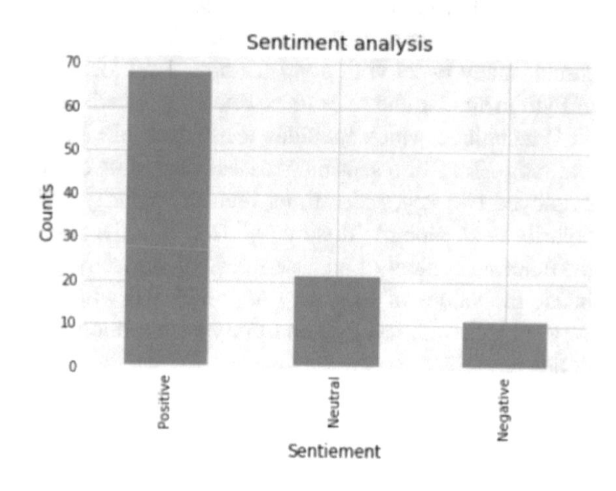

Fig. 4 Bar plot of evaluation

6 Outcome

The outcome of the project can be shown with bar plot which represents highest rate. Figure 4 represents bar plot of sentiment evaluation of the COVID-19 vaccines. Among all the tweets within side the data, approximately 1/2 of them had been distinctive via way of means of TextBlob as nice in sentiment (polarity > 0.0), with the opposite 1/2 of such as 75% impartial charged tweets (polarity = s 0.0) and 25% negatively charged tweets (polarity < 0.0).

7 Comparison Report

In terms of accuracy, we can say from the results that the Naive Bayes set of regulations has finished especially nicely as compared to the alternative algorithms (Fig. 5).

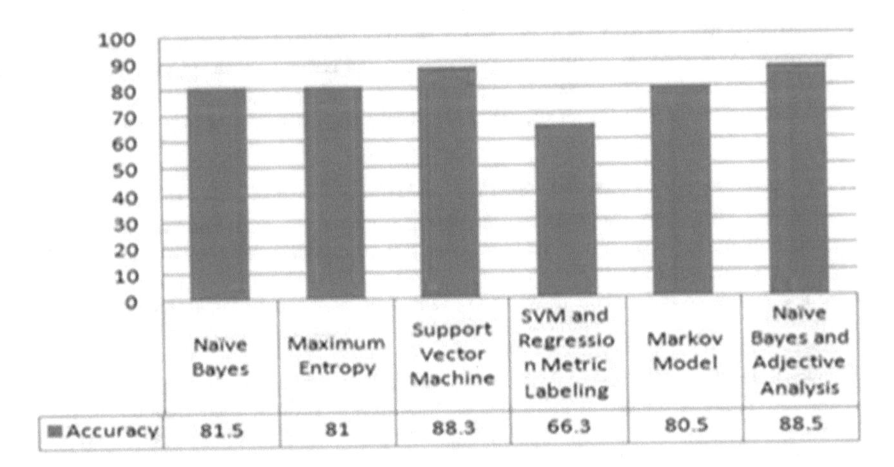

Fig. 5 Graph representing accuracy of numerous fashions

8 Conclusion

Twitter sentiment analysis is a thrilling manner to mirror attention where natural processing can be applied and results in approximately for textual content. It can be implemented in all accounts of media style assessment and may be sometimes, for marketing and other usage. Using this feedback of analysis utility in machine learning and Python is not always hard challenge, and it is a way to the days where it is simple to use and modernly available for all usage libraries. The application is basically for easy clarification of how this type of utility works.

Future Enhancement In future works, upon changing the dataset, the identical model can be knowledgeable to count on emotion, ethnicity, etc. Create an android/IOS app rather than a net web website online to be more on hand to the character. The system can be implemented using a cloud that may keep a massive amount of statistics for comparison.

References

1. Ahuja R, Chug A, Kohli S, Gupta S, Ahuja P (2019) The impact of features extraction on the sentiment analysis. Proc Comput Sci 152:341–348. https://doi.org/10.1016/j.procs.2019.05.008
2. Kharde V, Sonawane S (2016) Sentiment analysis of twitter data: a survey of techniques. Int J Comput Appl 139:5–15. https://doi.org/10.5120/ijca2016908625
3. Barbosa L, Feng J (2010) Robust sentiment detection on twitter from biased and noisy data. COLING 2010: Poster Volume, pp 36–44
4. Du J, Preston, Sharice S, Hanxiao S, Ross C, Rachel B, Julie S, Lara A, Muhammad T, Tao C (2021) Using machine learning–based approaches for the detection and classification of human papillomavirus vaccine misinformation: infodemiology study of reddit discussions. J Med Internet Res 23:e26478. https://doi.org/10.2196/26478

5. Bandi A, Fellah A (2019) Socio-analyzer: a sentiment analysis using social media data. In: Sixty-seven proceedings of twenty eighth international conference on software engineering and data engineering, vol 64, pp 61
6. Bifet, Frank E (2010) Sentiment knowledge discovery in twitter streaming data. In: Proceedings of the thirteenth international conference on discovery science. Springer, Berlin, Germany, pp 1–15
7. Venkataramaih MKA, Achar NA (2020) Twitter sentiment analysis using aspect-based bidirectional gated recurrent unit with self-attention mechanism. Int J Intel Eng Syst 13(5):97–110
8. Mohan Kumar AV, Nandkumar AN (2020) A survey on challenges and research opportunities in opinion mining. SN Comput Sci 1:1–6
9. Mohan Kumar AV, Nandakumar AN (2019) Sentiment analysis using robust hierarchical clustering algorithm for opinion mining on movie reviews-based applications. Int J Innov Technol Explor Eng (IJITEE), vol 8(8):452–457
10. Joyce B, Deng J (2017) Sentiment analysis of tweets for the 2016 US presidential election. In: IEEE MIT undergraduate research technology conference (URTC). IEEE, Cambridge, MA
11. Hermanto DT et al (2018) Twitter social media sentiment analysis in tourist destinations using algorithms naive bayes classifier. J Phys Conf Ser 1140 012037
12. Messias J, Diniz JP, Soares E, Ferreira M, Araujo M, Bastos L, Miranda M, Benevenuto F (2016) Towards sentiment analysis for mobile devices, pp 89-No12
13. Go RB, Huang L (2009) Twitter sentiment classification using distant supervision. Stanford University, Technical Paper, pp 456–876
14. Suppala K, Narasingarao (2019) Sentiment evaluation the usage of Naive Bayes classifiers. In: World wide magazine of progressive era and exploring engineering, vol 145(345)
15. Parikh R, Movassate M (2009) Sentiment analysis of user-generated twitter updates the usage of various classification techniques. In: CS224N final report
16. Pak A, Paroubek P (2010) Twitter as a corpus for sentiment analysis and opinion mining. In: Proceedings of the seventh conference on international language resources and evaluation, pp 1320–1326
17. Korkontzelos I, Nikfarjam A, Shardlow M, Ananiadou SS, Gonzalez GH (2016) Analysis of sentiment evaluation on extracting damaging drug reactions from tweets and discussion board posts. J Biomed Inform 62:148–158

Extraction of Buildings from Images Based on Segmentation

C. E. Ranjitha, Viswanath Kapinaiah, and T. N. Rishith Sadashiv

Abstract Extracting buildings and different structures from satellite/aerial images is an important task for updating maps and GIS databases. Structure, context and spectral data are used to extract buildings and other portions. To achieve accurate results for extraction by object segmentation, we use different segmentation approaches. In the first approach, we use K-means clustering to determine the different regions. To extract the buildings, texture segmentation approach with a conditional threshold value is used. Next, the image is separated into the building and non-building areas using the thresholding which is effective at allocating the bimodal histogram distribution. Finally, we use morphological opening and closing operations to extract bright foreground objects, and the closed building footprints are obtained using the marker-controlled watershed segmentation technique. The results are obtained by superimposing the segmented images with the original images to highlight the extraction. The results obtained from this approach are better in terms of accuracy.

Keywords K-means clustering · Texture segmentation · Thresholding · Watershed segmentation · Building extraction

1 Introduction

To obtain information about objects on the earth's surface, satellite and aerial images play an important role as they provide important features. The primary focus of many applications is on identifying objects and targets in aerial images. In various applications, extracting/detecting buildings from satellite/aerial images pose a significant

C. E. Ranjitha (✉) · V. Kapinaiah
Department of Telecommunication Engineering, Siddaganga Institute of Technology, Tumakuru, India
e-mail: ranjithabhagvath@gmail.com

T. N. Rishith Sadashiv
Department of Electronics and Communication Engineering, Siddaganga Institute of Technology, Tumakuru, India

A. Kumar et al. (eds.), *Proceedings of the International Conference on Cognitive and Intelligent Computing*, Cognitive Science and Technology,
https://doi.org/10.1007/978-981-19-2350-0_42

challenge. Typical applications include search, navigation, rescue and defence applications etc. Some interesting examples are urban monitoring, land use analysis, inspection and undertaking of GIS database, digital maps production and routing planning [1]. Building detection in aerial images is considered a difficult task because there are so many other objects in the image, such as vegetation, water bodies and roads. On the other hand, detecting buildings from aerial or satellite images based on rooftops is a difficult task. Due to the sun angle, the colour, structure and brightness of one location's image differ from that of another. In urban applications, building footprint extraction is very important for detecting buildings. Building detection from urban areas has been a focus of research in computer vision. Different algorithms and methods for building detection have been developed with the availability of high-resolution satellite images. Automatic detection of buildings with shadows in aerial images using multi thresholding methods has been proposed in [2]. Researchers have developed a way to identify building pixels from non-building pixels in an image, the proposed method consists of two stages: the first is the generation of a new image index; and the second is the threshold histogram of the index image [3].

The edge detection mechanism and FCM classifier are employed to detect and extract the geometric features of urban buildings by the authors in [4]. Spectral, structural and contextual information using a mathematical morphology-based context model as well as supervised machine learning [5]. Marker-controlled watershed algorithm and local Radon algorithm can extract different alignments of building footprint from satellite images [6]. The object-based technique uses the K-means algorithm and shape parameters as discussed in [7]. The marker controlled watershed transforms for the low level segmentation of single resolution and MR images are presented in [8]. The proposed technique can be divided into three main steps pre-processing, colour segmentation and feature extraction. All derived features were then fed to a neural classifier to detect building objects [9]. The automatic building detection technique [9] separates buildings from trees more effectively. An improved detector uses texture information from LIDAR and ortho imagery to identify the structure's shape [10], with three stages. In the first stage, using shadow evidence and fuzzy landscapes, extract information that is only related to building regions. In the second stage, image is automatically classified into four categories: building, shadow, vegetation and others. The final stage characterizes the regions belonging to buildings and roads [11]. The proposed building detection approach involves two steps: First is to convert RGB to CIE LUV colour space, and the second is to find buildings using thresholding method as presented in [12]. The authors presented the extraction of roads and river portions from satellite images in [13, 14]. Authors in [15] presented an approach using the fusion of regional and line segment features for urban area extraction from high spatial resolution remote sensing images. They proposed a technique with two levels of spatial information, i.e. microscopic (local spatial units) and macroscopic spatial extents. The work [16] proposes the extraction of buildings from high- and middle-resolution satellite images using machine learning methods. Different types of images and classifiers differ in weaknesses and strengths, depending on situations at hand. The resultant classified pixels were subsequently fused to produce final urban extraction. In this work, we present an approach for extracting buildings using segmentation algorithms. The results obtained from satellite images show that this approach provides more than 85% accuracy.

2 The Building Extraction

To obtain an image that may be used for further processing, pre-processing is required. Median filtering is commonly used in digital image processing because it reduces impulsive noise and also preserves edges in certain situations. Initial building prints are provided by an ensemble of supervised learning techniques [5]. In the preprocessing stage, colour images were enhanced using filtering, colour segmentation, and histogram-based contrast adjustments. The enhanced image is next subjected to the feature extraction process, which uses the Gray-Level Co-Occurrence Matrix (GLCM) and Single Value Decomposition (SVD) techniques as presented in [9]. In our work, we used a building print that provides the result of an unsupervised learning technique. This image is employed by the texture segmentation. In texture segmentation, we use entropy with that kernel to show the high entropy region and the low entropy region. On the resulting image, Otsu's thresholding is applied. This can separate the high entropy region from the low entropy region. Later, we plot the histogram to show the high and low entropy regions for separation. The proposed approach is presented in the Fig. 1.

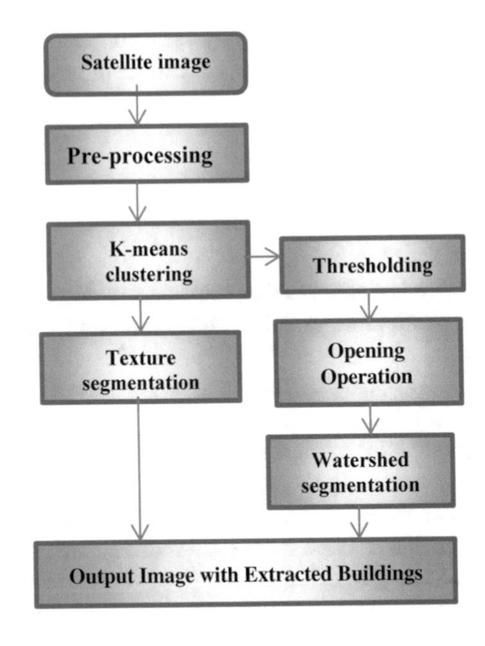

Fig. 1 The proposed approach for the extraction of buildings from images

2.1 K-Mean Clustering

The K-means algorithm is an unsupervised clustering technique that divides a given set of data into K number of clusters. It selects K centroids at random in the first phase and iteratively partitions each pixel to one of the closest centroids in the second phase, based on Euclidean distance. Here, $K = 1, 2 \ldots, n$ is a user-defined value that indicates how many pre-defined classes are there. The segmented image has been classified into six categories in this work. Basic urban features such as roads, vegetation, buildings, barren land, shadows and water area are divided into different classes (For example, $K = 6$). Different classes of building objects have been classified, such as varying rooftop reflectance and to differentiate buildings from other adjacent areas, each class requires additional processing.

2.2 Texture Segmentation

After the different objects have been identified, texture segmentation is preferred for grouping the similar objects. Texture segmentation is the process of dividing an image into regions with different textures and similar groups of pixels. To extract the building regions, a texture segmentation approach using a conditional threshold value has been used.

Entropy is used for quantify the texture, as it provides information about the local variability of the intensity values of pixels in an image. Hence, the local entropy can be used to characterize the texture of an image.

2.3 Thresholding

Image thresholding is a simple type of image segmentation in which a threshold value is applied to the pixel intensity to produce a binary image. Binary images have a pixel value of 0 for background and 1 for the foreground as building, whereas original image has 0–255 pixel values.

The Otsu's method is used to perform automatic image thresholding. The single intensity threshold that divides pixels into two categories: foreground and background, in the most basic form. Otsu's thresholding is a simple and iterative method for increasing the separation of results by iterating through all possible threshold values based on the histogram of the image. We used this approach for shadow detection that is based on this thresholding technique. As a result, the image is segmented according to different thresholds values using the Otsu method.

2.4 Morphological Operations

Using morphological approaches such as opening and closing operations are used to remove noisy pixels. Also, opening and closing are two operations that are used to restore an eroded image defined by a kernel. The term "opening" refers to the process of restoring or recovering the original image to the greatest extent possible. "Closing" is used to remove small gaps in the image that have been obtained. We use morphological operations erosion and dilation to remove the non-buildings object, with the structuring element equal to the average area of the buildings included in the original image.

2.5 Marker-Controlled Watershed Segmentation

Watershed segmentation is a mathematical morphology segmentation method. The watershed algorithm consists of two steps: The first is a sorting procedure, and the second is a submerging procedure. Sort each pixel on the grey level from low to high, then submerge from low to high using local minimum value. The watershed segmentation algorithm effectively extracts the object's continuous and closed boundary.

Extraction of Building Footprints:

A method for extracting building footprints is based on a marker-controlled watershed segmentation algorithm. The object's continuous and closed boundary is effectively segmented by the watershed algorithm as follows:

Step 1: Convert the colour image to a greyscale image.

Step 2: For satellite images, the gradient magnitude of the greyscale image is considered as segmented function. Then, by marking the foreground and background objects, a better segmentation results can be obtained.

Step 3: The mathematical morphology methods are used to mark the foreground objects.

Step 4: Thresholding is used for segmentation to calculate the background markers.

Step 5: Finally, the texture segmentation and the watershed segmented results are used together to highlight the extracted building portions in the satellite images.

Extracting a building footprint based on a marker-controlled watershed segmentation algorithm has the advantage of extracting the object's continuous and closed boundary. The gradient information is used from the building detection results, as well as the information between the building and the background.

Algorithm for the extraction of structures

Input: Satellite Image.
Output: Superimposed Image with highlighted portions of the structure/buildings
i. Apply a non-linear filter to remove impulsive noise.
ii. Initialize the number of clusters (Choose K) depending on the required portions of extraction.
iii. Group the input data into different clusters based on distance metrics using K-mean clustering.
iv. Pass the K-means clustered image through the texture segmentation.
v. Perform Otsu based thresholding.
vi. Perform morphological operations such as opening and closing for removing certain pixels and adding the pixel as defined by kernel.
vii. Perform watershed segmentation.
viii. Superimpose the texture segmented and watershed segmented images to obtain the highlighted extracted portions of the structures/buildings.

3 Results

In this section, the results obtained for building extraction are presented. The performance analysis using evaluation metrics such as accuracy and specificity are used to validate the proposed approach. The satellite image dataset of building has been obtained from resources quick bird, Google Earth, https://project.inria.fr/aerialimagelabeling/ and local city dataset can be obtained https://github.com/rishithSadashiv/road_extraction/tree/master/dataset. The accuracy evaluation strategy is a pixel-based measurement that divides all pixels in an image into four groups. True positive, true negative, false positive and false negative are denoted by TP, TN, FP and FN, respectively. They are basic statistical indices that can be evaluated based on segmentation results. The accuracy and specificity parameters gives that the correctness of the segmentation. The range of value of the score is from 0 to 1 and in percentage, 0-100%. The higher score, the better the image segmentation results. The original image is used as ground truth image to evaluate the scores. The results of this implementation are shown in Figs. 2, 3, 4 and 5 using the proposed approach.

Accuracy: The ratio of the correctly segmented area to the actual area, given as:

$$\text{Accuracy} = (\text{TP} + \text{TN})/(\text{TP} + \text{FN} + \text{FP} + \text{TN})$$

Specificity: It refers to the proportion of those who do not have the condition that received a negative result on this test and given as:

$$\text{Specificity} = \text{TN}/(\text{TN} + \text{FP})$$

Table 1 shows the values obtained after extraction of buildings from images as proposed in this work. One can observe that the building detection provides higher

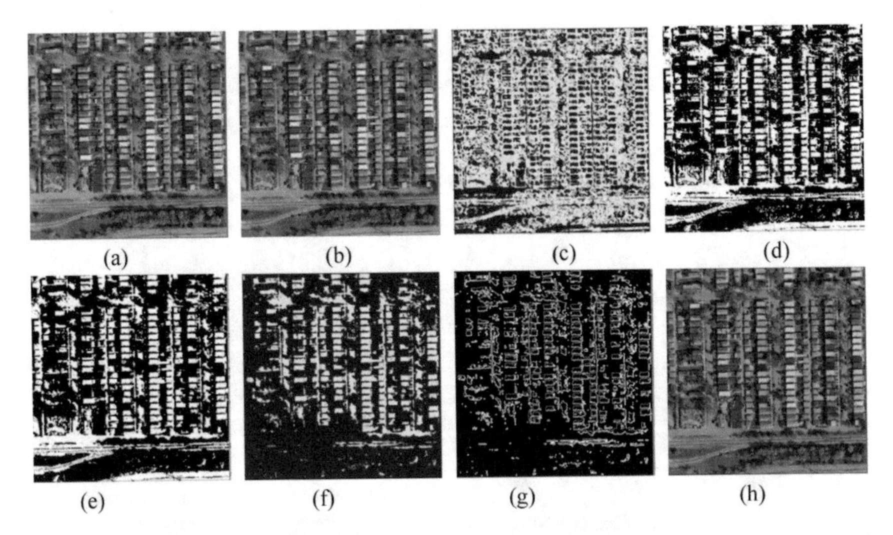

Fig. 2 Results of building extraction from satellite images on Chicago study area using $K = 6$. **a**) original images, **b**) K-mean clustered images, **c**) texture segmentation, **d**) outs thresholding, **e**) image opening, **f**) watershed sure background, **g**) watershed cluster and **h**) superimposition of the segmentation techniques

precision, which reveals the algorithm provides 83.5% accuracy in the detection of building rooftops.

4 Conclusion

In this paper, detection of buildings based on unsupervised classification approach and image segmentation has been presented. The experimentation results using this approach can detect buildings of various shapes and colours. Satellite data is important for updating maps and GIS databases. Structure, context and spectral data are used to extract buildings from satellite images. To achieve good segmentation, we use segmentation based on K-means clustering to determine the different regions. Then, to extract the buildings, we use a texture segmentation approach with a conditional thresholding. Finally, the resulting image is superimposed and highlighted with segmented portions in the images provided as output. The results show that the approch can provide more the 80% accuracy and specificity upto 87%. The results can be improved further by the use of supervised classification techniques. These results can be effectively used for the planning and development of smart city and villages.

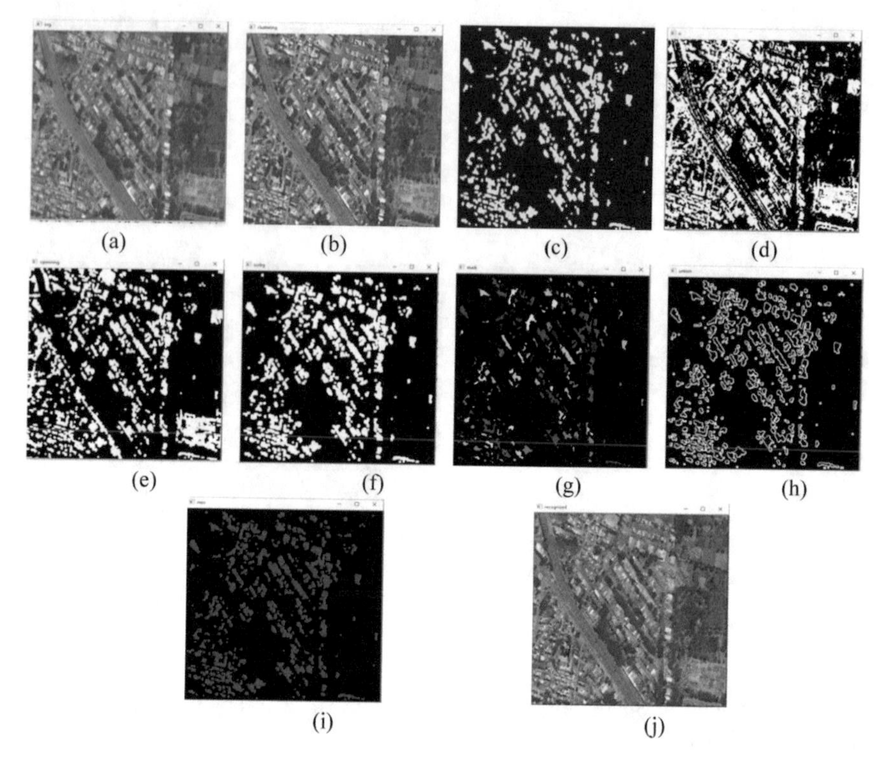

Fig. 3 Results of structures extraction from satellite images on Tumkur APMC study area, $K = 6$. **a)** original image **b)** K-mean clustered images, **c)** texture segmentation **d)** Otsu's thresholding **e)** Image after opening **f)** Watershed-sure background **g)** Watershed cluster **h)** Marker-control watershed segmentation using colour labelling **i)** Segmented image and **j)** Superimposition of the segmentation techniques

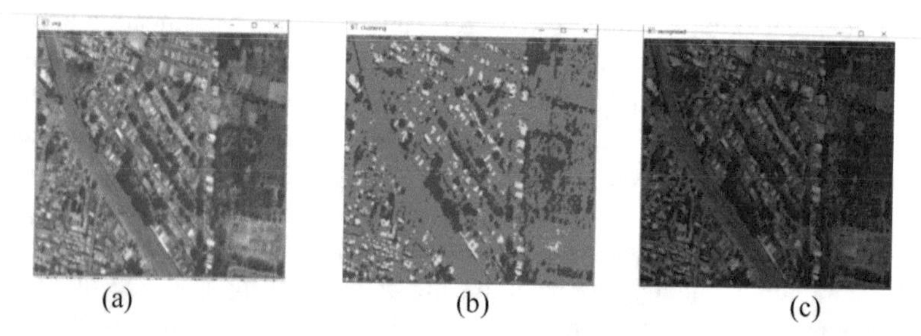

Fig. 4 Results of vegetation area extraction images on Tumkur City $K = 3$. **a)** original input images **b)** K-mean clustered images and **c)** superimposition of the segmentation techniques

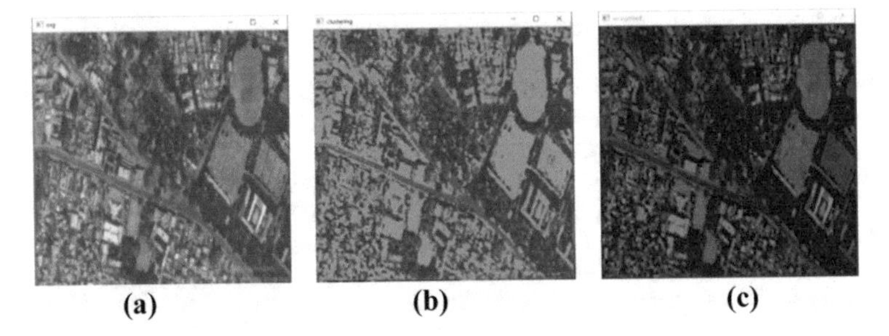

Fig. 5 Results of soil/playground area extraction from Tumkur City, $K = 4$. **a)** original input images **b)** K-means clustered image and **c)** highlighted and extracted portions

Table 1 Performance score after the extraction of the building structures for the image

Segmentation Techniques	Accuracy	Specificity
K-means	*0.8657*	*0.9259*
Texture	0.4031	0.3921
Watershed	0.7475	0.8604
Resulting images by the proposed approach	*0.835*	*0.8708*

References

1. Akçay HG, Aksoy S (2010) Building detection using directional spatial constraints. In: IEEE international geoscience and remote sensing symposium, pp 1932–1935. https://doi.org/10.1109/IGARSS.2010.5652842
2. Ghandour AJ, Jezzini AA (2019) Building shadow detection based on multi-thresholding segmentation. Signal Image Video Process (SIViP) 13:349–357. https://doi.org/10.1007/s11760-018-1363-0
3. Mostafa F, Mostafa Y, Nokrashy M, Yousef M (2021) Building extraction from worldview-2 images using invariant color models. https://www.researchgate.net/publication/350043287_Building_Extraction_from_WorldView-2_Images_Using_Invariant_Color_Models
4. Avudaiamma R, Dayana S, Prabhu R, Swarnalatha A (2018) Automatic building extraction from VHR satellite image. In: International conference on current trends towards converging technologies (ICCTCT), pp 1–6. https://doi.org/10.1109/ICCTCT.2018.8551119
5. Manandhar P, Aung Z, Marpu PR (2017) Segmentation based building detection in high resolution satellite images. In: IEEE international geoscience and remote sensing symposium (IGARSS), pp 3783–3786. https://doi.org/10.1109/IGARSS.2017.8127823
6. Yan L, Li P (2014) Building footprints extraction methods based on marker-controlled watershed segmentation and local radon transformation. Inf Technol J 13:1903–1907. https://doi.org/10.3923/itj.2014.1903.1907
7. Gavankar N, Ghosh SK (2018) Object based building footprint detection from high resolution multispectral satellite image using K-means clustering algorithm and shape parameters. Geocarto Int 626–643. https://doi.org/10.1080/10106049.2018.1425736
8. Gaetano R, Masi G, Poggi G, Verdoliva L, Scarpa G (2015) Marker-controlled watershed-based segmentation of multiresolution remote sensing images. IEEE Trans Geosci Remote Sens 53(6):2987–3004. https://doi.org/10.1109/TGRS.2014.2367129
9. Tanschotsrinon C, Phimoltares S, Lursinsap C (2013) An autonomic building detection method based on texture analysis, color segmentation, and neural classification. In: 5th international

conference on knowledge and smart technology (KST), pp 162–167. https://doi.org/10.1109/KST.2013.6512807
10. Awrangjeb M, Zhang C, Fraser CS (2011) Improved building detection using texture information. Int Arch Photogram Remote Sens Spatial Inf Sci XXXVIII-3/W22:143–148. https://doi.org/10.5194/isprsarchives-XXXVIII-3-W22-143-2011
11. Ok AÖ Automated detection of buildings and roads in urban areas from VHR satellite images. J Geodesy Geoinform 29–38. https://doi.org/10.9733/jgg.090315.1
12. Rahimzadeganasl A, Sertel E (2017) Automatic building detection based on CIE LUV color space using very high resolution Pleiades images. In: 25th signal processing and communications applications conference (SIU), pp 1–4. https://doi.org/10.1109/SIU.2017.7960711
13. Ranjitha CE, Viswanath Kapinaiah, Rishith Sadashiv TN (2021) Road extraction from satellite images based on K-Means clustering. In: International conference on big data, machine learning and IOT (ICBMI-2021), Kolkata, pp 31–35. ISBN: 978-93-90150-28-1
14. Deepika Rani GM, Viswanath Kapinaiah (2017) Extraction of river from satellite images. In: 2nd IEEE international conference on recent trends in electronics, information & communication technology (RTEICT), pp 226–230. https://doi.org/10.1109/RTEICT.2017.8256591
15. Zhang Q, Huang X, Zhang G (2017) Urban area extraction by regional and line segment feature fusion and urban morphology analysis. Remote Sensing 9(7):663. https://doi.org/10.3390/rs9070663
16. Puttinaovarat S, Horkaew P (2017) Urban areas extraction from multi sensor data based on machine learning and data fusion. Pattern Recognit Image Anal 27(2):326–337. https://doi.org/10.1134/S1054661816040131

Sentiment Analysis on Twitter Data

A. V. Mohan Kumar, M. Suhas, and Noah Fedrich

Abstract With the appearance of online innovation and its development, the Web presently contains an enormous measure of information for Web clients, just as a lot of information being produced. The Web has advanced into a stage for Internet learning, thought trade, and assessment sharing. Individuals utilize long-range informal communication locales, like Twitter, Facebook, and Google+ to impart and communicate their insights on different points, take an interest in conversations with assorted networks, and send messages everywhere on the world. In the subject of opinion analysis of Twitter information, a great deal of work has been finished. This examination centers on conclusion analysis of Twitter information, which is valuable for breaking down data in tweets where assessments are communicated. We use a variety of AI techniques, such as Naive Bayes and support vector machines, to investigate Twitter data streams.

Keywords Twitter · Sentiment analysis (SA) · Naive Bayes (NB) · Support vector machine (SVM)

1 Introduction

Individuals today communicate their views and sentiments in a new way thanks to the Internet. It is currently cultivated primarily through blog sections, online conversations, item audit sites, online media, and other comparison methods. Many people utilize social media platforms such as Facebook, Twitter, and Google Plus to communicate their emotions and perspectives on their daily life. Through Internet networks, we obtain intelligent media, where purchasers educate and influence others through gatherings. Web-based media is creating a huge volume of slant-rich information as tweets, announcements, blog entries, remarks, surveys, and so forth. Additionally, online media gives a chance to organizations by providing a platform for them to

A. V. Mohan Kumar (✉) · M. Suhas · N. Fedrich
Department of Information Science & Engineering, Don Bosco Institute of Technology, Kumbalgodu, Bengaluru, Karnataka, India
e-mail: avmohan@dbit.co.in

A. Kumar et al. (eds.), *Proceedings of the International Conference on Cognitive and Intelligent Computing*, Cognitive Science and Technology,
https://doi.org/10.1007/978-981-19-2350-0_43

connect with their customers in order to promote themselves. The majority of people prefer client-generated material to that found on the Internet.

The measure of content produced by clients is excessively tremendous for an ordinary client to examine. So, there is a need to mechanize this, different slant analysis methods are broadly utilized. Printed data recovery methodologies are essentially worried about handling, looking, and deciphering the authentic data accessible. Despite the fact that realities are evenhanded, there are some printed segments that address abstract characteristics. Opinion feelings, appraisals, perspectives, and feelings are the most widely recognized substance in sentiment analysis (SA). Because of the enormous development of accessible data on Web sources like Web sites and interpersonal organizations, it presents various testing chances for growing new applications.

In this examination, we look at Twitter, a famous microblog, and foster models for sorting "tweets" into positive, negative, and impartial mindset. Models are made for two arrangement errands: a parallel assignment of isolating positive and negative disposition and negative classes, similar to a three-way order task of positive, negative, and unbiased feeling classes. Because of this, component-based model, we utilize a portion of the characteristics proposed in past research.

2 Analysis of Emotions

Sentiment analysis is a technique for extracting mentalities, assessments, viewpoints, and emotions from text, sound, tweets, and other sources via natural language processing (NLP). The term "feeling analysis" refers to the process of classifying printed sentiments into categories such as "positive," "negative," and "nonpartisan." It is also referred to as subjectivity analysis, assessment mining, or appraisal extraction.

Albeit the terms assessment, feeling, view, and conviction are now and then utilized reciprocally, there are a few differentiations.

Opinion: A conclusion that is debatable (because different experts have different opinions).

- **VIEW**: Subjective opinion
- **BELIEF**: Deliberate acceptance and intellectual assent
- **SENTIMENT**: Opinion representing one's feelings.

An example for terminologies for sentiment analysis is as given below,

<SENTENCE> = the story of the movie was weak and boring.

<OPINION HOLDER> = <author>

<OBJECT> = <movie>

<FEATURE> = <story>

<OPINION > = <weak> <boring>

<POLARITY> = <negative>

An opinion can be represented mathematically as a quintuple (o, f, so, h, t), where

h = opinion holder;

t = time when the opinion is given;

o = object;

f = feature of the object o;

so = orientation or polarity of the opinion on feature f of object o;

h = opinion holder;

Object: A person, event, product, organization, or topic can all be considered an object.

Feature: An attribute (or a component) of the object for which an evaluation is performed.

Opinion orientation or polarity: An opinion's orientation on a feature f indicates whether it is positive, negative, or neutral.

Opinion holder: The person, company, or institution who expresses an opinion is known as the holder of that opinion. Figure 1 represents the architecture of sentiment analysis process.

3 Literature Survey

Numerous specialists recently did considerable research on the topic of "End Analysis on Twitter," which was originally produced for a combination plan that categorizes endings or reviews as good or bad.

Pak and Paroubek recommended three categories for tweets: sensible, positive, and negative. They collected tweets to generate a Twitter corpus. Tweets on social occasions using the Twitter API and typically clarifying those tweets with emoticons. They developed an inclination classifier using that corpus that is subject to the multinomial Naive Bayes algorithm. N-grams and POS-marks are two examples of features employed in this methodology. Due to the fact that the planning set they used contained just tweets with emoticons, it was less useful [1].

To group Parikh and Movassate combined a Naive Bayes and a maximum entropy algorithm to generate tweets. According to the researchers, Naive Bayes classifiers outperformed the maximum entropy model [2].

Go and L. Huang supplied a remote oversight strategy to evaluate analysis for Twitter data, in which their readiness data were composed of tweets containing emoticons that served as noisy imprints. They construct models using Naive Bayes, Maxent, and support vector machines (SVMs) [3]. In their component space, unigrams,

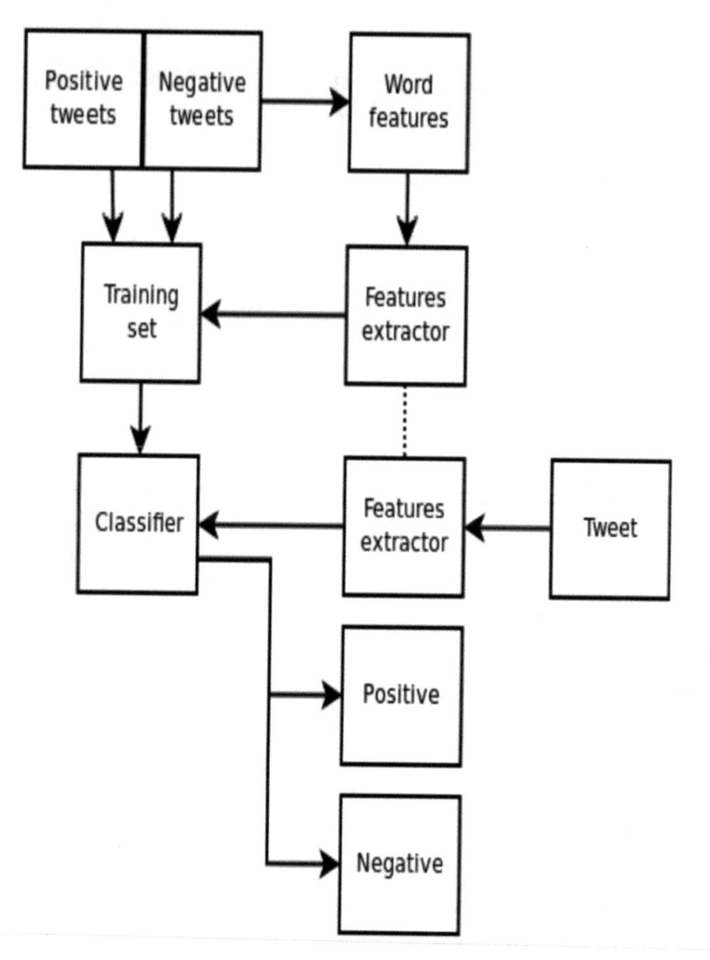

Fig. 1 Sentiment analysis architecture

bigrams, and POS were all present. They discovered that SVMs performed better than other models and that unigrams were more persuasive as features.

Barbosa et al. developed a two-phase strategy for tailoring assumption analysis. They solicited practical or emotional tweets, and then rated the theoretical tweets in the second round. In the part space, retweets, hashtags, affiliation, emphasis, and interposition markings were all used, and they were all identifiable by features such as before the word limit and the word limit and POS [4].

Bifet and Frank utilized Twitter streaming data provided by Firehouse API, which provided all messages sent and received by each client in a transparent and continuous fashion. They experimented with multinomial honest Bayes, stochastic tendency diving, and the Hoeffding tree [5]. They concluded that the SGD-based model outperformed the others when used with an appropriate learning rate.

Agarwal et al. pushed for a three-step process for categorizing evaluations as favorable, negative, or impartial. They investigated several models, including a unigram model, a segment-based model, and a tree part-based model. In the tree component-based method, they tended to tweet as trees. They concluded that characteristics connecting words' previous farthest point with their syntactic structures (pos) markings are normally substantial and play a big function in the request task [6]. The model based on tree fragments outperformed the other two models.

Davidov et al. provided an approach for limiting Twitter users' use of hash tags in tweets using highlight, single words, n-grams, and models as component types, which are then consolidated into a single component vector for feeling request [7]. They assigned assumption names using the k-nearest neighbor technique after creating a segment vector for each model in the readiness and test sets.

Po-Wei Liang et al. Twitter data that were acquired via the Twitter API are classified into four groups based on their preparation information (camera, movie, portable). Certain, negative, and speculative data are the three sorts of data. Suggestions have been culled from tweets. We employed the unigram Naive Bayes model, with the Naive Bayes algorithm working on the freedom presupposition. Additionally, they employed mutual information and chi square component extraction filters to eliminate superfluous highlights [8]. Finally, a tweet's direction can be anticipated as positive or negative.

Pablo and his companions' al. offered a variety of distinct Naive Bayes classifiers for identifying English tweets' extremes. Baseline classifiers (prepared to classify tweets as certain, negative, and nonpartisan) and binary classifiers (prepared to classify tweets as certain, negative, and nonpartisan) are two unique forms of Naive Bayes classifiers. Unbiased tweets are omitted). Classifiers evaluated lemmas (objects, action words, modifiers, and intensifiers), polarity lexicons, and multiword from different sources, as well as valence shifters [9].

Turney et al. utilized a sack-of-words technique for hypothesis analysis, in which no connections between words were considered and an archive was viewed as a collection of words. To establish the presumption for the entire archive, the sensations associated with each word were resolved and then merged via various aggregation capabilities [10].

Kamps et al., based on a variety of parameters, identified the passionate content of a word using the lexical dataset WorldNet. They devised a metric for WorldNet distance and quantified the semantic intensity of modifiers [11].

Xia et al. utilizing a sentiment classification outfit system designed with the merging of numerous capabilities and characterization methodologies. They conducted their analysis utilizing three base classifiers and two distinct sorts of capabilities (Data from parts of conversation and word relations) (Naive Bayes, maximum entropy, and support vector machines) [12]. They defined the assumptions in a variety of ways, including fixed mix, weighted mix, and meta-classifier mix, and obtained novel schemes [13].

Luoet discussed the difficulties and efficient methods for grading Twitter tweets. Spam and quickly changing language make recovering from a Twitter testing task challenging [14].

Nandakumar AN and Mohan Kumar AV, "A Survey on Challenges and Research Opportunities in Opinion Mining" on film critiques-primarily based totally packages via way of means which in details explain about all the aspects of sentimental analysis and taking different journals and discussed about the advantages and disadvantages and other aspects in which the sentimental analysis can be further refined and work can be carried out [15].

Mohan Kumar AV and Nandakumar AN, "Sentiment Analysis Using Robust Hierarchical Clustering Algorithm for Opinion Mining On movie Reviews-based Applications" in used natural language (NLP) primarily based totally technique to beautify the sentiment category via way of means of including semantics in function vectors [16] the usage of ensemble techniques for category wherein The person does now no longer want to learn in the way to use the interface become a benefit and reliability stays trouble [17].

4 Methodology

To develop a system, certain methodologies have been designed. It follows as:

4.1 Preprocessing of the Datasets

A tweet contains a lot of sentiments regarding facts that are transmitted in different ways by different clients. The Twitter dataset used in this study has now been divided into two classes: negative and positive extreme, making it easier to identify the impact of various highlights also during opinion analysis of the data. Extremely crude information is extremely susceptible to inconsistency and excess. After focuses, tweet preprocessing is included.

- Remove each and every URLs (e.g., www.xyz.com), hash labels (e.g., #subject), and targets (e.g., @username).
- Correct the spellings; there will be a number of reused characters to deal with.
- Replace all of the emojis with their estimated value.
- Remove all accents, pictures, and digits.
- Remove any words that serve as a stop sign.
- Increase the number of acronyms (we can utilize an abbreviation word reference).
- Remove any tweets that aren't in English.

4.2 Feature Extraction

The preprocessed dataset exhibits a number of properties that are unique. We remove qualities from the handled dataset by employing the element extraction technique. Later on, this component is used to interpret sentences' positive and negative extremes, which is necessary for utilizing models to determine individual suppositions.

Artificial intelligence procedures need that the most critical portions of text or archived data be addressed. These distinguishing characteristics are referred to as highlight vectors, and they are utilized during the arranging process. The following is a list of some of the models that have been discussed in the literature:

1. Words and Frequencies:

   Unigrams, bigrams, and n-gram models with their recurrence considers are underlined. To more easily illustrate this component, researchers have looked into using word presence rather than frequencies.
2. Grammatical forms Tags:

   Subjectivity and assessment can be indicated by grammatical forms such as descriptive words, modifiers, and a few groups of action words and things. Parsing or reliance trees can be used to generate syntactic reliance designs.
3. Words and Phrases for Evaluation:

   Aside from explicit words, other idioms and phrases that convey assumptions can be highlighted.

   Cost others an arm and a leg, for example.
4. Terms and their Position:

   The location of a term within a piece of content can have an effect on how much the term stands out in the overall assessment of the piece.
5. Invalidation is a critical yet difficult-to-understand component. The presence of nullification frequently modifies the extremity of the assessment. (For example, I am bothered.)
6. Punctuation:

   Many scientists use syntactic examples like collocations as highlights to understand subjectivity models.

4.3 *Training*

Supervised learning is a good way to deal with grouping concerns. The classifier might be prepared to deliver better expectations for obscure information later on.

4.4 *Classification*

The support vector machine investigations the information, characterizes the choice limits, and figures in input space utilizing portions. The information is made up of two m-vector configurations. Then, for each piece of information handled as a vector, a class is assigned. Following that, we detect an unconnected gap between the two classes. The distance defines the classifier's edge, which raises the edge and decreases reluctant selections. SVM additionally upholds characterization and relapse which are valuable for measurable learning hypothesis, and it likewise helps perceiving the variables decisively that should be considered, to comprehend it effective.

5 Technologies Used

5.1 The Natural Language Toolkit

The natural language toolkit (NLTK) is a Python framework for creating computers that deal with human language data in order to apply quantified conventional language (NLP).

It includes tokenization, parsing, depiction, stemming, naming, and semantic reasoning text planning libraries. It moreover incorporates graphical representations and test illuminating assortments, as well as a cook book and a book that teaches the concepts underlying the fundamental language taking care of duties that NLTK keeps up.

5.2 TextBlob

To get back to the slant analysis part of our undertaking, we can utilize our instinct that a few tweets are educational and others are obstinate to partition the bigger talk into discrete arrangements of tweets with comparable quantitative traits. In TextBlob, the feeling analysis is prepared using two main measurements: extremity and subjectivity. The extremity score is a number between −1.0 and 1.0 that reflects an assertion's or section's enthusiastic charge. Subjectivity is a numeric value between 0.0 and 1.0, with 0.0 indicating the most objective and 1.0 indicating the most emotional. It is a Python bundle that does a scope of NLP capacities utilizing a basic interface. It is natural and easy to use, making it appropriate for newbies. It is based on the shoulders of NLTK and another library named Pattern.

5.3 Twitter API

Programming interface permits you to get to assets just accessible on the worker. The Twitter API takes into account extraordinary and progressed automatic admittance to Twitter. It tends to be utilized to look at, gain from, and draw in with Tweets, direct messages, individuals, and other significant Twitter assets. For Web sites, Twitter permits you to bring live Twitter discussions on your Web page or application.

6 Tools Used

6.1 Django Server

Text Blob uses two main measurements to prepare its feeling analysis: extremity and subjectivity. The extreme score is a number between −1.0 and 1.0 that reflects an assertion's or section's eager charge. Subjectivity is measured on a scale of 0.0 to 1.0, with 0.0 representing the most objective and 1.0 representing the most emotive.

Django may be used to build virtually any style of Web site, from content management frameworks and wikis to informal communities and news sites. It is compatible with any customer-side structure and can provide content in nearly any format (including HTML, RSS channels, JSON, and XML). Django was used to develop the Web site you are currently viewing.

Inside, it very well might be changed to utilize different segments depending on the situation, while giving alternatives to for all intents and purposes any capacity you may need (e.g., different significant datasets, templating motors, and so forth).

6.2 PyCharm

PyCharm is the most well-known Python IDE, with astonishing highlights including code fulfillment and examination, an incredible debugger, and backing for Web programming and an assortment of systems. PyCharm was worked by Jet Brains, a Czech firm that spends significant time being developed conditions for Web improvement dialects, JavaScript, and PHP are two examples. PyCharm provides its clients and designers with the absolute greatest highlights in the angles below.

- Advanced debugging
- Coding completion and inspection Django and Flask are examples of Web development frameworks that are supported.

7 Conclusion

In this project, in this task, we target giving a prepared tweet information base to the representation of outsider applications. Text analysis centers around handling tweets to remove data from the crude information of the tweet, which can profit the application as far as ease of use and investigating text-designed information.

Likewise, we intend to serve a prepared twitter tweet information base to frontend outsider perception applications. Text analysis zeroed in on handling the tweets to separate data from the crude information of tweet, which can profit the application in extending more data to the client. The purpose of a Twitter tilt analysis is to learn about the public's reactions to a tweet/hashtag. There is information provided, such as a username or a hashtag. We discussed preprocessing and data recovery strategies for tweets sent via Twitter. Similarly, we learned about the method of directed learning: Text categorization using a support vector machine that may be used to find the extreme scarcely of any unimportant element and inadequate example vector. The presentation of SVM can be assessed utilizing exactness and review.

We give a review and analysis of existing assessment mining systems that combine AI with cross-space and cross-lingual methodologies, as well as specific assessment metrics, in this work. SVM and credulous are two techniques to artificial intelligence. Bayes has the highest precision and can be called a standard learning technique, according to research findings. We can concentrate our efforts on the study of merging AI technology with an assessment dictionary strategy in order to increase the precision of feeling order and adaptability to a range of settings and dialects.

Taking everything into account, we have met the approaches of conceivable information extraction through information treatment to give thorough data about person-to-person communication information.

References

1. Pak A, Paroubek P (2010) Twitter as a corpus for sentiment analysis and opinion mining. In:Twitter as a corpus for sentiment analysis and opinion mining, proceedings of the seventh conference on international language resources and evaluation, pp 1320–1326
2. Parikh R, Movassate M (2009) Sentiment analysis of user- generated twitter updates using various classification techniques, CS224N Final Report
3. Go RB, Huang L (2009) Twitter sentiment classification using distant supervision. Stanford University, Technical Paper
4. Barbosa L, Feng J Robust sentiment detection on twitter from biased and noisy data. COLING 2010: poster volume, pp 36–44
5. Bifet, Frank E (2010) Sentiment knowledge discovery in twitter streaming data. In: Proceedings of the 13th international conference on discovery science. Springer, Berlin, pp 1–15
6. Agarwal BX, Vovsha I, Rambow O, Passonneau R (2011) Sentiment analysis of twitter data. In: Proceedings of the ACL 2011 workshop on languages in social media, pp 30–38
7. Davidov D, Rappoport A (2010) Enhanced sentiment learning using twitter hashtags and smileys. Coling 2010: poster volume. Beijing, pp 241–249
8. Liang P-W, Dai B-R Opinion mining on social media data. In: IEEE 14th international conference on mobile data management, Milan, June 3–6, 2013, pp 91–96. ISBN: 978-1-494673-6068-5, http://doi.ieeecomputersociety.org/https://doi.org/10.1109/MDM.2013
9. Gamallo P, Garcia M Citius: a Naive-Bayes strategy for sentiment analysis on English tweets. In: 8th international workshop on semantic evaluation (SemEval 2014), Dublin, Aug 23–24 2014, pp 171–175
10. Neethu MS, Rajashree R (2013) Sentiment analysis in twitter using machine learning techniques. In: 4th ICCCNT 2013, at Tiruchengode. IEEE—31661

11. Turney PD (2002) Thumbs up or thumbs down: semantic orientation applied to unsupervised categorization of reviews. In: Association for computational linguistics, 40th annual meeting, pp 417–424
12. Kamps J, Marx M, Mokken RJ, De Rijke M (2004) Using wordnet to measure semantic orientations of adjectives
13. Xia R, Zong C, Li S (2011) Ensemble of feature sets and classification algorithms for sentiment classification. Inf Sci Int J 181(6):1138–1152
14. Luo Z, Osborne M, Wang T An effective approach to tweets opinion retrieval. Spr J WWW, Dec 2013. https://doi.org/10.1007/s11280-013-0268-7
15. Venkataramaiah MKA, Achar NA (2020) Twitter sentiment analysis using aspect-based bidirectional gated recurrent unit with self-attention mechanism. Int J Intel Eng Syst 13(5):97–110
16. Mohan Kumar AV, Nandkumar AN (2020) A survey on challenges and research opportunities in opinion mining. SN Comput Sci 1:1–6
17. Mohan Kumar AV, Nandakumar AN (2019) Sentiment analysis using robust hierarchical clustering algorithm for opinion mining on movie reviews-based applications. Int J Innov Technol Explor Eng (IJITEE) 8(8):452–457

Detection of Pervasive Developmental Disorder (PDD) Using Naïve Bayes Neural Network Algorithm

K. R. Nataraj, R. Rakshitha, Trupti, and K. Monika

Abstract Autism is defined as neuro developmental disorder which disturbs the behavioral condition and human communication level. This is a situation which is linked with the compound disorder of the brain which can lead to extensive difference in communication, interactivity, and social behavior of victim. We can apply various machine learning algorithms on autism dataset and predict autism at early stage with reducing cost for testing. This paper enables us to realize best neural network algorithm which helps us predict the pervasive developmental disorder using Naïve Bayes neural network algorithm with accuracy and reducing both time and cost for prediction. And this paper also gives the comparison between machine learning algorithms and helps us to know which algorithm gives better output with more accuracy.

Keywords Naïve Bayes · Linear regression · Validation dataset · Splitting samples · Test dataset

1 Introduction

Pervasive developmental disorder is defined as neurological development condition that has some trouble related to social conveying, oration, interaction, and how they involve themselves in world on all sides. Learning though signs is not much effective and easy, consultant also need for estimating the expressions of victims or patients to know the capability of victim to think, act, play, social activation, interactions, and communication. PDD victims can be found 1 in 68 persons as for the survey last few years. Initial detection of PDD will improve the psychological health of the victim. Machine learning procedure is used in PDD database to determine invisible feature and to erect analytical version for identification of chances that has been considered in this process or model [1]. Data categorization is most significant technique used in the machine learning the behavioral condition in the PDD in grown-up source

K. R. Nataraj (✉) · R. Rakshitha · Trupti · K. Monika
Don Bosco Institute of Technology, Bengaluru, India
e-mail: director.research@dbit.co.in

A. Kumar et al. (eds.), *Proceedings of the International Conference on Cognitive and Intelligent Computing*, Cognitive Science and Technology,
https://doi.org/10.1007/978-981-19-2350-0_44

have been categorized using neural Bayes, k-nearest neighbor, and linear regression and differentiate using precision measure [2]. The outcome specifies that NB has the highest precision. The database essential for this research has collected by the various surveys, conference accomplished by the specialists after the examination on PDD affected victims. Naive Bayes is known to predict PDD victims, and this process contributes an appropriate prediction of 96% found on all the properties on Naïve Bayes, J48, decision tree, k-nearest neighbor, and linear regression were bid on PDD of victim to determine the condition of victim [3].

2 Outline of Autism

Pervasive developmental disorder (PDD) which disturbs a person's way of conveying message, social interaction, and learning abilities. Diagnosis of PDD could be carried out at every single stage, its indications usually seen in initial 2 years of birth and cultivate along with time [4, 5]. Autism affected people face disparate types of dares, for example, complications with attentiveness, knowledge infirmities, psychological health difficulties such as nervousness, unhappiness, quickness, and sensual glitches.

In existing days, pervasive developmental disorder (PDD) is gaining its momentum exponentially. Apart detecting, autism using machine learning is very difficult job. Including the development of machine learning, autism could be projected at initial stage. In the projected reproach, Naïve Bayes algorithm plays a vital role forecasting pervasive developmental disorder [6]. Naïve Bayes is supervised learning algorithm built on the Bayes theorem.

3 Methodology

Figure 1 shows the model architecture where autism is predicted; first process is to collect and filter data then choosing a training model; next step is to evaluate the model, then tuning parameters, after tuning predictions are done.

Figure 2 shows block diagram where input or data are selected, processing of data and splitting of data, validation of data

1. **INPUT or DATA**: Any information or data sent to a model for processing is called as input, and they are assembled based on the user synchronal, where IoT system consumes using sensor data.
2. **PREPROCESSING**: This is a stage where the source can be organized as suitable for building and training the model. A source composing includes:
3. **SPLITING of SAMPLES: Data analyzing**: It can be used to impute the missing data, smoothing the noise, correcting the data, and identify the outliers.
4. **TRAINING DATASET**: The training dataset is a database used to fit the parameter for the process of PDD detection. The training dataset is the one which is

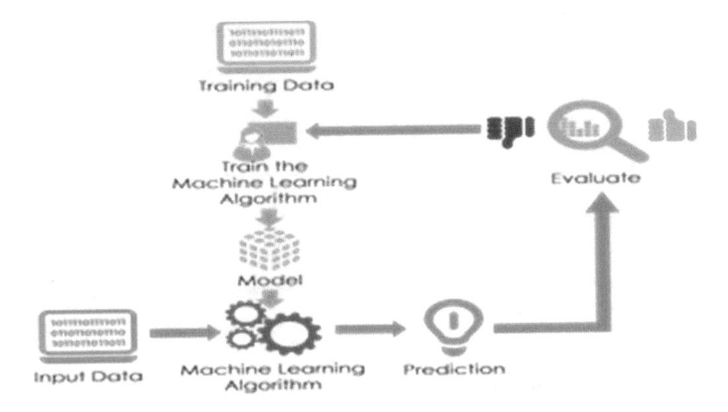

Fig. 1 Model architecture

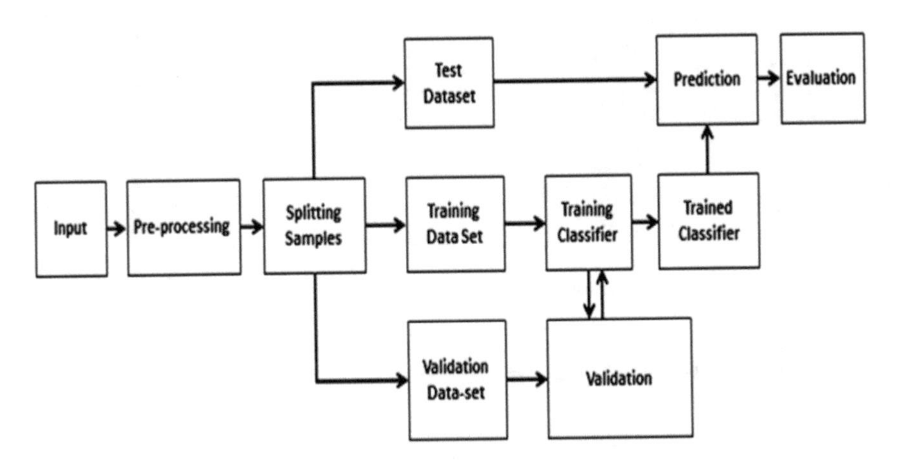

Fig. 2 Block diagram

used to train an algorithm to understand and how to apply concept such as neural networks to grasp and produce result for the PDD model. It incorporates both input data and anticipated output. The databank through which the computer learns how to process information.

5. **VALIDATION DATASET**: Validation datasets are datasets which are familiar to tune the hyper-parameters of a classifier. Also, named as "dev set." A hyper-parameter of neutral network includes hidden units in each layer.

   To keep away from over fitting, the framework is essential to balance, and it is compulsory to have validation dataset inclusion of training and test dataset.
6. **TEST DATASET** (Fig. 3):

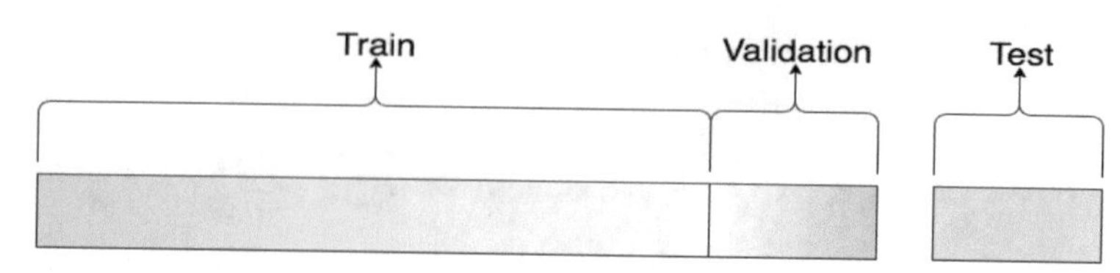

Fig. 3 Splitting of training dataset, validation dataset, and test dataset

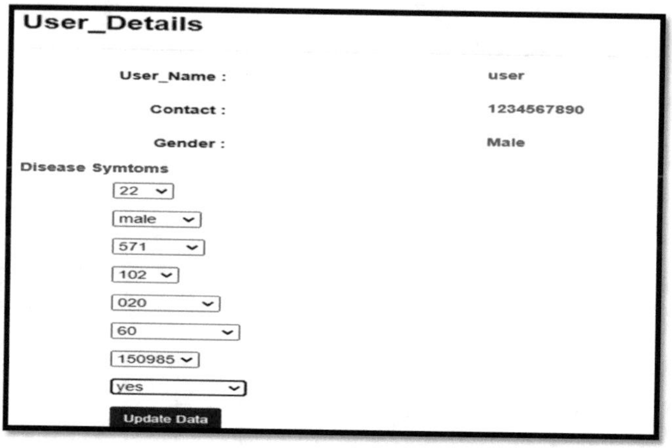

User_Details

User_Name : user

Contact : 1234567890

Gender : Male

Disease Symtoms

22

male

571

102

020

60

150985

yes

Update Data

Fig. 4 Disease symptom of patient 1

4 Simulation Results and Analysis

See Figs. 4, 5, 6 and Table 1.

5 Discussion

The performing of the suggested system of ours was put under test with the two-type dataset; one for children dataset and other for adolescent's dataset were chosen from UCI repository.

The results show clearly that the NB algorithm achieved the highest precision and the lowest error in case of children's dataset and adolescents dataset rate than the LR and KNN algorithm in detecting PDD disease (Tables 2 and 3).

Precision means the measure of quality. Recall is defined measure of quantity. F-measure is defined as the weighted harmonic mean of its precision and recall. MAE is defined as mean absolute error; it measures the average magnitude of the errors in a set of prediction. RMSE is defined as root mean squared error, and it is the standard deviation of the errors when a prediction is made on a dataset (Table 4).

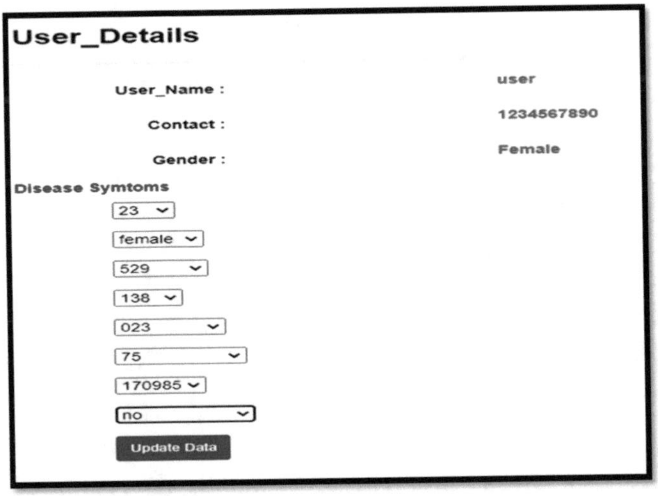

Fig. 5 Disease symptom of patient 2

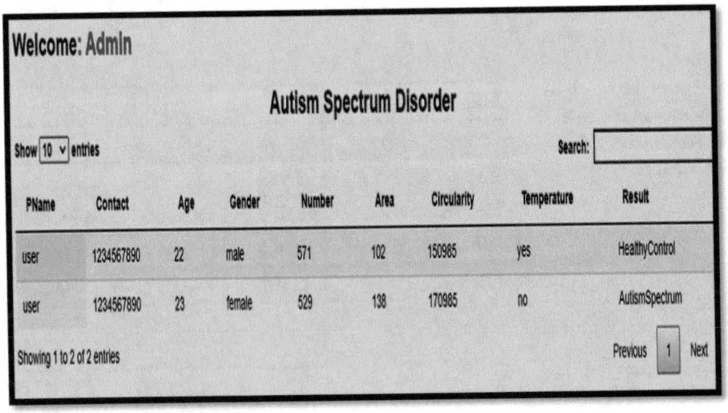

Fig. 6 Result of both patients 1 and 2

6 Conclusion

This project helps to learn machine learning classification technique and to detect the prediction of pervasive developmental disorder in grown-up. For the analysis of pervasive developmental disorder, the UCI autism disorder dataset was used.

From the outcomes which were analyzed using J48 classifier, Naïve Bayes, KNN, LR, and Bayesian network and on comparison of all these, we can predict that Naïve Bayes algorithm has the highest classifying precision than J48 LR, KNN, and

Table 1 Statistical summary of ABIDE preprocessed data

Class	Healthy control	Autism spectrum disorder
Number	571	529
Sex (M/F	479/99	465/64
Age (Years)	17.102 ± 7.762	17.082 ± 8.428
CC[2]_area (mm^2)	596.654 ± 102.93	596.908 ± 110.134
CC_perimeter (m)	196.405 ± 6.353	198.102 ± 17.265
CC_length (m)	70.583 ± 5.342	70.711 ± 5.671
CC_cicularity	0.194 ± 0.020	0.191 ± 0.023
W1[b] (Rostrum) (m)	20.753 ± 14.264	25.899 ± 10.809
W2[c] (genu) (m)	128.789 ± 32.13	128.855 ± 33.704
W3[d] (anterior body) (m)	91.088 ± 19.212	91.734 ± 20.302
W4[e] (mid-body) (m)	69.705 ± 13.351	69.345 ± 13.796
W5[f] (posterior body) (m)	59.007 ± 11.698	59.454 ± 12.501
W6[g] (isthmus) (m)	51.834 ± 12.519	52.137 ± 13.313
W7[h] (splemium) (m)	175.471 ± 32.35	174.483 ± 34.562
Brain volume (mm^3)	1,482,428.866 ± 150,985.32	1,504,247.415 ± 170,357.180

Table 2 Shows the result of children dataset in the term of precision, recall, F-measure, MAE, and RMSE

Measures	NB	LR (30)	KNN (30)
Precision (%)	98.975	95.216	88.822
Recall (%)	98.972	95.205	88.698
F-measure (%)	98.972	95.206	88.675
MAE	0.044	0.047	0.137
RMSE	0.109	0.217	0.304

Table 3 Shows the adolescents dataset. As important, as it is to find the right model, it is equally important to establish which models may not be the best choice

Measures	NB	LR (30)	KNN (30)
Precision (%)	98.136	95.229	94.413
Recall (%)	98.076	95.192	94.230
F-measure (%)	98.067	95.202	94.717
MAE	0.044	0.048	0.120
RMSE	0.126	0.219	0.284

Bayesian. But, during the analyzed result, the Naïve Bayes also performs better than other algorithms.

The goal of this exploration is to anticipate pervasive developmental disorder (PDD) in victim with help of machine learning algorithm and also plays an important role which enhances diagnostic timing and accuracy. This investigation compares and highlights the success of feature selection algorithm via recursive feature elimination

Table 4 Comparison between Naïve Bayes classifier, J48 classifier, and Bayesian network

Test split (%)	J48		Naïve Bayes		Bayesian network	
	Precision	Recall	Precision	Recall	Precision	Recall
10	89.1	88.8	96.5	96.5	91.8	91.6
20	90.7	90.6	96.1	96.1	95.8	95.7
30	89.5	88.6	94.7	94.7	93.5	93.5
40	91.8	91.9	94.3	94.3	94.3	94.2
50	91.3	91.2	95.1	95.2	94.8	94.8
60	91	91.1	94.3	94.3	95.4	95.5
70	91.3	91.5	95.3	95.3	96.2	96.2
80	92.1	92.2	95	95	97.9	97.9
90	98.6	98.6	94.8	94.3	98.6	98.6

(RFE) and correlation feature selection (CFS). This algorithm also proves to be efficient and effective to predict the result.

References

1. Bone D, Bishop SL, Black MP, Goodwin MS, Lord C, Narayanan SS (2016) Use of machine learning to improve autism screening and diagnostic instruments: effectiveness, efficiency, and multi-instrument fusion. J Child Psychol Psychiatr 57
2. Langley P, Iba W, Thompson (1992) An analysis of Bayesian classifiers. In: National conference on artificial intelligence, pp 223–228
3. Wall DP, Dally R, Leyster R, Jung J-Y, DeLuca TF (2012) Use of artificial intelligence to shorten the behavioural diagnosis of autism. PloS One 7(8):e 43855
4. Frith U, Happé F (2005) Pervasive developmental disorder. Current Biol 15(19):R786–R790
5. Thabtah F (2017) Pervasive developmental disorder screening: machine learning adaptation and dsm-5fulfillment. In: Proceedings of the 1st international conference on medical and health informatics. ACM
6. Jalaja Jayalakshmi V, Geetha V, Vivek R (2019) Classification of pervasive developmental disorder data using machine learning techniques, vol 8, Issue 6S. ISSN: 2249-8958
7. Werner E, Dawson G, Munson J, Osterling J (2005) Variation in early developmental course in autism and its relation with behavioral outcome at 3–4 years of age. Autism Dev Disorder 35(3):337–350

Face Mask Detection Alert System Using CNN

Akshay Kagwade, Rupali Kamathe, and Vandana Hanchate

Abstract In the time of 2019–20, a new virus surprised the world and there was war kind of situation where each individual facing a need to face COVID-19 virus. The virus spreads with air and cough and sneeze. Thus, as a primary concern, everyone needs to wear face mask and hand sanitizer. The problem starts when after sometime people won't follow the rules and the difficulties gets increase. The mandatory part is everyone should wear face mask everywhere. The schools, private institutes, colleges, and offices should follow this rule strictly. There should be a person at the doorstep of office who will check that each person is wearing mask or not. This manual job can be automated. Here comes the machine learning and artificial intelligence into picture. The machine learning can help the system to identify if that person is wearing mask or not. Furthermore, sometimes data got from survey is stale and that could cause wrong information about the severity of COVID-19 in particular city and area. This model can then help the authorities to manage the proper discipline and can help in planning distribution of masks, medicines, and vaccines in that city. This is nothing but real-time object detection, and we are detecting the mask as our preferable object. Then to get alert for the same an email will get send to the administration with his/her face as an attachment of that email if the person is not wearing mask. The real-time object detection techniques are now evolving in great amount with time. The existing algorithms are not that much up to the level with accuracy. This causes mis confusion for the machine and can predict wrong information with the available data. When it comes to real-time algorithm, CNN is best of them.

Keywords CNN · Face mask detection · Real-time object detection · Machine learning

A. Kagwade (✉) · R. Kamathe · V. Hanchate
E&TC Dept. PES Modern College of Engineering, Pune, India
e-mail: akshaykago@gmail.com; iamakshaykagwade@gmail.com

R. Kamathe
e-mail: hodentc@moderncoe.edu.in

V. Hanchate
e-mail: vandana.hanchate@moderncoe.edu.in

A. Kumar et al. (eds.), *Proceedings of the International Conference on Cognitive and Intelligent Computing*, Cognitive Science and Technology,
https://doi.org/10.1007/978-981-19-2350-0_45

1 Introduction

The unprecedented outbreak of 2019 novel coronavirus that happened in the world, termed as COVID-19 by the World health organization (WHO) has placed numerous governments around the world in a precarious position. Earlier only China witnessed the COVID-19 outbreak, where now it has become a great concern for virtually every country of the world. The unavailability of resources to endure the COVID-19 outbreak combined with the fear of overburdened healthcare systems impacted most of the countries, and forced into partial or complete lockdown.

The number of laboratory-confirmed coronavirus cases has been increasing at an alarming rate throughout the planet, with reportedly quite three million confirmed cases as of 30 April 2020 [1]. Adding to those woes, numerous false reports, misinformation, and unsolicited fears with regard to coronavirus, are being circulated regularly since the outbreak of the COVID-19. In response to such acts, we draw on various reliable sources to present a thorough review of all the most aspects associated with the COVID-19 pandemic. In addition to the direct health implications related to the outbreak of COVID-19, this study highlights its impact on the worldwide economy. In drawing things to an in depth, we explore the utilization of technologies like the web of Things (IoT), unmanned aerial vehicles (UAVs), blockchain, AI (AI), and 5G, among others, to assist mitigate the impact of COVID-19 outbreak.

Preventive measures to scale back the probabilities of infection include wearing a mask publicly, avoiding crowded places, keeping distance from others, ventilating indoor spaces, washing hands with soap and water often and for a minimum of 20 seconds, practising good respiratory hygiene, and avoiding touching the eyes, nose, or mouth with unwashed hands.

The WHO and therefore the US CDC recommend individuals wear non-medical face coverings publicly settings where there is an increased risk of transmission and where social distancing measures are difficult to take care of. But, if the mask includes an exhalation valve, a wearer that is infected (maybe without having noticed that, and asymptomatic) would transmit the virus outwards through it, despite any certification they can have. So, the masks with exhalation valve are not for the infected wearers and are not reliable to prevent the pandemic during a large scale.

The govt. made mandatory to wear masks at public places, and then only one is allowed to go out. For both who are infected and who are taking care of them both have to wear the masks compulsory. Figure 1 show person is wearing mask to get protection from COVID-19 virus. Proper hand hygiene after any cough or sneeze is encouraged. Healthcare professionals interacting directly with people who have COVID-19 are advised to use respirators at least as protective as NIOSH-certified N95 or equivalent, in addition to other personal protective equipment.

The problem is one cannot monitor if people are wearing masks or not. There are many lacking fields while doing this job manually. In case of offices, schools, colleges, and other public places, there one person is needed to monitor if people are actually wearing masks are not. This process is error prone and manual. This process can be automated using machine learning. Machine can be trained to identify if the

Fig. 1 Person wearing face mask to get protection from corona virus

person is wearing mask or not. This method also sends a mail to the respective given email id. With the help of this, one can monitor people in the respective organization.

2 Experimental/Methodology

2.1 Related Work

The COVID-19 outbreak was serious than the people assumed in early stages. The virus spread across the world with great speed and affected more than 200 countries got affected by it. The WHO provided guidelines to protect from coronavirus [2]. These guidelines include use of mask as this mask can enter human body through nose [3]. The large organizations such as schools, colleges, offices, and malls should take care of these guidelines. They should check the guidelines are getting followed by the people there. Such places already have CCTV cameras. There this system can easily automated using convolutional neural network by training the machine.

In the early stages, CNN was mainly used for image classification. Dr. Shailendra Narayan Singh successfully classified vehicle and another object detection using 295 image datasets with YOLO architecture [4]. The network is trained using the train faster R-CNN object detector function from the neural network toolbox. In the proposed approach, there is rigorous process of identifying all spaces of object location in image. However, if no regions are identified in the first stage of algorithm, then there is no need to further go to the second step of approach. Xiong Changzhen used three types of datasets to train three different models [5]. They have achieved 99% accuracy in detecting Chinese traffic signals using R-CNN model. They used datasets from German traffic sign detection benchmark (GTSDB) dataset, Germany traffic sign recognition (GTSRB) dataset, and Belgium traffic sign (Belgium TS) dataset.

Myeong Ah Cho, Tae-young Chung, Hyeongmin Lee, and Sangyoun Lee used new technique where they have introduced N-RPN and RPN [6]. Negative learning is used by their pre-trained model VGG-16. They have trained model to learn and identify hard examples and for that they are separating foreground and background. Zhipeng Deng, Hao Sun tried to identify objects from Google's satellite images of earth [7]. They used $650 \times 6 = 3900$ NWPU VHR-10 dataset from google earth. They tried to identify objects like, e.g.—airplane, ship, and grounds. They trained the model by augmenting the dataset, turning images by 90 degrees. Their model contains two steps, first for augmenting the data and second for detector training. This method is efficient as the dataset size got increased using the same images. There are two objectives of idea used in this paper. First is get more accurate localization for small objects. Second eliminate background region proposals when there are many small objects exists. It consists of four models—feature fusion module, region objectness network, region proposal network, and ROIs wise classification network.

Wenjie Guan proposed an algorithm that is used for small object detection from an image like birds or cars [8]. They have used faster R-CNN model with feature fusion model. Region proposal method based on ACF model. First it passes the input image into several convolutional layers and maze pooling layers to extract feature maps. Then a re-join of interest (RoI) pooling layer is utilized to pool the feature maps of each region proposal into a fixed length feature vector. They have compared three models—region free SSD, YOLOv2 model, and region-based model. Their proposed model achieves 8.8 and 8% higher average precision over that of faster R-CNN (Resnet-101) on UA-DATRAC and BSBDV 2017.

Dong and Wang proposed an approach for region proposal of pedestrian detection [9]. They used INRIA dataset in which there are 1805 images, PKU dataset where there are 4900 images with 1280×720 resolution and 1804 images of ETH dataset. For training the dataset, they have used five layers of CNN. Their model contains RoI pooling layer, Relu layer, and max pooling layer. They have used AlexNet architecture for vehicle detection, traffic surveillance, and intelligent transport system. Region proposals are extracted from images using cascade object detector.

Karim et al. analyzed a model for detection of vehicles from satellite images [10]. These small objects are difficult to identify in images. They used AlexNet architecture. They used dataset of VEDAI where 1024×1024 images were present. They have claimed over 90% accuracy using this R-CNN model.

Li et al. proposed an algorithm for same object retrieval [11]. The pre-trained faster R-CNN is fine-tuned by using a specific image dataset so that the confidence score can identify an object proposal to the object level rather than the classification level. SOR faster R-CNN technique is used in this paper. The publishers are checking the confidence score based on the image and also ranking them based on cosine distance. They have used coke cans images (500 images) for training the R-CNN model.

Based on the literature survey, the application of face mask detection is nothing but a binary classification system where determination of person is wearing mask or not is to be done. The binary classification is typically achieved by supervised learning [12]. Neural Networks are used in supervised methods for better results. Either to predict (regression) something or in classification [13]. Convolutional neural networks (CNN) are widely employed in image classifications and other object recognition applications.

The CNN method has obtained more accurate result than the YOLO architecture for the similar kind of application based on face mask detection [14, 15]. The proposed method also uses the CNN where layers are placed in a sequential order to provide the effective results.

2.2 *Methodology*

Figure 2 is block diagram which displays the exact idea that is proposed. Here, camera is most important part where the data will get processed. The following blocks are mentioned in Fig. 2.

The person will wear a mask or will not wear a mask. This is our subject. The machine will take input from this as a source and return the output. The machine which is taking the input from the camera and gives output by processing on it. The machine is designed such a way that it will detect if the person is wearing mask or not. Controller with CNN is the main data center where we keep our trained model and which will detect the images. The images will get to it by camera. The machine will process on those images and then gives output. The output will only contain if person is wearing mask or not.

2.2.1 Dataset

Datasets are a set of instances that each one shares a standard attribute. The examples of these data types are numerical data, categorical data, time series data, and text data.

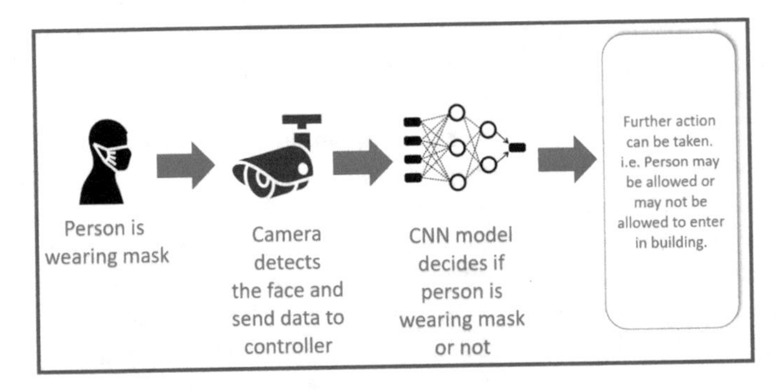

Fig. 2 Block diagram of proposed work

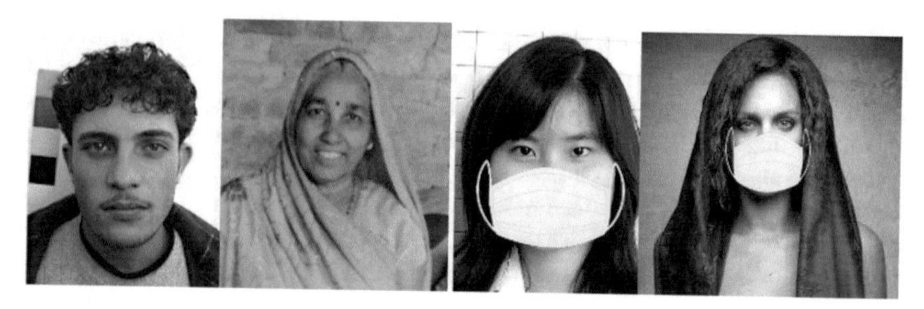

Fig. 3 Sample images from datasets [14]

In our case, the dataset will be categorical dataset as we have only two categories do define i.e. wearing the masks or not wearing the masks. Categorical data is sorted by defining characteristics. This can include gender, social class, ethnicity, hometown, the industry person works in, or a variety of other labels. We have used two labels here, with masked human faces and without mask human faces.

Here, we are using dataset provided by pyimage search [16]. They have provided the dataset for particularly this topic. The dataset contains total 1376 images. Out of which there are 686 images are of faces without masks, and remaining 690 are of faces with masks. These images with masks are added in augmented reality fashion. Here, the image with face is covered with the mask which is made augementally.

Figure 3 shows sample images from dataset obtained. These images are RGB images. The image contains various sizes and shapes. There is broad variety in the subjects i.e. people which are using masks. It is mixed dataset with female and male images. There are also some images of children.

2.2.2 Preprocessing of Images

Before starting the training, we need to preprocess the images. Pre-processing is required tasks for cleaning the data and making it suitable for a machine learning model which also increases the accuracy and efficiency of a machine learning model.

The data is first collected in 224 × 224 pixel resolution, so that all the images will be of uniform shape. The data is then normalized. Standardization is a useful technique to rework attributes with a normal distribution and differing means and standard deviations to a typical normal distribution with a mean of 0. The data is labeled whether it is with mask image or without mask image. The labeled data is used for understanding of machine learning model while training.

We can transform our data employing a binary threshold. All values above the edge are marked 1, and every one adequate to or below are marked as 0. This is called binarizing data or thresholding data. It can be useful when one has probabilities that one wants to make crisp values. It is also useful when feature engineering and one wants to add new features that indicate something meaningful.

The dataset is then increased using the ImageDataGenerator() function provided by tensor flow-keras [17]. This can increase the particular set of data by adding rotation angle, zooming in/out, shifting, and flipping the images. The updated dataset is then divided for 80% for training and 20% for testing purpose. This 80% is then provided to training model.

2.2.3 Training the Model

MobileNet-v2 is a CNN architecture model for image classification and mobile vision. There are other models also but what makes MobileNet-v2 special is, it has very little computation power to run or apply transfer learning [18]. This makes it an ideal fit for mobile devices, embedded systems, and computers without GPU or low computational efficiency with compromising significantly with the accuracy of the results. It is also best suited for web browsers as browsers have limitations over computation, graphic processing, and storage [19].

MobileNet-v2 for mobile and embedded vision applications is proposed, which are supported a streamlined architecture that uses depth-wise separable convolutions to create lightweight deep neural networks. Two simple global hyper-parameters that efficiently tradeoff between latency and accuracy are introduced. The core layer of Mobile Net-v2 is depth-wise separable filters, named as depth-wise separable convolution. The network structure is another factor to spice up the performance. Finally, the width and determination are often tuned to tradeoff between latency and accuracy.

Depth-wise separable convolutions which is a form of factorized convolutions that factorize a standard convolution into a depth-wise convolution and a $1 \times 11 \times 1$ convolution called a pointwise convolution. In MobileNet-v2, the depth-wise convolution applies a single filter to each input channel. The pointwise convolution then applies a $1 \times 11 \times 1$ convolution to combine the outputs of the depth-wise convolution.

Figure 4 illustrates the two phases of mask detection process. Here, in first phase face mask detector model is trained and saved in serialized format in hard disk. In second phase, the same pre-trained mask detector model is used to face mask detection. The camera detects the images and from that image, face region is identified with the help of OpenCV caffe model. The caffe model is more precise, efficient in order to detect human faces [20, 21]. These ROIs then passed to model where it detects if the person is wearing mask or not. Based on the results, the machine displays the output with a rectangle around the human face and labeled with proper comment if mask is present or not.

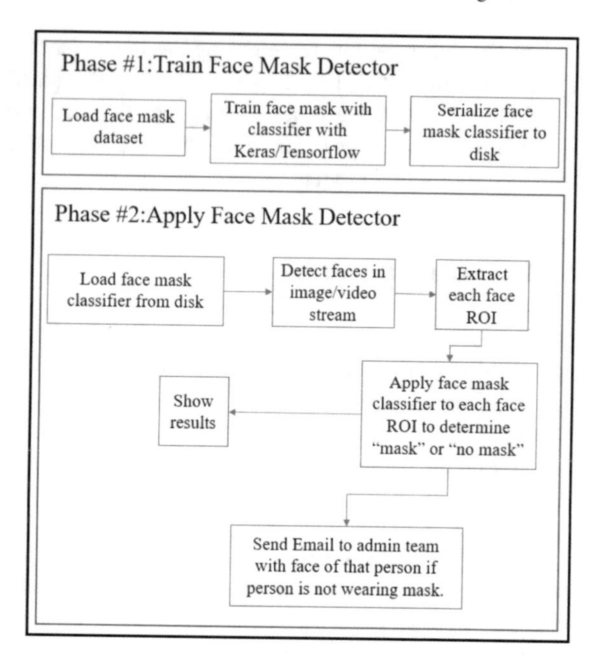

Fig. 4 Software system design in two phases

2.3 *Architecture of Model*

Figure 5 shows the architectural diagram of machine learning model. Pre-processing of data contains rearranging images into 224 × 224 pixels. Then there is average pooling layer of 7 × 7 grid. This 7 × 7 grid will do average and pool the value for that grid in image. Convolutional networks may include local or global pooling layers to

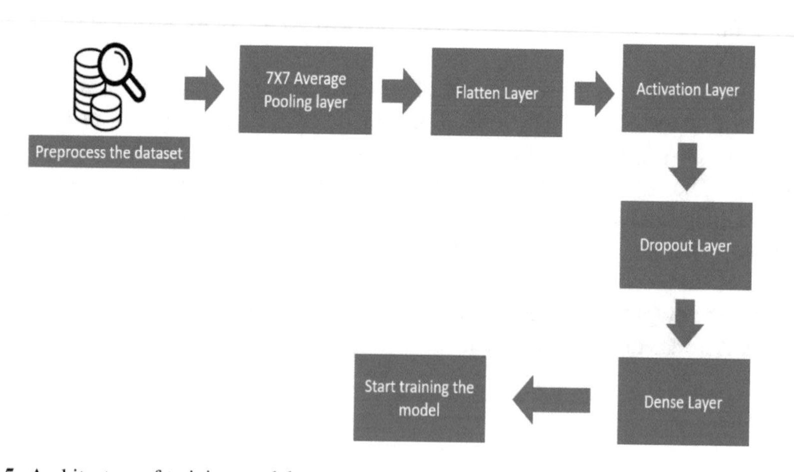

Fig. 5 Architecture of training model

streamline the underlying computation. Flattening a tensor means to remove all of the dimensions except for one. We are using Relu activation function.

As such, a wider network, e.g. more nodes, may be required when using dropout. Because the outputs of a layer under dropout are randomly subsampled, it has the effect of reducing the capacity or thinning the network during training. Each neuron in a layer receives an input from all the neurons present in the previous layer—thus, they are densely connected. In other words, the dense layer is a fully connected layer, meaning all the neurons in a layer are connected to those in the next layer.

3 Results and Discussion

Currently, the model is implemented with results with the parameters displayed in Table 1.

The batch size depends upon the processing power of machine (PC). In our case, Intel i5 8th generation processor is used and the RAM of PC is 8 GB. Table 2 is obtained after learning the model.

Here, the accuracy is 91%. This is a great accuracy score for a binary classification algorithm. This is actually more than the YoLo architecture that already exists [4].

The graph in Fig. 6 shows that the accuracy starts increasing on second epoch, and loss is decreasing after that. In each epoch, accuracy got increased and reached to almost 1 at the last 20th epoch. When the accuracy line is being stable, it indicates that there is no need for more iteration for increasing the model accuracy.

Table 1 Parameters used for training the model

Name of parameter	Value
Base model used	MobileNet_v2
No. of epochs	20
Batch size	32
Image pixel size	224 × 224

Table 2 Results after training the model

	Precision	Recall	F1-score	Support
With_mask	0.98	0.83	0.90	384
Without_mask	0.85	0.98	0.91	386
Accuracy			0.91	770
Macro avg	0.92	0.90	0.90	770
Weighted avg	0.92	0.91	0.90	770

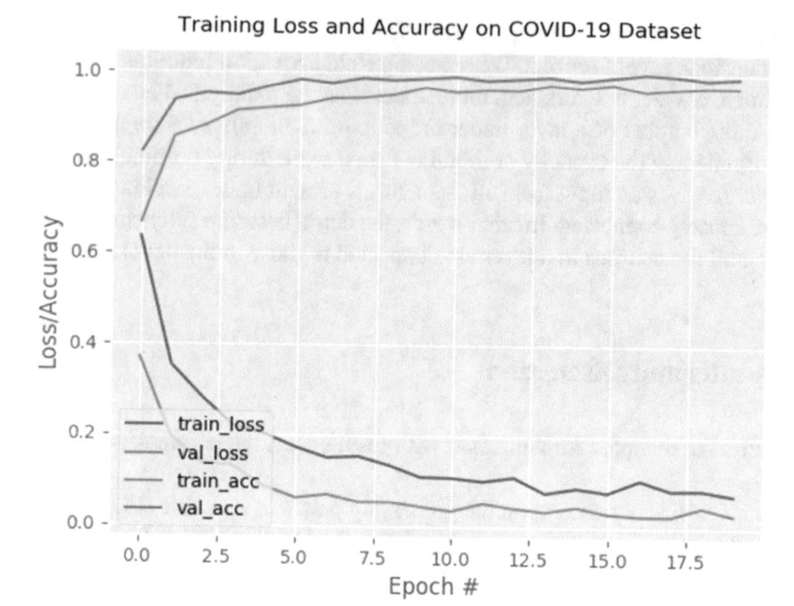

Fig. 6 Graph to identify losses and accuracies while training the model

Figures 7 and 8 display the result obtained from the system. Machine displays if that person is wearing mask or not. For result purpose, these images are obtained from images and not from live video.

Figure 9 shows that if the person is not wearing mask is detected then that face

Fig. 7 Detection of person is wearing mask

Fig. 8 Detection of person is not wearing mask

Fig. 9 Alert received on E-mail with the face of the person

is converted to image and an automated mail is get sent to the respective email. The subject and content of email can be changed. Based on the images, organization can take further take required action to manage proper guidelines to prevent COVID-19 spread.

4 Conclusion

There are many ways to define a model and train it. But while training, there are multiple layers that one should concern about. According to comparison of papers, the pre-defined model of MobileNet is more efficient and useful for binary classifications. The specific case taken into consideration in this paper is binary classification as there will be only two possibilities, person may wear a mask or may not.

The mask detector model predicts the images very well. The accuracy of 91% is great. This project focuses on machine learning and CNN. This project makes to study on CNN and machine learning algorithms. The model is processing the images with good accuracy. Better results in accuracy can be obtained by choosing a vast variety dataset. Such dataset should contain images of people wearing mask wrongly (wearing mask on chin). Different masks with designs and colors are also should be part of dataset to match the real-world problem accurately.

References

1. Chamola V, Hassija V, Gupta V, Guizani M (2020) A comprehensive review of the COVID-19 pandemic and the role of IoT, drones, AI, blockchain, and 5G in managing its impact. IEEE Access 8:1
2. https://www.who.int/emergencies/diseases/novel-coronavirus-2019/advice-for-public
3. Pereira LJ, Pereira CV, Murata RM, Pardi V (2020) Biological and social aspects of coronavirus disease 2019 (COVID-19) related to oral health, Sciflow Brazil
4. Abbas SM, Singh SN Region-based object detection and classification using faster R-CNN. In: International conference on computational intelligence and communication technology, CICT 2018, Ghaziabad
5. Changzhen X, Cong W, Weixin M, Yanmei S (2016) A traffic sign detection algorithm based on deep convolutional neural network. In: IEEE international conference on signal and image processing, Beijing
6. Cho MA, Chung T, Lee H, Lee S (2019) N-RPN: hard example learning for region proposal networks. School of Electrical and Electronic Engineering, Yonsei University, Republic of Korea fmaycho0305, tato0220, minimonia, IEEE, Taipei, Taiwan
7. Deng Z, Sun H, Zhou S, Zhau J, Lei L, Zou H Fast multiclass object detection in optical remote sensing image using region based convolution neural network. IEEE conference IGARSS, 2017, Fort Worth, Texas
8. Guan W, Zou YX, Zhou X (2018) Multi-scale object detection with feature fusion and region object ness network. In: IEEE international conference on signal and image processing, ADSPLAB/Intelligent Lab, School of ECE, Peking University, Shenzhen

9. Dong P, Wang W Better region proposals for pedestrian detection with R-CNN. In: IEEE international conference on signal and image processing, 2016, School of Electronic and Computer Engineering, Shenzhen Graduate School, Peking University
10. Karim S, Zhang Y, Yin S, Asif MR (2018) An efficient region proposal method for optical remote sensing imagery. In: IEEE conference paper, Valencia
11. Li H, Huang Y, Zhang Z (2015) An improved faster R-CNN for same object retrieval. IEEE J Latex Class Files 14:2
12. Japkowicz N (2001) Supervised versus unsupervised binary-learning by feedforward neural networks. Kluwer Academic Publishers, Springer
13. Yamashita R, Nishio M, Do RKG, Togashi K (2018) Convolutional neural networks: an overview and application in radiology. Insights into Imaging Springer 9:611–629
14. Susanto S, Putra FA, Analia R, Suciningtyas IKLN (2020) The face mask detection for preventing the spread of COVID-19 at Politeknik Negeri Batam. In: 3rd international conference on applied engineering (ICAE), Batam
15. Das A, Ansari MW, Basak R (2020) Covid-19 face mask detection using tensorflow, Keras and OpenCV. In: IEEE 17th India council international conference (INDICON), New Delhi
16. https://www.pyimagesearch.com
17. https://keras.io/api/preprocessing/image/
18. Liu X, Jia Z, Hou X, Fu M (2019) Real-time marine animal images classification by embedded system based on mobilenet and transfer learning. In: IEEE conference, OCEANS 2019, Marseille
19. Sanjaya SA, Rakhmawan SA (2020) Face mask detection using MobileNetV2 in the era of COVID-19 pandemic. In: IEEE conference, ICDABI, Sakheer
20. Dang K, Sharma S (2017) Review and comparison of face detection algorithms, CSE Department, ASET Amity University Noida. In: 2017 7th international conference on cloud computing, data science & engineering—confluence
21. Ren Z, Yang S, Zou F, Yang F, Luan C, Li K (2017) A face tracking framework based on convolutional neural networks and Kalman filter. Huazhong University of Science and Technology, Wuhan, China, ICSESS, School of Computer Science and Technology

Content Summarization: Journey Toward Word Clouds

Suwarna Gothane

Abstract Generating word cloud is an illustration demonstration of the occurrence of terminology within document. It is used to imagine the frequency of words in textbook, passage, manuscript, content, scientific data, speeches, and website tags. Sources of unstructured data varies from Emails, Text files considering Word processing, spreadsheets, PDF files, Websites, Social Media, Media such as images, video, audio, Mobile data, communication including live chat, collaboration software, IM, Customer-generated content, Raw data from marketing research, Books, magazines, and newspapers, Medical records. Word cloud are recognized as content clouds describe works simply with a concept maximum number of occurrences of a word associated with different data sources visualizes superior in the word cloud. It is an assortment or group of words represented in dissimilar sizes. More important words are those which display bolder and bigger in size. It means word appear maximum number of times in input document. It generates informative words and ideal way to generate excellent data visualization aid to our brain. It is useful for business and decision-making process to summarize information. Here, we discussed various online tool used for generation of word cloud and applied online tool for generation of word cloud. For security purpose, we also used python code implementation to generate word cloud.

Keywords World cloud · Python

1 Introduction

Word clouds generate image aid at rate of recurrence of tokens for the document data. It is helpful to identify frequency of words within structured and unstructured data sources. It is popular mean for business to perform data analysis. It considers important and informative information with catchy words. Word clouds are an incredible

S. Gothane (✉)
Department of Computer Science and Engineering, CMR Technical Campus, UGC Autonomous, Hyderabad, India
e-mail: suwarnagothane.cse@cmrtc.ac.in

A. Kumar et al. (eds.), *Proceedings of the International Conference on Cognitive and Intelligent Computing*, Cognitive Science and Technology,
https://doi.org/10.1007/978-981-19-2350-0_46

tool for data visualization. It is applicable to important speeches, seminar paper titles, movie title, or letters for summarization information in form of word cloud. With variety, velocity, and large volume of data, there is tremendous concentration and need to apply weighted list to visualize data to make it applicable for articulation of important information.

There are many online word cloud generators accessible which process data using inbuilt algorithm based on repetition of words. There is gap identified with limited performance of online generators, they can process only English characters.

Some of programming languages like Python and R generate word clouds, which need a skilled person and can obtain clouds with different colors, shapes, sizes, width, and height.

2 Literature Survey

Arnaboldi et al., presented Wormicloud tool that summarizes scientific articles into graphical form. Tool found useful for researchers and biocurators. Developed tool Wormicloud is modified for the Caenorhabditis elegans literature. It is found helpful to perform full-text searches using Textpresso with enhanced performance and get combined with WormBase [1]. Padmanandam et al., proposed work on dynamic data input with speech recognition enabled feature to guarantee semantic coherence and spatial stability of the words. Author analyzed text on vast collection of documents and voice commands [2].

Bafna et al., worked on large number of textual multiple documents using NLP on managing Marathi language and summarize the Marathi corpus. Author proved approach as a scalable, consistent, and noticed entropy and precision calculation results with robustness [3]. Calle-Alonso et al., worked for social network-based e-learning platform based on neurodidactics. Author implemented word clouds for students and teachers for course development, and evaluation process [4]. Basha et al., performed automatic summarization using nearest neighbor classifier approach. Author used parameters for sentence similarity frequency of the terms in single sentence and the similarity of that sentence to other sentences. Sentences are then ranked with threshold values and summarized [5].

Li et al., worked on cities geo-tagged textural data. Author provided a visual design combing metro map and wordles. The wordles are generated from keywords extracted from reviews, descriptions, and are embedded into the subareas based on their geographical locations. Results proved efficient and visual analysis was implemented on the restaurants data in Shanghai, China [6]. Hollander et al., studied parsimonious language model to construct word clouds on the proceedings of the European Parliament. Author applied stemming, stop words, N-gram, and multilingualism model, and implemented multiple word clouds using parsimonious language models or simple term frequencies [7].

Wu et al., introduced method to create semantic-preserving word clouds by leveraging tailored seam carving. Proposed method optimized word cloud layout and provide visual analytics [8]. Lohmann et al. worked on contrastive analysis of multiple texts on various texts examples. Concentric loud approach combines the words from numerous text documents into a single visualization turns beneficial [9].

Cui and Wu et al., proposed a layout approach on loopholes of existing methods and solved issues with dynamic word cloud. Algorithm depends on geometry meshes and an adaptive force-directed model with parameters semantic coherence and spatial stability and generated dynamic cloud [10]. Thomas Gottron proposed a plan of increasing performance of web documents with visualization technique from tag cloud of Web 2.0. Approach highlighted significant terms and traced relevancy [11].

Frances Miley and Andrew Read examined and summarized student usage for word cloud automated sources. Author performed study on student usage and collected student feedback and traced word clouds importance with consistency in learning process. They noticed it is helpful in working environment [12]. Yuping Jin focused on graphical user interface software to produce word cloud using Python programming language and conveyed its importance [13].

3 Approach to Generate Word Cloud

3.1 Use Optimize Data Set

Using online cloud generator with copy and paste textual data does not generate actual results. So, an important text-based and optimized dataset for context is needed. Several data analysts are capable on working data sources and convert into an optimized data set. Professional focus on the source data to be usable and actionable that find out attractive and appropriate content.

3.2 Use a Word Cloud Generator Tool

After correct and appropriate dataset, it needs to process through a word cloud generator tool. Businesses exercise on word cloud generator tool to routinely update data in database. Some of the online resources and available application to perform summarization include:

I. Wordle is a Java tool for generating "word clouds". It gives greater importance to words based on frequency where the clouds with added style of different fonts, layouts, and color schemes are generated.

II. TagCrowd is a web app supports representation of word frequencies. It provides option to paste text, upload, or enter the URL.

III. Tagxedo transforms terms into a word cloud, and it also uses images to create a custom shape.

IV. WordArt is an online word cloud art creator that generates wonderful and exclusive word cloud art with various customizable options for words, shapes, fonts, colors, layouts, etc.
V. ToCloudis a free word cloud generator based on measure of word frequency that generates a quick solution for page optimization.
VI. WordItOut is an open word tool available with control and custom settings options.
VII. Wordclouds.com is a liberated tool for cloud creator applicable for PC, tablet, or smartphone. It works with options to paste contents, upload, or URL. It supports cloud creation with shapes, themes, colors, and fonts, edit the word list, enables cloud size and gap size, supports clickable word clouds with links (image map), save the image and share.
VIII. Word Cloud Generator proposed by Jason Davies uses JavaScript and facilitate functionality for scale, word orientation, font, and the count of words from input contents for consideration.
IX. Vizzlo Word Cloud Generator is a non-free online data visualization means but supports high end word cloud generation capabilities.
X. Word Cloud Maker is a superior opensource tool allows user to input an image and create word cloud art in several formats such as vector svg, png, jpg, jpeg, and pdf.
XI. Google Docs facilates add-on option to generate word clouds.
XII. Infogram Word Clouds is an online chart maker used to design infographics, presentations, reports, etc.
XIII. WordSift helps educators to administer the hassle of words and educational language.
XIV. MonkeyLearn AI Word Cloud Generator uses AI to generate word cloud.

3.3 Export the Word Cloud

After creation of word cloud using online tools, we can save it in our system folders with download option and further refer it.

4 Implementation

4.1 Using Online Word Cloud Generator Tool

We consider paragraph on COVID-19 transmission and protective measures guideline and traced result of occurrence of word.

With online tool word frequency relevancy is shown in Table 1.

Figure 1 shows obtain word cloud generated by online tool.

Table 1 Word frequency relevancy

Word	Count	Relevance
Bent elbow	1	0.952
Touched object	1	0.952
Hand	2	0.952
Protective measures	1	0.952
New habits	1	0.952
Simple precautions	1	0.952
Close contact	1	0.952
Covid-19 transmission	1	0.952
Crowded place	1	0.952
Social gathering	1	0.952
Person	2	0.952
Joint responsibility	1	0.952
Someone	1	0.476
Surface	1	0.476
Mouth	1	0.476
Limit	1	0.476
Cough	1	0.476
Eye	1	0.476
Avoid	2	0.476
Disease	1	0.476
Nose	1	0.476
Time	1	0.476
Others	1	0.476

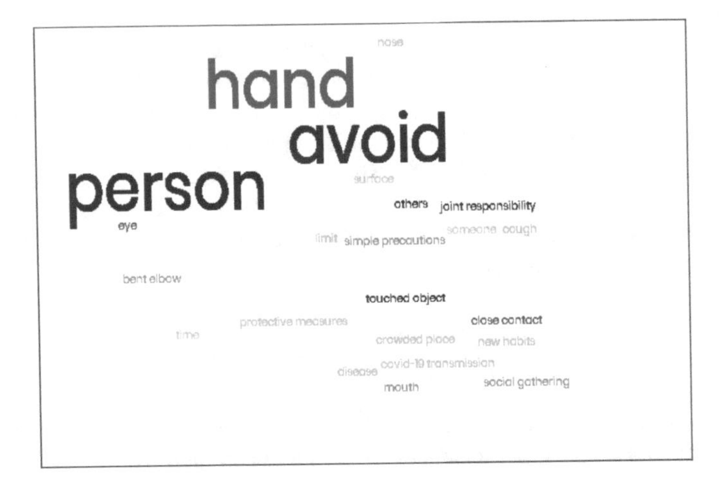

Fig. 1 Word cloud generated by online tool

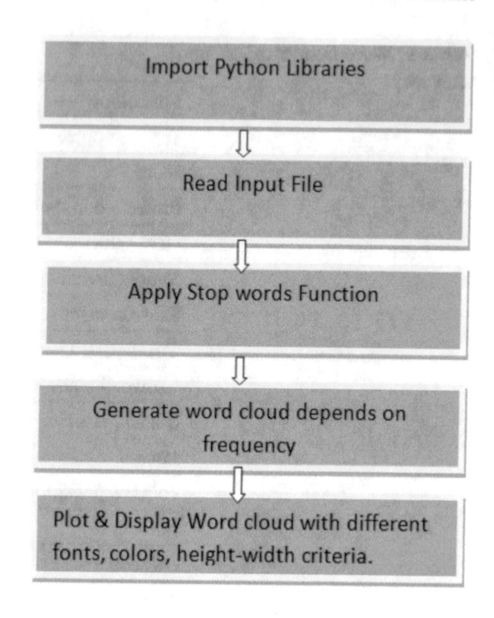

Fig. 2 Procedure for world cloud generation

4.2 Generation Word Cloud Using Python Language

We have noticed gaps about security precaution. The online tools are freely available for creating a word cloud and saves data in implicit collection. Compiled and analyzed data losses secret for industry, and it is extremely responsible for sectors like healthcare, banking, etc., to loss secrecy. So, appropriate means are required to maintain confidentiality. So, creation of word cloud from scratch is the solution. Here, we have used Python language for generation of word cloud. Python is a vital and progressive language.

Procedure for world cloud generation is shown Fig. 2.

We can create word cloud with the following equation criteria.

Word cloud of categories like weblogs, count to weblog entries are assigned to a group. For smaller frequencies, font sizes are assigned and for larger values, a scaling is applicable. In a linear normalization, The weight t_i of a descriptor is mapped to a size scale of 1 through f, where $t_{\min}$ and $t_{\max}$ are specifying the range of available weights. $F_{\max}$ species max. fontsize, t_i indicates count, tmin indicates min. count, tmax indicates max. count, and si shows display fontsize [6].

$$s_i = \left\lceil \frac{f_{\max} \cdot (t_i - t_{\min})}{t_{\max} - t_{\min}} \right\rceil \text{for } t_i > t_{\min}; \text{else } s_i = 1$$

For bigger ranges of values, a logarithmic representation is useful. It includes text parsing and filtering out unsupportive tags like common words, numbers, and punctuation.

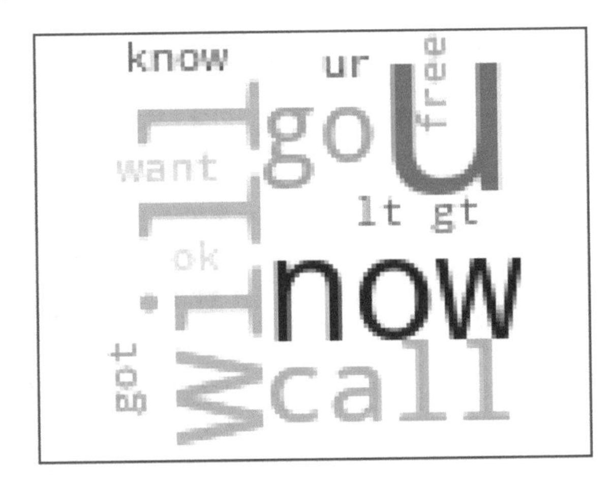

Fig. 3 Word cloud of dataset

Fig. 4 Word cloud spam classification

We processed spam dataset obtain from machine learning repository using Python and generated result to obtain word cloud. Word cloud for entire dataset is shown in Fig. 3.

Word cloud output of spam classification is shown in Fig. 4.

Word cloud output of ham classification is shown in Fig. 5.

5 Conclusion and Future Scope

We use apply word cloud for various types of structure as well unstructured data to summarize information. With online tools, word clouds are effortless to generate, engaging, and widely used. But, word clouds associated with data science work, journalists or scientific data lacks in security with online tools and shows deprived

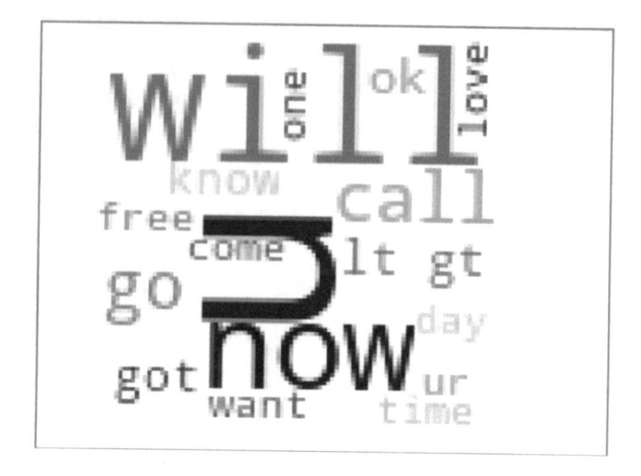

Fig. 5 Word cloud ham classification

to tackle relative values and to analyze. The field of information visualization needs to visualize the contents of textWords based on similar concept or associated with same topic need to form one group.

Separate groups of words from one another in category by assigning the same color to words with single concept. It makes use of principles from perceptual psychology. In the future work approach, word zones are easier to interpret. Visually grouped layouts become a valuable mean in time-constrained category understanding tasks compared to ungrouped layouts. Visual alignment is effective by separating categories via whitespace or by color distinction. Highly developed data visualization approach that turns important analytics into charts, graphs, and infographics will be effective in the future.

References

1. Arnaboldi V, Cho J, Sternberg PW (2021) Wormicloud: a new text summarization tool based on word clouds to explore the C. elegans literature. Database
2. Padmanandam K, Sai Priya VDS, Bheri LHV, Sruthi K (2021) A speech recognized dynamic word cloud visualization for text summarization. In: 6th international conference on inventive computation technologies (ICICT), proceeding to IEEE Xplore: 26 February 2021, IEEE
3. Bafna PB, Kumar J, Saini R (2020) Marathi text analysis using unsupervised learning and word cloud. Int J Eng Adv Technol (IJEAT) 9(3). ISSN: 2249-8958
4. Calle-Alonso F, Botón-Fernández V, Sánchez-Gómez JM, Vega-Rodríguez MA, Pérez CJ, de la Matal D (2019) Word clouds as a learning analytic tool for the cooperative e-learning platform neuro K. In: Proceedings of the 10th international conference on computer supported education (CSEDU 2018), pp 508–513. ISBN: 978-989-758-291-2. Copyright c©2019 by SCITEPRESS—Science and Technology Publications, Lda
5. Rahamat Basha S, Keziya Rani J, Prasad Yadav JJC (2019) A novel summarization-based approach for feature reduction enhancing text classification accuracy. Eng Technol Appl Sci Res 9(6):5001–5005
6. Li C, Dong X, Yuan X (2018) Metro-wordle: an interactive visualization for urban text distributions based on wordle. Visual Inform 2(1):50–59

7. de Hollander G, Marx M Summarization of meetings using word clouds. In: CSI international symposium on computer science and software engineering (CSSE). IEEE, 29 July 2011
8. Wu Y, Provan T Semantic-preserving word clouds by seam carving. ComputSci Comput Graph Forum. https://doi.org/10.1111/j.1467-8659.2011.01923.x. Corpus ID: 3242976, 1 June 2011
9. Lohmann S, Heimerl F, Bopp F, Burch M Concentri cloud: word cloud visualization for multiple text documents. In: Conference: 2015 19th international conference on information visualization, July 2015. https://doi.org/10.1109/iV.2015.30
10. Cui W, Wu Y, Liu S, Wei F, Zhou MX, Qu H (2010) Context-preserving, dynamic word cloud, visualization, Published by the IEEE Computer Society November/December
11. Gottron T (2009) Document word clouds: visualising web documents as tag clouds to aid users in relevance decisions
12. Miley F, Read A (2011) Using word clouds to develop proactive learners. J Scholarship Teach Learn 11(2)
13. Jin Y (2016) Development of word cloud generator software based on python. In: 13th global congress on manufacturing and management, GCMM

Optimization of Adaptive Queuing System with Priority-Based Scheduling in Underwater Sensor Networks

T. Aruna Jacintha and T. Jaya

Abstract Underwater sensor networks (UWSNs) have gained immense attention amongst researchers due to its advanced technologies. Underwater wireless communication (UWC) is important for monitoring marine life, water pollution, oil and gas rig exploration, natural disaster surveillance, navy tactical operations for coastal security and underwater environment monitoring. In this light, the broad adoption of UWC has become a critical topic of research in order to envision many military and commercial applications that have a rising interest in exploring the underwater environment for a variety of purposes. As signals in UWSN are sent by acoustic signals, and a major feature of UWSN is that it is far from other sensor networks. However, the elevated delay in UWSNs can lead to spatial–temporal uncertainty and thereby made transmission scheduling a complex task. To solve this issue, a novel hybrid optimization-driven technique is devised for priority-based scheduling in UWSN-based on queuing technique. Here, the first step is the simulation of UWSN. Thereafter, the optimal paths are discovered for transmitting the signals attained from the sensor nodes. The optimal routing is performed using a fractional gravitational search algorithm (FGSA). Once the optimal routes are obtained, the obtained signals are transmitted to BS, wherein the priority-based scheduling is performed. The priority-based scheduling is performed with a newly devised lion-based shuffled shepherd optimization algorithm (Lion-SS) and fitness factor. The proposed Lion-SS is newly devised considering lion algorithm and shuffled shepherd optimization algorithm. In addition, fitness is obtained by energy, delay and distance. Thus, the fitness factor and proposed Lion-SS are utilized for priority-based scheduling in UWSN.

Keywords Queuing · Underwater sensor network · Scheduling · Priority · Energy efficient routing

T. Aruna Jacintha · T. Jaya (✉)
Department of Electronics and Communication Engineering, School of Engineering, Vels Institute of Science Technology & Advanced Studies (VISTAS), Chennai, Tamil Nadu, India
e-mail: jaya.se@velsuniv.ac.in

T. Aruna Jacintha
e-mail: aruna11jacintha@gmail.com

A. Kumar et al. (eds.), *Proceedings of the International Conference on Cognitive and Intelligent Computing*, Cognitive Science and Technology,
https://doi.org/10.1007/978-981-19-2350-0_47

1 Introduction

UWSN is a powerful technology that exploited the ocean and is considered an emerging domain in the upcoming years. The UWSN contains unmanned vehicles and sensors. The sensors are linked with an acoustic transmitter that relays the data to a buoy [1]. The sensors acquire the features of seawater like density, acidity, and conductivity. The underwater sensors differed from other sensors in terms of communication method and cost. As underwater poses, distinct features and sensor devices are built for averting saltwater corrosion. The communication occurs amongst the sensors using sound waves [2]. In past years, the UWSN has gained huge interest due to its advanced technologies. However, the volatile platform made the acoustic signals suffer from issues like low data rate, low propagation speed and limited bandwidth [3]. These features made the UWSN a challenging domain. Thus, medium access control (MAC) protocols are devised that allowed multiple users to share a common medium effective manner. Thus, huge efforts are taken to devise a suitable protocol for precise UWSN [4].

Even though wireless protocols have been studied widely for more years, but still underwater techniques face several issues. In general, acoustic communication contains several issues like less latency, limited bandwidth and so on. Meanwhile, the medium access protocols (MAC) like frequency division multiple access (FDMA) are not apposite for underwater networks because of narrow bandwidth and susceptibility of limited band systems to fading [5]. Moreover, the time division multiple access (TDMA) needs accurate synchronization of time and huge guard time. In addition, the orthogonal MAC techniques poses scalability issues in which some nodes join and leave the network [6]. In recent days the scheduling of packet is considered as an imperative network for avoiding the rate of delay. Because of the huge delay, the queuing packets can lead to a high dropout rate. While sending the data from source to target node, all packets must be in a queue for transmitting data using medium availability. If a channel is busy, then the queuing length size becomes high. In addition, the high queue length may cause a delay in transmitting data to the target node [7].

This paper devises a novel technique for UWSN. Here, the preliminary step is the UWSN simulation. Thereafter, the routing is performed using FGSA to generate optimal paths for data transmission. Furthermore, priority-based scheduling is done using the considered data. The priority-based scheduling is done using the proposed Lion-SS, which is newly devised by combining the Lion algorithm and SSOA. The major contribution of the paper is proposed Lion-SS for priority scheduling. The priority scheduling is done using proposed Lion-SS, which is newly devised by combining Lion algorithm and SSOA. Here, the fitness is newly devised considering.

2 Motivations

This section explains the classical techniques along with its challenges. The challenges are considered as a motivation to design a novel optimization-driven adaptive queuing system with priority-based scheduling.

2.1 Literature Review

Sampath and Subashini [2] developed cluster-based MAC protocol, namely multilevel scheduling MAC (MLS-MAC) protocol for priority-based scheduling UWSN. Here, the clustering was done wherein the cluster head for transmitting the packet and helps to elevate network lifetime. For reinforcing the efficiency of the channel, the multilevel scheduling was initiated using queues based on applications set by the cluster head. Sivasubramaniam and Senniappan [8] developed enhanced core stateless fair queuing (ECSFQ) using a different scheduler. At first, the priority scheduler was adapted for flows that access edge routers. If real-time, then the packets will provide elevated priority, else less priority. Zhuo et al. [4] devises delay and queue aware adaptive scheduling-based medium access control (DQA-MAC) protocol for UWSN. The method combined handshaking packets, adaptive scheduling transmission for enhancing the network performance. Here, the time of data transmission was scheduled using propagation delays and queues. Ng et al. [6] devises multiple access collision avoidance (MACA) for packet scheduling in UWSN. Here, the technique examined state transition rules, backoff algorithm, and packet forwarding strategy. In addition, the performance was computed by simulating multi-hop underwater networks. Kumar et al. [9] devised a technique based on queuing theory for wireless networks. Here, the dynamic bandwidth algorithm was utilizing for shunning the probability of packets clash. In addition, the multi-sink WSN and multi-OLT PON were examined with computer simulations. Padmavathy, C. and Jayashree [7] developed a packet scheduling technique for avoiding the drop rate and delay in a wireless network. Here, the enhanced delay sensitive data packet scheduling (EDSP) algorithm was devised for initiating proper scheduling, which was based on weight calculation, medium selection and priority scheduling. Sundarasekar et al. [10] this high-end research offers the Adaptive energy aware quality of service (AEA-QoS) method, which combines a discrete temporal stochastic control process with deep learning approaches for UWSAN. The algorithm has been validated using traditional state-of-the-art methods, and the results show that the effective in terms of lower network overhead and propagation delay, as well as higher throughput and lower energy consumption for each reliable packet transmission. Faheem et al. [11] state that it selects the optimum forwarding relay node for data transmission and prevents route loops and packet losses. The suggested protocol outperforms existing routing protocols in terms of data delivery ratio, total network throughput, end-to-end delay and energy efficiency.

Khalil and Saeed [12], an effective relay placement strategy for relays in MI-based IoUT networks is performed. The relay placement problem is first formulated as a non-convex optimization problem. Then, to solve the non-convex function, global optimization strategy based on reverse convex programming (RCP) was performed. Ali et al. [13] For data transfer in an underwater environment, acoustic, optical and RF wireless carriers have been considered. The Internet of Undersea Things (IoUT) and next-generation (5G) networks have a significant influence on UWC because they support data throughput, connection and energy efficiency improvements. This study focuses on future developing UWC methodologies, as well as improving existing work with the use of a 5G network. This survey provides a complete assessment of existing UWC approaches, as well as potential future paths and recommendations for enabling next-generation underwater wireless networking systems.

2.2 *Challenges*

i. UWSN has acquired huge focus in recent days, and it uses various types of oceanic applications. However, this technique should handle energy constraints and propagation delay for initiating effective underwater communication [2].
ii. The issues of CSFQ involve the incapability for evaluating fairness wherein huge traffic flows are contained wherein the traffic can be short and bursty [8].
iii. In prior works, the UWSN has acquired huge attention due to large applications in marine exploration. However, the unstable ocean platforms made the underwater acoustic signals propagation to suffer from severe issues like less bandwidth, less rate of data and less speed of propagation [4].
iv. In recent days, the scheduling of packet is most imperative in a wireless network for avoiding drop rate and delay. Because of the higher queue length, it generates more delay for transmitting data to the destination node [7].

3 System Model

Assume UWSN as graph $G = (H, I)$, in which $H = \{k_1, k_2, \ldots, k_a, \ldots, k_d\}$ signifies a set of node and d represent total nodes contained in UWSN. In this model, I signifies set of a path which links two nodes, and is expressed as $I = \{g_1, g_2, \ldots, g_h, \ldots, g_e\}$. Consider Q signifies source node that transmits data to target node R. Every node requires high energy for sending data between nodes. While transmitting data from the source to other nodes, the distance is termed as the main hypothesis. Moreover, the topology of a network is vulnerable to attack because of the existence of vulnerable nodes. Thus, it is essential to devise a model using energy and mobility for effective data transmission in USWN (Fig. 1).

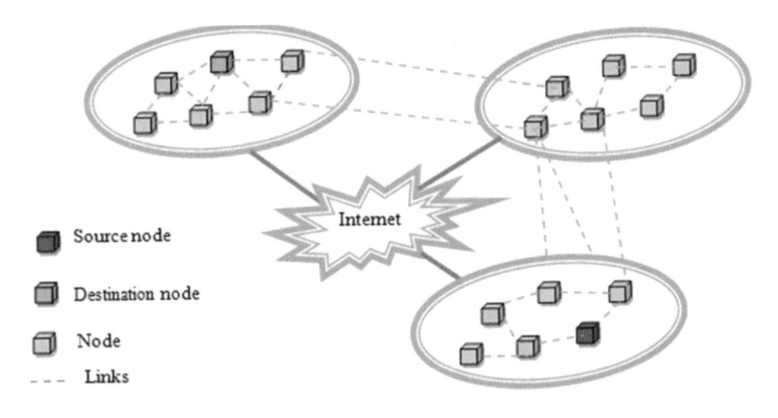

Fig. 1 System model of UWSN

3.1 Mobility Model

The mobility model is utilized for examining the movement of sensor nodes and for computing change in positionof node considering acceleration, velocity and position in a particular instant. Here, the mobility model is employed for discovering the efficiency of routing and is utilized for beginning movement amongst nodes. Imagine node a and b resides in position (u_1, v_1) and (u_2, v_2) in a particular time. The nodes a and b changes with various velocities in a specific time t. Hence, the Euclidean distance employed for node mobility a and b in particular path h is expressedas,

$$D_{(a,b,h)} = \sqrt{(u_1 - u_2)^2 + (v_1 - v_2)^2} \tag{1}$$

where, a and b signifies two nodes contained in path h, (u_1, v_1) and (u_2, v_2) express position of node a and b at particular time t.

3.2 Energy Model

The energy model of UWSN is considered forsending data amongst nodes. UWSN containsdistributed mobile nodes that operate batteries. Consider initial energy of nodes is, ε_0 that specifies batteries which is non-rechargeable. When data receives transmitter, some amount of energy is misplacedbased on communication distance. The network transmission based on protocol and energy dissipation is because of presence of power amplifier and radio electronics accessible in transmitter. Hence, it is noted that energy dissipates at time of data transmission based on distance and node nature. The energy model is defined as, If P comprises full battery energy of a node then remaining battery energy of node a at time t is given as $\varepsilon_a^{\text{rem}}(t)$ and expressed as,

$$\varepsilon_a^{\text{rem}}(t) = \varepsilon_a^{\text{rem}}(t-1) - \varepsilon_a^{P} \times n(t-1,t) - \varepsilon_a^{r} \times n(t-1,t) \tag{2}$$

where ε_a^{P} refers amount of energy needed to send data, n signifiescount of bits transmitted from t to $t-1$ and ε_a^{r} signifies energy required to receive data. In time $t = 0$, the remaining energy in node is full that is $\varepsilon_a^{\text{rem}}(t) = P$. The remaining energy is expressed as,

$$\text{energy ratio} = \frac{\varepsilon_a^{\text{rem}}(t)}{P} \tag{3}$$

where remaining energy resides in $0 \leq \varepsilon_a^{\text{rem}} \leq 1$.

4 Proposed Adaptive-HGSO for Priority-Based Scheduling Using Queuing

A novel hybrid optimization-driven technique is devised for priority-based scheduling in underwater sensor networks. Here, the first step is the simulation of UWSN. Thereafter, the optimal paths are discovered for transmitting the signals attained from the sensor nodes. The optimal routing is performed using a fractional gravitational search algorithm (FGSA). Once the optimal routes are obtained, the obtained signals are transmitted to BS, wherein the priority-based scheduling is performed. The priority-based scheduling is performed using a newly devised lion-based shuffled shepherd optimization algorithm (Lion-SS) and fitness factor. The proposed Lion-SS is newly devised considering lion algorithm and shuffled shepherd optimization algorithm. In addition, fitness is obtained by energy, delay and distance. Thus, the fitness factor and proposed Lion-SS are utilized for priority-based scheduling in UWSN (Fig. 2).

4.1 Routing Using FGSA

In an IoT network, routing through the effective path is a challenging task due to the exhausted energy of the battery. Here, the energy issues of nodes are evaluated through optimal path using the FGSA algorithm [14]. Thus, the FGSA routing algorithm is formed by updating the fractional theory with GSA. The purpose of selecting the optimal path is to diminish the energy and delay that increase the lifespan. The GSA algorithm has some disadvantages, like less convergence speed, high computational cost and so on. Thus, the mathematical expression for GSA is enhanced by including the fractional concept (FC) in the update equation of GSA. During transmission, the data is to be transmitted through the optimal path in order to reduce the power usage of node. The agent location is updated through fractional calculus, which is described by,

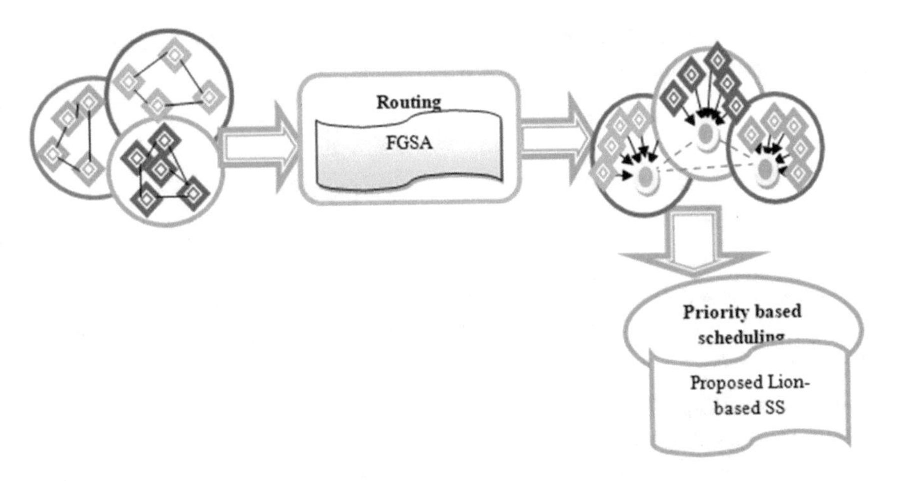

Fig. 2 Proposed lion-based SS for priority-based scheduling based on queuing

$$R_p^s(k+1) = R_p^s(k) + t_p^s(k+1) \tag{4}$$

where the term $R_p^s(k)$ depicts theposition of pth agent in sth cluster at time, and the velocity update is depicted as $t_p^s(k+1)$.

The above Eq. (9) can be modified as,

$$R_p^s(k+1) - R_p^s(k) = t_p^s(k+1) \tag{5}$$

From Eq. (10), the left side term $R_p^s(k+1) - R_p^s(k)$ depicts the discrete version of derivative order $\chi = 1$, then the equation is changed as,

$$J^\chi\left[R_p^s(k+1)\right] = t_p^s(k+1) \tag{6}$$

The degree of discrete derivative is assigned to real number $0 \leq \chi \leq 1$. Consequently, Eq. (10) can be modified with two differential derivatives, which are modelled as,

$$R_p^s(k+1) - \chi R_p^s(k) - \frac{1}{2}\chi R_p^s(k-1) = t_p^s(k+1) \tag{7}$$

$$R_p^s(k+1) = \chi R_p^s(k) + \frac{1}{2}\chi R_p^s(k-1) + t_p^s(k+1) \tag{8}$$

where $R_p^s(k+1)$ signifies position of agent p in sth dimension at time $k+1$, the position of pth agent at the present iteration k is depicted as $R_p^s(k)$, the term $R_p^s(k-1)$ denotes the position of agent in preceding iteration, and the velocity computed through GSA algorithm is depicted as $t_p^s(k+1)$. Thus, the suitable optimal paths are chosen using the FGSA for transmitting the information. The transmitted information through optimal multipath is considered as Q.

4.2 Priority-Based Scheduling Using Proposed Lion-SS in UWSN

The priority-based scheduling is carried out using the proposed Lion-SS in UWSN, considering the fitness factor. The solution encoding, fitness and developed Lion-SS is described below.

Solution encoding. The encoding of the solution is done by representing the optimal solution using the newly devised optimization. The solution of Lion-SS represents the count of nodes with which optimal node is selected using the maximal value of fitness. The solution encoding of Lion-SS is illustrated in Fig. 3. Here, t signifies prioritized nodes and h symbolize total nodes. Thus, the total prioritized node is less than the total nodes.

Fitness function. The fitness is computedwith certain parameters that involve delay, energy and distance with nodes. The fitness using lion-based SS is employed for discovering priority node. The fitness is evaluated as,

$$\text{fitness} = \frac{1}{N}[E + X + D] \tag{9}$$

where N signifies normalization factor, E signifies energy consumption, D is distance and X indicate delay.

Proposed Lion-based SS optimization algorithm. The optimal application migration is performed using the proposed Lion-SS, which is obtained by integrating the benefits of both SSOA and LOA. SSOA [15] is inspired by the mimicking shepherd's behaviour while operating each community. The algorithm offers a trade-off between exploitation and exploration ability. In addition, the SSOA is effective in solving engineering problems and discover an optimal solution with less parameter. The technique is simple to implement and can bypass local optima. LOA [16] is motivated from unique features of the lion, which include territorial takeover, territorial defence, laggardness exploitation and pride. The LOA helps to solve global

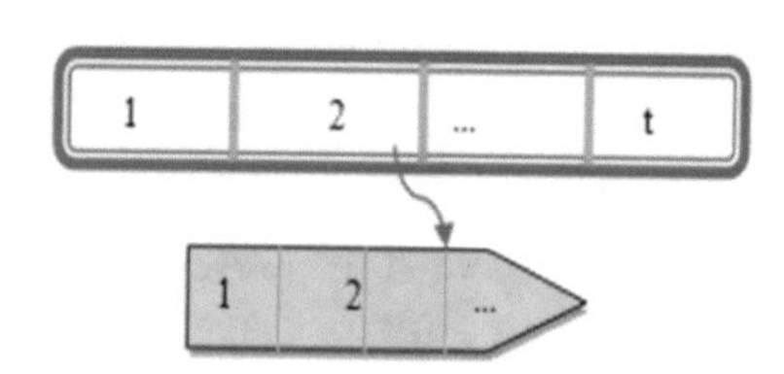

Fig. 3 Solution encoding

optimization issues and minimize errors. The technique prevents premature convergence. Thus, the incorporation of LOA and SSOA helps to offer improved performance and provides global optimal solutions. The algorithmic steps of the proposed Lion-SS are given as.

Step 1: Initialization. The preliminary step represent solution initialization expressed as, S considering total j solutions, such that $1 \leq i \leq j$

$$S = \{S_1, S_2, \ldots, S_i, \ldots, S_j\} \tag{10}$$

where j is total solution, and S_i signifies ith solution.

Step 2: Determination of fitness. The fitness is already defined with Eq. (14).

Step 3: Evaluation of update equation. The SSOA provides a balance between exploration and exploitation phases which subsequently helps to acquire global optimal solution. As per SSOA algorithm [15], the temple solution vector of each sheep is given as,

$$S_k^{\text{temple}} = S_k^{\text{old}} + \text{Step size} \tag{11}$$

where S_k^{old} signifies old temple solution vector, and $Step\ size$ indicate constant value.
It can be rewritten as,

$$S_k^{h+1} = S_k^h + \alpha\text{Rand}\left(S_f^h - S_k^h\right) + \beta\text{Rand}\left(S_e^h - S_k^h\right) \tag{12}$$

where, α is an exploration and β signifies exploitation parameters, Rand symbolize random numbers, S_k^h symbolize solution vector of a shepherd, S_f^h represent solution vector of a selected horse and S_e^h symbolize solution vector of selected sheep.

$$S_k^{h+1} = S_k^h + \alpha\text{Rand}S_f^h - \alpha\text{Rand}S_k^h + \beta\text{Rand}S_e^h - \beta\text{Rand}S_k^h \tag{13}$$

$$S_k^{h+1} = S_k^h(1 - \alpha\text{Rand} - \beta\text{Rand}) + \alpha\text{Rand}S_f^h + \beta\text{Rand}S_e^h \tag{14}$$

LOA [16] stores the best solution obtained so far and is mathematically formulated as,

$$S_k^{h+1} = S_k^h + 2D\text{Rand}(0, 1)\{N_1\} + H(-1, 1)\tan\theta J\{N_2\} \tag{15}$$

where S_k^h signifies present female lion position, D symbolize distance amongst female lion position and chosen point, $\{N_1\}$ signifies vector that starts point represent the previous position of female lion and $\{N_2\}$ is perpendicular to $\{N_1\}$ and $\text{Rand}(0, 1)$

symbolize random number between 0 and 1, and $H(-1,1)$ refers random number between -1 and 1.

$$S_k^h = S_k^{h+1} - 2D\text{Rand}(0,1)\{N_1\} - H(-1,1)\tan\theta J\{N_2\} \tag{16}$$

Substitute Eq. (21) in Eq. (19),

$$\begin{aligned} S_k^{h+1} &= \left(S_k^{h+1} - 2D\text{Rand}(0,1)\{N_1\} - H(-1,1)\tan\theta J\{N_2\}\right) \\ &(1 - \alpha\text{Rand} - \beta\text{Rand}) + \alpha\text{Rand}S_f^h + \beta\text{Rand}S_e^h \end{aligned} \tag{17}$$

$$\begin{aligned} S_k^{h+1} = {} & S_k^{h+1}(1 - \alpha\text{Rand} - \beta\text{Rand}) - (2D\text{Rand}(0,1)\{N_1\} \\ & + H(-1,1)\tan\theta J\{N_2\})(1 - \alpha\text{Rand} - \beta\text{Rand}) + \alpha\text{Rand}S_f^h + \beta\text{Rand}S_e^h \end{aligned} \tag{18}$$

$$\begin{aligned} S_k^{h+1} - S_k^{h+1}(1 - \alpha\text{Rand} - \beta\text{Rand}) = {} & -(2D\text{Rand}(0,1)\{N_1\} \\ & + H(-1,1)\tan\theta J\{N_2\}) \\ & (1 - \alpha\text{Rand} - \beta\text{Rand}) \\ & + \alpha\text{Rand}S_f^h + \beta\text{Rand}S_e^h \end{aligned} \tag{19}$$

$$\begin{aligned} S_k^{h+1}(1 - 1 + \alpha\text{Rand} + \beta\text{Rand}) = {} & -(2D\text{Rand}(0,1)\{N_1\} + H(-1,1)\tan\theta J\{N_2\}) \\ & (1 - \alpha\text{Rand} - \beta\text{Rand}) + \alpha\text{Rand}S_f^h + \beta\text{Rand}S_e^h \end{aligned} \tag{20}$$

$$\begin{aligned} S_k^{h+1}(\alpha\text{Rand} + \beta\text{Rand}) = {} & -(2D\text{Rand}(0,1)\{N_1\} + H(-1,1)\tan\theta J\{N_2\}) \\ & (1 - \alpha\text{Rand} - \beta\text{Rand}) + \alpha\text{Rand}S_f^h + \beta\text{Rand}S_e^h \end{aligned} \tag{21}$$

$$S_k^{h+1} = \frac{1}{(\alpha+\beta)\text{Rand}} \left[\begin{array}{l} (\alpha S_f^h + \beta S_e^h)\text{Rand} - \begin{pmatrix} 2D\text{Rand}(0,1)\{N_1\} \\ + H(-1,1)\tan\theta J\{N_2\} \end{pmatrix} \\ (1 - \alpha\text{Rand} - \beta\text{Rand}) \end{array} \right] \tag{22}$$

Step 4: Re-evaluate fitness for update solutions. The fitness of update solutions is re-evaluated wherein best solution is acquired for optimal application migration.

Step 5: Terminate. The optimum solutions are derived in an iterative manner until the maximum iterations is reached. Table 1 defines the pseudocode of developed Lion-SS.

Table 1 Pseudocode of developed Lion-SS

Input: Solution set S
Output: Best solution S^*
Begin
Initialize algorithmic agents
Evaluate fitness using Eq. (9)
Compute step size and temple solution vector
Update solution vector and merge herds
Update other algorithmic parameters
If $h < h_{max}$
Update solution using Eq. (22)
End if
Return S^*
End

5 Results and Discussion

This section gives a broad analysis of the evaluation process of proposed system through a fine tool Aquasim. It is extensive of NS-2 tool specially designed for underwater network communications. The performance metrics like delay, PDR, throughput and energy of MLS-MAC, EVFSQ and proposed Lion-SS are compared based on the simulation output. The priority-based queuing for scheduling of packets is performed by proposed Lion-SS optimization. By the proposed Lion-SS system, the delay time is reduced, loss of packets has been reduced, packet delivered is enached, throughput is increased and energy consumed is less.

5.1 Experimental Results

The implementation of the developed technique is performed in Aquasim extensive of NS-2 with 50 nodes. The developed technique is implemented on PC using Ubuntu OS, Intel i3 core processor and 2 GB RAM. The simulation set-up of the developed model is displayed in Fig. 4. In this model, the data is sent wherein the nodes are sent in a certain range of transmission. Here, the network contains 50 nodes which transmits the date in certain range. The red colour node represents the source nodes and green reprents the destination nodes and other nodes are used for transmission of data.

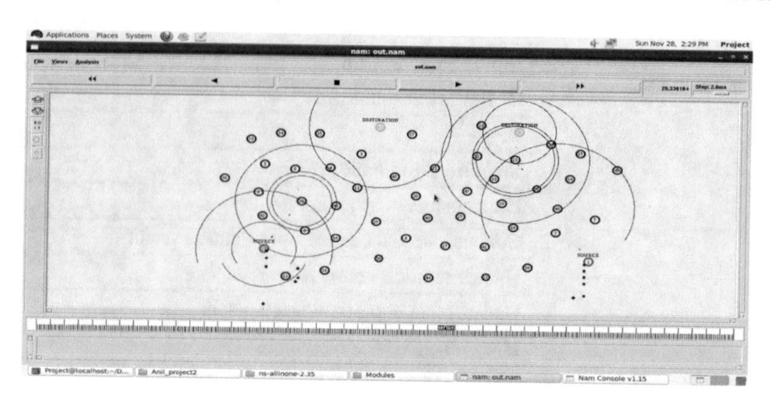

Fig. 4 Simulation results of developed Lion-SS model

5.2 *Performance Analysis*

This section gives a brief analysis of the evaluation techqunices of proposed Lion-SS system compared with MLS-MAC and ECSFQ. The performance metrics like delay, PDR, throughput, and energy for the simulation time of 50 s with 50 nodes.

Figure 5 represents the assessment using delay by varying the simulation time for 50 s. The graph is compared between MLS-MAC, ECSFQ and Proposed Lion-SS. The delay is calculated for time interval of 50 s. In MLS-MAC the time delay for 50 s is 0.13 s, for ECSFQ the time delay for 50 s is 0.08, whereas for proposed Lion-SS the time delay for 50 s is 0.02 s. By comparing these three methods, it is noted that the proposed Lion-SS delay is less. This is due to scheduling of packets using priority-based queuing techquies which helps to reduce the delay time.

Figure 6 represents the assessment of techniques based on PDR for the simulation time. The graph is compared between MLS-MAC, ECSFQ and proposed Lion-SS. The packet delivery ratio is calculated for the time interval of 50 s. The packet delivery ratio in MLS-MAC is 39% for 50 s, in ECSFQ is 49% for 50 s, and whereas in proposed Lion-SS it is 51% for 50 s. The packet delivery ratio in proposed Lion-SS

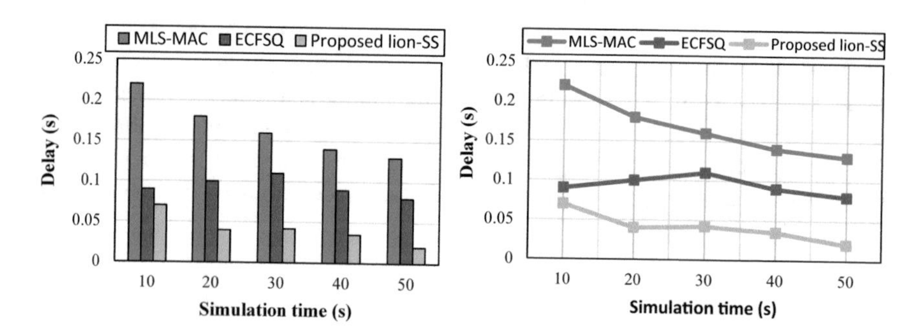

Fig. 5 Analysis based on delay

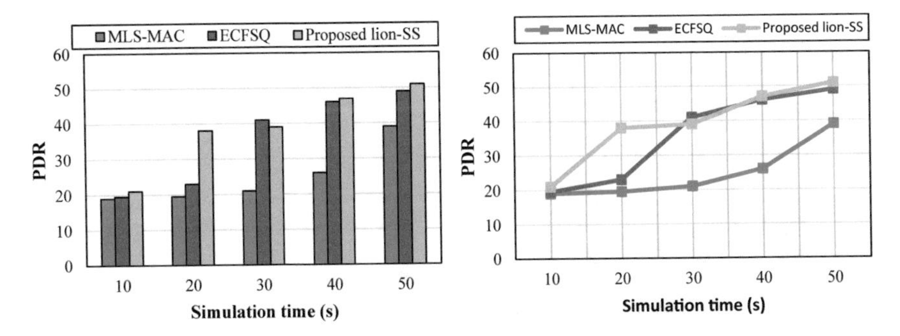

Fig. 6 Analysis based on PDR

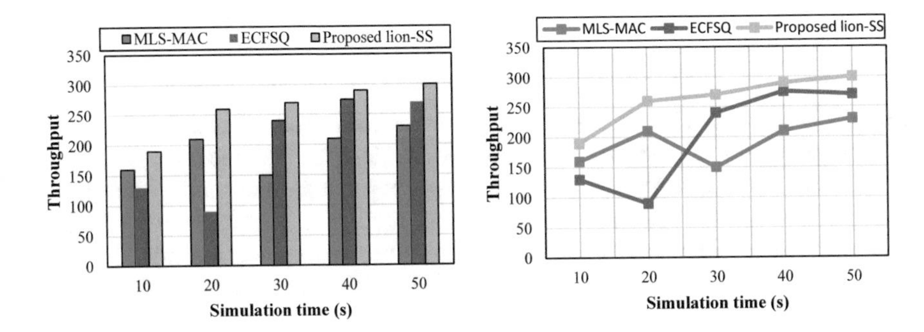

Fig. 7 Analysis based on throughput

is high as compare to MLS-MAC and ECSFQ. This is due to the incorporation of LOA and SSOA which helps to reduce the loss of packets and enchance the delivery of packects.

Figure 7 represents the assessment of techniques based on throughput for the simulation time. Here, the graph is compared between MLS-MAC, ECSFQ and proposed Lion-SS. The throughput is calculated for the time interval of 50 s. The throughput in MLS-MAC is 230% for 50 s, in ECSFQ is 270% for 50 s and whereas in proposed Lion-SS it is 300% for 50 s. The throughput in proposed Lion-SS is high as compare to MLS-MAC and ECSFQ. This due to the incorporation of LOA and SSOA which helps in increase of throughput.

Figure 8 represents the assessment of techniques based on energy for the simulation time. Here, the graph is compared between MLS-MAC, ECSFQ and proposed Lion-SS. The energy is calculated for the time interval of 50 s. The energy comsumed by MLS-MAC is 11.6 J for 50 s, in ECSFQ is 11.28 J for 50 s and whereas in proposed Lion-SS the energy is 9.95 J for 50 s. When compared to MLS-MAC and ECSFQ, the energycomsumed proposed Lion-SS is less. This due to the priority-based queuing technique which helps in reduction of energy in the node.

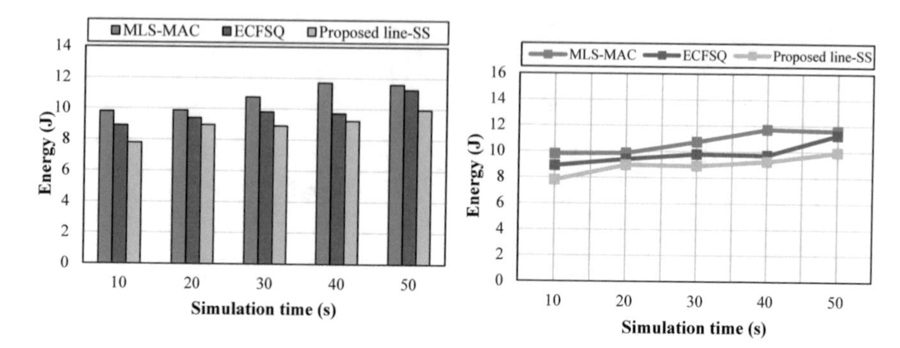

Fig. 8 Analysis based on energy

6 Conclusion

In this paper, a novel hybrid optimization-driven technique is devised for priority-based scheduling in underwater sensor networks. Here, the first step is the simulation of UWSN. Thereafter, the optimal paths are discovered for transmitting the signals attained from the sensor nodes. The optimal routing is performed using FGSA. Once the optimal routes are obtained, the obtained signals are transmitted to BS, wherein the priority scheduling is performed. The priority scheduling is done using newly devised Lion-SS and fitness factor. The proposed Lion-SS is newly devised considering Lion algorithm and SSOA. In addition, fitness is obtained by energy, delay and distance. Thus, the fitness factor and proposed Lion-SS are utilized for priority-based scheduling in UWSN.

References

1. Ram RSD, Vijayalakshmi P, Padmaja V, Jaya T, Rajendran V (2019) Research of channel capacity of acoustic communication network for oceanographic applications. Int J Innov Technol Explor Eng 9:3016–3019
2. Sampath SL, Subashini S (2018) A multilevel scheduling MAC protocol for underwater acoustic sensor networks. Int J Commun Netw Inf Secur 10:437–445
3. Ram RSD, Vijayalakshmi P, Padmaja V, Jaya T, Rajendran V (2019) Channel modelling of underwater communication network with AUVS. Int J Innov Technol Explor Eng 9:3012–3015
4. Zhuo X, Qu F, Yang H, Wei Y, Wu Y, Li J (2019) Delay and queue aware adaptive scheduling-based MAC protocol for underwater acoustic sensor networks. IEEE Access 7:56263–56275
5. Jaya T, Jacintha TA, Paranidharan KS, Sathyapriya, Rajendran V (2016) An optimum algorithm to improve BER using CDMA technique in underwater communication for BPSK modulation. Int J Res Adv Eng Technol 2:94–97
6. Ng HH, Soh WS, Motani M (2008) MACA-U: a media access protocol for underwater acoustic networks. In: IEEE GLOBECOM 2008–2008 IEEE global communication conference 1–5, IEEE
7. Padmavathy C, Jayashree LS (2017) An enhanced delay sensitive data packet scheduling algorithm to maximizing the network lifetime. Wirel Pers Commun 94:213–2227

8. Sivasubramaniam N, Senniappan P (2014) Enhanced core stateless fair queuing with multiple queue priority scheduler. Int Arab J Inf Technol 11:159–167
9. Kumar L, Sharma V, Singh A (2017) Bidirectional multi-optical line terminals incorporated converged WSN-PON network using M/M/1 queuing. Opt Fiber Technol 39:78–86
10. Sundarasekar R, Shakeel PM, Baskar S, Kadry S, Mastorakis G, Constandinos XM, Samuel RDJ, Vivekananda G (2019) Adaptive energy aware quality of service for reliable data transfer in under water acoustic sensor networks. IEEE access 7:80093–80103
11. Faheem M, Tuna G, Gungor VC (2019) x.: LRP: link quality-aware queue-based spectral clustering routing protocol for underwater acoustic sensor networks. Int J Commun Syst 30(12):e3257
12. Khalil, Amin R, Saeed N (2020) Optimal relay placement in magnetic induction-based internet of underwater things. IEEE Sens J 21:821–828
13. Ali MF, Jayakody DNK, Chursin YA, Affes S, Dmitry S (2020) Recent advances and future directions on underwater wireless communications. Arch Comput Methods Eng 27(5):1379–1412
14. Dhumane AV, Prasad RS (2019) Multi-objective fractional gravitational search algorithm for energy efficient routing in IoT. Wirel Netw 25:399–413
15. Kaveh A, Zaerreza A (2020) Shuffled shepherd optimization method: a new meta-heuristic algorithm. Eng Comput
16. Yazdani M, Jolai F (2016) Lion optimization algorithm (LOA): a nature-inspired metaheuristic algorithm. J Comput Des Eng 3:24–36

Performance Analysis of Leach, Leach Centralised and Leach Modified of Wireless Sensor Network

Zoya Akhtar, Himani Garg, and Sanjay K. Singh

Abstract Research in wireless sensor networks is booming exponentially, and this is due to the many factors that make WSNs a field of interest, one such factor being that it can be deployed in the remote areas for monitoring purposes. We require such technologies even more now because of the impact of climate changes. A recent report by the United Nations named 'Climate Change 2021' has issued code red for humanity as climate change is going to effect the human life to a great extend in the upcoming years. By deploying a wireless sensor network, we can monitor our surroundings as well as remote areas. There are many routing protocols used in wireless sensor network, and the most basic one is the Leach routing protocol. In this paper, we will study the performance analysis of leach, leach-C and Mod-Leach based on various parameters such as throughput, energy dissipation and network lifetime with respect to number of rounds in order to analyse which routing algorithm is more scalable when the number of rounds increases.

Keywords Wireless sensor network · Sensor nodes · Routing protocols · Performance analysis

1 Introduction

A wireless sensor network is nothing but a network that delivers the sense data to the base station wirelessly.

Now, as easy as it may sound but there are various factors that need to be taken into consideration before deploying a wireless sensor network. The main part of a wireless network is the selection of the routing algorithm which is going to be used for the functioning of the network. On the other hand, the sensor nodes used in the wireless network come with some limitations of its own although the working of a sensor is simply to sense the date of its surroundings but as the times passes, its battery

Z. Akhtar (✉) · H. Garg · S. K. Singh
Department of Electronics and communication technology of ABES Engineering College, Ghaziabad, Uttar Pradesh 201016, India
e-mail: zoyaakhtar17.10@gmail.com

A. Kumar et al. (eds.), *Proceedings of the International Conference on Cognitive and Intelligent Computing*, Cognitive Science and Technology,
https://doi.org/10.1007/978-981-19-2350-0_48

will eventually drive out. Routing protocols will help in consuming the energy of the battery efficiently so that lifetime of the wireless sensor can be prolonged. The routing protocols can be judged based on its energy dissipation rate, its alive and dead sensor nodes and its delivering rate of the packets to base station with increasing rounds in the wireless sensor network. There are various advantages of the wireless sensor networks if proper selection of the components is taken. One of the main advantages of WSN is with IoT, i.e. Internet of Things. The IoT is a powerful technology that authorises the communication and manages the association with other electronic devices over the network. With the IoT, communication between different electronic devices has significantly become much easy. However, using the IoT with the WSN should be done careful keeping both of them at a same level. The wireless sensor network can be used in the patient monitoring for any health purposes, it can be done by patient being at home, and its data can be sent wirelessly to his/her doctor directly so that if there is any critical change in his/her health, it can be immediately addressed without further delay and the treatment can be provided as soon as possible. Also, these wireless sensor networks can be easily deployed and any modification can easily be implemented in the network. We can even relocate the deployed sensor network if it is required at any time, and it will not affect the wireless sensor network functionality. It can also provide the monitoring of the environmental factors such as fire detection, landslides, green house effects and other environmental conditions that need to be monitoring. Its sensor nodes can be deployed in the remote areas randomly, manually or by an aircraft, and its data can be monitored wirelessly and regularly. In our times, this is one of the most important areas which require monitoring as the climate change has began to show us its impact with severe fires in the forests increasing landslides, floods and what not. Its effects are so severe now that is in some places the people has migrated permanently from their homes as their home is completely destroyed [1–3, 13, 14, 15]

2 Methods and Materials

Let us study the main routing protocols involved in this paper and briefly elaborate them (Fig. 1).

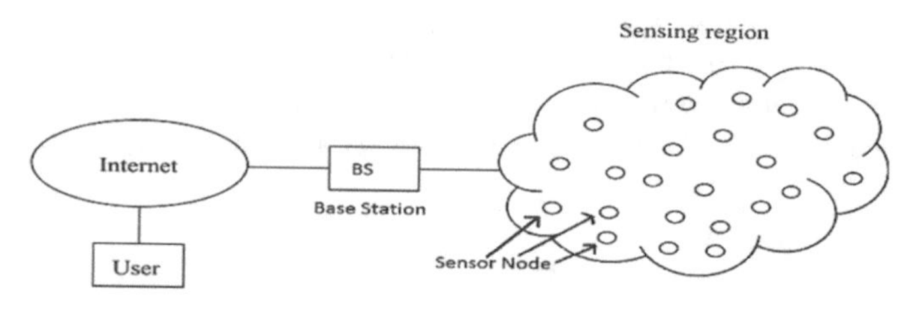

Fig. 1 A basic WSN

1. Leach routing protocol.
2. Leach centralised routing protocol.
3. Leach modified routing protocol.

2.1 LEACH Routing Protocol [4, 12]

Leach algorithm is the very basic energy efficient, and it stands for low energy adaptive clustering hierarchy. This is known to be the foremost routing protocol and the very first energy efficient routing protocol based on clustering. It is a hierarchical clustering protocol and is a distributed algorithm. In leach protocol, after the formation of the groups called the clusters from each cluster one sensor node will be selected to be the cluster head for the cluster based on the probability, the remaining sensor node present in cluster will forward their sensed information to its cluster head, and then this cluster head will be in contact with the sink node or base station. In this protocol, the overall energy of the wireless sensor network is assigned to every sensor node in the network periodically and this results in reduction of the consumption of energy and leading expansion in the lifespan of the wireless sensor network. The cluster head in the wireless sensor network is periodically rotated amid the sensor nodes for the balancing of the load. The periodic rotation of these sensor nodes inside the wireless sensor network is done in a way that a random number is chosen by the sensor nodes which lies between 0 and 1. For the selection criteria, if the random number is less than the threshold value, i.e. $t(n)$, then this sensor node will be elected as the cluster head for this round and the rest of the sensor nodes will join these cluster heads based on which is nearest to them. After the current round ends, this selection of the cluster heads will again take place.

The threshold value can be given as:

$$t(n) = P_o / (1 - P_o) * (R \mod 1 / P_o) \text{ if } [n\, e\, G],$$
$$0, \qquad \text{otherwise.}$$

where

P_o desired % of cluster head in the sensor nodes,
R current round no.
G set of sensor nodes that are not elected as cluster heads in the last $1/P_o$.

The working of Leach routing protocol is divided into two rounds, and every round consists of two phases.

2.1.1 Setup Phase

In this, sensor nodes are grouped together into clusters and election of cluster head takes place.

2.1.2 Steady Phase

The data is sensed by the sensor nodes, and the delivery to base station takes place (Fig. 2).

2.2 LEACH Centralised Routing Protocol [5, 7, 10, 11]

The basic leach routing protocol can be studied as the aggregated architecture design which is implemented on a distributed network system, in which the sensor nodes elect the cluster heads solely and the network is formed with different clusters, and in each cluster, there is a cluster head. Now a group of scholars including the Heinzelman has put forward this aggregated architecture design into a centralised network system instead of the distributed network system; hence, the name is leach centralised or Leach-C. It is basically a leach version with a centralised base station approach. In the leach-C, all the rounds for the election of the cluster heads that will take place the sink node should have the information of remaining energies of every

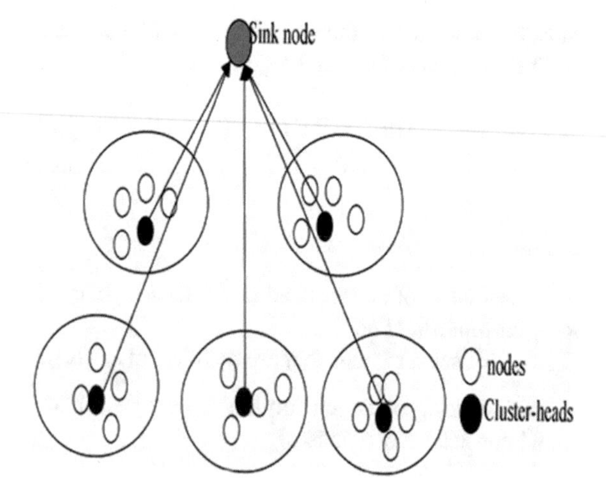

Fig. 2 Leach network structure

single sensor node with the location of individual sensor nodes. And on the basis of which the sink node is in the authority to select based on a particular method as to which of the sensor node will be elected as the new cluster head and also which sensor nodes will be grouped together for different cluster formation. In this way, some of the drawbacks of the leach protocol can be reduced by using this leach centralised protocol. Although it cannot be said that leach centralised is superior to the leach routing protocol. The major difference between the leach and leach-C is that the leach centralised is the sink node stationed at centre approach and control the setup phase. Every single sensor node will transmit its info to the sink node, and in return, the base station is owing to decide which one will be elected as cluster head and which sensor nodes are going to be grouped together. Then, the elected cluster heads (CH) will transfer this info to every sensor node. All of this process will require additional energy which will influence the performance of the routing algorithm.

Leach centralised also consists of two phases.

2.2.1 Setup Phase

It uses the centralised approach for the election of cluster heads and formation of clusters.

2.2.2 Steady State

It is same as leach routing protocol (Fig. 3).

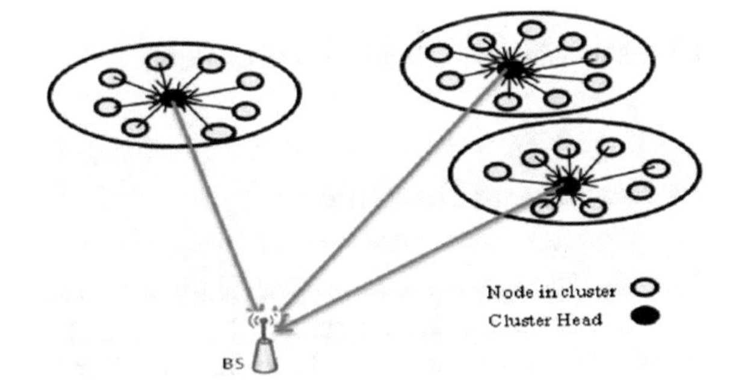

Fig. 3 Leach-C network structure

2.3 *Leach Modified (Mod-Leach) Routing Protocol [6, 8, 9]*

It is another variant of the leach routing protocol, and it only differs in the criteria of selection of the cluster heads (CH) in routing protocol in the wireless sensor network. In this, with the round this routing protocol will swap the cluster head (CH) and it will not allow it to be elected for the upcoming $1/p$ rounds. This modified protocol of leach initiates a new cluster head election scheme. In this scheme, it allows the elected cluster heads (CH) to be working for the upcoming rounds if it still has sufficient energy after its previous round and its residual energy is higher when compared with the threshold value $t(n)$. This approach will definitely result in saving the energy as it is eliminating the need to change the cluster heads (CH) after every round in the routing protocol but this should be kept in mind if the cluster head energy is less than threshold value, then it will update the cluster head with new sensor node having energy higher than the threshold value $t(n)$.

Modified leach protocol consists of two phases.

2.3.1 Setup Phase

The new introduced scheme for cluster heads, i.e. the selection of the cluster heads, should be done in a optimal way so that the need of changing the cluster heads after every round can be reduced. This can be done by selecting the sensor nodes as cluster heads (CH) who are closer to the sink node, and the sensor nodes should join the clusters based on their residual energy and position.

2.3.2 Steady State

It remains same as the leach routing protocol.

3 Results and Discussion

From the Figs. 4, 5 and 6, we can conclude that the remaining energy of the wireless sensor network in these routing protocols, i.e. Leach, Leach-C, Mod-Leach, is drained completely at round 600. Mod-Leach performs better in comparison to Leach and Leach-C algorithms as its total energy lasts up to 650 rounds. Another important thing to note here in Fig. 4 is that the Leach-C performs better if the rounds are limited to 100 in comparison to Leach and Leach-C and still has higher energy left at the end of 100 rounds (Table 1).

In Fig. 7, we can conclude that when the rounds are limited to 100 then Leach and Mod-Leach perform almost similar in terms of delivery of the packets while Leach-C is delivering more packets in the initial rounds but afterwards its delivery of the packets is less in relation to Leach and Mod-Leach. In Fig. 8, we see that up to 300 rounds leach and Mod-Leach deliver almost same amount of packets to the base station, and as the rounds increase, Mod-Leach delivers the highest amount of packets to the sink node while the performance of Leach-C is also better. In Fig. 9, we see that after 1000 rounds the routing protocol delivery becomes constant with Mod-Leach delivering highest number of packets.

Leach-C performs better in Fig. 10 with less number of dead nodes in the wireless sensor network. Now, in Fig. 11 Mod-Leach has the maximum number of sensor nodes alive in comparison to Leach and Leach-C. For a large wireless sensor network, Mod-Leach is better in terms of alive sensor nodes over the time (Fig. 12).

4 Conclusion

In this paper, we studied three routing protocols or two variants of Leach, i.e. Leach-C and Mod-Leach in comparison to the traditional Leach algorithm. We have plotted the graphs comparing the packet delivery to sink node with no of rounds, no of alive nodes with no of rounds and remaining energy as the rounds increase in the wireless sensor network. At last, we conclude that the Leach, Leach-C and Mod-Leach can be useful in small wireless sensor network as its energy drains completely after the rounds increase by 600. Out of all of these protocols, Mod-Leach performs better in

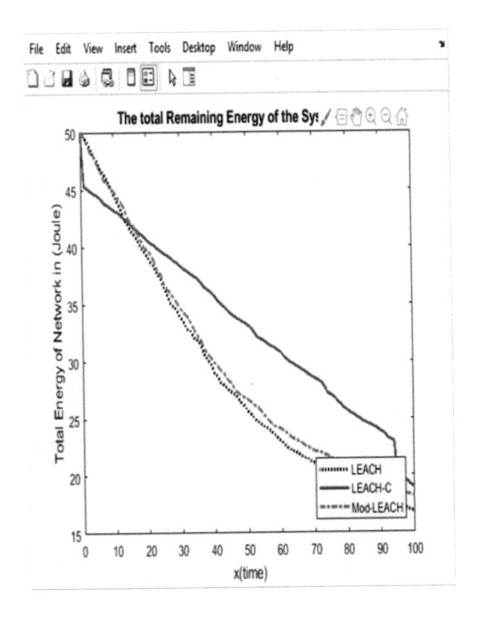

Fig. 4 100 rounds versus total energy

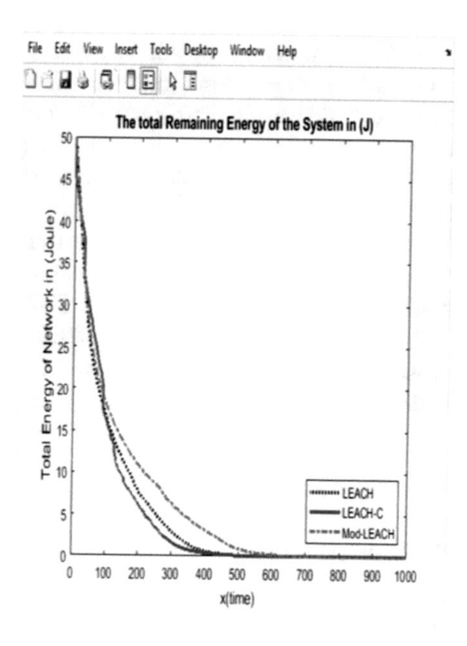

Fig. 5 1000 rounds versus total energy

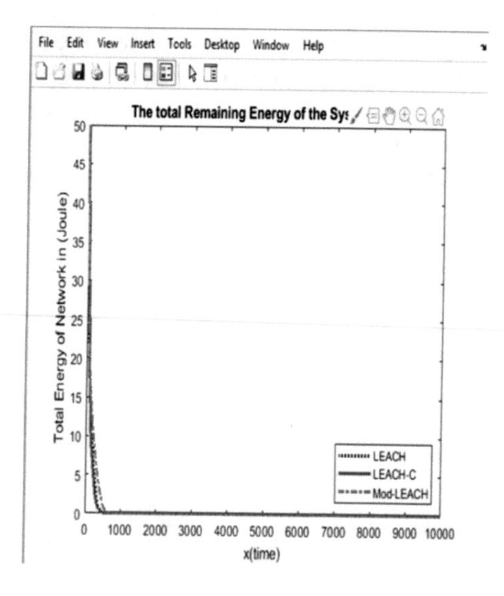

Fig. 6 10,000 rounds versus total energy

all aspects. As for the future, we should look for the new methods in the selection of the cluster heads in the wireless sensor network.

Table 1 Network parameters of the proposed wireless sensor network

Parameters of network	Values
Size of network	100×100 m^2
Number of sensor nodes	100
Routing algorithm used	Leach, leach centralised and leach modified
Primary energy Eo	1 J/node
Energy transmitted, i.e. ETX	50 nJ/bit
Energy received, i.e. ERX	50 nJ/bit
Amplified energy for short distance, i.e. EFS	10 pJ/bit/m^2
Amplified energy for long distance, i.e. EMP	0.0013 pJ/bit/m^2
Aggregated data energy, i.e. EDA	5 nJ/bit
Probability of the cluster head selection, i.e. *P*	0.05
No. of rounds (r)	100, 1000, 10,000

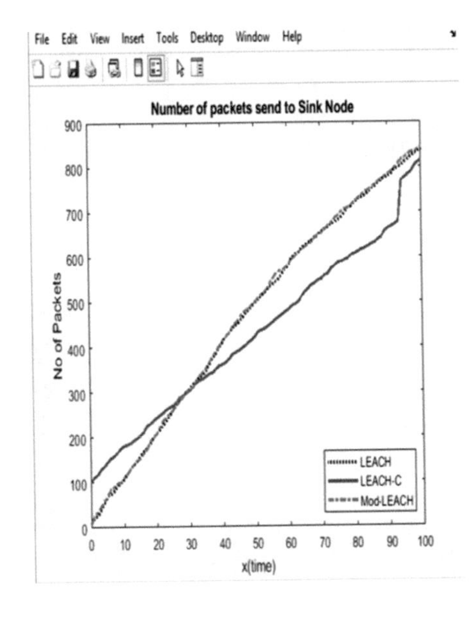

Fig. 7 100 rounds versus packets delivered

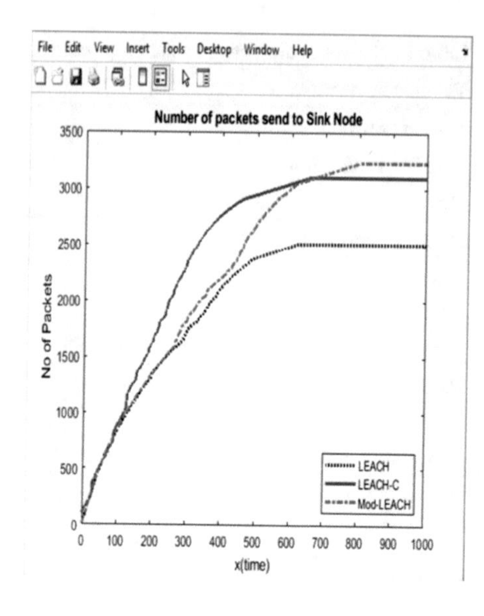

Fig. 8 1000 round versus packets delivered

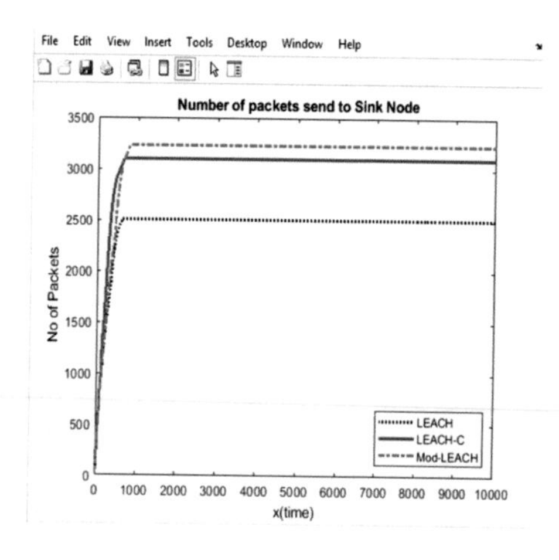

Fig. 9 10,000 rounds versus no of packets delivered

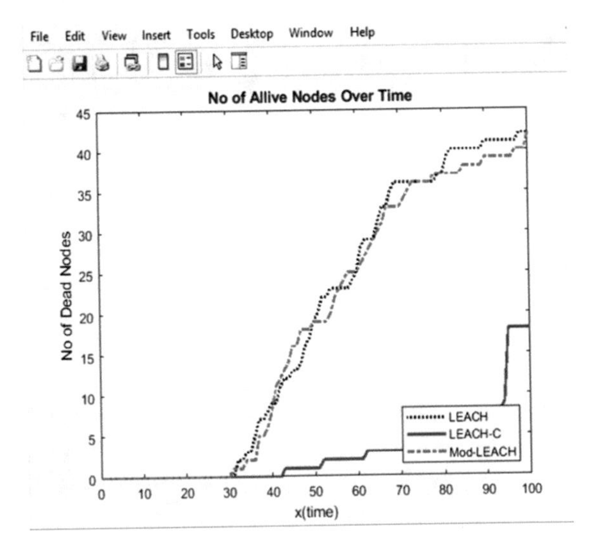

Fig. 10 100 rounds versus no of dead nodes

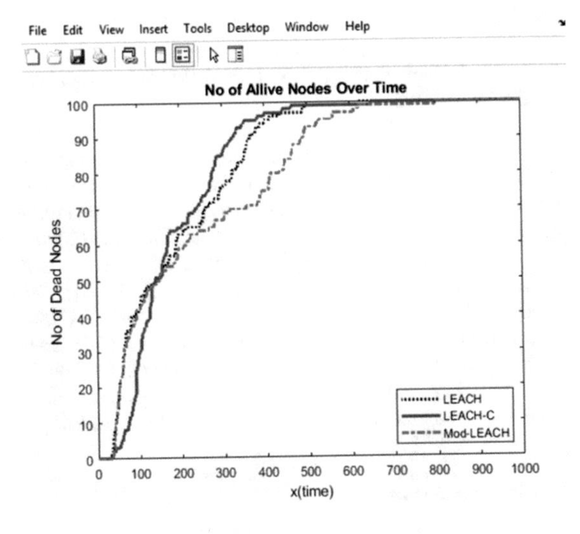

Fig. 11 1000 rounds vs no of dead nodes

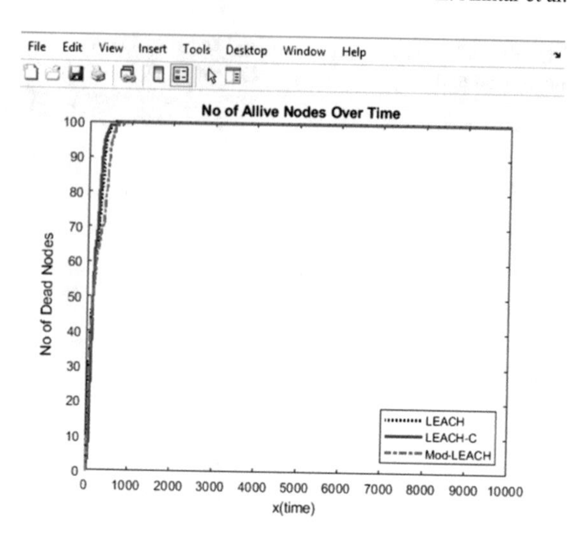

Fig. 12 10,000rounds versus no of dead nodes

References

1. Limin S, Jianzhong L, Yu C (2005) Wireless sensor networks. Tsinghua University press, Beijing, pp 3–24
2. QiangFeng J, Manivannan D (2004) Routing protocols for sensor networks. In: Presented at consumer communications and networking conference, CCNC 2004. First IEEE
3. Al-Karaki JN, Kamal AE (2004) Routing techniques in wireless sensor networks: a survey. IEEE Wirel Commun 11(6):6–28. https://doi.org/10.1109/MWC.2004.1368893
4. Heinzelman W, Chandrakasan A, Balakrishnan H (2000) LEACH: energy efficient communication protocol for wireless microsensor networks. In: Proceedings of Hawai international conference on system science. Maui, Hawaii, pp 3005–3014
5. Heinzelman W, Chandrakasan A, Balakrish-nan H (2002) An application-specific protocol architecture for wireless microsensor networks. IEEE Trans Wirel Commun 1:660–670
6. Mahmood D, Javaid N, Mahmood S, Qureshi S, Memon AM, Zaman T (2013) MODLEACH: a variant of LEACH for WSNs. In: 2013 eighth international conference on broadband and wireless computing, communication and applications, pp 158–163. https://doi.org/10.1109/BWCCA.2013.34
7. Romero MA et al (2006) Wireless sensor network. Mondragon Goi Eskola Politeknikoa, Mondragon Unibertsitatea
8. Ching HJM (2006) Wireless sensor network: localization and routing
9. Nissar N (2011) Secure routing in WSN. Al Akhawayn University Press
10. Li Y et al (2008) Wireless sensor networks and applications. Springer
11. Nayak A, Stojmenović I (2010) Wireless sensor and actuator networks: algorithms and protocols for scalable coordination and data communication. Wiley
12. Mahgoub I, Ilyas M (2006) Sensor network protocols. CRC Taylor & Francis
13. Sarkar A, Senthil Murugan T Routing protocols for wireless sensor networks: what the literature says? Alexandria Eng J Elsevier, 31 Aug. 2016. https://www.sciencedirect.com/science/article/pii/S1110016816302125
14. Ketshabetswe LK et al Communication protocols for wireless sensor networks: a survey and comparison. Heliyon, Elsevier, 21 May 2019. https://www.ncbi.nlm.nih.gov/pmc/articles/PMC6531673/
15. Wireless Sensor Networks. https://www.mdpi.com/topics/Wireless_Sensor_Networks

A Dialogflow-Based Chatbot for Karnataka Tourism

N. M. Madhu Manjunath and S. Ravindra

Abstract Imagine an invisible robot living within the Internet asking you questions. A Chatbot is a computer software developed to simulate communication with human users over the Internet. A Chatbot is a conversational agent that engages users in natural language interactions. The most basic communication model is a database of questions and replies, as well as the background history of the dialogs and the name of the related communication issue. With so many applications available now, Chatbots are becoming increasingly important for research and practice. The fundamental methodologies and technologies behind a tourist Chatbot allow individuals to textually converse with the goal of booking hotels, arranging excursions, and inquiring for interesting places to visit. So, development of Chatbot on tourism using Dialogflow consists of Karnataka Tourism Website, their user can get the information about the tourism places of Karnataka, they can get the sufficient services like transportation, accommodation, and so on, sign-in page helps the user to book a hotels, resorts, and they contact the organizer in any time to sort out the queries. The website Chatbot is there to help the user, about the places, foods, mode of transportation, etc.

Keywords Chatbot · Tourism · AI · Dialogflow · NLU

1 Introduction

With so many uses, Chatbots are becoming increasingly important in today's research. They are adaptable, and applications were used in many industries. Chatbots may suggest hotels, activities, sights, and arrangements of traveling completely in the industry of tourism. In order to download the map of navigation so want to make arrange of cities, vacation, users are needed to check many websites and

N. M. Madhu Manjunath (✉) · S. Ravindra
CSE, Jawaharlal Nehru New College of Engineering, Shivamogga, Karnataka, India
e-mail: madhumanjunath10@gmail.com

S. Ravindra
e-mail: ravindra@jnnce.ac.in

A. Kumar et al. (eds.), *Proceedings of the International Conference on Cognitive and Intelligent Computing*, Cognitive Science and Technology,
https://doi.org/10.1007/978-981-19-2350-0_49

Fig. 1 Image of travel (*Source* Google)

various applications on phones, Internet, and it is more complicated. This study presents "Clever Guidance," a Chatbot that is designed for mobile apps, to address this issue. It imitates a natural language conversation with users. It also allows for two-way communication and serves as an individual point of view for all customer responds. The Chatbot's final evaluation revealed that it can understand words and human language. Furthermore, customer wants an interactive action that is efficient, and that responses are sent quickly. The customer will be pleased and would like the suggestion of being able to convey their demands in their own language.

2 Ease of Use

2.1 Tourism and Travel Sector

Travel has grown in popularity as a popular economic and social activity around the world, owing to its capacity to generate jobs and generate new company ideas. Attractive travel sector of destination is maintained by technological things in a smart country for the supply and demand of value, pleasure, and the best experiences of the wealth and tourism organization. These travel departments assist in the development of new tourist attractions and their execution. The known phases were associated with smartphones in order to change the travel experience with an open field for special services in the travel sector. As can be seen in Fig.1, it shows the image of travel so that many people are increasingly interested in traveling, and our tourism department will receive cash for the development of the sites [1].

2.2 Chatbot

A Chatbot is a computer program that mimics human dialog using voice and text. Chatterbot is a short name for an AI that may be embedded and used in a variety of applications. It is capable of having discussions with users, such as clients, in order to keep them interested in a specific platform through conversation. A well-designed Chatbot will be able to:

a. Analyze conversation data to learn about the types of queries that different types of people might ask.
b. During the "training" time, try to examine the responses to those questions.
c. Learn the context using machine learning (ML) and natural language processing (NLP) to get better answers to inquiries in the future.

Chatbots can operate as a mediator to communicate with users by sending a conversation, allowing for mutual communication on websites and apps, as well as assisting business people in establishing new relationships with clients. These are bots, which are a type of automated client communication system. Bots that are simple and easy to use are employed in messaging programs like Skype and Messenger, as well as in food ordering apps such as Zomato, Swiggy, Dominos. The major goal is to automate things like flight confirmation and shopping for items. Smart AI technology allows the bot to react to user's questions in a roomier manner. Chatbots served a variety of tasks, but the most common was customer service.

2.3 *Artificial Intelligence*

Artificial intelligence (AI) refers to the simulation of human intelligence by robots that have been computer programmed to think and act like humans [2]. The word is used to describe machines that display human characteristics such as problem solving and learning. The labor deemed to constitute "intelligence" is more often removed from the AI definition as the computer becomes more capable, which is known as the effect of AI. "AI is whatever hasn't been done yet," according to Tesler's theorem [3]. The incidence, which is defined as a characteristic of optical frequency that recognizes the banned items as AI, has become an everyday activity in technological domains. Nowadays, modern capabilities are widely recognized for achieving high levels of comprehension in mind-tuning games such as Go and Chess, automatic car operating systems, intelligent routing delivery networks, and war simulations [4].

AI was found as an academic discipline in 1955. In those years, users experienced optimistic waves that were following the disappointment of fund loss, which is actually called "Winter of AI." It is followed with approaches of different levels, renewed funding, success. Next, AlphaGo was a successful diffusion player of a professional Go in the year of 2015. AI attracts global world attractions. Its research is divided into subfields where often communication fails with others. Machine learning, robotics, artificial neural networks (ANN), and logic are examples of subfields where a technique must be targeted. Subfields are social factors such as specific research work and specific institutions [5].

2.4 *Related Works*

2.4.1 Chatbot in Python

Kumar et al. in [6]—They created a Chatbot using Python programming language. It must answer queries for textual inputs. They used flask framework and also workflow to be implemented. Here, Python is the main platform; there they use AIML and LSA for text separation. Also, they used APIs like government services, news, weather, sports. Chatbot must answer based on the reality [6].

2.4.2 A Tool of Conversation: Chatbot

Dahiya in [7]—The Chatbot is designed to counterfeit communication for today's smart ground of spoken and text. Here, Chatbot is talking with the help of pattern matching. So that, however or whatever the sentence form the Chatbot has ability to understand. They designed a Chatbot keeping facts in the mind which are: OS selection, software selection, Chatbot creation, chat preparation, pattern matching, simple, conversational, and entertaining. Design of fundamental is dialog box creation, database creation. The description of modulus is Chatbot, random, adding of texts, in array. In the end, it knows that it is simple, easy knowledge, and some commercial products were recently improved and emerging in the same basis of Chatbot [7].

2.4.3 The First Chatbot of a Tourism/Hospitality Journal: Editor's Impressions

Ivanov in [8, 2], The hospitality of tourism was founded in the year 2017. Their architecture is using chat fuel, and it consists of AI rules, links, and blocks. The impact of 56 users of Chatbot is of communication. It consists of block buttons. Chatbots have limited editions of intended applications that must be dictated. Because of this, the bot remains unanswerable, so the user gets frustration also sometimes. They trained in the English language, and sometimes, it may be difficult when the user asks questions in other languages. The journal's online system is to simulate to direct the user. They update the Chatbot to interact properly [8].

3 System Architecture

The architecture of an AI Chatbot for tourism has been demonstrated here. When a user (tourist) first visits the website, a Chatbot will appear to assist the visitor. That Chatbot provides assistance to tourists so that they can learn more about their destinations. The tourist then responds to the Chatbot, requesting details. Both the

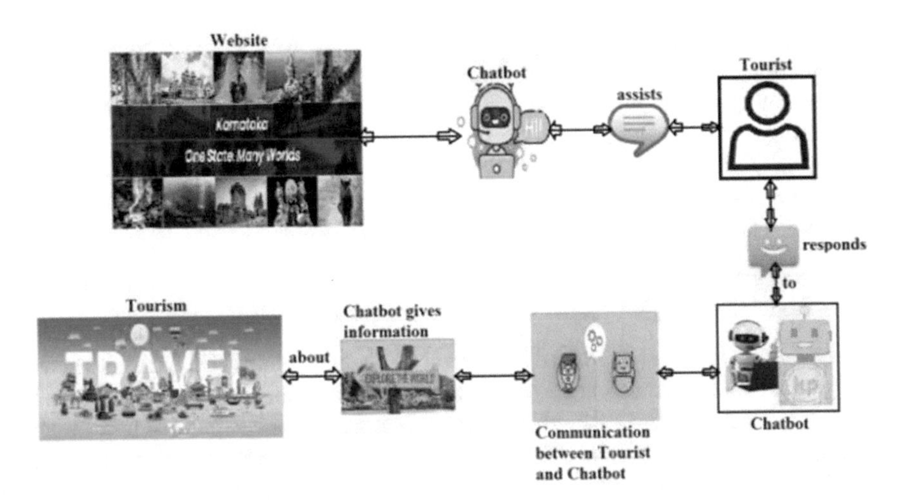

Fig. 2 System architecture

Chabot and the tourist engage in communication in this way. Following the talk, the Chatbot will learn about the tourist attraction and provide information. Finally, tourists learn about the location, food, and lodging options, among other things. In Fig. 2, system architecture is shown.

4 Design and Implementation

Front-end Design (Client-side development): Front-end development is a process to create type of user who sees when they open a web application, including the design, content, interaction. HTML, CSS, and JavaScript are used to do this. HTML stands for Hypertext Markup Language.

HTML: stands as Hypertext Markup Language is a markup language that is used to create content that will be displayed on a website. Website browsers receive HTML documents from a server and convert them to multimedia pages. The World Wide Web Consortium (W3C) is in charge of maintaining HTML and CSS standards. It is a collection of important terms called "Tags" that enable you to write a document that will be displayed on the Internet using a browser. It is a platform-agnostic language that may be used on any platform, including Windows, Linux, and Macintosh. To show a document on the web, it is necessary to use HTML tags to mark up the various sections of the content (headings, paragraphs, tables, and so on).

CSS: stands for Cascading Style Sheets and is used to describe the presentation of an HTML document. It is primarily used to style HTML sheets, which means HTML will write the program and content, but CSS will give the page style and design, making it look more appealing and wonderful. This is intended to distinguish content from presentation, which includes fonts, colors, and layout.

The ability to present a markup sheet in several forms of methods, which are print, voice, and screen, is made possible by the separation of format and content. The priority of stated to style rule over a rule must correspond with an element gave rise to this cascading term. The priority scheme's cascade effect is predicted. CSS is sometimes known as text/CSS.

Back-end Design (Server-side development): In back-end design, here, we are using the Dialogflow and natural language understanding (NLU).

Dialogflow: It is a platform of natural language understanding (NLU), and it makes design easy and understandable to the user interface in the mobile app, websites, bots, devices, responsive systems, and so on. Using this, we can provide new things of engaging with the users about the product. It can analyze in so many types like input from the user in the text or voice. It will make respond within the couple of seconds through text or synthetic speech.

Dialogflow is user-friendly and easy to use. Some of the versions of API in this are:

V2: It is stable and available generally for all.
V2Beta1: It is used to share the upcoming features of beta.

NLU: It is known as natural language understanding, and it is an AI that uses software to interpret any type of unstructured data and text. It can translate the text into computer language to human language to understand. It is also considered as artificial intelligence hard problem. Some of the applications of NLU are translation of machine, automatic reasoning, answering to the questions, activation of voice, analysis of large content, categorization of texts, archiving.

Implementation: Here, we are implementing the project through website for tourists. By using the flowchart, we have shown the implementation properly. A flowchart is the type of representation of an algorithm and the workflow. This shows the steps within the various kinds of boxes with the order of connected arrows. It is mainly used to analyze, design, documentation, and managing process in the variety of fields. In Fig. 3, we can see the flowchart. Firstly, the initial process is to start, there Karnataka Tourism website will open. There we can see the Chatbot, it will initiate the user, the user sends/requests the input query, and Chatbot responses the output to the user. It fetches from the natural language understanding (NLU)-based Dialogflow, and it has stored the data. From this, the Chatbot divided has trained and untrained Chatbots. Here, trained Chatbot is nothing but the data already given to the Chatbot, it gives more accuracy and proper reply, the untrained Chatbot not gives the sufficient data to answer the queries, and its accuracy is less. It is the structure of overall flowchart.

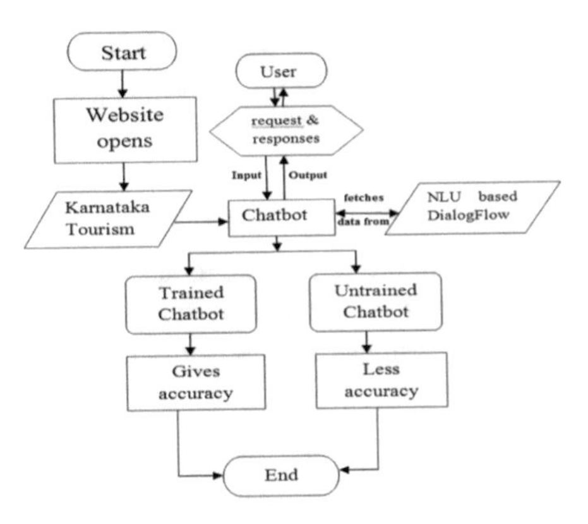

Fig. 3 Flowchart of Chatbot

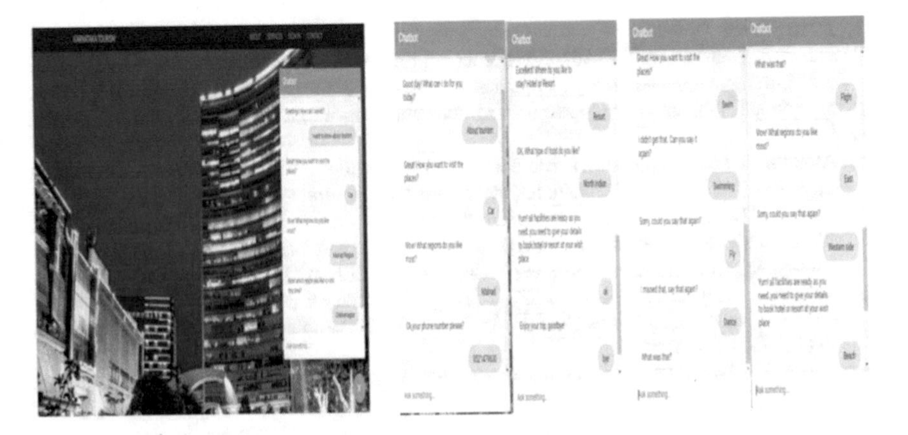

Fig. 4 Left: output, right: trained and untrained Chatbots

5 Results and Analysis

Figure 4 showing the result on the left side is the output and the right side is the analysis of trained and untrained Chatbots as shown.

6 Conclusion and Future Scope

This Chatbot is mostly for travelers to identify tourist destinations. The bot was created to assist anyone looking for the name of a location. This undoubtedly gives information on the locations. This saves time by opening the website directly, and the bot will provide information on places, including a brief description and the name of the location, which is useful for tourists. When the website is opened, it provides information, and the Chatbot communicates in a user-friendly manner. The Chatbot was created with crucial information in mind that is dependent on the needs of travelers.

The Chatbot implementation can be extended to

- Registration Services: The future requirements for user registration. As a result, they will be able to see all of the tourist attractions.
- Email Services: When a person registers, they will receive an email confirmation to their registered email address.

References

1. Main Source: Google
2. Ivanov S, Webster C (2017) Adoption of robots, artificial intelligence and service automation by travel, tourism and hospitality companies—a cost-benefit analysis. In: International scientific conference "contemporary tourism—traditions and innovations", 19–21 October 2017, Sofia University
3. Alotaibi R, Ali A, Alharthi H, Almehamadi K AI chatbot for tourism recommendations. A Case Study in the City of Jeddah. https://doi.org/10.3991/ijim.v14i19.17201
4. Kumar MN, LingaChandar PC, Venkatesh Prasad A, Sumangali K (2016) Android based educational chatbot for visually impaired people. In: International conference on computational intelligence and computing research (ICCIC), IEEE. IEEE, pp 1–4
5. Kasinathan V, Wahab MHA, Idrus SZS, Mustapha A, Yuen KZ (2020) AIRA chatbot for travel: case study of AirAsia. J Phys Conf Ser 1529 022101. IOP Publishing. https://doi.org/10.1088/1742-6596/1529/2/022101
6. Kumar A, Meena PK, Panda D, Sangeetha (2019) Chatbot in python. Int Res J Eng Technol (IRJET) 06(11). www.irjet.net
7. Dahiya M (2017) A tool of conversation: chatbot. Article in Int J Comput Sci Eng
8. Ivanov S The first chatbot of a tourism/hospitality journal: editor's impressions. Professor and Vice Rector (Research), Varna University of Management, 13A Oborishte str., 9000 Varna, Bulgaria. Editor-in-chief. Eur J Tour Res
9. Khan F, Shrigowri VB, Shwetha KM, Ayesha S, Vinutha K (2018) Artificial intelligence travel bot. In: Conference proceedings, ICRTT
10. Ukpabi D, Aslam B, Karjaluoto H (2019) Chatbot adoption in tourism services : a conceptual exploration. In: Ivanov S, Webster C (eds) Robots, artificial intelligence, and service automation in travel, tourism and hospitality. Emerald Publishing Limited, pp 105–121. https://doi.org/10.1108/978-1-78756-687-320191006

Smart Street Lighting Using Piezoelectric Technique

D. S. Dharshan Odeyar, K. R. Roopa, M. D. Arun, K. M. Chandu, and H. M. Hemanth

Abstract The natural sources used for powering functions are restrained assets and getting diminished day through way of day as the demand for it is rising. In creating countries, amount of generated electrical electricity is unable to keep up with the demand, and moreover, there is shortage of uncooked substances for producing the energy. In worldwide areas like India, 1/5th of electricity consumption is by way of avenue lighting. The widespread avenue lights are on the other hand designed in accordance with ancient necessities of reliability. Because of this, large quantity of electrical energy is wasted and it locations a lot of stress on the natural sources used for producing electricity. Alternative sources are now explored to prepare for the future dearth of regular power sources. A correct designed electricity surroundings nice avenue slight gadget have to permit website online website site visitors and pedestrian to trip at night with extraordinary visibility in safety and remedy even as decreasing electricity consumption and cost. The necessary purpose of our venture is to make use of the electricity generated as the end result of motion of automobiles on road to manipulate the avenue lighting fixtures and thereby creating their effectivity and additionally automating their process.

Keywords Arduino UNO · Piezoelectric sensor · Boost regulator · LDR sensor · UDM sensor · Smoke detector · Buzzer · Power LED

1 Introduction

Smart Street lighting pursuits at growing the effectively usage of street lights with the aid of automating their control, as and when, required, barring using any exterior supply. The automobiles moving on the street tends to vibration due to the piezoelectric cloth positioned under the road due to deformation, precipitated via the pressure of car passing [1]. The Piezoelectricity is a powered charge that gathers in positive strong substances because of mechanical pressure on it. The electrical vitality

D. S. Dharshan Odeyar (✉) · K. R. Roopa · M. D. Arun · K. M. Chandu · H. M. Hemanth
Don Bosco Institute of Technology, Bangalore, India
e-mail: dharshanodeyar@gmail.com

A. Kumar et al. (eds.), *Proceedings of the International Conference on Cognitive and Intelligent Computing*, Cognitive Science and Technology,
https://doi.org/10.1007/978-981-19-2350-0_50

produced from one Piezo element is really little and not more, at this point valuable for all intents and purposes; along these lines, we have a variety of piezoelectric transducers [2]. Considering the mammoth amount of piezo exhibits and huge pressure used with the guide of substantial vehicles, the electrical vitality produced increments [3]. The power produced from these transducers is comparably redressed and managed the use of solidarity gathering circuit. Presently, this immediate power is not utilized at this point straightforwardly, anyway the power created for the duration of the day is pared in batteries [4]. Henceforth, the combined measure of power put away in the battery is over the top adequate for driving of road lights. There is furthermore a computerization circuit which controls the street lights, in understanding whether it is day or evening time and also changes the power of road lights on the establishment of thickness of vehicles, at some random time. Street lights will be off during the daytime and will turn on mechanically during nighttime [5]. At night, avenue lights will glow with excessive depth, if there is a truthful amount of traffic; else avenue lights will glow at low depth. In addition to this [6], street lights can be turned on by vehicle movement during the nighttime by using a UDM sensor or an Infrared sensor, if there is no movement of vehicle street lights are turned off or can be made them to glow at low intensity.

2 Methodology

Mechanical energy is wasted during the vehicle movement on the roads, and this energy is converted into a useful form by using a Piezoelectric material. Piezoelectric sensors are embedded under the road, and due to the movement of vehicles, electrical energy is generated. Generated electrical energy is rectified, regulated, and stored in batteries as power supply. Required components are chosen to make a smart system. All the information ports are associated with the Arduino board. Stored energy in the battery is used to provide supply to Arduino UNO board. Power supply circuit provides the sufficient energy to light up the street lights and other sensors connected to Arduino UNO. When vehicle movement is detected based on the intensity level, street lights are turned ON (Fig. 1).

3 Background Theory

The existing Street Lighting system has many limitations. First of all, controlling of road lights is completely done manually and it is very difficult as we know, as it involves human interplay and that may also motivate blunders in control [7]. We have seen that street lights are sometimes ON even during the daytime, now and then street lights will not be on even when the darkness has settled in. If road lights are on, even during daylight, large quantity of energy is wasted [2]. Also, the existing street light gadget makes use of sodium vapor lamps [6]. Sodium vapor lamps consume an

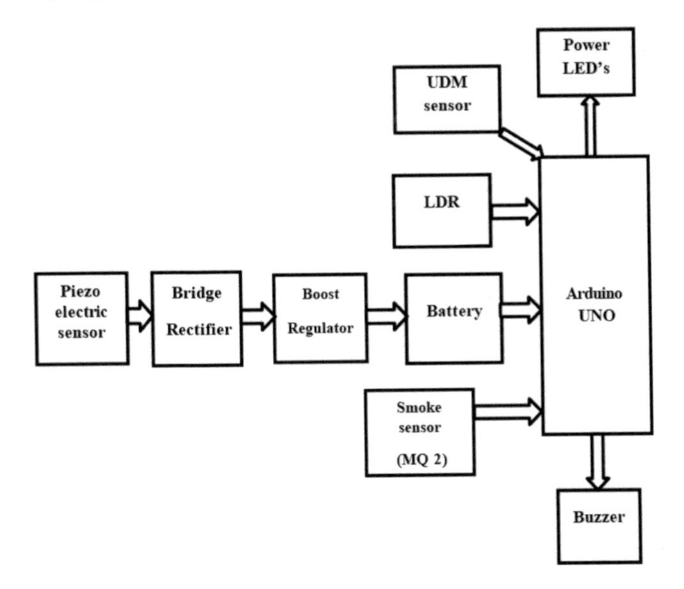

Fig. 1 Block diagram of the overall working methodology

awful lot large energy and additionally warm up, thereby leading to high electricity dissipation. During rainy season, due to cloud formation it is very challenging to produce the electrical energy by making use of solar energy, to produce electricity that is used for street lights.

4 Software Implementation

4.1 Installing Arduino IDE

Arduino IDE is Arduino's open-source programming tool. An IDE comprises all the basic hardware for programming improvement. We utilize Arduino board, the Arduino IDE to alter the source code and afterward transfer the code to the board. Arduino IDE is accessible for Windows and Mac. For putting in Arduino IDE, observe some guided steps. This is a just one-time procedure.

4.2 Selecting Board and Port

The board is chosen as Tools > Board > Arduino Uno, the port in which your board appears is chosen as Tools > Port > COMX, and here in this case it is COM3. The window will appear as proven after choosing the board and port.

4.3 *Verifying the Sketch*

The Arduino Sketch written in the Arduino IDE windows is then validated via using the Verify option. This will check for the compilation errors. If there is an error, then go to the error window and discover the error and debug it as a consequence and again verify it. If no error is discovered and compilation is successful, go to the subsequent step.

5 Results

See Table 1.

6 Conclusion

Our prototype will assist in harvesting energy from the robotically excited piezoelectric sensors and is viewed as a doable choice to change the modern-day power components for electricity-constrained embedded systems. It also eliminates the present day sodium vapor road lamps with better energy LED. It also avoids human interaction of controlling street lights. This prototype also alerts about pollution level, if the smoke from the automobiles crosses the threshold stage in the atmosphere (Fig. 2).

7 Future Scope

Circuit comprising of a road light on the divider having the working machine set up underneath the road. The plates appeared on the road are piezoelectric plates, used to produce electrical power due to the movement of car and the vitality be utilized for taking strolls the air sanitization and refreshing machine, the machine is moreover fueled by methods for the significant air conditioning gracefully.

Table 1 Outputs generated from piezoelectric sensors

S. No.	Approximate input weight (kg)	Approximate generated output voltage (V)
1	50	7
2	60	9
3	70	11
4	>75	>13

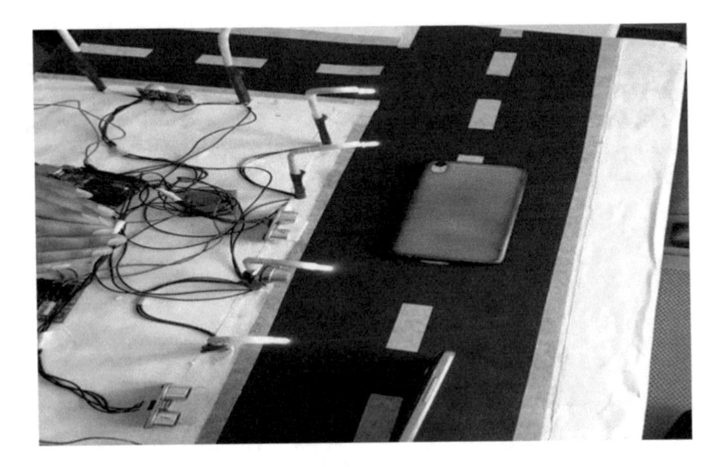

Fig. 2 Proposed model

The future enhancement of this venture is to format eco-friendly easier approachable model for the purifying of main air pollution like SO_2, NO_2, and CO_2. The attractions of the contamination are to be executed through a pull siphon allowing the cleaning gadget to get an adequate measure of attractions head to simple the dirtied air. This gadget will be set up over road lights where there are high air contamination levels. In the filtration chamber, all the parts of the dirtied air come and get broken down in the refined water. For PM (Particulate Matter) likewise with the guide of reality, they are substantial in nature, they settle down in the refined water and it flows internal in the chamber because of their weight. SO_2 make their individual acids, CO_2 does rise to, and after some time, the NO_2 which has been ended in the attractive spot makes their course inside the chamber and makes their separate acids. And these acids are dissolved in the distilled water by making a way out for the purified air.

References

1. Rudrawar O, Daga S, Chadha JR, Kulkami PS (2018) Smart street lighting system with light intensity control using power electronics. In: Technologies for smart-city energy security and power. Bhubaneswar, pp 1–5
2. Roy B, Acharya A, Roy TK, Kuila S, Datta J (2018) A smart street-light intensity optimizer. Emerging trends in electronic devices and computational techniques, Kolkata, pp 1–4
3. Vaghela M, Shah H, Jayswal H et al (2017) Arduino based auto streetlight intensity controller. In: Invention rapid embedded systems 2013, vol 3, pp 1–4
4. Abinaya R, Varsha V, Hariharan K (2017) An intelligent street light system based on piezoelectric sensor network. In: 4th international conference on electronics and communication systems (ICECS). Coimbatore, pp 138–142
5. du Toit P, Kruger C, Hancke GP, Ramotsoela TD (2017) Smart street lights usingpower line communication. IEEE AFRICON, Cape Town, pp 1581–1586

6. Kashiwao T, Izadgoshasb I, Lim Y, Deguchi M (2016) Optimization of rectifier circuits for a vibration energy harvesting system using a macro-fiber composite piezo electric element. Microelectron J 54:109–115
7. Gao S, Wu X, Ma H, Robertson J, Nathan A (2017) Ultrathin multifunctional graphene-PVDF layers for multidimensional touch interactivity for flexible displays. ACS Appl Mater Interfaces 9(22):18410–18416

Non-invasive Wearable Device for Monitoring Chronic Disease

S. Babitha, H. N. Chandan Kumar, Disha Manjunatha, and S. Chaithanya

Abstract Healthcare systems have stated innovating new devices to monitor patients vital parameters such as heart rate and oxygen level. The main purpose of the device designed is to monitor the old-age patients and inform doctors and beloved ones as soon as possible. Therefore, we are proposing a naive project to avoid such sudden death rates by monitoring patient's health which uses sensor technology and Internet of Things that helps to communicate to the loved ones in case of emergency. The proposed system uses set of sensors for monitoring patient's health. All the sensors are connected to the Raspberry Pi microcontroller to monitor the patient's health; also, LCD display is interfaced with microcontroller on which we can see the readings of specific sensors by using wireless sensing node we can save the patient's data through cloud. Therefore, in case of any abnormal changes in patient heart rate, amount of carbon dioxide, or excessive sweat, or variation in temperature or variation in pressure an indication is given through the buzzer using Internet of Things. This system allows us to track live data analysis of patient over the Internet work and provides medication at the earliest.

Keywords Please LCD · Sweat sensor · Raspberry Pi

1 Introduction

In medical industries, monitoring the vital health parameters continuously is a major problem for diseases, especially in case of chronic diseases. A disease is said to be chronic when it lasts for more than three months and may get worse over time. In early methods, doctors used to play a major role in checking up the patients' health. But, this procedure demands a lot of time for several steps like registration, taking appointment, followed by checkup. And also reports are not provided immediately to the patients; a lack of time is wasted here. This lengthy process makes the working people to forget the checkups or postpone it. Hence, this modern approach of using

S. Babitha (✉) · H. N. C. Kumar · D. Manjunatha · S. Chaithanya
Don Bosco Institute of Technology, Bengaluru, India
e-mail: babithagi@gmail.com

A. Kumar et al. (eds.), *Proceedings of the International Conference on Cognitive and Intelligent Computing*, Cognitive Science and Technology,
https://doi.org/10.1007/978-981-19-2350-0_51

wearable device will help to reduce time consumption in the process. By using recent technologies, we are able to monitor the physiological parameters in patients' body. Nowadays, wireless technology is growing rapidly for the need of upholding various sectors. And IOT is grasping almost of industrial area, especially in automation and control. For better health care, biomedical has become one of the recent trends. IoT can not only be used in hospitals but also can be used for personal health caring facilities. Through smartphone, it creates communication between a patient and medical personnel; this will collect, analyze, and transfer the required readings in the form of data for frequent review by a doctor. In the field of challenges, innovation, and research, medical scientists are trying from many years to improve health services and create happiness in human lives. The contribution of medical scientists in the field of medical is very important, and it cannot be neglected. From yesterday's basics only, new automotive structures have obtained root ideas. Also, chronic diseases can be detected at the earliest with this technology.

2 Literature Survey

The paper, 'Faster ways to flexible electronics', was published by MS Lavine on June 2015. Through this survey, we got to know that 40 million deaths will occur by chronic disease every year. This constitutes about 70% of global death. And four main chronic deaths are caused by the following diseases: diabetes, respiratory disease, cardiovascular disease, cancer [1]. The death rates for each disease are 1.6 million for diabetes, 3.9 million for respiratory disease, 8.8 million for cancer, and 17.7 million for cardiovascular disease [2]. The paper, 'Systematic review of Parkinson's disease', was published by Peter and Michael on April 2015. Parkinson's disease is a condition in which the person will not be able to stand or walk for long duration. And when reaches its extreme level, a gait will be formed on the legs, which makes the person to fall. Through this survey, it was got to know that the use of data support along with wearable sensors can help to cure the disease for some extent [3]. But, further, high-quality research is needed for better understanding [4]. The paper, 'Lifting the burden of chronic disease', was published by Sara Kreindler on May 2008. Self-management support was given more importance in this. And provide what patients cannot easily acquire, not just information. Target to patients with the greatest need and integrated into primary care. The appeals done by the mass media did not succeed in changing behavior [8]. By using well-researched and well-targeted social marketing, we may get positive results [5]. The paper, 'Chronic disease prevention and control', was published by Stephane Verguet in 2012.

Most of the global deaths like stroke, cancer, and heart disease will occur in middle-income and low-income countries. The main purpose of this paper is to identify the priorities for control of these chronic diseases. Findings of the disease control priorities and framework project are drawn by this paper [6]. The paper, 'Survey of chronic diseases and health monitoring using big data: A Survey', focuses on the full cycles of the big data processing, which includes security issues in big data, big data tools and algorithm, big data visualization, medical big data processing [6]. The paper, 'Chronic disease prevention and control', published by Stephane Verguet identifies priorities for control of these chronic diseases as an input into the Copenhagen consensus effort for 2012 [7]. This paper draws on the frameworks and findings of the disease control priorities project [3].

3 Methodology

Continuous monitoring is required for chronic disease care which may cause certain limitations in the patient's daily activities. This major health problem is mainly caused due to smoking, lack of exercise, and not maintaining balanced diet of the individuals. The chronic disease cannot be identified in the initial stage; this can be identified only when the symptoms are shown which severe enough to recognize the disease. Also, poor people who will be suffering from chronic disease should have to spend a lot of money for the treatment and also should visit the hospital frequently to monitor their health condition. By the continuous monitoring of patient's health, it is possible to detect the chronic disease as early as possible. The use of wearable device will provide a change to the traditional chronic disease care. In this proposed system, the following blocks are implemented (Fig. 1).

As shown in the proposed system, the controller used is the Raspberry Pi, and each sensor, i.e., heart rate sensor, CO_2 sensor, sweat sensor, temperature sensor and the pressure sensor are connected to the controller where the heart rate sensor monitors the variations in pulse level which is associated to cardiovascular disease.

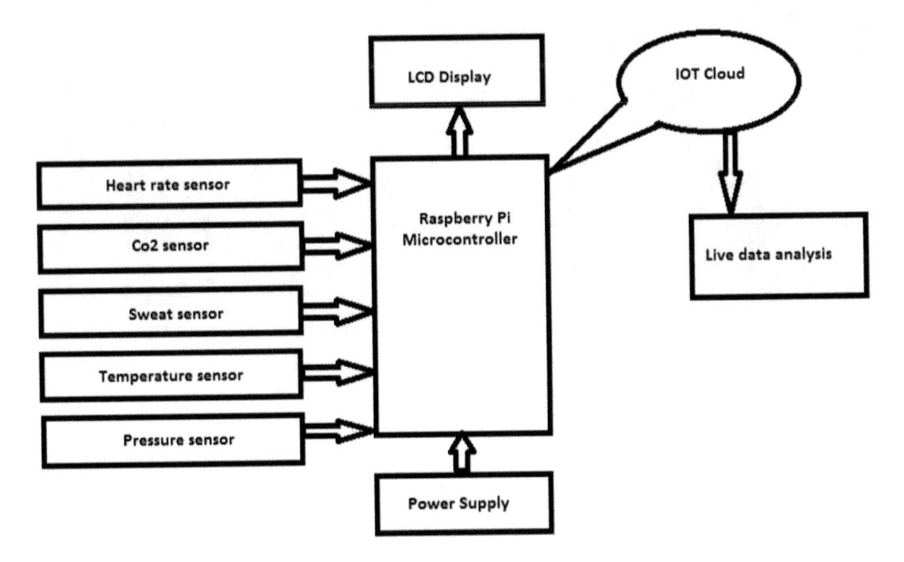

Fig. 1 Proposed system

The CO_2 sensor is used to monitor the amount of carbon dioxide present in lungs while respiration which is associated to asthma and other respiratory diseases. The sweat sensor is used to monitor the excess sweat present in the body which is associated to hyperhidrosis. Temperature sensor is used to monitor the temperature of the body associated to chronic fever. Pressure sensor is used to monitor the pressure present in patient's legs which is associated to neuropathic ulceration; then, the controller is given to the power supply of 5 V. The LCD which is of 16*2 display which is used to visualize the reading of each sensor and the variations of reading may lead to beep sound from the buzzer indicating alert, and hence, the patients data are stored in cloud so that continuous monitoring of patients health can be done, and the software used for overall is virtual network computing (VNC) is a software that allows us to perform the activities of keyboard and mouse by some other computer also, by providing the controller ID. But, the same Wi-Fi connection should be provided together to both system as well as the controller. And VNC is platform independent which uses remote frame buffer protocol (RFB). This RFB protocol works at frame buffer level, and it is applicable to all windowing systems and applications.

The detailed explication of each sensor working is shown in Fig. 2.

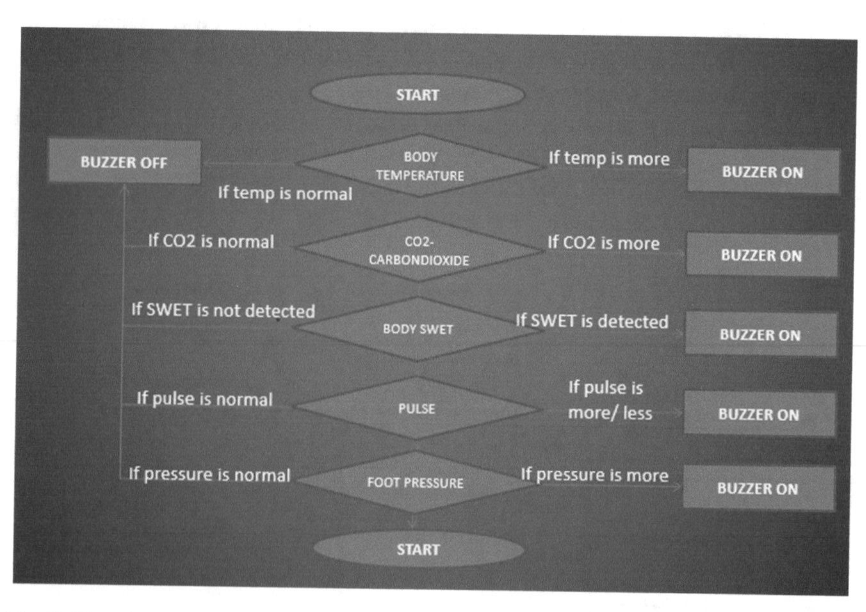

Fig. 2 Flowchart

4 Results and Discussion

Continuous monitoring of patients health is done. The sensor readings are displayed in LCD as well as in the computer screen. The variations in the readings may lead to an indication from the buzzer. This may lead to early detection of chronic disease affected patients, and hence, the patient can be treated in the early stage itself.

A. Prototype

Figure 3 model is a protype of the proposed system. It involves all the sensors explained.

B. Temperature Sensor

DHT11 temperature sensor features a temperature sensor complex with a calibrated digital signal output (Fig. 4).

C. Co_2 sensor

MQ-135 gas sensor senses gases like ammonia nitrogen, oxygen, alcohols, aromatic compounds, sulfide and smoke. The conductivity of the sensor increases as the concentration of polluting gas increases (Fig. 5).

Fig. 3 Results

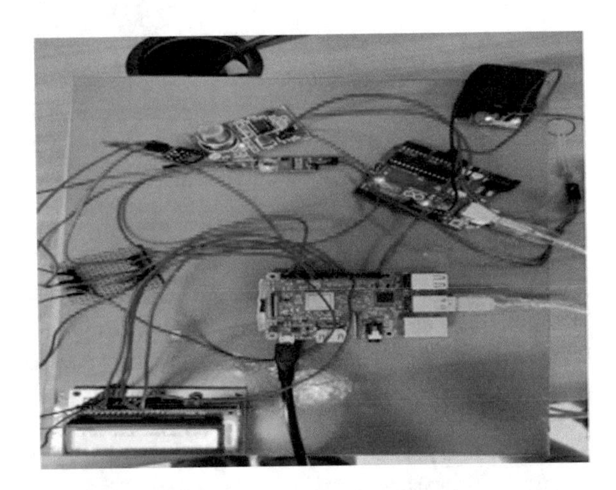

Fig. 4 Temperature sensor

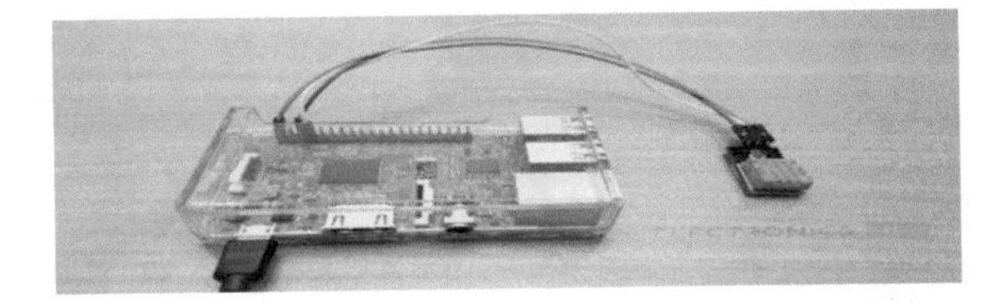

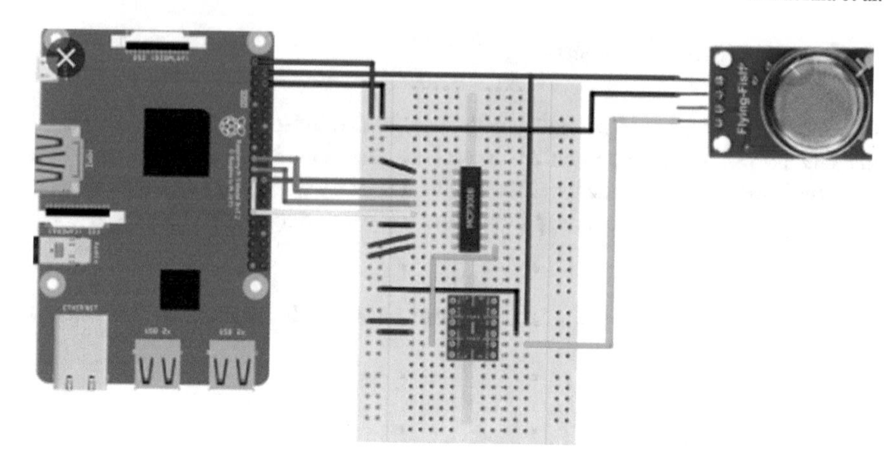

Fig. 5 Implementation of CO_2 sensor

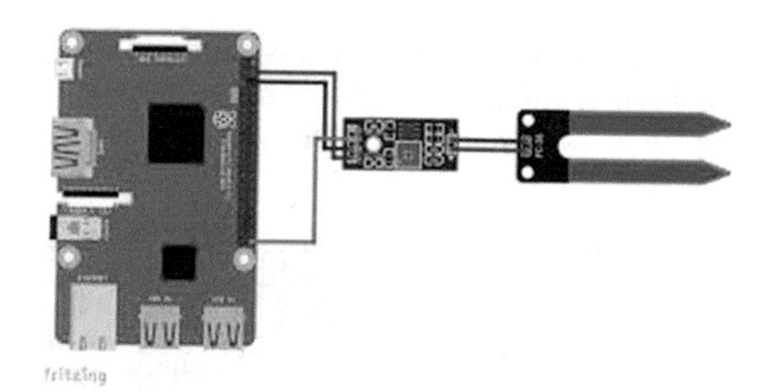

Fig. 6 Implementation of sweat sensor

D. Sweat Sensor

Sweat sensors is non-invasive technique used to provide hydration parameters of the human body. The testing of human sweat is in reference to the eccrine sweat gland which in contrast to the apocrine sweat gland has a lower composition of oils (Fig. 6).

E. Pulse/Heart Rate Sensor

The heart rate sensor senses the heart pulses and monitors hear rate; if there are variations in heart rate, the buzzer will send a signal (Fig. 7).

5 Conclusion

By developing this wearable sensor, early detection of patients affected is possible and hence lead to early monitoring, and the medication can be provided to patients. Implementation of this project can help the poor people who cannot afford for expensive treatments. In the future work, a technological advances in miniaturization and

Fig. 7 Implementation of heart rate sensor

nano-technology will help in the progress of wireless communication which allows for the further development of miniaturized devices, and this also gives us the flexibility to integrate the miniaturized devices with clothes and even implant in the human body. Self-monitoring helps to make it feasible to monitor disease in all situations.

References

1. Yao S, Swetha P, Zhu Y (2018) Nanomaterial-enabled wearable sensors for healthcare. Adv Healthc Mater 7(1):1700889
2. Ramirez-Bautista JA, Huerta-Ruelas JA, Chaparro-Cardenas SL, Hernandez-Zavala A (2017) A review in detection and monitoring gait disorders using in-shoe plantar measurement systems. IEEE Rev Biomed Eng 10:299–309
3. Konvalina G, Haick H (2014) Sensors for breath testing: from nanomaterials to comprehensive disease detection. Acc Chem Res 47(1):66–76
4. Koo JH et al (2017) Wearable electrocardiogram monitor using carbon nanotube electronics and color-tunable organic light-emitting diodes. ACS Nano 11(10):10032–10041
5. Gao W et al (2016) Fully integrated wearable sensor arrays for multiplexed in situ perspiration analysis. Nature 529(7587):509–514
6. Lavine MS (2015) Faster ways to flexible electronics. Science 348(6239):1103–1103
7. Teng XF, Zhang YT, Poon CC, Bonato P (2008) Wearable medical systems for p-Health. IEEE Rev Biomed Eng 1:62–74

FRCNN Algorithm Strategy with Image Cropping for Dental Radiographs

B. Naveen and M. Nandeesh

Abstract Location of cavity is crucial for the conclusion and treatment drafting of the dental sickness, which has been influencing a truly enormous populace all through the world. The objective of this work is to detection, cropping, extract enamel from the tooth with the Faster R-CNN network directing point by point examinations about the personality of the dental infection. The datasets, strategies, and results pointed to check dental treatment attributes employ periapical dental X-ray pictures. In this, we propose an attention toward FRCNN algorithmic strategy technique with the remedial grip to crop the ROIs for additionally handling toward clinical evaluation. In the proposed work, Faster R-CNN carry out outstandingly fine in regards to teeth detection and cropping, this found the location of teeth precisely with a high estimation of IOU with ground truth boxes, as well as good precision and review. The results give the model output of the cropped image detection boxes which covers enamel part of the teeth. The coordinates of predicated bounding box and classification accuracy are also verified.

Keywords Dental X-ray · Image segmentation · Dental caries · FRCNN · RPN

1 Introduction

Dental X-rays (radiographs) are the images where the dental specialist can roughly examine the apex of the root of the teeth. By the utilization of these X-rays with low degrees of radiation, specialist can identify and distinguish issues related to cavities, root, and affected part of the teeth. The main intention of radiograph segmentation is to localize the area of every tooth in a given X-ray image [1].

This radiographic division accordingly for instance causes the dental specialist to distinguish/recognize dental caries and periodontal infection which are the basic dental illnesses across world [2]. Today, dental caries moreover referenced as cavity or depressions is one of the premier normal and boundless steady infections which

B. Naveen (✉) · M. Nandeesh
BGS Institute of Technology, Adichunchanagiri University, B. G. Nagara, Karnataka, India
e-mail: naveenb@bgsit.ac.in

A. Kumar et al. (eds.), *Proceedings of the International Conference on Cognitive and Intelligent Computing*, Cognitive Science and Technology,
https://doi.org/10.1007/978-981-19-2350-0_52

are chief preventable. At the point, where the individual consume nourishments, microbes available in the teeth isolates and produce acids that can genuinely harm the hard tissues. This results in the development of dental caries (cavities).

Typically, cavity is frequently spotted on two explicit regions of the teeth mainly occlusal caries and the proximal caries [3]. It structures on the most noteworthy piece of the tooth where food particles are open direct contact with the teeth and entomb proximal caries, which are cavity that structure stucks between the teeth. These are the main two areas where microorganisms root and represent a danger to oral cleanliness. In the occasion where the teeth and encompassing zones aren't thought about effectively, the microorganisms force to convert sugar from the food what we eat and transfer to acid as a waste. The acid produced dematerializes the surface on our teeth and structure minuscule openings, the main phase of depression. Since polish starts deter down, the tooth lose the capacity to emphasize the calcium and phosphate structures. Normally, through spit properties as expected, corrosive enters into the tooth and decimates it from the back to front.

The creation of analytic quality radiographs relies upon numerous elements including appropriate film situating, appropriate X-ray exposure, and proper film processing techniques [4]. A slip-up in any of those elements will end in yet ideal or non-analytic radiographic picture. In this way, the point is to flexibly a demonstrative quality picture through segmentation which could empower the dental specialist to spot numerous conditions which will in any case stay undetected. Segmentation and form extraction are applied to remove highlights for programmed distinguishing proof of dental radiographs. Preceding dental caries location and confinement, it is essential to separate individual tooth tests from the X-ray. This extricated tooth tests will be gone through neural organization and later for caries location and characterization. A programmed editing for holding a necessary segment of fragmented picture dental X-ray pictures is proposed to keep away from the runtime of predisposing. Inside the limit of picture division approach, a few post methodologies, similar to veneer and bury proximal district trimming, ought to be performed.

In this paper, we first use calculations essentially lessen the outstanding task at hand of human specialists and can remove certain highlights that are hard for people to perceive. We propose a robotized editing strategy dependent on dental radiograph picture characterization, which coordinates clinical master's insight, with convolution neural organization and RPN-based learning [5, 6]. The trial dataset incorporates 148 explained tooth tests of periapical dental X-ray pictures. As most of information pictures in our investigation incorporate periapical dental X-ray pictures [7]. In dental X-ray, veneer locale is the main zone as beginning phase of hole initially happens from finish. In this way, extraction of tooth with the dominant part bit of lacquer is our objective. We propose a profound learning calculation, Faster R-CNN with district proposition network over conventional picture preparing methods to perform tooth division from a dental X-ray picture [8].

2 Materials and Methods

For this examination, dental X-ray pictures of 150 patients were collected from a dental center. The X-ray images were further cropped to a limit to remove unnecessary noise from the X-ray. The cropped image is then given to the deep learning algorithm (FRCNN) is show in Fig. 1 [8]. During our research, we tried preprocessing the dental image with image processing techniques like histogram equalization, contrast-limited histogram equalization, and blurring of the image. The image processing techniques however did not help in improving the feature quality [9].

The FRCNN algorithm works in two phases. Phase I of the algorithm works as a feature extractor. The convolution neural network extracts the features from a dental X-ray, this extracted features then act as an input to the regional proposal network [10]. Regional proposal network then draws anchor boxes over the feature map. The number of anchors produced by RPN was 32 of kernel size 2*2, 1*2, and 2*1. In order to determine the closest anchor box matching the ground truth box, RPN uses a voting process where each anchor in given a score based of the features shared with the ground truth box.

The anchor having maximum score is considered as the best anchor box overlapping the ground truth box. From the iteration, the region proposal network learns the features predict as the closest anchor box. Thus, minimizing error and improving the accuracy of the closest anchor box match with the ground truth box. Dental X-rays assembled for the experiment around 150 images were cropped as far as possible to decrease the noise from the dental pictures.

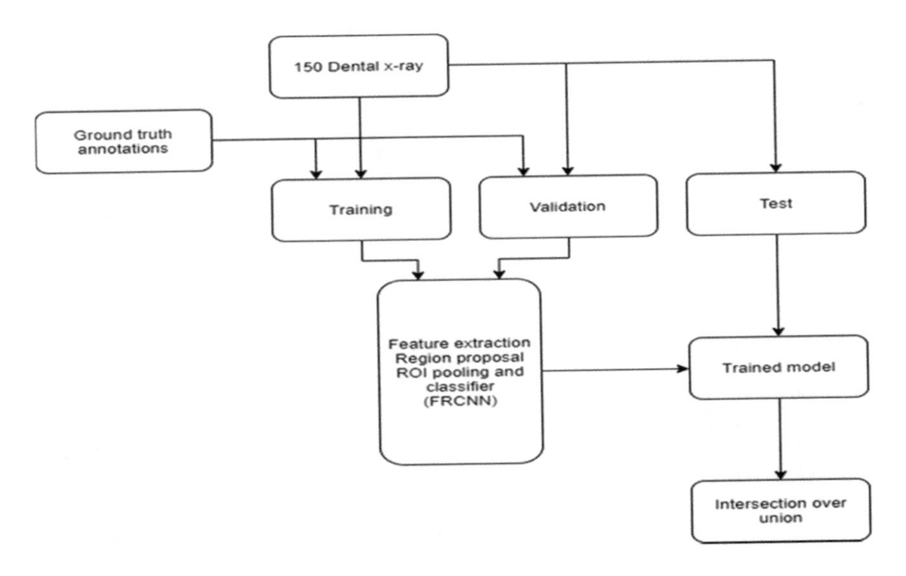

Fig. 1 Image data and ground truth annotations

Annotations of the ground truth box for each tooth are done manually [11]. Annotations of 150 dental X-rays were verified. Each ground truth value contains coordinates of a bounding box rectangle ×1, y1, ×2, y2. CNN uses a sliding window mechanism to convolve over the input image and generate feature maps shown in Fig. 2.

Neural network model construction, training, validation, network architecture, convolution neural network act as feature extractor and generate feature maps which are further passed to a region proposal network appeared in Figs. 3 and 4 [12, 13].

The RPN shapes two yield per anchor: anchor class: single of two categories: forefront or background. The FF class suggests that nearby possible an item within that box.

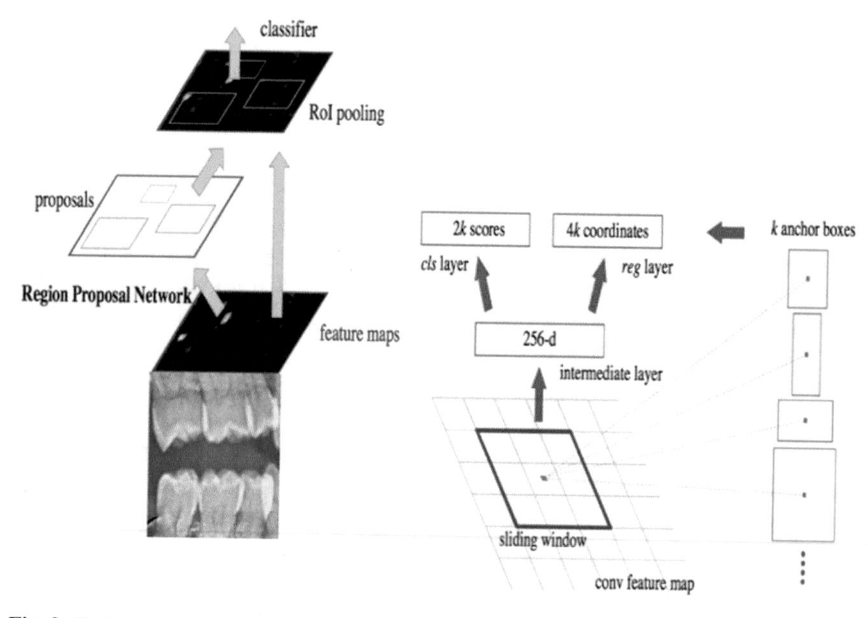

Fig. 2 Feature extraction

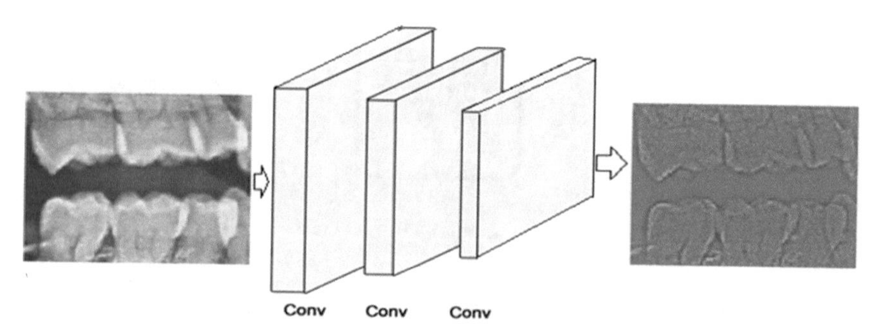

Fig. 3 Region proposal network

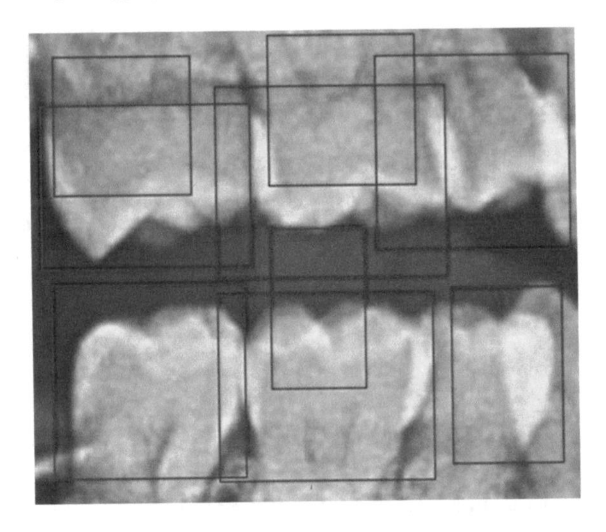

Fig. 4 Region proposal network generates anchor boxes on the feature maps

Bounding Box filtering: A forefront anchor (furthermore named positive anchor) probably won't be coaxial totally more the target. As a result, the RPN assesse a delta (% change in x, y, breadth, altitude) to process the anchor box to suit the object better. Employing the RPN expectations, crop the head anchors so as to probably going to have targets and process their zone with size. In the event that few anchors cover excessively, we retain the each with largest forefront closer view score and dispose of the rest (attributes to as Non_max Suppression) evident in Fig. 5. After that we have the last suggestions (ROI) that we move to the following level.

This level is executed on the region of interests (ROIs) recommended by the RPN. In addition, a lot equivalent to the RPN, it makes couple yields for single ROI group. The group of the item in ROI, rather than the RPN, which has couple of groups (FG/BG), this system be wider and has the capable to group locales to explicit groups (person, vehicle, chair, etc.). It can also develop a background group, which builds the ROI to be rejected.

Bouncing box filtering: As like how it is happened in the RPN, moreover, its inspiration is to additionally filter the position and size of the bounding box to enclose the item.

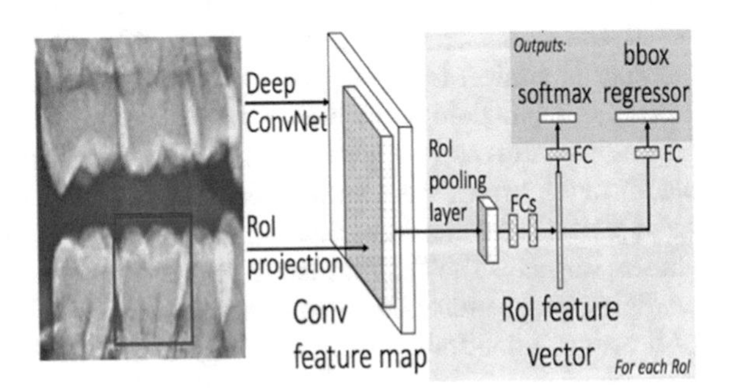

Fig. 5 ROI classifier and bounding box regressor

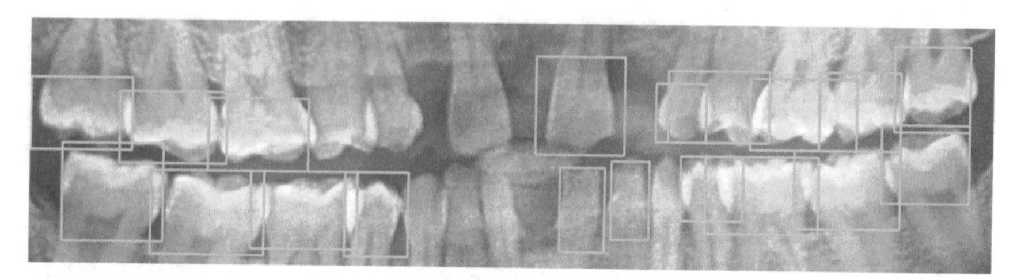

Fig. 6 Example of cropped image using FRCNN algorithm

3 Results

Figure 6 shows the model output of the cropped image detection boxes which covers enamel part of the teeth. The coordinates of predicated bounding box and classification accuracy are specified below.

Predicted bounding box and classification accuracy

[('Tooth', 99.94919300079346), ('Tooth', 99.929678440094), ('Tooth', 99.81977343559265), ('Tooth', 99.72871541976929), ('Tooth', 99.64505434036255), ('Tooth',99.64144229888916), ('Tooth', 99.51287508010864), ('Tooth', 99.34404492378235), ('Tooth', 9.16219711303711), ('Tooth', 97.75535464286804), ('Tooth', 96.69321179389954), ('Tooth', 5.71217894554138), ('Tooth', 95.28568983078003), ('Tooth', 94.92764472961426), ('Tooth', 94.4614052772522), ('Tooth', 93.35577487945557), ('Tooth', 92.22386479377747), ('Tooth', 8.05482387542725), ('Tooth', 87.7770185470581)]

4 Discussion

Many of investigations are centered on picture segmentation and classification. Twain conventional digital image processing algorithms as well as deep learning techniques are in big demand. In image segmentation zone, conventional image segmentation techniques be generally utilized during experimental images involve region-based strategies, edge tracing techniques, and integration techniques. Pleasing deep learning to clinical image segmentation is a natural theme newly with the widest variety in methodology, including the progress of unique CNN-based segmentation architectures and the wider function of RNNs. For the clinical image categories, nearby multiple attempts to utilize deep learning, particularly CNNs. Clinical picture arrangement examines have done in human eye, chest, cancer, brain, and different clinical impacts. The dataset picture classification cases and techniques dependent on digital image processing are utilized more widely. In summary, there is no state-of-the-art solution for dental image diagnosis in the literature, but there are related research efforts on image processing techniques and examination utilizing digital image processing techniques, conventional categories techniques, and deep neural networks. Proposed system is equipped for restoring numerous tooth segments dependent on an educator understudy structure. In this structure, it is prepared to evaluate candidate anchor boxes, and the scores got from each anchor box are then used to manage the preparation of the understudy. Finally, by training and validation, the input dataset was spit by 70/30 rule where 70% of the data for training and validation and 30% data used for testing. The network was prepared for 100 epochs with a loss threshold of 0.001.

5 Conclusion

In the proposed work, Faster R-CNN carries out outstandingly fine in regards to teeth detection and cropping; this found the location of teeth precisely with a high estimation of IOU with ground truth boxes, as well as good precision and review. This extracted tooth segment can further be utilized to extract enamel from the tooth. With as low as 150 tooth samples, the Faster R-CNN network can locate the region of the teeth on the dental X-ray image. The tooth can be cropped using the predicted bounding box. Cropped tooth can be then used in the further work to classify a tooth image as cavity or no cavity.

References

1. Nandeesh M, Naveen B (2020) A literature review on carries detection and classification in dental radiographs. Industr Eng J 13
2. Lin PL, Lai YH, Huang PW (2010) An effective classification and numbering system for dental bitewing radiographs using teeth region and contour information. Pattern Recogn 43:1380–1392
3. Yöntemleri DHGÇT (2016) Recent methods for diagnosis of dental caries in dentistry, Vol 1. Adnan Menderes University Faculty of Dentistry, Department of Pediatric Dentistry, Aydın, Turkey, pp 29–36
4. Said EH, Nassar DEM, Fahmy G, Ammar HH (2006) Teeth segmentation in digitized dental X-ray films using mathematical morphology. IEEE Trans Inf Forensics Secur 1:178–189
5. Milletari F, Navab N, Ahmadi SA (2016) V-net: fully convolutional neural networks for volumetric medical image segmentation. In: 4th international conference on IEEE, pp 565–571
6. Ronneberger O, Fischer P, Brox T (2015) U-net: convolutional networks for biomedical image segmentation. In: International conference on medical image computing and computer-assisted intervention. Springer, pp 234–241
7. Rad AE, Rahim MSM, Norouzi A (2013) Digital dental x-ray image segmentation and feature extraction. Indonesian J Electr Eng Comp Sci 11:3109–3114
8. Girshick (2015) Fast R-CNN. In: IEEE international conference on computer vision (ICCV), pp 1440–1448
9. Gonsalves PP (2017) Diagnosis of dental cavities using image processing. Int J Comp Appl 180:28–32
10. Yadav N, Binay U (2017) Comparative study of object detection algorithms. Int Res J Eng Technol (IRJET) 04:586–591

11. Chen H, Zhang K, Lyu P, Li H, Zhang L (2019) A deep learning approach to automatic teeth detection and numbering based on object detection in dental periapical flms
12. Esteva A, Kuprel B, Novoa RA, Ko J, Swetter SM, Blau HM, Thrun S (2017) Dermatologist-level classification of skin cancer with deep neural networks, Vol 542, pp 115–118
13. Chen H, Jain AK (2004) Tooth contour extraction for matching dental radiographs. In: 17th international conference on IEEE, Vol 3, pp 522–525
14. Wang C-W, Huang C-T, Lee J-H, Li C-H, Chang S-W, Siao M-J, Lai T-M, Ibragimov B, Vrtovec T, Ronneberger O et al (2016) A benchmark for comparison of dental radiography analysis algorithms. Med Image Anal 31:63–76
15. Leea J-H, Kima D-H, Jeonga S-N, Choib S-H (2018) Detection and diagnosis of dental caries using a deep learning-based convolutional neural network algorithm. J Dentistry 77:106–111
16. Ren S, He K, Girshick R, Sun J (2017) Faster R-CNN: towards real-time object detection with region proposal networks. IEEE Trans Pattern Anal Mach Intell 39:1137–1149
17. Raghu S, Naveen B (2019) A survey on QBIC system for ECG reports. Industr Eng J 12(12)
18. Raghu Kumar BS, Naveen B (2020) Analysis on CBIR system for ECG reports. Int J Eng Adv Technol (IJEAT) 9(5):586–592
19. Raghu Kumar BS, Naveen B (2020) Comparison analysis of feature extraction from the ECG graph reports. In: Proceedings of online international conference on smart modernistic in electronics and communication (ICSMEC), p 95. ISBN No. 978-93-80831-43-5
20. Naveen B, Rekha K, Nataraj K (2017) FPGA implementation of image splitting and enlarging. In: 2017 international conference on recent advances in electronics and communication technology (ICRAECT), Bangalore, India, 2017, pp 265–269

Adapting the Effect of Impulse Noise in Broadband Powerline Communication

D. Smitha Gayathri and K. R. Usha Rani

Abstract Powerline communication (PLC) has seen an extraordinary advancement since the 1920s; the spot earlier data transmission rates were to the extent of a few kbps and improved to a few Mbps. As the advancement has been made, quick improvement in machine loads related to the powerline has extended, inducing electromagnetic interference (EMI)/noise that taints correspondence execution. Standard help strategy cannot be applied clearly to reduce fuss from PLC authority, as the uproar that exits in powerline correspondence is impressive from the conventional correspondence system. In this endeavor, estimation and assessment of disruption in the repeat extent of 150–30 MHz are explored. From the evaluation, it is understood that the impulsive noise is a primary disturbance in PLC that is usually created by the operation of high-frequency electronic equipment or by transient switching, which began with the affiliation and separation of electric, mechanical, etc. The execution of correspondence in the OFDM powerline communication system is restricted by asynchronous impulsive noise and periodic impulsive noise. To ease this asynchronous impulsive noise and experience its sparsity in time-space, deficient Bayesian learning (SBL) method is applied to measure and remove the disturbance to achieve signal-to-noise ratio (SNR) increases of up to 9 dB.

Keywords Synchronous impulsive noise · Asynchronous impulsive noise · PLC · OFDM · Deficient Bayesian learning

---

D. Smitha Gayathri (✉)
ECE Department, RNS Institute of Technology, Bengaluru 560074, India
e-mail: smithasunil5@gmail.com

K. R. Usha Rani
ECE Department, RVCE, Bengaluru 560059, India

A. Kumar et al. (eds.), *Proceedings of the International Conference on Cognitive and Intelligent Computing*, Cognitive Science and Technology,
https://doi.org/10.1007/978-981-19-2350-0_53

1 Introduction

Powerline correspondence is a sensibly ongoing innovation. "Powerline correspondence" signifies an innovation that assists with moving information at narrowband or broadband through powerlines working with different development tweak innovations. This continues toward, as of now, existing powerlines for the transmission of information [1]. Utilizing the previously existing air conditioning powerlines as a medium to move the data information has assisted with associating the houses with an exceptionally rapid system passage with no need for new wiring establishment.

A powerline correspondence framework incorporates signals sent and received over a general nuclear family and a mechanical 50 Hz current passing on the powerline. In progressing time, PLC has furthermore been a bit of robotization of home and frameworks organization which goes under the overview of acclaimed propels. This is because powerline is a tolerably productive and more enthusiastic correspondence channel used worldwide. When in doubt, an essential propelled correspondence includes an encoder and a modulator on the transmission part and a decoder and a demodulator on getting some segment of the correspondence system. In any case, the two-way correspondence system uses a modem. In all cases, a powerline modem involves an encoder, a decoder, a transmission medium, and different coupling circuits are required for better confirmation, disengagement, and impedance planning. These modems also have various applications.

The progression of the PLC framework for broadband applications requires vast data on the correspondence channel characteristics and the common concern of focus that may affect the correspondence over the ideal channel. PLC, anyway, offers a fierce channel for correspondence signal, as the fundamental reason for existing is to send electric power at 50 Hz or 60 Hz. Clamor, multipath, blurring, and constriction are the significant properties of powerline correspondence and must be seriously considered while structuring a PLC framework. Notwithstanding these boundaries, arbitrary asynchronous noise described with brief lengths, and high amplitudes are called attention to as one of the significant insufficient boundaries that brings down the presentation of PLC system.

Even repeat division multiplexing (OFDM) is the generally utilized change strategy for PLC and has been viewed as the change plot for broadband PLC and is being used as the rule plan in this work. This is considering how OFDM decreases the impacts of multipath and gives high power against specific obscuring. It is likewise critical in the creativity of noise and works better than single-carrier transition strategies. If an OFDM picture is influenced by inspiration disturbance, the impact is spread over various subcarriers because of the recipient's discrete Fourier transform (DFT). Like this, the entirety of the simultaneously sent correspondence picture is barely affected by the incident rushed noise.

To achieve strong outcomes, proper channels and noise models are used in this paper. This zone will analyze a strong control technique that incorporates sparse Bayesian learning. The critical noise circumstance explored in this paper is careless

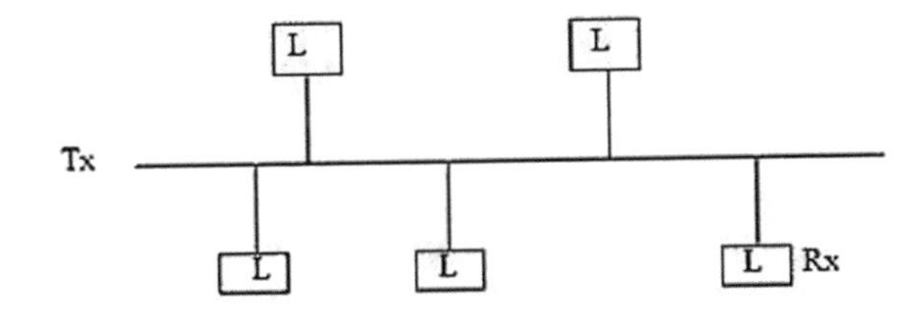

Fig. 1 Transmission line in the power grid

noise, where sudden noise is masterminded into composed impulsive noise and asynchronous impulsive noise. Results show that impulsive noise can genuinely affect the bit-botch pace of OFDM-based PLC frameworks.

I. Channel Modeling

In powerline correspondence, the spread of information signals does not follow a single way or unipath. Yet, they follow multipath following a model on an elementary level comparable to faraway signs attracted with cell communication. The power structure is a solitary focal transmission line with shooting stems completing near the end clients' place, as shown in Fig. 1. T_x is the explanation behind transmission (service supplier), and R_x is the inspiration driving receiver (automated meter/client/different machines) [2].

Signal force and BER of the got signal depend on transit followed and the length of the way or way. Multipath engendering is also liable for delay in PLC, which is given by

$$\tau_i = \frac{d_i\sqrt{\varepsilon_r}}{C_0} = \frac{d_i}{v_P} \tag{1}$$

d_i is the length of way, C_0 is light speed, and E is dielectric consistent of protecting material.

$$H(f) = \sum_{i=1}^{N} g_i A(f, d_i)\overline{e} j 2\Pi f \tau_i \tag{2}$$

$H(f)$ is the repeat response of the channel between two center interests. When the cross-section sort out makes colossal and complex, it could be isolated into sub-channels for lone assessment. An (f, d_i) are cable losses which could be in the form of heat or signal leakage etc. f is the frequency operation, g_i is weight factor which is genuinely identifying with several reflections and way followed

$$|g_i| \leq 1 \tag{3}$$

The estimations of and $A(f, d_i)$ are resolved tentatively. Because of the above-given factors, a scientific model of multipath PLC is proposed:

$$H(f) = \sum_{i=1}^{N} |g_i(f)| e^{\psi g_i(f)} e^{-(a_0 + a_1 f^k) di} e^{-j2\pi f \tau_i} \quad (4)$$

In light of broad examination on trial information, $A(f, d_i)$ can be approximated scientific recipe for lessening factor (α)

$$\alpha(f) = a_0 + a_1 . f^k \quad (5)$$

Utilizing $A(f, d_i)$ in (f) gives the channel model for the PLC transmission line

$$H(f) = \sum_{i=1}^{N} |g_i(f)| e^{\psi g_i(f)} e^{-(a_0 + a_1 f^k) d_i} e^{-j2\pi f \left(\frac{d_i}{v_p}\right)} \quad (6)$$

Noise Modeling in PLC

This research examines the available PLC models of hurried noise (asynchronous impulsive noise and synchronous impulsive noise). The noise relief system proposed is nonparametric; these models are remarkable for propagation in order to examine the intensity of the approach proposed.

(A) Modeling of Asynchronous Impulsive Noise

This work shows the time-space property of asynchronous impulsive noise. In broadband electrical cable correspondence, synchronous noise estimations at higher recurrence groups are found from several kHz to 20 MHz, where asynchronous noise is predominant. The prompt plentifulness of the noise is clarified utilizing different examinations that exactly fit the noise information to Gaussian noise and Middleton Class A noise [3]. For the comfort of dialogue in later areas, we rapidly define the Middleton Class A and the Gaussian mix as follows:

(1) **Gaussian noise Model**

A flying variable Z has a Gaussian noise dissipation when its probability density function is a weighted accumulation of different Gaussian noises, i.e., in which $N(z; 0, \mu k)$ is Gaussian pdf with zero mean, and changes μk, and μk is the k-th Gaussian part's combining probability.

(2) **Middleton Class A Model**

A covering factor defines the Middleton Class A model and foundation to noise power proportion Ω ($A \in [11 - 2, 1]$ and $\Omega \in [11 - 6, 1]$) by and large. Time-area noise follows reproduced from Gaussian blend, and Class A models are delineated in Figs. 2 and 3, respectively.

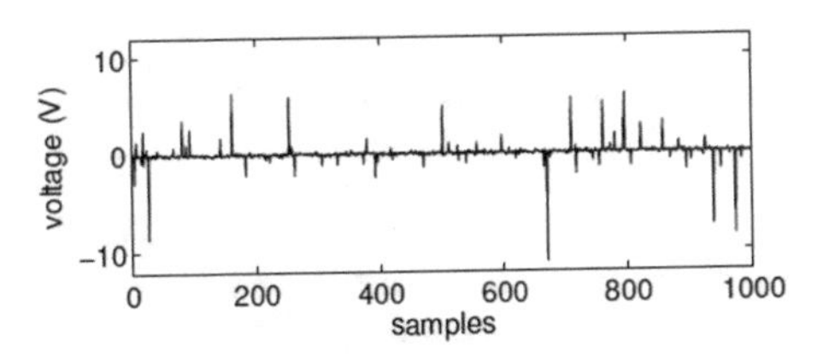

Fig. 2 Synchronous noise and $\gamma = [1, 100, 1000]$ reproduced from a Gaussian blend conveyance with $x = [0.9, 0.07, 0.03]$

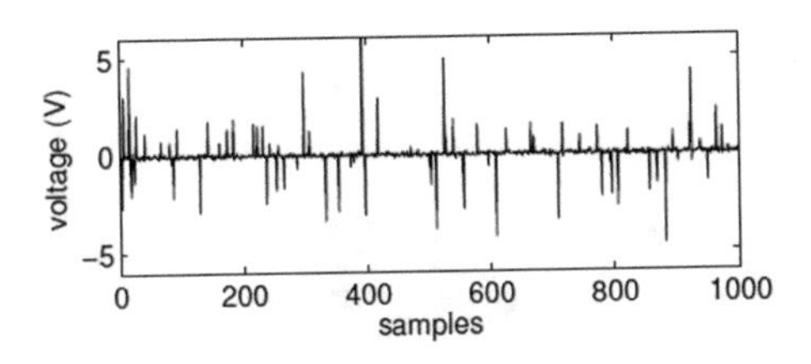

Fig. 3 Asynchronous impulsive noise recreated from *a* middle tone class A dissemination with $A = 0.2$, $\Omega = 0.02$, and the pdf shortened to the initial 11 blend segment

(B) Synchronous Impulsive Noise

The racket in the NB-PLC involved can be allotted in three classes: establishment upheaval, fuss due to narrowband impediment, and tactless noise. Notwithstanding the route in the broadband powerline communication channel, the most dangerous noise is the asynchronous impulsive noise, accomplished by an accomplice or isolating electrical gadgets on the network; in the narrowband powerline communication case, the regular noise section is the cyclo-fixed drive racket.

The narrowband deterrent is made by supporters in the longwave run, what is more by exchanging mode energy supplies with development frequencies above 21 kHz. Such a fuss changes logically in time, and the transmission capacity of a tight-band interferer can be considered to follow an even dispersal in [6, 21 kHz]. To review the impact of such bang on our reenactment framework, we have thought that the interferer has an inside repeat (f_c) around the point of convergence of the OFDM transporter signals used for data transmission. In this way, the most unfavorable situation is thought of. It is considered that f_c is evenly distributed in [f_c − 21 kHz, f_c + 21 kHz], where f_c is given by

$$f_c = f_{u1} + f_{u2}/2 \tag{7}$$

The boundaries f_{u1} and f_{u2} represent the first and terminating transporter signals used to communicate data bits separately, and they are characterized as follows

$$f_{u1} = f_{tr1} + N\,\text{null} * \Delta f \tag{8}$$

$$f_{u2} = f_{tr1} + (N\,\text{null} + N\,\text{carriers} - 1) * \Delta f \tag{9}$$

where f_{tr1} speaks to the essential transporter recurrence of the OFDM image, Δf is the transporter dispersing, null is the measure of zero transporters associated with the OFDM image before the leading transporter to transfer information, and N carriers is the hard and fast number of OFDM. To complete the image, the foundation commotion is exhibited as AWGN.

2 Noise Mitigation

2.1 *Asynchronous Mitigation of Noise*

In BB PLC, as with remote frames, synchronous impulsive noise, for example, emerges off the cell frameworks and the cuff. Earlier processes in abrupt asynchronous noise reduction include parametric techniques that recognize a certain measurable concussion model and usually test the limitations of the reality model during the preparedness phase. Such calculations consolidate methodology for pre-isolation, zero, and cutting strategies, MMSE picture-by-picture pointers and iterative decoders [4].

From late on, energy was generated to produce nonparametric denoising techniques that misuse the insufficient driving noise structure in time. The packed detection methods were used in particular to assess noise disruption from the tones of the received signal (i.e., tones that are not transmitted by data or pilots). However, the introduction of the figure is affected by constraints explicitly adapted to the level of noise inside the OFDM picture and the noise level of the foundation [5].

In order to increase execution and strength, CS-based calculation is expanded to include the sparse Bayesian learning (SBL) approach.

2.2 *Synchronous Noise Mitigation*

When everything is said and said, boundary estimates are inherently more troublesome in synchronous impulsive noise than those in abrupt synchronous noise. It is a consequence of the essential increase to the boundaries and henceforth, the degrees of opportunity for the non-needless association of time-space to become sometimes noisy. The correct measurement of these limits normally requires a lot of data, i.e., in different cycles, including basic overhead planning. However, what is more, there is normally no notable memory in current powerline communication modems. In this research, it was found that the cyclo-range, i.e., the difference between the Fourier

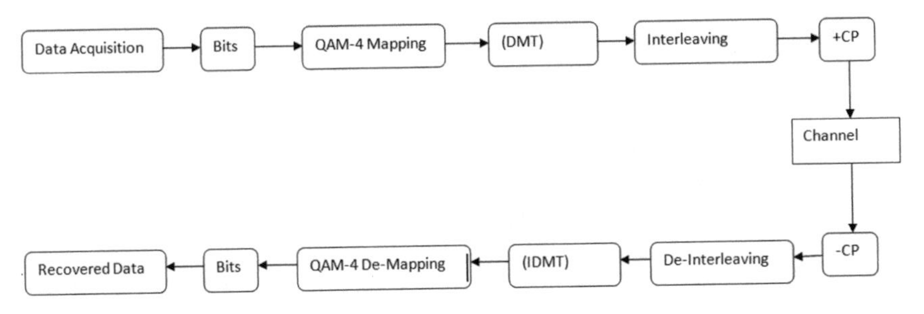

Fig. 4 Square outline of an OFDM framework comprising a transmitter, a PLC channel, and a collector

auto-affiliate work and the second-request cyclostationary process, comprises consonant pinnacles, and this may, therefore, be used for the finding and excavation of this type of system [2]. The second-request noise estimates revealed a direct minimal average square error that reflects area equalizers for single-transporter orthogonal frequency multiplexing (OFDM) architectures. SBL was first proposed and introduced by Wipf and Rao for sparse recovery signals. In general, SBL is a Bayesian learning strategy to dealing with the direct repetition problem.

The largest a posteriori (guide) of w is the average μw behind. Because of the sparse propellant aspect of the previous noise, most parts of α and so μw are made to beeline for zero, thus providing an improper measurement of w. It has proven that SBL has changed the quality of other compact detection computations like premise pursuit and FOCUSS, as the perfect overall is the most sparse course of action, all the excellent local game plans are insufficient and the amount of optima closely reduced (Fig. 4).

2.3 Estimation Using Null and Pilot Tones

This approach requires frequency estimates to pick the p-DOA for the transmitted p-signals to minimize interference from other signals on each DOA. Our instance suffers from lowest interference from others to project any IN sequence interlinked with the generated subspace into the temporal domain. The frequency evaluation function might be specified (3). Since there were found p in the number of interleaved IN sequences, α values of p_i matching to the major function values were found, and α values were viewed as vector α μ of p_i in length. Consequently, all observed IN interleaved sequences are supported by IN support.

2.4 Estimation Using All Tones

The number of S segments in this support estimate is dependent on being more than the interlocking p IN sequences. In general, previous knowledge about the IN width might be used, but the width to estimate the number of S that should be used is not known in advance [6]. It is also expected, that all S-section subcarriers are used for noise estimation in LS as null subcarriers. In short, with fewer segments or fewer items in each segment, the accuracy of the IN support estimate will be reduced. Consequently, the support of the IN candidate is increased by a factor of two based on the frequency evaluation role. For example, when the preliminary estimate of p_1 is p_1, the first two p_1 α values are selected that match the most significant estimate function. If an OFDM symbol has less IN interference, the primary p_1 is predicted to be smaller and more accurate. Therefore, we choose only twice the size of a supporting candidate. Conversely, it is anticipated that the preliminary p_1 estimate will be more significant and less precise when IN interferes severely. The support for IN estimations is twice the original size, such that no real IN support is overlooked.

3 Results and Discussion

To go at this figure, we replicate a complex baseband multiplexing system of the orthogonal frequency division [7]. We utilize the sparse learning estimate as a whole as the sequential figure of the SBL is indistinguishable. We experiment with our proven figure in a multifarious structure and complexity with their performance at bit error rates and the compressed recognized estimates in asynchronous impulsive noise. To look at the demonstration of the mentioned invariable frameworks to the variable ones, we likewise execute the two minimum mean square estimation markers, since they are flawless in the minimum mean square estimation sense among other varying strategies, for example, resetting and cutting all estimations are imitated with and without the convolutional code (Figs. 5, 6 and 7).

4 Conclusion

This paper presents three approaches for refining correspondence execution in a synchronous and asynchronous impulsive noise of the orthogonal frequency division multiplexing line communication systems. We implement sparse learning (SBL) structures to measure the rash agitating impact of the got data by watching info either on the invalid and reference subcarriers or above all subcarriers. Beneath sudden incidental change, the technique involves handling a period space interleaving orthogonal frequency division multiplexing handset formation to divide long blast impacts the length of different orthogonal frequency division multiplexing pictures

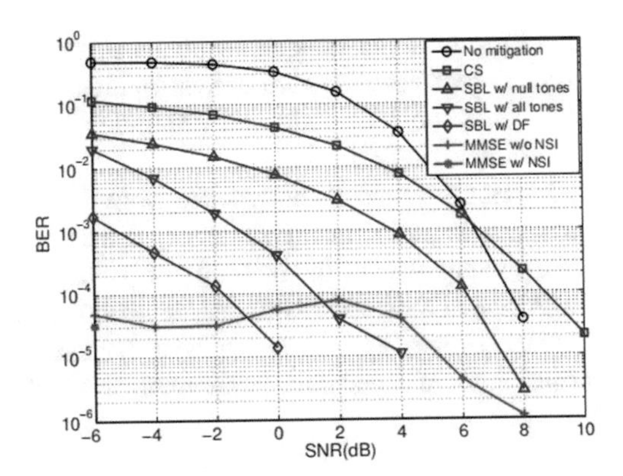

Fig. 5 BER execution of the proposed calculations in Gaussian blend displayed asynchronous impulsive noise, in examination with the customary

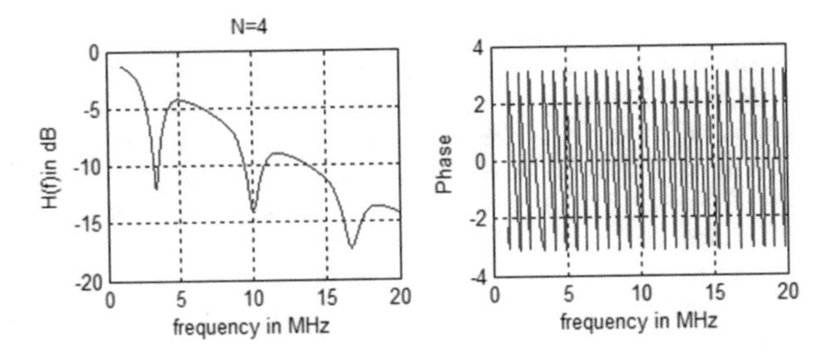

Fig. 6 PLC channel model gain and phase plot, $N = 4$

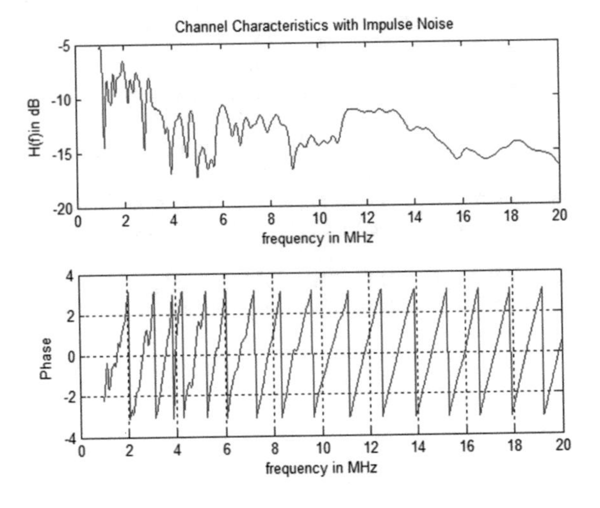

Fig. 7 PLC channel model gain and phase plot, $N = 15$

into short impacts, and like this apply the deficient learning systems. All structures remain unchanged; i.e., no predefined system or system breakpoint data is needed. We grasp the intended count subject to odd motivation commotion and rare drive clack from various measurable models.

This paper reviews smart grid and its emergence from the existing power grid while enlisting its advantages and future challenges. Specific industry standards claim promising data throughputs over the last mile of powerline communication. Homeplug Powerline Alliance Standards is a leading group specializing in home PLC products and services. Homeplug releases series of standards named Homeplug 1.0, turbo, AV, offering physical layer data throughput of 14, 85 and 200 Mbps, respectively, within a home network.

References

1. Berger LT, Schwager A, Escudero-Garzs JJ (2013) Power line communications for smart grid applications. JECE 2013:1–16
2. Li B, Tucci M (2015) Member, IEEE and Marco Raugi "impulsive noise mitigation with interleaving based on MUSIC in power line communication. J Latex Class Files 14(8)
3. Ndo G, Labeau F, Kassouf M (2013) A Markov-Middleton model for bursty impulsive noise: modeling and receiver design. IEEE Trans Power Del 28(4):2317–2325
4. Lin J, Nassar M, Evans BL (2011) Non-parametric impulsive noisemitigation in OFDM systems using sparse Bayesian learning. In: Proceedings of the IEEE conference on global communication
5. Shrestha D, Tonello A, Mestre X, Payar M (2016) Simultaneous cancellation of narrow band interference and impulsive noise in PLC systems. In: IEEE international conference on smart grid communications, Sydney, NSW, 2016, pp 326–331
6. Al-Naffouri TY, Quadeer AA, Caire G (2014) Impulse noise estimation and removal for OFDM systems. IEEE Trans Commun 62(3):976–989
7. Zimmermann M, Dostert K (2002) A multipath model for the powerline channel. IEEE Trans Commun 50(4):553–559
8. Anastasiadou D, Antonakopoulos T (2005) Multipath characterization of indoor power-line networks. IEEE Trans Power Del 20(1):90–99
9. Shrestha D, Mestre X, Payar M (2016) Asynchronous impulsive noise mitigation based on subspace support estimation for PLC systems. Int Symp Power Line Commun Appl Bottrop 1–6
10. Smith Gayathri D, Usha Rani K (2019) Adapting the effect of asynchronous impulsive noise using interleaving based on compound signal classification in broad Plc. Int J Innov Technol Explor Eng 9(2)

Technological Perspective on Precision Medicine in the Context of Big Data—A Review

C. Mallika and S. Selvamuthukumaran

Abstract This paper focuses on the recent developments in precision medicine (PM) and health informatics in the context of big data. PM aims to integrate a complete dataset gathered continually about a patient's lifestyle, surrounding conditions influence on an individual, and genetics to propel scientific knowledge in these areas of diagnosis, intervention, drug management, and assure provision of suitable treatments. The proliferation of biomedical big data will impose many issues in terms of data security, privacy, and governance. Resolving these issues will discover the efficient and secure application of data in the medical sector. This article explores the prior research works and future openings associated with big data for precision medicine and delineating some of the key challenges that need to be resolved. This paper also illustrates how advanced computational platforms and tools could be employed for converting billions of datasets into diagnostics and accurate decisions for precision medicine with improved clinical quality.

Keywords Big data · Data integration · Precision medicine · Data analytics · Processing pipeline · Omics data

1 Introduction

In the current epoch, the torrents of data generated by sources including computational health informatics, medical information systems, clinical decision support systems, therapeutic Websites, computer-aided diagnosis, and electronic health records rise rapidly from terabytes (10^{12}) to zettabytes (10^{21}) or yottabyte (10^{24}) scale. Therefore, it is not surprising that big data and healthcare informatics are

C. Mallika (✉)
Department of Master of Computer Application, E.G.S. Pillay Engineering College, Nagapattinam, India
e-mail: cmallikachinna@gmail.com

S. Selvamuthukumaran
Department of Computer Applications, A.V.C. College of Engineering, Mannampandal, India
e-mail: smksmk@gmail.com

A. Kumar et al. (eds.), *Proceedings of the International Conference on Cognitive and Intelligent Computing*, Cognitive Science and Technology,
https://doi.org/10.1007/978-981-19-2350-0_54

jointly studied. Big data in the clinical system is irresistible not only because of its massive amount of data in an unprecedented scale but also on account of complexity, diversified data formats, the rich context of data being produced, and the pace at which it must be manipulated. According to common search engines, the usage frequency of big data is doubling every year. About 90% of the data in the world today have been produced in the past two years. It is also expected that the developments of the Internet of things (IoT) will produce twice the amount of data over every twelve hours instead of every twelve months. The velocity of data production is escalating across various platforms; by 2025, omics data, which are playing a vital role in medical and healthcare informatics, will create the uppermost volume of data [1]. The exponential growth in data creation needs an equally essential ability for data collection, storage, retrieval, and analytics; nevertheless, this information technology system has not developed as fast a pace—really.

Big healthcare data refer to clinical datasets that are very large, very fast, and very difficult for medical service contributors to analyze and apply further procedures or extract results using traditional data processing applications. The proliferation of electronic health records (EHRs) of individuals and assimilation of social media, various omic technologies with eHealth, mHealth, telehealth, and smart health technologies reconciling with behavioral science, clinical genomics, and socio-economic status have led to the evolution of novel infrastructures and the transformation of healthcare systems for the PM and improved health services. Big data promise more PM for individuals with superior diagnostic accuracy, and psychoanalysis combined to their characteristics and needs, genetics, environmental risk, drug discovery, and accurate disease phenotype. This assurance comes from the information gathered from sources such as bio-specimens, sensor informatics, medical images, and EHRs, from which disease-specific parameters, the pattern of interest among the sources, and relations can be statistically identified and employed to determine therapies appropriate to the patient [2]. The acquisition of these data into networks affords greater knowledge of health and disease management with the objective of providing the appropriate therapies at the right time to the right patient.

From a classification perspective, big data are widely categorized into three main types: *structured, unstructured,* and *semi-structured*. Structured data have a high level of structured format, which makes it conform to depict within the rows and columns of a data matrix. They are usually stored in relational databases in an ordered manner. Since structured data are represented by a data matrix in a database, it is easy to enter, store, query with a variety of search algorithms, and evaluate by means of computers. Basically, semi-structured data are the structured data but it is unorganized. Data from telehealth technology, wireless sensor devices are examples of semi-structured data in the medical industry. In contradiction of structured data, unstructured data have no definite association, and it may have its own organization, it does not amenable to data matrix and is very difficult to store in databases. Physicians' notes and medical images are examples of unstructured data. While these data converted into a structured format, some information is certainly lost. It has been expected that about 80% of the data in the medical system is unstructured [3]. The heterogeneity of these data makes data collection and analysis more complex.

Big data are described according to its five important characteristics (5Vs): *volume, velocity, variety, veracity*, and *value*. Obviously, the *volume* of data is a significant feature of big data. In fact, the US healthcare organizations alone already produced 150 exabytes (10^{18}) [4], and this data will go on increase in future. The countries with the highest populations and promising economies such as China and India will be dealing with zettabytes and yottabytes. *Velocity* denotes how often data updates, or to rise over time [5]. Data that update in real-time, or near real-time, have provided additional value as it can aid investigators and medical practitioners to make clinical decisions that provide strategic benefits. Velocity is a mainly imperative attribute of population-based datasets as gathering and manipulating data in near real-time can provide better knowledge about infectious diseases and outbreaks.

Veracity is also a significant factor for big data, for instance, personal health records (PHRs) may have cryptic notes, abbreviations, and typographical errors. Ambulatory readings may not be accurate due to less reliable, unmanaged circumstances related to clinical quality measures, which are derived by skilled physicians. The use of impulsive and uncontrolled information can initiate erroneous decisions as its context is not always identified properly. Real *value* to both medical applications and patients can only be achieved if issues in big data analytics can be tackled in a consistent manner. The total estimated healthcare expenditure in the US by 2022 will create 19.9% of the gross domestic product (GDP) [6] while the estimated healthcare contribution of the GDP in the UK is anticipated to reach 6.4% by 2021 [7]. In these respects, if utilized correctly, big data can be valuable resources that can offer important insights for evading unnecessary therapies, reducing adverse effects of drugs, decreasing healthcare expenses, and finally delivering much more efficient and effective healthcare services.

In this article, we review major developments that happened over recent years in constructing clinical infrastructures, particularly in precision medicine to organize data collection, data management (i.e., storage and sharing) and data analytics (i.e., analyzing, searching, and decision support system) as a complete clinical solution for the provision of enhanced medical services. We begin by explaining about big data and its characteristics. In Sect. 2, we discussed about precision medicine. Sect. 3 discusses some of the impacts and potential of big data in health and medical sciences. In Sect. 3, the processing pipeline of PM is provided. Section 4 depicts the technical challenges facing big data in the precision medicine model. Finally, we conclude this study in Sect. 5.

2 Precision Medicine

According to US Food and Drug Administration (FDA), precision medicine is defined as "the tailoring of medical treatment to the individual characteristics, needs, and preference of a patient during all stages of care, including prevention, diagnosis, treatment, and follow-up" [2]. In precision medicine, extensive clinical laboratory testing is implemented for deciding proper and optimal treatments depends on the context of a patient's genetic information. The extraordinary breadth, depth, and volume of data, which can now be composed drove US President Obama's proposal of the Precision Medicine Initiative in 2015 [8] and will eventually facilitate the objective of PM by individuals surrounded by a cloud of data. In this model, a patient's distinctive characteristics are employed to combine therapy in a particular way that might be more sophisticated than the typical approaches. For instance, nowadays cardiologists apply an algorithm that for a patient envisages the occurrence of a myocardial infarction within 10 years derived from body weight, arterial pressure, blood lipid analysis reports, smoking status, and individual as well as family medical records. PM can be employed in the analysis and prevention of disease, including cancer, due to developments in next-generation sequencing (NGS), liquid biopsy methods, high-throughput functional screening, computational techniques, and analytical methods.

3 Big Data Framework for Precision Medicine

The computational framework of big data in precision medicine comprises of a sequence of phases. Every phase in this pipeline plays a significant role in depicting competent and precious results of big data analytics. The gathering, storing, and sharing phases produce suitable input for the following phases (i.e., analyzing, searching, decision support), where different tools are employed to determine meaningful patterns for making apt and useful decisions. The subsequent sections will explain this processing pipeline in detail.

3.1 Data Collection

The detection and applications of useful patterns in datasets have aided to bring remarkable achievements in several organizations. Nevertheless, the utilization of analytics to enhance results in the precision medicine has not acquired much more attention owing to the complexity of gathering voluminous biomedical data, which is the very first phase of the pipeline. Data in precision medicine are categorized by several different sources (from both external and internal sources), different structures (e.g., database records, flat files, etc.), and diverse sites (either physically distributed devices or multiple sites) [9]. Taking the potential benefits of big data means gathering large size of data produced in various dynamic formats rapidly. Still, there are some problems such as scalability and latency. High scalability and low latency are highly preferred characteristic for stream processing. Apache Hadoop combined with MapReduce model and other data integration software could enable parallel processing environment to provide efficient data collection. Of late, Hadoop Spark [10], a descendant model that is more prevailing and flexible than MapReduce model, is receiving increasing attention owing to its iterative computation, queries, lower-latency, and real-time processing.

3.2 Data Storing

Novel technologies and the explosive growth of EHRs are generating large amounts of digital clinical data with increased complexity, from different medical images to clinical notes, laboratory reports, and so on. It is important to deliver valuable outcomes for economical storage and management. More particularly, there are many different significant characteristics for a preferred PM data storage and management [11]:

- Scalability: In order to lodge massive datasets, the storage medium ought to be incrementally scalable.
- Availability: High availability is preferable for medical researchers and clinicians to retrieve records rapidly, securely, and reliably anywhere, anytime. Reliable and fast access to patient's data saves precious time for medical professionals and allows them to create responses and provide immediate care, which can decide the difference between life and death.
- Cost-effective: Storing and processing big data could be redundant and intricate. Good storage architecture should decrease the complexity and cost of data storage. It should give security to the data without degrading its performance.

As stated before, medical images and sensors are the main sources of PM. In order to meet the unusual requirements of these data, picture archiving and communication system (PACS) technology is introduced, which has the benefits of cost-effective storage and suitable retrieval [12]. With the explosive growth of big data analytics in PM, several corporate companies have designed enterprise-level solutions for data storing and management. For instance, NetApp [13] introduces PACS solutions and EHR for decreasing complexity in storage management. Furthermore, Intel Distribution for Apache Hadoop platform gives distributed statistical models for storing, preprocessing, formatting, and normalizing unstructured data, which are normally time-consuming and hard to store and process. It realizes scalability by manipulating each record individually as well as in parallel.

3.3 Data Sharing

After collecting healthcare data, secure data sharing among researchers and medical professionals across medical industries over institutional, regional, or even nationwide is problematic [13]. There are several issues related to data sharing in the clinical system, for example: (i) the informal utilization of various data structures and methods in various platforms and systems; (ii) the promise of the controlled data sharing through secure systems. Recently, big data have been omnipresent due to technological advances in the cloud platform. By assuring server management, storage capacity, bandwidth utilities, and network power, cloud technology will facilitate synchronization of data storage with medical instruments so that the data being produced automatically streams into storage medium for facilitating data sharing. PACS-on-demand system is used to enable the sharing of healthcare information among different entities. Through storing and sharing healthcare data with the cloud, it decreases the necessity to invest in IT capacity and enabling competent as well as secure cooperation with radiology professionals and allied practices. This system also facilitates information sharing among healthcare providers and other stakeholders, which frequently have different data formats that are unable to organize to make proper clinical decisions. The researchers also propose many secure and sensitive information sharing methods on big data domains, which will help organizations to provide cost-effective PM as well as value-added healthcare services.

3.4 Data Analyzing

Big data instigated an explosion in the implementation of more extensive data mining approaches in the healthcare system. This section will discuss the important steps of data analyzing phase.

3.4.1 Preprocessing

Original medical data are skewed, heterogeneous, and noisy in nature. Therefore, there is no feasibility to apply an analytical procedure to the unprocessed data directly owing to its complexity and heterogeneity. Hence, to enhance data quality and make them for further analysis, data cleaning, missing data interpolation, synchronization, normalization, and data formatting processes are important. An archetypal preprocessing algorithm in the healthcare system generally comprises of following steps based on the type of data format and data sources [14]:

- Data cleaning: This step used to eliminate noise from raw data. The raw data may comprise of frequency noise and artifacts. Usually, thresholding techniques are applied to eliminate incompliant data [15], and electronic filtering techniques are frequently employed to eliminate frequency noise in biosignals [16].
- Interpolation: Missing data may be due to the intentional causes (e.g., transportation of patients) or unintentional causes (e.g., sensor malfunctions). Single missing data due to sensor malfunctions are interpolated by its prior and subsequent values [15].
- Synchronization: Sensor signals are gathered at various data rates with timestamps depending upon their internal clocks [14].The data from multiple sensors have to

be aligned in time. Therefore, it is essential to make some valid assumptions and develop alignment procedures to achieve synchronicity across the sensor clocks.

- Normalization: Normalization is necessary to manage the disparity in the data recording procedure [14]. For instance, a daily heart rate may denote average rate per day or heart rate in a particular period. Moreover, the normalization is used to convert the raw dataset into an equivalent code set by using standardized format.
- Formatting: All analytical methods need data in a particular structure. Thus, there is a need to convert the raw clinical data into a machine-readable form. For instance, the data ought to be stored in the attribute-relation file format (ARFF) and the data characteristics should be defined in an identifiable way in the WEKA tool [17].

3.4.2 Feature Selection

The process of selecting and extracting a subgroup of significant and appropriate features from a huge dataset is known feature selection or extraction. Generally, raw clinical data may have redundant features, which can be converted into a small set of appropriate features by means of dimensionality reduction methods. The implementation of feature selection algorithms for multifaceted and massive data is statistically costly, and they have space and time limitations. High-dimensional data in the healthcare system may cause low accuracy and inefficiency. To resolve this problem, several investigators propose different feature selection algorithms (FSA) in the medical system [18]. Recently, a number of FSA handling big data have been proposed. As compared to traditional FSA, online feature extraction is more appropriate for voluminous datasets in such a manner that every feature is processed upon its arrival, and every time, the best feature set is selected [19]. Tan et al. [20] also suggest a novel FSA for large datasets. This approach first converts the problem into a convex semi-infinite programming (SIP) problem. Then, a new feature generating machine (FGM) is used to resolve this issue. The FGM continually selects the most appropriate features by means of multiple kernel learning (MKL) sub-problem.

3.4.3 Machine Learning

At present, machine learning techniques (MLTs) have been employed in various domains such as autonomous vehicle, audio processing, and detection of fraudulent credit card activity. MLT came into focus of big data analytics due to their important aptitude to integrate heterogeneous and large-scale clinical datasets, one of the primary issues in PM. Usually, MLT is divided into two classes [21]: (i) predictive or supervised learning and (ii) descriptive or unsupervised learning. Predictive methods, such as classification and regression, use historical data (training data). They map input training data samples with known class labels. The prediction rules are derived from training dataset to calculate hidden class labels, and then, it is used in the classification of the new patient's data. The descriptive methods, such as dimensional reduction and clustering, aim to discover the structure of unlabeled samples based on their similarity. The descriptive model is learned by revealing unknown patterns in the input and arranging the data samples into meaningful subsets.

Classification is a predictive learning method for finding out the type of input data based on the known training samples. It classifies data samples by ordering them in a tree. Decision tree-based classification method has been effectively implemented in precision medicine models [22]. Clustering is an unsupervised learning approach used to categorize a group of datasets into clusters (subgroups) in such a way that the datasets in a cluster are more similar to each other related to those in other clusters [23]. There are three different clustering algorithms as (i) *partitioning algorithms* divide the input datasets into clusters in a way that each object is found in only one cluster [24], (ii) *hierarchical algorithms* develop a hierarchy called as a dendrogram [25]. Bottom-up (agglomerative) [26] and the top-down (divisive) [27] are two common approaches for hierarchical clustering. (iii) *Density-based clustering* manages dynamic and multifaceted inputs and determines outliers and random shapes particularly in medical images. But, it is slow in big data environment. Hitherto, several investigations have been conducted on MLTs to delineate the pros and cons of MLTs in medical sciences [22]. It is noteworthy that there is an important overlap amid MLT and mining techniques as both are used in big data analytics.

3.5 Searching

Medical professionals utilize analyzed data to extract meaningful patterns of interest, which helps them to achieve better treatment through more precise diagnosis. Usually, mining techniques are used to search and explore big data to identify systematic and consistent patterns. The heterogeneous nature of data is the main challenge in big data mining. Text mining is a helpful information retrieval technique broadly employed in the clinical system. The main problem in text mining is that the physician's notes are unstructured. In order to resolve this problem, different methods including natural language processing [28], content-based image retrieval [29], and composite anchor graph hashing algorithm [30] are used in the literature.

3.6 Decision Support System

Clinical decision support system (CDSS) is a system for increasing the trait of clinical services. It assists patients and medical professionals to take better decisions. CDSS has made great contributions to reduce healthcare errors and improve clinical care [31]. Medical systems can leverage new techniques to provide better CDSS. At present, CDSS is a vital issue and important since it increases the effectiveness of the clinical system. Using big data frameworks, the issues in CDSS have been resolved, and dynamic knowledge-based algorithms have been developed to implement more sophisticated frameworks for the CDSS.

4 Research Issues in Big Data Related to Precision Medicine

Medical industries are not exposed much with the highly developed information technological aspects. They implement the same protocols for years. This state of mind has formed a rigid culture, and it is too hard to bring novel practices in clinical systems. The precision medicine also poses new challenges for clinicians, policy-makers, researchers as well as the public, as it establishes a new type of personal data in the medical system. Traditional methods for data storage, management, retrieval, and analysis are not enough for the zettabytes of biological data produced every year [32]. Furthermore, as datasets become more heterogeneous and complex, advanced distributed database management system and processing techniques are required to make the data valuable. In addition, result reproducibility and data sharing policies continue to be vigorously debated issues. Most of the clinical systems, for example, emergency units, diagnosis sections, case management systems, and clinical laboratories, etc., all have got their own approaches for gathering clinical data. Even today most of them use paper charts, paper folders and fax to share the data.

- *Data Heterogeneity*: Due to the proliferation of EHR, the integration of old and modernized systems is becoming a difficult endeavor. It is also difficult to store heterogeneous data into a standardized storage system.
- *Legacy Infrastructures*: Present clinical systems are comprised of legacy systems which cannot cope up with the highly developed technologies. This issue is resolved with the help of appropriate middleware which makes the compatibility between legacy systems and advanced technologies before processing the data.
- *Poor data quality*: In the clinical system, another problem is a deficiency of nursing facility and high quality of service with low cost. Manual data entries have bit error, and hard to retrieve different data types. Erroneous, missing, or inadequate data provide useless or wrong healthcare information. When this wrong information is linked with a clinical system, it may cause severe irrecoverable consequences. In order to make the right decision at the right time, data quality is imperative.
- *Data privacy and security*: Protecting and preserving the patient's sensitive information are a big challenge as it transmits over the unsecured channel. Hackers may alter these data which results in misdiagnosis, or imprecise evaluation of illnesses causes wrong treatment and consequently increases the death rate. Transferring medical information through IoT devices is vulnerable to security issues.
- *Data storage*: The conventional storage system is difficult to realize in big data analytics; so the cloud computing architecture such as Amazon EC2, Amazon S2, and Elastic Block Store provides a solution to the big data analytics by delivering boundless storage space with appropriate fault tolerance techniques. Besides

storage, an additional issue is transmitting big data to cloud system with high speed and low cost. Hence, the major issues are information storage, retrieval, and mining useful patterns by utilizing several algorithms.

5 Conclusion

Precision medicine plays an important role in the present and future healthcare system. This article presents an overview of major developments that happened over recent years in constructing clinical infrastructures particularly, in precision medicine to organize data collection, management, and analytics as complete clinical outcomes for the provision of enhanced medical services. This paper also illustrates how advanced computational platforms and tools could be employed for converting billions of datasets into diagnostics and accurate decisions for precision medicine with improved clinical quality. We have summarized the characteristics of big data in the context of precision medicine. We discussed about impacts and potential of big data in medical sciences. The data processing pipeline of big data for precision medicine is discussed. The healthcare informatics will definitely benefit from the big biomedical data in the near future.

References

1. Stephens ZD, Lee SY, Faghri F et al (2015) Big data: astronomical or genomical? PLoS Biol 13:e1002195
2. Aiyegbusi OL, Macpherson K, Elston L et al (2020) Patient and public perspectives on cell and gene therapies: a systematic review. Nat Commun 11:6265. https://doi.org/10.1038/s41467-020-20096-1
3. Kong HJ (2019) Managing unstructured big data in healthcare system. Healthcare Inform Res 25(1):1–2. https://doi.org/10.4258/hir.2019.25.1.1
4. Pastorino R, De Vito C, Migliara G, Glocker K, Binenbaum I, Ricciardi W, Boccia S (2019) Benefits and challenges of big data in healthcare: an overview of the European initiatives. Eur J Public Health 29(3):23–27. https://doi.org/10.1093/eurpub/ckz168
5. Bazzaz Abkenar S, Haghi Kashani M, Mahdipour E, Jameii SM (2021) Big data analytics meets social media: a systematic review of techniques, open issues, and future directions. Telematics Inform 57:101517. https://doi.org/10.1016/j.tele.2020.101517
6. Dash S, Shakyawar SK, Sharma M et al (2019) Big data in healthcare: management, analysis and future prospects. J Big Data 6:54. https://doi.org/10.1186/s40537-019-0217-0
7. Public Expenditure Statistical Analyses, HM Treasury, London, U.K., 2012
8. Precision Medicine Initiative. The White House; 2015. Accessed 17 Dec 2015
9. León A, Pastor O (2021) Enhancing precision medicine: a big data-driven approach for the management of genomic data. Big Data Res 26:100253. https://doi.org/10.1016/j.bdr.2021.100253
10. Matei Zaharia, Mosharaf Chowdhury, Michael J. Franklin, Scott Shenker, and Ion Stoica. 2010. Spark: Cluster computing with working sets. In: Proceedings of the 2nd USENIX conference on hot topics in cloud computing, pp 10–11

11. Lysaght T, Ballantyne A, Toh HJ, Lau A, Ong S, Schaefer O, Shiraishi M, van den Boom W, Xafis V, Tai ES (2021) Trust and trade-offs in sharing data for precision medicine: a national survey of Singapore. J Personalized Med 11(9):921. https://doi.org/10.3390/jpm11090921
12. Khaleel HH, Rahmat ROK, Zamrin DM (2019) Components and implementation of a picture archiving and communication system in a prototype application. Rep Med Imaging 12:1–8. https://doi.org/10.2147/RMI.S179268
13. NetApp (2011) http://www.netapp.com/us/solutions/industry/healthcare/
14. Crowne A (2014) Preparing the healthcare industry to capture the full potential of big data. http:// sparkblog.emc.com/2014/06/preparing-healthcare-industry-capture-full-potential-big-data/. Retrieved 25 Feb 2015
15. Hassler A, Menasalvas E, García-García F et al (2019) Importance of medical data preprocessing in predictive modeling and risk factor discovery for the frailty syndrome. BMC Med Inform Decis Mak 19:33. https://doi.org/10.1186/s12911-019-0747-6
16. Apiletti D, Baralis E, Bruno G, Cerquitelli T (2009) Real-time analysis of physiological data to support medical applications. IEEE Trans Inform Technol Biomed 13(3):313–321
17. Frantzidis CA et al (2010) On the classification of emotional biosignals evoked while viewing affective pictures: An integrated data-mining-based approach for healthcare applications. IEEE Trans Inf Technol Biomed 14(2):309–318
18. Hall M, Frank E, Holmes G, Pfahringer B, Reutemann P, Witten IH (2009) The WEKA data mining software: an update. ACM SIGKDD Explor Newsletter 11(1):10–18
19. Hegde SK, Mundada MR (2021) Hybrid generative regression-based deep intelligence to predict the risk of chronic disease. Int J Intell Comput Cybern ahead-of-print No. ahead-of-print. https://doi.org/10.1108/IJICC-06-2021-0103
20. Li HG, Wu X, Li Z, Ding W (2013) Online group feature selection from feature streams. In: 27th AAAI conference on artificial intelligence. Citeseer, pp 1627–1628
21. Tan M, Tsang IW, Wang L (2014) Towards ultrahigh dimensional feature selection for big data. J Mach Learn Res 15(1):1371–1429
22. Sarker IH (2021) Machine learning: algorithms, real-world applications and research directions. SN Comput Sci 2:160. https://doi.org/10.1007/s42979-021-00592-x
23. Shaikhina T, Lowe D, Daga S, Briggs D, Higgins R, Khovanova N (2019) Decision tree and random forest models for outcome prediction in antibody incompatible kidney transplantation. Biomed Signal Process Control 52:456–462. https://doi.org/10.1016/j.bspc.2017.01.012
24. Alashwal H, El Halaby M, Crouse JJ, Abdalla A, Moustafa AA (2019) The application of unsupervised clustering methods to Alzheimer's disease. Front Comput Neurosci 13:31. https://doi.org/10.3389/fncom.2019.00031
25. Zolfaghar K, Meadem N, Teredesai A, Basu Roy S, Chin SC, Muckian B (2013) Big data solutions for predicting risk-of-readmission for congestive heart failure patients. In: 2013 IEEE international conference on big data, pp 64–71
26. Nithya N, Duraiswamy K, Gomathy P (2013) A survey on clustering techniques in medical diagnosis. Int J Comp Sci Trends Technol (IJCST) 1(2):17–23
27. Belciug S (2009) Patients length of stay grouping using the hierarchical clustering algorithm. Ann Univ Craiova-Math Comp Sci Ser 36(2):79–84
28. Chipman H, Tibshirani R (2006) Hybrid hierarchical clustering with applications to microarray data. Biostatistics 7(2):286–301
29. Hameed IM, Abdulhussain SH, Mahmmod BM, Pham DT (Ed) (2021) Content-based image retrieval: a review of recent trends. Cogent Engineering 8:1. doi: https://doi.org/10.1080/23311916.2021.1927469
30. Cinar Akakin H, Gurcan MN (2012) Content-based microscopic image retrieval system for multi-image queries. IEEE Trans Inform Technol Biomed 16(4):758–769
31. Mallika C, Selvamuthukumaran S (2021) A hybrid crow search and grey wolf optimization technique for enhanced medical data classification in diabetes diagnosis system. Int J Comput Intell Syst 14:1–18, Impact factor: 1.736

32. Mallika C, Selvamuthukumaran S (2021) Hybrid online model for predicting diabetes mellitus. Intell Autom Soft Comput 31(3):1873–1885, Impact factor: 1.647
33. Liu W, Wang J, Kumar S, Chang SF (2011) Hashing with graphs. In: Proceedings of the 28th international conference on machine learning (ICML'11), pp 1–8

Fundamental and Technical Challenges in Software Testing and Their Applicability in Small Software Companies

Irfan Ahmad Khan and Dipti Kumari

Abstract Software testing is the final step of any product before the deployment or market launch. Through testing, we can tune the requirements of the client with the final product. Testing helps us to know the gap between the user's requirement and the product developed by the programmers. Testing is a costly procedure and is used to authenticate new software. Fundamental and technical are the main challenges in which we need to verify and validate the outcome of the software objective. The key objective of this paper is to check the ground reality of the testing challenges (Fundamental and Technical) and their applicability in small software development companies. After analyzing and studying both types of challenges, we surveyed to find the ground reality of these two categories. We contacted a total of 50 software development companies, all these companies working for small projects and belongs to ten different states of India. Many reforms like technical and administrative training, infrastructure, working in the specified area of development and re-engineering and re-defining the business & working environment and most important competent team are the main needs of this sector to overcome with all these issues.

Keywords Technical challenges · Fundamental challenges · Software testing · Small software companies

1 Introduction

Quality control is widely used in industrial practice by adopting different stages and types of testing. To handle testing, there are several tools and testing techniques are available which ultimately enhance the quality of the software product. Testing is

---

I. A. Khan (✉)
Research Scholar, Department of Computer Application, RKDF University, Ranchi, India
e-mail: irfan16143@gmail.com

D. Kumari
Assistant Professor, Department of Computer Science and Engineering, RKDF University, Ranchi, India
e-mail: dipti5679@gmail.com

A. Kumar et al. (eds.), *Proceedings of the International Conference on Cognitive and Intelligent Computing*, Cognitive Science and Technology,
https://doi.org/10.1007/978-981-19-2350-0_55

the method of analyzing a program to discover errors. Once testing is completed positively, it will eliminate all the faults from the product. For the excellent performance of the product, software should be zero error. Testing is a method, to assess the capability of a software application with a commitment to discover whether the program met the user requirements or not and to recognize the faults to confirm that the product is faultless to develop an excellent program. As mentioned above that this is the final step of the complete developmental cycle. Every phase of SDLC plays an important and prominent role, but due to the last step, it becomes more valuable. Some reasons are listed below which make testing more significant in the entire software development process.

- User satisfaction
- Product excellency
- Cost efficiency
- Security concern.

If we summarized the entire detail of testing, we can say testing is the process of verification and validation, and testing is important for its scalability.

2 Problem Statement

Testing is the final and one of the important phases in SDLC before the market release. Testing is applicable for the scalability of the software. During this phase, what are the contribution of fundamental and technical challenges and how it has been handled has been discussed in the paper.

3 Research Objectives

The key objective of this paper is to check the ground reality of the testing challenges (Fundamental and Technical) and their applicability in small software development companies.

4 Methodology

Testing is a process through that we can scale the quality. Fundamental and technical are the main challenges in which we need to verify and validate the outcome of the software objective.

5 Elements of Testing

Test strategy, testing plan, test cases, test data, and testing environment are the basic elements needed for the successful testing of software [1]. These five fundamentals play the main role in the entire testing if any of these are missing or not considered during the testing your result could be different than your main objectives.

- Test Strategy: The strategy of the test lets you know what type and amount of testing you needed.
- Test Plan: To execute the test strategy, you need to make a plan.
- Test Case: It is a detailed example that must be prepared in advance to meet the requirements.
- Test Data: While executing a test case, you need to consider both types of data, i.e., input data and DBMS data.
- Test Environment: Environment factors in the entire developmental phase coming under the category of non-technical factors [2]. A comfortable and friendly environment plays an important role that needs to carry during the entire phase. This eventually needed for a better outcome.

6 Aim of Software Testing

The main aim of any software testing is to find glitches and repair them to expand the quality of the product. Normally, testing of any software product consumes a handsome percentage of the entire budget. The testing aim can be categorized into following [3]:

- Product Demo: It shows processes under distinct conditions and displays that software is ready for use.
- Prevention: It gives evidence to avoid or decrease the number of faults, simplify system performance, recognize ways to escape the threat and future glitches.
- Investigation: It determines flaws, errors, and absences. It also outlines system abilities, boundaries, quality of modules.
- Quality Improvement: We can reduce faults and increase product quality by performing actual testing.

7 Fundamental Challenges

In this section, we will focus on the different fundamental challenges which need to be in consideration when we move toward the testing phase of the SDLC. If we know the challenges in advance, we can make proper arrangements to counter the issue. Fundamental challenges are those points that do not include the technical side; it is the logical side. According to "Kaner," fundamental challenges can be categorized into four major groups [4]. These groups and their details are listed below.

7.1 *Complete Testing is not Possible*

Almost, it is impossible to go with complete testing or extensive testing because:

- Sometimes, it is not possible to generate all probable implementation situations of the system. This becomes more important when the performance of the program depends on real data like weather altitude and temperature and so on.
- The strategy matters may be too difficult for an entire test. For example, maybe static variables are used by the programmer for controlling the execution of the program.
- The entire program may have several situations which act in different states, and timing is important for the different inputs. Input value and time need to be in consideration.
- The input program of a domain may be too large for use in the testing system.

7.2 *Process Myths*

The software industry is comparatively new although it is having some myths and fallacies which end with many mystifying nodes on the client as well as the entire development team. These assumptions are unsafe. The main intention to mention these myths in this paper is to let the people know some more about the fundamental challenges which are associated with the non-technical side. Some of the process myths are listed below.

- Following specific SDLC model for development.
- Collection of client requirements at the beginning of the project and stay with the same throughout the development time.
- Connecting the client's behavior with the requirements of the client.
- Fixing the bugs on any specific level and never changing that level to avoid any other bugs.
- According to Software Development Academy London, U.K., some more myths are listed below while keeping in the consideration of process myths [5].
- Big team size giving better result.
- Offsite developers' performance is not better.
- The development of software is a direct and expected process.
- Programmers are only associated with code writing.
- The process of development should be fast.
- Adding any new attribute to the program is an easy task.
- If the development team working dedicatedly, all will go well.

- Modular testing is not required.
- Releasing the product means the project is closed.

7.3 *Multiple Mission*

In most cases, companies are working on multiple missions, and it is very casual in today's competitive environment. Different hierarchy levels of the company working for different challenges these challenges are not unique in some scenarios. For the better performance and good result of his own, the particular employee is focusing on his desk work. But, when these all things are combined, the required result is not generated. The main reason behind this is multiple foci. Multiple foci are not the good practice. Different projects demand and are developed by the different stakeholders. All have distinguished parameters for their own and based on client parameters Concern Company developing software. There is no uniformity in the same type of software structure. This is another reason behind the multiple missions because everyone worried about his client and his assigned task. A huge conflict of interest, preference, and need is noticed in this area. Analyst and programmer sometimes trying to resolve the conflicts of different stakeholders, sometimes it works, but most of the time it remains throughout the project. Some common and repeated missions are listed Table 1 with word cloud visualization.

Table 1 Some common and repeated missions

Find flaws 
Product release
Technical budgets
Assess conformance
Use of the product
Measure quality
Verify correctness
Product outcome
Assure quality

7.4 *Weak Programmers*

Term weak programming does not mean the person who was involved in software development. A software programmer is the one who develops the program as per the specification, but when we are talking about the testing, if the same programmer is testing his self-developed program, in this case, he will be considered a weak programmer. In most cases, small companies are doing the same. The one who developed the program cannot do software testing effectively. For software testing dedicated Software Development Engineering in Test (SDET), a professional is required. SDET designs the testing framework based on the software specification and test the developed program with cull capacity and will check every parameter for a better outcome. Some of the key roles of SDET are listed below.

- Testing framework for multiple applications as per the requirement specification.
- Examine client issues.
- Must develop different test scenarios.
- SDET must work in coordination with the deployment process.
- These responsibilities may differ from company to company. Generally, a simple software tester or programmer who develops the entire source code is not suitable for software testing. In most of the case, companies are giving this task to the tester. Table 2 what is the difference between tester, and SDET is mentioned.

With the above discussion and after knowing the role and importance of SDTE, we can understand the need and availability of SDTE in a company is much important. Manual tester or software developers when handling the entire testing phases, they were tagged as a weak programmer. And it is one of the big challenges for the company to come out of this image.

Table 2 Comparative analysis between tester and SDET

SDTE	Manual tester
• Identifies the complete structure from start to end	Partial information about the project
• SDET can take part in test mechanisms and tool expansion	Not an expert in test mechanisms or frameworks
• Recognize the specification and procedures of the product	No such knowledge is expected from QA professionals. Quality assurance professionals cannot handle these parts
• Extremely capable specialized with the knowledge of development and testing	The manual tester worked only in making and implementing the test circumstances

8 Technical Challenges

Software testing is a classy procedure that is used to authenticate new products. Imperative means of excellence management are broadly used in developmental practice. It is predictable that 80% cost of software is consumed on identifying and protective faults [6]. Various studies and data are being collected for testing and its challenges; in this section, we will focus on the technical challenges of the software testing on the company as well client end. While studying and analyzing this area of software testing, different types of technical challenges have been observed which are decreasing the progress of the company. These challenges are purely technical, and it is associated with both parties (Client and Development Company) which straight or ultimately disturb the efficiency of software business.

8.1 Limited Source Challenges

Limited resource for any company is the biggest challenge, and this issue exists at almost every level. Especially in small companies, while visiting some of the organizations, I personally observed the limitation of resources which results in the entire development phase and delay in project completion. In my observation, I found a lack of technical setup as well as lack of human resources too, which caused multiple focusing of the same employee on a different level. After discussing this issue with some of the administrative authorities, they replied: Due to fewer projects and keen competition in the market, we are not in a position to manage a luxury setup. Sometimes, employees left the job in the middle of the project which results in a major effect on the development as well as testing. Hardware equipment, license copy of the development, and testing tool are the other resource challenges also observed in my physical visit of the companies.

8.2 Technological Updating Challenges

Now a day, frequently, technological updating occurs; if development and testing do not synchronize with the updated trends in this scenario, every step of the SDLC will be affected; if testing with upgraded tolls not done will surely faceless error shoot and a bad release to the client, it will cause a negative impact on company image as well company will lose client recommendations. So, proper migration of technological updating is highly recommended.

8.3 Challenge with Project Size

There are different models in SDLC, and the selection of the model is depending on the size of the project. But, in most cases, development companies follow the same type of model for all types of projects. This is also one of the challenges for maintaining the quality of the projects. In case the development is too big, then also it is hard to succeed. In the same way, if the project is small, again, less significance is given to the project, which resulting bad quality. If projects are developed in such types of situations, this is a big challenge to manage in a testing phase because project size and the model selection are the prominent parameters for the entire development which are supposed to be synchronized according to the defined parameters [7].

8.4 QA and QC Team Availability Challenge

Quality assurance (QA) and quality check (QC) are of the main aspects that are very closely associated with testing. Software requirement specification (SRS) is one of the documents which is very closely related to these QA and QC. Quality assurance is the process that is going on parallel with the development and QC performed after the completion of the development. Now, what are the challenges faced by a company here? In most small software development companies, there is no separate team for QC; in this case, software companies need third-party support for the entire quality check process. This is a big cooperation task and challenging because software SRC is written by a software analyst, reviewed by the project manager, and implemented by the developer, and now, a third party who/which is new for this project involved in this process. Another scenario is important to discuss here which we observed in our studies, i.e., sometimes, the client is not ready for QC with the same development company; client needs a third-party QC report for the final approval. In this case, the company also needs to take help from the third party for QC even after having its own QA and QC team. Normally, the second scenario is not found in small software companies, but on a large scale, it is very common. In either case, software development companies have to face this issue, and it is a big challenge to coordinate with the QC team for a better outcome.

8.5 Comfortable Relation of Developer and QA as a Challenge

In our study, we found a notable challenge of developer and quality assurance relationships. As mentioned in the above heading that development and quality assurance is a parallel process, which means both are coordinating on each step to finalize the task. Now, if the quality assurance person is not in deep into the practice with his job just to support his relation and not objecting to every small point at the end, it will become a big bug. Due to multiple pressures, many times, the developer wanted to finish his work from his table. So, here again, the higher authority of the company needs to observe this point very carefully and need to handle this comfortable relation challenge of both the parties very smartly.

8.6 Time Frame Challenges

While visiting different software development companies, we found there is no extra time for testing. In most cases, whatever, time is committed by the development company is for the development only. Due to cut competition in the market, development companies are not taking much time to deliver the product; in this case, there is no separate slot for the testing; this is a big challenge for all the small development companies to manage the testing in the given time itself. If the client is the first-timer, the analysis part consumes a major portion of the entire project in this scenario development phase also not getting a standard time slot. In our entire study, we found time is one of the major issues for almost all small development companies. In most cases, a committed time slot is not applied. So, time frame is a big challenge for testing because of no separate slot allotted in the entire development.

8.7 Testing Structure Challenges

As we discussed above, testing is a complex process, and it is monitored and implemented on various levels. For better understanding, we can consider the entire structure in the following three levels [8]:

- Testing stage(s) of the software
- Testing tool(s) of the software
- Testing technologies of the software.

Testing of software is generally defined in different stages of development. In every testing stage, different types of tools are used for conducting analysis. Consistent testing mechanisms like average situation data, reference executions, test actions, and test situations (manual and automated) deliver the technical base for industrial testing tools. Once the software starts working on its respective site, maintenance starts where software practically repairs, based on real data and circumstances. Throughout this process, most of the steps of testing will be repeated multiple times. For the software development company, this complete sequence of structure is a big challenge. To implement these, all on-site companies are facing the issue of staff change, technology updates, and tools availability. Sometimes, client location change is also a big challenge which creates a big issue between the clients and development companies.

8.8 *Wrong Tool Challenges*

Testing tools are one of the important phenomena. If the applied tool is technically not good that will not give you a better outcome. A good process and organizational management are equally important for the testing. In our personal experience, we found in many developments companies' wrong tools were selected. Proper research and complete documentation should be needed before applying a testing tool; otherwise, in the middle of the test, if it needs to change the tool, it gives lots of financial burden on the company as well as the time frame also gets disturbed. Before buying any tool (especially the first time), expert opinion and discussion with the testing team are highly recommended. If any of the known persons, insider or outsider of the company know about the respective tolls, must discuss and then take the decision. The selection of the right testing tools is one of the big challenges for development companies on every level [9].

9 Survey Report and Result Discussion

After analyzing and studying both types of challenges, we surveyed to find the ground reality of these two categories. In this section, we are presenting a data collection report in the form of a table and also discussing results and their applicability in small software development companies.

We contacted a total of 50 software development companies, all these companies working for small projects and belong to ten different states of India. The names of these respective states are Uttar Pradesh, Bihar, Gujrat, Goa, Maharashtra, Karnataka, Madhya Pradesh, Chhattisgarh, Telangana, and Jharkhand. Many companies were not agreeing to answer any types of documentation, but we had an oral discussion. Most of the companies are even not familiar with the name of different types of challenges, but they are handling many issues during the entire development, and these issues are almost the same as we categorize in our study. More than 90% agreed with the issues and their challenges, and some of them seriously working and trying to find the solution to these challenges and some of them passing out the issue to the higher-level management because of the lack of resources and less skilled for QA and QC. One more point we want to mention here about the QA and QC team, almost, every small software development company doesn't have a separate team for this task. That is why multiple mission challenges of the fundamental category are highly dominated. Some of the companies agreed to give documentation but did not agree to mention the company name by keeping into consideration of our commitment. Survey data are mentioned in Tables 3 and 4.

As data mentioned in Table 5, defining the ground reality of the small software development companies, most of the companies are facing both challenges of testing; many of them are not aware of the technical terms of the challenges, but in practice,

Table 3 Software development company (SDC)

Name of the company	Fundamental challenges				Technical challenges							
	1	2	3	4	1	2	3	4	5	6	7	8
SDC-1	Y	Y	Y	N	Y	Y	Y	Y	N	Y	Y	Y
SDC-2	Y	Y	N	Y	Y	Y	N	Y	Y	Y	Y	Y
SDC-3	Y	N	Y	Y	Y	Y	Y	Y	N	Y	Y	Y
SDC-4	Y	Y	Y	Y	Y	Y	Y	Y	Y	Y	Y	Y
SDC-5	N	Y	Y	Y	Y	Y	N	Y	Y	Y	Y	Y
SDC-6	N	Y	Y	Y	Y	Y	N	Y	Y	Y	Y	Y
SDC-7	Y	Y	Y	Y	Y	Y	Y	Y	Y	Y	Y	Y
SDC-8	Y	Y	N	Y	Y	Y	Y	Y	Y	Y	Y	Y
SDC-9	Y	Y	Y	Y	Y	Y	N	Y	Y	Y	Y	Y
SDC-10	Y	Y	N	Y	Y	Y	Y	Y	Y	Y	Y	Y
SDC-11	Y	Y	Y	Y	Y	N	Y	Y	Y	Y	Y	Y
SDC-12	Y	Y	Y	Y	Y	Y	N	Y	Y	Y	Y	Y
SDC-13	Y	Y	Y	Y	Y	N	Y	Y	Y	Y	Y	Y
SDC-14	Y	Y	N	Y	Y	Y	N	Y	Y	Y	Y	Y
SDC-15	Y	Y	Y	N	Y	Y	Y	Y	Y	Y	Y	Y

Table 4 List of fundamental challenges

Fundamental challenges	1	Complete testing is not possible
	2	Process myths
	3	Multiple mission
	4	Weak programmers

Table 5 List of technical challenges

Technical challenges	1	Limited source challenges
	2	Technological updating challenges
	3	Challenge with project size:
	4	QA and QC team availability challenge
	5	Comfortable relation of developer and QA as challenge
	6	Time frame challenges
	7	Testing structure challenges
	8	Wrong tool challenges

they are facing these challenges. Because of the less skill and lack of infrastructure, these challenges are not countered in a professional way.

Our main objectives of this paper is to check the ground reality of both types of testing challenges in small software development companies and let the people know to identify these issue and create a proper counter plan in a professional way to overcome with all challenges. So, these companies can also come in the counting of developing quality software, and these practices will help them to create a good reputation in the software development field.

10 Conclusions

In our entire studies, we found a very clear picture of the fundamental and technical challenges and their existence and major effects in the small software development sector. We also observe that these sectors are not much aware of these challenges, and neither they have a proper plan to overcome them. Some of the points mentioned here are denoting the major loss of quality as well as reputation in the market.

- Customer un-satisfaction
- Financial and social threatening
- Loss of projects
- Destruction of programming ethics
- Confusion of client necessities.

Many reforms like technical and administrative training, infrastructure, working in the specified area of development and re-engineering and re-defining the business and working environment and most important competent team are the main needs of this sector to overcome with all these issues. The main strength of this sector is the experience, based on experiencing many of them running their firm in the absence of high class technical and experimental knowledge.

References

1. Singh SK, Singh A (2012) Software testing. Vandana Publications
2. Khan IA, Kumari D (2021) Factors influencing the software development process in small scale industries. Design Eng 4089–4098
3. Tuteja M, Dubey G (2012) A research study on importance of testing and quality assurance in software development life cycle (SDLC) models. Int J Soft Comput Eng (IJSCE) 2(3):251–257
4. Kaner C (2003) Fundamental challenges in software testing. Colloquium at Butler University
5. https://sdacademy.dev/10-myths-about-software-development/
6. Silva LS (2010) Automated object-oriented software testing using genetic algorithms and static-analysis (Doctoral dissertation, Master's thesis, Swiss Federal Institute of Technology Zurich, 2010. ga10aLucasSerpaSilva⇒http://se. inf. ethz. ch/old/projects/lucas_silva/report. pdf)
7. Naik K (2008) Software testing and quality assurance-theory and practice. Department of electrical and computer engineering-university of waterloo

8. Planning S (2002) The economic impacts of inadequate infrastructure for software testing. National Institute of Standards and Technology
9. Rice RW, CSQA C, Solutions LRC (2003) Surviving the top ten challenges of software test automation. CrossTalk: J Defense Softw Eng 26–29

Greenhouse Service Implemented on Raspberry Pi-3

K. R. Nataraj, S. Sheela, K. R. Rekha, and Santosh M. Nejakar

Abstract Agriculture is more than just cows and plow. The main aim of the greenhouse monitoring method used in agriculture because of low-cost system to obtain the good yield in the plants irrespective of the climatic condition, rain, and soil condition. It keeps on monitoring continuously and keeps on updating the data (environmental parameters). The rain falls which is unequal distributed leads to the scarcity of water and it creates difficult situation for the farmers. It requires some advanced method for the water irrigation depending on any type of weather condition, soil type, and variety of crops, by greenhouse we can manage the development of the agriculture. We use sensor for the sensing (temperature, humidity in soil, and light intensity). Raspberry Pi and the Internet of Thing (IOT), by this method we can monitor from anywhere based on correct precipitation, efficient transmission, and intelligent synthesis. This research focuses on the automatic monitoring the temperature, soil condition, humidity, and light intensity. Here, we use Raspberry Pi which is programmed and used as the central hub to manage the sensors. This is the web-based climatic condition monitoring all the parameter for the growth of the crops. Later the application is created which is used for the interaction with the greenhouse controller.

Keywords Raspberry Pi-3 · SD card · LDR sensor · DHT11 sensor · DS18B20 · Soil moisture sensor · Arduino ADC driver · Linux OS · Python

1 Introduction

Agriculture is the main occupation of India, and it is the backbone of Indian economy, because of the economic conditions the old conventional method is still followed in

K. R. Nataraj (✉) · S. Sheela · S. M. Nejakar
Don Bosco Institute of Technology, Bengaluru, India
e-mail: director.research@dbit.co.in

K. R. Rekha
SJB Institute of Technology, Bengaluru, India

A. Kumar et al. (eds.), *Proceedings of the International Conference on Cognitive and Intelligent Computing*, Cognitive Science and Technology,
https://doi.org/10.1007/978-981-19-2350-0_56

India, which led to the lagging of the agriculture without the knowledge of the modern techniques. During the last 50 years has demonstrated the strong correlation between agricultural growth. In order to improve the economy and the yield, modern technique has to be taken to improve in the economic prosperity productivity, profitability, and sustainability of our major farming system. Fastest growing sector is the greenhouse industry [1]. It is one of the oldest techniques, but here in modern method, we use sensors and continuously monitor the crops which will finally give an output product in a bulk amount, and it gives the organic food. Here, in this method, we use controlled environmental area to grow plants. In modern technique of greenhouse, we can grow plants in extreme cold, cloudy, and windy climate. Negative impact on the plants takes when high temperatures and a high humidity environmental factors are extremely present inside the greenhouse, so controlling the environment is the essential factor for the plant growth good and healthy.

The project main idea is to design and build a greenhouse controller by using live sensor readings, and it is capable of displaying the system status to the respective owner it is also capable to maintain the environmental parameter. Raspberry Pi acts as a controller where it receives input from a variety of a sensors, it also controls motor, light, and other actuators used in greenhouse [2]. To control all the parameters, the programs are written on the Raspberry Pi and to exchange data from one another that is from the electronic circuits, sensors, software, and network connection; we use IOT where IOT is the network of physical thing embedded with sensors software, etc. The Raspberry Pi is the real-time remote control which helps in greenhouse monitoring system to track environmental parameters [3]. It acts as a server which is a low-cost ARM powered Linux-based computer, and it communicates with clients with Wi-Fi module which is inbuilt in Pi [4]. The greenhouse is capable of creating its own environment according to the development of the crops and the climatic condition required for the crops. By manipulating environmental condition as per crops requirement, we can cultivate many varieties of crops at a time this is the main advantage of greenhouse. By this method, we have an advantage of higher crop yield, prolonged production period, better quality, and organic method (less use of chemicals) [5, 6].

2 Related Work

Agriculture is not a simple job in India; it plays the vital role where it is considered as the backbone of the country because it is the oldest and the most important occupation where the major part of the economy is dependent on it, therefore, this paper main focus is to use the new technology which is automated and used to make all the aspects more efficient and useful for the agriculture [7]. By the survey's done it clear that agriculture industry is far behind in the new technologies, modern techniques which is efficient of cost where the problem faced in the agriculture industry can be overridden by this method of modern greenhouse proposed where the controlling capacity of the factors and the monitoring of the greenhouse and the other sectors of the agriculture industry can be done. With the implementation of this method, the production rate is increased with the increase in the economy of the country where it is affordable by everyone and it is efficient of cost [8]. This paper is mainly based on the automation of the agriculture which has the presence of the low-cost embedded system with the use of GSM and WEB modern technology, where these technologies are used for the monitoring of the sensors and to get the status of the sensors where there is a presence of the sensors like temperature, soil, moisture sensors, etc. These sensors are used to detect the temperature, water level where these are the important factors for the agriculture and these sensed data is send by the SMS to the web page these web pages contain the data of the sensor's status [9]. The monitoring is done continuously by the micro-controller, and with the GPRS method, the information at the remote location can be viewed and helps in the monitoring process [10].

3 Methodology

3.1 Hardware Components

3.1.1 Humidity Sensor (DHT11)

The humidity sensor DHT11 is a four-signal pin package with the spacing of 0.1 that are used to sense the humidity, it has the property where it is highly reliable and best stability for long-term. The humidity sensor is incorporated for a resistive measurement part that is associate with execution of 8-bit micro-controller, which has the worthful quality, response rapidly, anti-interference capacity, and cost efficient. Each element of DHT11 must be calibrated strictly which has an extreme accuracy on humidity calibration. The coefficients of the calibration are stored, and the stored calibration are in the format of programs in the OTP memory. This OTP memory is useful for the sensor's decreasing process of the internal signal. The signal-wise serial interface makes system integration quick and easy and up to 20-m sign transmission. The humidity sensor has the presence of two electrodes, and these two electrodes have substrate in between which are moisture holding substrates. When there is changes in the substrate conductivity as humidity keeps changing and the electrodes transforms satisfy the substrates transforms between them. These resistance changes are calculated by the ready IC which is used to be read by the GPIO pins of Raspberry Pi. With the accuracy of 5%, it is good for the humidity of 20–80% of readings. Temperature readings have accuracy of ±2 °C where temperature is of 0–50 °C and sampling rate must not be more than 1Hz every second once.

3.1.2 Soil Moisture Sensor

By using the soil moisture sensor (SMS) in the landscape form of irrigation which leads to the water saving process also. The soil water content (SWC) which is used to detect the soil content, and the detection is done by the SMS. The transmission of the data takes place in two methods that is by wire and wireless method from and to the server and various viewpoint of the batteries are used in farming which are contrasted. The capacitance sensor which is two sides folded having the capacity to check the moisture in the soil, when there is a change in the moisture of the soil it brings to the notice and the process of checking the moisture of the soil is known as the permittivity of the soil as the moisture content in the soil keeps on changing and this is the inter digital device.

3.1.3 Temperature Sensor DB18B20

The DS18B20 is a digital thermometer having the description in general where it has measurement of temperature from 9-bit to 12-bit Celsius with the alarm function which is user-programmable of lower trigger points and upper trigger points having the non-volatile nature. Here, the communication is done over the wire which is 1-wire bus where the requirement is only single data line and it is grounded with the microprocessor which is a central for establishing the commutation and additionally from the data line the power is derived directly known as "parasite power" where the power supply present externally is eliminated. In each DS18B20, it has the presence of 64-bit unique serial code allowing the multiple DS18B20s to work on the same and single 1-wire bus, thus there is a use of one simple microprocessor. There is an advantages here that is monitoring systems for the, temperature present internally of the buildings or present in the machineries or the equipment's, it also has the environmental controlling HVAC which helps in the monitoring of the process and has the controlling capacity of the system.

3.1.4 Fan and Water Pump

The greenhouse constructed uses a transparent acrylic sheet having various cut-outs, and there is a presence of two fans which are placed diagonally in opposite direction, where one of the fans works as the exhaust and the fan in the other end of the diagonal work as the cooler as shown in Fig. 1. The plants are supplied with the water as per the requirement of the respective crop and also helps to save the water.

3.1.5 Raspberry Pi

The open source format was formed during the Raspberry Pi-3 B model forming the get-go which means that Linux board inserted with the entire gatherings which is

Fig. 1 12 V DC supply fan and water pump

a defector, there is a presence of the BCM2837B0 which is the Broadcom present along with the Cortex-A53 which is a 64-bit and the SoC present at the 1.4 GHz, LPDDR2 of 1 GB SDRAM present with 2.4 GHz and the IEEE802.11. b/g/n/ac of 5 GHz which is a LAN and it is wireless. There is a presence of the BLE Ethernet which has the gigabit over USB 2.0 (maximum throughput 300Mbps) 4 × USB 2.0 ports which has the extension of the 40 header pins of the GPIO. HDMI of the 1 × full size gives the images which are high defined, and there is a presence of the MIPI DSI display port along with the MIPI CSI camera port for capturing the pictures. There is a presence of the output which is a stereo of four poles and the pot of video H.264 which is also an output for the receiver. The temperature of operating is 0–50 °C. It has the OpenGL ES 1.1, H.264 encode (1080p30), MPEG-4 decodes (1080p30), for the purpose of the data storing and the operation loading system there is a presence of the 2.0 graphics micro SD card, the ADC of 5 V/2.5 through a micro USB connector and the DC of 5 V connected to the Ethernet (PoE) in enabled condition which means separate PoE HAT is the requirement and connected through the header power of the GPIO.

3.1.6 Software Part

There is the presence of the sensors like the temperature, humidity, to find the light intensity and to find the water content in the soil where the soil moisture sensors are used. The software is basically separated in two parts where the first part has the soil moisture and the sensors to find the intensity of the light. The Raspberry pi is used which basically retrieve the data from the sensors later the required data or the bench mark data will be used to compare with the input data based on the result obtained from the comparisons made the supporting environmental factors are switched on or off this process is done by following the method of passing the signal, these signal are passed to the relay board, based on the data received from the system calculation is done weather the requirement is maximum or minimum where the control panel is used which is not achievable.

4 Implementation

In the greenhouse method, the water pump is placed externally along with the Raspberry Pi in the greenhouse, there is the presence of the LED lights on the roof top. Water plays the vital role in the plant development where the plants are supplied with the water from the tanks which store the water. The sensors are placed in certain position for the good working, and the soil moisture sensors are placed beside the crops and with the sensors like temperature and humidity are located internally in the greenhouse. This Raspberry Pi placed externally will take care of all the needed requirements until the passing of the optimal required values set and the data values are updated in the web pages which are used to display in the respective web sites. Here, there is the presence of the GPIO pins where these pins use Python language for programming where these pins are used for the input and the output connections given to the sensors and all the rest of the components used in the greenhouse method as shown in Fig. 2.

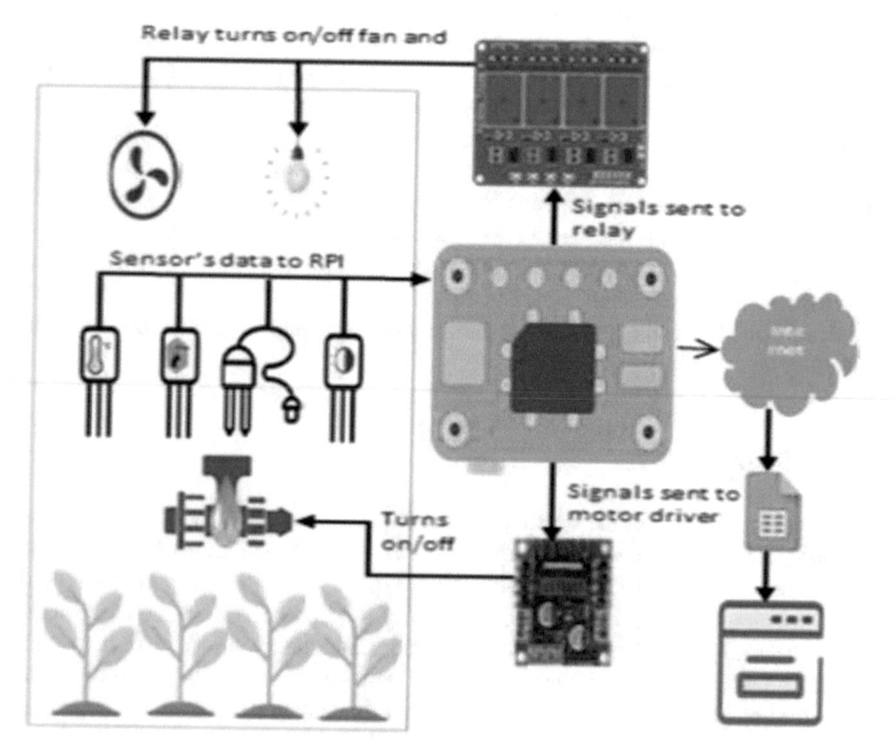

Fig. 2 Automated greenhouse architecture

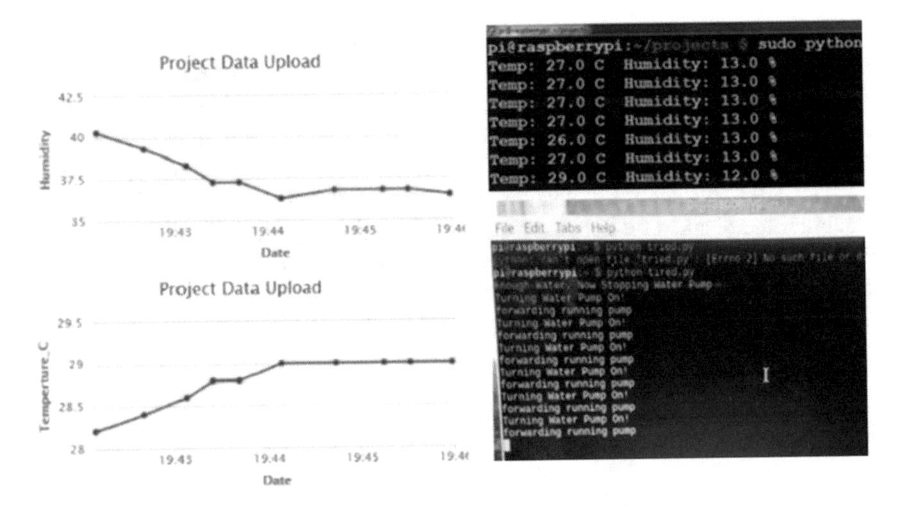

Fig. 3 Results

5 Results

After we get the data from the graphs; it is compared with the standard values when the situation requires the temperature more it is taken care and the requirement of temperature is less also being taken care as per the requirement in the same method; the humidity is taken care and the water requirement for the growth is also taken care where the conservation of water is also plays an advantaged rule; finally, the fan is on and off as per the requirement of the situations. The situations can be handled in the computers of else in the mobile phones with internet connection irrespective of the locations. Temperature and humidity are monitored continuously to turn on/off water pump as shown in Fig. 3.

6 Conclusion

IT is mainly a real-time project where the temperature, soil, and climatic conditions are measured; the software is accessed by any one so it form an open source, and the hardware used is a real time which looks after the crops requirements and notifies the authorized persons which gives a technical support for the automatize development of the plants which helps in the increment of the economy of the country simultaneously. All the objectives mention above have met the automated greenhouse which makes the good development in the agriculture fields. In further days, food becomes the most difficult parts because of the growing population and following old method. The process of growing the crops is the difficult task because of no proper water distribution and climatic condition. Many of the plants have become rare and difficult for the available because of the environmental condition, no knowledge about the plants and mainly due to the global warming where this global warming has become the bigger threat. The future generation has to take care of the development of the agriculture practices irrespective of the climatic conditions, and modern techniques must be implemented for the drastic development of the agriculture field which will not affect the other animals, humans, and other living things. The automatic machineries are taken for the observation of the greenhouse when any changes occur it will be updated in the web pages and this change will be monitored without affecting the crops. The decision is taken care by the human. By following this method, there is increment in the economy of the agro-based industries.

References

1. Elijah O, Member S, Rahman TA (2018) An overview of internet of things (IoT) and data analytics in agriculture: benefits and challenges. IEEE Internet Things J 5(5):3758–3773
2. Kodali RK, Jain V, Karagwal S (2016) IoT based smart greenhouse. In: IEEE region 10 humanitarian technology conference (R10-HTC), pp 1–6
3. Li R, Lin K (2014). Smart greenhouse: a real-time mobile intelligent monitoring system based on WSN. In: International wireless communications and mobile computing conference (IWCMC), pp 1152–1156
4. Meghana KC, Nataraj KR (2016) IOT based intelligent bin for smart cities. Int J Rec Innov Trends Comput Commun 4(5):225–229
5. Sheela S, Shivaram KR (2016) International journal of innovative research in science, engineering and technology proceedings of international conference on "low cost alert system for monitoring the wildlife from entering the human populated areas using IOT devices, India, Special Issue, Vol 5, 2016, pp 128–132
6. Sheela S, Shivaram KR (2016) International journal of innovative research in science, engineering and technology proceedings of international conference on "low cost automation for sorting of objects on conveyor belt. IJIRSET, India, Special Issue Vol 5, pp 195–200
7. Nejakar SM, Pruthvi S (2013) Wireless infrared remote controller for multiple home appliances. Int J Electr Electron Res (IJEER) 2:25–35
8. Bin Jafar I, Raihana K, Bhowmik S, Rahman Shakil S (2014) Wireless monitoring system and controlling software for smart greenhouse management. In: IEEE 3rd international conference on informatics, electronics and vision, Vol 7, pp 61–65
9. Singh H, Pravanda S, Ranjan S, Singla D (2015) Remote sensing in green house monitoring system. Int J Electron Commun Eng (Ssrg-Ijece)-Eees
10. Enokela JA, Othoigbe TO (2015) An automated greenhouse control system using arduino prototyping platform. Australian J Eng Res

Defect Detection Based on Image Recognition

D. Chandraprakash, Nikhitha Gaddam, and M. Narayana

Abstract Corrosion and flaws are two of the most common activities in the corporate world. Industrial materials include periodic flaws that are difficult to detect, even by trained human inspectors, throughout the manufacturing process. This technique assesses commercial materials in real-time at a consistent rate. The intended system's most important approach is to notice the flaws using digital images of the fabric victimization image processing tools. The goal of this research is to develop a method for detecting defects that is both reasonable and thorough. To eliminate the subjective features of manual examination, an automatic scrutiny system exploitation picture method will be designed and implemented. There are a few procedures to follow to spot defects in empty materials. First and foremost, the system should use a camera to capture images of objects. When using a particular image recognition and analysis approach, the images must be compared to templates to discover common flaws such as uneven shape, unlabeled, and various defects. When compared to existing approaches, the suggested system can resist any predicament and is more efficient and swifter.

Keywords Template matching · Defect detection · Face recognition

D. Chandraprakash (✉)
KG REDDY College of Engineering, Hyderabad, India
e-mail: chandraprakash.d@gmail.com; dcprakash99@gmail.com

N. Gaddam
ECE Department, Vardhaman College of Engineering, Shamshabad, Hyderabad, India

M. Narayana
ECE Department, Anurag University, Hyderabad, India
e-mail: narayanaece@anurag.edu.in

A. Kumar et al. (eds.), *Proceedings of the International Conference on Cognitive and Intelligent Computing*, Cognitive Science and Technology,
https://doi.org/10.1007/978-981-19-2350-0_57

1 Introduction

Defects are one of the most common activities in the industrial sector. Periodic flaws in materials are difficult to detect by human inspection. The goal of this proposed approach is to locate the flaws in the digital picture exploitation image processing technology. Model matching is a computer vision approach for identifying a matching sub-image of a target picture. This approach is widely used in areas such as police investigation, automobile pursuit, robotics, medical imaging, and production. Model matching techniques are usually divided into two groups: one that supports the extended bar chart methodology, and another that supports the feature extraction methodology.

1.1 Object Detection

Object detection is a branch of computer vision and image processing that investigates occurrences of linguistic items of a certain type (such as people, buildings, and automobiles) in digital images and videos. Face detection and pedestrian detection are two well-studied object detection areas. Object detection has a wide range of applications in the computer world. Picture processing is an important application area in which algorithms are used to locate and extract several desirable sections or forms (features) of a digitized image or video stream. It is extremely important in the field of optical character recognition.

1.2 Feature Extraction

Feature extraction in machine learning, pattern recognition, and image processing starts with a set of measured data and builds derived values that are supposed to be informative and non-redundant, easing the next learning and generalization steps and in some cases, resulting in better human interpretations. The term "feature extraction" is used to describe the process of reducing the size of a set of spatial properties. When an algorithmic program's input data are just too large to analyze and are considered to be redundant (for example, a comparable measurement in feet and meters, or the repetitiveness of photographs supplied as pixels), it is frequently rebuilt into a smaller set of possibilities (also named a feature vector). Determinative, a feature selection, refers to a group of the first selections. Because the selected alternatives are anticipated to include the relevant data from the input file, the stated job is frequently accomplished by utilizing this simplified depiction rather than the entire initial dataset.

2 Literature Works

2.1 *Real-time Inspection System for Printed Circuit Boards*

When compared to current approaches, the PCB testing system presented in this study will detect flaws such as wire breakage and tension while being significantly faster. The intended examination method is based on denotative matching between the hold on the reference picture and the examination (observed) image. In denotative matching, block matching is used to determine the arrangement. To reduce the complexity of the process, we usually implement the proposed algorithmic rule utilizing one instruction multiple data (SIMD) direction, which is referred to as SSE2 [1, 2].

2.2 *Using Spin Images for Efficient Object Recognition in Cluttered 3Dscenes*

Present a 3D shape-based visual perception system for simultaneous identification of many items in settings with litter and occlusion in this project. Matching surfaces by matching points using the spin image representation is the basis for recognition. The spinning image is a form descriptor at the knowledge level that is used to match surfaces represented as surface meshes. We offer a compression theme for spin images that results in cost-effective multiple visual perceptions, which we demonstrate with data demonstrating the simultaneous recognition of multiple items from a library of twenty models [3, 4].

2.3 *A Simple Guidance Template-Based Defect Detection Method for Strip Steel Surfaces*

Quality control has become an important part of the commercial production inspection system. It was mostly done by delicate inspectors in the previous years, which was both time-consuming and punishing. Furthermore, a human vision-based inspection cannot fulfill the demands of speed, dependability, and strength. Throughout this article, we will mostly This research was funded in part by the National Science Foundation of China under Grant 61,403,119 and in part by the Hopei Science Foundation specializes in computer vision-based inspection techniques that are increasingly being used in manufacturing lines [5, 6].

2.4 Automated Defect Detection in Uniform and Structured Fabrics Using Gabor Filters and PCA

This work provides an algorithmic approach that has been evaluated in the TILDA picture collection for texture defect identification in uniform and structured materials. The proposed strategy is divided into two sections: a feature extraction portion that uses a sophisticated bilateral Gabor filter bank and principal component analysis (PCA), and a defect diagnosis section that uses the geometer norm of choices and compares it to cloth-sort specific characteristics. Our research is done on a patch-by-pattern basis, rather than pixel-by-pixel [7, 8].

2.5 Performance Evaluation of Full Search Equivalent Pattern Matching Algorithms

Signal processing, computer vision, and image and video processing all use pattern matching. Complete search equivalent methods speed up the pattern matching approach while producing the same results as the full search. The goal of this study is to analyze and compare progressive methods for full search equivalent pattern matching [9, 10].

2.6 Improved Normalized Cross-Correlation for Defect Detection in Printed Circuit Boards

In automatic printed circuit board (PCB) inspection, the suggested system is a modified low-complexity NCC technique for locating missing integrated circuits (ICs). In the proposed technique, 2D sub-images are transformed into 1D feature descriptors throughout the matching process. To construct the feature descriptors, the running sub-images are scanned vertically and horizontally, enhanced with spatial statistical data, and then converted using the discrete cosine transform (DCT). The suggested technique outperforms the classic NCC and other systems based on 1D representations, according to experimental findings of template matching on a 32-image dataset of PC and mobile phone PCBs. Signal-to-noise ratios also demonstrate the suggested technique's resilience against noise [11].

2.7 Automated Solder Joint Inspection System Using Optical 3D Image Detection

An automated technique for visually evaluating SMD solder connections has been created (surface mounted devices). The device can analyze fine pitch components as small as 0.3 mm pitch quad flat packages (QFPs). To get exact 3D pictures of solder junctions, a novel image detecting approach was devised. A confocal microscopy concept is utilized; however, many sensors are used to detect reflected light at multiple focusing points at the same time. Secondary reflection and dead angles have no effect on the system. Using the detected 3D pictures, the warp in a printed circuit (PC) board surface is computed in real time, and the board height to be detected in subsequent regions is forecasted. Using newly developed defect detection algorithms, real-time automated focusing control is then done; the system can distinguish leads, pads, and solder filets from the discovered pictures [12].

2.8 Automatic Inspection System for Printed Circuit Boards

The goal of this work is to highlight issues and solutions in the field of automated visual inspection of printed circuit boards (PCBs). When it comes to accurately recognizing PCB layouts, vertical and diagonal lighting are helpful. An algorithm is developed that compares local aspects of the patterns to be examined with those of the pattern to be referred. There is also a description of a technology-based inspection system [13].

2.9 Individual Stable Space: An Approach to Face Recognition Under Uncontrolled Conditions

Face recognition under uncontrolled situations is the topic of this article. The individual stable space (ISS) is the key, as it exclusively represents personal features. The ISNN neural network is presented as a way to map a raw facial picture into the ISS. Following that, three ISS-based algorithms for FR under uncontrolled settings are developed. The photos supplied into these algorithms are not restricted in any way. Furthermore, unlike many other FR approaches, they do not need any additional training data, such as the view angle [14].

2.9.1 Occlusion Resistant Face Detection and Recognition System

This work describes a convolutional neural network that was trained to increase face identification accuracy while also capturing facial traits. The suggested solution solves the problem of occlusion when the face is present. For the inputted picture, the face detection network calculates all of the face regions and facial landmarks. After that, the face is aligned using a facial landmark and entered into a face recognition network for identification [15].

3 Proposed Design

3.1 Proposed Fast Template Matching Algorithm

The current template matching design methodology we present uses a correlation technique coupled with the Hadoop framework. One of the limits of the traditional models described is that this technique is not mastered. Here, the restriction on matching several items at a time has been removed. Template matching entails matching a template image, i.e., the input image, with a targeted object that is the polar opposite of the source image. Figure 1 illustrates an example of a model definition. In this case, we consider the area unit of the supply and the template images, then map them using a matching algorithmic program. This produces the target picture, which is displayed as the output. The supply and template pictures for processing are already in the Hadoop distributed filing system. In addition, the output picture, including the identified target objects, is saved on the HDFS. To simplify the procedure, our original and template pictures are regenerated into greyscale images. The area units of these greyscale photos are compared using the window method at the same time. It has now achieved the highest matching matrix value, which is used to designate the object picture in a rectangular box or bounding box. The sliding window approach is used to match every grid of the source picture with the template image

Fig. 1 Image of printed circuit boards (PCB)

in the template matching mode. A matrix price is calculated for each grid. We look at the associated value among all of these matrix prices and consider them to be the existence of the target object. Finally, within a bounding box displaying the result, the mapped item in the source picture is designated as the target object. Figure 3 depicts the template matching algorithmic formula in its entirety. The algorithm for template matching is described below. Algorithm:

Step 1: The first step is to open the original picture.
Step 2: Open the template image and load it.
Step 3: Incorporate the correlation method.
Step 4: Convert the original and example images to grayscale images.
Step 5: Calculate the picture dimensions. If the size of the template image is more than the size of the actual image, use the template image; otherwise, use the correlation technique.
Step 6: Calculate the measure of similarity.

6.1: Images are divided into $n \times n$ blocks.
6.2: Using the window to calculate the similarity measure for each block.
6.3: The matrix kind of similarity metric is sliding.
6.4: A similar match is chosen.

If the correlation value is more than 0.75, the template image was also discovered. Otherwise, the template picture will not be matched.

Step 7: Displays the template picture that matches the bounding box.
Step 8: Stop.

This arrangement template matching the algorithmic rule's main convictions is as follows:

To begin, we compute the centroid bounding circle of the template to the maximum radius of r. Second, using the rr size matrix, split the entire picture into blocks. Finally, compare the color histograms of each block with the template's maximum centroid bounding circle. If the affinity surpasses the edge, we will be looking at this block as a candidate location for the template centroid. The template matching procedure starts right away. This algorithm is not at all like the prior template matching algorithm. Our algorithmic approach chooses the centroid point as the template's reference point in most cases, utilizing the common higher left purpose. Finally, using the NCC algorithmic rule with the centroid point for each block, we will achieve the best match (Fig. 2).

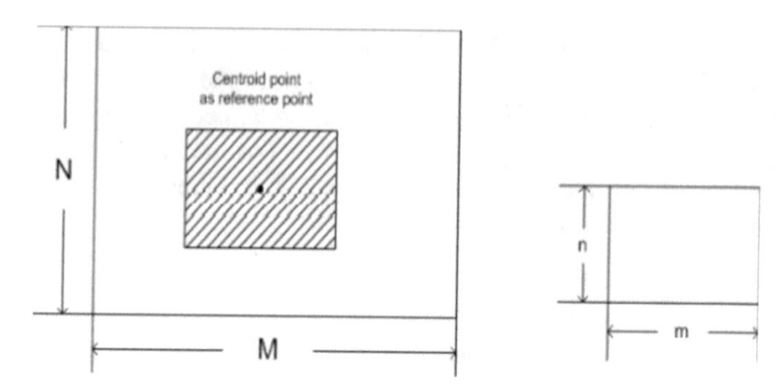

Fig. 2 Template matching using centroid point as reference point

3.2 *Principle of Centroid Bounding*

The idea of centroid bounding is illustrated in Fig. 4. When splitting a picture into blocks, a formula for each block can be used. The color histogram should be calculated. The details will be discussed in the next section. The template's centroid point can shift inside the possible block that the template is in. Third, using the NCC formula with the centroid bound is entirely different from using the ancient NCC formula, where the template moves with the centroid point. Between the top-left point of the preset template and the centroid point, an offset is determined (Fig. 3).

$$NCC(x,y) = \frac{\sum_{i=1}^{B_m}\sum_{j=1}^{B_n} I(x+i-offx, y+j-offy).T(i,j)}{\sqrt{\sum_{i=1}^{B_m}\sum_{j=1}^{B_n} I(x+i-offx, y+j-offy)^2}\sqrt{\sum_{i=1}^{B_m}\sum_{j=1}^{B_n} T(i,j)^2}}$$

Fig. 3 NCC formula

Fig. 4 Cap template image

Where *offx* and *offy* are the *x* and *y* coordinate offsets between the centroid point and the upper left point, respectively. The block that NCC is calculating is denoted by B_{m} and B_{n}. The suggested algorithm is usually summarized in the block diagram (Figs. 5, and 6).

Fig. 5 Original bottle image

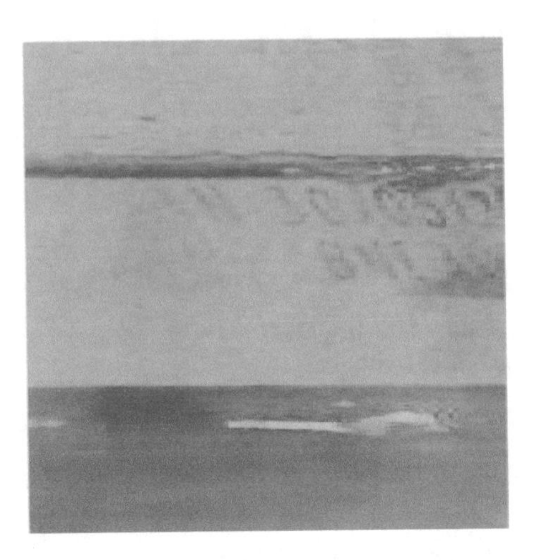

Fig. 6 Volume template

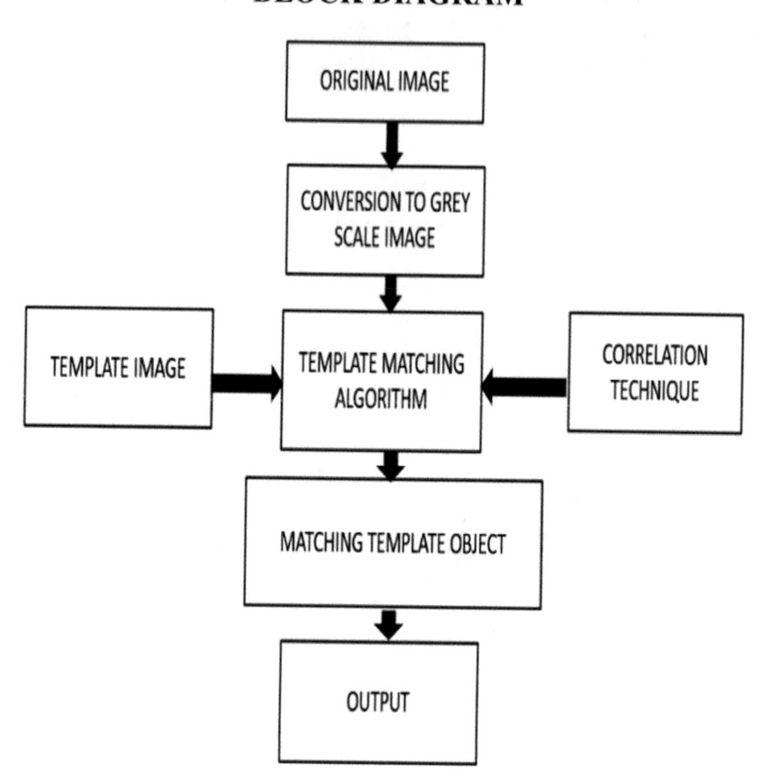

4 Result

The proposed work produced the following results:

1. It detects the position of the bottle top.
2. It determines the bottle's volume.

As a result, if a flaw is discovered, it shows an output message that says:

- Cap missing or in the wrong place, in case of incorrect position of the cap.
- Inadequate volume, in case of irregular volume of liquid observed.

Otherwise, it produces a rectangular border around the matched template if the template is correctly matched. This technique is discovered to be more efficient and generates faster results in comparison with existing methods (Fig. 7).

Fig. 7 Output images of all the possible instances

5 Conclusion

We present a model matching technique with centroid bounds for pattern recognition from an image in this project. In comparison with the model matching techniques that are now available on the market, the findings of this algorithm can provide us with a cost-effective and high-accuracy model matching approach. The suggested algorithmic program, we believe, has a wide range of applications, such as object identification and picture retrieval.

6 Future Scope

Most computer and robot vision systems may require the capacity to detect objects. The most recent research in this area has made significant progress in numerous ways. The capacity to locate things is useful in a variety of applications, including self-driving automobiles, where it allows the vehicle to see pedestrians, bikers, road signs, and other vehicles, as well as security, where intrusive people may be spotted. Apart from centroid bounding, we believe other choices will be extracted to help future work reduce template matching computation time. This algorithm is useful in detecting faults in any type of object, so it must be defect-free. If it has defects, it will not function, and every value that might be wasted will be preserved. This algorithmic rule will become increasingly important in future to overcome object faults caused by flaws.

References

1. Sackin D (1999) A study of the embedded microdevices for infrastructure monitoring. MS Thesis, Carnegie Mellon University, Pittsburgh, PA
2. Kim NH, Pyun JY, Choi KS, Choi BD, Ko SJ (2001) Real-time inspection system for printed circuit boards. In: IEEE international syposium on industrial electronics (ISIE), Pusan, Korea, Vol 1, pp 166–177
3. Akinci B, Garrett J, Patton M (2002) A vision for active project control using advanced sensors and integrated project models. In: Specialty conference on fully integrated and automated project processes, Virginia Tech, January 23–25, 2002, ASCE, pp 386–397
4. Johnson AE, Hebert M (1999) Using spin images for efficient object recognition in cluttered 3D scenes. IEEE Trans Patt Anal Mach Intell 21(5):433–449
5. Li X, Tso SK, Guan X-P, Huang Q (2006) improving automatic detection of defects in castings by applying wavelet technique. IEEE Trans Ind Electron 53:1927–1934
6. Wang H, Zhang J, Tian Y, Chen H, Sun H, Liu K (2019) A simple guidance template-based defect detection method for strip steel surfaces. IEEE Trans Industr Inf 15(5):2798–2809. https://doi.org/10.1109/TII.2018.2887145
7. Kumar S, Ansari MD, Gunjan VK, Solanki VK (2020) On classification of BMD images using machine learning (ANN) algorithm. In: ICDSMLA 2019. Springer, Singapore, pp 1590–1599
8. He FQ, Wang W, Chen ZC (2006) Automatic visual inspection for leather manufacture. Key Eng Mater 326–328:469–472

9. Gunjan VK, Shaik F, Venkatesh C, Amarnath M (2017) Artifacts correction in MRI images. In: Computational methods in molecular imaging technologies. Springer, Singapore, pp 9–28
10. Duan W, Kuester F, Gaudiot J-L, Hammami O (2008) Automatic object image alignment using Fourier descriptors. Image Vis Comput 26:1196–1206
11. Shaik F, Sharma AK, Ahmed SM, Gunjan VK, Naik C (2016) An improved model for analysis of Diabetic Retinopathy related imagery. Indian J Sci Technol 9:44
12. Matsuyama Y, et al (1996) Automated solder joint inspection system using optical 3-D image detection. In: Proceedings 3rd IEEE workshop on applications of computer vision. WACV'96, 1996, pp 116–122. doi: https://doi.org/10.1109/ACV.1996.572014
13. Hara Y, Akiyama N, Karasaki K (1983) Automatic inspection system for printed circuit boards. IEEE Trans Patt Anal Mach Intell PAMI-5(6):623–630. doi: https://doi.org/10.1109/TPAMI.1983.4767453
14. Geng X, Zhou Z, Smith-Miles K (2008) Individual stable space: an approach to face recognition under uncontrolled conditions. IEEE Trans Neural Netw 19(8):1354–1368. https://doi.org/10.1109/TNN.2008.2000275
15. Tsai AC, Ou YY, Wu WC, Wang JF (2020) Occlusion resistant face detection and recognition system. In: 2020 8th international conference on orange technology (ICOT), 2020, pp 1–4. doi: https://doi.org/10.1109/ICOT51877.2020.9468767
16. Shaik AS, Karsh RK, Suresh M, Gunjan VK (2022) LWT-DCT based image hashing for tampering localization via blind geometric correction. In: ICDSMLA 2020. Springer, Singapore, pp 1651–1663

Load Balancing in Cloud Computing

Shivangi Khare, Uday Chourasia, and Anjna Jayant Deen

Abstract Nowadays, many organizations are using cloud-based applications and platforms because of its on-demand service and quick response. The biggest concern of cloud computing is overloading of the system for any individual, group or organization. So, nowadays, load balancing is gaining popularity, its algorithms and solutions are also getting better day by day. This review article gives a brief overview that how the traditional load balancing was done, and the updates made in it with time. This paper also provides comparison table of the major research papers published on load balancing, various algorithms of load balancing, context on which load balancing is used like round robin, throttled, honey bee, max min, min-min, OLB, etc. Also, cloud service providers in CC who deal with powerful computational structures which are entirely determined on usage and offer their solutions in a stable manner. In addition, this research article provides extensive list of areas for research under load balancing and its future scopes.

Keywords Cloud computing · Load balancing · Cloud platforms · Cloud services · Cloud applications

1 Introduction

1.1 Cloud Computing

Cloud computing is very new and effective technology that is used widely in today's world. Cloud computing is the network technology that provides on-demand services to its users. It provides hardware as well as software applications with platforms and testing tools. Cloud computing is the distribution of computer services that includes

S. Khare (✉) · U. Chourasia · A. J. Deen
University Institute of Technology, RGPV, Bhopal (M.P.), India
e-mail: Shivangikhare27@gmail.com

U. Chourasia
e-mail: udaychourasia@rgtu.net

A. Kumar et al. (eds.), *Proceedings of the International Conference on Cognitive and Intelligent Computing*, Cognitive Science and Technology,
https://doi.org/10.1007/978-981-19-2350-0_58

intelligence, databases, servers, networking, storage, etc. It provides data at very low cost, and users can access resources any time whenever they need.

1.2 Cloud Service Model

Software as a service—It is an on-demand software service in which applications are delivered by service providers. It is hosted on remote server that is managed from central location.

Platform as a service—It is a kind of platform on which applications are being developed, tested, run and managed by the programmers. It provides an ability to "auto scale".

Infrastructure as a service—It is also known as hardware as a service. It is cost-effective that means it provides user to avoid cost and complexity of managing and purchasing the physical servers.

1.3 Load Balancing

Load balancing is the contribution of work among different machines or nodes ensuring that no computing device will be overloaded. It enables organization to control or manage applications demand by giving out resources among servers, networks or computers. It is used to circulate the traffic that is on cloud server in order to maintain the system stability. Also, it maintains different parameters of cloud server, i.e. it increases response time, execution time and therefore, it increases the overall performance of the cloud (Fig. 1).

1.4 Parameters of Load Balancing

Overhead—The overhead supports an extra cost of merging the algorithms.

Scalability—Scalability is the kind of parameter by which load of the system is managed or handled. It used to increase or decrease the resource as per need.

Fault tolerance—Whenever the failure occurs in some of cloud components, then fault tolerance helps to continue the operation.

Response time—The time which is taken by the server to process any task.

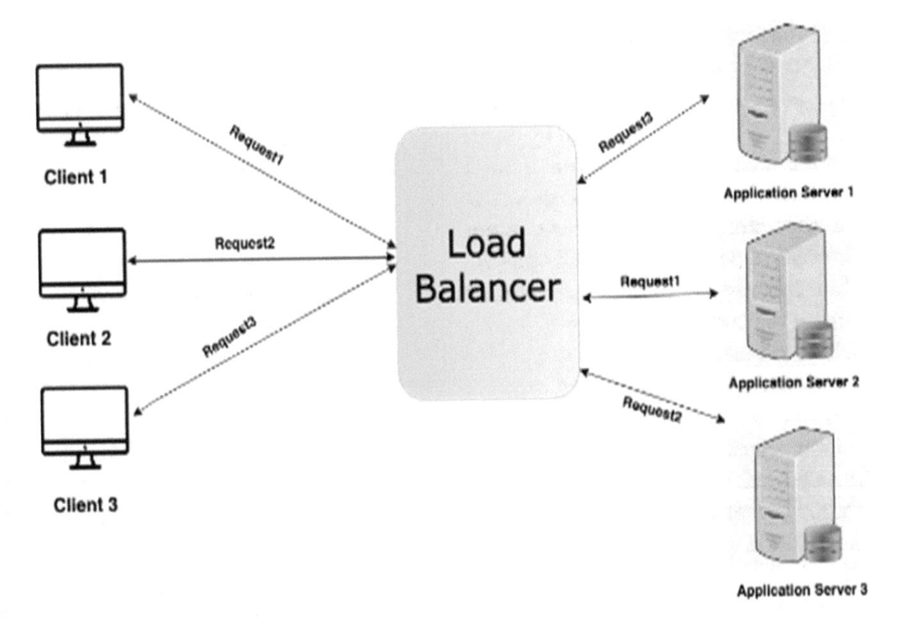

Fig. 1 Load balancing phenomena

2 Literature Survey

In throttled algorithm, Violetta N. Volkova suggested that the load balancer controls a table of virtual machine indexes, as well as their location whether it is reachable or occupied. In the given algorithm, the load balancer controls an index table of a virtual machines and their specific positions. Data centre needs to ask for virtual machine allocation for desired load balancer. Whenever implementing order of consumer, the load balancer gives-1 to the data centre if efficient virtual machine does not get recognized. So the consumer first requests the balancer to pick an optimal virtual machine for implementing essential activities [1].

Dina and Fatima proposed storing and retrieving data on Internet rather than hard drive of computer. The components taken part in cloud computing are clients, datacentres and distributed servers. It means to divide the workload among different nodes equally so that no single node will be overloaded. In this, we introduced an architecture and algorithm which will upgrade the load balancing problems by reducing the response time. We have introduced the amplified version of subsist modulate balancing proposal for cloud computing greedy algorithm of balancing the load. To check the performance of presented approach, we have used the cloud analyst simulator [2].

Scheduling of task that depends on LB-Y. Fang proposed a two-level scheduling of task technique depends on to meet dynamic demands of customers which acquire high resource deployment. It receives load balancer by first mapping task to virtual machines and then upgrading the performance, response time and resource deployment of the cloud environment [3].

Priyanka Sagwan and Manmohan Sharma proposed the fundamental part of operating system is multiprogramming. It is a strategy for operating various procedures all in the memory. We have rolled out a progress in round robin algorithm so that the bringing out of CPU can be moved forward and compared it with original round robin scheduling and exhibit the consequences of why our model is systematic and well organized than the original one [4].

R. Achar proposed distributed intra cloud load balancing algorithm to balance and compare depends on sampling to reach balanced solution. Cost of running virtual machine on each of the host was determined and it ensures that virtual machine locates from physical machine with respect to larger cost to those with low and high was difference of cost [5].

Gao proposed a multi-objective virtual machine placement by ant colony algorithm system. This exerts to carry out Pareto solution which at same time reduces resource power consumption and wastage. This will use the value of pheromone concentration of a virtual machine sets on the server for resolution. Requirements of a virtual machine put on the server is inversely proportionate to the amount of resource wastage and utilization associated with the placement plan [6].

Accessibility at one of the most important aspects is environmental systems. Load balancing is used in it to ensure Internet availability by distributing traffic across many data centres of computer hardware, mitigating recourse limitations and software failures. It talks about the load balancing and after reveals a case study on accessibility of system depends upon a typical management of database of hospital solution [7].

Balancing the load under cloud environment has basic effect on the performance. This object defines a best load balancer model for the public cloud, which is founded on cloud division theory and provides a switch mechanism that involves selecting multiple strategies for different circumstances. To improve efficiency in the cloud context, the system combines game theory to the computation offloading technique. [8].

Michael D proposed a graph technique to express and economic networking. It has both complex and simpler connections between platforms. They employed graph method to build some static and random fabrication networks. Here gives detailed information on these aspects of economic communication networks along a broad range of industries [9].

Bhathiya has proposed two virtual machine load balancing strategies have been deployed for load balancing in cloud server farms at the very same time. The very first algorithm is the automatic inspection base station, which divides the platform's load consistently. The throttled is the second algorithm that makes sure that only of a predefined jobs should be located to individual virtual machine at any certain given time. The major issue happens whenever the hardware layout of VMs is different and it fabricates the undergoing load and over load conditions in VMs [10].

Pham proposed a latest community depends on search algorithm known as bees algorithm. This algorithm emulates the food foddering behaviour of flocks of the honey bees. In this bees locality, spy bees are casting for exploring food origin by moving casually from one origin to other. After when they got success in finding food origin with standard, threshold the spy bees back to swarm and dance on the floor. Bees, those having the most fitness's are taken as picked bees and locations visited by those bees are considered for neighbourhood explore [11].

Li et al. A dynamic min-min and list scheduling approach is being used as a resource provisioning strategy. This technique initiated for infrastructure as a service layer, and works depend on the job response times. To boost virtual machine, migration technique, we are putting forward a better load balancing algorithm that is stimulated from direction of the human's hormone system that is known as endocrine algorithm. This algorithm authorized the overloaded virtual machines to get off their additional jobs to suitable virtual machines through communications that is stimulated from feedbacking technique between hormones. There are benefits of this technique [12].

Sukhpreet Kaur et al. presented an improved genetic algorithm (IGA) for assigning virtual machines to users' tasks. The goal of this proposed strategy is to increase resource consumption while cutting task completion costs and spending as few effort as possible [13].

According to Sheetal Karki et al. data is kept in an unified VMs known as the cloud, and cloud service companies are responsible for sending the offerings to end clients [13].

Cloud technology is one of the areas in computation that is changing fast at the present. Companies and individuals demand vital services, and CC delivers them in the shape of IaaS, SaaS and PaaS. The Internet has made it easier to develop apps as well as provide capabilities via the web [14].

3 Comparison Tables

See Table 1.

4 Conclusion and Future Work

After having a deep exploration of all mentioned research papers evaluated in this article, it is observed that load balancing had abundant limitations. So, in recent times, many other algorithms came like flexible load sharing algorithm, dynamic load balancing, firefly algorithm, etc., which enhance the overload problem to some extent.

Then also the load balancing problem is at risk to high order because if system become overloaded then the system may be crashed, fault tolerance and other parameters. So, it can be enhanced further with some better modern algorithm or combination of two or more algorithms which can give result for improvement of balancing

Table 1 Comparative study among previous work on the basis of different parameters and performances

Parameter	Security	Cost-effectiveness	Scalability and elasticity	Execution time	Performance	Algorithm
Violetta et al.	Intermediate	Low	High	Intermediate	high	Throttled algorithm
Dina et al.	Low	High	High	Low	Intermediate	Greedy load balancing algorithm
LB-Y. Fang et al.	Intermediate	Intermediate	Low	High	Intermediate	Two-level task scheduling algorithm
Priyanka et al.	High	Intermediate	Intermediate	High	Intermediate	Round robin algorithm
R. Achar et al.	Low	Intermediate	High	High	High	Distributed intra cloud load balancing algorithm
Gao et al.	High	Intermediate	Intermediate	Intermediate	High	Ant colony system algorithm
G. Xu et al.	Intermediate	Low	High	Intermediate	High	Cloud partitioning for public cloud
R. kanniga et al.	Low	High	High	Low	Intermediate	The game theory to the load balancing algorithm
Michael D et al.	Intermediate	Intermediate	Low	High	Intermediate	Graph technique
Bhathiya et al.	High	Intermediate	intermediate	High	Intermediate	Two virtual machine load balancing algorithms
Pham et al.	Low	Intermediate	High	High	High	Bees Algorithm
Li et al.	High	Intermediate	Intermediate	Intermediate	High	Token-ring algorithm

the load. This research concentrated on the problems of cloud computing and their challenges. It is a computer technology that gives client support every time. For load balancing, this paper narrated out the effective algorithms.

Keeping this data on an open network naturally creates concerns about balancing load and its parameters. We simulate the approach on a network with over 50 nodes installed on edge computing or fog edge. The results of the experiments show that the suggested approach is effective, detecting dropping workload, data packets received on the network.

Acknowledgements I take this opportunity to gratefully acknowledge the assistant of my college faculty guides Prof. Uday Chourasia and Dr. Anjna Deen, who had faith in me.

I am also very much grateful to UIT-RGPV, DoCSE for motivating the students.

References

1. Shahid MA, Islam N, Alam MM, Su'Ud MM, Musa S (2021) A comprehensive study of load balancing approaches in the cloud computing environment and a novel fault tolerance approach. IEEE Access 8:130500–130526. doi: https://doi.org/10.1109/ACCESS.2020.3009184
2. Altayeb DF, Mustafa FA, History M (2019) Analysis on load balancing algorithms implementation on cloud computing, Vol 6, no. 02, pp 32–36
3. Al Nuaimi K, Mohamed N, Al Nuaimi M, Al-Jaroodi J (2012) A survey of load balancing in cloud computing: challenges and algorithms.In: Proceedings of the IEEE 2nd symposium networking cloud computer applications, NCCA 2012, pp 137–142. doi: https://doi.org/10.1109/NCCA.2012.29
4. Sangwan P, Sharma M, Kumar A (2017) Improved round robin scheduling in cloud computing. Adv Comput Sci Technol 10(4):639–644
5. Mohamed Shameem P, Shaji RS (2013) A methodological survey on load balancing techniques in cloud computing. Int J Eng Technol 5(5):3801–3812. doi: https://doi.org/10.9790/0661-18215561
6. Thakur A, Goraya MS (2017) A taxonomic survey on load balancing in cloud. J Netw Comput Appl 98(February):43–57. https://doi.org/10.1016/j.jnca.2017.08.020
7. Chaczko Z, Mahadevan V, Aslanzadeh S, Mcdermid C (2011) Availability and load balancing in cloud computing
8. Xu G, Pang J, Fu X (2013) A load balancing model based on cloud partitioning for the public cloud. Tsinghua Sci Technol 18(1):34–39. https://doi.org/10.1109/TST.2013.6449405
9. Kanniga Devi R, Murugaboopathi G, Vijayakumar P (2017) A graph-based mathematical model for an efficient load balancing and fault tolerance in cloud computing. In: Proceedings of the 2017 2nd international conference on recent trends challenges computer model ICRTCCM 2017, pp 136–140. doi: https://doi.org/10.1109/ICRTCCM.2017.25
10. Soni G, Kalra M (2014) A novel approach for load balancing in cloud data center. In: Souvenir 2014 IEEE international advanced computing conference on IACC 2014, no. February, pp 807–812, 2014. doi: https://doi.org/10.1109/IAdCC.2014.6779427
11. T. Nadu (2014) Nature inspired preemptive task scheduling for. No. 978, pp 0–5, 2014
12. Aslanzadeh S, Chaczko Z (2015) Load balancing optimization in cloud computing: applying endocrine-particale swarm optimization. IEEE Int Conf Electro Inf Technol 165–169. doi: https://doi.org/10.1109/EIT.2015.7293424

13. Kalpana, Shanbhog M (2019) Load balancing in cloud computing with enhanced genetic algorithm. Int J Recent Technol Eng 8(2, Special Issue 6):926–930. doi: https://doi.org/10.35940/ijrte.B1176.0782S619
14. Amal Murayki Alruwaili BAA, Humayun M, Jhanjhi NZ (2021) Proposing a load balancing algorithm for cloud computing applications. J Phys Conf Ser 1. doi: https://doi.org/10.1088/17426596/1979

Embedded System based Smart Gloves for Soldiers

T. R. Lakshmidevi, Darshan H. Ray, Raunak Raj, P. V. Darshan, and T. S. Hitesh

Abstract The method which involves the conversion of those hand signals into respective voice signals that can be transmitted via wireless modules. The Hand Talk glove is a normal cloth driving glove fitted with electrode sensors. The sensors output a stream of data that varies with degree of bend made by the fingers. Electrode sensors are sensors that change in resistance depending on the amount of bend on the sensor. They convert the change in bend to electrical resistance—the more the bend, the more the resistance value. The output from the sensor is converted to digital and processed by using Arduino and then it responds in the voice using speaker. In this project we have used microcontroller, Wi-Fi module and voice kit to produce the voice signal. Arduino UNO, DAC, wireless speaker, tactile sensors, flex sensors, strength grant, and Voice IC (APR 9600) are the hardware components used. The importance of embedded systems is growing continuously. Exponentially increasing computing power (Moores law) connectivity and convergence of technology have resulted in hardware/software systems being embedded within everyday products and places. Already 90% of computing devices are in embedded systems and not in PCs. The growth rate of Embedded System is more than 10% per annum and it is forecasted there will be over 40 billion devices worldwide by 2020.The value added to the final product by embedded software is often orders of magnitude higher than the cost of the embedded devices themselves. The aim of this project is to develop smart gloves which are useful for our defense forces in the borders and in the surgical strikes.

Keywords Arduino UNO · Wireless speaker · Embedded C-language · Flex sensor

T. R. Lakshmidevi (✉) · D. H. Ray · R. Raj · P. V. Darshan · T. S. Hitesh
Department of Electronics and Communication Engineering, Don Bosco Institute of Technology, Bengaluru, India
e-mail: laxmikala.devi7@gmail.com

A. Kumar et al. (eds.), *Proceedings of the International Conference on Cognitive and Intelligent Computing*, Cognitive Science and Technology,
https://doi.org/10.1007/978-981-19-2350-0_59

1 Introduction

In Militia, there will be a strain at collaborating with team mates while undertaking pursuit maneuver and exploring. The military use hand signs even so it will not be acknowledged by others due to little prominence of sunny through dark, creation of haze. It is unfeasible to use voice gesture with high intensity, meanwhile it will support the others invaders or impostors. The signal conveyed can be seized as well so it is not benign for contact. From now, they practice tactical gestures. In order to over whelm the drawbacks of this in life-threatening zones we practice this technique for factual period communiqué.

In command to overwhelmed the glitches in these dangerous situations, gesticulation-meticulous instructions that are programmed can be taken which can be deciphered at the recipient and transformed to vocal sound communications for essential accomplishment through actual time communiqué [1]. The equitable of this scheme is to change the structure that can translate gesture into encoded digital data that can be transmitted through Wi-Fi module and deciphered at the recipient side and transformed into vocal sound messages. On a quantity of curve of the instrument will alter the given exchange of curve to voltaic resistant more the curve, greater the resistant data.

Productivity from instrument has been changed for cardinal data is communicated through radio communication network. In recipient, this data is related from predetermined information, will be kept in recollection. Contingent of equated data produces vocal sound gesture form particular feed in signals. Later, individual recipient obtains a vocal sound note.

2 Proposed Method

The Hand Exchange gauntlet is static with flex sensors and free-fall sensor. These devices will track record the dissimilar data for diverse hand gesticulations. Free-Fall sensors examines situation of the palm with esteem to three-dimensional axes. These devices produce move of records that differ by diploma of curve thru with an aid of the limbs [2]. Flex sensors are instruments that alternate in resistant relying In this project design, structured modular plan idea is adopted and the device is by and large composed of a double microcontroller, LCD, MP3 module, headphones, Wi-Fi module, flex, and accelerometer sensors. The microcontroller located at the center varieties the manage unit at the transmitter side. Embedded within the microcontroller is an application that helps the microcontroller to take motion based on the inputs provided.

3 Block Map and Overall Explanation

3.1 Hardware Description

(i) **Microcontroller: AT-mega-2560**

This Arduino AT-Mega-2560 is a microcontroller panel built on the AT-mega-2560 (Datasheet). It contains 56 digital input/output pin numbers, 18 analogy input, 4-UARTs pin, a 16-MHz crystal oscillator, a Universal Serial Bus (USB) connection, a specific power-jack, an In-Circuit Serial Programming (ICSP) header, and a given reset button for resetting operation (Figs. 1 and 2).

The Mega2560 panel can be automated with the Arduino Software (IDE). The Mega2560 is power-driven via the Universal Serial Bus (USB) linking or through a peripheral power source. The power basis is designated robotically [3].

The AT-mega 2560 is a small-slung Complementary Metal Oxide Semiconductor (CMOS) 8-bit microcontroller built entirely on the Advanced Virtual RISC (AVR)

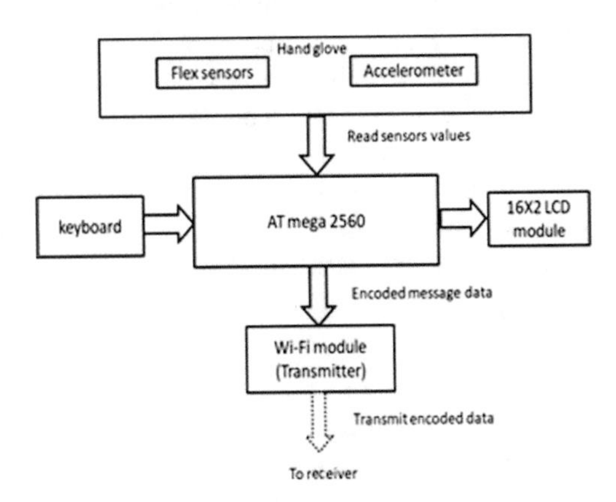

Fig. 1 Block map of transmitter side

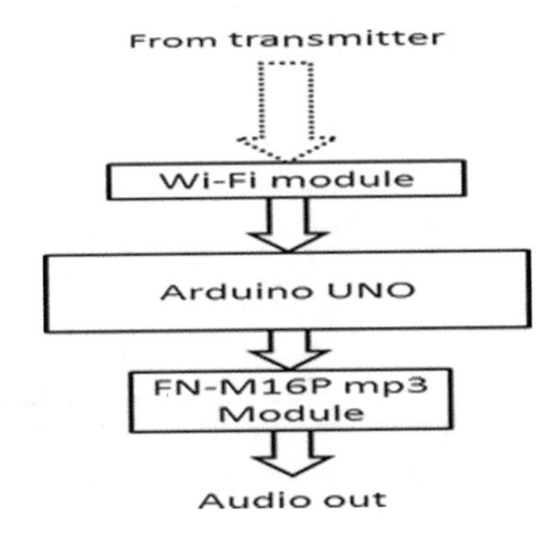

Fig. 2 Block map of receiver side

improved Reduced Instruction Set Computer (RISC) planning. Can performing authoritative guidelines from only timer, this AT-Mega 2560 attains quantities drawing near 1 MIPS per MHz letting this machine intended by enhance electricity ingesting as opposed to dispensation speediness.

(ii) **Arduino UNO**

This Arduino Uno is a microcontroller boarding mainly built from AT-mega-328. It has 12 digital input/output pin (from which 8 should be used as Pulse Width Modulation (PWM) outputs), 8 analogy input pin, a 14 MHz crystal oscillator, a Universal Serial Bus (USB) connection, a strength knave, an In-Circuit Serial Programming (ICSP) header, and a retune pin.

This comprises the entirely wished to aid the microcontroller clearly attach the device through a USB cable or supply with an AC to DC device or battery-operated. This UNO varies as of all previous panels in which it does not make use of FTDI USB to sequential driven mark. In its place, it facets the AT- mega 8U2 compiled as a USB to serial changer. This Arduino Uno should power-driven through varies way of the Universal Serial Bus (USB) linking or by an exterior electric source. This electricity supply can be chosen robotically [5]. Exterior control can derive both by AC to DC connector or free style. This connector can be connected by a 2.2 mm center positive wadding in to the panel's electric jack. This battery-operated should be inserted in ground and input voltage (Vin) pin to influence connector. This panel should operate through an external provider of 5–21 V. Voltage is more than 8 V, yet, 4 V pin can additionally furnish least 5 V and panel can also have unhinged. When voltage is greater than 11 V, the voltage controller might also burn and harm the panel. The suggested choice is 6–11 V.

(iii) **Sensors**

A device is a gadget that senses and replies to nearly of enter by the corporeal atmosphere. By precise contribution should be sunlit heat, wave, moistness, compression, or anyone of an incredible no of different ecological miracles. This yield was normally a gesture that was transformed to humanoid-legible shown by the instrument place or communicated by electronic means concluded by a system for interpretation or additional dispensation [4].

(a) Flex Sensors

Flex device is the instrument used to change its resistant reliant upon by quantity of curvature by instrument. Flex devices is analogy voltage divider. This resistor effort is adjustable analogy resistor. Exclusive flex device is carbon renitent fundamentals inside skinny stretchy substratum. Furthermore carbon resources fewer resistant. As soon as the substratum was determined the instrument harvests resistant yield comparative by twist range. When flex device attains incredible aspect ratio of the tinny bendy substratum. As soon as the substratum is twisted, the instrument harvests a resistant yield linked by the turn ambit was proven in Fig. 3. Minor the radius, greater the resistant worth [6].

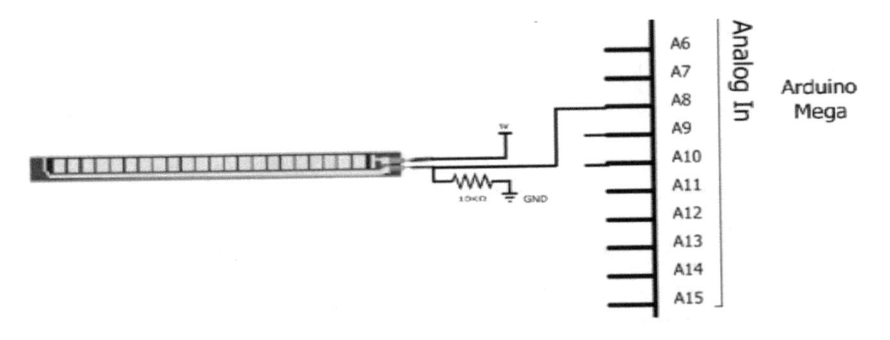

Fig. 3 Flex sensor

(b) ADXL335 Accelerometer

They are typically used in one of three modes:

(i) Initial measurement is velocity.
(ii) Initial measurement of position.
(iii) As a sensor of inclination, tilt or orientation in two or three-dimensions.

A free-fall device was a detecting thing was used to compute speeding up; rushing is the rate of alteration of rapidity by recognize to stint. It has magnitude and direction so it is a vector quantity. Free -fall device estimates in device of g–a-g is the quickening quantity for gravity which is equal to 9.8 m/s^2. Free-fall device consume advanced from a humble aquatic pipe with a mid-air bubble that confirmed the route of the hastening to a combined course that will be positioned on a circuit panel. Free-fall device that can amount: sensations, shockwaves, slope, effects and movement by a thing. A free-fall device is a sensor that used to measures acceleration forces. Practically a free-fall device works through checked frame of a coil (Fig. 4).

(iv) **4 x 4 Keyboard**

Interface 4 × 4 matrix keypad to an Arduino board is the essential aspect of this project. Most of the electronics units use them as person inputs. Knowing how to join a keypad to a microcontroller like Arduino is very precious for building industrial

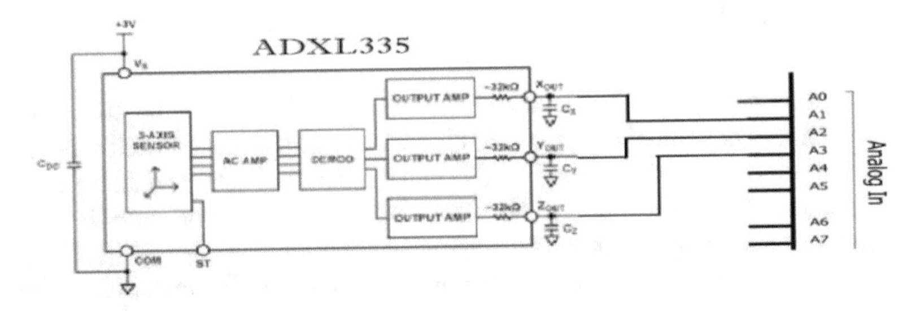

Fig. 4 Free-fall sensor

products. Whenever a key is pressed corresponding action will be seen on the LCD screen. Arduino is very precious for constructing commercial products. Whenever a key is pressed corresponding motion will be seen on the LCD screen (Fig. 5).

(v) **L.C.D Display**

A liquid crystal demonstration (L.C.D) is a plane pane demonstration, digital visible show, built totally on Liquid Quartz Knowledge. A liquid crystal display comprises a collection through small sections that will be operated by current data. Liquid quartzes do not release sunlit straight alternatively used mild tempering methods. LCD's are used in an extensive variety of tenders, such as PC displays, TV, gadget panels, plane arena presentations, indication. Those are mutual from patron gadgets similar to cinematic companies, gamble strategies, timepieces, wristwatches, adding machine, and phones. This is a by electronic means tempered optical ruse thru upon a little no of sections occupied per liquid quartzes are organized in forward-facing of a sunlit foundation (backlight) or reverberator to yield pictures in color or indeterminate (Fig. 6).

(vi) **ESP-8266 Wi-Fi Module**

The ESP-8266i Wi-Fi Unit is a self-sufficient System on Chip (S.O.C) with combined T.C.P/I.P decorum heap that can use at all microcontroller right of entry to your Wi-Fi system. The ESP-8266i is successful for mutually web holding a software and unburdening all Wi-Fi schmoozing task from additional tender workstation. To each ESP-8266 unit* derives cybernetic* with an AT command* set firmware*, meaning*, you can in reality hook* this up to your Arduino* gadget and get about as a whole lot of Wi-Fi* capability as a Wi-Fi Safeguard* offers. The ESP8266 module* is an extremely* within your means board* with a huge*, and ever community (Fig. 7).

(vii) **FN-M-16-P Module**

FN-M-16-P is minor and cheap MP4 unit with basic productivity at the orator. The unit are rummage-sale as a separate unit along with attached battery-operated, orator, and thrust keys or castoff in mixture by an Arduino UNO or slightly further by receiver/transmitter competences (Fig. 8).

3.2 *Software Requirements*

This Arduino Integrated Development Environment (IDE) can be used multi-platform that can be printed in purpose of C and C++. By the use to inscribe and sync plans to Arduino well-suited panel, by the aid of mediator nuclei and another seller progress panel. This base code for the I.D.E can be unconfined beneath the G.N.U Universal Civic Authorization, kind.

This Arduino I.D.E provisions the languages C and C++ by means of superior rubrics of code configuring. Thus, Arduino I.D.E provide a software collection since

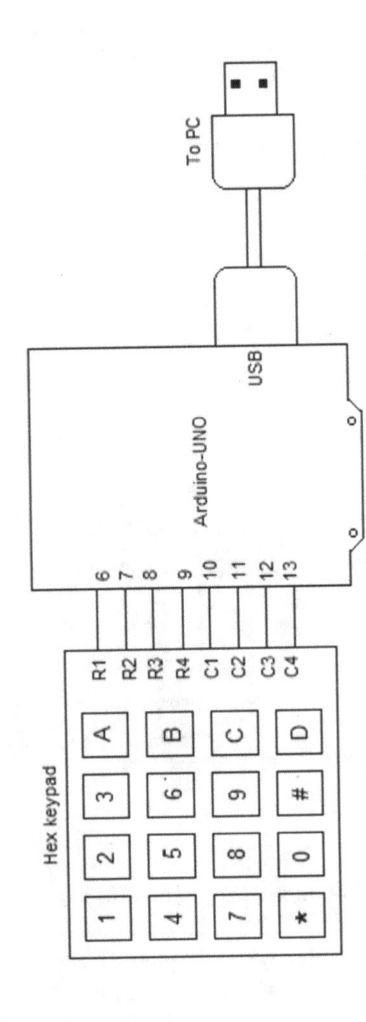

Fig. 5 4 × 4 keyboard interface with Arduino UNO

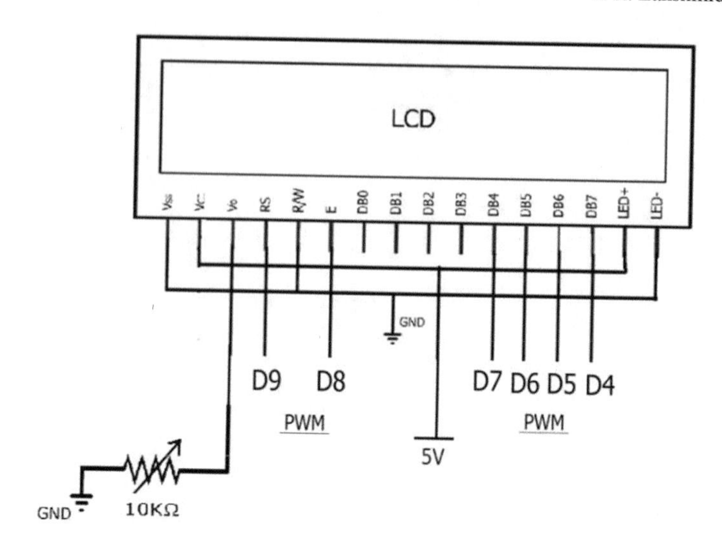

Fig. 6 L.C.D interface with Arduino AT-mega 2560

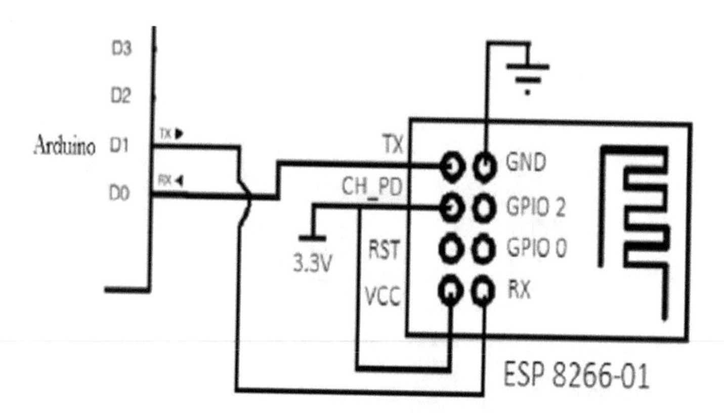

Fig. 7 ESP8266*-01 Wi-Fi module*

the cabling venture, which offers numerous public effort and trials. Programmer-noted code solitary necessitates two elementary purposes, from preliminary the draught and core program circlet that can assemble and allied with a program remnant core in to a viable cyclical execution database through the G.N.U toolkit, likewise encompassed by the I.D.E dispersal. So, Arduino I.D.E hires the plug-in Arduino to adapt the viable code in to a manuscript in hexadecimal encrypting that is encumbered plug-in in the panel's firmware. Though, Arduino is castoff as the sync utensil to flashy the programmer code on to certified Arduino panel.

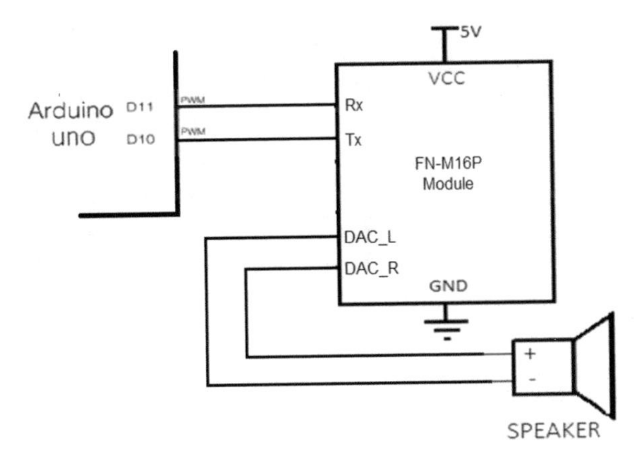

Fig. 8 FN-M-16-P module

3.3 *Algorithm*

TRANSMITTER SIDE

Step 1: Switch on the device.
Step 2: Enter your passkey.
Step 3: If passkey is correct allow user to access the system, else go to Step 2.
Step 4: Read values from hand glove.
Step 5: Compare flex sensor value with the pre-defined range of data along with accelerometer.
Step 6: The both values are read and assigned to a character.
Step 7: If the values are not read show the reading process On LCD display.
Step 8: Send the assigned character through Wi-Fi unit to the recipient side.

RECEIVER SIDE

Step 1: Receiver Wi-Fi unit accepts assigned the character by the transmitter side.
Step 2: The assigned character is compared with pre-defined character.
Step 3: The corresponding audio will be played from the mp3 player. Some Common Mistakes.

3.4 *Project Outcome*

In this paper was to help army personnel. For these we have made some gloves that will soldiers to communicate harsh environments through hand signals. When the soldiers preform the hand signals the Arduino AT- mega 2560 circuit in the gloves will convert the hand signals in to voice signals that will be communicated to other fellow soldiers via earpieces that they are wearing and the signals will also be encrypted. So, any kind of hacking on to signals will be very difficult. This project will help the soldiers to perform more efficiently as a team and in an environment, where they are

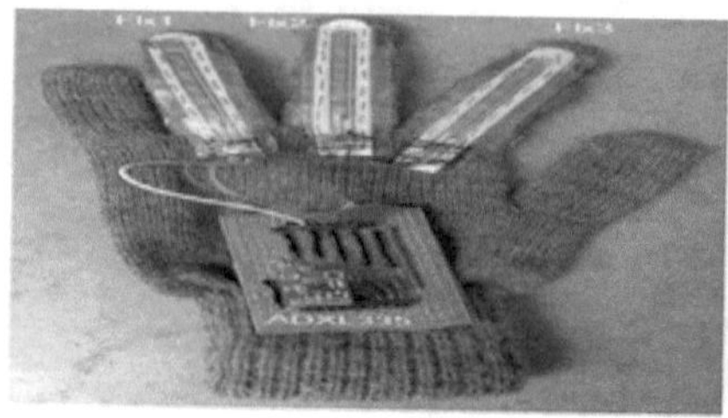

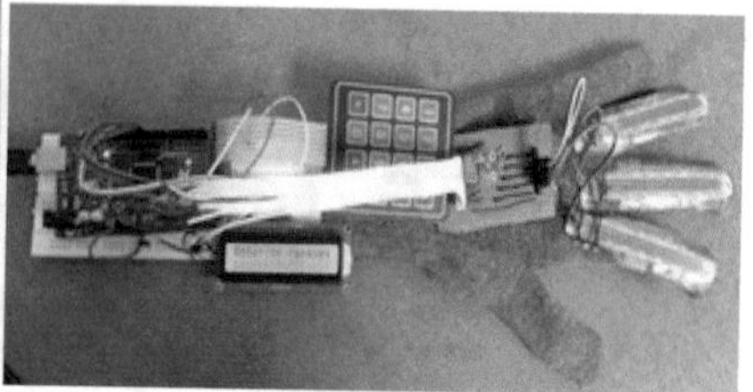

Fig. 9 Gloves

not able to see each other clearly. They will also monitor the health of soldiers by using hearts sensors and BP sensors and if something goes wrong with one of the soldiers the other fellow soldiers will be alerted (Fig. 9).

4 Conclusion

Smart glove is very useful in many ways. We can use this device to communicate with control station in noisy environment. From highly précised devices, it would be easy to recover the wounded person. We will upload the measured health conditions to cloud through the IOT. To monitor the Soldiers health conditions in the war field. Android tender can use for exhibiting signal notes.

References

1. Mehdi SA, Khan YN (2002) Sign language recognition using sensor gloves. In: Proceedings of the 9th international conference on neural information processing, 2002. ICONIP '02 2002, Vol 5, pp 2204–2206
2. Preetham C, Ramakrishnan G, Kumar S, Tamse A, Krishnapura N (2013) Hand talk-implementation of a gesture recognizing glove. In: India educators' conference (TIIEC), 2013 Texas instruments, Bangalore, 2013, pp 328–331
3. Rajam PS, Balakrishnan G (2011) Real time Indian sign language recognition system to aid deaf-dumb people. In: 2011 IEEE 13th international conference on communication technology, Jinan, 2011, pp 737–742
4. Deo N, Rangesh A, Trivedi M (2016) In-vehicle hand gesture recognition using hidden Markov models. In: 2016 IEEE 19th international conference on intelligent transportation systems (ITSC), Rio de Janeiro, Brazil, 2016, pp 2179–2184
5. More SP, Sattar A (2016) Hand gesture recognition system using image processing. In: 2016 international conference on electrical, electronics, and optimization techniques (ICEEOT), Chennai, 2016, pp 671–675
6. Harish N, Poonguzhali S (2015) Design and development of hand gesture recognition system for speech impaired people. In: 2015 international conference on industrial instrumentation and control (ICIC), Pune, 2015, pp 1129–1133

Computer-Aided Diagnosis Mechanism for Melanoma Skin Cancer Detection Using Radial Basis Function Network

Sreedhar Burada, B. E. Manjunath Swamy, and M. Sunil Kumar

Abstract Skin cancer is one of the most common form of cancers among humans. Malignant melanoma is an assertive type of the skin cancer, and its roots are from the melanocytes, located at epidermis of the skin. Detecting the skin cancer at early stage helps dermatologists for accelerating their diagnosis. The proposed model is a computer-aided diagnosis method for detecting the melanoma skin cancer using radial basis function network. Initially, this model converts the color image into gray scale image and after that applies the filters like median, which remove the noise and other articulated objects. Next, it detects the lesion segments using segmentation methods and then extracts the features. Finally, the radial basis function network classifies the segmented image, either malignant melanoma or benign cancer. This paper discusses the results of the sample images and performance of the proposed model.

Keywords Median filter · Segmentation · Dermatology · Melanoma · Radial basis function

1 Introduction

The occurrence of skin cancer in all over the world has increased steadily over the past decades, because of changes in the pattern of sun exposure of the population. Mainly skin cancer is classified into malignant melanoma and benign. Malignant melanoma is an assertive type of the skin cancer, and the root of this skin cancer is from melanocytes, which is in the epidermis of the skin. Malignant melanoma tumors are often cancerous. The death rate of melanoma skin care is dramatically increased from the last few decades. According to the American Cancer Society, the statistics

S. Burada (✉) · B. E. M. Swamy
Don Bosco Institute of Technology, Bengaluru, Karnataka, India
e-mail: sreedharburada1@gmail.com

M. S. Kumar
Sree Vidyanikethan Engineering College, Tirupati, Andhra Pradesh, India

A. Kumar et al. (eds.), *Proceedings of the International Conference on Cognitive and Intelligent Computing*, Cognitive Science and Technology,
https://doi.org/10.1007/978-981-19-2350-0_60

of melanoma skin cancer in the year 2020, over 100,350 new melanoma skin cases have been estimated [1]. Among these, 60,350 are the male cases and 43,070 are the female cases [2]. It is becoming a dangerous issue in present days. The dermatologist can diagnosis the skin cancer based on his or her medical experience; sometimes, it is hard to diagnosis; and in those cases, dermatologists can prefer the biopsy.

If the result of the biopsy is negative, it will cost to the patients. So that an automaton computerized diagnosis mechanism is required. The automated diagnosis helps dermatologists for detecting the melanoma skin cancer at early stage. It helps to reduce the morbidity and treatment cost [1, 2]. The dermatological images are classified based on the ABCD rule which is a well-organized standard form for the skin cancer medical image classification [3, 4]. ABCD rule [5] is having many standards features: A (asymmetry), B (border), C (color), and D (Diameter). One additional feature is also added that is E (evolution), and the rule is ABCDE [2]. Many computer-aided diagnosis mechanisms have been proposed by many researchers. Such mechanisms could accelerate the diagnosis for the dermatologists. The objective of the computer-aided diagnosis mechanism is to store the images with the diagnostic information for further investigation and create new methods to diagnosis of melanoma skin cancer. The CAD of melanoma skin cancer improves the accuracy of diagnosis and also consistency of interpretation of image by dermatologists using the computer as a guide. The performance of this system has been evaluated with the opinion of dermatologists on both normal and abnormal images. This paper aims to build a computer-aided diagnosis mechanism for classifying the skin lesion into malignant melanoma or no-malignant melanoma using radial basis function neural network. In this paper, first phase introduces about image preprocessing like removing the noise, hair removal, etc. The second phase discusses about proposed methodology that is, how the features are extracted, and using radial basis function network, how it is going to classify the melanoma skin cancer. The third phase explains the obtained results, and the fourth phase suggests the conclusion.

2 Proposed CAD for Detecting Skin Cancer

The proposed computer-aided diagnosis mechanism is explained by a block diagram of Fig. 1, which comprises preprocessing, segmentation, feature extraction, and classification.

The preprocessing phase helps to convert the color image into gray scale image and removes the noise and some unwanted artifacts like hair-occluded information. Hair removal algorithm [6] which removes the hair-occluded information and helps to increase the accuracy of the dermoscopy images [7]. Noises are commonly present in images, and it can be removed using filters. Generally, filters are classified into 1. linear filter and 2. nonlinear filter. In the case of an image having uniform noise distribution, linear filters are used. Linear filters are most of the time idle. If the image noise distribution is random, in that case nonlinear filters are used. Pepper and salt images are coming under nonlinear filtering. The segmentation phase extracts the

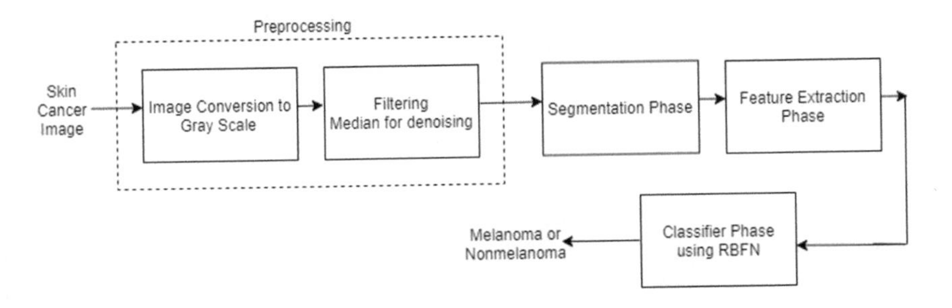

Fig. 1 Block diagram for RBFN classifier

lesion of cancerous from the skin background image [8–11]. The computer-aided diagnosis mechanism is built with the 29 features, which comprises basic shape—5, asymmetry—1, border irregularity—5, color—7, and texture feature—11 [4].

The classifiers process the cancer image features and classify [5] whether it is melanoma skin cancer and non-melanoma skin cancer. The proposed methodology uses radial base function for classification of melanoma skin cancer [12–14]. Radial base function network does work on nonlinear transformation, over the input vectors, before the input vectors are fed for classification [13]. By using such nonlinear transformation, it is possible to convert nonlinearly separable problems into linearly separable problems.

The RBF network works on increasing the dimensionality of the feature vectors. Because it has been found that, if we increase the dimensionality of the feature vectors then converting the non-separable problems into linearly separable problems. It is very difficult to classify the lower dimension problems like curve-fitting problems. It may be possible that they will be linearly separable in higher dimensional space. And more and more if we increase the dimension, so that it can able to convert non-separable into separable problems. RBF network is a fully connected layers, namely hidden and output layers as shown in Fig. 2. An RBF network can be viewed as a curve-fitting problem in a high-dimensionality space [14].

The training phase is always searching for best fit to the training data in higher dimensional search space. The input layer neurons are directly fed to the hidden layer neurons which is called as radial basis function. The nodes in the output layer perform linear combination of the outputs of the hidden layer nodes. Output layer has a linear classifier that is classification is done only at output layer. Whereas, at the hidden layer only perform the radial basis function which given the output values between the range 0–1, that is a transformation of the input feature vector which is skin cancer image to a hidden space. RBF has two levels of training, and the first level of training is training of the hidden layer nodes. The training and testing process gives the design of the proposed model [12].

The purpose of training the hidden layer is to find out what should be the center of the node. The output *i*th hidden neuron is expressed in the form of a nonlinear Gaussian distribution function which can be written as

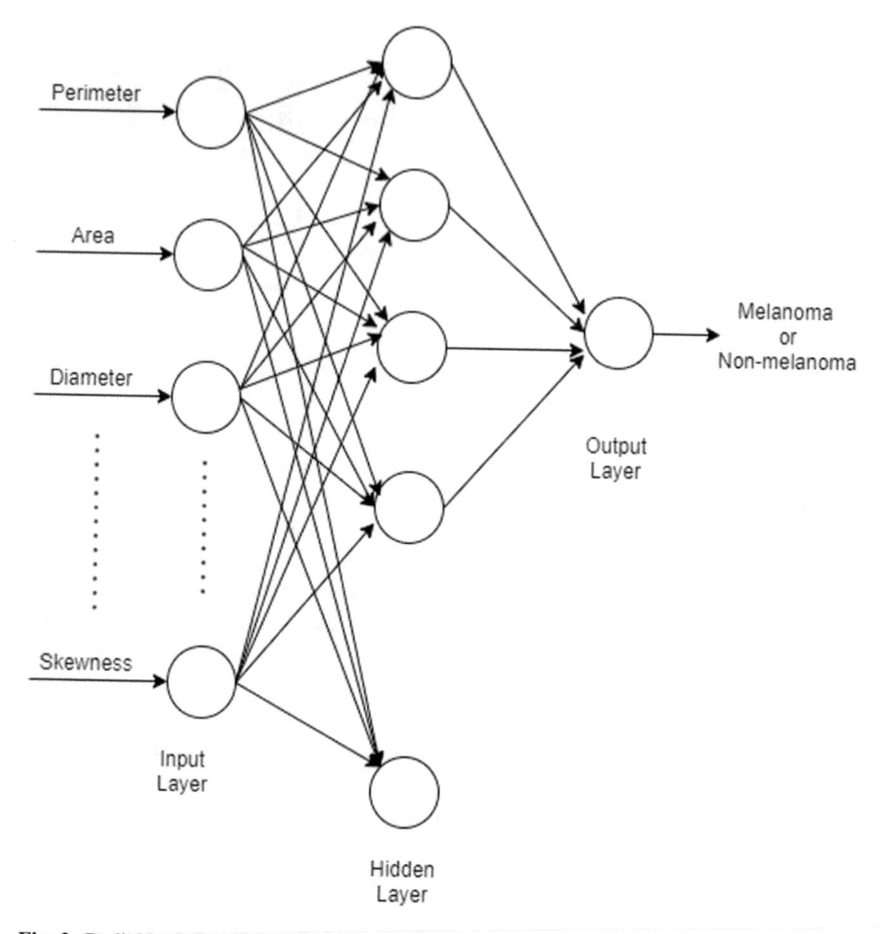

Fig. 2 Radial basis function network architecture

$$Ø(X_i) = e^{\left[-\frac{r^2}{2\sigma^2}\right]} \tag{1}$$

where σ indicates that what is the spread of radial basis function, so that it is also trained.

r is the distance between the input node X_i and center of the hidden layer node C_i, and r can be written as

$$r = \|X - C\| \tag{2}$$

Each neuron in the output layer has a linear transformation function that sums the weighted outputs of all hidden neurons connected to that output neuron. The output OUT (Y_j) of jth neuron is

$$\text{OUT}(Y_j) = \sum_{j=1}^{N} w_j \emptyset(X_j) \tag{3}$$

The objective of this configuration is simple mathematical representations and lowered computational efforts. The proposed network model is trained with the known training data set which comprises to extract skin cancer image features and target function. Once training has done at the hidden layer nodes, the proposed network can be ready to perform classification of new skin cancer image. The input data set X and the target output Y vectors can be represented as

$$\{X \leftrightarrow Y\} = \{f_1, f_2, f_3, \ldots, f_{29} \leftrightarrow \text{Class}\} \tag{4}$$

3 Results Discussion

The malignant melanoma and benign are taken from [14, 15] in tailoring the proposed computer-aided diagnosis mechanism. The detailed results of the malignant melanoma and benign skin cancer lesion have been discussed in this section. Figure 3 consists of two sample images, while Fig. 4 consists of the converted gray scale images and median filter sample images [3]. It has been observed from the tables that the thin hairs and other noises are effective by using median filter and hair removal algorithms. Figure 5 shows the edge detection, which is obtained by the segmentation process of melanoma skin cancer lesion region. Out of 29 features, 11 features were marked as texture feature and 18 were considered as color and shape features. These features are evaluated for all skin lesion segments in both positive and negative classes. Table 1 lists all the evaluated features of the sample malignant melanoma and benign skin lesions.

The training and testing data sets are created by considering the extracted features as input vectors and assign a value for benign as 0.1 and for melanoma as 0.9 which is a target vector. The output value of the radial basis function network is closest to

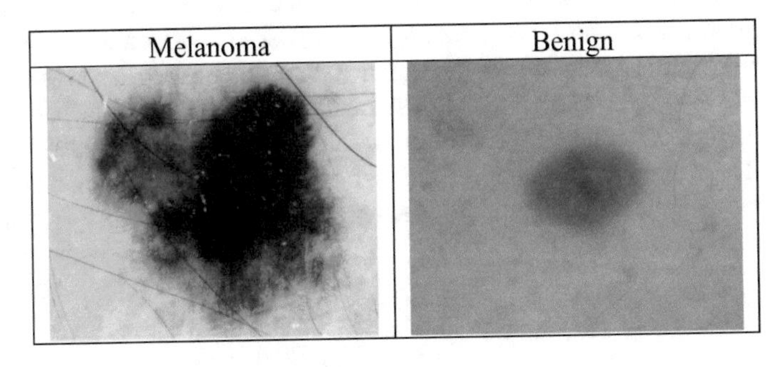

Fig. 3 Sample dermoscopic images

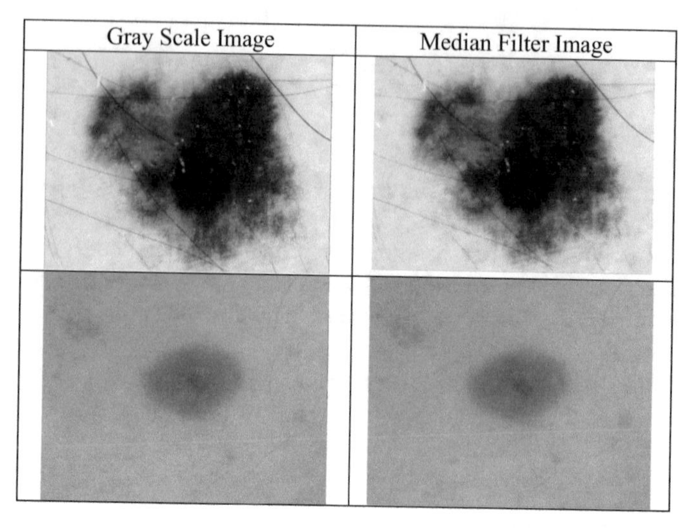

Fig. 4 Preprocessing the dermoscopic images

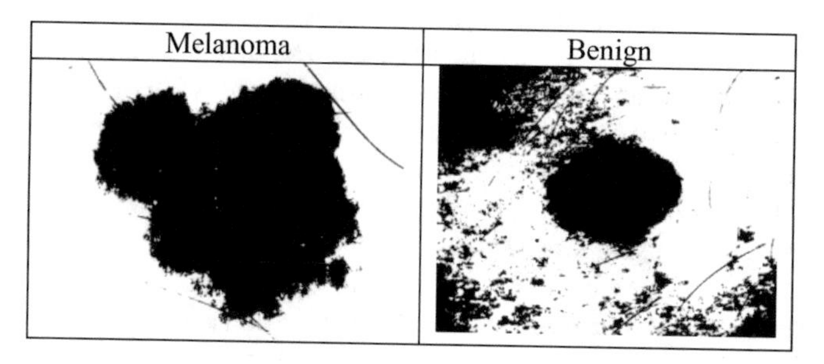

Fig. 5 Segmentation of sample dermoscopic images

the 0.9 vector value which will be treated as malignant melanoma. Generally, 80% of the input data set will be considered as training data set and remaining 20% of the data set will be considered as the testing data set.

After training, the radial basis function network classifier is ready for classification [12, 16] of skin cancer lesion images. The proposed network model gets an error result, because it fails to identify an abnormality. A confusion matrix consists of information about original and predicated classifications. This matrix is usually created for studying the performance of classifier, and in this, we consider the RBFN classifier [17, 18]. The common quantitative measurements of the confusion matrix, such as accuracy, sensitivity, and specificity, are calculated in this paper by the following equations.

$$\text{Accuracy} = \frac{(\text{TP} + \text{TN})}{(\text{TP} + \text{TN} + \text{FP} + \text{FN})} \quad (5)$$

Table 1 Extracted image features of sample dermoscopic images

	Melanoma	Benign
Area	1.3557e+04	3.2000e+03
Asymmetry index	1.0671e−01	1.1588e−01
Color score	2.0000e+00	2.0000e+00
Color count	2.1000e+01	2.8400e+02
	2.0000e+00	0.0000e+00
	0.0000e+00	0.0000e+00
	2.9690e+03	5.4000e+01
	1.3450e+03	1.5040e+03
	0.0000e+00	0.0000e+00
Contrast	1.8209e−01	3.4531e−01
Correlation	8.9493e−01	7.9220e−01
CI	1.1000e+01	1.5164e+00
Diameter	1.3138e+02	6.3831e+01
Energy	4.3649e−01	2.2591e−01
Entropy	6.1937e+00	6.4097e+00
Greatest diameter	1.3909e+02	6.71129e+01
Homogeneity	9.1227e−01	8.3816e−01
IrA	1.0098e−01	7.7168e−02
IrB	9.8421e+00	3.6785e+00
IrC	9.3643e−01	2.7544e−01
IrD	1.2084e+01	4.6763e+00
Kurtosis	8.0192e+00	1.0234e+01
Mean	7.8580e−01	7.0420e−01
Perimeter	1.3690e+03	2.4694e+02
RMS	7.9274e−01	7.1127e−01
Smallest diameter	1.2701e+02	6.2453e+01
Standard deviation	1.1543e−01	1.0724e−01
Skewness	−2.2136e+0	−2.2512e+0
Variance	1.0085e−02	9.7871e−03

$$\text{Sensitivity} = \frac{\text{TP}}{(\text{TP} + \text{FN})} \tag{6}$$

$$\text{Specificity} = \frac{\text{TN}}{(\text{TN} + \text{FP})} \tag{7}$$

Table 2 shows the confusion matrix. It is constructed from the results of test images. Table 3 describes the performance metrics of the proposed classifier with

Table 2 Confusion matrix

	Predicted classes	
	Melanoma	Benign
Melanoma	13	1
Benign	3	19

Table 3 Compare with existing methods

Ref.	No. of selected features	Classifier	Accuracy (%)	Sens. (%)	Spec. (%)
[3]	21	KNN	73	87	92
[8]		PSNR	96.26		
[9]	17	SRM DTEA RAC SDP		65.74 82.67 45.67 80.25	79.28 79.20 5.10 92.18
[10]	31	Fusion		83.6	81.7
[19]		Chan-Vese active contour method	94	78.5	99
Proposed method	29	RBFN	88.89	92.86	86.36

those of the existing work of the [3, 8]. The analysis of abovementioned table information indicates the superior performance of the proposed classifier through achieving 94.44% of accuracy, 97.85% of sensitivity, and 90.91% of specificity, which are much higher than those of the existing methods. From the below Table 2 it has clearly concludes that from the above discussion the computer-aided diagnosis mechanism using radial basis function network performs better in view of classifying the skin lesions into melanoma and non-melanoma skin cancer.

4 Conclusion

Radial basis function network-based classifier has been proposed for helping the dermatologists in diagnosing the skin cancer. This model has been built to eliminate artifacts such as noises and hair removing on the image, segment the cancerous, extract features that represent the characteristics of skin lesions, and classify into malignant melanoma or benign. The results on the two images, that is one belongs to malignant melanoma and another to benign, have clearly exhibited the superior performance of the proposed computer-aided diagnosis mechanism in terms of accuracy, sensitivity, and specificity. The proposed system performance can be improved further by considering the statistical features through k-means, SVM, and wavelet transform.

References

1. Cancer Facts and Figures (2020) American Cancer Society. https://www.cancer.org/content/dam/cancer-org/research/cancerfacts-and-statistics/annualcancer-facts-and-figures/2020/cancer-factsand-figures-2020.pdf. Accessed 8 Jan 2020
2. Sreedhar B, Swamy MBE, Kumar MS () A comparative study of melanoma skin cancer detection in traditional and current image processing techniques. In: 2020 4th international conference on I-SMAC (IoT in social, mobile, analytics and cloud) (I-SMAC), 2020, pp 654–658. doi: https://doi.org/10.1109/I-SMAC49090.2020.9243501
3. Ganster H, Pinz P, Rohrer R, Wildling E, Binder M, Kittler H (2001) Automated melanoma recognition. IEEE Trans Med Imaging 20(3):233–239. https://doi.org/10.1109/42.918473
4. Green A, Martin N, Pfitzner J, O'Rourke M, Knight N (1994) Computer image analysis in the diagnosis of melanoma. J Am Acad Dermatol 31(6):958–964. ISSN 0190-9622, https://doi.org/10.1016/S0190-9622(94)70264-0
5. Abbasi NR, Shaw HM, Rigel DS et al (2004) Early diagnosis of cutaneous melanoma: revisiting the ABCD criteria. JAMA 292(22):2771–2776. https://doi.org/10.1001/jama.292.22.2771
6. Abbas Q, Garcia IF, Emre Celebi M, Ahmad W (2013) A feature-preserving hair removal algorithm for dermoscopy images. Skin Res Technol 19:e27–e36. https://doi.org/10.1111/j.1600-0846.2011.00603.x
7. Huang A, Kwan S, Chang W, Liu M, Chi M, Chen G (2013) A robust hair segmentation and removal approach for clinical images of skin lesions. In: 2013 35th annual international conference of the IEEE engineering in medicine and biology society (EMBC), pp 3315–3318. doi: https://doi.org/10.1109/EMBC.2013.6610250
8. Al-abayechi AAA, Logeswaran R, Guo X, Tan W (2013) Lesion border detection in dermoscopy images using bilateral filter. In: 2013 IEEE international conference on signal and image processing applications, pp 365–368. doi: https://doi.org/10.1109/ICSIPA.2013.6708034
9. Abbas Q, Celebi ME, Fondón García I, Rashid M (2011) Lesion border detection in dermoscopy images using dynamic programming. Skin Res Technol 17:91–100. https://doi.org/10.1111/j.1600-0846.2010.00472.x
10. Emre Celebi M, Wen Q, Hwang S, Iyatomi H, Schaefer G (2013) Lesion Border detection in dermoscopy images using ensembles of thresholding methods. Skin Res Technol 19:e252–e258. https://doi.org/10.1111/j.1600-0846.2012.00636.x
11. Emre Celebi M, Iyatomi H, Schaefer G, Stoecker WV (2009) Lesion border detection in dermoscopy images. Comp Med Imaging Graph 33(2):148–153. ISSN 0895-6111, https://doi.org/10.1016/j.compmedimag.2008.11.002
12. Mengistu AD, Alemayehu DM (2015) Computer vision for skin cancer diagnosis and recognition using RBF and SOM. Int J Image Process (IJIP) 9.6:311–319
13. Przystalski K, et al (2010) Semantic analysis of skin lesions using radial basis function neural networks. In: 3rd international conference on human system interaction. IEEE
14. Jayapal P, et al (2014) Skin Lesion classification using hybrid spatial features and radial basis network. Skin 3.3
15. Garnavi R, Aldeen M, Bailey J (2012) Computer-aided diagnosis of melanoma using border-and wavelet-based texture analysis. IEEE Trans Inf Technol Biomed 16(6):1239–1252
16. Elgamal M (2013) Automatic skin cancer images classification. (IJACSA) Int J Adv Comp Sci Appl 4.3:287–294
17. Emre Celebi M, Kingravi HA, Uddin B, Iyatomi H, Alp Aslandogan Y, Stoecker WV, Moss RH (2007) A methodological approach to the classification of dermoscopy images. Comp Med Imaging Graph 31(6):362–373. ISSN 0895-6111, https://doi.org/10.1016/j.compmedimag.2007.01.003

18. Denton WE, Duller AWG, Fish PJ (1995) Boundary detection for skin lesions: an edge focusing algorithm, pp 399–403
19. Meskini E, Helfroush MS, Kazemi K, Sepaskhah M (2018) A new algorithm for skin lesion border detection in dermoscopy images. J Biomed Phys Eng 8(1):117–126. Published 1 Mar 2018

Personalized Ranking Mechanism Using Yandex Dataset on Machine Learning Approaches

B. Sangamithra, B. E. Manjunath Swamy, and M. Sunil Kumar

Abstract Web service technology is extensively utilized as data source by every client. As the quantity of Web information develops quickly, search engines are capable to retrieve data based on the customer preferences. Users now rely on the Internet to meet their information needs, but current search engines often return a long list of results despite using sophisticated document indexing algorithms, many of which were not constantly applicable to the needs of the customer. Because a customer has a precise aim in mind when looking for data, personalized exploration will deliver outcomes that precisely match the user's plan and purpose. Query-based investigate is commonly used by businesses to assist customers in finding information and products on their Websites. We look at how to rank a collection of outcomes returned in reply for a query in the most efficient way possible. Based on a customer search and click record, we propose a personalized ranking mechanism. We present our Yandex personalized Web search challenge solution. The goal of this challenge was to personalize top-N document rankings for a group of test users using historical search logs.

Keywords Personalization · Information retrieval · Web search · Yandex · Ranking

1 Introduction

With the World Wide Web, you can gain access to a vast amount of information that is stored in a variety of locations around the world in a short amount of time. As the amount of data on the Internet increases, different Internet users need different information. This has created several problems, including making it more difficult to

B. Sangamithra (✉) · B. E. Manjunath Swamy
Department of CSE, Don Bosco Institute of Technology, Bengaluru, Karnataka, India
e-mail: mithra197@gmail.com

M. Sunil Kumar
Department of CSE, Sree Vidyanikethan Engineering College, Tirupati, Andhra Pradesh, India

A. Kumar et al. (eds.), *Proceedings of the International Conference on Cognitive and Intelligent Computing*, Cognitive Science and Technology,
https://doi.org/10.1007/978-981-19-2350-0_61

find relevant information, to extract helpful information and to learn about customers. In order for a search engine to be competent to retrieve specific information when a customer presents a keyword in order to locate exact data, search engines will be capable to gather data in the manner that the customer has specified as appropriate. However, the search engine returns a list of pages that have been ranked according to their similarity to the query. Due to the absence of many relevant terms from queries, as well as the presence of ambiguous words, the results may not always be consistent with the interests of users. In future, approximately, half of the documents retrieved will not be associated with the desired search result, according to the results of the study [1].

As a result of the small form factor of portable devices, the communication among customer and portable devices is severely restricted when it comes to mobile search. The ability of a mobile search engine to recognize and understand the needs of its users, as well as the delivery of highly relevant information to them, are both critical requirements for a mobile search engine. This reduces the number of interactions between the user and the search interface. One method of resolving the issue is through personalized search. An individual user's preferences can be catered to through the use of user profiles that contain information about that user's interests, which is accomplished through personalized re-ranking of investigation outcomes acquired from general search engines by a personalized search middleware. User profiles are important in the personalization process because they are used to re-rank investigated results, which helps to make the process more efficient. Therefore, they must be trained on a regular basis using the search activities of the user, which must be done on a regular basis. In order to model users' content preferences based on their clicking and browsing behaviors, numerous personalization methods have been planned [2–5]. These techniques are based on the investigation of customer's clicking and browsing history. Because locality data are so significant in portable investigation, our paper proposes that user profiles should include the customer locality preference plus their contented choice and other preferences, in addition to their other preferences. customer contented and position preference (i.e., "multi-facets") are being captured and stored in an ontology depended, multiple faceted (OMF) customer profiling approach, which is being reviewed and improved for the purpose of developing an adapted search engine for portable clients. The common procedure of personalized search is depicted in Fig. 1, which is comprised of 02 key actions: (1) re-ranking and (2) revise of profile.

- **Re-ranking**: The search results are retrieved from the backend investigation engines whenever a customer submits a search query (e.g., MSN search). The search results were shared and re-graded in line with the customer report, which was created based on the customer's previous search history and is unique to each user.
- **Profile Revision**: Following the recovery of query items from backend Web investigation tools, the substance and area ideas (i.e., significant terms and expresses) and their connections are extracted online from the indexed lists and stored separately in the substance philosophy and area philosophy datasets as content

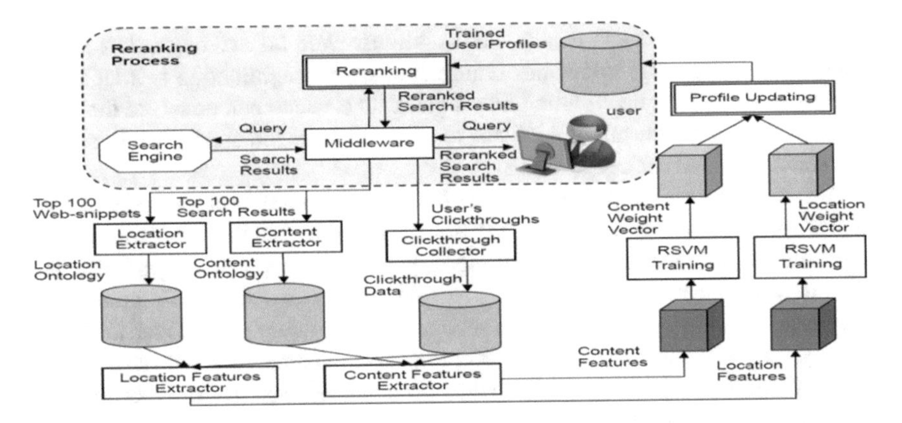

Fig. 1 General process of personalization search

ontology and area cosmology. At whatever point a client taps on an output, the clicked outcome, just as the related substance and area ideas, are saved in the client's navigate information. RSVM [6] preparing is then used to produce a substance load vector and an area weight vector for re-ranking the client's list items, which are then utilized related to the substance and area ontology's, just as navigate information.

Website personalization algorithms enhance the Web investigation practice by utilizing an personality information, such as the area of concern, favorites, query record (including the search terms used), browser history, and other information. They mine the outcomes that are appropriate to that entity based on the factors they consider [7]. The rest of this paper is organized as follows. Section 2 focuses on related work, and Sect. 3 focuses on the proposed methodology of the work. Section 4 compares our work with existing works, and Sect. 5 concludes the work.

2 Related Work

This section presents some of the most recent research findings in the fields of personalized Web search, recommender systems, and Web mining. The results of personalized search are affected by the information gathered from customer, whether absolutely or openly, according to research. As Morris [8] discovered, combining the data of several people with similar interests can help to improve the personalized nature of Web investigation outcomes. A difficult task is identifying closely connected set of public from the information that will be utilized for personalization purposes. The user significance mold proposed by jones can be used to create a user interest profile, which is described in detail in [9, 10].

Web history aids in the delivery of more personalized search results based on the search terms entered and the Websites visited. Early on, there may not be much of a difference in your search results, but as you use Web history more, they should improve over time. In most major search engines, hit-highlighting is a technique used to assist users in determining how significant a search outcome is to their data needs. A highlighted result in the list is one that contains the user's search terms in the name, fragment, or URL [11, 12].

A system was developed by Dou et al. [13] to compare profile-dependent and click-dependent hunt approaches. The researchers discovered that profile-dependent approach, in both extended and little term perspective, are more significant than click-depended methods in terms of increasing search performance. Reference [14] examines how entropy will be worn to answer a few key query's about Web investigation and how it be able to be worn to develop search outcomes. It was discovered that personalized backoff is more effective than personalized backoff alone, according to the findings of this study. When we do not have enough information about a specific user, we should move on to bigger and bigger sets of related customers. Jones et al. [10] affirm that people use a wide range of methods to manage Web information in order to re-contact and recycle it in future.

The empirical literature on marketing personalization, for starters, has a connection to this topic. Personalization increases profits by 7.6%, according to Sangamithra et al. [15], who distributed a persuasive paper measuring the advantages of balanced valuing utilizing buy history information. Also, Kumar and Harika [16] track down that content-situated mail can build navigate rates by up to 62%, and Harika et al. [17] track down that custom advancements can support their benefits through "installed expenses" or social issues. Then again, a few late papers have moved the re-visitation of personalization in an assortment of settings, from exposure to inn positioning in movement locales. Our paper contrasts from them on two primary measurements: scalability and predictive performance. In these previous papers, which are applied to small(er) datasets, the methodologies used (e.g., Markov chain Monte Carlo) are limited in scalability and are not built to maximize prediction performance. Our method, on the other hand, employs techniques such as automatic feature selection and can handle very huge datasets without losing performance or speed.

The use of experimental data on ten subjects inferred user interests from browsing history [18]. Teerevan et al. and Elheickh off et al. illustrate a return to personalization in various explicit search varieties concerning uncertain inquiries and a distinctive investigation session. According to Beurtnnett et al., durable record is helpful untimely in the gathering, while immediate history is useful afterward in the gathering. Three major dissimilarities exist. First, unlike previous works, we have the limitation of not knowing the context when it comes to data. Because we cannot use ideas like topic similarities across queries in our algorithm because of the data limitation, our task becomes significantly more difficult. Secondly, the previous papers use much smaller sets of data than we do, helping to avoid the scalability problems we have. Thirdly, we present a latest position of substantial results in personalization returns on heterogeneity. In particular, none of the previous papers calculate the return to personalization based on the user history length, the previous query performance or the user intentions of the query.

3 Methodology and Evaluation

3.1 Dataset

The data are open to the public as part of Kaggle's personalized Web search challenge. Customer ids, queries, query words, URLs, spheres, URL grade, ticks, and time stamps were extracted from Yandex logs. IDs, rather than actual data, are used to make user information unidentifiable. For faster experimentation and emphases of the learning calculations, we used an example from this dataset. We haphazardly chose 19568 special clients, then, at that point chosen every one of their solicitations and afterward partitioned the Kaggle information into a bunch of trains (first 12 days) and tests (after that 2 days). We do not utilize the test information of Yandex, as it does not have the data needed to ascertain NDCG. We have also detached queries from both the train and test datasets that do not have any click data. Some basic characteristics of our sampled dataset are shown in Table 1 and the graphs in Figs. 2, 3, and 4.

The Yandex-provided dataset's search results were not personalized. Depending on the customer, investigation is recorded. We have been working to rank the search results again. We used implicit click relevance because we did not have enough

Table 1 Sample dataset

Feature	Train	Examination
Days	12	2
Session count	68,975	8978
Exclusive customer	16,589	5689
Inquiry total	89,689	14,666
Exclusive conditions	97,865	98,698
Exclusive provinces	198,569	49,800
Exclusive URLs	802,689	16,589
Ticks total	189,678	24,896
Total records	1,464,066	219,318

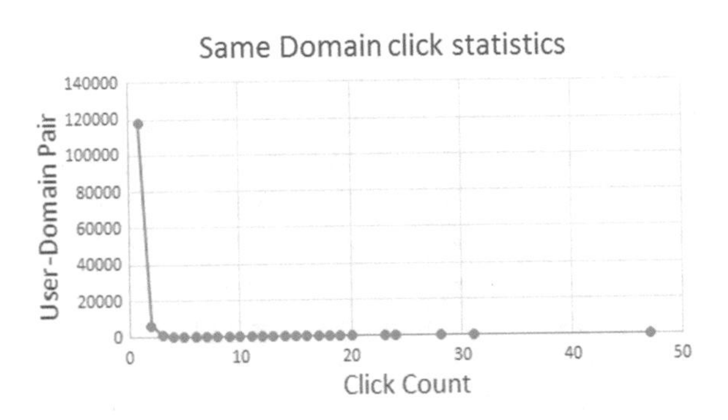

Fig. 2 Domain click statistics

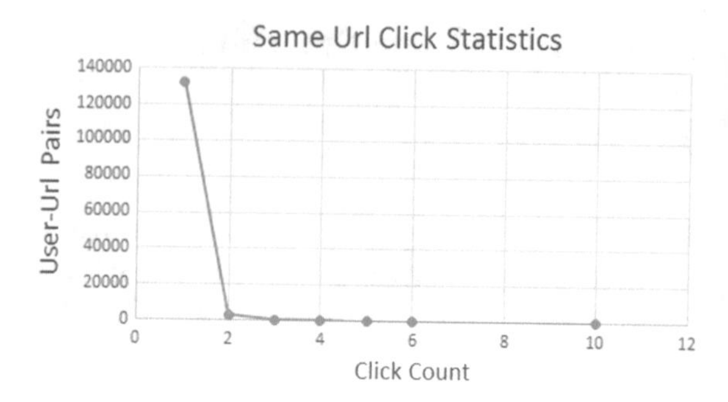

Fig. 3 URL click statistics

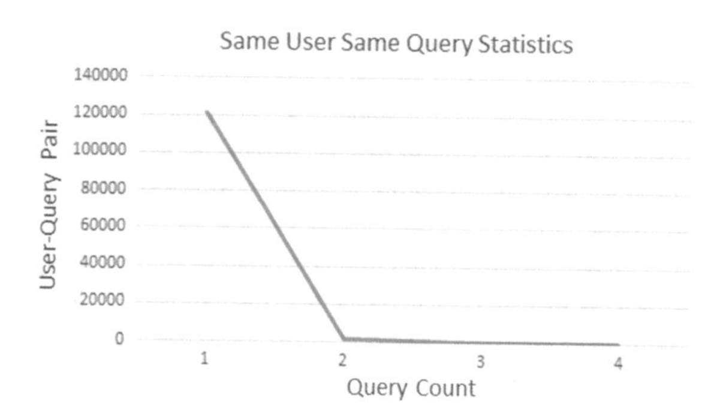

Fig. 4 User query statistics

information to make explicit judgments on relevance. We use clicks as a proxy for what the user believes to be important. Yandex also anonymizes user data by revealing only the session, user, query, URL, and domain IDs, rather than any other information. Therefore, we are unable to make use of any URL contented, query contented, or personal user summary information, as numerous works suggest. The characteristics to record individual distortions depended on the user's exploration and records as well as training a ranking system to re-rank the results were among the projects we worked on.

3.2 Data Labeling

URLs can be skipped, clicked, or missed for a specific session and query. All URLs under the previous ticked URL were skipped, and every URLs over it were not clicked. We are fascinated in URLs that have been clicked. As previously stated, we lack the human judgment data required to assign a relevance score to URLs. We divide the clicked URLs into three categories based on the number of clicks and the quantity of time used up on every one.

- High Relevance: The URL was ticked and more than 300 units were spent on it.
- Relevant: The time spent on a URL is between 50 and 300 units.

- Irrelevant: Click a URL and spend less than 50 units on it, or if you miss/skip URLs.

The quantity of time used up is considered based on the timestamp data contained in the dataset. In order to calculate the time spent, the time stamp for each click and query in relation to a session must be provided. The time spent is then calculated by subtracting the time distinction among the ticks. The time used up on a click is assigned 300 units, if it is the last activity in a session and it can be classified as satisfied.

3.3 *Normalization of Query*

The quantity of time exhausted is computed using the dataset's timestamp data. The timestamps for a click and a query are provided in relation to a session, and the time spent is calculated by subtracting the timestamp difference between the two clicks. If this tick was the last movement in a meeting, the time used up is assigned 250 pieces, because it will be classified as happy.

4 Data Features

This section covers the dataset's features that we derived from it. These characteristics are used in our model's training and testing.

4.1 *Navigational Query*

When it comes to navigational queries, personalized search should not be used. For instance, the question "Facebook" demonstrates that individuals would consistently go to the site facebook.com to join. While deciding navigational questions, we utilize the inquiry entropy highlight, which estimates the variety in clicked results across people. Coming up next is the equation for computing click entropy:

$$\text{Tick entropy}(b) = -\sum a(Nx|b) * \log 2(a(Ny|b))$$

$p(Nx|b)$ is the chance that URL $\times$ was ticked subsequent inquiry b

$p(Ny|b)$ = (amount of times that URL is ticked for particular inquiry)/(sum of URLs get ticked for that inquiry).

An increase in tick entropy indicates that a lot of pages were ticked (in response to the query, rather than a decrease in click entropy) a decrease in tick entropy specifies that only a few pages were ticked (in response to that inquiry).

4.2 User Preferences

Users' previous domain preferences: Some customers expressed a favorite for specific elements for a specific topic in the past, and this is still the case today. For instance, some users prefer shopping on Amazon, and others prefer shopping on eBay. According to train data, the same user clicked on 8282 domains (out of a total of 126,448 clicked domains).

- Customer's previous favorite for URLs: Calculate the number of times a search is handled repeatedly (Fig. 3). In accordance with the train data, 4563 URLs (among 268,976 ticked URLs) were ticked more than once by the similar user.
- The customer's previous favorite on the similar query: Query-specific URL and domain preferences are expressed by the user. For example, a user may prefer the result of a technical query from stack overflow, whereas a user may prefer the result of a movie query from IMDb. In the same way, the user's preference for URLs may change depending on the query type.

In total, 1540 queries (from a total of 122797 total queries) were submitted by the same user more than once, according to data from the train system.

Using every predicate compute the 03 characteristics that match to 03 URL categorizations for each of the 03 URLs:

- Higher significant prospect: (total higher significant URLs for a predicate + 2)/(total URLs for a predicate + 4)
- Lower significant prospect: (total lower significant URLs for a predicate + 2)
- Predicate-related probability = (total significant URLs for a predicate plus one)/(whole URLs for a predicate) + 4
- Inappropriate probability: The sum of all irrelevant URLs for a predicate multiplied by one-third of the total number of URLs for a predicate + 4. We use Laplace leveling to deal with the sparseness of the information in this case.

Query-Based Feature

URL re-positioning likewise relies upon inquiry equivocalness. For catching question uncertainty, we have included two highlights:

- query length : more modest the length greater likelihood of vagueness.
- query normal situation in meeting: later in the meeting implies less questionable as client continues refining inquiry.

5 Results and Comparison

Our approach to this problem has been to take a step-by-step approach. Regression or classification methods are typically used in conjunction with point-wise algorithms. This was accomplished through the categorization of the examination URL's into one of three categories: 0 (miss/inappropriate/skip), 1 (appropriate), and 2 (appropriate/skip) (highly relevant). When developing our model to categorize URLs into the categories listed above, we used the subsequent algorithms to train it:

- Random forest
- Gradient boosted trees.

These algorithms were implemented using the Scikit-Python library, which we found to be very useful (Table 2; Fig. 5).

The following parameters were used in the random forest algorithm:

Table 2 Comparison of different dataset

Model	Zero entropy restriction (do not use personalization where query entropy is zero)	Baseline NDCG score on test data	Predicted NDCG score on train data	Baseline NDCG on train data	Predicted NDCG on train data
Gradient boosting	No constraint	0.9268	0.8356	0.9153	0.8798
Gradient boosting	With constraint	0.9268	0.83659	0.9153	0.8536
Random forest	No constraint	0.9268	0.8365	0.9153	0.9268
Random forest	With constraint	0.9268	0.8309	0.9153	0.9256

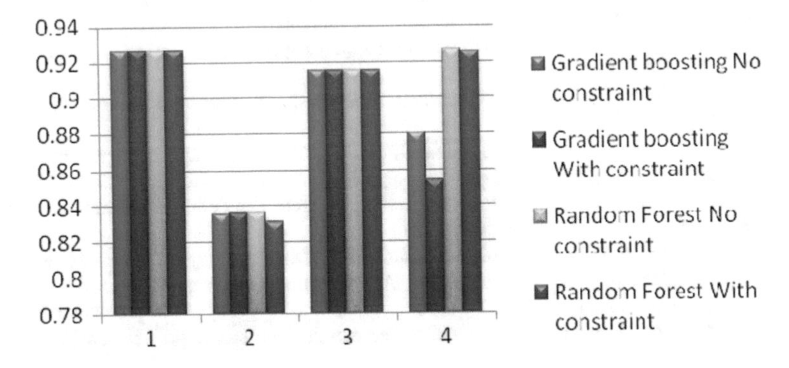

Fig. 5 Result analysis

n_estimators=20, max_depth=Nil, min_samples_split=4, random_state=None, n_jobs=-2

Gradient boosted trees have the following parameters, which we used in our experiments: n_estimators=20, learning_rate=1, max_depth=2, random_state=1

6 Conclusion

In this work, we examined the current state of research in the area of Web search personalization. This research demonstrates that there is a great deal of variation in the techniques utilized for personalized search. A choice of personalization strategies and personalized search methods have proposed and developed, but it has been discovered that none of them are optimal. The collection of users' requirements, so that results can be recovered based on the favorites of the users, has potential for development in future. By examining the results, we have discovered that personalized Web search does not perform uniformly well in all situations under consideration. In certain cases, it is a significant improvement over normal Web search, while in others, it has little effect and may even have a detrimental influence on query performance. Our future work will be focused on personalized Web search, which will make use of neural networks to improve the user-related search experience.

References

1. Smyth B, Balfe E (2006) Anonymous personalization in collaborative web search. Inf Retrieval 9(2):165–190
2. Agichtein E, Brill E, Dumais S (2018) Improving web search ranking by incorporating user behavior information. In: Proceedings of ACM SIGIR conference
3. Joachims T (2019) Optimizing search engines using clickthrough data. In: Proceedings of ACM SIGKDD conference
4. Ng W, Deng L, Lee DL (2019) Mining user preference using spy voting for search engine personalization. ACM TOIT 7(4)
5. Tan Q, Chai X, Ng W, Lee D (2018) Applying co-training to clickthrough data for search engine adaptation. In: Proceedings of DASFAA conference
6. Leung KW-T, Ng W, Lee DL (2018) Personalized concept-based clustering of search engine queries. IEEE TKDE 20(11)
7. Liu B, Lee WS, Yu PS, Li X (2017) Partially supervised classification of text documents. In: Proceedings of ICML conference
8. Morris MR, Teevan J, Bush S (2008) Enhancing collaborative web search with personalization: Groupization, smart splitting, and group hit-highlighting. In: Proceedings of CSCW '08
9. Li Z, Xia S, Niu Q, Xia Z (2007) Research on the user interest modeling of personalized search engine. Wuhan Univ J Nat Sci 12(5):893–896
10. Jones W, Dumais S, Bruce H (2002) Once found, what then? a study of "keeping" behaviors in the personal use of web information. Proc Am Soc Inform Sci Technol 39(1):391–402
11. Teevan J, Morris MR, Bush S (2009) Discovering and using groups to improve personalized search. In: Proceedings of the second ACM international conference on web search and data mining 2009 Feb 9, pp 15–24
12. Morris MR (2008) A survey of collaborative web search practices. In: Proceedings of CHI '08, pp 1657–1660

13. Dou Z, Song R, Wen J-R (2007) A large-scale evaluation and analysis of personalized search strategies. In: Proceedings of the 16th international conference on World Wide Web
14. Mei Q, Church K (2008) Entropy of search logs: how hard is search? With personalization? With backoff?" In: Proceedings of WSDM '18
15. Sangamithra B, Swamy BM, Kumar MS (2021) A comparative study on a privacy protection in personalized web search. Mater Today: Proc
16. Kumar MS, Harika A (2020) Extraction and classification of non-functional requirements from text files: a supervised learning approach. Psychol Educ J 57(9):4120–4123
17. Harika A, Kumar MS, Natarajan VA, Kallam S (2021) Business process reengineering: issues and challenges. In: Proceedings of second international conference on smart energy and communication. Springer, Singapore, pp 363–382
18. Sangamithra B, Neelima P, Kumar MS (2017) A memetic algorithm for multi objective vehicle routing problem with time windows. In: 2017 IEEE international conference on electrical, instrumentation and communication engineering (ICEICE). IEEE, New York, pp 1–8

Leading Platforms in Robotic Process Automation: Review

Pranav Desai, Shreyas Joshi, Yesha Desai, Naman Kothari, and Dattatray Sawant

Abstract This paper talks about robot process automation (RPA) its background and emergence, leading companies in the field of robotic process automation (RPA) their differences, specialties, and different orchestrator technologies used by different leading companies. How processes are automated and implemented using these leading platforms for RPA and their results and effectiveness.

Keywords Robotic process automation (RPA) · UiPath · Automation Anywhere · Blue Prism · RPA orchestrator

1 Introduction

The growth in population in today's world has practically resulted in an increase in all needs. To keep up with the rising demand, businesses and organizations want their personnel to perform increasingly more tasks. Organizations in all sectors of industries are attempting to discover a means to raise their pace of work, as well as efficiency and quality, as a result of increased competition. As a result, many businesses are turning to automation. Automation does not necessitate a large number of people and also requires less effort from users, from a human aspect. Automation is capable of doing everyday repetitive activities with near-perfect efficiency every time and at a rate quicker than humans.

P. Desai · S. Joshi · Y. Desai · N. Kothari
Department of Computer Engineering, SVKM's NMIMS Mukesh Patel School of Technology Management and Engineering, Mumbai 400056, India

D. Sawant (✉)
Department of Mechatronics Engineering, SVKM's NMIMS Mukesh Patel School of Technology Management and Engineering, Mumbai 400056, India
e-mail: dattatray.sawant@nmims.edu

A. Kumar et al. (eds.), *Proceedings of the International Conference on Cognitive and Intelligent Computing*, Cognitive Science and Technology,
https://doi.org/10.1007/978-981-19-2350-0_62

1.1 Definition

Robotic process automation (RPA) is a technology that allows anybody to set up computer software, or a "robot," to mimic and combine the activities of a human engaging with digital systems to complete a business process [1].

The IRPA-AI Institute provides one of the most up-to-date, clear, and comprehensive definitions of robotics process automation (RPA) [1]. RPA is defined as "the application of technology allowing employees in a company to configure computer software or a "robot" to capture and interpret existing applications for processing a transaction, manipulating data, triggering responses, and communicating with other digital systems" [1]. RPA robots use the user the UI to capture data and manipulate applications just like humans do.

1.2 Background and Emergence

To understand the history of RPA, we must look at three important technologies: screen scraping, workflow automation, and management tools, as well as AI. RPA's major ancestors are these three. RPA was born out of the need to bring these three together and enhance their capabilities. RPA employs a variety of technologies now, but these three technologies and concepts remain the most significant. Workflow automation and management technologies, as well as artificial intelligence (AI), have been around since roughly 2000, 1920, and 1956, respectively. The phrase robotic process automation first appeared in the early 2000s.

1.3 Current Form

The basic concepts of RPA have not changed, however, new technology has dramatically enhanced RPA's capabilities. With the help of software robots, we can now easily automate simple tasks, and as predicted previously, automation on a larger scale in various organizations and factories has already begun.

2 Literature Review

2.1 Leading Platforms

UiPath Studio, Automation Anywhere, and Blue Prism are the three primary platforms for developing the RPA process. Currently, Blue Prism (34%) is the most popular platform, followed by Automation Anywhere (28%) and UiPath (26%)

Fig. 1 Popularity pie chart

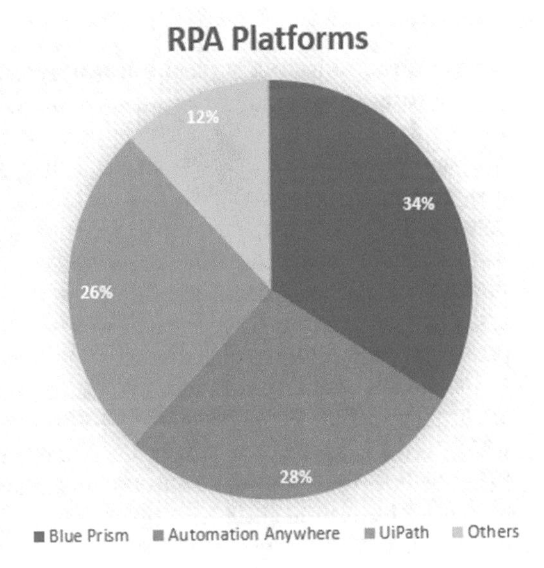

before all other platforms (12%). Although Blue Prism and Automation Anywhere are now the top two systems, UiPath might soon replace both since it offers a free community version of its RPA development platform, UiPath Studio, as well as several free learning programs. Apart from that, there are quite a few differences between the three platforms variety of which are shown below (Fig. 1).

- UiPath

  UiPath was founded in 2005, while it was still an outsourcing firm. They saw the need for robotic process automation (RPA) as a result of significant market demand and decided to begin developing an industry-standard platform for training and coordinating software robots [2].

  Their source code is integrated with a variety of products and enterprises throughout the world, performing tasks such as document management, call center, healthcare, finance, data extraction and migration, process automation, and API enablement [2].

- Automation Anywhere

  Automation Anywhere was founded back in 2003 and it started as Tethys Solutions [3]. The term "Tethys" is a Greek term that means Goddess of Water [3]. Automation Anywhere is a software company that employs software bots to complete processes and is a developer of RPA [3]. The product still looks like it was released in 2009 [3]. They also had a built-in ROI calculator. RPA's user interface is heavily inspired from here as the studio where the configuration of the workflow happens, and the control central which is used to establish the bots is loosely based on this. Automation Anywhere has an IQ Bot which is a combination of RPA and AI [3].

- Blue Prism

  Blue Prism is another leading company that provides RPA solutions and RPA tools [4]. It is based on Java and .NET framework and provides a drag and drop approach for bot designing [4]. The four main components of Blue Prism are process diagram, process studio, object studio, and application modeler [4]. Process diagrams are business workflows that are created by utilizing core programming concepts [4]. These graphical representations of workflows are used to create, analyze, modify, and scale business capabilities [4]. Process studio provides a platform to create process diagrams with various drag and drop activities [4]. Object studio is used to create visual basic objects which are used to communicate with other applications [4]. Application modeler is the functionality to create application models with Object Studio [4]. This exposes the UI elements of a target application to Blue Prism program [4]. Blue Prism introduces connected-RPA which works in association with artificial intelligence and cognitive capabilities [4]. Blue Prism also provides control room for analyzing bot activities and audit trails [4]. It has also integrated cross-platform support for many other AI and cloud technologies [4]. Work queues are used for workload management for managing the execution of multiple bots simultaneously [4]. Blue Prism also supports intelligent surface automation, multi-language interface support, customized dashboard, robot screen capture, etc., to name some [4].

2.2 *Specialties and Differences*

Specialties:

UiPath

- The foundations of UiPath, which uses drag and drop functionality and is not reliant on software coding, are Workflow Automation and Screen Scraping. It also has the capability of recording. It includes recorders for Citrix environments, desktop apps, and terminal emulators, allowing for much faster and more precise automation creation.
- UiPath Test Suite has strong integration capabilities, allowing it to connect to a variety of ALM tools, including JIRA. UiPath benefits the organization by reducing the manual workforce, improving customer service, and working more efficiently.
- UiPath offers a complete solution for automating third-party applications, application integration, business IT processes, etc.
- The key components of RPA are the software bots that are designed to mimic humans to reduce the manual workforce

- The life cycle is simpler as it involves very few steps and is flexible in dealing with the clerical processes.

Automation Anywhere

- Enhanced the user experience and other features like multi-personal, guided navigation, reusable ML libraries, strong security, improved analytics, and cloud delivery [5].
- Automation Anywhere has three strategic strengths which that uses product portfolio, pricing, and innovation [5]
- IQBots, Bot Insight, and other major components are all available within the enterprise package and with A2019 are offered in three bundled solutions that are competitively priced [5]
- The Automation Anywhere no longer has a dedicated testing product suite like UiPath, and there are certain bots that can be customized around software testing. Previously, the testing suite was called Testing Anywhere and is still available in limited release to some customers, but you must fill out a special form to get a copy of the software that is considered a legacy on-prem solution [5].

Blue Prism

- Intelligent Surface Automation uses computer vision to compensate for system and color variations increasing productivity and effectiveness.
- Adaptive Positioning Technology automates application controls and fields regardless of their onscreen position to increase robot speed.
- Custom Digital Frontends integrate front-office agents and back-office automation to improve human–robot interactivity and productivity. Mitigate risks with a platform used by the world's most secure financial institutions.
- Published dashboards offer unprecedented insight by packaging and externalizing analytics data for visibility within 3rd party tools along with a new Analytics UI.
- Introduction of multilingual UIs expands usability far beyond our 33,000 registered users and developers worldwide.

Differences:

There are many differences between the 3, few of these are (Table 1).

3 Major Tools and Orchestrators

3.1 Major Tools

UiPath:

UiPath is based on VB.net and C#. It is a drag and drop platform that allows a developer to automate different processes using a low amount of coding. UiPath

Table 1 Comparative table for different platforms of RPA

Features/categories	Automation Anywhere	UiPath	Blue Prism
Product availability	A one-month trial is available in the enterprise edition whereas the community edition is available to use with only BotCreator rights. Audit log, management, and API features together with control room repository access are not available	The community edition is available for all to use but the bots created cannot be distributed. Enterprise edition is available on 60 days free trial with 1 orchestrator, 10Licenses for UiPath Studio, StudioX, Studio Pro, 10 Attended, 10 Unattended, 10Test, 2 AI Robots, 10 ActionCenter, 1 Insight	Provides one-month free trial of the product. It has a limitation of 15 processes and 1 digital worker. In the learning edition, the free license is given for 180 days, limiting 1 digital worker and 5 processes
Usability	UI is complex. More suited for people with proper coding knowledge and developers	UI is very simple and easy to use. Can be used by fresh users too	UI is simple and provides an easy generation of bots
Type of processes that can be automated	Can be used for back-office and front-office automation	Can be used for back-office and front-office automation	Can be used for only back-office automation
Recorder	The three types of recorders that can be used for desktop as well as Web applications are—Smart, screen, and web	UiPath has a robust set (5) of recorders namely—basic, Web, desktop, image, and Citrix. These make it easier to capture human actions to mimic them further	One has to create a process using drag and drop features as there are no recorders available
Cognitive capability	Medium cognitive capabilities	Medium cognitive capabilities	Low-cognitive capabilities
Coding requirement	Coding is not obligatory as it supports both recordings and drag-drop approaches	Coding is not obligatory as it supports both recordings and drag-drop approaches	It supports coding but it is not obligatory. One can use the process diagrams and in-built functionalities as they are handy and convenient

(continued)

Table 1 (continued)

Features/categories	Automation Anywhere	UiPath	Blue Prism
Pricing	Cloud starter 9000$ (Customisable) Approx 20,000$ annually	Customizable as per requirement Approx 18,000$ annually	Around 15,000$ annually
Reliability and security	High security is provided as Automation Anywhere provides a credential vault to save confidential user information which is encrypted strongly	Proper encryption has been incorporated keeping in mind that the credential manager is used to save user data that is sensitive and confidential	The sensitive information is secured as it is saved in the Blue Prism credential manager. Users can choose an algorithm to generate the key and where to save it
Encryption algorithm	RSA with 2048-bit master key is used for encryption AES-256 bit key is used for data encryption	Supports encryption algorithms like AES, DES, RC2, Rijndael, and TripleDES	Source code obfuscation for all codes reduces the risk of attacks or reverse engineering or patching. Cipher obfuscation is used for credential information
Certification	Available online	Available online	Available online
Clients	Google, Siemens, Cisco, Dell	PWC, Lufthansa, HP, DHL	O2, Walgreen, Heineken

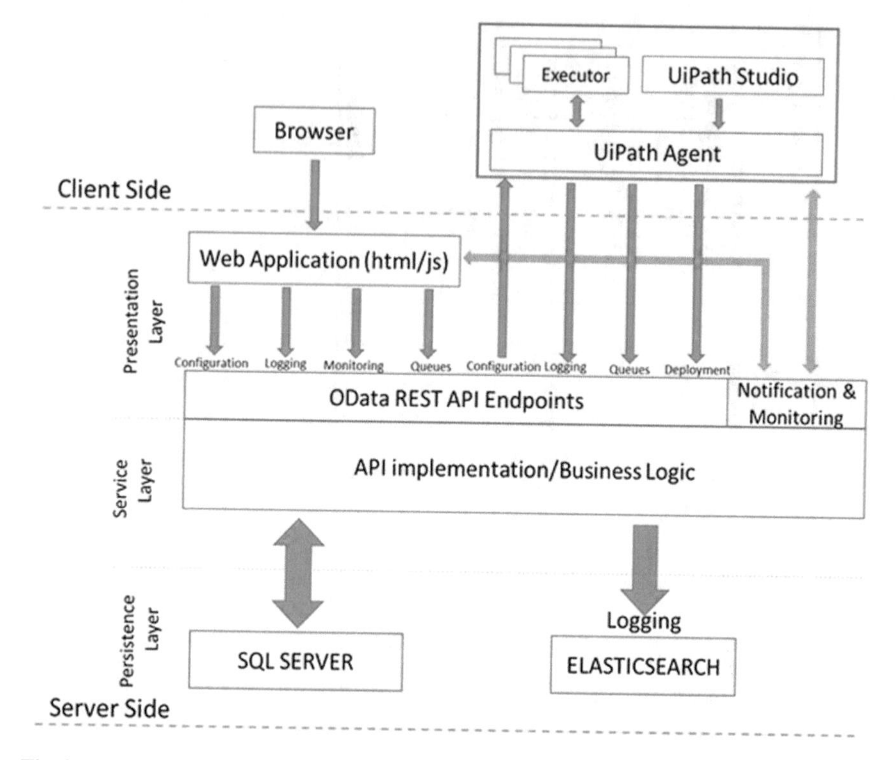

Fig. 2 Architecture of UiPath

provides UiPath studio as its development platform where you have the ability to use different pre-built templates, packages, etc.

It has 3 main elements of UiPath which make automation possible they are

1. UiPath Studio
   - It is a development software that allows both developers and business users the ability to build automated processes.
2. UiPath Orchestrator
   - It is a Web application created to enable developers to monitor and control processes and software robots.
3. Software Robots
   - The UiPath robot is the execution host that runs processes in the built-in UiPath Studio. They are of 2 types attended and unattended (Fig. 2).

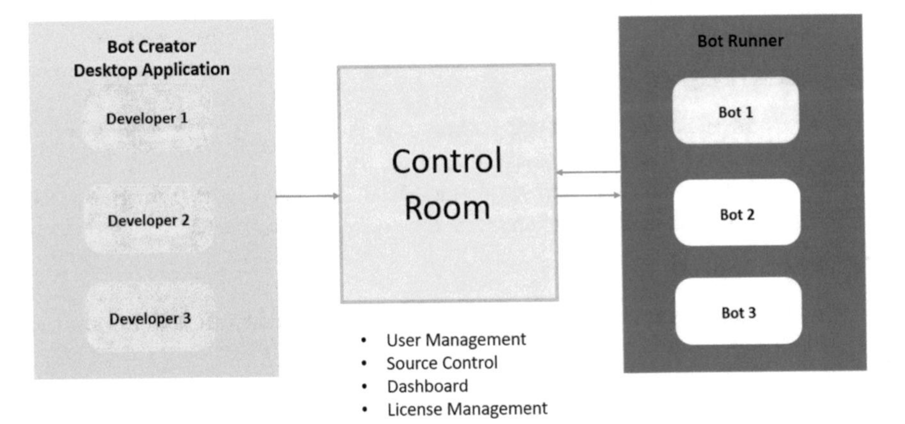

Fig. 3 Architecture of automation anywhere

Automation Anywhere

Automation Anywhere relies on C#. It is a Web-based management system that uses a control room to run the automated tasks. Developers use desktop-based applications to form bots. Their dev licenses are checked therewith configured within the room. On authentication, the code of the bots they create is stored within the room. Different developers may create individual tasks/bots. These bots might be merged and executed without delay.

Automation Anywhere architecture has 3 primary components

1. Control Room
    - It provides automated provisioning, orchestration, governance, and actionable analytics for enterprise-wide implementation.
2. Bot Creator
    - Bot Creators are simply used to create bots. These are desktop-based applications whose sole role is to upload or download bots and connect them to the control room
3. Bot Runner
    - The Bot Runner is the software machine that runs bots. When a bot is created using Bot Creator, then Bot Runners can run bots at scale (Fig. 3).

Blue Prism

Blue Prism relies on Java programming language and provides a simple-to-use interface with a visual designer. It permits us to use a flowchart like an interface with basic drag and drop functionalities for automating business processes step by step.

There are a total of four main components present in Blue Prism, such as:

1. Process Diagram

   - It consists of a different state. These states include data collection, decision, action, etc. These help to create a visual representation of the process flow. The execution of the program starts from the start of these diagrams. We choose a business object and then the corresponding action.

2. Process Studio

   - It contains a flow of the process or steps of the process. It is in a form of a flowchart.

3. Object Studio

   - Contains object diagrams that are similar to process diagrams but its main motive is to interact with external components. Has two default pages initialize, it is the first thing that executes can be used to keep global variable, etc. The second default page is cleanup this performs cleanup operation when an action is completed.

4. Application Modeler

   - An entity within object studio where we identify elements with the target application and subsequently perform action later (Fig. 4).

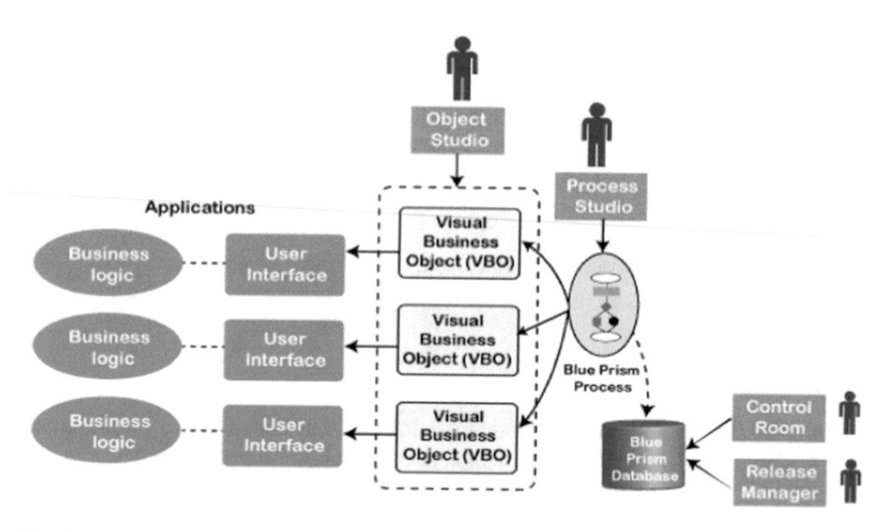

Fig. 4 Architecture of Blue Prism

3.2 *Orchestrator Technology*

Orchestrator is the main control center of automation. This is software that allows developers to save, start, and schedule software robots. It gives developers a place to analyze, report, track, and form of monitoring and controls over the automated processes and the software robots created to execute those bots. Each platform tries to provide a unique feature and user-friendly UI so that developers and control their automation more effectively.

Each platform deploys its own orchestrators.

UiPath

UiPath's orchestrator power comes from its capability of managing your entire robot fleet. Attended, Unattended, Nonproduction, Studio, or StudioX Robots—they can all be managed from this centralized point [6].

Code developed using UiPath studio is usually deployed in the orchestrator, and then, it can be assigned to a particular machine to run the robot [7]. Orchestrator provides various methods to run the robot. For a single instance of running a robot, one can go to the jobs section in the orchestrator and fire a job after selecting the required process [7].

Orchestrator Main Capabilities

- Processes—It contains all the automated processes that are uploaded on orchestrator.
- Assets—It contains all the saved passwords and data that are used in different processes
- Triggers—Triggers enable you to execute jobs in a preplanned manner, at regular intervals (time triggers) or whenever new items are added to your queues (queue triggers). The triggers page enables you to create new triggers, manage existing ones, as well as instantly launch a job for any existing trigger.
- Queues—Ensures automatic workload distribution across robots
- Users—This contains the different users of that account.
- Jobs Status—Contains all the information about processes that are currently running, stopped, etc.
- Jobs History—Shows history and numbers of processes that are successfully completed, stopped, and failed. (Fig. 5).

Automation Anywhere

The control room is the central interface for all things bot management, global value creation, user management, deployment, and testing. The tabs you most likely have access to as a developer are

1 Activity: From the activity, tab users can see in-progress bot runs, set up, and modify schedules for unattended bot runs, view event trigger details for bots that execute from a trigger, as well as view historical details on bots that have previously executed. Should you be executing unattended bot runner tests or

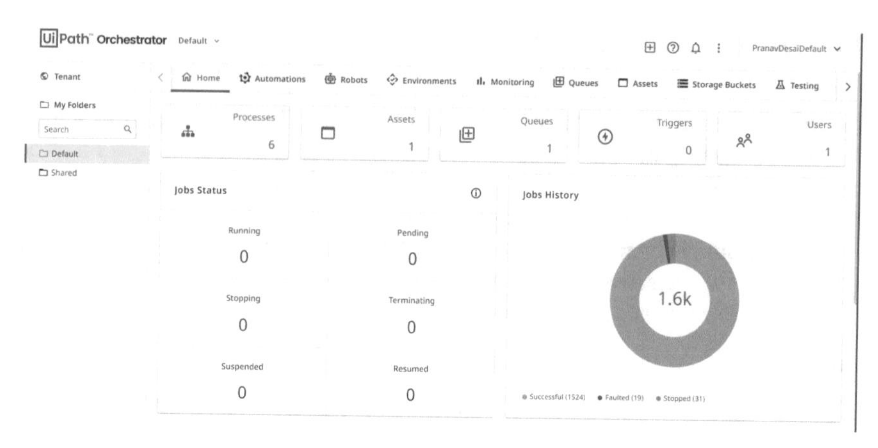

Fig. 5 Orchestrator of UiPath

reviewing production deployments—this will be the tab where you spend quite a bit of time.

2. Bots: From the bots tab (where bot builders spend a majority of their time)—we can access our private bot repository as well as view bots and bot components that have been checked into the public bot repository. Additionally, users have the capability to manage credentials (values that are securely stored in the control room for use by bots), manage global values (environment-specific values which are accessible to all bots), as well as the packages section where developers have the ability to manage and install custom packages.
3. Devices: The devices tab is where users come to set up devices that can be made available during development, testing, and production bot execution. It is on this tab that users can add or remove devices, upgrade devices, and validate that the Bot Agent on a particular device is able to successfully connect to the control room.
4. Workload: Workload management in Automation 360 provides users with the capability of managing bot workload by dividing work into logical segments which can be processed simultaneously across bot runners. The Workload tab of the control room allows administrators and developers to set up, manage, and modify workload queues that have been set up for groups of bot runners – including adding and removing devices to different workload pools.
5. Bot Store: The Bot Store tab allows users to directly install free or paid content from the Automation Anywhere Bot Store directly into their control room. With this capability, pre-developed bots and custom packages that are acquired via Bot Store can be instantly installed and used in your own environment to accelerate the bot building process (Fig. 6).

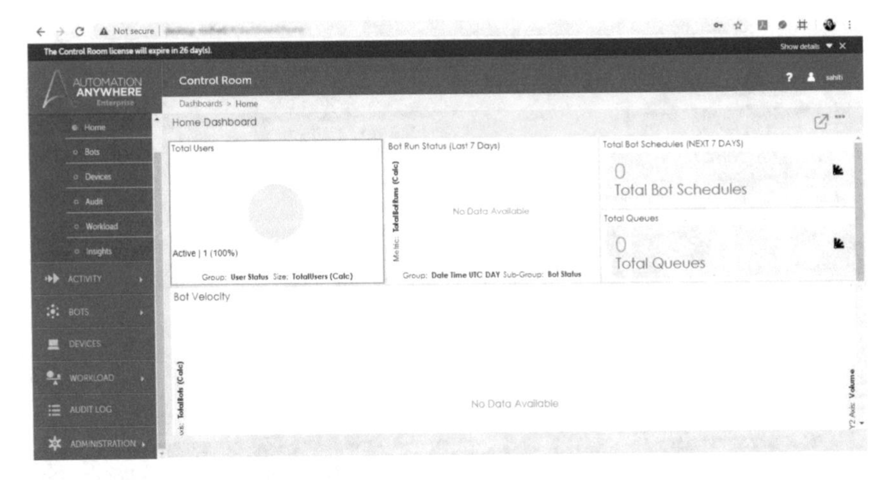

Fig. 6 Control room (orchestrator) of automation anywhere

Blue Prism

The control room is Blue Prism's command center: It provides a centralized administration console for controlling, monitoring, executing, and scheduling automated processes to the Digital Workforce similarly as an overall hub for resource management. Often underutilized, understanding the room is critical to reaching the total potential of your connected—RPA Digital Workforce.

- Scheduling allows for time-critical processes to be completed within defined SLAs even during peak volumes without the requirement to recruit for overtime or out-of-hours human workers.
- Digital workers will be allocated to certain processes or can be put to figure across all processes sort of a multiskilled worker conditional volumes this enables you to be reactive consistent with business needs.
- You have got access to any or all run data and run logs that may tell you exactly what actions have been completed by your Digital Workforce; this will be more transparent than actions completed by the human workforce and ensures full audibility of processes automated.
- It is a real-time view of what your Digital Workforce is currently performing on, at what speed, and how effective it is (Fig. 7).

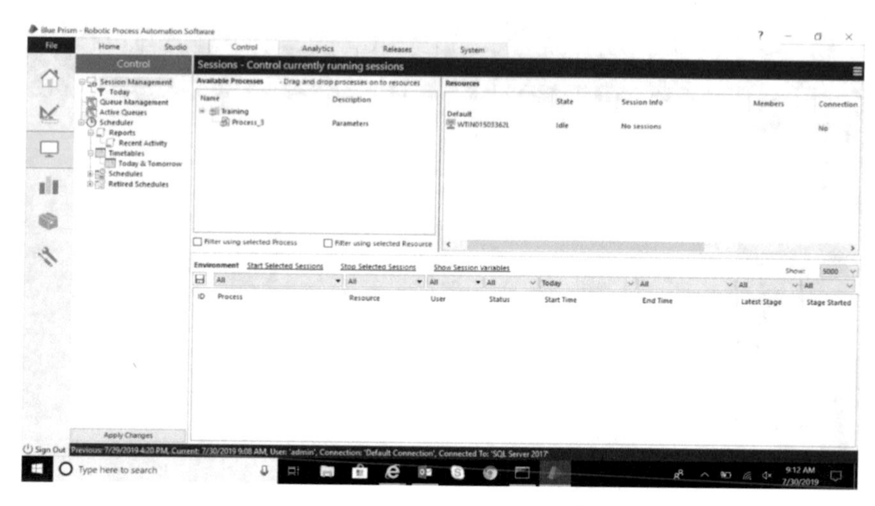

Fig. 7 Control room (orchestrator) of Blue Prism

4 Uses of Automation and Improvements That Can be Made

4.1 Uses of RPA in Various Domain

RPA has many benefits which make it one of the most in-demand and popular technology. A few of these benefits include:

- Increased Productivity

  The majority of robots created using the RPA technique are completely focused on a single task. In a matter of seconds, robots can calculate, click, and navigate through the screen. This time saving is not adequate to replace the labor with the robot. It does, however, make the worker more productive, which will help future cost avoidance. Business productivity will increase as a result of process automation. For e.g., RPA bot that permits a worker to create a monthly report in 20 min, which manually endures four hours.
- Increase Efficiency

  RPA software can work 24 h a day, 7 days a week, and 365 days a year without taking a break. It also does not take a vacation or become ill. In most cases, a single RPA robot can replace two to five full-time employees, if not more. Robots can complete the same amount of work in less time or more work at the same time as humans.
- Enhance Accuracy

  As software robots follow the particular activities/commands, they will not make errors normal employees do due to reasons like tiredness, lack of interest, etc.

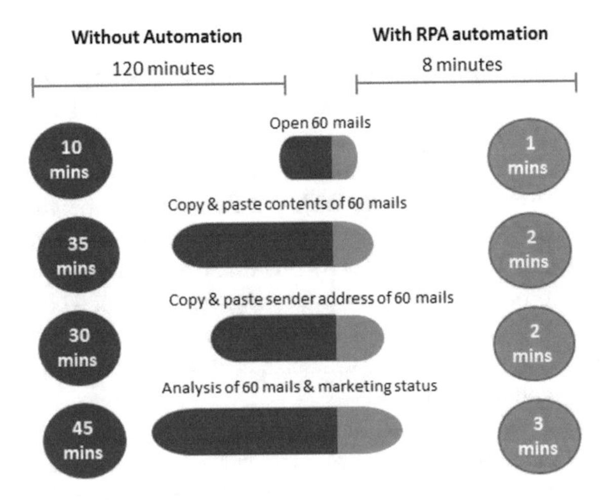

Fig. 8 Time required to do common tasks manually versus with RPA

- Increase Security

  Data accesses are well controlled and documented in any process that is automated. It functions on a granular level. As the bot performs only a single task, there is no worry of information leakage from one part to another.
- Boost Scalability Opportunities

  Robotic method automation offers flexibility to adapt what the instant demands in means that of the sort, and therefore, the variety of tasks required for any given objective. Robotic automation will support businesses to accommodate the chosen needs of bound objectives (Fig. 8).

RPA is in high demand in every sector as today's world needs technologies that can perform boring repetitive tasks faster, more reliably, and with greater consistency. A few of the industries where RPA is used are:

- *Banking*: There are many uses of RPA in the banking sector like Accounts Receivable, Customer Onboarding and Invoicing, Cash Application, Inventory Turnover, Logistics Management, Automatic Report Generation.
- *Finance and accounting*: In F&A processes, RPA reduces costs, improves efficiency, accuracy, and delivers significantly faster cycle times at more than 50% cost reduction. Processes that can be automated using RPA are procure-to-pay, order to cash, record to report, vendor management, incentive claims, collections, sales order management.
- *Retai:*: It can improve companies' finance and logistics processes across all its 5 centers. It can also help remove internal process bottlenecks to provide the best services and it can also create a human delivery center structure complemented by a virtual delivery center. It can perform tasks like report generation and data extraction, etc.
- *IT*: RPA delivers solutions to both enterprise computing and end-user computing. Approx. 30% of the time is spent on low-level, repetitive tasks. Processes that can

be automated using RPA enterprise computing processes: servers, data centers, security, infrastructure, end-user computing: any application used by human users: from Citrix to Excel, hardware or collaborative tools.

- *Data Analysis*: With large amounts of sales data to analyze and action every day, these processes demand intensive effort by employees as fast as possible. RPA provides full audits with real-time insights, helping retailers and suppliers carry out a more effective analysis to maximize sales opportunities.
- *HR*: RPA is a good fit for processes that require filling, capturing data, updating and processing requests, increasing efficiency and cost-effectiveness. These account for approx. 65% of HR rules-based processes. Processes that can be automated using RPA are data management, onboarding/offboarding, payroll, compliance, and reporting
- *Healthcare*: Max healthcare is one of India's top healthcare service providers. Around 1 crore pending payments were recovered. 65–75% of the time saved in CGHS and ECHS processing. Almost 50% of the turnaround time was reduced for claims processing.
- *Transportation*: DHL Global Forwarding Freight (DGFF) is a leading provider of ocean, overland, and air freight-forwarding services. After the pilot project was successful, the team started a center of excellence (CoE) to define the standards for RPA-based process automation such as the configuration and services for DGFF. If we talk of the numbers, then UiPath has helped DGFF achieve the following numbers 50% reduction in the total resources required to complete the process, enabled human employees to do such rewarding work, 300 robots are providing work equivalent to 300 full-time employees, who are presently working with better initiatives, has deployed 80+ robots is less than a year, achieve complete ROI in 1 month.
- *Telecommunication*: Sprint Corporation is one of the top telecommunication companies based in Overland Park, Kansas. It is the fourth-largest mobile network operator in the United States and offers a wide variety of wireline and wireless services. It has around 53.9 million customers and has an annual revenue of $33.6 Billion. More than 20,000 h are saved. 50+ automation developed. More than 50+ automation in the pipeline.
- *BPO*: MAXIMUS is a worldwide leading provider of government services. More than 2.5 million $ saved annually. 39 bots deployed in production and more than 10 are expected to be deployed in production by next year.
- *Education*: The University of Melbourne is located in the heart of Melbourne. This university has partnered with leading research centers and is one of Australia's oldest and most reputed institutions. It increases the efficiency of critical business processes. Boost staff engagement and reduce 10,000 h of manual force was reduced. Improve customer experience. 97% of throughput in processing supplier details. 22 processes are automated (Fig. 9).

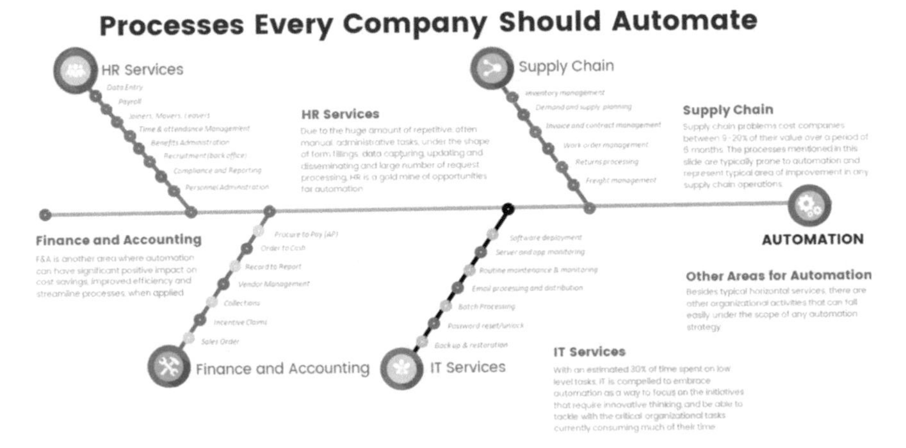

Fig. 9 Processes that can be automated in different departments of an organization

4.2 *Improvements that Can Be Done*

From the research that we have conducted, we found that there is no platform/software that can allow the user to select their preferred platform in combination with other platforms. For example, if an organization wants to use UiPath for one process and Automation Anywhere for another process currently there is no software that can allow a user to perform this function. So, adding this feature will be greatly useful for everyone involved as it will allow organizations that want to implement RPA ability to use the best feature of all three platforms.

5 Conclusion

Robotic process automation (RPA) is a great tool to automate processes that are boring and repetitive. RPA not only saves time and money by doing these tasks more effectively than a normal employee, but it also allows employees to focus on a more important task that requires their actual effort rather than them wasting effort on a task that can be completed by a few repetitive moves.

There are three major players in the field of RPA UiPath, Automation Anywhere, Blue Prism each have their own pros and cons depending on the need of the user they can select one suitable to them.

Automation Anywhere is built for mid and large-sized companies in the banking, finance, IT telecom, and healthcare sectors. The digital workforce of AA is focused on purchase, payment, and HR management makes it suitable for the banking, finance, and IT sectors. Blue Prism is created for large-sized companies. It provides strong support for back-office automation and thus makes it more suitable for media manufacturing companies and healthcare companies. UiPath is intended for both mid-sized and large-sized companies. It provides strong support for both front-office automation and back-office automation which makes it useful in various industries like human resources, insurance, and infrastructure.

References

1. Enríquez JG, Jiménez-Ramírez A, Domínguez-Mayo FJ, García-García JA (2020, February) Robotic process automation: a scientific and industrial systematic mapping study
2. Issac R, Muni R, Desai K (2018, February) Delineated analysis of robotic process automation tools. In: International conference on advances in electronics, computer and communications (ICAECC)
3. Desai D, Jain A, Naik D, Panchal N, Sawant D (2021) RPA: a business automation & reality. In: International conference on smart data intelligence (ICSMDI 2021)
4. Khan S (2020, November) Comparative analysis of RPA tools-UiPath, automation anywhere and blue prism. Int J Comput Sci Mob Appl
5. Andrade D (2020) Challenges of automated software testing with robotic process automation RPA—a comparative analysis of UiPath and automation anywhere
6. Nayak R, Vekhande V, Sheth B, Dhumal R, Patra P (2020, April) Product delivery optimization
7. Karn S, Kotecha R (2021, May) RPA-based Implementation of IoT

Comparison of Different Deep CNN Models for Leukemia Diagnosis

Chetna Agarwal and Virendra P. Vishwakarma

Abstract Deep learning has been gaining popularity as one of the most fascinating fields in the technology domain. Training deep learning models from scratch encounters several challenges in the real world. For one, it is not always possible to collect the necessary amount of data. In addition, one might not have access to high end hardware required for handling exceedingly large amounts of data. Transfer learning is a technique by which one can use the already trained models made available by the previous researchers and tweak them according to one's specific needs. Here, we present a comparison of several of these pretrained models in order to diagnose whether a patient is suffering from Leukemia, a disease which can be fatal if left undetected or untreated, based on microscopic images of patient's blood samples. The datasets used for performing said classification are the ALL IDB1 and ALL IDB2 datasets. Proposed classifier utilizes transfer learning in conjunction with fine tuning, a modification to the former which usually gives better results. Four of the most frequently applied pretrained models, namely VGG19, MobileNetV2, Resnet50, and InceptionV3, are used here and their accuracies are compared to see which one fits our problem statement the best. These models are compared on the basis of 4 different metrics, which are accuracy, precision, recall, and F1 score. The best model was obtained with InceptionV3, having an accuracy of 98.65%.

Keywords Leukemia · Deep learning · Transfer learning · Fine tuning

C. Agarwal (✉) · V. P. Vishwakarma
University School of Information and Communication Technology, Guru Gobind Singh Indraprastha University, Sector 16C, Dwarka, New Delhi 110078, India
e-mail: chetnaagarwal1997@gmail.com; chetnaagarwal_97@yahoo.com

V. P. Vishwakarma
e-mail: vpv@ipu.ac.in

A. Kumar et al. (eds.), *Proceedings of the International Conference on Cognitive and Intelligent Computing*, Cognitive Science and Technology,
https://doi.org/10.1007/978-981-19-2350-0_63

1 Introduction

Healthcare is one of the most exciting and salubrious domains of deep learning. Not all people have access to advanced healthcare facilities or can afford medical costs. In contrast, technology is one thing that is available to everyone these days regardless of a person's economic status or geographical location. What this research work aims to achieve is that healthcare facilities be made available to every individual at low cost by making use of technology. For this, we propose a deep learning model that can serve as an initial diagnosis for leukemia.

Leukemia, commonly called blood cancer, is an abnormal proliferation of blood forming cells in the body which encompasses multiple biologically distinct subgroups. Broadly, leukemia can be classified as myeloid or lymphoid in origin depending upon the predominantly affected cell type, and as acute or chronic depending upon the rate of cell growth. Each type of leukemia affects a specific demographic group and presents itself with its own set of symptoms and laboratory findings. Common symptoms include weight loss, fatigue, repeated infections, easy bruising, and bleeding from multiple body sites. Diagnosis of leukemia can be confirmed by blood tests, including estimation of the number of circulating blood cells and examination of their morphology. Cancerous blood cells have unique pathological features which can be easily evaluated on light microscopy, including changes in cell size, shape, and abnormalities in the cell nucleus (Fig. 1).

In this paper, these pathological features are extracted by deep learning methods to differentiate between cancerous and non-cancerous cells. For this, transfer learning and fine tuning techniques are utilized. Transfer learning is a technique used in deep learning wherein models whose weights and parameters have been optimized by the previous researchers for their specific problem statement are remodeled according to the current problem at hand [1–4]. The earlier layers of such models are reused, and custom layers are added at the end, whose weights are tuned. This way, it takes much less time and resources and at the same time, gives good results. It is also used when the amount of data one has is lesser than that required to train an effective model. There are many pretrained models available online that one can directly use in such projects. Fine tuning is a type of transfer learning technique which has been employed here to increase the accuracy even further. In this, the weights of pretrained models are taken as a starting point and modified up to some extent, along with the custom layers.

The paper is further divided into the following sections. In Sect. 2, a survey of work done by various researchers up until now is presented. Section 3 gives the complete end-to-end workflow of the proposed methodology, including dataset modeling, model initialization, training, evaluation, and inference. Section 4 explores the results obtained with complete analysis and discussion. Section 5 gives the conclusion and future work.

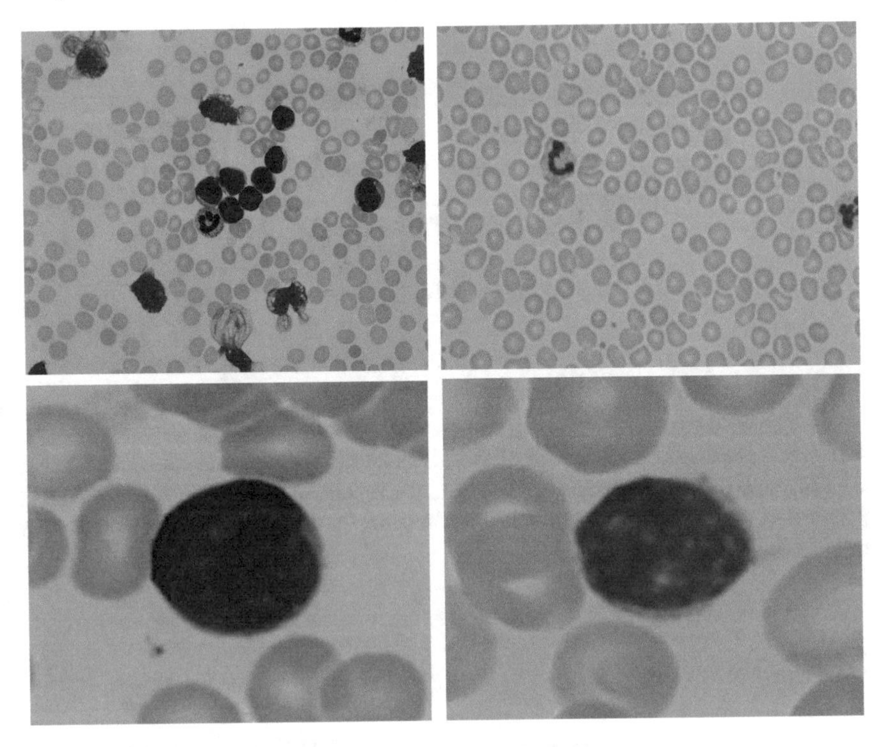

Fig. 1 Images of diseased and healthy samples from DB1 and DB2

2 Literature Survey

Researchers have previously tried and tested various classification methods for leukemia using machine learning, image processing, and deep learning methods. In [5], image segmentation techniques like k-means clustering, marker controlled watershed algorithm, and HSV color-based segmentation were used to extract morphological features from blood smear images, and classification was done using SVM classifier. In [6], image processing algorithms like image enhancement, image thresholding, mathematical morphology, and labeling were implemented on Leukemia blood smear images. In [7], a fuzzy system was proposed to classify leukemia into its subtypes. In [8], AlexNet, a pretrained model, was used with Leukemia images. In [9], a comparative study of machine learning algorithms like support vector machines, k-nearest neighbors, neural networks, and Naive Bayes was presented for diagnosis of leukemia and its subtypes. Genovese et al. [10] improved the sharpness of blood smear images before performing classification using CNNs. In [11], the chronological SCA algorithm was used to select optimal weights for deep CNN.

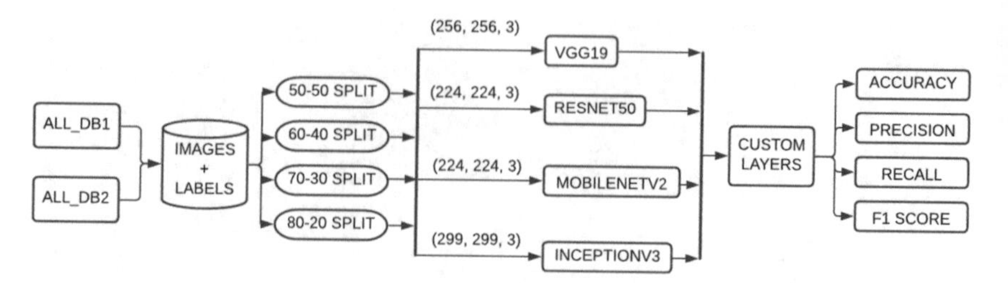

Fig. 2 End-to-end workflow of proposed methodology

3 Proposed Methodology

The paper presents a comparison of results obtained by applying transfer learning with fine tuning using 4 different pretrained models on the ALL DB1 and ALL DB2 databases [12–15] available in UCI Repository for Leukemia diagnosis as shown in the Fig. 2.

3.1 Dataset Modeling

The images from the 2 databases such as ALL DB1 and ALL DB2 were first combined to form a single image source, containing a total of 371 labeled images. They were then divided into training and testing sets using 4 different split percentages. The splits used were 80-20, 70-30, 60-40, and 50-50.

3.2 Model Initialization

For each of these splits, transfer learning and fine tuning techniques were applied with four different pretrained models and hyperparameters like batch size, optimizer used, learning rate, loss, number of frozen layers of pretrained models, etc., were tuned. The models used were VGG19, MobileNetV2, ResNet50, and InceptionV3. Custom layers were added at the end.

3.2.1 Pretrained Models

For our problem, we have chosen 4 of these and tested how well they perform for leukemia diagnosis problem. These 4 models are

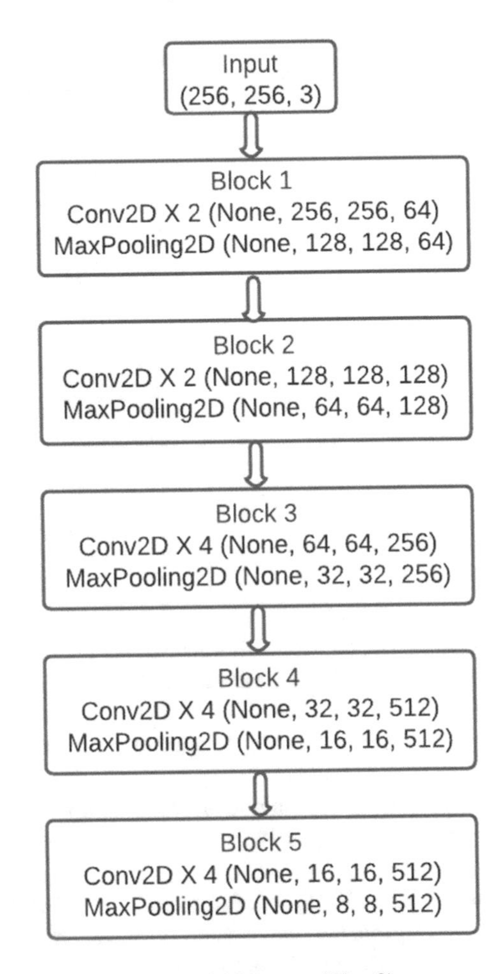

Fig. 3 VGG model architecture

- VGG19—created at Oxford University and consists of 19 layers (Fig. 3).
- MobileNetV2—created by Google and consists of consists of 53 layers.
- ResNet50—proposed by He et al. and consists of 50 layers. It introduced the concept of residual blocks and won the ImageNet competition in 2015.
- InceptionV3—is the culmination of many different architectures proposed by researchers over many years and consists of 48 layers.

3.2.2 Hyperparameter Tuning

- Binary cross entropy loss function was used. It compares the predicted probabilities to the actual class outputs, which are either zero or one, meaning it only works for binary classification. It then calculates the score that puts a penalty on the probabilities obtained based on how close or far away it is from the expected value.

Fig. 4 Values of hyperparameters used

Hyperparameter	*Value tuned*
Loss	Binary Cross Entropy loss
Learning Rate	0.0001
Rescaling factor	[1/255]
Optimizer	Adam
Dropout	0.5
No of frozen layers	100
Activation functions used	ReLu and Sigmoid

- Learning rate controls how quickly the parameters of the model are adapted to the problem at hand while training.
- Adam optimizer was used. The role optimizers play in training of deep learning models is to minimize the error function so as to maximize the efficiency of the resulting model.
- Dropout is a technique one can employ to reduce overfitting on a dataset, wherein several units of a deep learning model, i.e., neurons, are ignored at random during training.
- The first 100 layers of pretrained model were frozen, meaning their weights were not optimized, whereas all the layers after that, including all the added custom layers, were fine tuned.
- Activation function is a non-linear function one can use to transform the output of one layer before sending it over to the next layer. In custom layers, rectified linear unit (ReLU) and sigmoid activation functions were used (Fig. 4).

3.2.3 Custom Layers Added

The output from the pretrained models was fed into another deep learning architecture, which consisted of layers added manually. Several combinations of these layers were tried and tested to see which one would give good results (Fig. 5).

5 blocks of such layers were added, each consisting of a combination of the following layers:

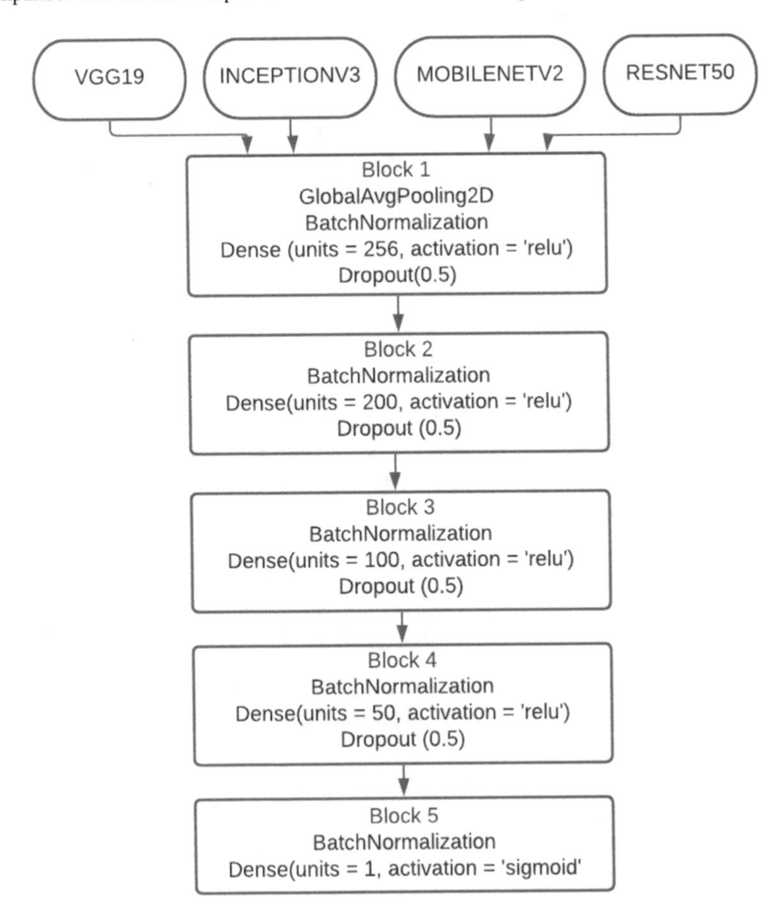

Fig. 5 Model head architecture

- Global Average Pooling Layer—This layer reduces the dimensions of the representation, hence reducing the number of parameters and computation time, while at the same time, retaining the maximum number of important features.
- Batch Normalization Layer—This layer is used to normalize outputs of each layer before feeding them into the next layer. This normalization allows every layer of the network to do more learning independently.
- Dense Layer—This layer consists of neuron units, which are connected densely to the layers before and after it, meaning all units of this layer are connected to all units of the previous/following layers. The activation functions used with these layers were ReLU and sigmoid.
- Dropout Layer—This layer is added to decrease the extent of overfitting on the training data.

3.3 Model Training

For various combinations of training and testing splits and models used, model was trained, and results were obtained. Four metrics were evaluated, i.e., precision, recall, F1 score, and accuracy. The model training was done using an NVidia T4 GPU, with a RAM of 12.69 GB and a disk space of 78 GB. Python 3 was used for implementation.

3.4 Model Evaluation and Inference

For comparing results of different models trained, the mean of these metrics for 20 epochs and 30 epochs was calculated. Also, the best accuracy obtained until 200 epochs was found out. The models were evaluated on the basis of 4 different metrics. The metrics used were

- Accuracy is useful when comparing the performance of the trained model across all classes and when all the classes have equal importance. It is defined as the ratio of the number of predictions made correctly to the total number of predictions made.
- Precision is defined as the ratio of the total number of positive samples which are correctly predicted to the total number of samples predicted as positive, either correctly or incorrectly.
- Recall is defined as the ratio of number of positive samples correctly predicted to the total number of positive samples.
- F1 score is calculated as the average of precision and recall. The best value for F1 score is at 1 and worst is at 0.

4 Results and Discussion

The accuracy, precision, recall, and F1 score are calculated separately for each of the pretrained models. Average of the values for these metrics until 20 and 30 epochs is calculated. Also, the best value for these metrics until 200 epochs is reported.

4.1 InceptionV3

See Figs. 6, 7, 8 and 9.

Fig. 6 InceptionV3 50-50

	Epochs = 20	Epochs = 30	Epochs = 200
Accuracy	52.58%	56.59%	88.04%
Loss	0.6849	0.6563	0.2627
Recall	73.42%	81.49%	100.00%
Precision	49.63%	52.56%	81.85%
F1 Score	55.05%	60.84%	89.63%

Fig. 7 InceptionV3 60-40

	Epochs = 20	Epochs = 30	Epochs = 200
Accuracy	65.47%	73.44%	97.97%
Loss	0.6210	0.5478	0.1104
Recall	41.84%	57.94%	98.61%
Precision	74.92%	79.17%	97.42%
F1 Score	49.44%	62.47%	97.93%

Fig. 8 InceptionV3 70-30

	Epochs = 20	Epochs = 30	Epochs = 200
Accuracy	64.14%	70.48%	98.20%
Loss	0.6455	0.6170	0.1566
Recall	69.85%	73.22%	96.93%
Precision	60.81%	68.97%	100.00%
F1 Score	62.43%	68.85%	98.40%

4.2 *MobileNetV2*

See Figs. 10, 11, 12 and 13.

4.3 *VGG19*

See Figs. 14, 15, 16 and 17.

Fig. 9 InceptionV3 80-20

	Epochs = 20	Epochs = 30	Epochs = 200
Accuracy	62.09%	69.68%	98.65%
Loss	0.5849	0.5112	0.0578
Recall	91.28%	94.19%	96.97%
Precision	56.88%	63.58%	100.00%
F1 Score	67.99%	74.22%	98.41%

4.4 ResNet50

See Figs. 18, 19, 20, and 21

Fig. 10 MobileNetV2 50-50

	Epochs = 20	Epochs = 30	Epochs = 200
Accuracy	48.56%	49.20%	91.85%
Loss	0.7199	0.7230	0.2506
Recall	87.53%	91.69%	100.00%
Precision	47.55%	48.20%	84.77%
F1 Score	60.38%	62.13%	91.42%

Fig. 11 MobileNetV2 60-40

	Epochs = 20	Epochs = 30	Epochs = 200
Accuracy	60.87%	62.13%	94.59%
Loss	0.6622	0.6484	0.1557
Recall	83.01%	88.58%	94.53%
Precision	57.55%	57.84%	94.88%
F1 Score	66.28%	68.63%	94.66%

Fig. 12 MobileNetV2 70-30

	Epochs = 20	Epochs = 30	Epochs = 200
Accuracy	55.09%	57.59%	94.59%
Loss	0.7040	0.6943	0.1949
Recall	44.75%	43.00%	96.00%
Precision	58.03%	62.15%	94.83%
F1 Score	47.88%	48.35%	95.14%

Fig. 13 MobileNetV2 80-20

	Epochs = 20	Epochs = 30	Epochs = 200
Accuracy	59.79%	61.44%	94.59%
Loss	0.6632	0.6492	0.1595
Recall	95.29%	92.65%	95.35%
Precision	55.85%	57.38%	93.33%
F1 Score	70.15%	70.34%	93.89%

Fig. 14 VGG19 50-50

	Epochs = 20	Epochs = 30	Epochs = 200
Accuracy	54.94%	59.85%	90.22%
Loss	0.6759	0.6646	0.2820
Recall	9.15%	31.90%	93.77%
Precision	32.63%	43.35%	86.79%
F1 Score	12.95%	31.65%	89.76%

Fig. 15 VGG19 60-40

	Epochs = 20	Epochs = 30	Epochs = 200
Accuracy	57.90%	59.07%	94.59%
Loss	0.6712	0.6590	0.1635
Recall	29.74%	27.10%	94.37%
Precision	75.89%	78.37%	94.70%
F1 Score	33.13%	33.03%	94.37%

Fig. 16 VGG19 70-30

	Epochs = 20	Epochs = 30	Epochs = 200
Accuracy	50.09%	49.90%	87.39%
Loss	0.7077	0.7128	0.2672
Recall	98.99%	99.33%	96.67%
Precision	49.79%	49.93%	77.99%
F1 Score	65.61%	65.89%	85.20%

Fig. 17 VGG19 80-20

	Epochs = 20	Epochs = 30	Epochs = 200
Accuracy	57.43%	63.55%	95.95%
Loss	0.6729	0.6574	0.1500
Recall	28.37%	36.47%	97.62%
Precision	56.31%	70.31%	95.56%
F1 Score	29.14%	41.62%	96.55%

Fig. 18 ResNet50 50-50

	Epochs = 20	Epochs = 30	Epochs = 200
Accuracy	49.13%	51.15%	82.61%
Loss	0.7050	0.7104	0.4014
Recall	65.00%	20.61%	75.23%
Precision	31.05%	11.83%	86.49%
F1 Score	41.47%	13.93%	78.60%

Fig. 19 ResNet50 60-40

	Epochs = 20	Epochs = 30	Epochs = 200
Accuracy	49.83%	50.60%	89.86%
Loss	0.7288	0.7221	0.2712
Recall	75.06%	78.92%	89.47%
Precision	37.86%	41.01%	90.59%
F1 Score	49.28%	52.79%	89.81%

Fig. 20 ResNet50 70-30

	Epochs = 20	Epochs = 30	Epochs = 200
Accuracy	50.36%	49.93%	87.39%
Loss	0.7183	0.7278	0.4129
Recall	60.00%	43.51%	100.00%
Precision	29.88%	22.21%	78.37%
F1 Score	39.48%	28.82%	87.32%

Fig. 21 ResNet50 80-20

	Epochs = 20	Epochs = 30	Epochs = 200
Accuracy	50%	50%	93.24%
Loss	0.7000	0.7408	0.3354
Recall	10.00%	6.67%	96.67%
Precision	50.05%	3.33%	89.29%
F1 Score	6.61%	4.41%	92.83%

Based on the results obtained up until first 200 epochs, we can see that InceptionV3 gives the best overall performance. VGG19 gives the second best accuracy, followed closely by MobileNetV2. ResNet50 gives the worst performance out of all the four pretrained models chosen.

At 80-20 split, InceptionV3 gives the best accuracy of 98.65%. Similarly, accuracy reached by VGG19 is 95.95%, by MobileNetV2 is 94.59% and by ResNet50 is 93.24%, ResNet50, despite having 50 layers and therefore being quite deep, gives the worst performance. This shows that skip connections, which are a feature of ResNet models, do not work well for our problem. On the other hand, InceptionV3, which stacks multiple kernels at the same level and is therefore able to extract both local and global features, works the best. VGG19 also works fairly well, despite the fact that it only has 19 layers and is not as deep as the rest of the networks considered here.

5 Conclusion and Future Work

In this paper, we can see that using CNN with pretrained models works very well for Leukemia diagnosis problem. In our case, the dataset taken was quite small. The performance of the proposed model could be increased if more number of images for training can be acquired.

Also, using iterative approach takes a long time for training models. Other methods could be utilized to decrease this training time.

References

1. Vishwakarma VP, Dalal S (2020) A novel non-linear modifier for adaptive illumination normalization for robust face recognition. Multimedia Tools Appl 79(17):11503–11529
2. Chand S, Vishwakarma VP (2021) Comparison of segmentation algorithms for Leukemia classification
3. Ahuja B, Vishwakarma VP (2021) Deterministic multi-kernel based extreme learning machine for pattern classification. Expert Syst Appl 115308
4. Dalal S, Vishwakarma VP (2021) Classification of ECG signals using multi-cumulants based evolutionary hybrid classifier. Sci Reports 11(1):1–25
5. Jagadev P, Virani HG (2017) Detection of Leukemia and its types using image processing and machine learning. In: 2017 International conference on trends in electronics and informatics (ICEI), pp 522–526. https://doi.org/10.1109/ICOEI.2017.8300983
6. Raje C, Rangole J (2014) Detection of Leukemia in microscopic images using image processing. In: 2014 International conference on communication and signal processing, pp 255–259. https://doi.org/10.1109/ICCSP.2014.6949840
7. Khosrosereshki MA, Menhaj MB (2017) A fuzzy based classifier for diagnosis of acute lymphoblastic Leukemia using blood smear image processing. In: 2017 5th Iranian joint congress on fuzzy and intelligent systems (CFIS), pp 13–18. https://doi.org/10.1109/CFIS.2017.8003589
8. Shafique S, Tehsin S (2018) Acute lymphoblastic Leukemia detection and classification of its subtypes using pretrained deep convolutional neural networks. Technol Cancer Res Treatment 17:1533033818802789
9. Maria IJ, Devi T, Ravi D (2020) Machine learning algorithms for diagnosis of Leukemia. IJSTR 9:267–270
10. Genovese A, Hosseini MS, Piuri V, Plataniotis KN, Scotti F (2021) Acute lymphoblastic Leukemia detection based on adaptive unsharpening and deep learning. In: ICASSP 2021—2021 IEEE international conference on acoustics, speech and signal processing (ICASSP), pp 1205–1209. https://doi.org/10.1109/ICASSP39728.2021.9414362
11. Jha KK, Dutta HS (2019) Mutual information based hybrid model and deep learning for acute lymphocytic Leukemia detection in single cell blood smear images. Comput Methods Programs Biomed 179:104987. https://doi.org/10.1016/j.cmpb.2019.104987
12. Labati RD, Piuri V, Scotti F (2011) All-IDB: the acute lymphoblastic Leukemia image database for image processing. In: 2011 18th IEEE international conference on image processing. IEEE, New York, pp 2045–2048
13. Scotti F (2006) Robust segmentation and measurements techniques of white cells in blood microscope images. In: 2006 IEEE instrumentation and measurement technology conference proceedings. IEEE, New York, pp 43–48
14. Scotti F (2005) Automatic morphological analysis for acute Leukemia identification in peripheral blood microscope images. In: CIMSA. 2005 IEEE international conference on computational intelligence for measurement systems and applications. IEEE, New York, pp 96–101
15. Piuri V, Scotti F (2004) Morphological classification of blood leucocytes by microscope images. In: 2004 IEEE international conference on computational intelligence for measurement systems and applications, 2004. CIMSA. IEEE, New York, pp 103–108

CNN-Based Car Damage Detection

G. Naganandini, Shubhodeep Adarak, and Prashanth Bagel

Abstract Identifying the damage in a vehicle is considered to be the highest priority task in redeeming insurance from insurance industries. Whenever an accident occurs the guilty must be identified and the charges must be fined on him, these activities are actually carried out by the insurance companies. In this work, we present a different approach of identifying the vehicle's damage by capturing 2D images, hence if the body of the vehicle gets deformed, the deprived driver of the vehicle need not have to wait for the insurance company's consent and calculation of the ensured amount, instead the driver himself can calculate the recover amount from the insurance company. This work contains the CNN-based approach to check for the intensity and severity of damage. A computer vision approach is used accurately to classify damage of a vehicle and facilitates in claiming the right amount at right time.

Keywords Claim · Ensured amount · CNN

1 Introduction

The most important and widely uttered word in today's scenario is about insurance. INSURANCE companies have been playing a major role since the vehicle came into existence. There is absolutely no doubt if we say that the insurance industry is currently the hot and happening industry. There was an insurance outlay rise in the 2015 and 2016, adhering to National Association of Insurance Commissioners. There was an average outflow in the 2016, and it was highest in New Jersey and New York. The number of accidents on road increased drastically and a heavy load was imbibed on insurance companies. Increase in road accidents gave rise to enormous pressure on insurance companies. The accuracy of the insurance companies was hindered due to this pressure.

G. Naganandini (✉) · S. Adarak · P. Bagel
Department of Information Science and Engineering, Don Bosco Institute of Technology, Bengaluru, India
e-mail: nandinigrao@dbit.co.in

A. Kumar et al. (eds.), *Proceedings of the International Conference on Cognitive and Intelligent Computing*, Cognitive Science and Technology,
https://doi.org/10.1007/978-981-19-2350-0_64

A car damage analysis system is in place, when a car becomes paralyzed in an accident, whoever is the guarantee they will bear all the damage expenses, however, there might be a wrong estimation [1]. The approximation involves blue-collar labors, mortal experts, and the time the event occurred to estimate the damage. An initial requirement for each types of bodies and insurance companies. A machine learning imbibed methods to develop training models will help to segregate the cars that are damaged and estimate the repair that is been looked for.

2 Related Work

Good results are got when used with machine learning applications. CNNs always give good performance if the computer vision task is involved, tasks such as object identification and detection [2]. CNN based method identifies the damage by assessing the structure. A deep learning-based mechanism is used for structural health monitoring (SHM) to recognize the repairment in the form of flaws on a multiple material [3]. Unsupervised learning is also used to identify the damage in the vehicle. A lot of models under CNN are trained using ImageNet, and these models are used to detect the damage in the car. Many publicly available ImageNets are there in order to maintain the image-based detection. This works better for some dataset.

3 Proposed Idea

We choose a CNN as our neural network variation since we are using 2D photos as data. CNNs are superior in this field. CNNs are also typically easy to use, as they have fewer parameters than other types of ANN: s and are typically simple to train. With five hidden layers, we picked a rather straightforward structure for our CNN. The individual weights are multiplied and added together to produce the final output which helps in making accurate network predictions [4]. Here, five separate modules (layers) in united to create our portal. Each layer is in charge of judging a distinct attribute. On the upload page, the user is first greeted. After you have uploaded an image, it will be sent to the layers for analysis (Figs. 1 and 2).

Image Upload: The extent of damage detection is done only when the vehicle image is uploaded, and further, it is taken care of in the further stages.
Stage-1: Only if the image received is a vehicle, then the further processing takes place.
Stage-2: This stage will identify the presence of damage. This assessed image will be either chosen or dropped to further layers based on the decision taken.
Stage-3: This stage detects the presence of damage at front or back end of a
Stage-4: The score of severity of damage on the basis of low, medium, and high are identified.

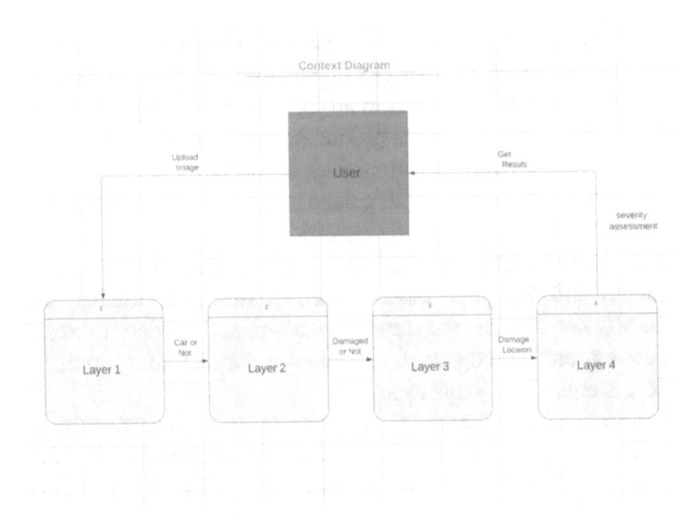

Fig. 1 Context diagram

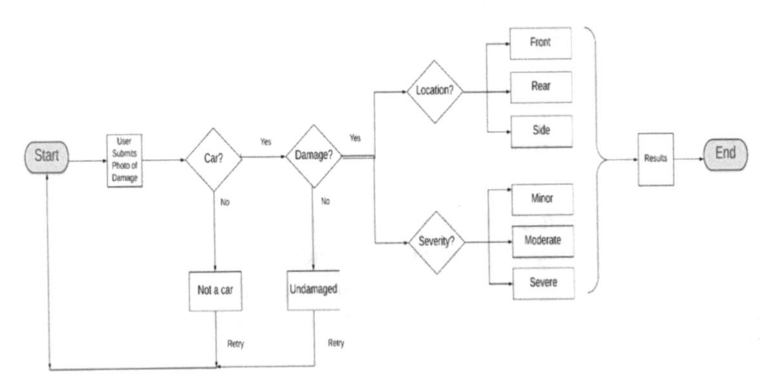

Fig. 2 Flow diagram

Stage-5: This stage combines all the previous stages outputs and consolidates it to produce the final output that will be shown to the user.

4 System Design

Designing of a system involves organizing the architecture, different components, different stages, and modules, its interfaces to satisfy the criteria of a system [4]. It is possible to think of system design as the application of systems theory to product development. All the identified design involves the combination of theory which are overlapped. The product development will be carried out only when it has a proper strategy to integrate design, manufacturing, and marketing, all these result in designing a process which defines and develops systems to meet the user requirement.

Designing of a system plays an important role in acknowledging an important role in the data processing sector in the early. The modular systems design has the ability which are enabled by the standardization of hardware and software in the 1990s. The importance of software engineering has gained its pace and runs on general platforms that has grown till date.

The ultimately used approaches for designing the computer systems are object-oriented analysis and design methodologies [5]. In object-oriented analysis and design, the UML has become the default standard. It is widely used for software systems modeling, and its popularity is achieved for becoming more popular for non-software systems and organizations [6].

One of the most crucial stages of the software development process is system design. The design's goal is to plan a solution to a problem that is outlined in the requirement specification. In other words, the design of the project is the initial step toward a solution [7]. The system's architecture is perhaps the most important aspect determining software quality. The goal of the design phase is to create the software's overall design [8]. Its goal is to determine which modules should be included in the system to efficiently meet all of the system's requirements.

5 System Implementation

The implementation is usually the integration of all the SDLC steps from the beginning till coding [9]. It incorporates planning, designing, manufacturing. Straight away deployment is done only when the product undergoes execution. The product builds in this way. For a particular specification or standard, there may be numerous implementations [10].

Implementation is a critical stage in the software development life cycle (SDLC). It includes all of the steps that are really involved in accessing new software or hardware up and running in its environment, including installation, configuration, testing, and any necessary adjustments. It comprises translating the design into a fully functional system and coding the system in a specific computer language. This phase of the system is performed with the mentality that anything is generated should be implemented if it satisfies the needs of the users, the system's aim and scope. During the implementation phase, the user's solution is developed [11].

Pseudocode is a representation of algorithm or program steps to execute each step. It generally has set of rules which has structure of a particular programming language and its readable by humans than machine, and it is easy understandable. Every statement is connected to each other, and this code is understandable by a lay man without any confusion in understanding. It helps the easy readability of a program, and it is one of the best approaches to understand the implementation. It acts as a mediator between the algorithm, flowchart, and a program. In industry, the role of pseudocode plays a major role in documentation [12, 13].

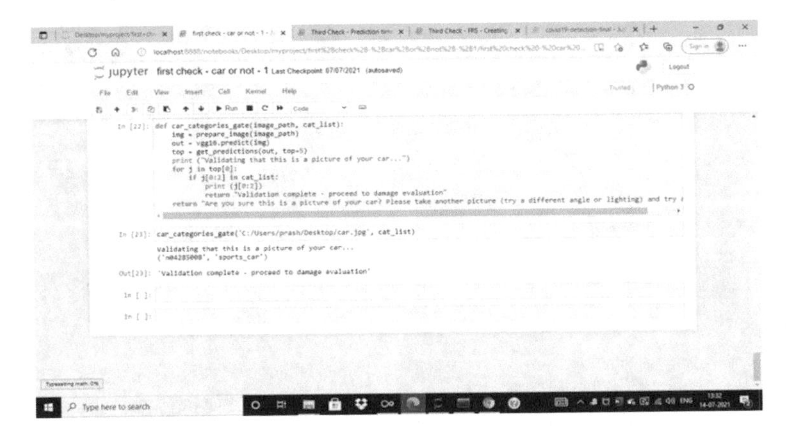

Fig. 3 Image preprocessing

Pseudocode:

Step-1: First user submits the photo.
Step-2: Identify it is a car picture or not.
Step-3: If it is not a car, then user upload the new photo.
Step-4: If it is a car, then identify it is damage or not
Step-5: If is not damage, the retry upload new photo.
Step-6: If it is damage, then identify the location-; front, rear, side.
Step-7: Identify the severity - minor, moderate, severe
Step-8: Result

6 Results

The first step is to upload the picture, then identify it as a car or not, if car then identify the presence or absence of damage [10]. The accuracy of identifying it as damage or not is 89%, then identifying the damaged part of the car rear, back, front accuracy is 69%, identify the level of damage a car as low, moderate, or high, the accuracy of the identified damage level is 66%, the final step gives the combined result of all the four-module the accuracy of the final step is 74%. They help the company to claim the insurance on car (Figs. 3, 4, 5, and 6).

7 Conclusion

This project put forth the idea of detecting the damage at the front and rear side of the car with an accuracy of 89%. Further, the car damage detection analysis done using CNN will improve the accuracy even further and will help the victim claim the

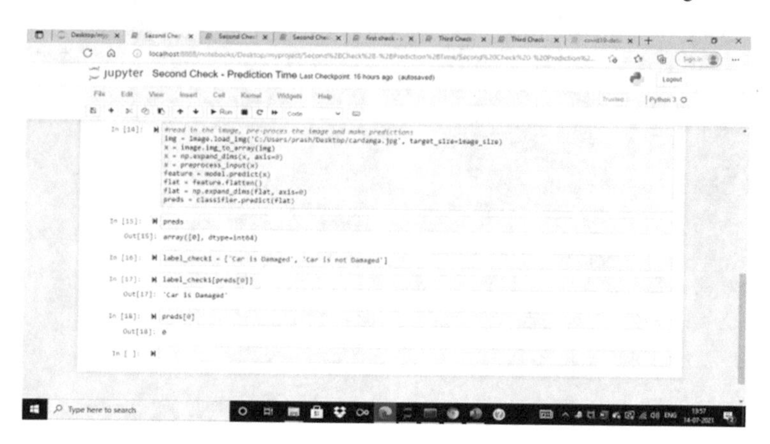

Fig. 4 Damage classification with labels

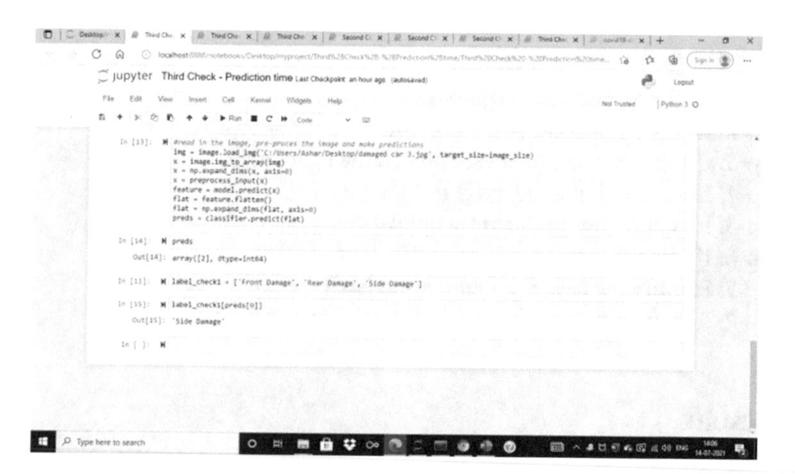

Fig. 5 Damage prediction

insurance effectively. False claims can be completely avoided. Overall, increasing the accuracy rate will help the victim or the deceased to claim insurance to the concerned company car damage detection's datasets availability is very rare, and the collection of these important datasets must be improved Though there is a good amount of accuracy in damage detection of the car that occurs during accidents, a hybrid model such as fuzzy neuro model can be used to increase the accuracy even further. This model can be embedded in a working humanoid robot which acts as a first aid providing and damage detection robot which comes to the aid of rescue of people and also victims.

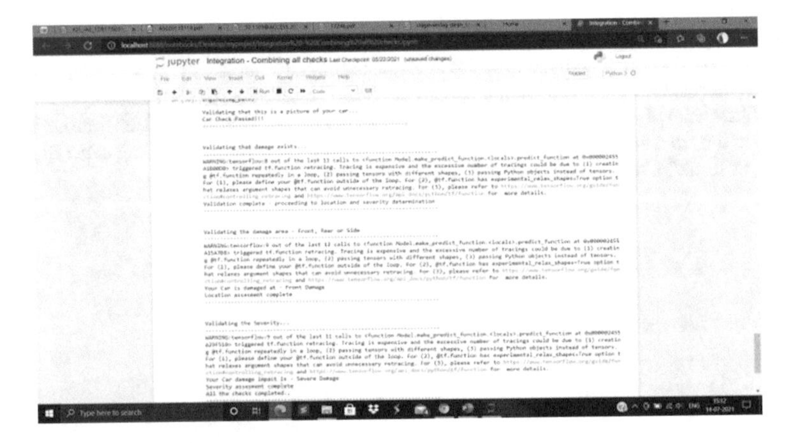

Fig. 6 Severity and validation

References

1. Zhang Q, Chang X, Bian SB (2020) Vehicle-damage-detection segmentation algorithm based on improved mask RCNN. IEEE Access 8:6997–7004. https://doi.org/10.1109/ACCESS.2020.2964055
2. Szegedy C, Vanhoucke V, Ioffe S, Shlens J, Wojna Z (2015) Rethinking the inception architecture for computer vision. arXiv preprint arXiv:1512.00567
3. Sharma A, Verma A, Gupta D (2019, November) Preventing car damage using CNN and computer vision. Int J Innov Technol Explor Eng (IJITEE) 9(1). ISSN: 2278-3075
4. Liu Y, Zhang X, Zhang B, Chen Z (2020) Deep network for road damage detection. In: IEEE international conference on big data (Big Data), pp 5572–5576. https://doi.org/10.1109/BigData50022.2020.9377991
5. Harshani WAR, Vidanage K (2017) Image processing based severity and cost prediction of damages in the vehicle body: a computational intelligence approach. In: National information technology conference (NITC), pp 18–21. https://doi.org/10.1109/NITC.2017.8285649
6. Patil K, Kulkarni M, Karande S (2017) Deep learning based car damage classification. https://doi.org/10.1109/ICMLA.2017.0-179
7. Guo P, Zhuang P, Guo Y (2020) Bayesian pan-sharpening with multiorder gradient-based deep network constraints. IEEE J Sel Topics Appl Earth Observ Remote Sens 13:950–962. https://doi.org/10.1109/JSTARS.2020.2975000
8. Soumalya Sarkar MGMRG, Reddy KK (2016) Deep learning for structural health monitoring: a damage characterization application. In: Annual conference of the prognostics and health management society
9. Goularas D, Kamis S (2019) Evaluation of deep learning techniques in sentiment analysis from twitter data. In: International conference on deep learning and machine learning in emerging applications (Deep-ML), pp 12–17. https://doi.org/10.1109/Deep-ML.2019.00011
10. Simonyan K, Zisserman A (2014) Very deep convolutional networks for large-scale image recognition. arXiv preprint ArXiv: 1409.1556
11. Lee KB, Shin HS (2019) An application of a deep learning algorithm for automatic detection of unexpected accidents under bad CCTV monitoring conditions in tunnels. In: 2019 International conference on deep learning and machine learning in emerging applications (Deep-ML), pp 7–11. https://doi.org/10.1109/Deep-ML.2019.00010

12. Huang Z-H, Wang C-M, Wu W-C, Jhang W-S (2020) Application of vehicle detection based on deep learning in headlight control. In: 2020 International symposium on computer, consumer and control (IS3C), pp 178–180. https://doi.org/10.1109/IS3C50286.2020.00053
13. Alfarrarjeh A, Trivedi D, Kim SH, Shahabi C (2018) A deep learning approach for road damage detection from smartphone images. In: IEEE international conference on big data (Big Data), pp 5201–5204. https://doi.org/10.1109/BigData.2018.8621899

A Study and Analysis of Energy Efficient Machine Learning Algorithm for Virtualized Sensor Cloud Infrastructure

Anasuya N. Jadagerimath, S. Prakash, and K. R. Nataraj

Abstract Today, world is moving towards digital economy; each and every application demands samples of data from real world to analyse every piece of information for better productivity. With the rapid reduction in the cost of sensors, in next five years, sensors can be deployed and used in all sectors and there will be seamless communications among all the applications. IoT devices will directly share data using open flow standards and connect to World Wide Web through Internet. Today, most of the data generated from physical sensors is data of specific applications which cannot be used as input for other domain applications. Many physical sensors deployed in one domain will only be used by that specific application. With the convergence of IoT and cloud computing, the data generated from physical sensors of different applications deployed at different scenarios can be imported on cloud environment which in turn can be used by all the applications. It is high time to address high usage of energy manageable data centres. It is time to identify and eliminate inefficiencies in delivering electric services to IT resources. This can be achieved by enhancing the efficiency of physical infrastructure, adopting efficient resource allocation and management algorithms. This paper focuses on energy management of IT equipments in virtualized sensor cloud data centre. This work emphasizes on minimizing energy consumption in virtualized sensor cloud infrastructure. We have implemented S-learning algorithm on sensor cloud environment and proved that the solution is able to handle dynamic requests. For dynamic request generation rate, the response time, violation of the SLA and the energy utilisation are measured. The algorithm is quite effective on three parameters: quick response time, reduced number of SLA violations and reduced energy consumption. The amount of energy conserved is measured and compared with other approaches. The model has built on

A. N. Jadagerimath (✉) · K. R. Nataraj
Don Bosco Institute of Technology, Bengaluru, Karnataka, India
e-mail: anasuyaprakash@dbit.co.in

K. R. Nataraj
e-mail: director.research@dbit.co.in

S. Prakash
AIT, CSE, Chandigarh University, Mohali, Punjab, India

A. Kumar et al. (eds.), *Proceedings of the International Conference on Cognitive and Intelligent Computing*, Cognitive Science and Technology,
https://doi.org/10.1007/978-981-19-2350-0_65

a test bed with dynamic data requests and measures the number of SLA violations in the system wherein the proposed scheduling method is associated with other methods and found this solution is more efficient.

Keywords IoT · WSN · Virtualization · Machine learning · SLA · M2M

1 Introduction

The wireless sensor network is a major domain of 3rd platform technologies. As per the Cisco report, 50 billion gadgets are associated with the Internet in near future. IoT empowers billions of everyday physical entities like automotive and sensors to be connected on World Wide Web that will transform the way we work, live and enjoy. IoT systems will improve efficiency, make new plans of action and produce new income streams. IoT emerged from M2M (machine to machine) communications; but in today's scenario, it is not just machine communicating with machines; it is almost everything on the world being connected through Internet. Many of these things are on mobility environment, and user can bring any device (BOYD) and seamlessly get connected to IoT ecosystem. IoT took shape in mid-2000; down the line, many sectors such as health care, manufacturing and retail started using IoT in their business to increase productivity. IoT technology roadmap is largely applicable in many horizontal and vertical applications. The range starts from security surveillance, medicine, transport, inventory monitoring, pollution monitoring and many more applications. IoT technology will directly touch day-to-day lives of common man by connecting objects to Internet. RFID and GPS will play major role in tracking and positioning of objects on IoT ecosystem. IoT refers to establishing communications and sensing capabilities to real-world physical objects. With the rapid reduction in the cost of sensors, in next few years, sensors can be deployed and used in all sectors and there will be seamless communications among all the applications. IoT devices will directly share data using networking protocols such as Wi-Fi, Bluetooth and radio networks and connect to World Wide Web through Internet.

2 Related Work

This section is about exhaustive literature survey on related research around sensor cloud infrastructure that constitutes IoT and cloud computing and virtualization [1, 2]. The summary of literature survey includes the information about few statistics on growth of WSN [3–6], IoT, few deployment scenarios of IoT, applications of IoT, cloud computing and mobile cloud computing and its advantages. Here, it is highlighted the impact of convergence of IoT and cloud computing that has led to sensor cloud infrastructure. Hence, it is elaborated on the components of sensor cloud infrastructure and its advantages. Then, we have touched on the issues and challenges in sensor cloud infrastructure. The issue of energy management in sensor cloud infrastructure is reflected in detail.

2.1 Sensor Network Infrastructure and IoT

IoT will profoundly have greater impact on the society and the lives of the common man. Its applications are already visible for home automation, health care, smart transportation, smart farming and smart cities. With the advancements in semiconductor and sensor technologies, IoT has taken new shape in transforming the communications of physical devices as shown in Fig. 1. The IoT is an arrangement of related "things" (which also joins people). The relationship will be between human people, human things and things-things (Fig. 2).

3 Existing System

Lan, K. T., "What's next? Sensor+Cloud?" [7]. Today, we are witnessing various intriguing advancements of technology. To begin with, there is an expanding appropriation of detecting advancements (e.g. RFID, cameras and cell phones) in numerous businesses. Second, the Web has turned into a resource of continuous data (e.g. through Web journals, informal communities, live discussions) for daily transactions occurring around us. The real data that is dynamic in nature is sourced from

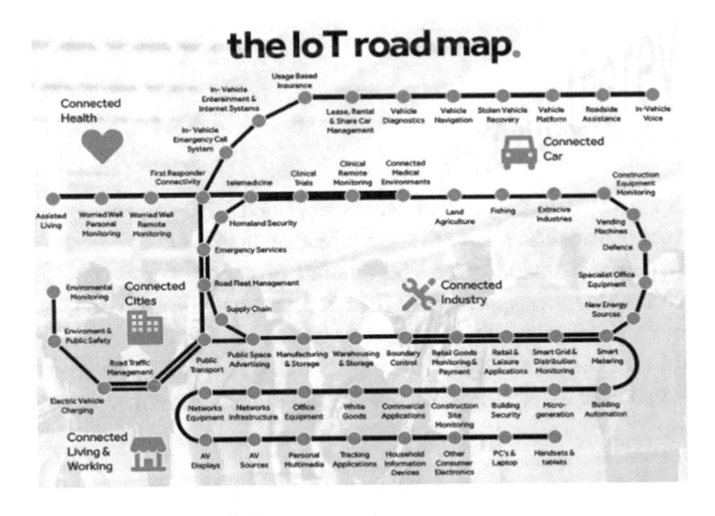

Fig. 1 Connecting all sectors on IoT

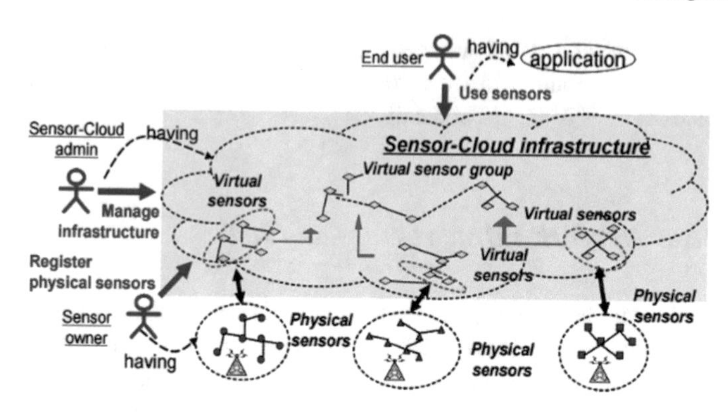

Fig. 2 Sensor cloud infrastructure

physical sensors [3]. The sensor data is deployed on cloud computing environment where in this data can be made available for all the stake holders who are in requirement of this data/information for their applications [8]. Zhang et al. [4] propose a novel architecture for sensor segregation that can easily scale up deployment of SNI. This architecture proposes many sink centres that participate in identifying and assembling the data from sensing devices. Every sink centre is accountable for social event data from the sensors inside a zone. These sink centres are mapped and associated with virtual sinks on cloud environment. These virtual sinks are highly secured on cloud environment. Authors claim that the proposed solution improves execution efficiency of SNI when integrated with cloud environment. The results show decreased packed transmission batch rate with reduced number of end-end skips. The authors propose a novel framework from a remote sensor sort out perspective that can enhance the execution of SNI. The proposed idea enhances the execution of SNI.

4 Problem Definition

The problem has been clearly and in detail elaborated on the objectives of the paper title. The two metrics used to verify the effectiveness of the approach are energy optimization and SLA violations. The amount of energy conserved is measured and compared with other approaches. SLA is indicator of the number of QOS [7] violations because of scheduling. Here, the test bed has been set with dynamic data requests and the number of SLA violations is measured in the system wherein the proposed scheduling algorithm is compared with other algorithms.

5 Proposed Algorithm

The proposed algorithm is modified SARSA [9] algorithm.

5.1 *Novelty of the Proposed Solution*

Present algorithms for resource management do not perform well in sensor cloud application scenarios because of unpredictable load and time varying load requirements on VM [1, 10]. Most of the existing schemes are based on threshold value that cause lot of VM migrations and increase the energy consumption with lower response time. The data centre workloads are dynamic in nature, and hence, the number of tasks and requirement of the resources for virtual machines are not known in advance. There is negligence amount of research done till date on dynamic sensor data input, and we need exclusive algorithms that are adaptable to dynamic nature of sensor cloud infrastructure. The novelty of this research work is that we have proposed and implemented S-learning algorithm that performs better for dynamic conditions of sensor cloud environment and proved that the solution is able to handle dynamic requests by considering the parameters of energy-efficient algorithms [11]. For dynamic request generation rate, the response time, SLA violation and the energy consumption are measured [12].

6 Methodology

6.1 *Machine Learning-Based Dynamic Scheduling of VMs in Virtualized Data Centre (VDC) for Sensor Cloud Infrastructure*

Sensor cloud is a dynamic environment. In this, the user requests for sensor data and the physical sensor feed to cloud is dynamic. The virtual sensors must be scheduled to handle the physical sensor feed and also the queries from user. Since the request is highly dynamic, the scheduling policy must be able to handle these dynamic requests and guarantee QOS in terms of response time for the user and process time of sensor data feed.

If we assume the number of requests arriving from user is *Nu,* the number of events arriving from the sensor is *Ns,* the response time deadline for each user requests is *Tud,* and the response time deadline for each event from sensor is *Tsd*, the scheduling must achieve the following

Choose the subset of *Nu* with *SNu* request in a time window *T* such that

$$T/SNu \Leftarrow \text{Tud for each request in } SNu \tag{1}$$

With the constraint to minimize *Nu-SNu*

Choose the subset of *Ns* with *SNs* request in a time window T such that

$$T/Sns \Leftarrow Tsd \text{ for each request in } SNs \tag{2}$$

With the constraint to minimize *Ns-SNs.*

The scheduling must be learning based in sense that it must be able to adjust itself to the data rate arrivals and fine tune its scheduling policy.

The energy consumption problem can be translated to number of idle or low CPU usage machine in the data centre.

If over a time T, if the number of idle, low CPU machine in data centre is $\{Nu1, Nu2, \ldots NuT\}$ then,

$$TE = \sum_{k=0}^{T} Nuk \tag{3}$$

And if the *TE* is minimized to low value as possible without comprising the conditions 1 and 2, that is the maximum energy that can be saved.

The proposed reinforcement machine learning technique is very much suitable for the dynamic task allocation. It is formalized as the reinforcement learning as follows. The defined values are state, action and reward of the environment with stochastic finite state machine. The details are shown below. State transition function

$$P(X(t)|X(t-1), A(t)) : P(Y(t)|X(t), A(t)) \tag{4}$$

Observation (output) function

$$E(R(t)|X(t), A(t)) \tag{6}$$

Reward function. And hence the state transition function is:

$$S(t) = f(S(t-1), Y(t), R(t), A(t)) \tag{7}$$

Policy/output function:

$$A(t) = pi(S(t)) \tag{8}$$

The agent finds the policy to increase the expected discounted rewards.

$$E[R_0 + gR_1 + g^2 R_2 + \cdots] = E \sum_{t=0}^{\text{nifty}} \gamma R_t \tag{9}$$

where $0 \Leftarrow \gamma \Leftarrow 1$ is a discount factor which models the fact future reward is worth less than immediate reward.

7 State Transition Function

Figure 3 depicts the state transition function. The special case is that $Y(t) = X(t)$, we say the world is completely recognizable, and the model turns into a Markov Decision Process (MDP). In the event that we know the model (i.e. the move and reward capacities), we can gauge for the ideal approach in about n^2 time utilizing strategy emphasis. Tragically, if the state is made out of k paired state factors, then $n = 2^k$, so this is much too moderate.

In the MDPs, ideal approach is to have an arrangement that augments the summation of future prizes. Accordingly, ideal strategy comprises a few activities which have a place with a limited arrangement of activities.

An operation of an agent is $s0$, $a0$, $r1$, $s1$, $a1$, $r2$, $s2$, $a2$, $r3$, $s3$, $a3$, $r4$, $s4$, ..., which is a sequence of state action-rewards, this shows that $s0$ state was the previous state with reward $r1$ and showing the present state as $s1$, then proceeded with the action $a1$ with $r2$ reward and completed the action $a2$ and so on (Table 1).

A novel algorithm has been proposed that is based on improving the SARSA algorithm for the case of sensor cloud requirements [8]. Our algorithm is the S-learning (enhanced SARSA) which makes use of fuzzy scoring for variable rewards instead of using the fixed reward/punishment values. The input for S-learning would be from the sensor network. S-learning algorithm can predict the load of each sensor based on the data arrival from sensors and the query on sensors. This prediction

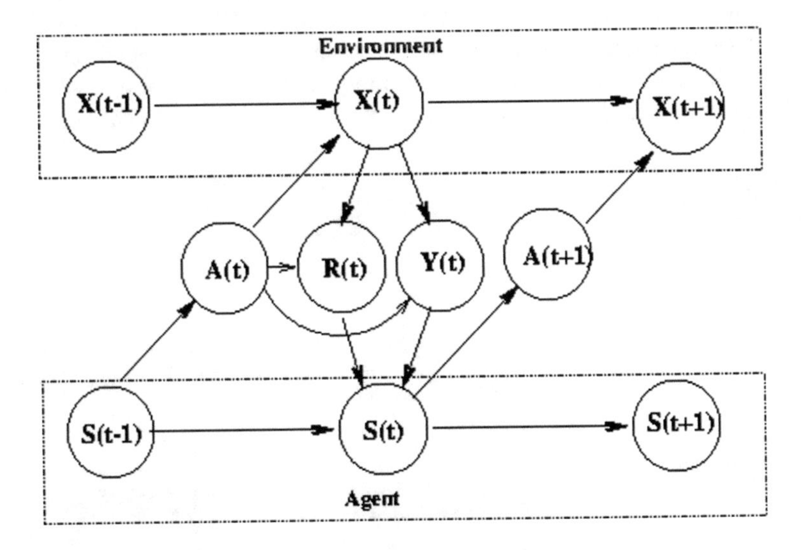

Fig. 3 State transition diagram

Table 1 State, action and rewards

State	Action	Reward
$s0$	$a0$	$r1$
$s1$	$a1$	$r2$
$s.$	$a.$	$r.$
Sn	an	$rn + 1$

would be done by using ARIMA model. The sensors are grouped based on load into 3 categories: Low, Medium and High. Using the RL, fuzzy score is calculated by the virtual sensor applications running on VM [5, 13] for each category based on network bandwidth consumption, CPU usage, the average response time and energy consumption at VM. The main job of scheduler is to make decision to allocate a VM to PE, migrate a VM to PE, schedule a sensor app to VM and migrate sensor app to VM. The reinforcement is the continuous process which is learned based on the change of state, action and rewards. The algorithm records the states (i.e. the conditions at VM) when score is highest and also remembers the states when the score is lost. The values are reset when the load on sensor network changes. Over the period of time, this state learning process is stabilized and the RL algorithm makes the decisions to move the states to points of high score. Since the RL algorithm is iterative and it will continue with the learning stage till it stabilizes, and once stabilized, it makes better decision. The linear output is generated for the ARIMA forecasting equation used for a stationary time series in which the predictors consist of lags of the dependent variable and/or lags of the forecast errors. This can be illustrated as follows (Fig. 4).

Sensor load prediction module: This will predict the load at sensor based on ARIMA model. Fuzzy scoring module: It gives the score to the decision made by scheduler while making a scheduling decision to allocate a VM to PM. Reinforcement learning (RL) module: It implements the S-learning algorithm to optimize the states where the highest score is given by fuzzy scoring. Thereby, it finds the best scheduling decisions. Reinforcement learning module stores the states, events and score in an

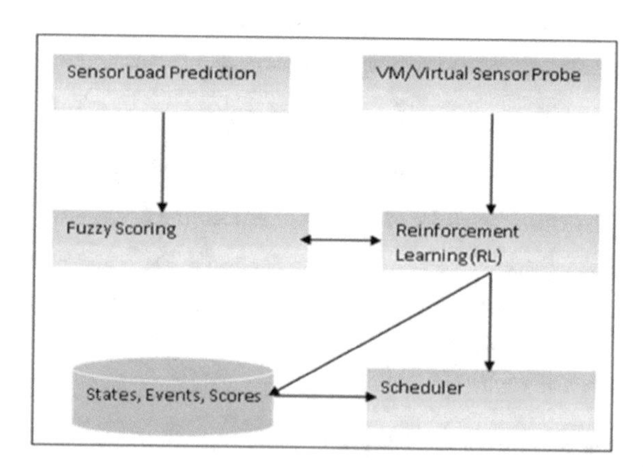

Fig. 4 Block diagram of S-learning algorithm

in-memory repository. Scheduler module: This module implements the scheduling algorithm [14] to allocate VM to PM. VM probe: This module probes the VM, finds the CPU utilization and reports to the reinforcement learning module. This can be illustrated as follows. Predicted value of Y is a constant and/or a weighted sum of one or more recent values of Y and/or a weighted sum of one or more recent values of the errors. Suppose the predictors consist only lagged values of Y, then it is a pure autoregressive ("self-regressed") model, which is just a special case of a regression model, and this could be fixed with standard regression software. The lag of one period of ARIMA model can be defined as a first-order autoregressive which could be written as ("AR (1)"), (LAG (Y, 1). The ARIMA model is not a linear regression model when the predictors have errors. This leads to the model predictions which are not linear functions of the coefficients, even though they are linear functions of the past data. And hence, the nonlinear optimization methods ("hill-climbing") can be used for the ARIMA model which has the lagged errors. The other models of ARIMA are shown below. ARIMA (p, d, q) model is a non-seasonal ARIMA model where p defines the number of autoregressive terms, d defines the number of non-seasonal differences needed for stationary, and q defines the number of lagged forecast errors in the prediction equation.

Hence, the forecasting equation is constructed as follows. In first step, y denotes the dth difference of Y, which means:

$$\text{If} \quad d = 0{:}yt = Yt \tag{6}$$

$$\text{If} \quad d = 1\ {:}yt = Yt - Yt - 1 \tag{7}$$

$$\begin{aligned} \text{If} \quad d = 2\ {:}yt &= (Yt - Yt - 1) - (Yt - 1 - Yt - 2) \\ &= Yt - 2Yt - 1 + Yt - 2 \end{aligned} \tag{8}$$

Hence,

$$\hat{y}t = \mu + \phi 1yt - 1 + \cdots + \phi pyt - p - \theta 1et - 1 - \cdots - \theta qet - q \tag{9}$$

The above equation shows the moving average parameters (θ's) with negative signs in the equation. The parameters can be represented as AR(1), AR(2), …, and MA(1), MA(2), … etc.

Figure 5 shows the proposed solution. The sensors in the sensor network detect events and route data through the gateway to the virtual sensor app. For each physical sensor, there is a virtual sensor app which runs on the cloud. Any events from that sensor or user query on that sensor are directed to the virtual sensor for execution. One or more virtual sensor app run on VM, and VM uses the physical resources for execution. User executes queries on the virtual sensor application, which maps it as query on physical sensors and returns the query result to the user. Existing algorithms for resource management do not perform well in sensor cloud application scenarios

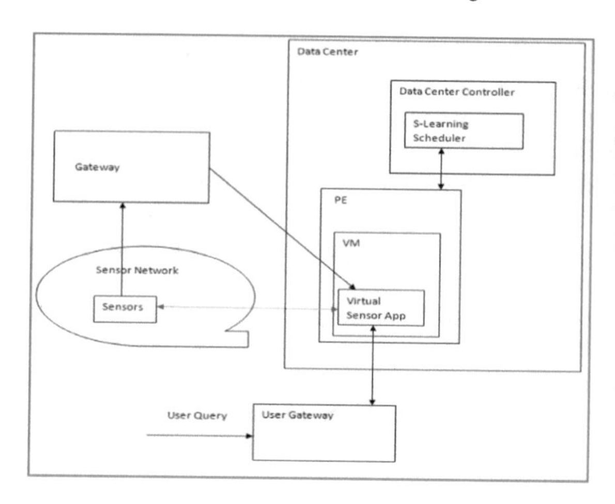

Fig. 5 Architecture of the proposed solution

because of highly unpredictable load and time varying load requirements on VM. Most of the existing schemes are based on threshold, cause lot of VM migrations and increase the energy consumption and lower response time. The data centre workloads are dynamic in nature, and hence, the number of tasks and requirement of the resources for virtual machines are not known in advance. The proposed reinforcement learning algorithm is very much suitable for the dynamic task allocation. This algorithm allows agent to act in the world so as to maximize the rewards based on the state of action.

8 Software Tools Used

We have used CloudSim [15] and NS2 tool for simulating our proposed solution. NS2 is used for sensor infrastructure, and the results of this simulation are imported to CloudSim simulator. The scheduler code was added to the CloudSim tool for the virtual sensor application scheduling. Following are the details of test bed environment.

8.1 CloudSim

Starting entries of CloudSim used Java as the discrete task activity engine that supports many functionalities such as handling of events, production of cloud structure components, correspondence among fragments and organization of the restoration clock. The CloudSim restoration layer offers interfaces for VMs, memory, stockpiling and information exchange limit. The real issues, for instance, provisioning of hosts to VMs, regulating application execution and checking dynamic structure state, are dealt by this layer. A cloud provider, who needs to consider the capability of different methodologies in assigning its hosts to VMs (VM provisioning), would need to complete this method at this layer. Such use ought to be conceivable via consequently opening up the VM provisioning facility. There is an distinctive refinement at this layer related to provisioning of hosts to VMs. As mentioned in QoS levels, VMs can accept the cloud host actions at the same time. This layer also revealed the functionalities that a cloud application planner can connect with performance of complex workload profiling and application execution. The top-most layer in the CloudSim stack is the user code that revealed central substances for hosts (number of machines, their specifications and so on), applications (number of endeavours and their necessities), VMs, number of customers and their applications and mediator arranging approaches (Fig. 6).

Here, user code is the main interface for the user to create cloud scenarios by specifying application specifications and user requirements, and then, user code is submitted to the data centre broker. User interface structures: This acts as an interface between the user and the CloudSim engine.VM services: This manages the execution of the tasks created as cloudlets on virtual machines and also maintains the details of VM task allocation and migrations. Cloud services: This allows us to define the required services like virtual machine provisioning, CPU allocation, memory allocation, storage allocation and bandwidth allocation. Cloud resources: Based on the user request and the policy, these resources are allocated to the users. In the same way, data canters are created in association with the cloud coordinator. Network: This is the core part of the simulator which can be assigned to the user based on the next configurations requirements.

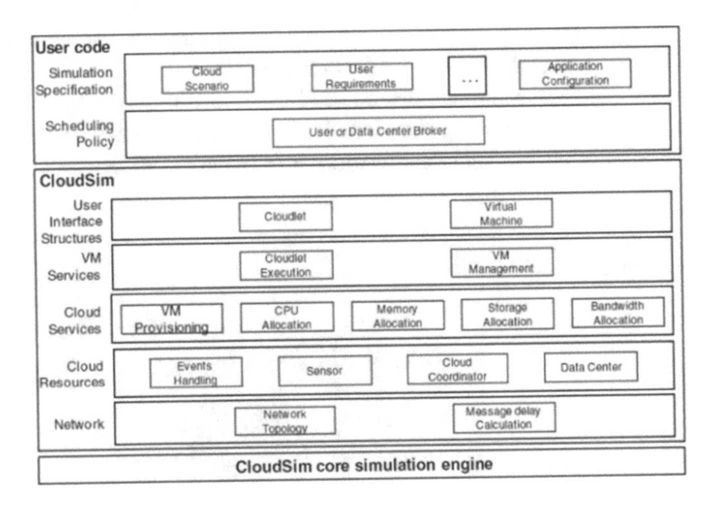

Fig. 6 CloudSim simulation engine

8.2 NS2

It is an object-oriented, discrete event-driven network simulator. It uses C++ programming and TCL scripting language. It simulates the sensor network concepts with the help of OTCL and NS2 libraries. The output is displayed with network animator.

8.3 Eclipse

Eclipse IDE is used in our development platform. Eclipse is formed generally in Java, and it is mainly used for development of Java applications. As CloudSim is built using java, Eclipse is found to be most suited IDE for this environment.

9 Results

The input variables for fuzzy scoring are based on the mentioned parameters, number of applications, the CPU usage, network bandwidth consumption, VM energy consumption and the average response time for virtual sensor applications.

All the input variable is characterized in three fuzzy values as HG (High), LW (Low) and MD (Medium). The output fuzzy score is also characterized to three fuzzy values as HG (High), LW (Low) and MD (Medium).

Figure 7 depicts fuzzy score for the number of applications. Figure 7 explains the fuzzy scoring probability in terms of specific input variables. The input variable number of application (NA) are fuzzified as shown below.

The fuzzy score for the applications ranges from 0 to 100. Fuzzy scoring will be Low if the fuzzified number of applications is in range 0–30. Fuzzy scoring will be Medium if the fuzzified number of applications is in range 25–80. Fuzzy scoring will

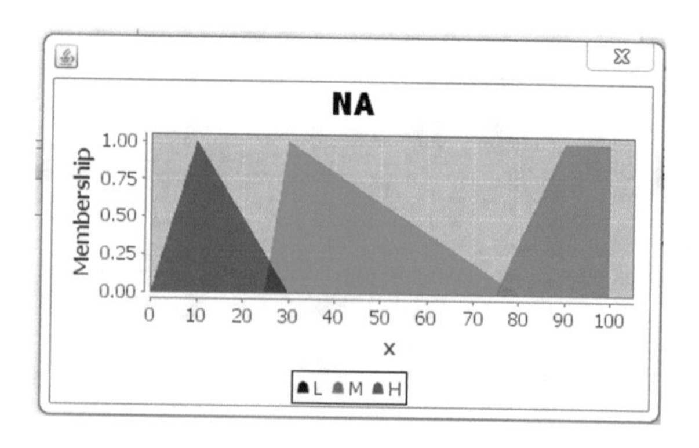

Fig. 7 Fuzzified values for number of applications

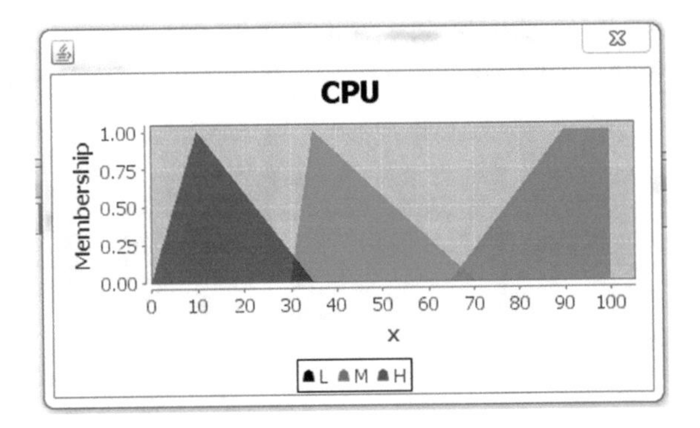

Fig. 8 Fuzzified value for CPU usage

be High if the fuzzified number of applications is in range 75–100. The fuzzy score for the CPU usage is shown below (Fig. 8).

Fuzzy scoring will be Low if the fuzzified value of CPU usage is in range 0–35. Fuzzy scoring will be Medium if the fuzzified value of CPU usage is in range 30–70. Fuzzy scoring will be High if the fuzzified value of CPU usage is in range 65–100. Fuzzy score for bandwidth consumption is showed in Fig. 9.

Figure 9 depicts the fuzzy score for network bandwidth consumption. Fuzzy scoring will be Low if the fuzzified value of bandwidth is in range 0–40. Fuzzy scoring will be Medium if the fuzzified value of bandwidth usage is in range 30–80. Fuzzy scoring will be High if the fuzzified value of bandwidth usage is in range 75–100. The fuzzy score for energy consumption is showed in Fig. 10.

Fuzzy scoring will be Low if the fuzzified value of energy consumption is in range 0–50. Fuzzy scoring will be Medium if the fuzzified value of energy consumption is in range 40–70. Fuzzy scoring will be High if the fuzzified value of energy consumption is in range 65–100. The fuzzy score for average response time is showed in Fig. 11.

Fuzzy scoring will be Low if the fuzzified value of average response time is in range 0–50. Fuzzy scoring will be Medium if the fuzzified value of average response time is in range 40–70. Fuzzy scoring will be High if the fuzzified value of average response time is in range 60–100. Defuzzification of fuzzy score is showed in Fig. 12.

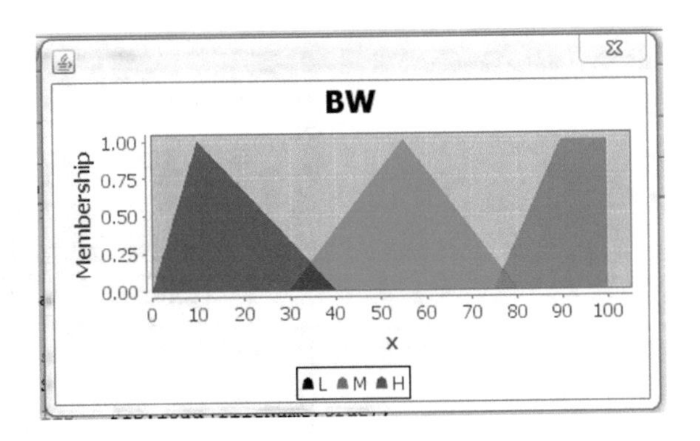

Fig. 9 Fuzzified value of bandwidth consumption

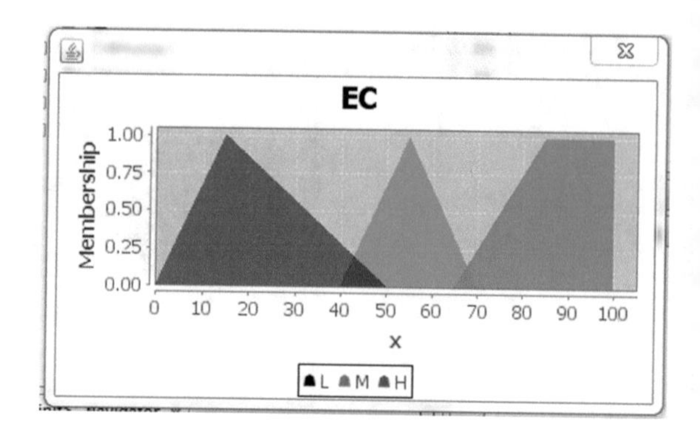

Fig. 10 Fuzzified value of energy consumption

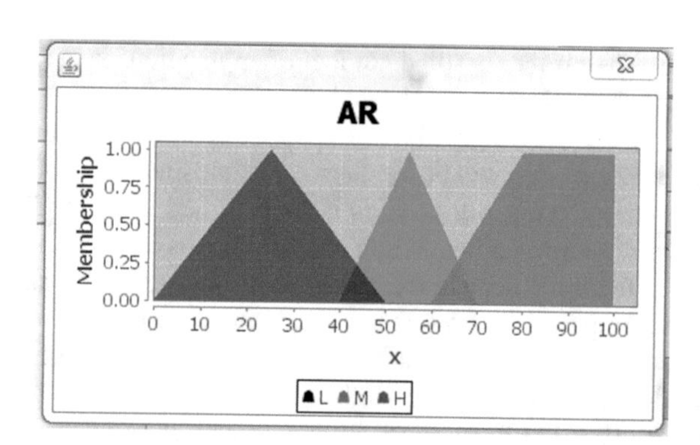

Fig. 11 Fuzzified value for average response time

Figure 12 depicts the defuzzication of fuzzy score, which is the process of producing a quantifiable result in crisp logic for a given fuzzy sets and corresponding membership degrees. It is typically needed in fuzzy control systems. Fuzzy scoring will be Low if the fuzzified value of fuzzy sets is in range 0–5. Fuzzy scoring will be Medium if the fuzzified value of fuzzy sets is in range 3–7. Fuzzy scoring will be High if the fuzzified value of fuzzy sets is in range 6–10. The details of the notations used are number of applications-NA, the CPU usage-CPU, network bandwidth

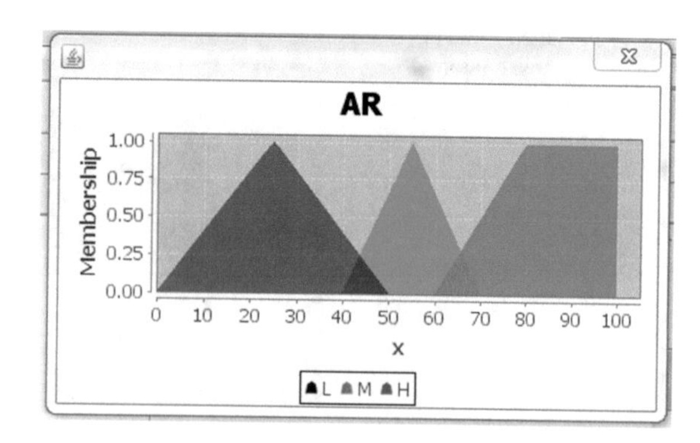

Fig. 12 Defuzzification of fuzzy score

Graph 1 Energy consumption versus response time

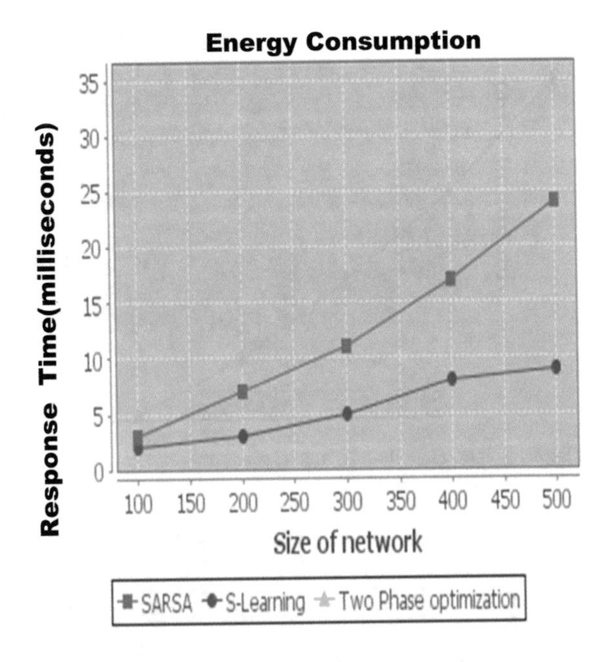

consumption-BW, energy consumption at VM-EC and the average response time for virtual sensor applications-AR. This gives the output of fuzzy score-FS. The parameters considered are shown below. Low = LW, Medium = MD and High = HG.

Energy consumption based on the size of the network

Graph 1 depicts energy consumption in data centre for different size of sensor network. The size of network is varied from 100 to 500 sensors; each sensor sends data for every sec. The simulation is done for 1 hour, and the energy consumption in cloud is measured and plotted below.

In Graph 1, X-axis is size of network and Y-axis is energy consumption in Joules. The comparison of energy consumption of SARSA, S-learning and two-phase optimization algorithms is depicted. It clearly reflects that the S-learning algorithm consumes lesser energy compared to other SARSA and two-phase optimization approaches.

Graph 2 shows the response time for the queries on sensors and is measured for different query arrival rate from 10 to 50 queries per millisecond, and the results of SARSA, S-learning and two-phase optimization are depicted. The X-axis shows the query rate, and response time is taken and displayed along the Y-axis. The comparative results show that the response time in S-learning is better than the SARSA and two-phase optimization approaches. The energy consumption for query execution rate for SARSA, S-learning and two-phase optimization generated from CloudSim is depicted.

Graph 3 shows the percentage of SLA violation with respect to the size of the

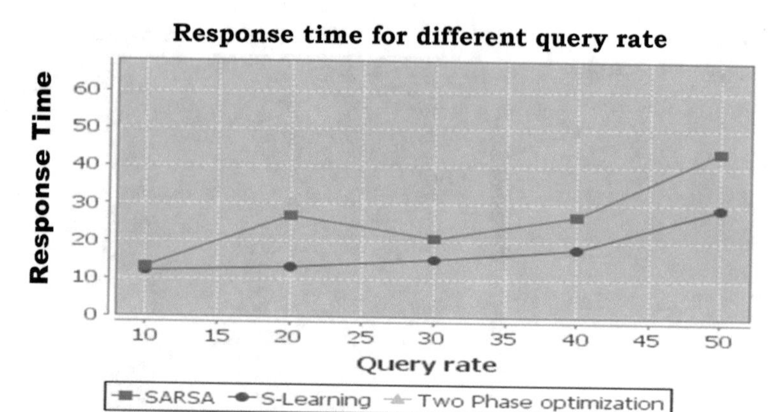

Graph 2 Response time versus response time

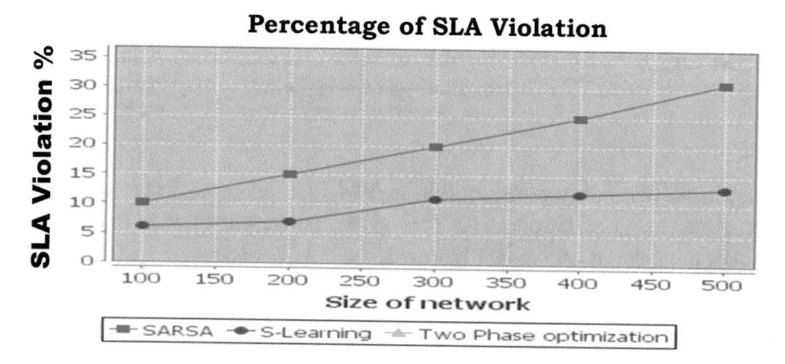

Graph 3 Size of network versus SLA violation

network. The X-axis represents the size of the network ranging from 100 to 500, and Y-axis depicts the percentage of SLA violation. It shows that when the number of nodes are considered till 500, the percentage of SLA violation is less in S-learning algorithm compared to SARSA and two-phase optimization algorithms. Hence, S-learning algorithm exhibits less energy consumption with respect to percentage of SLA violation.

Number of VM migration: The number of VM migration is measured for different size of sensor network. When the number of VM migration is less, the energy consumption is also less.

Graph 4 shows the number of virtual machine migration when different size of network is considered. The X-axis depicts the size of the network ranging from 100 to 500 nodes, and Y-axis shows the number of virtual machine migration ranging from 0 to 125. The two-phase optimization technique shows the highest migration rate compared to S-learning and SARSA algorithm, and this leads to the result that the energy consumption is also more. Comparing these technique, S-learning algorithm

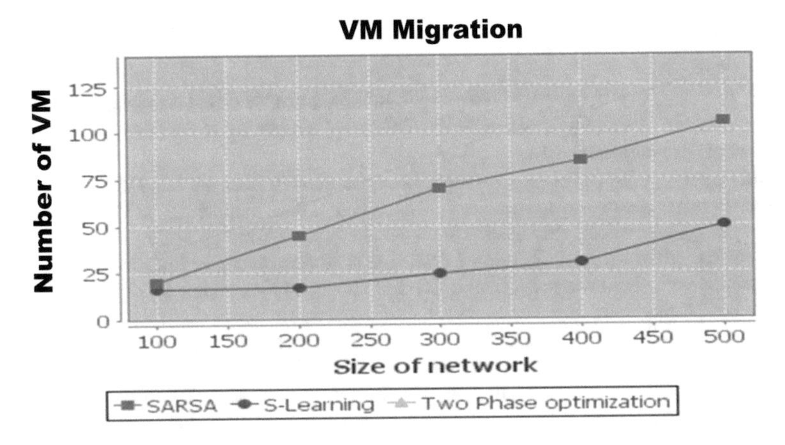

Graph 4 VM migration versus number of VM

has less number of virtual machine migration. Hence, the energy consumption is also less.

10 Conclusion

In this paper, we presented the machine learning algorithm implementation which performs better for dynamic conditions of sensor cloud environment. We have implemented the S-learning algorithm on sensor cloud environment and proved that the solution is able to handle dynamic requests. For dynamic request generation rate, we measured the response time, SLA violation and the energy consumption. For simulating the sensor cloud environment, CloudSim and NS2 simulators have been used.

11 Future Enhancement

In future, this concept can be carried out to implement a parallel multi-learning systems which continuously learn multiple instances of S-learning with different rule sets and the S-learning which achieves best results is taken as optimized. By this way, user can achieve even more reduction in response time, SLA violations and energy consumption.

References

1. Buyya R, Yeo CS, Venugopal S, Broberg J, Brandic I (2009) Cloud computing and emerging IT platforms: vision, hype, and reality for delivering computing as the 5th utility. Futur Gener Comput Syst 25(6):599–616
2. Zhu Q, Zhu J, Agrawal G (2010) Power-aware consolidation of scientific workflows in virtualized environments. In: Proceedings of the international conference on high performance, computing, networking, storage and analysis. IEEE, New York, pp 1–12
3. Shah SH, Khan FK, Ali W, Khan J (2013) A new framework to integrate wireless sensor networks with cloud computing. In: IEEE conference, pp 1–6. https://doi.org/10.1109/AERO.2013.6497359
4. Zhang P, Yan Z, Sun H (2013, January) A novel architecture based on cloud computing for wireless sensor network. In: Proceedings of the 2nd international conference on computer science and electronics engineering, pp 472–475. ISBN 978-90-78677-61-1
5. Ahmed K, Gregory M, Integrating wireless sensor networks with cloud computing. In: IEEE Seventh international conference on mobile Ad-hoc and sensor networks, pp 364–366
6. Jadagerimath AN, Basu A, Chitrapu SD (2014, September) Battery capacity management in wireless sensor network rechargeable sensor nodes. Int J Eng Comput Sci 3(9):8129–8135. ISSN: 2319-7242
7. Alamri A, Ansari WS, Hassan MM, Shamim Hossain M, Alelaiwi A, Anwar Hossain M (2013) A survey on sensor-cloud: architecture, applications, and approaches. Int J Distrib Sensor Networks 30(10), Article ID 917923, 18 p. ISSN: 15501477
8. Yuriyama M, Kushida T (2010, September) Sensor-cloud infrastructure physical sensor management with virtualized sensors on cloud computing. In: Proceedings of the 13th international conference on network-based information systems, pp 1–8, Japan
9. Corazza M, Sangalli A (2015) Q-learning and SARSA: a comparison between two intelligent stochastic control approaches for financial trading. Working Paper - Department of Economics Ca' Foscari University of Venice, ISSN: 1827-3580 No. 15/WP/2015
10. Sekhar J, Jeba G, Durga S (2012, November) A survey on energy efficient server consolidation through VM live migration. Int J Adv Eng Technol 5(1):515–525. ISSN: 2231-1963
11. Albers S (2010) Energy-efficient algorithms. Commun ACM 53(5):86–96
12. Katsoras et al (2011) A service framework for energy aware monitoring and VM management in clouds. In: Including special sections, advanced cloud monitoring systems & the fourth IEEE international conference on e-science
13. Mills K, Filliben J, Dabrowski C (2011) Comparing VM-placement algorithms for on-demand Clouds. In: Proceedings of the 3rd IEEE international conference on cloud computing technology and science (CloudCom), pp 91–98
14. Younge AJ, von Laszewski G, Wang L, Lopez-Alarcon S, Carithers W (2010) Efficient resource management for cloud computing environments. In: Proceedings of the international conference on green computing (GREENCOMP '10). IEEE Computer Society, Washington, DC, USA, pp 357–364, Aug 15–18, 2010
15. Buyya R, Beloglazov A, Abawajy J (2010) Energy-efficient management of data center resources for cloud computing: a vision, architectural elements, and open challenges. In: Proceedings of the international conference on parallel and distributed processing techniques and applications (PDPTA), pp 1–12

Classification of Flower Species Using Machine Learning Algorithm

N. J. Anasuya, B. N. Yashasvi, P. Puneeth Kumar, and B. S. Medha

Abstract Since its invention, the computer has begun to have an impact on our day-to-day lives. It enhances the grade of our existence by making them more convenient and productive. Allowing a machine to think and learn like a person is an intriguing concept. Machine learning is the process of allowing a computer to build learning skills on its own using pre-programmed knowledge. Pattern recognition can be compared to a system's ability to recognize many types of items. As a result, pattern recognition and machine learning are inextricably linked. The Iris flower is the subject of this paper. Iris has three separate classifications in its dataset: Setosa, Versicolor, and Virginica. These three separate types of Iris will be distinguished by the developed recognition mechanism. Iris is a genus of flowering plants that includes between 260 and 300 species. We are looking at three different species: Setosa, Versicolor, and Virginica. The goal is to determine the flower's species from its petal and sepal measurements. This work should be improved to classify different plant species. This makes it simple for agriculturists and gardeners to understand different plant types simply by providing measurements.

Keywords Setosa · Versicolor · Virginica · Pattern recognition · Machine learning

1 Introduction

Pattern recognition, a fundamental talent for personage but a difficulty for machines, has been used as a strong technique to build artificial intelligence using machine learning. Pattern recognition has set off an indispensable and vital approach in the discipline of artificial intelligence as computer technology has progressed. Pattern recognition can recognize letters, images, voices, and other things, as well as status, scope, and other abstractions. Since its invention, the computer has begun to have an impact on our daily lives. It enhances the standard of our lives by making them more convenient and productive. Allowing a machine to think and learn like a person is an

N. J. Anasuya (✉) · B. N. Yashasvi · P. Puneeth Kumar · B. S. Medha
Don Bosco Institute of Technology, Bengaluru, India
e-mail: anasuyaprakash@dbit.co.in

A. Kumar et al. (eds.), *Proceedings of the International Conference on Cognitive and Intelligent Computing*, Cognitive Science and Technology,
https://doi.org/10.1007/978-981-19-2350-0_66

intriguing concept. Machine learning is the process of allowing a computer to build learning skills on its own using pre-programmed knowledge. Pattern recognition can be compared to a computer that can recognize various plant species [1]. As a result, pattern recognition and machine learning go hand in hand. The flower classification [2] is the main objective of this paper. Setosa, Versicolor, and Virginica are the three classifications represented in the Iris data collection. These three types of Iris will be distinguished by the recognition mechanism that has been built. After the project is completed, the computer should be able to combine three separate Iris flower classifications into three categories. The entire machine learning operation should go seamlessly. Users do not need to inform the system which class the Iris is allied to; the computer is capable of recognizing all of them on its own. The ultimate goal of this effort is to provide a fundamental understanding of machine learning to anyone who reads this thesis. Even though they have never worked in this industry, they can see how the machine learning algorithm will grow in popularity and utility in the upcoming. Furthermore, the Iris identification case study will demonstrate how to use Scikit-learn software to implement machine learning. This study could be expanded to include the discovery of various plant species that will benefit gardeners and farmers.

2 Related Work

Classification techniques are commonly used to divide data into different categories. Different systems employ classification strategies to quickly determine the class and group to which it belongs [3]. For classification, a variety of algorithms are utilized [4]. There are two primary kinds of categorization algorithms. There are two varieties of classification algorithms: supervised and unsupervised. Artificial neural networks, learning vector quantization, decision tree induction, nearest neighbor classifier, support vector machine, regression trees, etc., are supervised classification algorithms. Expectation–maximization algorithm, vector quantization, generative topographic map, information bottleneck method, and K-means are unsupervised classification algorithms.

One of the most significant approaches in machine learning is classification. Data analysis is the primary function of machine learning [5]. Classification techniques include decision trees, Naive Bayes, back propagation, neural networks, artificial neural networks, multilayer perception, multiclass classification, support vector machine, and K-nearest neighbor, among others [6]. Three strategies are described in depth in this work. The Iris dataset is used in the implementation. For the implementation, the Scikit tool is used. This research primarily uses classification and regression techniques on the Iris dataset to uncover and analyze patterns utilizing the flower's sepal and petal sizes. In comparison with KNN [7] and logistic regression models, we discovered that the SVM classifier provides the best accuracy.

The action of finding user access patterns from a website's log is known as web usage mining. The web log usually accommodates unorganized, loud, and immaterial information. This data must go through a data preprocessing phase before it can be used for pattern mining and pattern survey [8]. Data preprocessing enhances data quality while simultaneously reducing the size of the web log file. Data collection, data cleansing, session recognition, user identification, and path completion are all steps in the data preprocessing process. This document discusses a number of data pre-processing strategies that can be used to prepare and put together raw data for mining and analysis.

3 Methodology

In the subject of artificial intelligence, machine learning plays a significant role. The AI system did not have a comprehensive learning aptness from the start of its development, therefore, the system as a whole was not ideal. When a computer encounters a problem, for example, it is unable to self-adjust. Furthermore, the computer is unable to collect and discover new information on its own. The program's inference requires more deduction than deduction. As a result, computers can only deduce truths that already exist. It lacks the ability to develop new logical theories, rules, and so on. Pattern identification, artificial intelligence, computer vision, data mining, text categorization, and more industries have already used in machine learning. Machine learning offers a novel advance towards to increasing the intelligence of machines. It also makes it easy to assist people in analyzing data from large datasets (Fig. 1).

Learning difficulties can be divided into two groupings: supervised and unsupervised learning:

Supervised Learning: It is a wide cast-off machine learning method that may be used in a variety of computer science domains. The system can create a learning model based on the training dataset in the supervised learning approach. The algorithm can be used by a computer to forecast or analyze new information, conforming to this learning model. A system or machine can determine the optimal solution and lower the error rate on its own by utilizing sophisticated algorithms. The most common applications of supervised learning are classification and regression. When a developer sends the computer several representatives for supervised learning, each representative is always associated with some categorization details. When a classifier performs recognitions for each pattern, the system will examine these representatives to gain learning occurrences, lowering the mistake rate. A distinct machine learning model is used by each classifier. A neutral network method and a decision tree learning algorithm, for example, are two dissimilar types of classifiers.

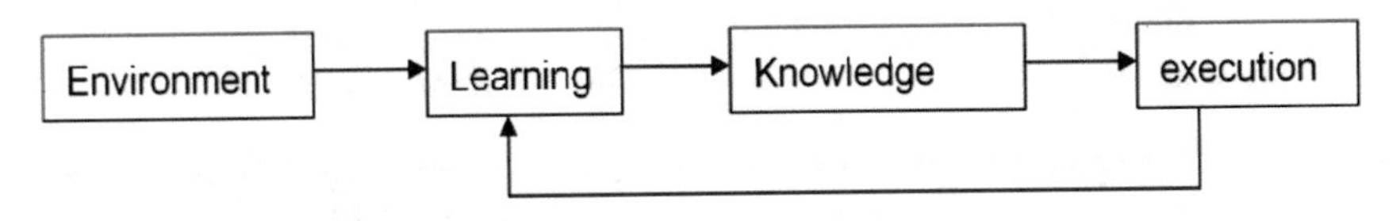

Fig. 1 Machine learning model

Unsupervised Learning: Original data can also be classified using unsupervised learning. In the unsupervised learning practice, the classifier's goal is to notice grouping details for unlabeled representatives. The objective of unsupervised learning is to allow the system to acquire a knowledge on its own. We do not instruct the machine on how to perform the task. The computer is supposed to analyze the representatives provided. In unsupervised learning, the system is unable to determine the optimum course of action and has no way of knowing whether or not the output is right. When the system accepts the original data, it may immediately locate the potential control within the details, which it will then apply to the new situation. This is where the divergence between supervised and unsupervised learning is made. This is where the distinction between supervised and unsupervised learning is made.

The following are the steps to properly prepare data for a machine learning model:

Step 1: Gathering the data

The term "gathering" or "collecting" refers to all of these data collections.

Step 2: Handling missing data

Real, unprocessed data is rarely usable right away. We will have to deal with data gathering failures, corrupt records, missing values, and a slew of other issues.

To convert the data into a format that can be further processed, we must first clean it.

Step 3: Deciding which key factors are important

The more data we offer to the model, the more expense.

If certain features have no bearing on output, they can be ignored.

Step 4: Splitting the data into training and testing sets

The well-known data splitting rule is 80–20% training and testing sets, respectively.

Cross validation is a crucial process for estimating the model's skill on new data.

K-Nearest Neighbor Algorithm:

The KNN method is a sort of supervised machine learning method that may be used to work out the two categorization and regression forecasting issues. However, in diligence, it is mostly utilized to work out and resolve categorization and prediction issues. The KNN forecasts the values of newly discovered data points using "feature correspondence," which hints that the newly discovered data point will be allocated value depending on how intimately it blends with the points in the training set. With the help of the steps below, we can understand how it works (Fig. 2).

Step 1: We require a dataset to execute and put in practice any algorithm. As an out-turn, we must stuff training and test data at the initial stage of KNN.

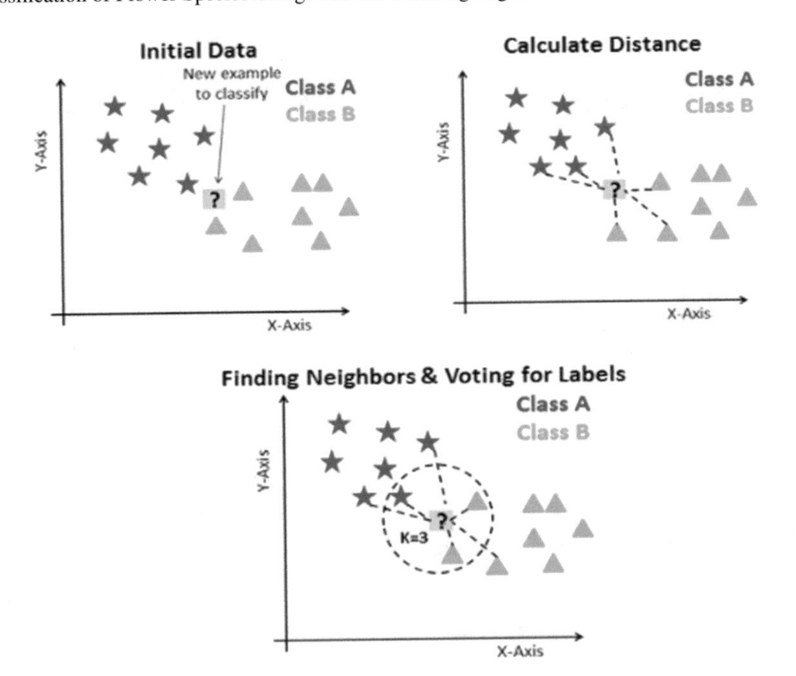

Fig. 2 K-nearest neighbor classification

Step 2: The value of K, the closest data points should be chosen. K being any integer.

Step 3: Perform the ensuing for every individual point in the test data:

- 3.1: Using any of the following techniques, compute the interspaced between test data and every individual row of training data: Euclidean, Manhattan, or Hamming distance. The Euclidean method is extensively utilized technique for calculating distance.
- 3.2: Classify them in its way up order depending on the distance value.
- 3.3: The furthest up K rows of the classified array will then be selected.
- 3.4: The test point will now be designated a class based on the mass common set of the rows.

Step 4: End

Logistic Regression:

This procedure is basically a supervised classification procedure. For a specified group of features (or inputs), *x*, the target variable (or output), *y*, can only take distinct values. Logistic regression is a regression model, converse to familiar kind presumption.

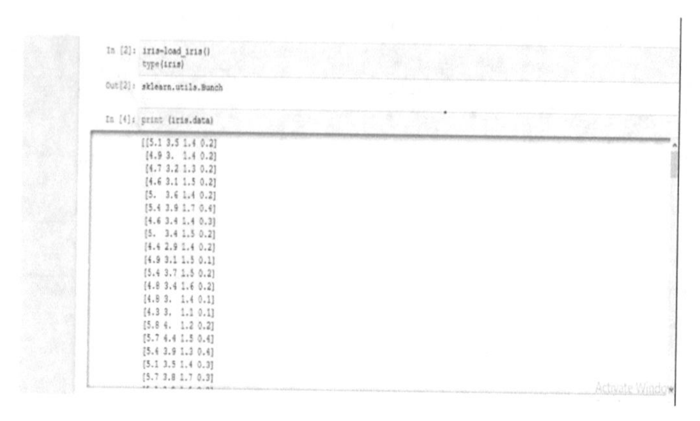

Fig. 3 Iris datasets

Hypothesis function for logistic regression:

$$\frac{p}{21 - p} = \exp(b_0 + b_1 x) \tag{1}$$

While training the model as mentioned in Eq. 1, we give:

x input training data
p labels to data
b_0 constant
b_1 coefficient of x and defines the steepness of the curve.

It suits the finest line to predict the value of p for a specified value of x when training the model. By working out the finest b_0 and b_1 values, the procedure obtains the finest regression fit line. We get the best fit line once we find the best b_0 and b_1 values. So when we are finally using our model for prediction, it will forecast the value of y for the input value of x.

4 Results

See Figs. 3, 4, 5, and 6.

5 Conclusion

With the expeditious growth of technology, AI has been put in various fields. Machine learning is the basic method for achieving AI. This contention discusses the working basis of machine learning, two distinct machine learning formats, and a machine

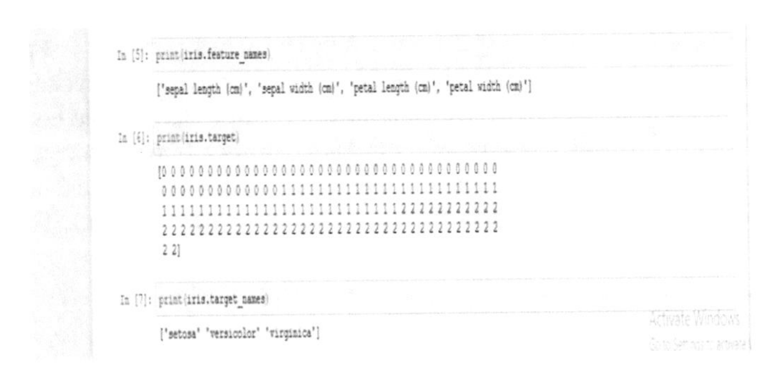

Fig. 4 Iris datasets

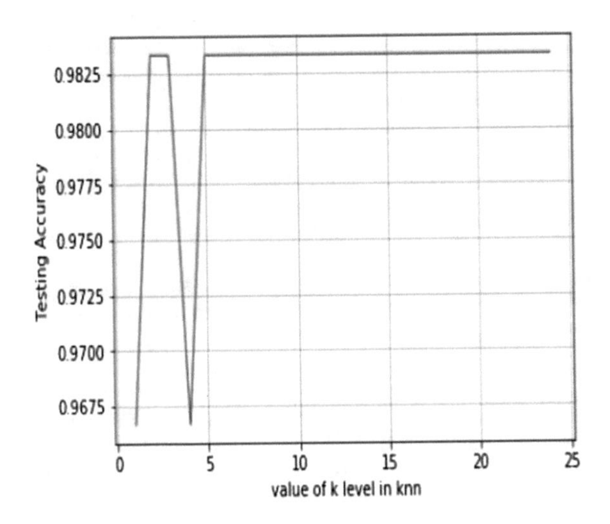

Fig. 5 Accuracy graph

```
In [128]: knn=KNeighborsClassifier(n_neighbors=5)
          knn.fit(X,y)
          pred = knn.predict([[5,3,1,1],[6.3, 3.3, 6, 2.5]])
          print(format(iris['target_names'][pred]))

          ['setosa' 'virginica']
```

Fig. 6 Output

learning application. A case study of Iris flower identification is also provided to demonstrate the workflow of machine learning in pattern identification. The definition of pattern recognition and how machine learning performs in pattern recognition have been presented in this case. The supervised learning method employs the K-nearest neighbor method, which is a fairly basic machine learning algorithm. The paper also demonstrates how to learn machine learning using the SciKit-learn package.

This project is a simple use case of supervised machine learning method. But this methodology can be used to build a system which detects different species of plants. Using a large data and more efficient classification algorithms in machine learning, one can build a system which can classify a variety of species of plants. This will be helpful for farmers, gardeners and also for forest department.

References

1. Kusumo BS, Heryana A, Mahendra O, Pardede HF (2018) Machine learning-based for automatic detection of corn-plant diseases using image processing. In: International conference on computer, control, informatics, and its applications (IC3INA), Tangerang, Indonesia, pp 93–97
2. Mete BR, Ensari T (2019) Flower classification with deep CNN and machine learning algorithms. In: 3rd international symposium on multidisciplinary studies and innovative technologies (ISMSIT), pp 1–5. doi: https://doi.org/10.1109/ISMSIT.2019.8932908
3. Gamit C, Swadas PB, Prajapati NB (2015) Literature review on flower classification. Int J Eng Res Technol (IJERT) 04(02):510–512
4. Hsu T, Lee C, Chen L (2011) An interactive flower image recognition system. Multimedia Tools Appl 53(1):53–73
5. Pinto JP, Kelur S, Shetty J (2018) Iris flower species identification using machine learning approach. In: 4th International conference for convergence in technology (I2CT), pp 1–4. doi: https://doi.org/10.1109/I2CT42659.2018.9057891
6. Liu Y, Tang F, Zhou D et al (2016) Flower classification via convolutional neural network. In: Proceedings of the IEEE international conference on functional-structural plant growth modeling, simulation, visualization and applications, Qingdao, China, pp 110–116
7. Prashar K, Talwar R, Kant C (2019) CNN based on overlapping pooling method and multi-layered learning with SVM & KNN for American Cotton Leaf Disease Recognition. In: International conference on automation, computational and technology management (ICACTM), London, United Kingdom, 2019, pp 330–333
8. Dwivedi SK, Rawat B (2015) A review paper on data preprocessing: a critical phase in web usage mining process. Int Conf Green Comp Inter Things (ICGCIoT) 2015:506–510. https://doi.org/10.1109/ICGCIoT.2015.7380517

Implementing Spot the Differences Game Using YOLO Algorithm

Charan Sai Kondapaneni, Modi Venkata Sai Teja, Rohith Kavuri, Deepika Tinnavalli, and Shahana Bano

Abstract The requirement for object detection has been expanding with computational force. Object identification is a technique that identifies the semantic class of the objects in the image or video. In this work, we talk about implementing an application that connects with the user and distinguishes the various objects present in two generally comparable images utilizing object recognition algorithms (YOLO algorithm). The user collaborates with the application through speech and recognizes the object. This execution has a precision of 70%. Object detection is one of the most demanding task in computer vision, and it involves in finding the type of the object and knowing the location in the image. Through object detection, we can achieve both object detection and tracking, whereas object tracking generally locks the moving object. However, YOLO is one if the top algorithm for object detection it gradually increased its accuracy with version updates one by one which helped it to become the best algorithm.

Keywords Object detection · YOLO · Speech recognition · Object classification

1 Introduction

Object detection is one of the most generally utilized tools in computer vision. The natural eye can recognize objects to a wide reach; however, carrying that ability to a computer is a difficult errand. Object detection is directly or indirectly utilized in many fields like self-driving vehicles, video surveillance, robust vision, and many more. The principle thought of object recognition is to recognize objects and arrange them likewise. Object detection began with customary algorithms like scale-invariant element change (SIFT) [1] which is found in the '90s and the performance of these

C. S. Kondapaneni (✉) · M. V. S. Teja · R. Kavuri · D. Tinnavalli · S. Bano
Department of CSE, Koneru Lakshmaiah Education Foundation, Vaddeswaram, AP, India
e-mail: kondapanenicharansai@gmail.com

S. Bano
e-mail: shahanabano@icloud.com

A. Kumar et al. (eds.), *Proceedings of the International Conference on Cognitive and Intelligent Computing*, Cognitive Science and Technology,
https://doi.org/10.1007/978-981-19-2350-0_67

sorts of calculations are not satisfactory, then the profound learning methods have arisen which gave goliath jump in the presentation and exactness. Our principal philosophy is to implement an application that distinguishes differences between two images utilizing YOLO algorithm. Simultaneously, it can recognize various objects in the image. There are a few algorithms for object recognition, these are isolated into two kinds like classification and regression [2]. CNN and RNN come out under groups under classification while YOLO groups to regression. Previously, classifiers are used to perform object detection, but by using YOLO algorithm object detection is framed as a regression problem.

CNN and RNN are comparatively slow as we have to run these algorithms every time for a selected region in images. But in YOLO, we can directly predict the bounding boxes and classes. YOLO is one of the pre-trained models. There are various versions of YOLO algorithm each having an advantage over the previous version [2].

1.1 Working of YOLO Algorithm

YOLO is the short term of "YOU ONLY LOOK ONCE," an object detection algorithm used to distinguish and identify different objects present in images in real time. As the name recommends, it runs just a single time through the whole image. To comprehend the working of YOLO, one should understand what it is predicting. YOLO algorithm predicts the class probabilities and bounding boxes of the objects present in the image.

The algorithm does not only find the region of interest; rather it isolates the whole image into networks or cells typically of size $S \times S$ [3]. Every cell is then responsible for recognizing K-bounding boxes (Fig. 1).

Bounding boxes can be portrayed utilizing four parameters, specifically center of the box (b_x, b_y), height of the box (b_h), width of the box (b_w) [2].

An object is considered to lie in a cell provided that its center coordinates line inside that cell. The height and width coordinates are constantly determined compared to the image (Fig. 2).

1.2 Non-Maximum Suppression

Objects sometimes tend to get detected more than one time, resulting in more than one bounding box. This can be resolved using non-maximum suppression, which suppresses the bounding boxes with non-maximum probabilities (Fig. 3). Hence, the bounding box with maximum class probability is used.

Fig. 1 Splitting of image into cells of size S×S

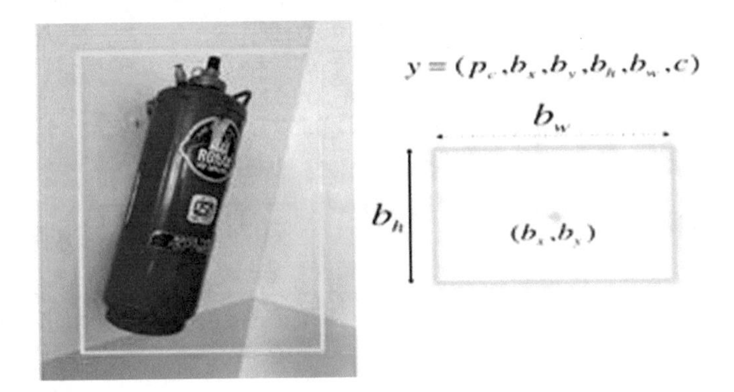

Fig. 2 Parameters relative to bounding box

1.3 YOLO Architecture

YOLO architecture is a combination of convolution and max pool layers (Fig. 4). The max pool layer picks the maximum element from the feature map which will be having more prominent features than the remaining. Whereas the convolution layer extracts all the features from an image among the available features the max pool layer picks the maximum element out of it.

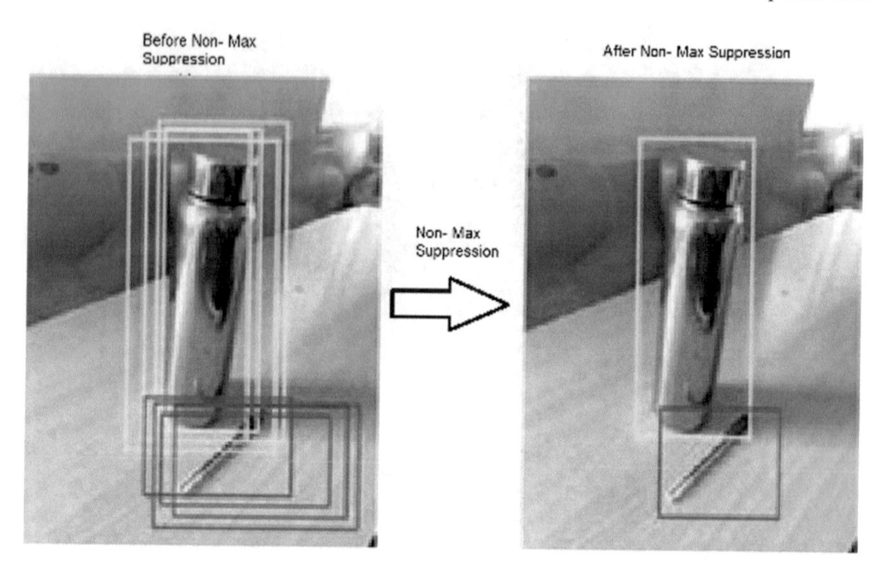

Fig. 3 Non-maximum suppression of image

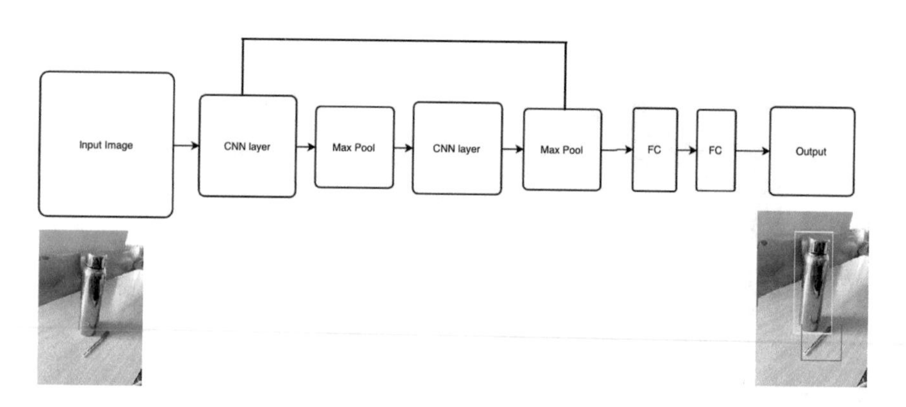

Fig. 4 YOLO architecture

1.4 Comparison of Various Versions of YOLO Algorithm

The comparison of various versions of YOLO algorithm has been summarized in Table 1.

Accuracy of various versions of the YOLO algorithm on images which contain objects on which these algorithms are trained has been shown in Table 2.

There are many metrics to evaluate the performance of these algorithms, and one of the popular metrics is average precision (AP) [4].

YOLO v3 is one of the most widely used object detection method, and it also uses k-means cluster method [5].

Table 1 Comparison of versions of YOLO algorithm

Parameter	YOLO v2	YOLO v3	YOLO v4
Pre-trained objects	80	80	80
Bounding boxes per cell (for 416 × 416 pixel image)	845	10,647	10,647
Frames per second	40–90	> 90	> 90
Time taken for identifying objects		439 ms/(16–17) objects	570 ms/ 16 objects
Framework	Darknet—19 19 convolutional layers and 5 max pool layers	Darknet—53 53 convolutional layers. For detection 53 more are added making 106 layers in total	
Small size object detection	Smaller objects are detected	Better than v2	

Table 2 Performance of versions of YOLO on images used for implementation

Version	Correct prediction	Wrong prediction	Accuracy (%)
YOLO v2	60	20	75.00
YOLO v3	65	15	81.25
YOLO v4	64	16	80.00

2 Procedure

The point of our work is to implement an application that recognizes the different objects present in two comparative images. The user communicates with the model through the speech module. This speech module is utilized to change speech over to message.

First, the system arbitrarily shows two comparable images with a couple of differences. It additionally distinguishes the objects present in the images, yet won't be shown to the user. Now the user recognizes the various objects present in the images. Then, at that point, by utilizing the speech module (speech to message converter), the user inputs the distinctive object present in the first image and not in the second image. In the event that the information given by the user is available in the rundown of objects distinguished by the model, then, at that point, the model will return "Right Answer" as output and draw a bounding box to indicate the area of the object. In the event that the info, given by the user isn't recognized by the model or then again assuming that information is absent in the main image, the model returns "Wrong Answer" as output and won't draw a bounding box.

2.1 Data for Implementation

The data utilized for testing and execution of the application are modified images that contain objects that are now pre-trained by the model. That is, few of the individual images of the objects that are pre-trained are consolidated together to frame one new image. Such produced images are kept in folders. Now, the model chooses two images arbitrarily and recognizes the distinction between these two images.

In the above images represented in Figures 5, 6, and 7, we collected different images of objects and combined them into one single image. Now, the model detects the objects present in this new image.

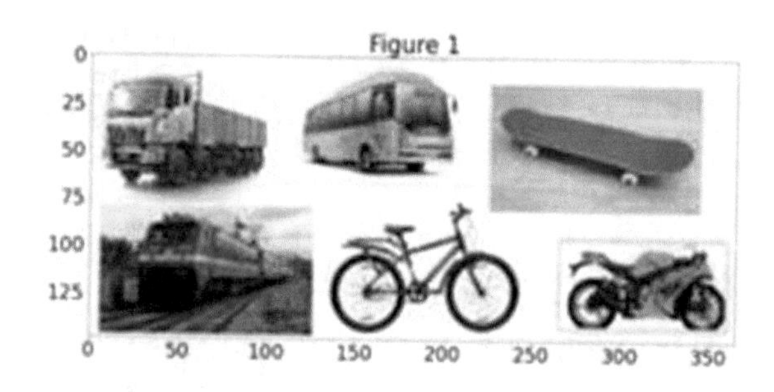

Fig. 5 Custom-made image 1

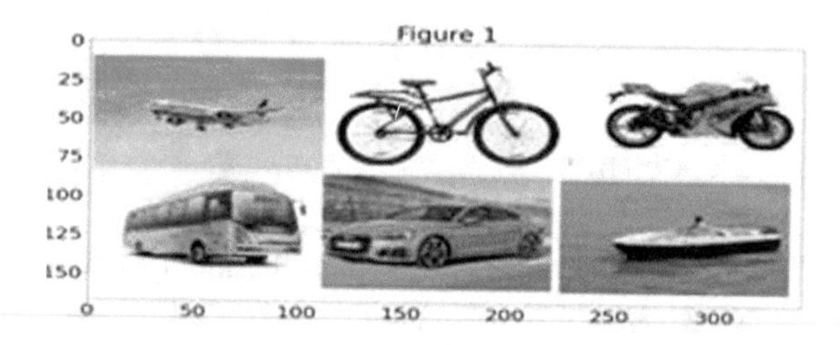

Fig. 6 Custom-made image 2

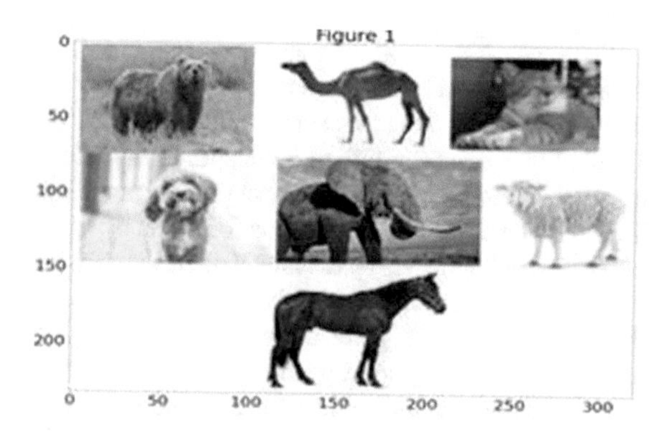

Fig. 7 Custom-made image 3

3 Pseudo Code

Step1: Start

Step2: Import YOLO v3 weights and configuration in the program

Step3: Declare an empty list to store the names of the objects which are present in the coco names files

Step4: Copying all the names in the coco file to classes file using a loop

Step5: Print the length of classes and all the names in the list

Step6: Read the two images that are from the local computer and store them in img and img1, respectively

Step7: S1-Start detecting the objects in the images one by one in both the images

Step8: S2-Print the name of the objects which have confidence values greater than 0.5

Step9: Run the S1 and S2 for the img1 too

Step10: Write a function that takes the object name as an input from the user. Store the input in the MyText variable

Step11: Use the MyText variable to find the object in both the images using a loop

Step12: Generate Label names and confidence values

Step13: If labelname==MyText and confidence value greater than 0.5 then print it is right answer

Step14: Else print the output as the wrong answer

Step15: Generate bounding boxes of the MyText value is equal to the label value

Step16: Generate bounding boxed if the MyText value is not equal and show which is the right answer for the user

Step17: Stop

Figure 8 represents the workflow representation of the whole process. The randomly generated images are displayed to the user. The user identifies the differences from the image.

The data flow diagram is represented in Fig. 9.

4 Results

The model selects these two images randomly and displays them to the user.

Input:

See Fig. 10.

Output:

See Figs. 11 and 12.

As shown, when the input given by the user is selective (Fig. 11), that is, it is present in the first image (Fig. 10a) and not in the second image (Fig. 10b), the model displays "Right Answer" and draws a bounding box around that object (Fig. 12).

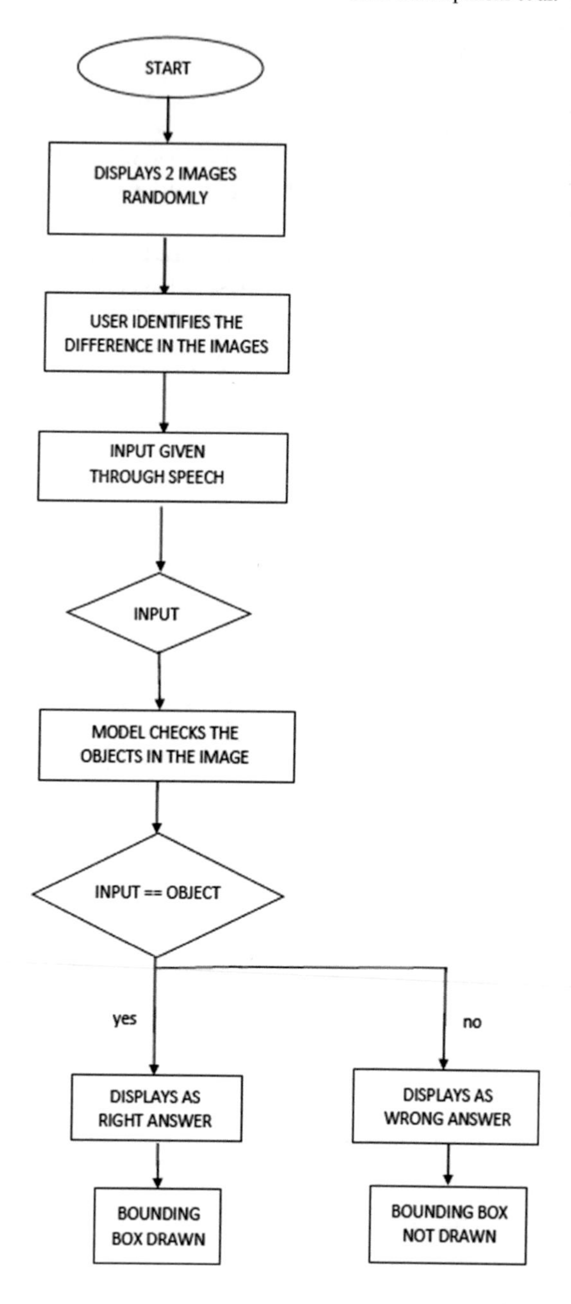

Fig. 8 Workflow representation

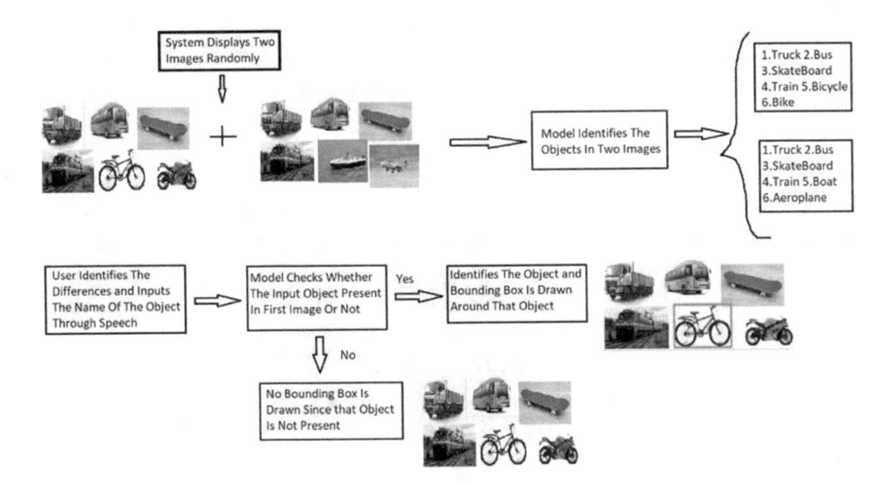

Fig. 9 Flow of the image in the implementation

Fig. 10 **a**, **b** Input images 1

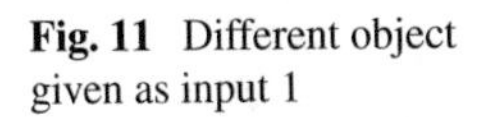

Fig. 11 Different object given as input 1

```
Detection_Of_Object("motorbike")
Right Answer
```

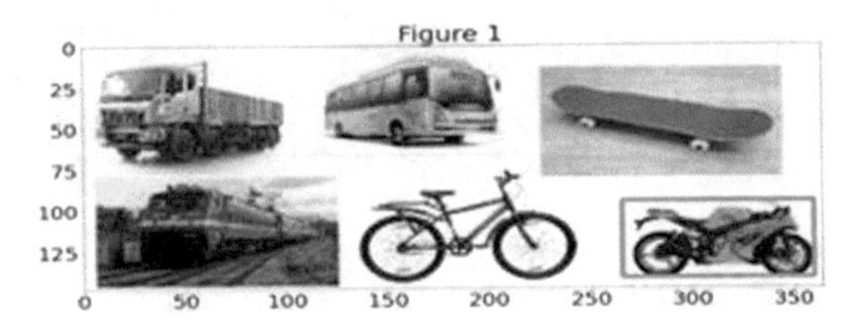

Fig. 12 Output image 1

As shown in Fig. 13, when the input given by the user is present is both the input images (Fig. 10a and b), the model displays "Wrong Answer" and does not draw a bounding box (Fig. 14).

Fig. 13 Common object given as input 1

```
Detection_Of_Object("train")

Wrong Answer
```

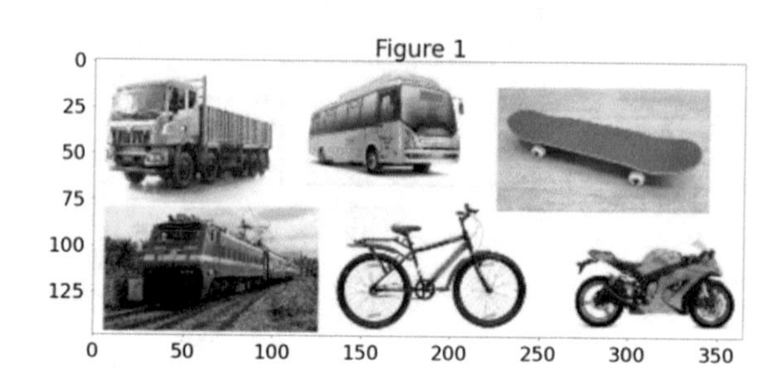

Fig. 14 Output image for input given in Fig. 13

Input:

See Figs. 15 and 16.

Two images (Fig. 15a and b) were randomly selected and displayed to the user by the model.

Output:

See Fig. 17.

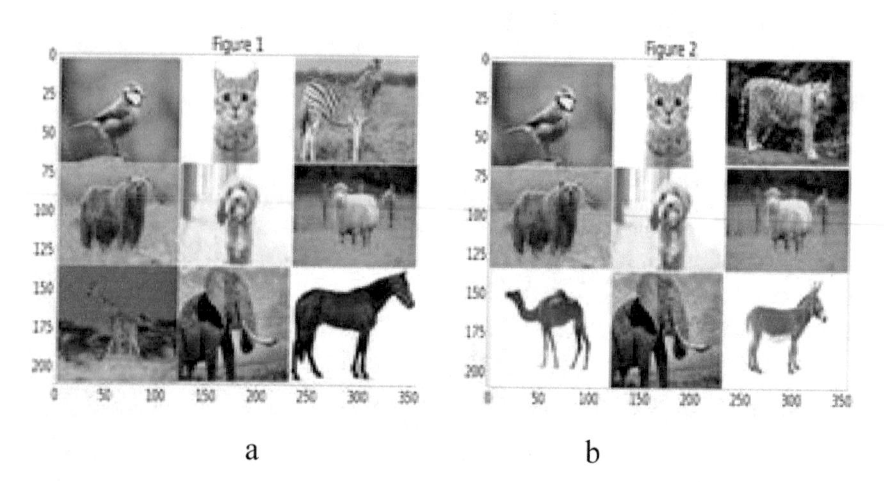

Fig. 15 **a**, **b** Input images 2

Fig. 16 Different object given as input 2

```
Detection_Of_Object("giraffe")

Right Answer
```

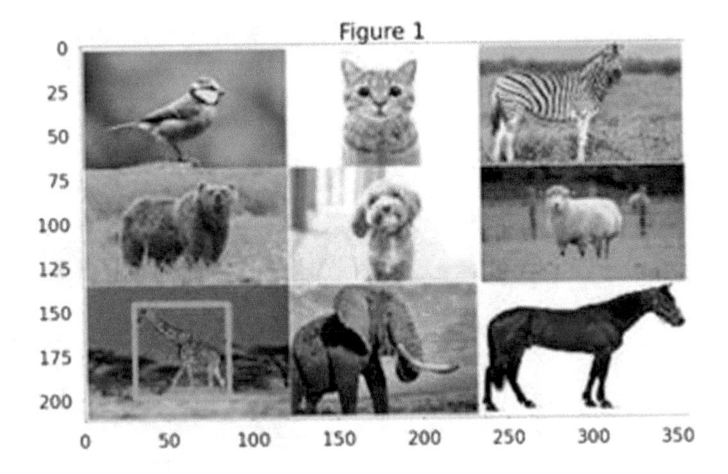

Fig. 17 Output image 2

The input given by the user (Fig. 16) has been identified, and a bounding box is drawn around that object (Fig. 17).

Input:

See Fig. 18.

Output:

See Figs. 19 and 20.

The input given by the user (Fig. 19) is present in both the images (Fig. 18a and b). Hence, no bounding box is drawn (Fig. 20).

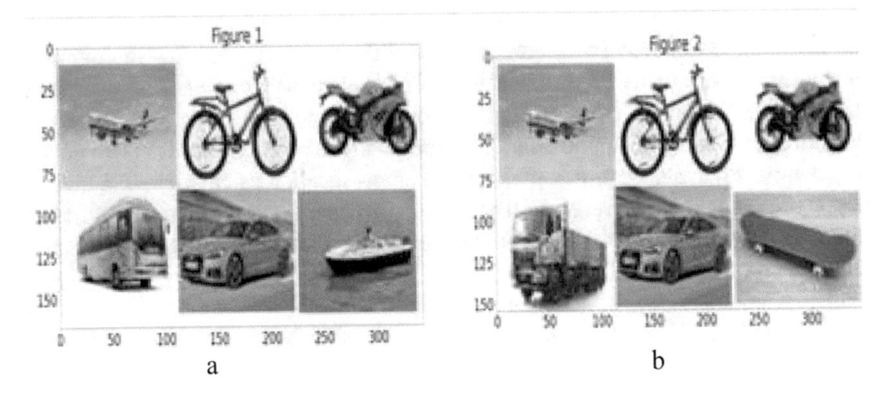

Fig. 18 **a**, **b** Input images 3

Fig. 19 Common object given as input

```
Detection_Of_Object("motorbike")

Wrong Answer
```

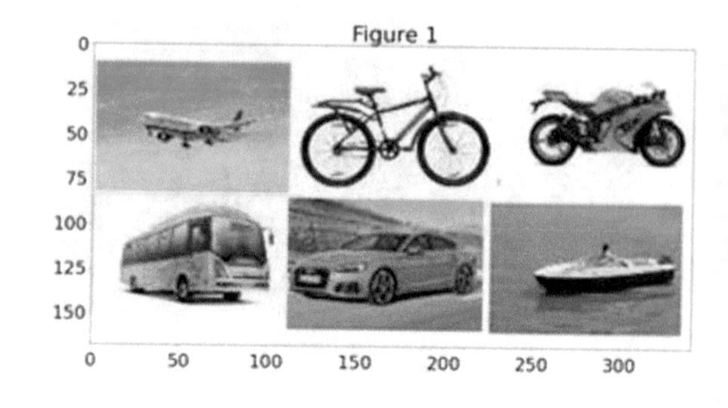

Fig. 20 Output image 3

5 Drawback

This application is confined to recognize just 80 pre-trained class objects. The fundamental disadvantage of this application is that it wrongly recognizes an object of undeveloped classes as one of the pre-trained class objects. For example, it recognizes an image of a donkey as a horse and an image of a white tiger as cat. To beat this downside, we can train the model with more custom datasets and classes.

6 Conclusion

Recognizing the objects in images is one of the crucial assignments in computer vision on account of its expansive scope of uses. This paper extends an outline of the working of the YOLO algorithm for object recognition. This paper likewise centers on executing an application for finding various objects present in two generally comparative images utilizing YOLO v3 object location calculation. The diverse object present in the main image will be represented by the bounding box. By utilizing non-max suppression work this model recognizes the bounding box with the greatest confidence values.

7 Future Scope

The application examined in this paper can be utilized for distinguishing the differences between two images. This can likewise be created to distinguish similitudes among images [6], and it additionally can be created as a gaming application for the

"Spot the Differences" game. This can additionally be created to recognize objects in crime location examination. YOLO v3 algorithm can be utilized in different fields to tackle some genuine issues like security, observing traffic parking assistance [7], port administration [8], and visually impaired people. This execution can additionally be scaled to distinguish different custom objects.

References

1. Kurra H, Guttikonda TM, Reddy GG, Bano S, Trinadh VB (2021) Detecting images similarity using SIFT. In: International conference on expert clouds and applications
2. Handalage U, Kugunandamurthy L (2021) Real-time object detection using YOLO: a review
3. Geethapriya S, Duraimurugan N, Chokkalingam SP (2019) Real-time object detection with YOLO. Int J Eng Adv Technol
4. Sun Z, Cao S, Yang Y, Kitani KM (2021) Rethinking transformer-based set prediction for object detection
5. Zhao L, Li S (2020) Object detection algorithm based on improved YOLO v3
6. Appana V, Guttikonda TM, Shree D, Bano S, Kurra HM (2021) Similarity score of two images using different measures. In: 6th international conference on inventive computation technologies
7. Xu Q, Lin R, Yue H, Huang H, Yang Y, Yao Z (2020) Research on small target detection in driving scenarios based on improved YOLO network
8. Li H, Deng L, Yang C, Liu J, Gu Z (2021) Enhanced YOLO v3 tiny network for real-time ship detection from visual image
9. Van der Weken D, Nachtegael M, Kerre EE (2002) An overview of similarity measures for images. In: IEEE international conference on acoustics, speech, and signal processing
10. Ahmad T, Ma Y, Yahya M, Ahmad B, Nazir S, Ul Haq A, Ali R (2020) Object detection through modified YOLO neural network
11. Masurekar O, Jadhav O, Kulkarni P, Patil S (2020) Real time object detection using YOLO v3. Int J Eng Adv Technol
12. Ibrahem H, Ahmad Salem AD, Kang HS (2021) Real-time weakly supervised object detection using center-of- features localization
13. Huang R, Pedoeem J, Chen C (2019) YOLO-LITE: a real- time object detection algorithm optimized for non-GPU computers
14. Padilla R, Netto SL, da Silva EAB (2020) A survey on performance metrics for object-detection algorithms
15. Yin Y, Li H, Fu W (2020) Faster-YOLO: an accurate and faster object detection method

Design and Implementation of Hand Gesture for Various Applications

K. R. Nataraj, S. Sheela, and K. R. Rekha

Abstract In human–computer interaction, hand gesture recognition is the forthcoming technology, making it easy to interact with computer and many research is performed to improvise this by implementing in several home appliance devices. People are using this insight in many ways in day-to-day life for colour detection, face recognition, hand recognition, etc. This paper presents six different gestures right click, left clicks, right, left, up, down based on the input accelerometer and flex sensor. This paper is based on latest MEMS technology, which provides specific output based on hand gesture.

Keywords MEMS · Hand gesture · Mouse control · Receiver · Transmitter · Zigbee module · Home automation

1 Introduction

The interaction of human–machine is increasing in our day today lives that has made user interface technology progressively more important. Physical gesture will greatly ease the process of interaction and enable humans to more naturally command computers. Gesture device can be designed for any part of the body motion or state but most commonly designed for face and hand. This gesture recognition enables the human to communicate and to interact with the machine without any mechanical devices. By using this concept, it is possible to point the hand towards the computer screen and by the moment of hand, the cursor will move in the same direction of hand motion. This could make conventional input devices such as mouse and controlling home appliances [1].

K. R. Nataraj · S. Sheela (✉)
Don Bosco Institute of Technology, Bengaluru, India
e-mail: sheelatcdbit@gmail.com

K. R. Rekha
SJB Institute of Technology, Bengaluru, India

A. Kumar et al. (eds.), *Proceedings of the International Conference on Cognitive and Intelligent Computing*, Cognitive Science and Technology,
https://doi.org/10.1007/978-981-19-2350-0_68

A gesture is type of non-verbal communication it is also called as non-vocal communication in which the body actions communicate with computer. The project mainly focuses on the advancement of a hand-operated computer mouse which consists of sensors placed in the gloves to determine the position of hand and to operate as the simple human mouse. The sensors which are fixed in the gloves detect the hand motion and drive the left or right displacement of the mouse. Here, the gesture detection is used by accelerometer for various corresponding orientation in two axes, here, the collection of output from accelerometer is done by microcontroller and the corresponding information is displayed on LCD screen. The accelerometer sensor is the main component here which controls the screen as well as the home appliance via Zigbee.

There are different types of gesture they are emblematic gestures, iconic gestures, metaphoric gestures, affect gestures, beat gestures. The hand gesture recognition system is c l assified as glove based or sensor based, vision based, and depth based. We are using glove-based hand gesture recognition in this project because in this sensor based recognition, to capture the position and motion of the hand gesture electronic gloves. Accelerometer detects various gestures, input is forwarded to microcontroller, which is connected to computer via USB using two Zigbee modules, one at transmitter and the other at receiver, and output is displayed on LCD screen.

2 Problem Definition

This project model controls the cursor movement of mouse and consists of several home applications. This project is developed to design a human control interaction (HCI) module so that this could be implemented on a hand glove, which will be helpful for the age old people, paralysed patients and physically challenged population to have a self-dependent live. HCI module is very useful and this technology reduces the gap between the computer and human. As the less educated or elderly people find difficulty in operating phones, laptops, or any other gadgets, this technology is easy to use, operate and maintain [2].

3 Proposed Architecture Design

The system is designed to measure the three-dimensional of the tasks, under-utilizing the available resources and restricts the expressiveness of application use. Nowadays acceleration of human hand motion with adequate accuracy and precision, the necessary bandwidth for normal human motion, and the amplitude range is required for the highest normal acceleration.

The system is designed to measure the 3-dimensional of the tasks, under-utilizing the available resources and restricts the expressiveness of application use. Nowadays acceleration of human hand motion with adequate accuracy and precision, the necessary bandwidth for normal human motion, and the amplitude range required for the highest normal acceleration. The main objective of this project is to control the cursor of the computer and the switch of the home from specified distance. This design developed a hand glove which helps people to operate the computer. By this hand gloves, we can operate the cursor without touching mouse and as well as using the same gloves we can control home appliances by shifting the switch. This is not only used in domestic application but also used in the medical field and in military for sending secret information [4].

The system will develop the executable Java application which reads data from USB and controls the cursor movement (right, left, up, down, right click, and left click) and all the direction cursor movement. Read data from USB which is connected Zigbee module to USB of system which receives data from other Zigbee where hand gesture movement is done. This accelerometer sensor is the main part of the transmitter which works on the earth gravity and its unit is defined as "g", i.e. approximately equals 9.81 m/s^2 (32.17 ft/s^2) as shown in Fig. 1. This works on the 5 V DC supply. Embedded is present in conductive particles in which sensor's one side is polymer ink printed. 30kohms of resistance is given by the particle ink when sensor is straight. And the resistance will increase form the conductive particle ink during the bent stage of sensor.

Home automation also performs the same operations as that of the cursor movement device, and it consists of minor difference, i.e. this device has two loads along with two relays. Here, load is nothing but a light, bulb, fan, phone, and any other device connected to the load. Receiver of this home automation is connected to the same transmitter of cursor controller as shown in Fig. 2.

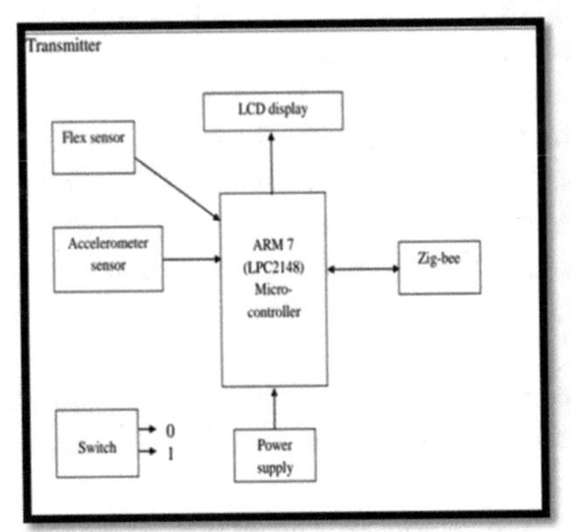

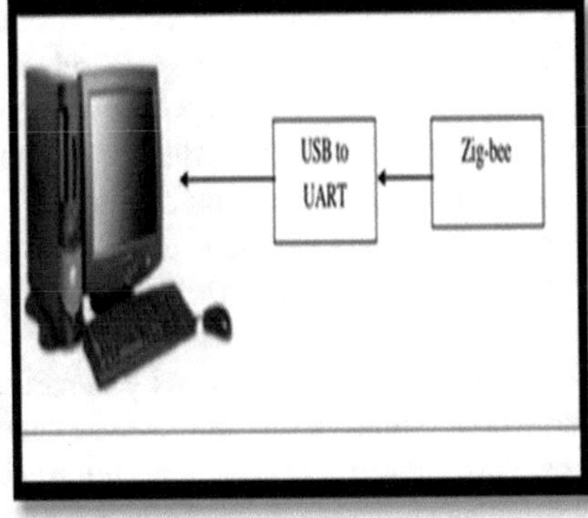

Fig. 1 Block diagram consisting of transmitter and receiver of cursor control

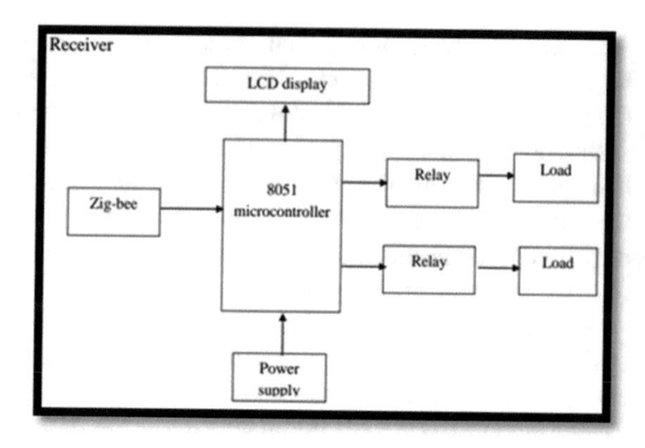

Fig. 2 Block diagram consisting of receiver of home automation

4 Implementation and Results

Figure 3 shows the implementation of cursor movement of the mouse, this consists of two axis planes in x and y direction, when the switch is ON, it performs home automation and when the switch is in OFF state, it controls the cursor movement of the mouse. Accelerometer performs up, down, right and left movement, and flex sensor performs right and left click. This flowchart shows when a peculiar hand gesture is performed the sensor senses the direction, according to the axis and respective output is obtained.

Figure 4 shows the implementation of the home automation, and this is operated when the switch is in ON state, any number of loads can be connected when hand gesture is in x-axis then a specific load performs ON and OFF operation and when hand gesture is recognized in y-axis another load performs its functions.

The overall hand gesture-based devices are shown in Fig. 7.0 overall two 12 V-2A DC power supply is used to power up the devices. Initially, the device will be OFF state. Once it is powered up, there is a sequence of activities that occur as mentioned below. Two devices are connected to the circuit; the audio will play when device is ON or OFF. Which means if the switch is in ON state, then the home automation takes place, if the switch is OFF, then cursor movement of mouse is operated, without touching the mouse as shown in Fig. 5.

Here, in the above picture as you can see the bulb glowing, which means home automation switch is ON and mouse controller switch is turned OFF. This audio speaker also informs whether the switch is in ON or OFF state. Several home automations can be performed by using this device for example, fan can be operated, lights can be turned on or off, mobile phones can be charged, etc. Two devices are connected to the circuit; the audio will play when device is ON or OFF. Which means if the switch is in ON state, then the home automation takes place, if the switch is OFF then cursor movement of mouse is operated, without touching the mouse as shown in Fig. 6. Several home automations can be performed by using this device for example, fan can be operated, lights can be turned on or off, mobile phones can be charged, etc. [6].

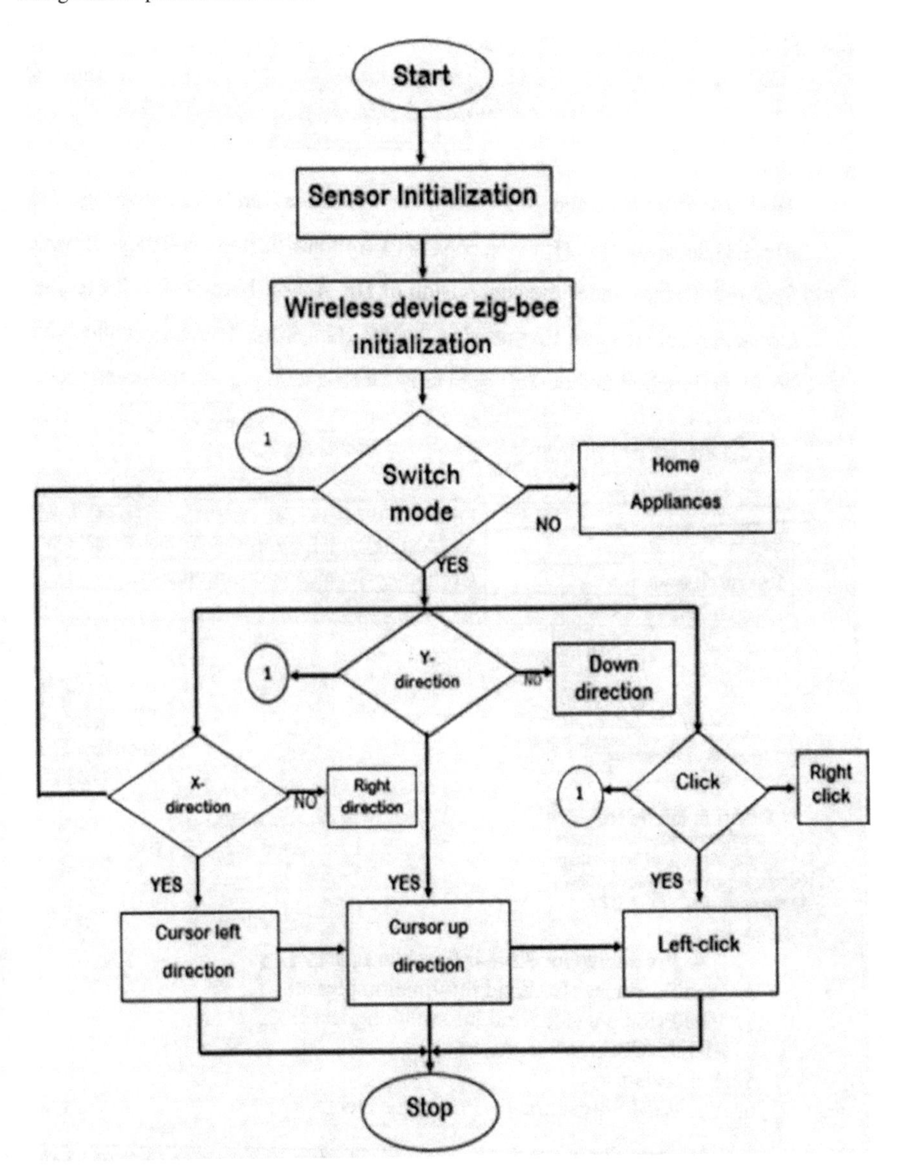

Fig. 3 Mouse cursor movement

5 Conclusion

With the help of several sign languages, which disabled patients use as a means of communication can be given as an input and as an output we can acquire proper speech which is easily understandable even by a person who does not understand their sign language. This can be made possible with the help of some additional devices

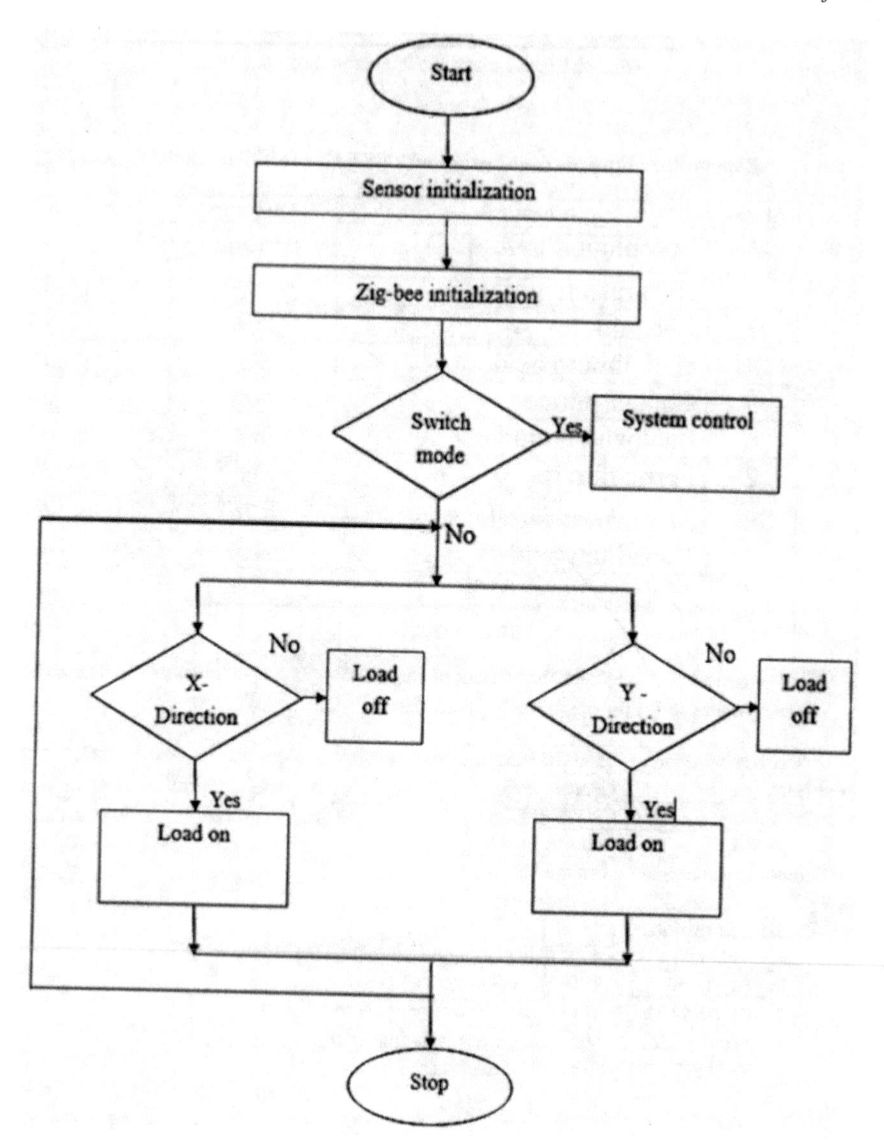

Fig. 4 Home automation

such as flex sensors, audio speaker, and microphone coupled with the accelerometer. This implementation is one such approach for gesture recognition and brings us one step closer to realize the goal of interfacing of humans with machines on their own natural things. This implementation can be used to develop electronic hand gloves, by storing the data on the gloves, the hand gestures are recognized. Similar to that of mouse, this can be implemented on keyboards which would be helpful for the disabled patients. This is used in biomedical field, for smart home automation, etc.

Fig. 5 Result on LCD screen for cursor movement of mouse

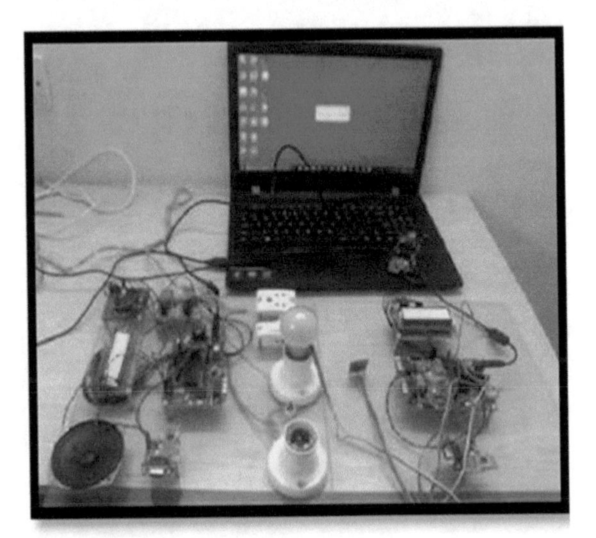

Fig. 6 Result on LCD screen of home automation

References

1. Sai Prasanth P, Gopalakrishnan A (2019) Enhancing user experience using hand—gesture control. Int J Innov Technol Explor Eng (IJITEE) 8(6):2278–3075
2. Ghute MS, Anjum M, Kamble KP (2020) Gesture based mouse control. In: Proceedings of the 2nd international conference on electronics, communication and aerospace technology
3. Chowdhury SR, Pathak S, Anto Praveena MD (2018) Gesture recognition based virtual mouse keyboard. In: Proceedings of the 4th international conference on trends in electronics and informatics

4. Xu R, Zhou S, Li WJ et al (2012) MEMS accelerometer based nonspecific—user hand gesture recognition. IEEE Sens J 12(5)
5. Zhou S, Dong Z, Li WJ, Kwong C (2008) Hand-written character recognition using MEMS motion sensing technology. In: Accepted to 2008 IEEE/ASME international conference on advanced mechatronics
6. Liu J, Wang Z, Zhong L (2009) Accelerometer-based personalized gesture recognition and its applications. In: 2009 IEEE international conference on pervasive computing and communication
7. Gonzalo PJ, Holgado –terriza Juan A (2015) Control of home devices based on hand gesture. In: 2015 IEEE 5th international conference on consumer electronics, Berlin (ICCE-BERLIN)
8. Panchal PB, Nayak VH (2015) A hand gesture based transceiver system for multi-application. In: 2015 2nd international conference on electronics and communication system (ICECS)

English to Hindi Machine Translation Using Sub-classed Model

K. H. Asha, Ayushya Jaiswal, and Ankit Anand

Abstract Natural language processing (NLP) includes a component called machine translation that tries to bridge the communication gap between different social groups. The principle of translation in a language translation system is simple: a system analysis, the pre-translated sentences in the database to determine a suitable translation of an input sentence. As a result, the more pre-translated sentences in the database, the more accurate the language translation system will be. Language translation is one area in which machines clearly fall short of human cognitive abilities. One of the traditional approaches to solving the challenge of machine translation is statistical machine translation. This strategy necessitates large datasets and works effectively on grammar structured language pairs with a similar grammar structure. Neural machine translation (NMT) has evolved as an alternative method of addressing the same problem in recent years. We investigate various setups for putting up a neural machine translation system for the Indian language, Hindi, in this study. For English to Hindi translation, we tested six alternative architecture combinations of NMT and compared our results to those of traditional machine translation techniques. In this work, we also discovered that NMT takes a small quantity of data for training and so provides excellent translation for tens of thousands of training sentences. The objective of the MT mechanism is to automatically translate between two different natural languages, with neural machine translation (NMT) attracting interest since it provides reasonable translation accuracy in the context of context analysis and fluent translation.

Keywords Machine translation · Example-based MT (EBMT) · Model

K. H. Asha (✉) · A. Jaiswal · A. Anand
Don Bosco Institute of Technology, Bengaluru, India
e-mail: asha.kh06@gmail.com

A. Kumar et al. (eds.), *Proceedings of the International Conference on Cognitive and Intelligent Computing*, Cognitive Science and Technology,
https://doi.org/10.1007/978-981-19-2350-0_69

1 Introduction

Language may be a medium, utilized by humans to act with one another and to practice their thoughts to every other. Over 121 languages are used in India by numerous societies and communities of individuals. It is virtually not possible to know each and every language spoken by a selected community of individuals. The majorities of Indians, particularly the remote villagers, do not perceive, read, or write English, so implementing an associate degree economical language translator is required. The basic plan behind the building of the MT system is to fill the gap between people knowing two altogether completely different languages and facilitate them to act with each other in such a way that with the help of a machine, the message of one person gets regenerate into a language which can be understood by completely different and vice-versa. A machine translation model takes a string/sentence of a particular language as input and generates a corresponding string in another language. In English to Hindi MT, the English human's language, i.e., language of the sentence to be translated is named because the supply language, and additionally, the Hindi language, i.e., language of the translated sentence is named the target language. There are diverse strategies of machine translations including rule-based MT, direct-based MT, transfer-based MT, and knowledge-based MT [4].

In rule-based MT, the system primarily retrieves linguistic information regarding the source and target language by changing the most semantic, morphological, and grammar regularities of each language, respectively. Then, the RBMT system converts the input sentences to the corresponding target language supported by the linguistic system collected during its training. In direct-based machine translation, the system returns word by word through the linguistic communication text, translating every word as it goes. The machine uses no intermediate structure; besides the direct-based MT will partner in a nursing in-intensity evaluation of the supply language after which increase it, they talk to attain a language unbiased illustration of this means that or reason among the supply and goal language. Each supply phrase is then immediately mapped onto a few focused-on phrases.

In transfer-based MT, the device operates over three degrees, i.e., analysis, transfer, and generation. First, the supply language textual content is parsed right into a supply language unique intermediate grammar structure. Then, linguistic guidelines unique to the language integrate redesign this instance right into a comparable instance in the goal language. Finally, the last goal language textual content is generated. Statistical system translation is a computational linguistics paradigm wherein translations are generated through the concept of implemented math fashions, i.e., statistical fashions whose parameters are derived from the evaluation of bilingual textual content corpora.

Hybrid-based MT takes the advantage of applied math and rule-based translation methodologies, and a brand new approach was developed, referred to as the hybrid-based approach, which has proven to own higher potency within the space of MT systems.

This paper aims to train the encoder-decoder model by first converting the input to the corresponding context vector and then getting the output text from the generated context vector. The model can be further run with speech-to-text and computer vision technologies to offer full sustainability. In this way, a completely independent experience can be delivered to users with some disabilities. The rest of the paper is organized as follows: Section 1 presents a study within the field and a few of the present systems, the proposed idea is presented in Sect. 2, and in Sect. 3, the conclusion is given [1].

2 Related Work

1. IBM Watson Language Translator

   Users can utilize IBM Watson Language Translator [5] to convert documents and text from one language to another while keeping the formatting. Users observe enhanced translation accuracy and faster speeds when using the latest neural machine translation algorithms. Watson Language Translator also allows you to create bespoke models for your business based on industry or geography.
2. Amazon Translate

   Amazon Translate is a machine gadget translation carrier that supplies fast, high quality, and less costly language translation. Amazon Translate permits you to without problems translate big volumes of textual content efficiently and to localize websites and programs for worldwide users.
3. Babylon

   Babylon computing device translator is to be had as a loose down load and affords translations in over seventy-seven languages. The translator works each online and offline, handing over immediate effects without ever having to alternate applications. With simply the easy click on of a mouse, your asked translation will immediately appear. Babylon's precise computing device changed into evolved the use of optical character recognition (OCR), a modern generation that permits the gadget to pick out each the time period you need to translate in addition to the textual content surrounding it. What this means is that the software program is capable of offer contextual translations with an unrivaled stage of accuracy.
4. Unbabel

   Unbabel eliminates language barriers, allowing businesses to thrive across cultures and regions. For fast, efficient, amazing translations that get wiser over time, the company's language operations platform combines excellent synthetic intelligence with human editors. Unbabel fits effortlessly into any channel, allowing marketers to provide consistent multilingual support from within their existing workflows, making it simple for businesses to expand into new markets and build customer trust in every corner of the globe. Unbabel integrates smoothly with the most popular CRMs and chat systems to provide translations within current workflows via virtual support channels such as chat, email, and FAQs. All of this is managed through the portal, which allows clients to alter translation flows, display critical data such as speed and quality, and assign different tasks in order to operationalize the use of many languages within their organization. Unbabel based in San Francisco works with major customer service divisions at companies like Facebook, Microsoft, Booking.com, and Uber to communicate effectively with customers all over the world, regardless of their native language.

3 Proposed Idea

The main goal is to examine the vast field of machine learning in order to identify the best strategies for application in this context. It is simple to pick one that is good for this purpose because there are so many algorithms and architecture to choose from, but machine translation is an area where robots plainly fall short of human talents. When dealing with a problem that needs to be solved utilizing deep learning approaches, in-depth analysis is required to overcome this limitation. Fortunately, there are numerous papers accessible, most of which focus on the theoretical elements of the solutions provided. The standard of the code becomes a priority when designing an application that needs to be run at the assembly level so that users may utilize it. Another goal is to create a program that is easily accessible, quick, and accurate for users in order to improve their experience and suit their needs. This entails investigating and putting to the test a variety of platforms and packages for constructing deep learning applications. With the growth of this discipline, a number of tools became available, all of which optimized the compiling and computational parts so that developers could concentrate on the algorithms themselves, making them higher and faster. At an equivalent time, integration must be considered additionally to performance. As long as this application is meant to be incorporated into a spread of larger projects, it should be created in such how that integration is straightforward, leading to an object-oriented approach. Within the approach for developing a machine translation model, there are a number of steps that must be completed. These steps are described in further sub-sections of this chapter [1, 2].

A. The Dataset

The dataset that is used for training and testing the model is a CSV file containing English sentences and their corresponding Hindi sentences. This dataset is specifically created to train the machine learning model for developing the machine translation model. It is a large dataset containing over 1 lakh of instances of English-Hindi sentences. This dataset contains two columns, namely English and Hindi columns, and as the name suggests, English column contains English sentences, and their respective Hindi sentences are stored in the Hindi column. For this paper, only 40,000 instances of the dataset are used; once the dataset is loaded, then further data preprocessing is to be performed.

B. Data Preprocessing

Data preprocessing may be a vital step in any machine learning activity. Preprocessing is very important as a result of it completely directs ML activities and makes the additional processes less complicated. To bring down all the sentences within the needed format that is accepted by the application, the preprocessing task starts initially by changing the given input sentence into lowercase, and then further areas, single quotes, punctuations, and different stop words are removed from the sentence. Once this is done, English and Hindi numbers are removed from the sentence. In the end, 'start_' and '_end' tokens are added at the start and end of the sentences. Once all the sentences come to a simpler form, tokenization is the next step that needs to be performed. Tokenization essentially splits sentences, paragraphs, or an entire text document into smaller units, i.e., individual words. Each of these individual words is called a token. TensorFlow offers a simple function named 'Tokenizer' to get the individual tokens. All these tokens are stored in a list of texts.

These tokens cannot be directly passed as a word to models, and words have to be converted to a numerical representation. 'texts_to_sequences' is a function that takes words as an input and returns their distinct numerical representations. In the end before passing the list to the model, every sequence should be of the identical length. 'pad_sequences' is a function which ensures that every one sequence in a list has an identical length. The list of source and target tokens is then split in the relation of 80:20, where 80 percent of the data is used for training purposes and the rest 20 percent is used for testing purposes [3].

C. The Encoder-Decoder Architecture

The encoder-decoder is a recurrent neural network designed to address sequence-to-sequence problems, popularly known as seq2seq. This architecture is widely used in machine translation applications [1]. This machine translation model works by first consuming the words of the source language sequentially, and then sequentially predicting the corresponding words in the target language. However, under the hood, it is actually two different models, an encoder and a decoder. Each of these models uses a recurrent neural network, i.e., in total, two RNN models are used to build the encoder-decoder model (Fig. 1).

The encoder and decoder models that we are creating for this study are sub-classed models. Model sub-classing is a fully adjustable object-oriented method of implementing the feedforward mechanism for our custom-designed deep neural network.

The functions __init__ and 'call' are the most significant functions in model sub-classing. Essentially, we define all of the tf.keras layers, as well as any custom implemented layers, in the '__init__' method and then call those layers depending on the network design in the 'call' method, which is used to accomplish feedforward propagation.

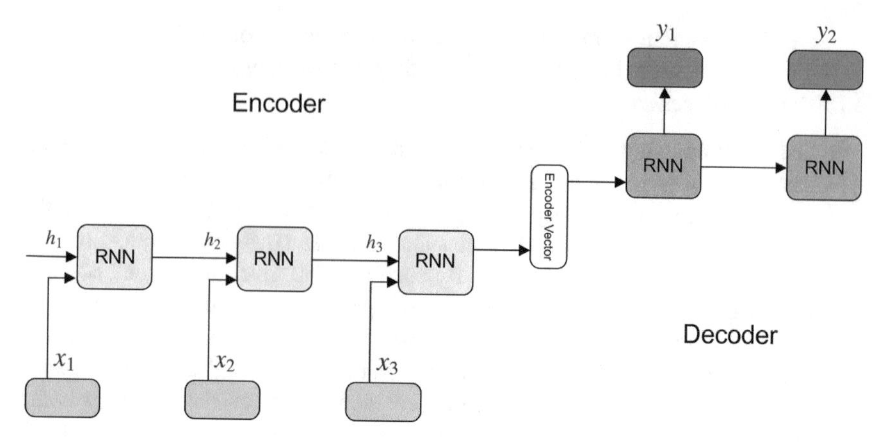

Fig. 1 Encoder-decoder architecture

In natural language processing, the attention mechanism outperformed the encoder-decoder-based neural machine translation system. Attention is a feature of the RNN that allows it to focus on specific parts of the input sequence when predicting a specific section of the output sequence, making learning easier and more accurate. The integration of attention mechanisms increased performance in a wide range of tasks, making it a necessary component of modern RNN networks.

1. The Encoder

   The encoder takes the numerical representation of English words as inputs and produces a corresponding representation of the inputs known as a context vector. Some important attributes of the dataset are the average number of words, the average length of the sentence, and the size of the vocabulary. These parameters are required to define the input layer of the encoder.

   A sentence is a time series input, which means that past words influence each word in the sentence. The encoder is a sequential model that goes from one input to the next while creating an output at each time step. At time step 1, the first word is processed, at time step 2, the second word is processed, and so on. In our translator, the encoder model is made up of GRU model which stands for gated recurrent unit (Fig. 2).

   The GRU layer absorbs the input, consumes the starting state, and outputs the output as the new state at each time step. This GRU layer continues in this manner until the sentences finish. The prior step's hidden state serves as a memory of what the model has previously observed. These hidden states are calculated using the GRU model's internal parameters, which are learned during model training. The final state of the GRU layer is a context vector, which will be sent to the decoder as inputs later.

Fig. 2 GRU in encoder

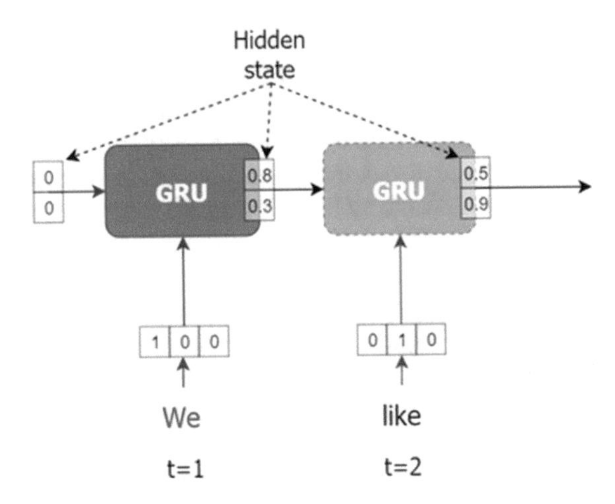

2. Bahdanau Attention

Bahdanau et al. proposed an interesting mechanism that learns to align and translate jointly. It is likewise referred to as additive interest because it plays a linear mixture of encoder states and decoder states.

Let us recognize the attention mechanism advised with the aid of using Bahdanau:

i. All hidden states of the encoder (ahead and backward) and the decoder are used to generate the context vector, not like how simply the final encoder hidden country is utilized in seq2seq without attention.
ii. The interest mechanism aligns the enter and output sequences, with an alignment rating parameterized with the aid of using a feed-ahead network. It allows being aware of the maximum applicable data within side the supply sequence.
iii. The version predicts a goal phrase primarily based totally on the context vectors related to the supply role and the formerly generated goal words.

3. The Decoder

The decoder receives the context vector and produces probabilistic predictions for each time step. At each time step, the phrase with the highest probability is chosen. The decoder creates continuous probabilistic outputs despite the encoder receiving just ones and zeros as inputs. The decoder is implemented identically to the encoder using the Keras GRU layer.

To get the output sentence, repeat the context vector N times, for example, to get a ten-word Hindi sentence, repeat the context vector 10 times. We use the repeat vector layer offered by Keras for this. We may use the repeat vector layer to repeat an input for a set amount of times. Using the repeat vector layer, we define the decoder input by repeating the encoder state for a specified number of times. Now we will put together the decoder GRU layer. All of the decoder's GRU layer outputs are required in decoder because each of those GRU outputs is ultimately used to forecast the proper Hindi word for each decoder location [4]. These models are called sequence-to-sequence models because they map a sequence that is an English sentence to another sentence, i.e., a Hindi sentence.

D. The Training Phase

The supply and goal datasets are entered into the encoder layer after the dataset has been preprocessed to put together the context vectors from the sentences. The decoder is then given the encoder output, the encoder hidden states, and the decoder enter (that is the begin token). The predictions and hidden states are again through the decoder. The hidden nation of the decoder is then exceeded returned into the model, and the loss is calculated by the use of the predictions. We use GRU and an attention mechanism in our version design. We additionally encompass a dense layer that serves as a hyperlink between the encoder and the decoder. We hired teacher forcing to decide the decoder's subsequent input.

The subsequent enter to the decoder is decided with the aid of using trainer forcing. The goal phrase is dispatched as the following enters to the decoder by the use of the trainer forcing technique. The gradient is calculated and implemented to the optimizer, and lower back propagated within side the ultimate step. Multiple epochs are used to educate an encoder and decoder version with attention. The time taken for every epoch is decided, in addition to the loss functions.

We are utilizing a dataset of size 40,000 for training our model in an 80:20 ratio, which means that we are using 80% of the data for training. We are also using GPU to conduct 20 epochs. The neural network takes about an hour to train for each epoch because we used GPU. As the dataset and the number of epochs grow larger, the model's prediction accuracy improves. However, there is a limit to how many epochs can be used; once this limit is reached, increasing the number of epochs will not improve accuracy.

The compared characteristic is pretty equal to the schooling loop, with the exception that we do now no longer appoint teacher forcing. The decoder gets its beyond predictions, in addition to the hidden country and encoder output, as input.

When the version predicts the stop token, it stops forecasting and saves the eye weights for on every occasion step.

E. Results

When we give the model some sentences to translate, at least six to seven sentences out of ten match the expected output, as seen in the diagram below. We receive some results that are similar to the expected result for the remaining sentences. We may conclude that our system is 70% accurate by comparing terms and memorizing all the changes. A more powerful machine is required for improved accuracy, as we will be able to train the model on a larger dataset and spend less time at more epochs (Fig. 3).

```
translate(u'Human contracted itself blind, malignant.')
Predicted translation: इंसानियत ने खुद से क़रार कर रखा है अंधेपन विद्वेष का

translate(u'They are talking about music.')
Predicted translation: वे संगीत के बारे में बात कर रहे हैं।

translate(u'That student loves History')
Predicted translation: जो कि समाज को इतिहास को जानता है

translate(u'English is a beautiful language')
Predicted translation: अंग्रेज़ी पढ़ाना एक लम्बी है

translate(u'But modernization actually brought communication')
Predicted translation: लेकिन आधुनिकीकरण वास्तव में संचार लाया

translate(u'I am going to work.')
Predicted translation: मैं पूरी काम कर रहा हूँ।

translate(u'You need to work smart.')
Predicted translation: तो आपको काम करना होगा

translate(u'Hello')
Predicted translation: नमस्कार।

translate(u'I live in Bangalore')
Predicted translation: मैं श्री उबिराजरा में रहता हूँ।

translate(u'it fires whenever the animal gets near')
Predicted translation: यह तरंग भेजती है जब भी जंतु
```

Fig. 3 Result of ten sentences

4 Conclusion

The machine constructed combines the fashions of recurrent neural network giving the very best accuracies inside ten epochs. It has been determined that taking more than one fashion at a time ended in a higher version giving better accuracy for the machine as a whole. It has additionally been determined that as person fashions, recurrent neural network with embedding offers the most accuracy within addition iterations, observed by bidirectional recurrent neural network. The least appearing person fashions encompass the primary recurrent neural network and the recurrent neural network. The version has to be selected smartly, and at the same time as combining, it should be taken care that the machine has to now no longer end result to over-fitting troubles and different overheads, as they will cause worse accuracy and an inefficient machine for gadget translation the usage of neural networks. According to the effects determined from the experiments done, bidirectional recurrent neural network with encoding set of rules offers an excessive accuracy in comparison to that of the given fashions of recurrent neural network. Single recurrent neural network fashions like the primary RNN, simplest RNN with encoding, simplest bidirectional RNN, and RNN with encoder and decoder offer lesser accuracy comparatively. The paintings may be prolonged for a bigger dataset consisting of the even larger textual content corpus, in order that the machine turns into best for translating each viable sentence as opposed to only a few sentences. Further, the viable mixture of fashions may be broadened into different fashions of RNN as well, thinking about all viable combos inside them, in order that the structures may be made in addition accurate.

References

1. Puscasiu A, Franca A, Gota D, Valean H (2020) Automated image captioning. In: 2020 IEEE international conference on automation, quality and testing, robotics (AQTR)
2. Saini S, Sahula V (2018) Neural machine translation for English to Hindi. In: 2018 4th international conference on information retrieval and knowledge management (CAMP)
3. Nangi U, NK Jain, Rastogi P, Malik R, Jain P (2021) Artificial neural network algorithms for the optimum solution of economic load dispatch problem. In: 2021 international conference on communication, control and information sciences (ICCISc)
4. Nair J, Amrutha Krishna K, Deetha R (2016) An efficient English to Hindi machine translation system using hybrid mechanisms. In: 2016 international conference on advances in computing, communications and informatics (ICACCI)
5. Li Y, Muthiah M, Routh A, Dorai C (2017) Cognitive computing in action to enhance invoice processing with customized language translation. In: 2017 IEEE international conference on cognitive computing (ICCC)

Spoof Invariant Assistive System for Physically Disabled People Using Facial Recognition

A. S. Chaithra, S. Aishwarya, and B. J. Shukrutha

Abstract Facial recognition is a tough and quickly growing topic in biometrics. A variety of advancements based on extraction of features and matching algorithms to recognize the face accurately and efficiently have been developed. Facial recognition techniques were proposed to develop a computationally efficient facial biometric application for visually challenged people. LBP is used for feature extraction and SVM classification to categorize photographs in this system as valid or invalid, which will help vision impaired people improve the lifestyle and safety of blinds. The system has been put to the test in the actual world. Actual customer feedback and printed images have been fantastic.

Keywords Wavelet transform · CSLBP · Support vector machine (SVM) · K-nearest neighbor · Feature extraction · Local binary pattern (LBP)

1 Introduction

Blindness is a condition that affects roughly 0.7% of the population in the world. To the current estimation, around one million people have some sort of vision impairment, with approximately thousands of people suffering from blindness as a result of the retinal diseases mentioned. There is the potential for spoofing attacks in facial recognition systems. When a person offers the camera with a photograph of the desired person rather than their own, this can happen. This is a significant issue because obtaining a photograph to be utilized later is pretty easy and straightforward. We believe that this is a critical issue for those with vision impairment. In this premises, the purpose is to develop, build, and validate an anti-spoofing and facial recognition architecture which can be integrated into a mobile application as well as a video door entry. We hope that by doing so, we will be able to provide an instrument or tool to the blind, deaf, and hard of hearing, as well as the visually impaired, that

A. S. Chaithra (✉) · S. Aishwarya · B. J. Shukrutha
Don Bosco Institute of Technology, Bangalore, India
e-mail: aschaithra@gmail.com; aschaithra@dbit.co.in

A. Kumar et al. (eds.), *Proceedings of the International Conference on Cognitive and Intelligent Computing*, Cognitive Science and Technology,
https://doi.org/10.1007/978-981-19-2350-0_70

will allow them to improve their quality of life by boosting their sense of security at home or when dealing with others [1].

2 Problem Statement

2.1 Existing System

Face recognition for visually impaired people has been researched using a variety of techniques. The work is summarized below, with the important features that drove the architecture proposed here highlighted for each of its development. Although meetings have mostly focused on what elements of the visual field which was captured by the device are significant to the subject, and hence, facial recognition technologies for the visually handicapped are presently available in mobile devices [2, 3].

2.2 Proposed System

The proposed system was tested with the users, surrounding us, tested on them with them, simulating the exact situation that could arise the situation where the video or photographs will be captured by some impaired person using a mobile device. Some of the contributions are as follows:

- The acquisition of the image is proposed.
- Second, we propose employing computer vision to detect faces.
- Third, we look at how face recognition and facial expression detection techniques like the Wavelet transform, support vector machine, CSLBP, and k- nearest neighbor work are all examples of local binary patterns.
- Finally, for visually challenged people, the estimated person's name and expression are transformed into speech [2, 6].

3 System Design

The idea for implementing this system design is to bridge the gap between the visually impaired person and a normal person for ease of communication between them and to avoid spoofing. The proposed project is built with the most trending programming language also used in major industries—Python. The system is provided with five unique and individual modules [4].

- **Facial Detection**: Viola and Jones proposed that the facial detector is often used to track the face throughout time. The face detector works well if the face is very frontal, but it fails when the rotations are up to 45° on both the horizontal and vertical axes. But, it sometimes does not track your face over time (Fig. 1).

- **Tracking and Facial Recognition**: The following steps are to track the face (tracking) and normalize the region of the face. The locations of faces in the image which correlate to rectangles were found by algorithm named Viola and Jones. This is not exact because there are pixels in the rectangle's corners that do not correspond to facial regions. As a result, a normalizing algorithm is required [5] (Fig. 2).
- **Extraction of characteristics through LBP**: Face attributes have been extracted using the LBP operator for use in both face identification and spoofing detection.

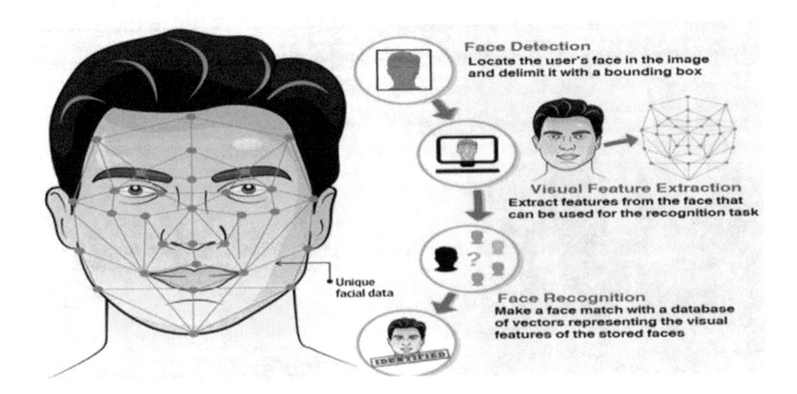

Fig. 1 Facial detection

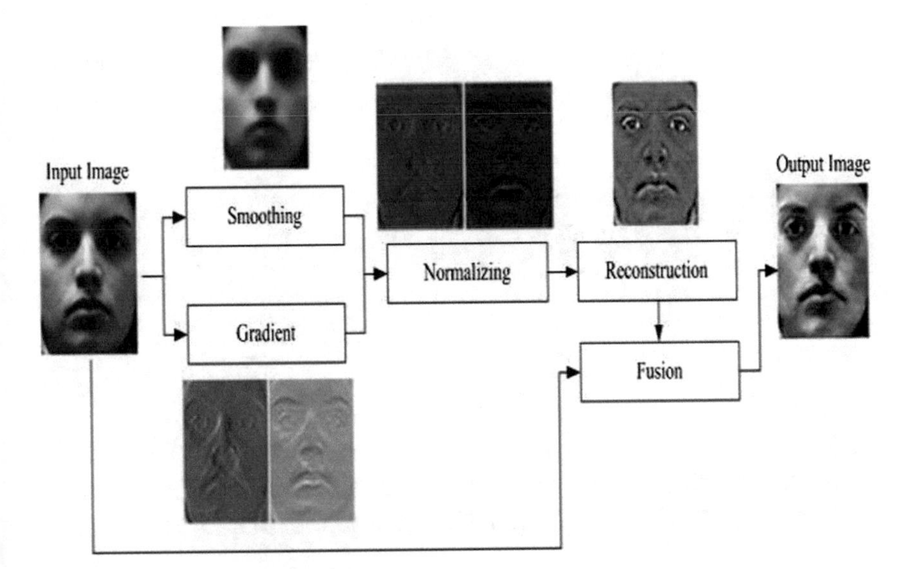

Fig. 2 Tracking, normalizing, and facial recognition

The LBP is a type of operator that is used to handle classification problems. In anything that has to do with texture categorization, it is a very powerful operator (Fig. 3).

- **Feature Extraction for Identification of Face and Spoofing Detection using the LBP operator**: Following a thorough evaluation, a review, and some preliminary tests, the LBP operator was found to perform well in recognition of face and spoofing detection. This is also a computationally straightforward operation. In order to construct a fast, reliable, and light solution, thus investigating which operators and descriptors may be employed to represent the region of face efficiently [8] (Fig. 4).
- **Audio Generator and Information Aggregator**: When analyzing a collection of N specific frames, the results are given to the use to reduce the information. The user checks that is as accurate as possible. As a result, the system analyzes

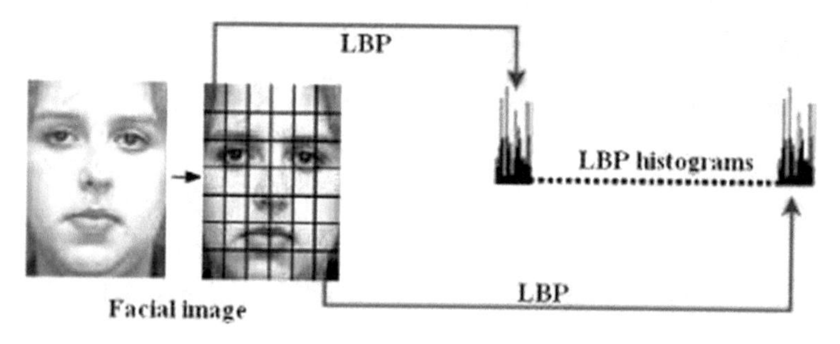

Fig. 3 Extraction of characteristics using LBP

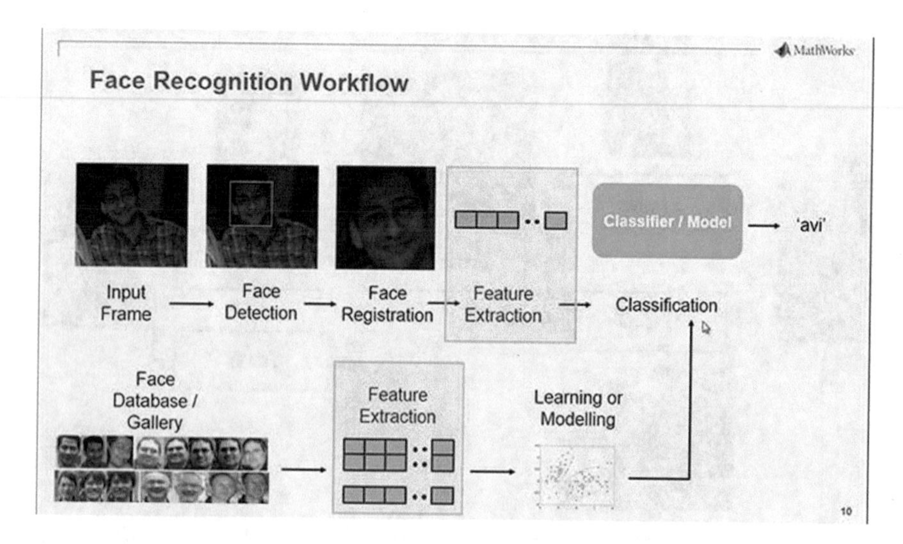

Fig. 4 Extraction of feature using LBP

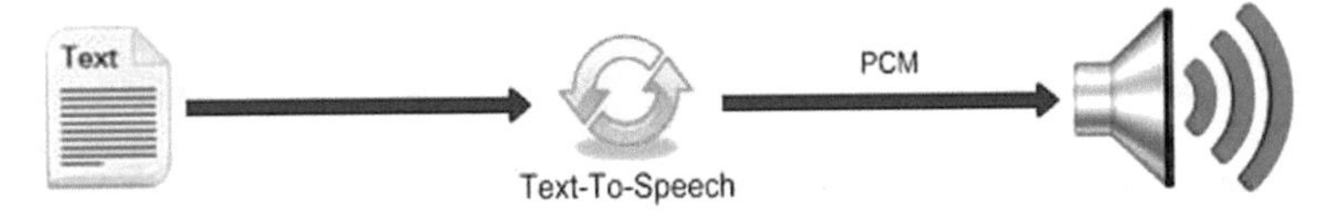

Fig. 5 Text to speech conversion

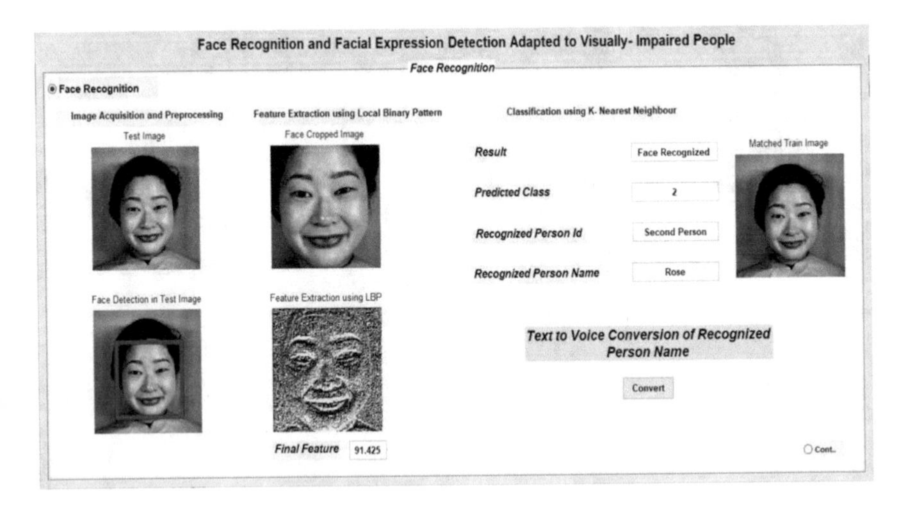

Fig. 6 Face recognition, extraction, and its classification

the preceding N frames. Then, the tracking algorithm detects the face presence to provide the essential information to the user to prevent spoofing [3] (Fig. 5).

4 Results

The proposed system also hands the role of facial recognition, emotion recognition detection, and transformation of text to speech.

Speech for blinds, text for deaf, dumb, and mute person. All these modules are available for single system general purpose as well as for complete visually impaired person. Below are the three major phases of detection of facial recognition and facial expression (Figs. 6, 7, and 8).

5 Conclusion

For person with visual impaired, the main aim is to provide the face recognition technique and spoofing detection. The bone and muscle of the design were to target,

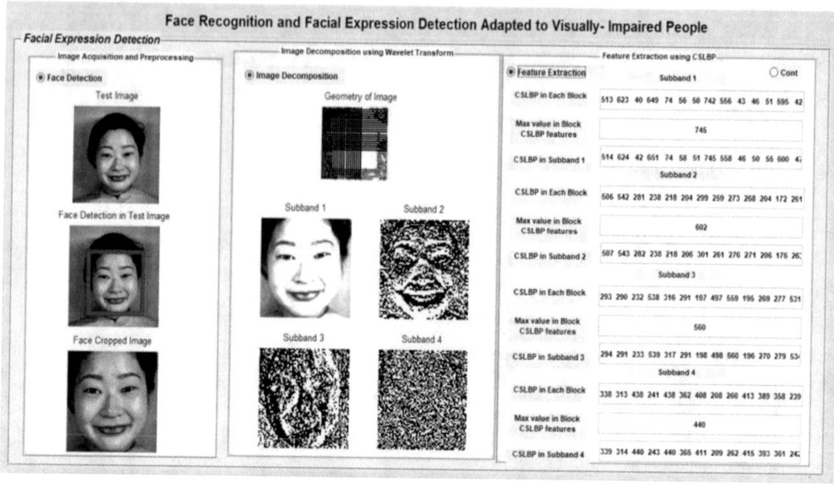

Fig. 7 Image acquisition, decomposition, and feature extraction using LBP

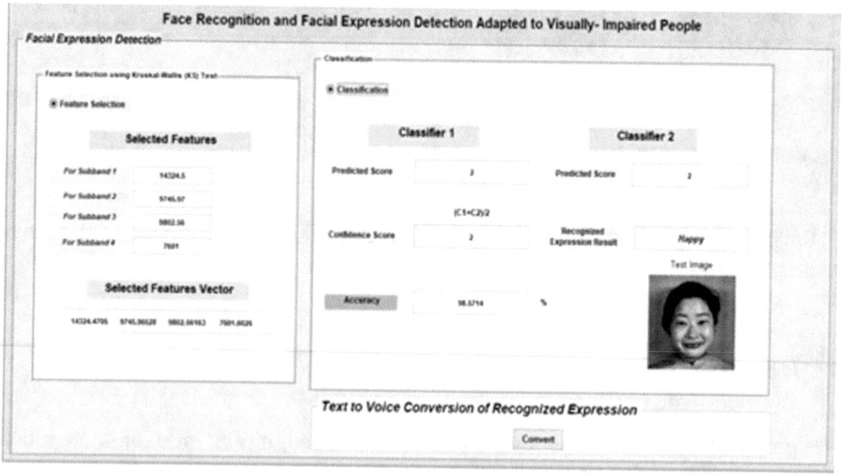

Fig. 8 Facial expression detection and text to voice conversion

the production of a stable system, which could yield great lightweight results with cutting-edge commutating powers, which may include digital tech such as mobile phone or gadgets. The system been developed is the most efficient to be used by any visually disabled people, thus been tested on different environment and people to get accurate results.

Some of the future enhancements of the system would be to largely work simultaneously in two components. The one of the two component is to make system more resilient to get results fast as possible. This is very much important to check how the preprocessing state of the system is proportional to the spoofing detection module. Secondary, word on the spoofing detection approach is enhancing the recognition rate. Finally, we believe that person who have visual impairments will benefit from the existing approach, which will enhance their overall quality of life.

Acknowledgements The development of the proposed system was extremely supported by Don Bosco Institute of Technology. We present gratitude to our guide Mrs. Chaithra A. S., Assistant Professor, Department of Information Science and Engineering, DBIT, co-heartedly provided extra support to make this project successful with team members—Aishwarya S. and Shukrutha B. J. We thank the department and management to provide the idea in coming up with this system and extremity in resources.

References

1. Isewon I (ed) (2014) Design and implementation of text to speech conversion for visually impaired people. Int J Appl Inf Syst (IJAIS)
2. Vision 2020: The right to sight (2011) World Health Organization (WHO) and International Agency for Blindness Prevention (IAPB). Action Plan 2006–2011, Eurostat web portal
3. Maidenbaum S, Hanassy S, Abboud S, Buchs G, Chebat DR, Levy-Tzedek S, Amedi A (2014) The, "EyeCane", a new electronic travel aid for the blind: technology, behavior and swift learning. Restor Neurol Neurosci 32(6):813–824
4. Kramer CKM, Hedin DS, Rolkosky DJ (2010) Smartphone based face recognition tool for the blind. In: Engineering in medicine and biology society (EMBC), annual international conference of the IEEE, pp 4538–4541
5. Balduzzi L, Fusco G, Odone F, Dini S, Mesiti M, Destrero A, Lovato A (2010) Low-cost face biometry for visually impaired users. In: Biometric measurements and systems for security and medical applications (BIOMS), IEEE workshop, pp 45–52
6. http://citeseerx.ist.psu.edu/viewdoc/download?doi=10.1.1.670.7344&rep=rep1&type=pdf---Feasible Camera-based functional & supporting text and Product Tag Studying from held things for Unsighted users
7. A Novel perspective for divulgence among unsighted, deafened and mute users
8. http://www.ijirset.com/upload/2014/icets/131_EC343.pdf-paper

Leaf Disease Detection, Classification, and Analysis in Agriculture Sector

S. R. Shankara Gowda, J. Pavithra, and S. Pravallika

Abstract The agricultural field contributes significantly to a country's revenue. Plants are vital because they provide a source of food for humans. The farmers usually employ manual agricultural methods. Leaf diseases, particularly those affecting the leaves, harm agricultural output quality and quantity. The human eye's perception is not quite as sharp as it is capable of noticing minute differences in an infected leaf. Sometimes plant diseases are identified late, resulting in financial losses. Farmers suffer losses, which have an impact on the state's and country's economies. Uneven backdrop during picture capture, segmentation, and classification of images are some of the problems in illness diagnosis and classification. Once diseases have been recognized based on their symptoms, control mechanisms can be used based on the features. According to the report, traditional methods and machine learning techniques are still inefficient. When compared with traditional methods, deep learning algorithms produced improved outcomes for illness detection and categorization.

Keywords Data collection · Data preprocessing · Machine learning (ML) · Deep learning (DL) · Disease detection · Convolutional neural network (CNN)

1 Introduction

Agriculture is the economic backbone of any country. Most of the farmers' desire to adopt modern agriculture, but they are unable to do for a variety of reasons, including a lack of knowledge about new technology, high technological costs. Machine learning-based approaches have shown to be effective in a variety of image processing applications in recent years. Learning that is based on artificial intelligence applications has shown positive results. Machine learning approaches teach the system to learn autonomously and improve its outcomes based on its own experiences. Plant diseases are difficult to control since their population varies depending on environmental conditions.

S. R. S. Gowda (✉) · J. Pavithra · S. Pravallika
Don Bosco Institute of Technology, Bengaluru, India
e-mail: shankargowda@gmail.com

A. Kumar et al. (eds.), *Proceedings of the International Conference on Cognitive and Intelligent Computing*, Cognitive Science and Technology,
https://doi.org/10.1007/978-981-19-2350-0_71

Farmers in developing nations employ a traditional approach that demands more labor and takes longer. It is also conceivable that manual detection or observation with the naked eye won't yield fruitful results. It has been noticed that the farmers are making use of pesticides to eliminate the effects of illness without identifying the specific diseases. In agriculture, farmers practice pesticides in an unrestricted manner, which can harm plant quality and human health.

Plant leaf disease recognition and its classification using deep learning (DL) and machine learning can assist farmers in identifying illnesses and taking appropriate measures to manage them. In comparison with standard image processing approaches, ML and DL algorithms for detecting plant leaf diseases are more accurate and take less time.

This project mainly focuses on plant diseases, gathering the datasets of diseased plant leaves using deep learning. A web-based system is developed and deployed where we can directly upload the image of the diseased leaf. Finally, the system predicts the appropriate results by analyzing the diseased leaf image.

2 Literature Survey

Keke Zhang, Qiufeng Wu, Anwang Liu, and Xiangyan Meng, can deep learning identify tomato leaf disease? Mentioned in their research that tomatoes are a widely grown commodity that is high in nutrients, has a distinct flavor, and has health benefits, thus they play a significant part in agricultural sector and commerce across the world. The commercial importance of tomatoes, which is essential to increase yield and product quality. There are eight types of diseases in tomatoes. As a result, real-time and accurate recognition technology is important. This article focuses on utilizing deep CNN and transfer learning to diagnose tomato leaf disease. The utilized networks are based on the AlexNet, GoogLeNet, and ResNet deep learning models that have been pre-trained. We first examined the performance of these deep networks using the stochastic gradient descent and Adam optimization methods, finding that ResNet using the stochastic gradient descent optimization technique achieves the greatest result with the best precision, 96.5% [1].

Muhammad Hammad Saleem, Johan Potgieter, and Khalid Mahmood Arif, plant disease detection and classification by deep learning mentioned in their research that plant diseases have an effect on the evolution of their own kind, hence early detection is critical. Several machine learning (ML) models are used to identify and classify plant illnesses, but with recent developments in a subset of ML, DL, this field of research has a great potential in terms of enhanced precision. To identify and categorize the symptoms of plant diseases, a variety of modified or developed DL architectures, and numerous visualization approaches are used. These architectures/techniques are evaluated using several performance measures [2].

This review described how to use deep learning to identify plant diseases. Furthermore, various visualization techniques/mappings were summarized to detect illness signs. Although substantial progress has been made in the previous three to four years, there are still research gaps, as mentioned below:

- Because the severity of plant diseases varies over time, deep learning models should be enhanced or adjusted to allow them to identify and categorize illnesses during their entire life cycle.
- The DL model/architecture should be effective across a variety of illumination situations; therefore, the datasets should include images collected in a variety of field scenarios in addition to the real world.
- To understand the parameters that influence the detection of plant leaf diseases, such as dataset classes, sizes, learning rate, illumination, and so on, complete research is necessary.

Sharada Prasanna Mohanty, David Hughes, and Marcel Salathe, using deep learning for image-based plant disease detection mentioned in their research that crop diseases are dangerous to food security, but due to a lack of infrastructure in many regions of the world, early diagnosis is challenging. Smartphone-assisted illness detection is now achievable thanks to a combination of rising worldwide smartphone adoption, recent breakthroughs in computer image made possible by DL. We trained a deep CNN to detect 14 different plant species and 26 types of illnesses using available dataset of 54,305 pictures of sick and non-diseased, i.e., healthy plant leaves gathered under precise settings (or absence thereof). On a test set, the trained model achieves an accuracy: 99.35%, demonstrating the viability of the method [3]. All of the previous experiments were carried out in this study using our branch of caffe, which is a fast, open-source deep learning framework. A regular instance of caffe may also be used to reproduce the fundamental findings, such as total accuracy.

Vijai Singh a, A.K. Misra, detection of plant leaf disease using image segmentation and soft computing techniques [4] mentioned in their research that, in today's globe, the agricultural landmass is more than just a source of food. Agriculture production is significantly vital to India's economy. As a result, in the area of agriculture, disease identification in plants is critical. The adoption of an automated disease detection approach is useful in detecting a plant disease in its early stages. For example, in the United States, a disease known as tiny leaf disease is a dangerous illness that affects pine trees. The afflicted tree's development is stunted, and it dies within 6 years. It influences Alabama and Georgia, as well as other parts of the southern United States. Early identification in such situations might have been beneficial.

3 System Architecture

3.1 Flow Diagram

See Fig. 1.

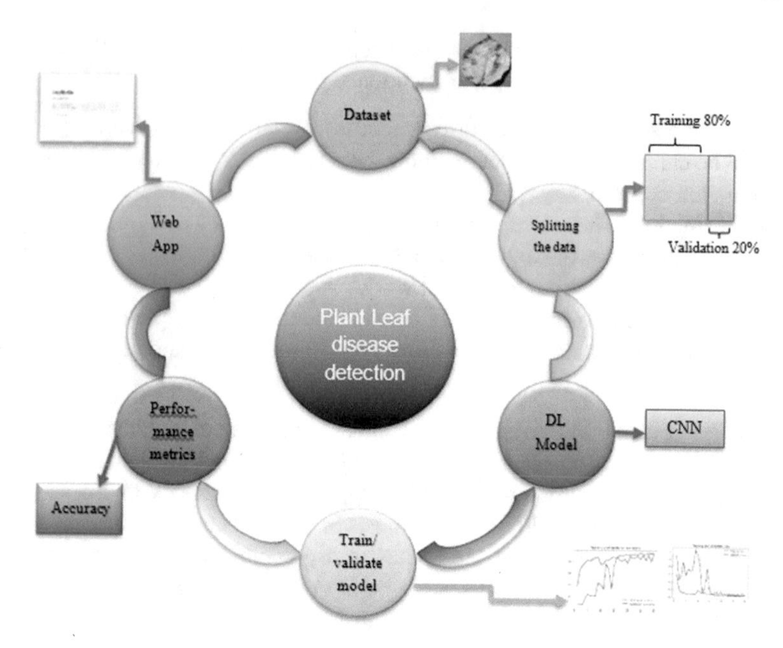

Fig. 1 Flow diagram

3.2 *Flow of the System*

1. Input leaf image.
2. Image preprocessing.
   The data points are scaled from [0–25] (the minimum and maximum RGB values of the images to the range [0–1]).
3. Data augmentation.
4. Apply the CNN model.
5. Classify the images into the type of disease.
6. Show the output.

4 Methodology

4.1 *Data Collection*

Images are downloaded from the Internet by using the keywords plant and the diseases names or also can be downloaded from the plant village datasets. The datasets of this work contain four different plants and 13 classes in total (Fig. 2).

Fig. 2 Random images of the datasets

4.2 Data Preprocessing

The images in the dataset may be in a variety of formats, quality levels, and resolutions. As a result, the images must be preprocessed; for example, images with a lower resolution or a dimension of lesser than 500 pixels will not be recognized valid for the dataset. To minimize training time, the rest of the images will be reduced to 256 * 256 pixels.

4.3 Data Augmentation

Data augmentation is a technique that performs shifts, random rotations, flips, sheers, and crops on our dataset. It allows us to use a smaller dataset and achieve high accuracy.

4.4 Proposed Algorithm

4.4.1 Convolutional Neural Network

A ConvNet/convolutional neural network (CNN) is a DL algorithm that can take input as image, allocate significance to various instances in the input image, and segregate between them. CNN model, when related to other classification methods, the size of preprocessing required by a CNN is considerably less (Fig. 3).

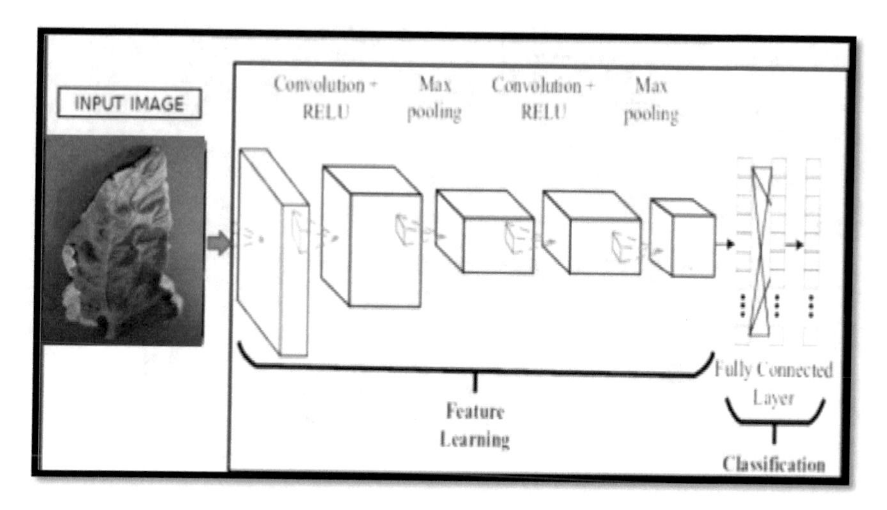

Fig. 3 CNN structure diagram

Image classification is the most frequent use for CNNs, such as recognizing satellite pictures that contain roads or classifying plant diseases, handwritten characters, and numerals. Other more common tasks, like feature extraction and signal processing, are well-suited to CNNs.

The three layers that form a CNN are:

1. Convolutional layer.
2. Fully connected layer.
3. Pooling layer.

The major benefit of CNN is that it recognizes significant characteristics without any human intervention.

4.5 *Training*

The deep convolutional neural network will be trained in this stage to create an image categorization model. After that, saturating nonlinearities will be replaced with rectified linear units (ReLU). This adaptive activation function will learn the parameters of rectifiers and increase accuracy at a low computing cost. We also apply the model with batch normalization, max pooling, and 0.25 (25%) dropout.

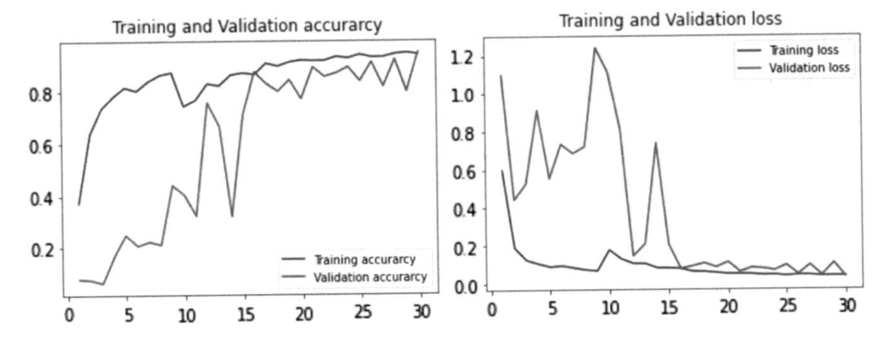

Fig. 4 Accuracy and loss rate

4.6 Testing

The test set for predicting whether a leaf is healthy or unhealthy, together with the disease's name, will be used to evaluate the classifier's performance in this phase.

4.7 Plotting a Graph

Using Matplotlib, graph is plotted for accuracy and loss rate (Fig. 4).

4.8 Evaluation Measures

Measures such as accuracy are computed for calculating the performance of the model. The accuracy obtained for this work is 94.61%.

5 Experimental Design

5.1 Software and Hardware Requirements

In this project, the development and testing process is implemented using, Python-based computer vision and deep learning libraries. This technique is implemented using Anaconda Python and libraries like OpenCV, Sklearn, Keras, Matplotlib, TensorFlow, and OS. GPUs will be used for training.

```
[INFO] Calculating model accuracy
17/17 [==============================] - 1s 33ms/step - loss: 0.0296 - accuracy: 0.9462
Test Accuracy: 94.61538195610046
```

Fig. 5 Accuracy of test

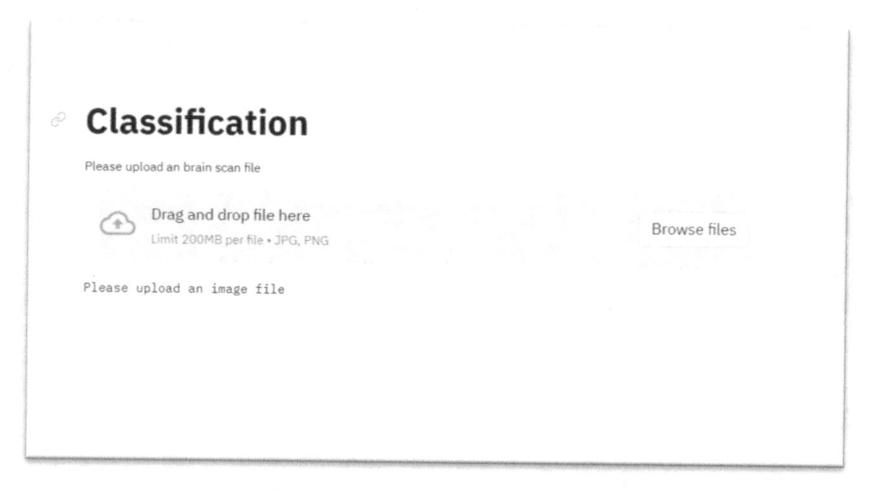

Fig. 6 Web-based system

6 Result

The results of the trained CNN model show that the accuracy of the model in plant disease detection is 94.61%. Fig. 5: shows the experimental result of the network model that has smart recognition accuracy for plant disease detection.

This project mainly focuses on plant diseases, gathering the datasets of diseased plant leaves using deep learning. A web-based system is developed and deployed where we can directly upload the image of the diseased leaf. Finally, the system predicts the appropriate results by analyzing the diseased leaf image. Figure 6 shows the web page of this model, and Fig. 7 shows the output of the uploaded image.

7 Conclusion

In this, the plant leaf disease detection method is developed using deep learning, which primarily aims at diseases in the plant leaf. By image preprocessing, disease detection, recognition, and classification, these techniques effectively determine the disease of the plant. The result displays that the accuracy of this technique is 94.61%.

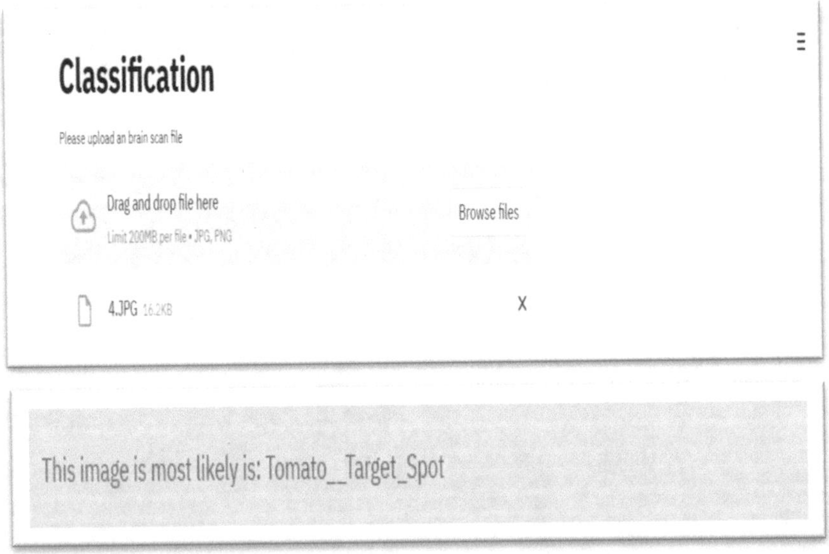

Fig. 7 Test result

A web-based system is developed and deployed where we can directly upload the image of the diseased leaf. Finally, the system predicts the appropriate results by analyzing the diseased leaf image.

8 Future Enhancement

The future enhancement of this project is that it can be deployed into a mobile app. By deploying the model into a mobile application, it makes the users use the application flexibly. Also, the datasets can be extended so that it doesn't affect the total accuracy.

Acknowledgements The development of the proposed system was extremely supported by Don Bosco Institute of Technology. We present gratitude to our guide Mr. Shankara Gowda S. R., Assistant Professor, Department of Information Science & Engineering, DBIT co-heartedly provided extra support to make this project successful with team members—Pavithra J & Pravallika S. We thank the department and management to provide the idea in coming up with this system and extremity in resources.

References

1. Zhang K, Wu Q, Liu A, Meng X. Can deep learning identify tomato leaf disease?
2. Saleem MH, Potgieter J, Arif KM. Plant disease detection and classification by deep learning

3. Mohanty SP, Hughes D, Salathé M. Using deep learning for image-based plant disease detection
4. Singh V, Misra AK. Detection of plant leaf diseases using image segmentation and soft computing techniques

An Effective Approach for Forgery Detection of Fake COVID-19 Report

K. R. Nataraj, B. R. Harshitha, and S. Vidya

Abstract With advancements in technology and image processing software, digital image forgery has become increasingly simple. However, because digital images, such as COVID-19 reports, are a common source of information, the authenticity of these digital reports has become a big concern. In recent days, it has been discovered that an increasing number of academics have begun to focus on the issue of digital report manipulation. A new deep learning-based digital picture forgery detection solution has been acquired. The mechanism is intended to ensure that the COVID-19 report on health care has not been amended or tampered with.

Keywords Deep learning · Image acquisition · Pre-processing · Feature extraction · Dissimilarity measure · Classification · DOG · SIFT · Convolutional neural network · Binary masking · Nearest neighbour matching

1 Introduction

In terms of facilities, the tending industry has recently witnessed a significant improvement. Individuals are now consulting doctors without physically visiting them. In recent years, telemedicine has gained high quality because of the actual fact that medical pictures will be sent across the web to speak patient's restorative state that makes medical analysis easy for medical workers. In telemedicine, a medical image will be altered over the web throughout the transmission method and might be misperceived and it loses its integrity. Commonly, forgery in medical information cannot be perceived by the oculus. Digital image forgery or we are able to say that meddling of digital pictures became one amongst the foremost issues. If medical information is leaked or manipulated, for example, the patient may be embarrassed or frustrated, whilst others may gain an unfair advantage. Information tampering with digital technology dates back to the late twentieth century. Currently, digital pictures

K. R. Nataraj (✉) · B. R. Harshitha · S. Vidya
Don Bosco Institute of Technology, Bangalore, India
e-mail: Director.research@dbit.co.in

A. Kumar et al. (eds.), *Proceedings of the International Conference on Cognitive and Intelligent Computing*, Cognitive Science and Technology,
https://doi.org/10.1007/978-981-19-2350-0_72

will be simply changed by affordable software system. However, digital information denotes one amongst the foremost common sources of data over the web. In recent times, digital information alteration has become easy because of quick digital software system growth, and victimisation totally different tools, information will be simply altered. Once the utilisation of digital information will increase, new software system tools square measure used for the manipulation of pictures that square measure being enforced within the framework. As a result, a system inside a sensible tending framework should be able to check whether or not medical data has been tampered with during transmission by hackers or intruders. The term of forgery detection is to be derived from the deep learning techniques. There square measure some ways through that the reports will be changed. The strategies for manipulating medical information will be roughly divided into two categories: active strategies and passive strategies. Active strategies square measure used for the protection of digital information that embodies digital watermarking and digital signatures. In these strategies, throughout information pre-processing, a watermark is embedded into information, or a signature is made throughout the information creation part. In distinction, victimisation passive strategies to switch information may be a nice challenge within the field of digital processing. Passive strategies embody information retouching, copy–move and information junction. Not like active strategies, in passive strategies, there is no information pre-processing. Tamper detection victimisation passive strategies are to analyse a raw image supported totally different linguistics and statistics of the image to localise the meddling. The classification of image modifying techniques is illustrated in Fig. 1. Many passive strategies are wont to modify digital pictures in telemedicine, like copy–move, image junction, and image retouching.

Copy–move forgery method is the most popular technique used in tampering with the actual image. A digital image is copied from one region and pasted on different locations. The suggested forgery detection method's main goal is to eliminate altered COVID-19 reports. DOG and SIFT descriptors are used to detect forgeries in digital images. The CNN classifier is applied to medical images to evaluate a forgery that is not visible by the naked eye. CNN aims to detect active forgery and identifies the digital watermarks present on the images. In this case of a forgery, the system seeks for the actual report and compares it with the tampered one and finds the end result [1].

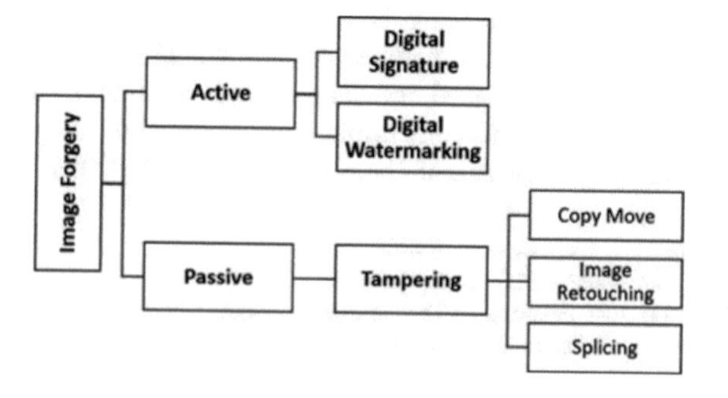

Fig. 1 Types of data forgeries

2 Image Forgery Detection

Different forgery detection strategies are given antecedent. However, most of them use non-intrusive security measures that are not gift in most of the everyday life's documents. An honest summary of various varieties of non-intrusive security measures is found. There are also approaches exploitation intrusive document feature. For detecting duplicated image portions, copy–move forgery detection is used. A non-intrusive search that compares the image to each cyclic shifted version of itself would be one direct solution to the existing problem. However, this strategy would be computationally difficult, and a picture of size $m \times n$ would require $(m \times n)$ steps. Also, the kind of search may not add the case wherever the derived space has undergone approximate modifications. There square measure different forgery detection strategies, where the author uses the auto-correlation properties. Only when the repeated sections comprised an excessive percentage of the image did this strategy prove to be effective. The block-matching technique is used in one of the approaches. The image is segmental into overlapping blocks first using this method. Rather than detecting the full rebuilt region, the goal here is to locate related picture blocks that have been derived and altered. It is worth noting that the derived portion would include numerous overlapping blocks, and because each block would be affected by the same amount of shift, the space between each repeated block combined would be the same. Therefore, the forgery created is given by saying there is square measure with quite a particular variety of comparable image blocks amongst identical distance and these blocks square measure connected to every different in order that they kind two regions of identical form. Image forgery detection is a vital technique in several PC vision applications like image fusion, image retrieval, object following, face recognition, amendment detection then on. Native feature descriptors, i.e. a way to find options and the way to explain them, play a basic and vital role in image forgery detection method that directly influences the accuracy and hardness of image recognition. This technique mainly focuses on the range of native feature descriptors as well as some theoretical analysis, mathematical models, and strategies or algorithms along-side their applications within the context of forgery detection. The goal of our project is to produce a regular care system for individuals with at-most security and caution wherever it keeps the intruders and different malpractices away. This will be achieved by planning an efficient rule for the detection of forgery on COVID-19 report [2].

3 Image Acquisition with Pre-processing

Image acquisition's key objective is to convert optical image into numerical data array that can then be produced on a computer. However, before any forgery can take place, the image must first be collected and scanned and then converted into a manageable entity. This process mainly consists of three major states: the system focuses on the energy. The focused energy is reflected from the object. Finally, there is a sensor that measures how much energy is being used. An appropriate lens is used to capture images. Before scanning the photographs, we employ a variety of lenses to capture them. If we need to scan an image, we can film in sensitive and avoids noise in the image. Images in black and white can be found in a variety of grayscales. These colours range from 0 to 255, with 0 indicating black and 255 indicating white, with intermediate values indicating different shades of black and white. Grayscale refers to the spectrum of neutral tonal values from black to white. The presence of colour images is also considered where they are made up of coloured pixels. We prefer to use grayscale photographs despite the fact that colour can capture a considerably wider range of values than grayscale. The red, green, blue, i.e. RGB, are the most commonly used colours in this system. The amount of charges that are later converted to digital form by using an analogue or digital converter is largely dependent on the hardware used, where the reflected light energy from the object is captured and converted into electrons, which are then spread across the internal sensor chip, which is similar to a 2D array of cells. This digital image can now be enhanced, restored, segmented and subjected to various alterations. Pre-processing is a common operation used on images at the lowest level of abstraction, where the input and output images are both very bright. The original data is scanned and then processed, yielding a higher-intensity image represented by the image function values in-matrix [3]. Pre-processing is used to improve image data by removing any unnecessary distortions or enhancements. Although geometric modifications such as rotation, scaling, and translation of images are significant in further processing, only a few picture attributes are important and are segregated. These methods are segregated based on the pixel size neighbour to capture the nearest neighbour point with regard to the initial start point. Each category deals with pixel brightness alterations and specifies geometric transformations. Image restoration and image enhancements can be accomplished out via pre-processing procedures that are applied to the processed image pixel's local neighbourhood. This requires knowledge of the complete image. There are N number of methods where the classification of image pre-processing is differently processed on the image for enhancement, covering pixel brightness transformations and image restoration. This method is mostly utilised when there is a lot of repetition in an image. Neighbour pixels belonging to the same item in real images have fundamentally same or comparable illumination values; thus, if an inaccurate pixel is discovered, it is reinstated as the regular value of the nearest surrounding pixels [4] (Fig. 2).

4 Feature Extraction

Feature extraction is a crucial phase image processing that entails turning data into numerical form that is analysed by maintaining integrity of novel data set. It produces enhanced outcome when compared to other solutions. This is performed either physically or automatically, as seen below: identifying and characterising the features

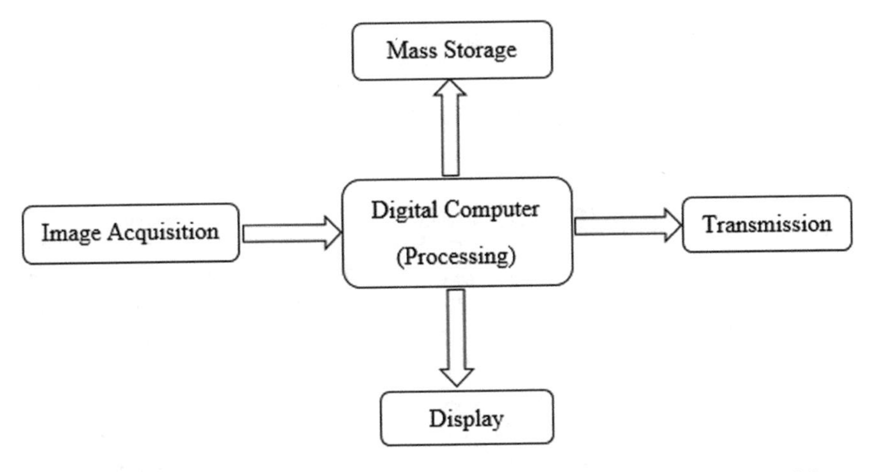

Fig. 2 A simple method to demonstrate how image processing is implemented in the workflow

that are crucial to the situation at hand, as well as designing a method for extracting those features, are all part of manual feature extraction. Knowing the context or domain might aid in making informed selections about which characteristics might be valuable. Researchers have spent years developing feature extraction algorithms for photographs, signals and text. When the user needs to go swiftly from input data to create deep learning algorithms, this technique comes in handy. Automated feature extraction is exemplified via wavelet scattering. The earliest layers of deep networks employing the upgraded deep learning technique have mostly replaced feature extraction—but mostly for picture data. As one of the most critical processes, feature extraction remains the first hurdle that must be overcome before successful predictive models can be built. Advanced approaches, such as wavelet scattering and deep neural networks, have emerged from signals to automatically extract features. They are mostly used to extract features from provided data and also to minimise the dimensionality of the data. Wavelet scattering networks are computer-assisted systems that extract low-variance features from a data set. This method generates data representations with minimum changes inside a class while maintaining the classes' discriminability. When there is no large data to begin with, wavelet scattering works perfectly. The extraction of unique key-points from the actual set of information presented is what feature extraction for picture data is all about. Initially, specialised identification, extraction and feature matching algorithms were used to complete this approach. Deep learning is currently widely utilised in image and video scrutiny, and it is recognised where it accepts raw set of data as input and skips the phase. Image registration, object detection and classification, and content-based image retrieval are all there are several computer vision applications that require an effective approach for representing unique features, through the initial layers of a deep network or explicitly through the use of well-established systems [5].

5 Dissimilarity Measures—SIFT and CNN

Dissimilarity measure is one of the update procedures that is carried out after key-point extraction. The initial step in this procedure is that provides the dissimilarities present in the data on comparison with the actual or original data set. Finally, the relationship between the presented technique and the projection to doubly stochastic matrices is highlighted. We can also learn how to compute the CDM efficiently for large data sets here. Local descriptors are used to characterise the n number of data sets. To extract the unique key-points, the SIFT descriptor is employed. Principal component analysis is used to reduce 128-dimensional SIFT descriptors to 36-dimensional vectors. Scale-invariant feature transform (SIFT) is a feature-based manner-obtaining necessary distinct point from images, and it extracts feature points, where valid comparison amongst various angles of an entity but also detects interest point locations. The SIFT characteristics are unaffected by image orientation or size, and they allow accurate matching over a wide variety of related biased, changes in 3D viewpoint, noise accumulation and illuminated deviations. SIFT characteristics are initially retrieved and saved in a database from a set of reference photographs. Each feature from the input image is compared, and the testing candidate matches features grounded on Euclidean distance of their feature vectors. This process will actually support the image forensics investigation where SIFT descriptor features have been proposed and used. When a forged or tampered or suspected photograph is given as input, it can indeed extract all the possible key-point matches, if a certain region has been changed with intensity and furthermore, it determines the geometric transformation like scale and rotation applied on image to perform tampering on the image [6].

Convolutional neural network (CNN) is one of the most widely used technique in automation of image classification systems. As a result, features in the top layer of the CNN utilised for classification may not collect enough data that is considered relevant information to forecast image correctness and produce the intended output in many cases. When compared to features from the top layer, characteristics from the lower layer carry more discriminative information in some instances. As a result, using features for a specific layer just for the purpose of grouping sems appears to be a method that does not take advantage of CNN's actual learned potential, while also exploiting the discriminant power to its greatest extent. This intrinsic ability necessitates the merging of available features from many layers [7]. To continue the flow of work, the proposed approach of merging features from many layers in specified CNN models is proposed. The CNN models are trained using sample photographs, which are then utilised to extract features from many layers, and the test images are compared and classification performed to obtain the final result. The suggested fusion method is tested using the CIFÂR-10, NORB and SVHŃ image ordering benchmark data sets. In the vast majority of cases, the suggested method outperforms the existing models, with reported performance of 0.38 per cent, 3.22 per cent and 0.13 per cent, respectively [8] (Fig. 3).

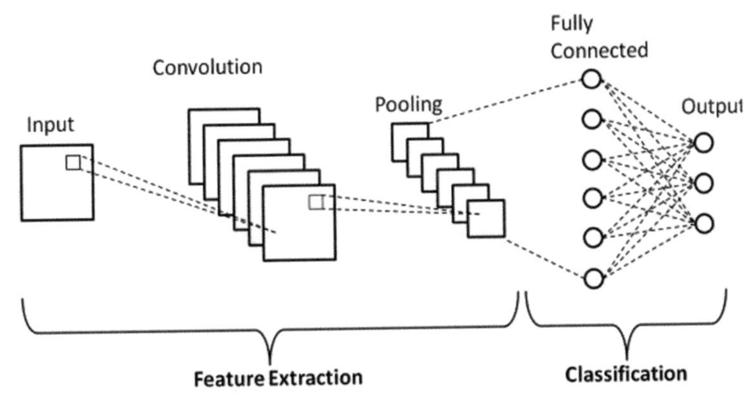

Fig. 3 Working flow of CNN technique where it provides the result on performing the classification

6 Proposed Framework

This section makes a case for the projected work for detection forgery on report. The projected work consists of five modules: image acquisition, pre-processing, key purpose extraction, dissimilarity live and classification. Figure shows the pair of proposed work stages.

Image Acquisition: The action of retrieving a picture from a source is commonly referred to as image acquisition in image processing. Because the picture non-inheritable is completely raw and noise-free, playing acting image capture in image processing is usually the first step in the workflow sequence. Initially, the scanned copy of the COVID report is given because of the input within the acquisition step. The image here could be an electronic health record (HER) image which provides decent quantity of text knowledge and provides smart quality of picture element.

Pre-processing: In pre-processing, the unwilling distortions area unit suppressed or image options area unit is increased. Here segmentation algorithmic rule is meted out that facilitates to nullify the unwanted knowledge and stores the particular knowledge from the input info given. The colour image is reborn into grey image and has similar pel distribution, then the tampered region was extracted from background mistreatment binary masking in order to avoid the computation time and memory house. If identifies the distinctive key-points with terribly minute description, in and of itself that identical letter within the text is classified unambiguously supported checking whether or not it is in graphic symbol or character so key-points area unit known [9].

Feature Extraction: In feature extraction, the unique key-points on the COVID-19 image area unit detected mistreatment DOG descriptor so SIFT descriptor can select key-points to extract it. These descriptors have been shown to vanquish to alternative detector and descriptor combos in several applications. If any noise is found within the image, it is reduced by Wiener filter so processed through the SIFT descriptor [10].

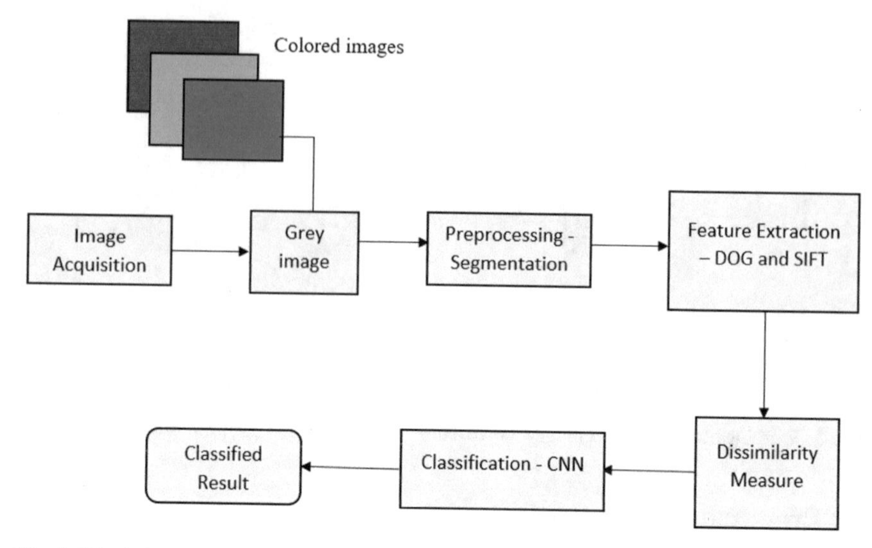

Fig. 4 Block diagram

$$\begin{aligned} D(x, y, \sigma) &= (G(x, y, k\sigma) - G(x, y, \sigma)) * I(x, y) \\ &= L(x, y, k\sigma) - L(x, y, \sigma) \end{aligned}$$

Dissimilarity Measure: Unique key-points are extracted based on the distance ratio, a decision with respect to whether the key-point consideration; i.e. closest neighbour key-points are matched. Geometric verification is used for matching the unique key-points generated from the process.

$$NN = (X\max - X\min) * (Y\max - Y\min).$$

Classification: The classification process is carried out on the key-points that are extracted. CNN classifier is applied over the features, and the classification is done by saying whether the report is tampered or not (Fig. 4).

7 System Design

The process of modelling a system's appearance, components, modules, interfaces and data in order to suit specific needs. The use of systems theory in development may even be seen as a product of system design. The disciplines of system analysis, systems design and systems engineering may be combined with theory. If the broader topic development "blends the aspect of mercantilism, design and manufacturing into one approach to development", previously fashionable the act of discussing mercantilism knowledge and preparing factory-made products. Systems scheme is therefore the methodology of shaping and developing schemes to satisfy the given wants of the user. The increasing significance of package running on generic platforms has multiplied the discipline of package engineering. Object-oriented analysis and style methods became the foremost wide used methods for laptop computer systems style. The UML has become the standard language in object-oriented analysis and magnificence. It is widely used for modelling package systems and is a lot of and a lot of used for high bobbing up with non-software systems and organisations. This is one in every of the foremost very important phases of package development methodology. The aim of set upcoming look is to set up the solution of a drag given by the requirement documentation. In different words, the first step in resolution is that the design of the project. The planning of the system is maybe the foremost vital issue poignant the quality of the package. The target of the planning half is to produce overall style of the package. It aims to figure out the modules that got to be inside the system to fulfil all the system wants in economical manner. Its goal provides representation of utility of system connections. This indicates which unit tasks are executed by which actor. The actor's involvement within the system is graphically portrayed (Fig. 5).

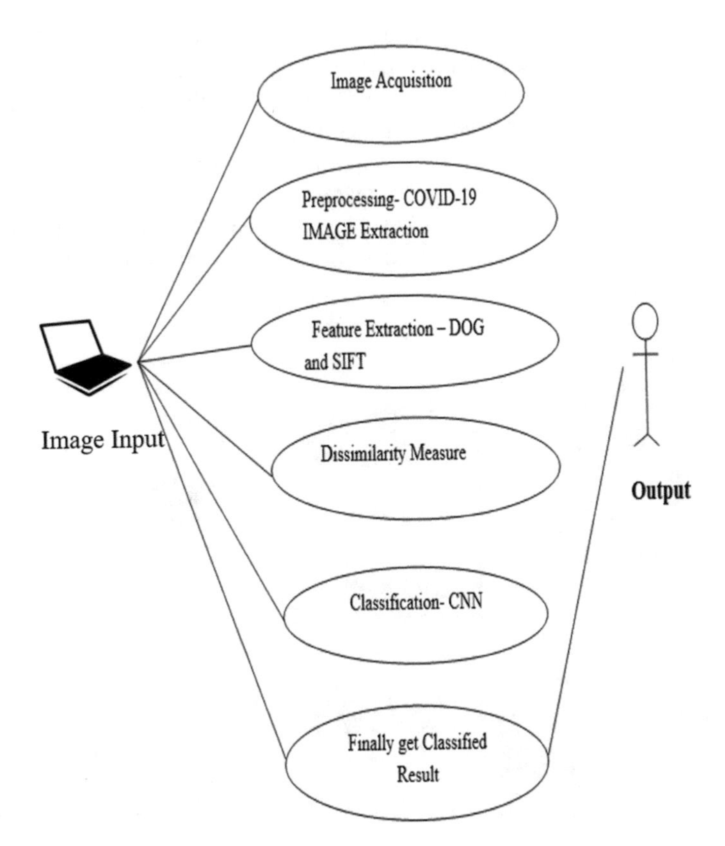

Fig. 5 Use case diagram

8 Advanced Level Design

The diagrammatic representation, given below of varied modules designed so as to urge an efficient approach for the detection of false COVID-19 report, gives the output saying, whether the report is tampered or not. Here it represents how the process in the project is executed where all the below-mentioned modules are carried out to get the result (Fig. 6).

Convolutional neural networks (CNNs) are a type of deep neural network architecture developed for specialised applications such as image classification. It is mostly used to categorise the results. Convolutional neural networks have been presented as a way to decrease the number of parameters and tailor the specification to vision tasks particularly. CNNs are a subset of deep neural networks that are frequently used to analyse visual data. This algorithm compares the properties of the original image to determine the final result. The input image is split, and then image pooling is performed, as can be seen. The results are achieved after the completion. There are three types of layers in a systematic natural network:

- Input Layer: This is the layer where we supply our model with data. This layer has the same amount of neurons as our data's total number of attributes.
- Hidden Layer: The hidden layer receives the data inflow from the input layer. There are often numerous hidden layers, reliant on our model and the amount of data we have. Each hidden layer can have a varied number of neurons, which should be greater than the number of features.
- Output Layer: The output of the hidden layer is then input into a logistic function such as sigmoid, which turns the output of each class into a possibility mark for each class (Fig. 7).

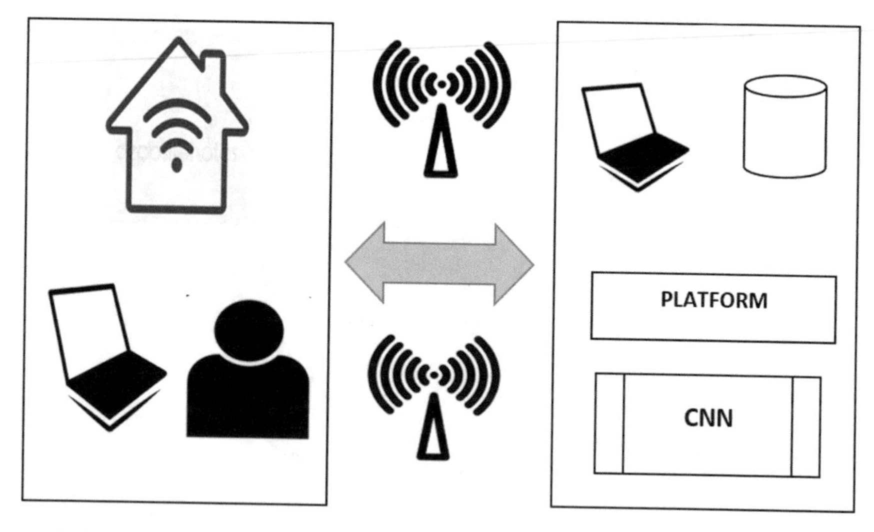

Fig. 6 Proposed smart framework of the designed system

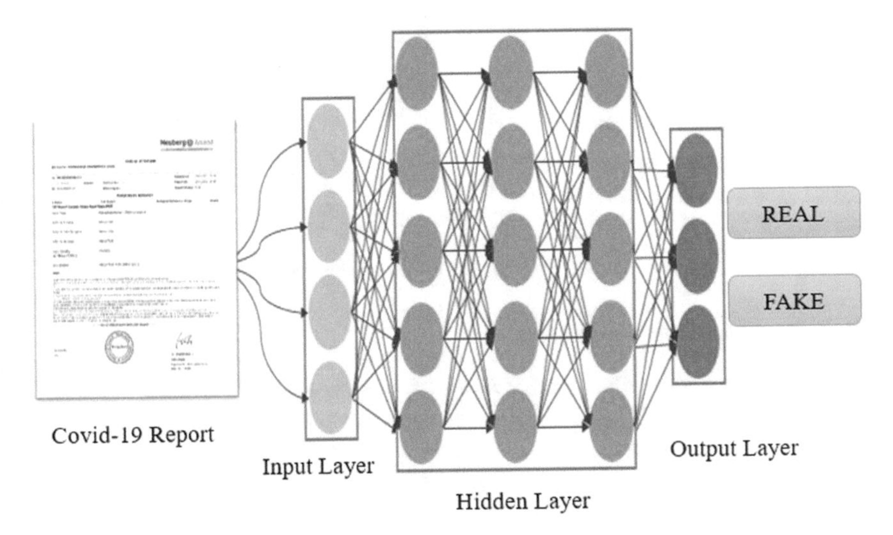

Fig. 7 Example of CNN classification

Nearest Neighbour: The closest neighbour chooses the most effective immediate position next to chosen key-point. This offers the similarities between the take a look at knowledge and train knowledge. Ensuing nearest neighbour graph for sites is whose vertex is of degree one and in degree at the most half-dozen. The nearest neighbour search locates the k-nearest neighbours or all neighbours inside such as distance to question knowledge points, supported the required distance metric. To classify associate unknown instance painted by some feature vectors as a degree within the feature area, the k-NN classifier calculates the distances between the purpose and points within the coaching knowledge set. Usually, the geometrician distance is employed because the distance metric. Nearest neighbour formula primarily returns the coaching example that is at the smallest amount distance from the given take a look at sample. K-nearest neighbour returns k (a positive integer) coaching examples a minimum of distance from given take a look at sample. This section explains the projected work for detection forgery in scanned document relying on component primarily based sort. Figure shows the proposed work stages. Allow us to think about associate example with A because the key-point: begin at A. opt for the closest neighbour at every unvisited vertex. Come to A once visiting all different vertices (Figs. 8 and 9).

$$\text{Euclidean function} = \sqrt{\sum_{i=1}^{k} (x_i - y_i)^2}$$

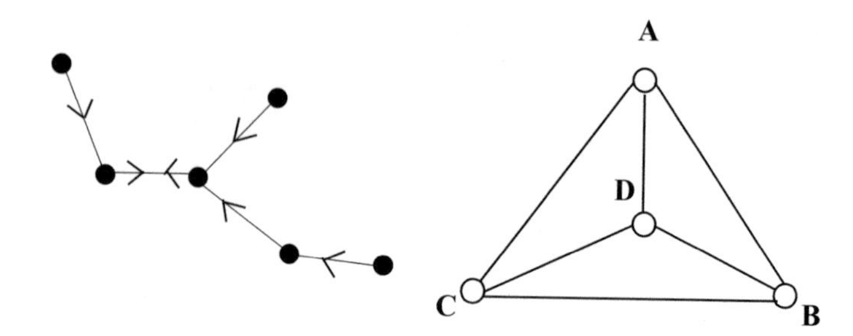

Fig. 8 Nearest neighbour working states

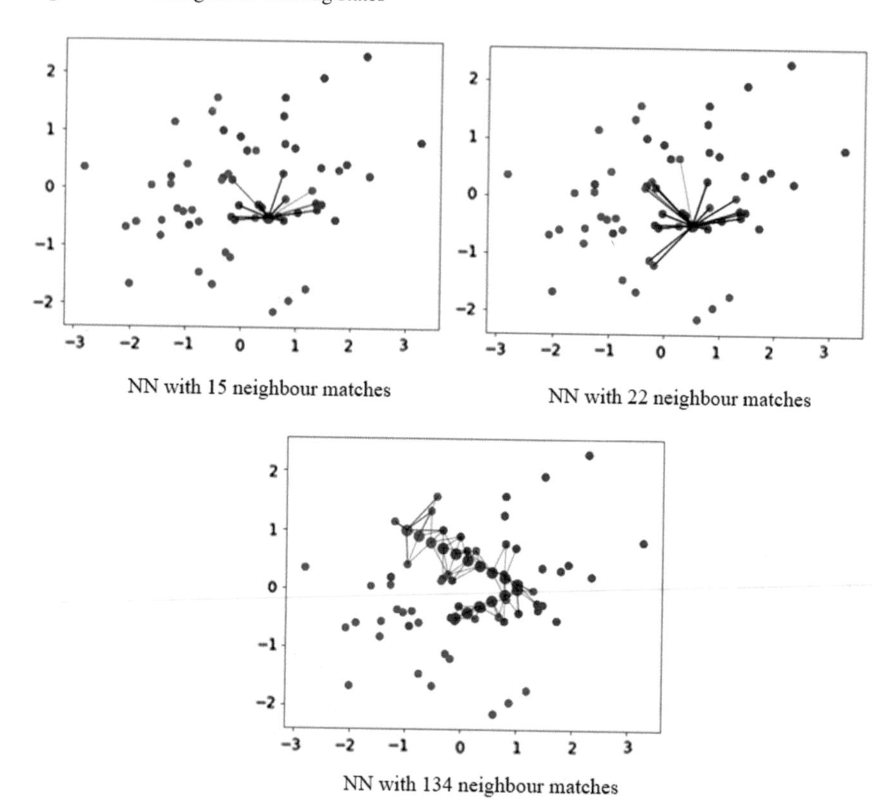

NN with 15 neighbour matches
NN with 22 neighbour matches
NN with 134 neighbour matches

Fig. 9 In the above-graph representations, we can see how the nearest neighbour matching works. To start with 15 neighbour immediate position are matched with the chosen point. In addition to this, further 22 points are matched. Finally to conclude more than 134 matches are made with the chosen point

9 Result of the Proposed Work

The process of implementing the system and its implementation was done using (MATLAB program for programming and design using the graphical user interface) by working on a data set that had been created personally. The data set contains original official documents and tampered official documents. The result of system work explained the experiments that applied on proposed work for calculation and discussed the results of scanned document (Figs. 10, 11 and 12).

The results of system work explained the experiments that applied on planned work for prediction and discuss the results of scanned COVID-19 report. So as to comprehensively value the performance of our planned CNN-based forgery detection, we tend to specialise in pre-processing effectiveness, binary masking, image handicraft or compression, key purpose extraction, difference live and eventually classifying the result. First, we will describe the info utilised in our analysis. In our analysis, we tend to haphazardly choose the image and train it at the start by

Neuberg
MOLECULAR BIOLOGY
2019 Novel Corona Virus Real Time PCR
NEGATIVE
NEGATIVE FOR SARS CoV 2

Fig. 10 Binary masking

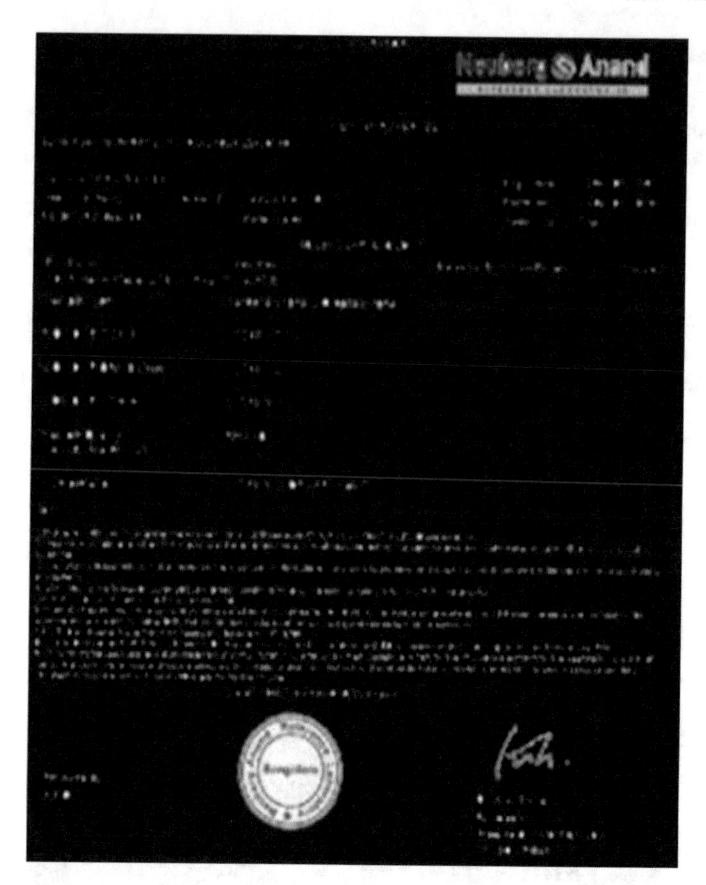

Fig. 11 Image stitching

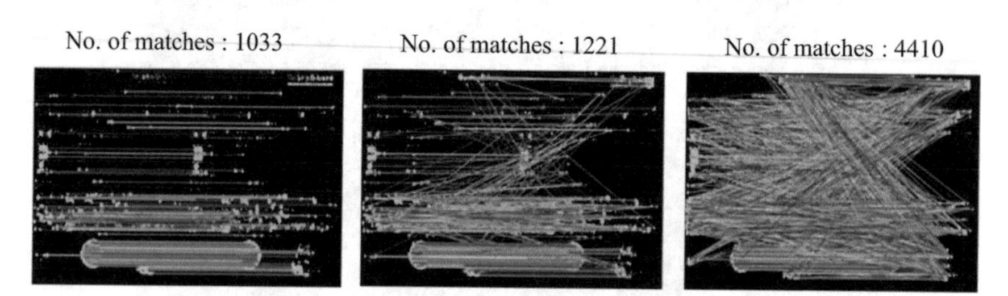

Fig. 12 Nearest neighbour matching with the descriptor associated with the key-point in terms of Euclidean distance

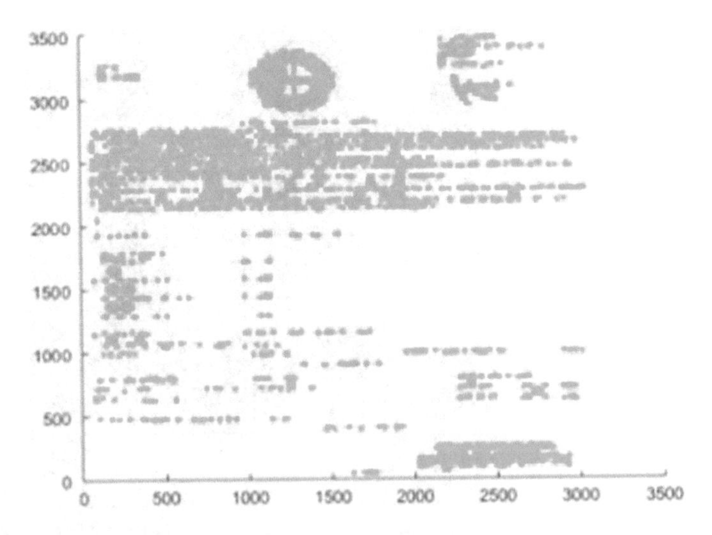

Fig. 13 Representation of graph of report

pre-processing. Area unit going to extract the distinctive key-points that are used for matching it with check image to predict whether or not the report is original or tampered. Finally, it is planned to validate supported the trained modules. It ought to be noted that we tend to use an equivalent solid data set in our experiments for honest comparison (Figs. 13 and 14).

10 Conclusion

The major goal is to create a system for the healthcare framework that verifies that the COVID-19 report on health care has not been amended or tampered with. The field of medicinal imaging forgery finding requires more kindness in order to restore patients' belief and prevent embarrassment. The next generation of network technology will provide enormous computational power as well as pervasive service. Active and passive digital data forgeries are divided into two kinds. The proposed methodology is assessed by comparing it to several COVID-19 data sets. We retrieved the unique key spots in this paper using feature extraction, which aids in the localisation of tampered images. Using nearest neighbour matching, the resulting unique descriptors are compared to the trained data set. We used CNN to classify the results, and we got some really promising results. For training, some authentic COVID-19 report photographs are required for each sort of COVID-19 image. Finally, the required result is produced, indicating whether or not the image has been tampered.

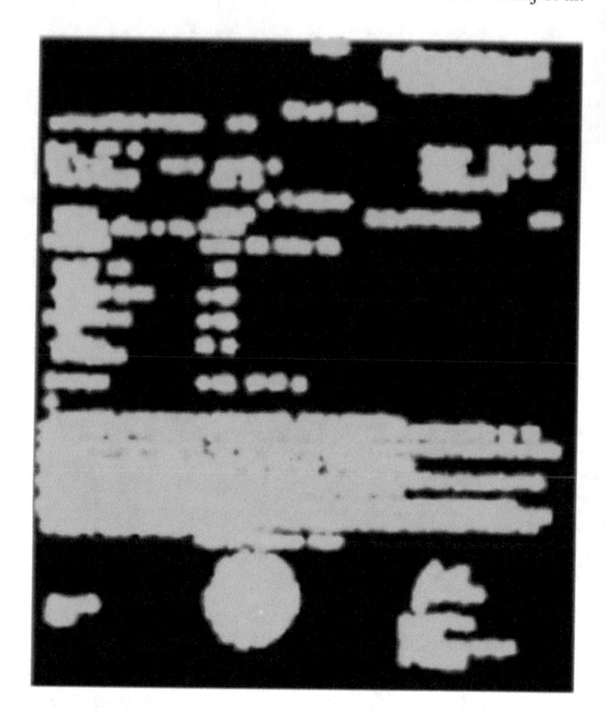

Fig. 14 Binary masking of image

Acknowledgements The development of the proposed system was extremely supported by Don Bosco Institute of Technology. We present gratitude to our guide Mrs. Navya Shridhar C. S., Assistant Professor, Department of Information Science and Engineering, DBIT co-heartedly provided extra support to make this project successful with team members—Harshitha B. R. and Vidya S. We thank the department and management to provide the idea in coming up with this system and extremity in resources.

References

1. Ulutas G et al (Dec 2017) Medical image tamper detection based on passive image authentication. J Digital Imaging 30(6)
2. Muhammad G et al (May 2014) Image forgery detection using steerable pyramid transform and local binary pattern. Mach Vis Appl 25(4)
3. Zhao X et al (2011) Detecting digital image splicing in Chroma spaces. Proceedings international workshop on digital watermarking. pp 12–22
4. Yadav A, Yadav P (2009) Digital image processing. University Science Press
5. Mitto R, Wang F, Wang S, Jiang X, Dudley JT (May 2017) Deep learning for healthcare: review, opportunities and challenges. Brief Bioinf 6. https://doi.org/10.1093/bib/bbx044
6. Dong J, Wang W (2011) CASIA tampered image detection evaluation (TIDE) database, v1.0, and v2.0
7. Solanas et al (Aug 2014) Smart health: a context aware health paradigm within smart cities. IEEE Commun Mag 52(8):74–81

8. Krizhevsky A, Sutskever I, Hinton GE (2012) Image net classification with deep convolutional neural networks. In: Advances in neural information processing systems (NIPS). pp 1097–1105
9. Mendonca AM, Campilho A (2006) Segmentation of retinal blood vessels by combining the detection of centerlines and morphological reconstruction. IEEE Trans Med Imaging 25(9):1200–1213
10. Chen J et al (Sept 2010) WLD: a robust local image descriptor. IEEE Trans Pattern Anal Mach Intell 32(9)

Applying Ensemble Techniques of Machine Learning to Predict Heart Disease

K. Shilpa and T. Adilakshmi

Abstract Heart disease is a leading cause of death in the world today. It is a serious task in the field of health data investigation to forecast cardiovascular disease. ML (machine learning) has shown to be effective in assisting in making judgments and result forecasts from the massive volume of statistics created by the medical industry. ML methods are forecast for cardiovascular disease. We have proposed an ensemble classification technique that results in enhanced accuracy of the cardiovascular illness classification. It applied ADA boosting, bagging, and stacking to attain improved accuracy. The proposed modal is introduced using different groupings of features and certain identified techniques of classification. The proposed modal is presented using various mixtures of attributes, and techniques of classification are considered. We have taken datasets from the UCI dataset with 918 instances and 14 attributes. The aim of research is to discover the best classification algorithm with maximum accuracy. In this, Weka tool is applied for accuracy prediction. We proposed an ensemble technique with 100% accuracy while compared to Bayesian Network (Bayes NET) and IBK Instance based for K-Nearest Neighbor (IBK). We proposed an ensemble technique, which has given 100% accuracy when compared to Bayes NET and IBK. The Bayes NET algorithm came in second place with 98.801% accuracy. Third is that IBK has given 98.692% accuracy. Moreover, the time taken to build a model is less when compared to other algorithms. This result will help in selecting the best classification algorithm for heart disease prediction and can be used for detection and treatment.

Keywords Heart Disease Prediction · Machine Learning · Ensemble Technique · Ada Boosting · Bagging · Stacking · Bayesian Network (Bayes NET) · Instance Based for K-Nearest Neighbor (IBK) · Weka 3.8.5

K. Shilpa (✉)
Department of CSE, UCE, OU, Hyderabad, India
e-mail: shilpamtech555@gmail.com

T. Adilakshmi
Department of CSE, Vasavi College of Engineering, Hyderabad, India
e-mail: t_adilakshmi@staff.vce.ac.in

A. Kumar et al. (eds.), *Proceedings of the International Conference on Cognitive and Intelligent Computing*, Cognitive Science and Technology,
https://doi.org/10.1007/978-981-19-2350-0_73

1 Introduction

Machine Learning (ML) consists of various algorithms that improve spontaneously over experience and use of data. ML is a part of AI. The ML algorithm uses training data to construct a modal based on sample data, with the aim of creating estimates or results without any obvious programming.

ML algorithms apply in a variety of applications, such as email filtering, medicine, computer vision, and even speech recognition, where constructing conventional algorithms to execute desired tasks will be challenging.

The heart is a very essential organ in the human body. Heart disease may take the lives of many people. Chronic heart failure (CHF) affects a large number of people. In recent years, medical utensils have been developed to accurately identify heart disease. There are certain important signs that confirm the physical structure of a person, whether the person is normal or abnormal: 1. High blood pressure, 2. Temperature, 3. The pulse rate, and 4. Rate of respiration. The pulse rate gives the heartbeat. Chest movements depend on the respiratory rate. Blood pressure is another important signal that gives the maximum and minimum pressure. ML techniques find a solution to aforementioned problem by identifying the heart disease accurately.

1.1 Motivation

- To early detection of heart disease prediction and increased survival rates,
- Because the right diagnosis leads to the right treatment.

1.2 Objectives

- To identify the best algorithm for disease prediction,
- To accurate prediction of the classification algorithm,
- To analyze the performance and accuracy of machine learning algorithms.

2 Literature Work

Beulah Christalin Latha, S. Carolin Jeeva: This paper tests a technique called ensemble classification. This technique enhances the performance and correctness of weak algorithms through the merging of multiple classifiers. The purpose of the paper is to improve the weak classification accuracy algorithms and the execution of the algorithms with health datasets for early stage prediction. They have shown results of ensemble techniques to improve weak classifiers' accuracy and shown acceptable performance in detecting the risk of heart disease. They demonstrated that ensemble classification can achieve a maximum of 7% weak classifier accuracy. They further enhanced performance with the technique of feature selection [1].

Xiao-Yan Gao, Abdelmegeid Amin Ali, Hassan Shaban Hassan, and Eman Anwar: This paper employed ensemble techniques to improve the performance of heart disease prediction. They used two feature extraction methods, such as PCA and LDA, to choose essential features. As a result, they have shown best performance through ensemble learning using the bagging method through a decision tree as a result [2].

Vardhan Shorewala: He has focused on the heart disease of coronary risk prediction. They have used predictive techniques on base of standard such as precision recall and roc curve. These classifiers like KNN, naïve Bayes, and logistic binary classification are associated with ensemble techniques. Bagging technique increased mean accuracy of 96 percentage, and boosted technique increased mean accuracy of 73.4 percentage but AUC score 0.73. The KNN, RF, and SVM using stacked model showed maximum actual with accuracy of 75.1 percentage. Furthermore, the attainment of models has been evaluated using K-Fold cross-validation [3].

Yar Muhammad, Muhammad Tahir, Maqsood Hayat, and KilTo Chong: They have used many ML algorithms to remove irrelevant and noisy data. It is predicted that the proposed system will be beneficial and obliging for the physician to detect heart disease accurately and effectively [4].

Bayu Adhi Tama, Sun Im, and Seungchul Lee: They have used a new CHD technique. They have used a stacked design to predict the class label of 3 ensemble learners: Gradient boosting, extreme gradient boosting, and random forest. They have used several datasets of heart disease to evaluate detection [5].

3 Methodology

In this paper, ensemble techniques like Ada boosting, Bagging, and Stacking are used, and Bayesian network and IBK machine learning algorithms are used on heart disease prediction to find out better accuracy.

Ensemble technique: It is ML technique, which associates various techniques to construct the best possible predictive model.

Bayesian network: A Bayesian network evaluates the network variable from training and performance find out against the validation set as an outcome.

IBK: It specifies the number of nearest neighbor's and applies when classifying a test instance, and the outcome is found out by vote (Fig. 1).

In this research paper, first we choose the UCI dataset, and then pre-processing is applied. Next, we applied attribute section classifier for feature selection to remove misleading attributes. Next, we applied Bayesian network and IBK and ensemble techniques with vote (Ada boosting, bagging, and stacking) combined to get better accuracy.

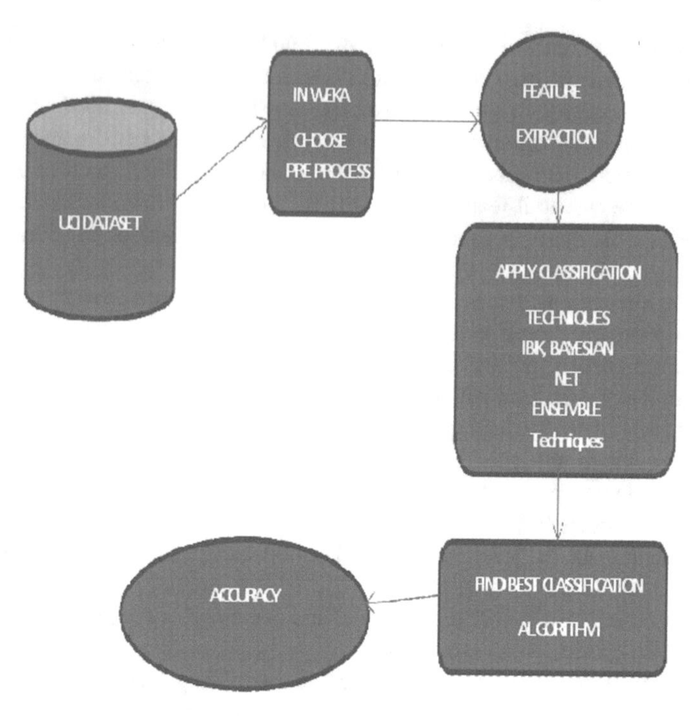

Fig. 1 Workflow of heart disease prediction

3.1 Heart Disease Prediction Dataset

The data used in this research are provided by the UCI Machine Learning repository located in Heart Disease dataset. 5 Databases: Hungary, Cleveland, Switzerland, stat log, and VA Long Beach are combined, having 918 instances, 1 class (yes/no), and 14 attributes. (Fig. 2) Class distribution: yes = 1 means have heart disease, no = 0 means not have disease. Here, we have taken YES = 508, NO = 410.

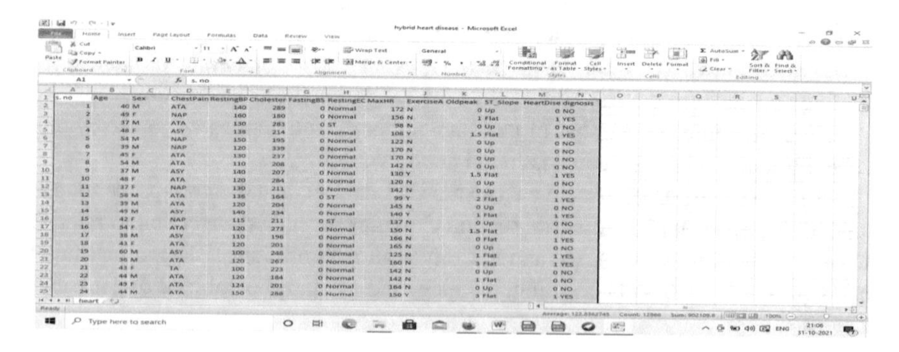

s. no	Age	Sex	ChestPain	RestingBP	Cholester	FastingBS	RestingEC	MaxHR	ExerciseA	Oldpeak	ST_Slope	HeartDise	dignosis
1	40	M	ATA	140	289	0	Normal	172	N	0	Up	0	NO
2	49	F	NAP	160	180	0	Normal	156	N	1	Flat	1	YES
3	37	M	ATA	130	283	0	ST	98	N	0	Up	0	NO
4	48	F	ASY	138	214	0	Normal	108	Y	1.5	Flat	1	YES
5	54	M	NAP	150	195	0	Normal	122	N	0	Up	0	NO
6	39	M	NAP	120	339	0	Normal	170	N	0	Up	0	NO
7	45	F	ATA	130	237	0	Normal	170	N	0	Up	0	NO
8	54	M	ATA	110	208	0	Normal	142	N	0	Up	0	NO
9	37	M	ASY	140	207	0	Normal	130	Y	1.5	Flat	1	YES
10	48	F	ATA	120	284	0	Normal	120	N	0	Up	0	NO
11	37	F	NAP	130	211	0	Normal	142	N	0	Up	0	NO
12	58	M	ATA	136	164	0	ST	99	Y	2	Flat	1	YES
13	39	M	ATA	120	204	0	Normal	145	N	0	Up	0	NO
14	49	M	ASY	140	234	0	Normal	140	Y	1	Flat	1	YES
15	42	F	NAP	115	211	0	ST	137	N	0	Up	0	NO
16	54	F	ATA	120	273	0	Normal	150	N	1.5	Flat	0	NO
17	38	M	ASY	110	196	0	Normal	166	N	0	Flat	1	YES
18	43	F	ATA	120	201	0	Normal	165	N	0	Up	0	NO
19	60	M	ASY	100	248	0	Normal	125	N	1	Flat	1	YES
20	36	M	ATA	120	267	0	Normal	160	N	3	Flat	1	YES
21	43	F	TA	100	223	0	Normal	142	N	0	Up	0	NO
22	44	M	ATA	120	184	0	Normal	142	N	1	Flat	0	NO
23	49	F	ATA	124	201	0	Normal	164	N	0	Up	0	NO
24	44	M	ATA	150	288	0	Normal	150	Y	3	Flat	1	YES

Fig. 2 Heart Disease Dataset from UCI Machine Learning repository

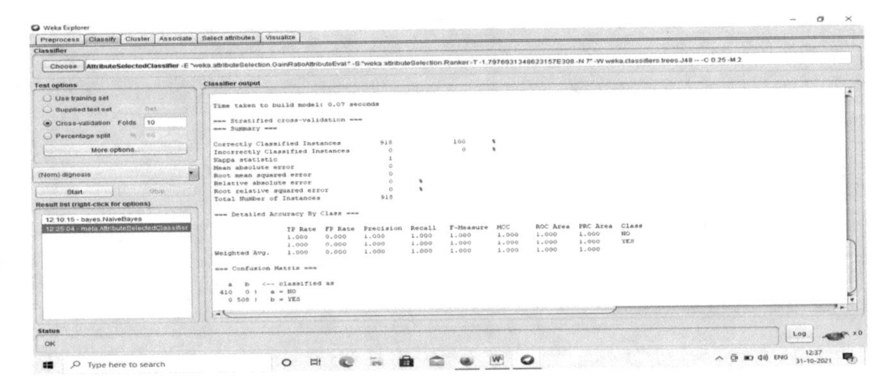

Fig. 3 Attribute selection classify

4 Results and Discussion

We have used Weka for experimentation.

4.1 Experimental Tool

In this, we have used WEKA version 3.8.5. The tool which is preferred in research to investigate medical datasets and estimate the accuracy and performance of DM techniques practical to these sets. The particular ML methods are taken attributes for analyses. Additionally, the measures of model performance are presented which are the source for the comparison of methods' effectiveness and accuracy. To conclude, the visualization of each algorithm's accuracy is shown for medical datasets. This is built on personal experience with WEKA environment.

Figures 3, 4, 5, and 6.

Table 1 shows time taken to build model, ranking, performance, and accuracy. The best algorithm in terms of accuracy is ensemble with 100% compared to Bayesian Net and Instance Based for K-Nearest Neighbor IBK algorithm. However, overall performance was better when compared to other algorithms, next second place taken by Bayesian Net algorithm with 98.8017% Accuracy more over time taken to build model is less when compared to other algorithm. Third is that IBK is given 98.692% accuracy (Figs. 7 and 8, Tables 2 and 3).

$$\text{Accuracy} = \frac{\text{Total Correct Predictions}}{\text{Total Number of Instances}}$$

$$\text{Precision} = \frac{\text{True Positive}}{\text{True Positive} + \text{False Positive}}$$

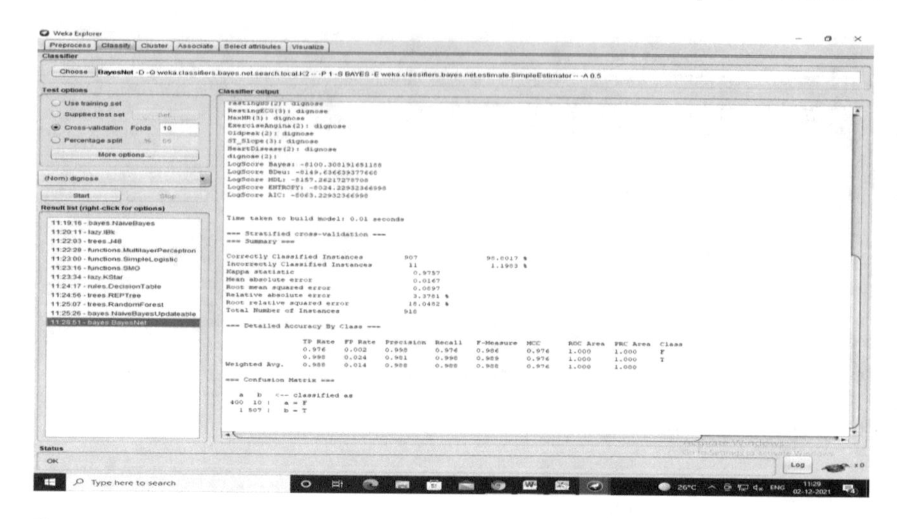

Fig. 4 Bayesian net algorithm

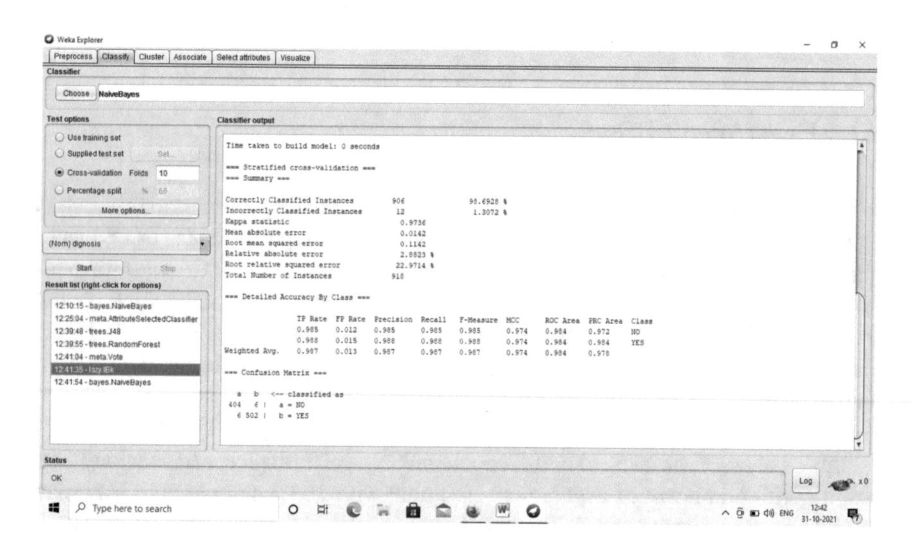

Fig. 5 Instance based for K-nearest neighbor (IBK)

$$\text{Recall} = \frac{\text{True Positive}}{\text{True Positive} + \text{False Positive}}$$

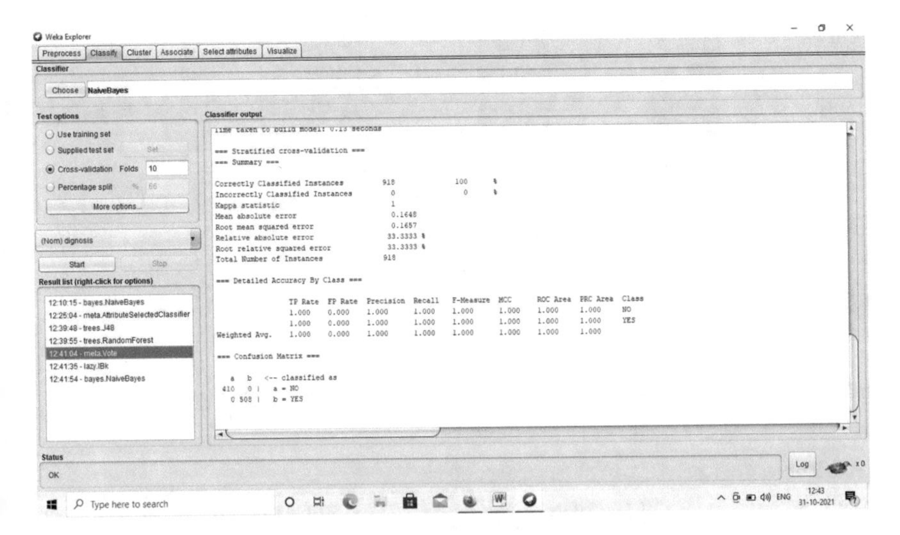

Fig. 6 Ensemble technique vote (Ada boosting, bagging, and stacking)

Table 1 Training and simulation error

	Bayesian net	IBK	ENSEMBLE
Time taken to build model	0.01	0.0	0.13
kappa statistic	0.9757	0.9736	1.00
Mean absolute error	0.0167	0.1142	0.1648
Root mean squared error	0.0897	0.1142	0.1657
Relative absolute error	3.3781	2.88	33.33
Root relative squared error	18.0482	22.97	33.33
Total number of instances	918	918	918
Accuracy	**98.8017**	**98.692**	**100%**

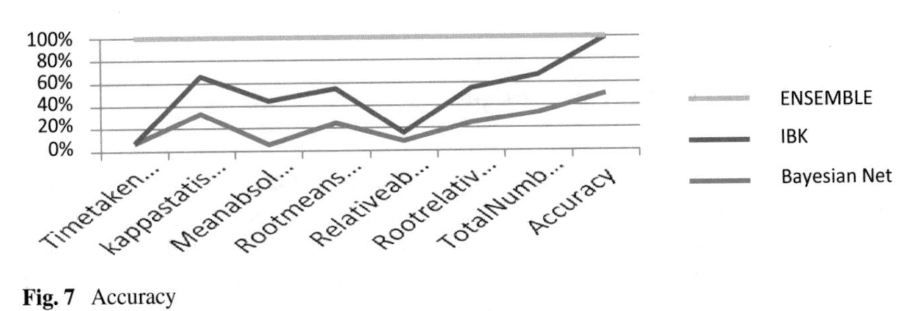

Fig. 7 Accuracy

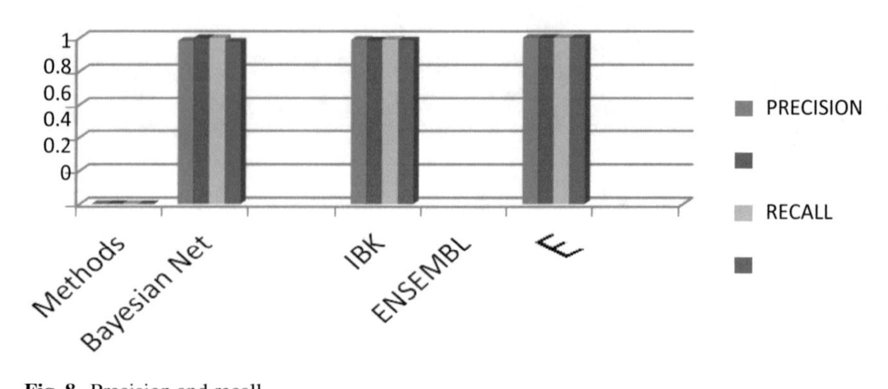

Fig. 8 Precision and recall

Table 2 Precision and recall of different algorithms

Methods	PRECISION		RECALL	
	YES (%)	NO (%)	YES (%)	NO (%)
Bayesian Net	98.1	99.8	99.8	97.6
IBK	98.8	98.5	98.8	98.5
ENSEMBL E	100	100	100	100

Table 3 Confusion matrix of different algorithms

	A = NO	B = YES
Bayesian Net	400	10
	1	507
IBK	404	6
	6	502
ENSEMBLE	410	0
	0	508

5 Conclusion and Future Scope

This paper proposed an ensemble technique that enhances the performance and accuracy of cardiovascular disease diagnostics. In this, Ada boosting, Bagging, and stacking combined by vote are used as ensemble and Bayesian net and IBK to attain improved Performance and accuracy. We have taken datasets from the UCI Dataset with 918 instances and 14 attributes. The main aim is to discover the best classification algorithm with maximum accuracy. In this, we used the Weka tool for accuracy prediction. This proposed model uses an ensemble technique, and it has given 100% accuracy when compared to Bayesian Net and IBK. The next second place was taken by the Bayesian Net algorithm with 98.8017% accuracy. Third is that IBK has given 98.692% accuracy. Moreover, the time taken to build a model is less when compared to other algorithms. This result will help in selecting the best classification algorithm for heart disease prediction and can be used for detection and treatment. I want to use the real-time dataset in the future for my research work.

References

1. Latha CBC, Jeeva SC (2019) Improving the accuracy of prediction of heart disease risk based on ensemble classification techniques. ELSEVIER
2. Gao XY, Ali AA, Hassan HS, Anwar EM. Improving the accuracy for analyzing heart diseases prediction based on the ensemble method
3. Shorewala V. Early detection of coronary heart disease using ensemble techniques
4. Muhammad Y, Tahir M, Hayat M, Chong KT. Early and accurate detection and diagnosis of heart disease using intelligent computational model
5. Tama BA, Im S, Lee S. Improving an intelligent detection system for coronary heart disease using a two-tier classifier

Integrated Image Recommendation Based on Category and Color Utilizing CNN and VGG-16

R. Bhuvanya and M. Kavitha

Abstract In earlier days, image retrieval is based on text whereas due to the improvement in technology, and this paper aims at deep learning (DL) techniques to produce integrated image recommendations. As the recommendation system has become inevitable in e-commerce sites, the proposed work is based on recommending the matching color of the same category product. Initially, the preprocessing of the dataset is carried out by applying lemmatization and stop words removal then by applying term frequency-inverse document frequency (TF-IDF), the top 15 frequent categories of the product are identified which is considered for further processing. Then, the Web scraping technique is applied to retrieve the images from the product URL. To identify the color of the product, the color histogram approach is incorporated. The fetched images are fed into the DL techniques of convolutional neural network (CNN) and VGG-16 to recommend the matching color of the same category product. The evaluation metrics of precision, recall, F1score, and accuracy are applied to evaluate the implemented model. From the experimented results, it is justified that the VGG-16 technique scores better than CNN.

Keywords Color histogram · Convolutional neural network · Image retrieval · TF-IDF · VGG-16

1 Introduction

As the volume of data grows rapidly, it is essential to improve the quality of recommendations. Hence, the data scientists have replaced the traditional machine learning (ML) algorithms with deep learning (DL) models. In the past few years, deep learning techniques have perceived great success in computer vision applications. Recently,

---

R. Bhuvanya (✉) · M. Kavitha
Vel Tech Rangarajan Dr. Sagunthala R&D Institute of Science and Technology, Chennai, India
e-mail: bhuvanyaraghunathan@gmail.com

M. Kavitha
e-mail: kavitha@veltech.edu.in

A. Kumar et al. (eds.), *Proceedings of the International Conference on Cognitive and Intelligent Computing*, Cognitive Science and Technology,
https://doi.org/10.1007/978-981-19-2350-0_74

the DL models have started their influence in the field of recommender systems (RS) as it achieves high-quality recommendations. Before diving into the techniques of DL, this introductory session explains the basic concept of RS. Recommender system analyzes the user and item profile [1] and helps to discover new items by reducing the searching time of the user. It is usually classified into collaborative, content-based, and hybrid filtering techniques [2]. Collaborative filtering generates the recommendation list by considering the user-item historical interactions, browser history, and user ratings whereas content-based filtering [3] generates the recommendation list by considering the features of items such as color, shape, text, etc. The hybrid filtering produces the recommendation by integrating one or more filtering techniques [4]. The category of recommendation can be non-personalized, semi-personalized, and personalized [5]. If the recommendation list is generated based on the relevance of the most popular category, it is considered a non-personalized system. If the recommendation is generated based on a group, for instance, students, professionals, old aged people, etc., then it is a semi-personalized recommendation. At this level, the same category of people will get the same type of recommendations. The last level of personalized recommendation provides the items by analyzing the current user of the system. The generated list will be made only to the specific user. This recommendation can be applied to various application domains such as e-commerce, healthcare, tourism, and education. The famous e-commerce site Amazon.com has launched its recommendation engine two decades ago and millions of users have benefitted from the RS, by discovering unknown items [6, 7].

Deep learning is a subset of ML which is a subset of artificial intelligence (AI) [8]. With the ability to work with a large set of data along with good computational power, DL algorithms can able to self-learn the hidden patterns to make predictions and recommendations. The difference between the prediction and recommendation is the predictions help to quantify the items whereas the recommendations help the users to discover the unknown items. With the increase of research publications based on deep learning-based recommendations, it is evident that the DL models have proved their efficiency. All the social media sites and e-commerce sites have implemented the DL models for recommending items. For instance, YouTube, e-Bay uses deep neural network (DNN) models, and Spotify, Netflix uses convolutional neural network. When the neural network is applied, it can process complex user-item interaction patterns. Also, due to the advancement of deep learning models, it is possible to build hybrid models for recommending items. The deep learning models proved their efficiency in the tasks of both supervised and unsupervised ML. As the field is teeming with new ideas, DL models have become more successful in image processing and speech recognition techniques, e-commerce, healthcare, advertising, entertainment, etc. The major reasons for implementing DL models in RS are non-linear transformation, representation learning, sequence modeling, and flexibility [9].

Non-linear Transformation: The traditional matrix factorization technique combines the user and item latent factors linearly. But the neural networks (NN) are capable of approximate the function by applying various activation functions.

Representation Learning: The real-world applications usually contain a large amount of descriptive information about user-item interactions. To get a better quality recommendation, it is essential to make use of this information effectively. When the deep learning model is incorporated, it learns the features from the raw data. Also, it can make use of text, image, audio, and video to build a recommendation model.

Sequence Modeling: In the past decade, DNN has exposed its efficient results in natural language processing and speech recognition techniques. The deep learning techniques of Recurrent Neural Network (RNN) and Convolutional Neural Network (CNN) work effectively in the above-mentioned tasks. RNN is a feedforward network, in which the information moves in the forward direction. It consists of input, hidden, and output layers. RNN achieves this task by the presence of internal memory. Because of the internal memory, the RNN can make the prediction easier, hence it is useful for the time series data, sentiment analysis. CNN works on the image by taking a segment also known as filters. Sequence modeling has become important in session-based recommendations. Hence, it can be concluded that the deep learning models are extremely useful for the pattern mining task.

Flexibility: Deep learning has the framework of Tensorflow, PyTorch, DeepLearning4j, Keras, ONNX, Caffe, etc. Hence, it provides an easy way to combine various neural network models which makes the task of prediction and recommendation efficient.

This paper focuses on color and category-based image retrieval [10, 11] by implementing convolutional neural network (CNN) and VGG-16. Initially, the category of the product needs to be identified, which is done by applying natural language processing techniques. Then, the color of the user interested product is identified through color histogram which is then fed into CNN and VGG-16 to retrieve the same color products from the dataset, which will reduce the browsing time of the user.

1.1 Organization of the Paper

The paper is organized as follows. Section 1 introduced the significance of the work, Sect. 2 describes the techniques covered under deep learning, and Sect. 3 describes the proposed methodologies of CNN and VGG-16. Finally, Sect. 4 explains the evaluation metrics followed by the conclusion.

2 State of the Art

Deep learning is a subset of machine learning, which can be thought of as a subset of machine learning. It is a field devoted to computer algorithms that learn and evolve on their own [12]. In contrast to machine learning, which employs simpler concepts, deep learning employs artificial neural networks, which are designed to mimic how humans think and learn. This section summarizes the most popular deep learning approaches.

Multilayer Perceptron

It is also known as a fully connected neural network which consists of three layers known as the input layer, one or more hidden layers(s), and output layer. As it is known as a fully connected network, each neuron is connected with every other neuron. It is extremely applicable to the tasks of classification and regression. Here, the training of data will be accomplished through backpropagation [13]. The input layer receives data and passes it on to the hidden layer. Weights are assigned to each input at random by the links between the two layers. After the weights have been multiplied separately, a bias is added to each input. To the activation function, the weighted sum is passed. The activation function can be linear or non-linear, and if the linear function is used, then it is termed a single-layer perceptron. The activation function is applied at the output layer to generate an output. The above-mentioned process is feedforward NN. Observing the loss function of output, the previous weights are adjusted and backpropagated to minimize the error.

Convolutional Neural Network

It is a type of neural network that has shown to be useful in image recognition and classification applications. It is also known as the convolution net, and it is made up of three layers: convolution, pooling, and fully linked layers [9]. CNN works in a different way than normal neural networks. The height, breadth, and depth of the layers in CNNs are all determined in three dimensions. The neurons in one hidden layer only communicate with a subset of the neurons in the other layer, rather than all of them. In addition, the result is grouped along the depth dimension into a single vector. The convolution and pooling layers do feature extraction, whereas the fully connected layer performs classification.

Recurrent Neural Network

A recurrent neural network (RNN) is considered as the multiple feedforward neural network, which transfers the information from one to the other. It has profound applications in speech recognition, stock market prediction, language translation, and image recognition [9]. The major drawback with the Naïve RNN is it tends to forget the old information also it suffers from the issue of vanishing gradient and exploding gradient. The problem with the feedforward network is the predictions are done based on the current input as it cannot remember the historical information. On the other hand, RNN predicts the output based on current and previously learned input. A recurrent neural network is based on the notion of preserving a layer's output and feeding it back to the input to predict the output. Hence, the nodes of a particular layer can remember the information of the past steps. The best example for RNN is Google's auto-filling words. The performance of RNN can be enhanced by extending the memory by implementing long short-term memory (LSTM) and gated recurrent unit (GRU). They have internal gates that can control information flow. These gates can learn which data in a sequence should be kept and which should be discarded.

This allows it to convey relevant information down a long chain of sequences to generate predictions.

Autoencoders

Autoencoders are a specific sort of feedforward network, with an input identical to the output. It compresses the input into a lower-dimensional code and then reconstructs the result. Encoder, code, and decoder are the three parts of an autoencoder. The encoder compresses the input and generates the code, which the decoder subsequently uses to reconstruct the input. With a non-linear activation function and multiple layers, an autoencoder can learn non-linear transformations. It does not need to learn dense layers. Bottleneck refers to the layer that exists between the encoder and the decoder, i.e., the code. This is a well-designed method for determining which aspects of observed data are important and which can be ignored.

Restricted Boltzmann Machine

RBMs are one of the most basic neural networks because they only have two layers: a visible layer and a hidden layer [14, 15]. In a forward pass, we feed our training data into the visible layer, and during backpropagation, we train weights and biases between them. Each hidden neuron's output is generated using an activation function like ReLU. Neurons in the same layer cannot communicate directly with one another. The connections will exist between the two layers. Since the communication is restricted, it is known as restricted Boltzmann machine.

Deep Reinforcement Learning

Deep reinforcement learning (DRL) revolutionizes artificial intelligence (AI) and is a step toward designing autonomous systems to better understand the visual world [16]. A machine is continuously learning through a sequence of trials and errors, making this technology perfect for dynamic, changing situations. Although reinforcement training was available for decades, it was paired with deep learning much more recently, which provided fascinating outcomes. The deep part of reinforcement is many (deep) layers of artificial neural networks that emulate the human brain structure.

Generative Adversarial Networks

GANs are deep learning algorithms that create new, training-like data instances [17]. GAN consists of two components: a generator and a discriminator. During the training data, the former generates false data and the latter can learn from the false information. GAN is used to generate realistic images, cartoons, human face photographs, and 3D objects.

3 Proposed Methodologies

As the work concentrates on both color and category-based image retrieval, the TF-IDF method is applied, as a preprocessing step. Then, the online scraping technique, also known as Web harvesting, is used to extract images from the product URL as the procedure focuses on color. Because there are 20,000 items in the dataset and each product can have 5 images, all of the images must be obtained from the URL. Web scraping is used to retrieve and download images from product URLs in the dataset. The extracted images are passed into convolutional neural network and VGG-16 to retrieve similar types of products. Section 4 incorporates the proof of retrieval images. Figure 1 displays the example of retrieval images. And Fig. 2 shows the product category tree.

3.1 Preprocessing

Initially, the case-folding also known as the lowercasing approach is used, which turns all letters to lowercases. To eliminate the punctuation marks and non-alphabetic characters, string manipulation and regular expression procedure are used. Furthermore, a lemmatization method is applied to fetch the meaningful category for the product. TF-IDF [18] is applied, to fetch the top and least 15 frequent categories from the dataset. TF-IDF is a feature extraction technique that presents the relative importance of a term. TF is the ratio of no. of times term '*t*' appears in the document to the total number of terms whereas IDF is the ratio of the number of documents to the number of documents that has a term '*t*' which will compute the logarithmic further. By applying the TF-IDF, the most frequent and least frequent product categories are identified. Only the top-15 frequent categories are considered for further process. The prediction of categories can be easily identified through the above-mentioned approach.

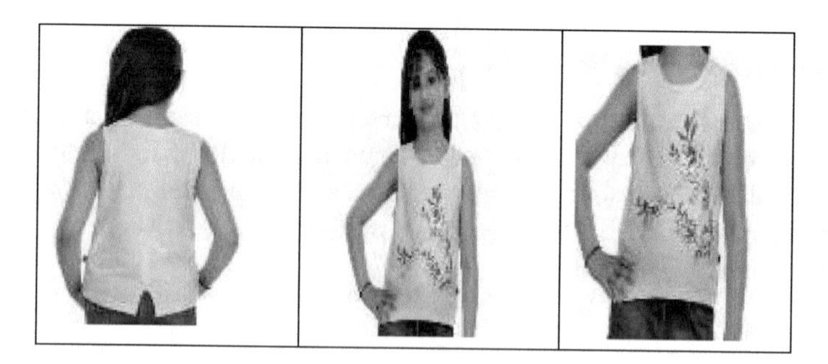

Fig. 1 Example of retrieval images

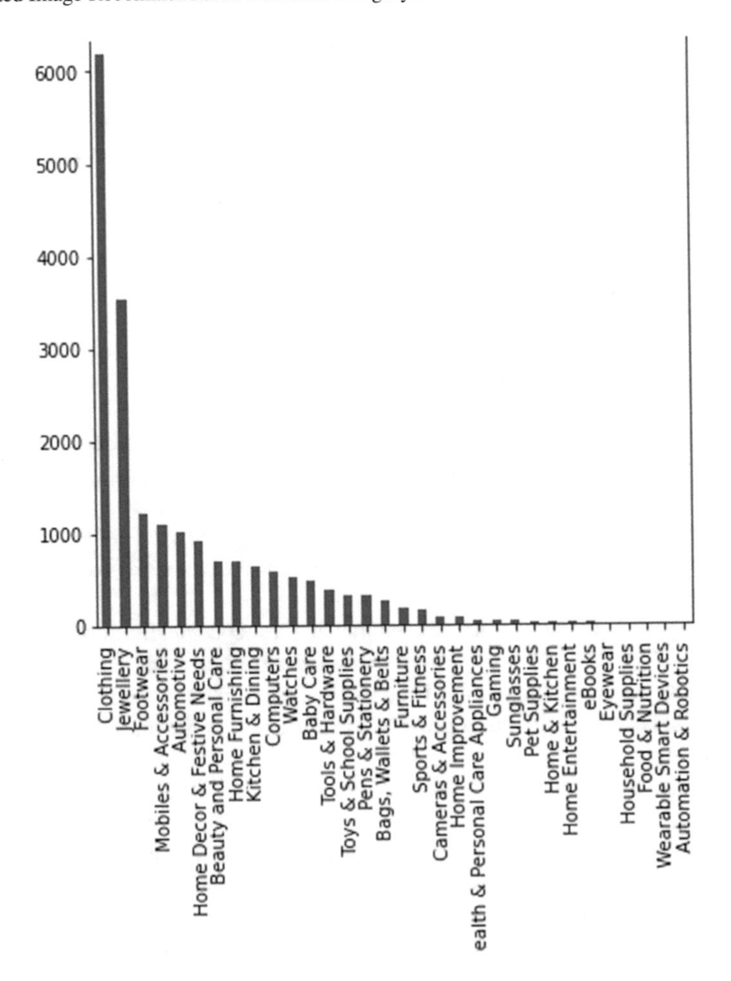

Fig. 2 Product category tree

3.2 *Color Identification*

The color feature is used to find the corresponding product based on a similar intensity of pixels. To extract color information, color histogram techniques, color coherence vectors, and color moments are often utilized. This paper utilizes the color histogram to identify the color of the product. A color histogram [19] is a graph that depicts the distribution of color composition in an image. In addition, the color histogram bar shows the number of pixels in each type of color in an image. A histogram's data are obtained by counting each color in the image. To identify the similarity matching between the images, the Euclidean distance is computed. The formula to compute the Euclidean distance is given below in Eq. (1).

$$D = \left((x_2 - x_1)^2 + (y_2 - y_1)^2\right)^{1/2} \quad (1)$$

The color of the retrieved images will be determined this way. Finally, the color identified through the color histogram will be passed into the DL algorithms, which will predict the matching image for the selected product.

3.3 Application of DL Techniques

3.3.1 Convolutional Neural Network

The reason behind the choice of CNN is, when compared to other classification methods, the amount of pre-processing required by a ConvNet is significantly less. It is also known as convolution nets, which have three types of layers known as convolution layer, pooling layer, and fully connected layer. When compared to regular neural networks, CNNs work differently. The architecture of CNN for the proposed work is depicted in Fig. 3.

The image gets differed from one another in their special structure. CNN could be used to derive a higher-level representation of image content. In addition to the color identification product in the E-commerce sites, various applications can be listed for the color identification process. For instance, in the agricultural industry, to determine the color of fruits [20] that are riped completely and to discard it, to identify the color of the light under different conditions, detecting and identifying the color of the traffic signal to maintain road safety. As the implemented work is incorporated in the images taken from the e-commerce site, this method can be useful to recommend similar color products [12] by observing the user's behavior.

In CNN, the overhead to preprocessing will be lesser. The basic elements of CNN are the convolutional layer, pooling layer, fully connected layer. The convolutional layer is responsible for feature extraction. The term 'convolution' is a mathematical

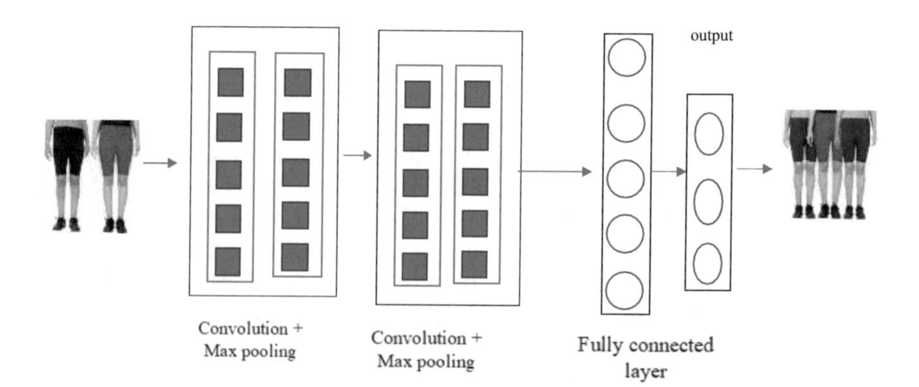

Fig. 3 Architecture of CNN

function that combines two functions and produces a third function using the merging technique. The images retrieved from the e-commerce sites are represented in terms of the pixel matrix. The input to the convolutional layer is the height, width, and number of channels in the image. As we have taken the RGB image, the number of the channel will be 3.

As the CNN will not consider the whole image of pixels, only the part of the pixels is supplied as input. For instance, instead of taking the whole pixel matrix, a small subset of matrix values will be taken. For instance, from the whole matrix, either 3 × 3 or 5 × 5 matrix values and the extracted feature map will be applied with the filter which is of the same size. The number of features chosen by the CNN is directly proportional to the training time. If the number of features is high, then the time taken for training will also be longer. Then, the process of activation function begins, here the 'ReLU' function is used, which will replace the non-negative value with zero. The pooling layer works similar to convolution, and there are various types of pooling available known as max pooling, min pooling, and average pooling. It slides the value obtained through convolution by considering the mentioned value. For instance, the max pooling is applied here, which will stride the max. Value from the pixel groups. The obtained values are then randomly passed to the fully connected layer. The final layer has a 'Softmax' function which will predict the recommendation for the query image by considering the category and color.

3.3.2 VGG-16

VGG-16 is a type of CNN model known as a very deep convolutional neural network. VGG-16 has a convolutional layer which is 3 × 3 size, and the max pooling layer has 2 × 2 size followed by the fully connected layers at the end. Hence, '16' denotes the number of layers that have weights [20]. The formation of convolution and max pool layer is consistent with the whole architecture. Figure 4 shows the architecture of VGG-16.

In 2014, Simonyan and Zisserman [21] presented VGG-16 architecture as a very deep convolutional network for large scale image recognition. A sequence of convolutional layers is followed by a max pooling layer in each VGG block. All convolutional layers have the same kernel size (3 × 3). After each block, a max pooling of size 2 × 2 with strides of 2 is used to halve the resolution. There are two completely linked hidden layers and one fully connected output layer in each VGG model. The dimensions of the input image change to 224 × 224 × 64 when it is passed through the first and second convolutional layers. The output is then sent to the max pooling layer with a stride of two.

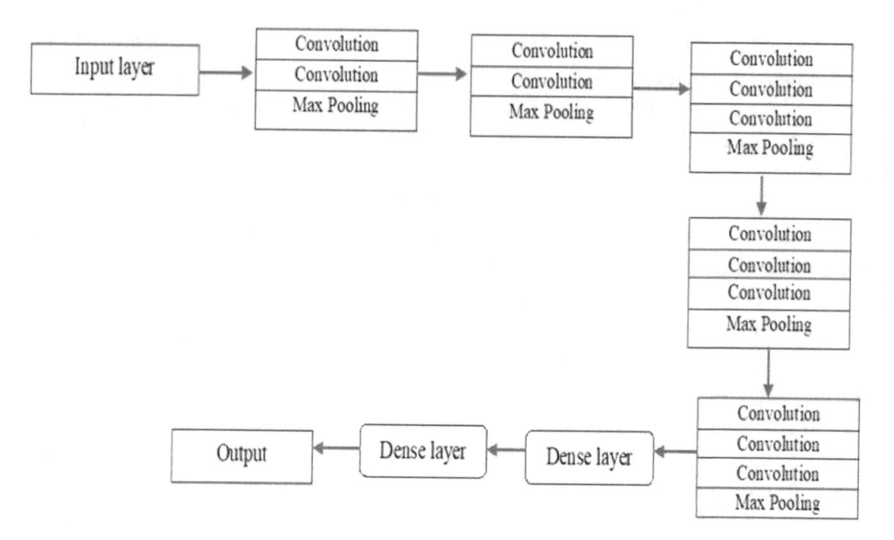

Fig. 4 Architecture of VGG-16

4 Performance and Evaluation

4.1 Datasets

The evaluation dataset is a pre-crawled dataset derived from data extracted from Flipkart.com, a leading e-commerce site. Product URL, product name, product category tree, pid, retail-price, reduced price, image, is FK advantage product, product description, product rating, overall rating, brand, and product specification are among the fields in this collection [22]. Figure 5 shows the process of preprocessing such as stopwords removal, lemmatization, etc. Figures 6, 7 describe the top and least 15 categories of items. Figure 8 depicts the color histogram of the query image, and finally, Fig. 9 shows the proof of image retrieval based on color and category.

4.2 Precision, Recall, F-Measure, and Accuracy

To evaluate the performance of CNN and VGG-16, the performance metrics [23] of precision, recall, F-measure, and accuracy are applied. Precision defines, 'how many selected items are relevant', and recall defines, 'how many relevant items are selected'. Accuracy defines the correctly predicted data to the total observations. The formula to calculate precision, recall, F-measure is mentioned below in Eqs. (2), (3), (4), and (5). Where TP, TN, FP, and FN denote true positive, true negative, false positive, and false negative, respectively.

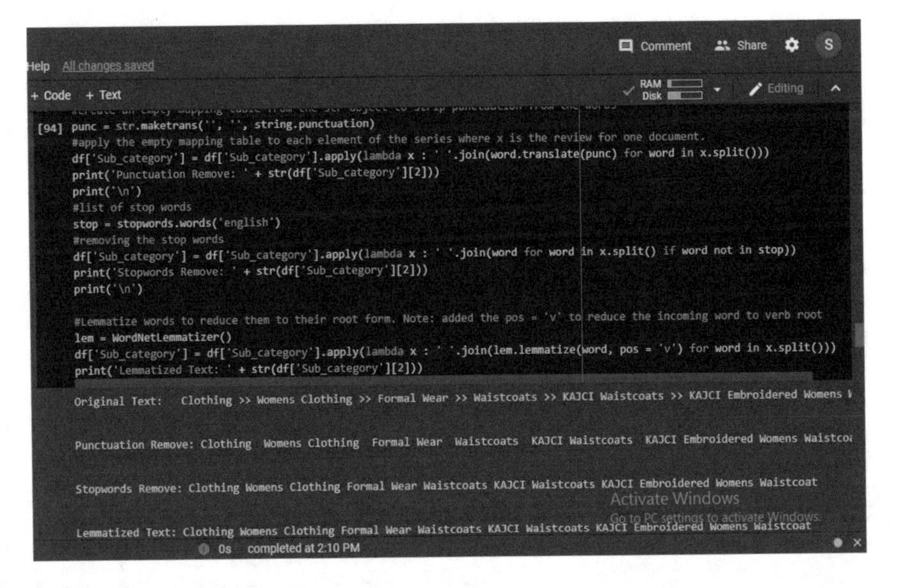

Fig. 5 Preprocessing the dataset

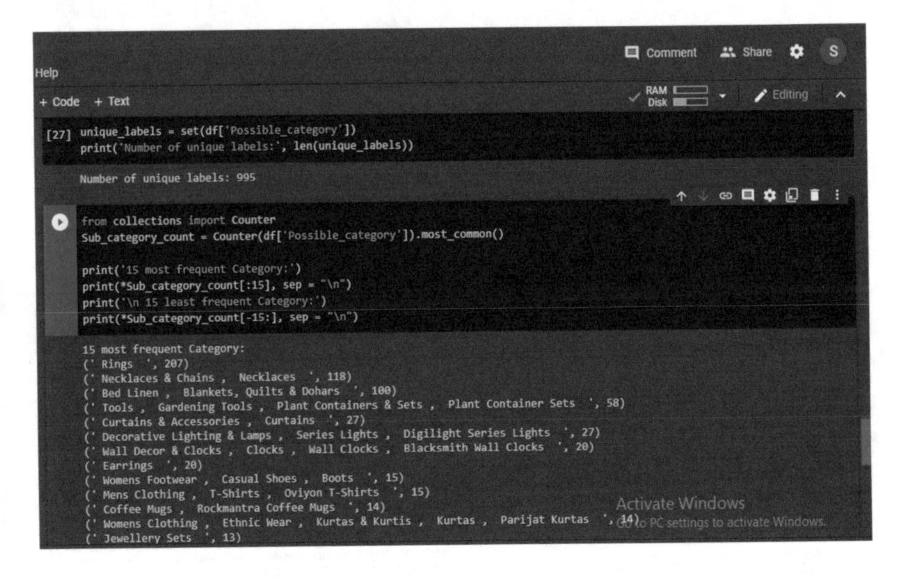

Fig. 6 Retrieving the top 15 frequent category

$$\text{Precision}(P) = \text{TP}/\text{TP} + \text{FP} \tag{2}$$

$$\text{Recall}(R) = \text{TP}/\text{TP} + \text{FN} \tag{3}$$

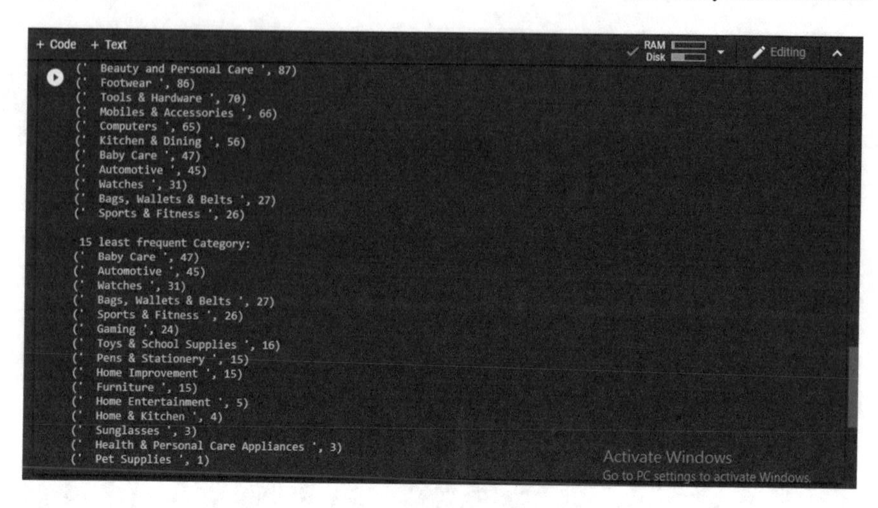

Fig. 7 Retrieving the least 15 frequent category

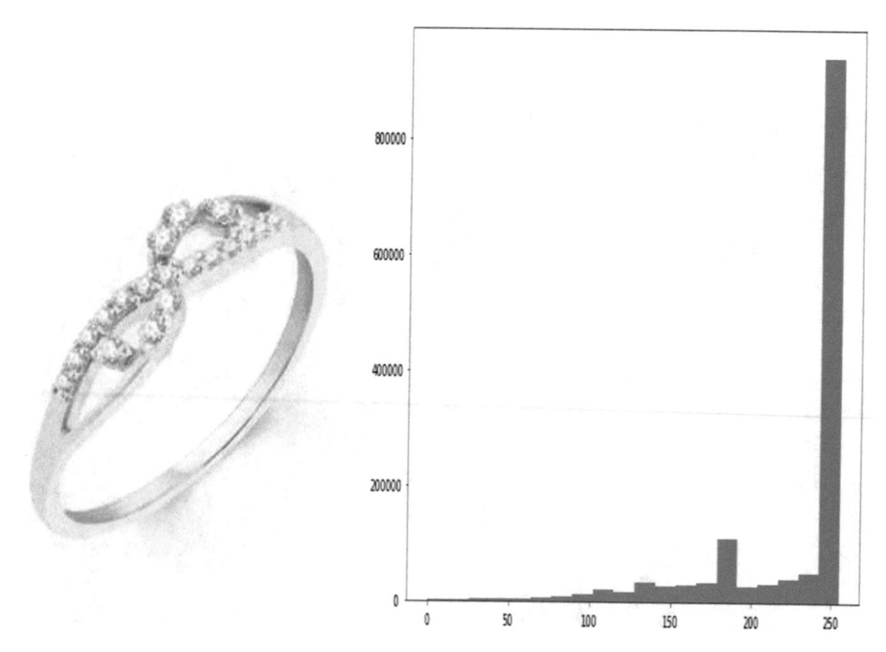

Fig. 8 Color histogram of the query image

$$\text{F-Measure} = 2 * (P * R)/(P + R) \quad (4)$$

$$\text{Accuracy} = \text{TP} + \text{TN}/\text{TP} + \text{FP} + \text{FN} + \text{TN} \quad (5)$$

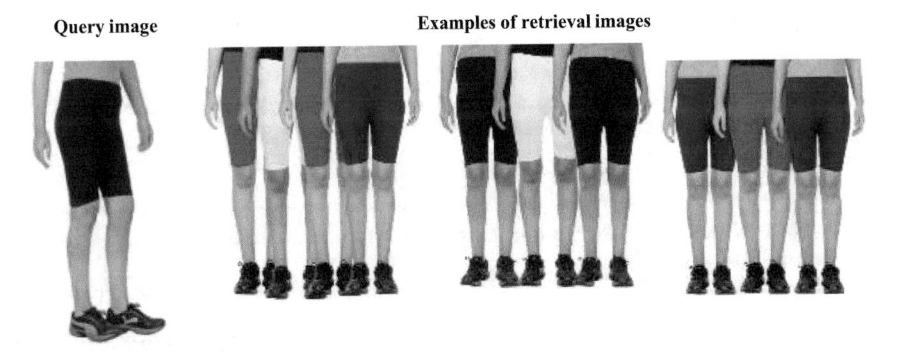

Fig. 9 Category and color-based retrieval

Table 1 Comparison of performance

Metrics/techniques	CNN	VGG-16
Precision	73.52	90.47
Recall	87.71	95
F-measure	0.8	0.92
Accuracy	85.32	91.25

Table 1 shows the obtained results by applying CNN and VGG-16. And Fig. 9 shows the similar product retrieved by applying the above technique.

From the obtained results, it can be concluded that the results of VGG-16 outperform the CNN.

5 Conclusion

This paper adopts the strategy of convolutional neural network and VGG-16 to fetch the color-based products. Also, 20,000 images are trained to predict the products based on color. Initially, the TF-IDF is applied to identify the category of the product. Then, the Web scraping technique retrieves all the images from the product URL, and it is fed into a color histogram, to identify the color of the product. From the color histogram, the color of the product is easily identified and it is passed into CNN and VGG-16 to fetch the matching images for the chosen product. The experimental results show a good prediction for the implemented models. The evaluation metrics of precision, recall, F-measure, and accuracy demonstrate that our system performs well with the trained model, and VGG-16 performs better with CNN. This paper concentrates on image retrieval in the domain of apparel, and further, it can be enhanced to various products available in the dataset. This image retrieval can be implemented on e-commerce sites to reduce the browsing time of the user.

References

1. Guan C, Qin S, Ling W, Ding G (2016) Apparel recommendation system evolution: an empirical review. Int J Clothing Sci Technol 28(6):854–879
2. Tewari AS (2020) Generating items recommendations by fusing content and user-item based collaborative filtering. Procedia Comput Sci 167:1934–1940
3. Pazzani M, Billsus D (2007) Content-based recommendation systems. In: Brusilovsky P, Kobsa A, Nejdl W (eds) The Adaptive web, lecture notes in computer science, vol 4321. Springer Berlin Heidelberg, pp 325–341
4. Chu W-T, Tsai Y-L (2017) A hybrid recommendation system considering visual information for predicting favorite restaurants. WWWJ 1–19
5. Wang Y-F, Chuang Y-L, Hsu M-H, Keh H-C (2004) A personalized recommender system for the cosmetic business. Expert Syst Appl 26(3):427–434
6. Blake MB (2017) Two decades of recommender systems at Amazon
7. McAuley J, Targett C, Shi Q, van den Hengel A (2015) Image-based recommendations on styles and substitutes. In: Proceedings 38th international ACM SIGIR conference research developmental information retrieval. pp 43–52
8. Goodfellow I, Bengio Y, Courville A (2016) Deep learning. MIT Press. http://www.deeplearningbook.org
9. Zhang S, Yao L, Sun A, Tay Y (2019) Deep learning-based recommender system: a survey and new perspectives. ACM Comput Surv 52(1):38, 5. https://doi.org/10.1145/3285029
10. Sabahi F, Ahmad MO, Swamy MNS (2016) An unsupervised learning-based method for content-based image retrieval using hopfield neural network. 2016 2nd international conference of signal processing and intelligent systems (ICSPIS). pp 1–5. https://doi.org/10.1109/ICSPIS.2016.7869882
11. Wei W, Wang Y (2019) Color image retrieval based on quaternion and deep features. IEEE Access 7:126430–126438
12. Deng L, Yu D (2014) Deep learning: methods and applications. Found Trends® Sig Process 7(3–4):197–387
13. Zhang P, Jia Y, Gao J, Song W, Leung H (2020) Short-term rainfall forecasting using multi-layer perceptron. IEEE Trans Big Data 6(1):93–106. https://doi.org/10.1109/TBDATA.2018.2871151
14. Salakhutdinov R, Mnih A, Hinton G (2007) Restricted Boltzmann machines for collaborative filtering. In: Proceedings of the 24th international conference on Machine learning, ICML, association for computing machinery. New York, NY, USA, pp 791–798
15. Liu X, Ouyang Y, Rong W, Xiong Z (2015) Item category aware conditional restricted Boltzmann machine based recommendation. In: International conference on neural information processing. Springer, pp 609–616
16. Arulkumaran K, Deisenroth MP, Brundage M, Bharath AA (Nov 2017) Deep reinforcement learning: a brief survey. IEEE Sig Proc Mag 34(6):26–38. https://doi.org/10.1109/MSP.2017.2743240
17. Cai X, Han J, Yang L (2018) Generative adversarial network based heterogeneous bibliographic network representation for personalized citation recommendation. In: AAAI
18. JingL-P, Huang H-K, Shi H-B (2002) Improved feature selection approach TFIDF in text mining. In: Proceedings of the international conference on machine learning and cybernetics, vol 2. pp 944–946
19. Juang C-F, Sun W-K, Chen G-C (2009) Object detection by color histogram-based fuzzy classifier with support vector learning. Neurocomputing 72(10–12):2464–2476
20. Yang Z, Yue J, Li Z, Zhu L (2018) Vegetable image retrieval with fine-tuning VGG model and image hash. IFAC-PapersOnLine 51(17):280–285

21. Simonyan K, Zisserman A (2015) Very deep convolutional networks for large-scale image recognition. CoRR arXiv:1409.1556
22. https://www.kaggle.com/PromptCloudHQ/flipkart-products
23. Gunawardana A, Shani G (2009) A survey of accuracy evaluation metrics of recommendation tasks. J Mach Learn Res 10:2935–2962

Development of Opto-Mechanical Method to Measure Liquid Volume, Density, and Level

Nityananda Hazarika, Ram Kishore Roy, Munmee Borah, Hidam Kumarjit Singh, and Tulshi Bezboruah

Abstract We demonstrate here an optical method, capable of measuring liquid volume, level, and density by using linear extension spring. This measurement method is based on linear extension of vertically suspended helical spring with a light source fixed on the top end and a light-dependent resistor on the bottom end of the spring, both being in same alignment. Loading liquid samples in a cylindrical container attached to the bottom end of the spring results in corresponding elongation in length of the spring, thereby changing the longitudinal distance between light source and light-dependent resistor. Light intensity is proportional to the square of the distance from the source; hence, resistance of the light-dependent resistor changes accordingly.

Keywords Light intensity · Linear extension spring · Longitudinal distance · Optical method · Volume measurement · Level measurement · Density measurement

1 Introduction

The precise measurement of liquid volume, level, and density plays a significant role in research fields and many industrial fields such as petrochemical, pharmaceutical, textile, food industries, etc. Over the passing years, numerous mechanical, electrical, and electronic approaches have been seen to be implemented in the measurement of these quantities [1]. Many unconventional and novel techniques have got profound research and development interest as reported in [2–7]. Most of the modern techniques are embedded with microcontroller/computer to obtain ergonomically perfect configuration and enhance accuracy. In reference, most common sensor used in the

N. Hazarika · R. K. Roy (✉) · M. Borah · H. K. Singh · T. Bezboruah
Department of Electronics and Communication Technology, Gauhati University, Guwahati, Assam 781014, India
e-mail: r.kore51guece@gmail.com

A. Kumar et al. (eds.), *Proceedings of the International Conference on Cognitive and Intelligent Computing*, Cognitive Science and Technology,
https://doi.org/10.1007/978-981-19-2350-0_75

measurement of liquid level, volume, and density is fiber optic [2–6]. Fiber optics-based sensing relies on the principle of bending loss [3], multimode interference effect [6], light intensity modulation [2], fiber long-period, and Bragg gratings [5].

In this paper, we have proposed an optical method for measuring liquid volume, level, and density by utilizing property of linear extension of spring and inverse square law of light for precise and accurate measurements.

2 The Proposed Method

The basic principle of the proposed method is based on inverse square law and Hooke's law. According to Hooke's law, within the elastic limit of the spring, elongation in length of extension spring (l_E) is directly proportional to the applied weight (W) as in Eq. (1).

$$l_E = kW \tag{1}$$

where k is the spring constant. The weight of the given liquid sample in the cylindrical container can be related with its volume (V), level (l), and density (ρ) as following equations as in [7].

$$W = mg = \rho V g = \varepsilon_V V \tag{2}$$

$$W = \rho g V = \rho g \alpha l = \varepsilon_l l \tag{3}$$

$$W = mg = \rho V g = \varepsilon_\rho \rho \tag{4}$$

where $\varepsilon_v = \rho g$, $\varepsilon_l = \rho g \alpha$, and $\varepsilon_\rho = gV$ are constant for a liquid sample and g is the acceleration due to gravity of a place. Equations (2)–(4) show that the elongation in the length of the spring changes with change in the volume, level, or density of the liquid in the given container.

The sensing principle of the proposed method is schematically shown in the Fig. 1a, b. The effective change in elongation length of the spring for a particular level or volume or density of liquid sample in the attached container is determined by the separation distance between the light source and the light-dependent resistor (LDR) attached at the two ends of the spring as shown in the Fig. 1a, b.

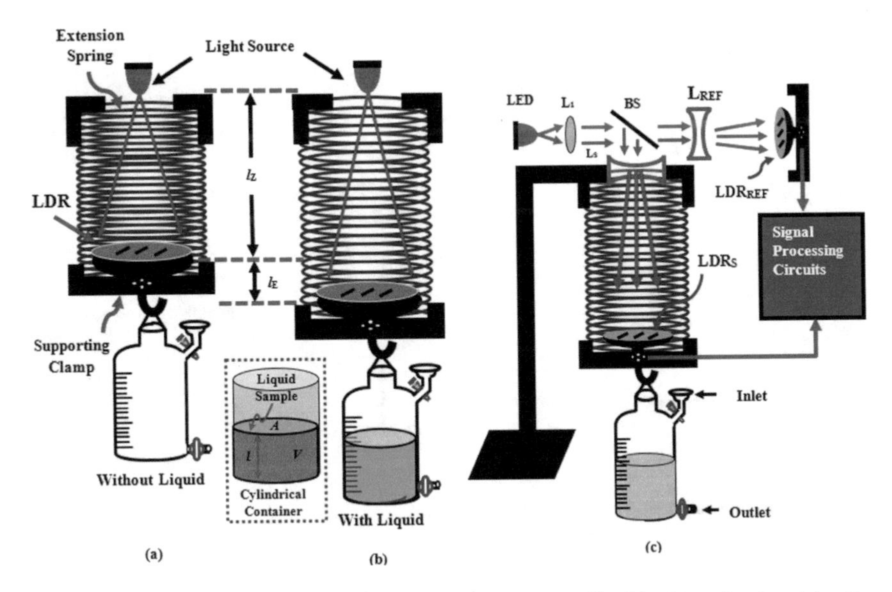

Fig. 1 Sensing principle and arrangement for the measurement of liquid volume, level, and density: **a** zero liquid state, **b** liquid state (inset: cylindrical container of inner cross-sectional area *A* filled with liquid sample of volume *V* at level *l*), **c** simplified schematic used for experimental arrangement

3 Experimental Arrangements

The experimental arrangement for the proposed measurement system is shown in Fig. 1c. A light-emitting diode (LED) is used as a primary light source. The diverging light rays from the light source get collimated by a lens L_1 and traverse through a 50:50 beam splitter to provide secondary light beams. The secondary light beams are once again diverged by using two identical objective lenses L_s and L_{REF}. These light beams are finally allowed to incident on reference light-dependent resistor (LDR_{REF}) and sensing light-dependent resistor (LDRs), as shown in Fig. 1c.

4 Signal Processing Method

The electronic arrangement for the processing of the optical signal includes two identical LDRs–one LDR is used to detect reference light beam represented as LDR_{REF}, while the other one is used to detect sensing light beam represented as LDRs. The two LDRs in the sensing arm and reference arm along with its pull up resistors *R* form a Wheatstone bridge circuit as shown in the Fig. 2. The bridge network translates the resistance of LDR_S (R_S) and LDR_{REF} (R_{REF}) into corresponding voltages V_S and $V_{REF,}$ respectively. An instrumentation amplifier (IA) (model: AD620) takes V_S and V_{REF} as inputs and generates a differential output voltage V_O with gain A_V. Finally,

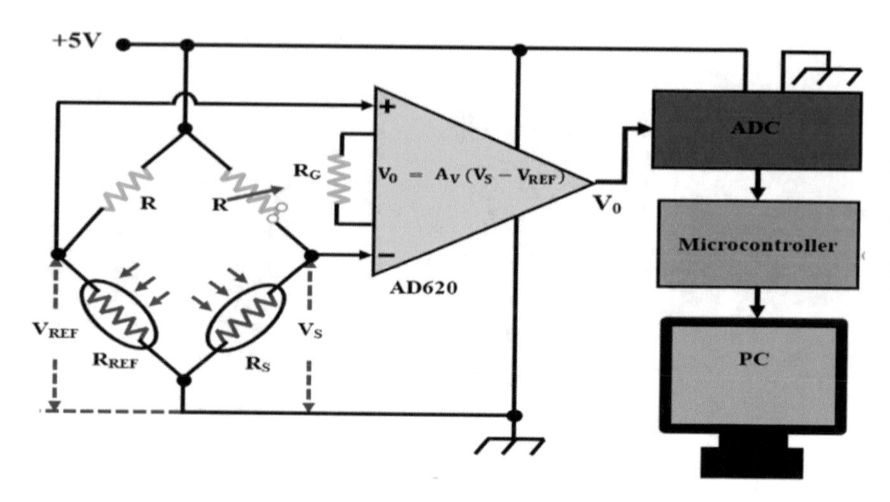

Fig. 2 Simplified schematic of signal processing unit [8]

the output of IA is processed by data logger based on microcontroller (μC) interfaced with analog to digital converter (ADC) circuit, serially connected to personal computer (PC).

The V_S is function of V, l, and ρ as R_S is function of V, l, and ρ. The amplified output voltage of IA with gain A_V is given by [8]:

$$V_O(V, l, \rho) = A_V(V_S - V_{REF}) \tag{5}$$

Here, V_O is also function of V, l, and ρ as V_S is function of V, l, and ρ. As the liquid volume, level, or density increases, it will cause corresponding changes in the output voltage.

5 Experimental Observations

The complete experimental setup is arranged on a vibration isolation optical table in a complete dark room at room temperature. For performing the experiments, three different liquid samples, viz. kerosene, water, and pure glycerine of known densities, have been chosen. At room temperature, standard densities of kersosene, water, and glycerine are 820.1 kg/m^3, 997.297 kg/m^3, and 1260 kg/m^3, respectively.

With the spring mounted in the system, we have measured the longitudinal distance between LED and LDR$_S$ by introducing liquid samples in the container. The volume limit for each sample is chosen in such a way that it does not exceed elastic limit of the spring. Then, elongation, l_E, in the length of spring is calculated by subtracting zero liquid state length (l_Z) from the length corresponding to a liquid state. The observed responses have been plotted in Fig. 3a, b.

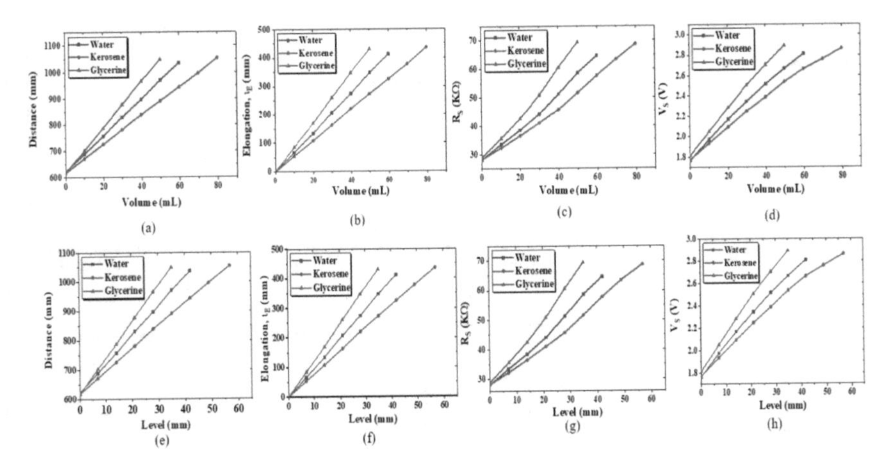

Fig. 3 Scattered plot showing: **a** liquid volume versus longitudinal distance, **b** liquid volume versus elongation in length (l_E) of spring, **c** liquid volume versus resistance R_S of LDR$_S$, **d** liquid volume versus voltage V_S across LDR$_S$, **e** liquid level versus longitudinal distance, **f** liquid level versus elongation in length (l_E) of the spring, **g** liquid level versus resistance R_S of LDR$_S$, and **h** liquid level versus voltage V_S across LDR$_S$ with standard deviation (SD) as error bar

Then, we have measured the changes in resistances of LDR$_S$ (R_S), voltages across LDR$_S$ (V_S), and output voltage V_O for the changes in the volume of each liquid samples repeatedly for 10 times. A standard digital multimeter is used to measure the value of R_S, V_S, and V_O. The observed responses, with standard deviation (SD) as error bar, are plotted in the Figs. 3c, d and 4a.

Similarly, the observed responses for the change in longitudinal distance, l_E, R_S, V_S, and V_O with the change in level for three liquid samples with SD as error bar are plotted in the Figs. 3e–h and 4b, respectively. Then, we have recorded observations for the change in the density of the liquid samples with corresponding change in $R_{S\rho}$, $V_{S\rho}$, and $V_{O\rho}$ at a constant volume and liquid level for the three liquid samples. The

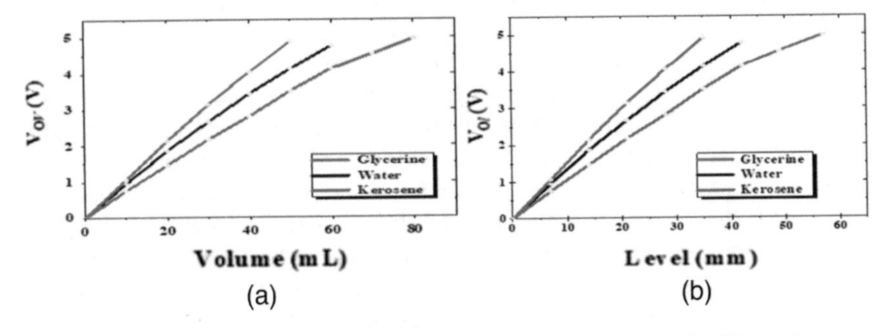

Fig. 4 Scattered plot showing change in **a** volume and **b** level of the three liquid samples versus change in output voltage V_O with SD as error bar

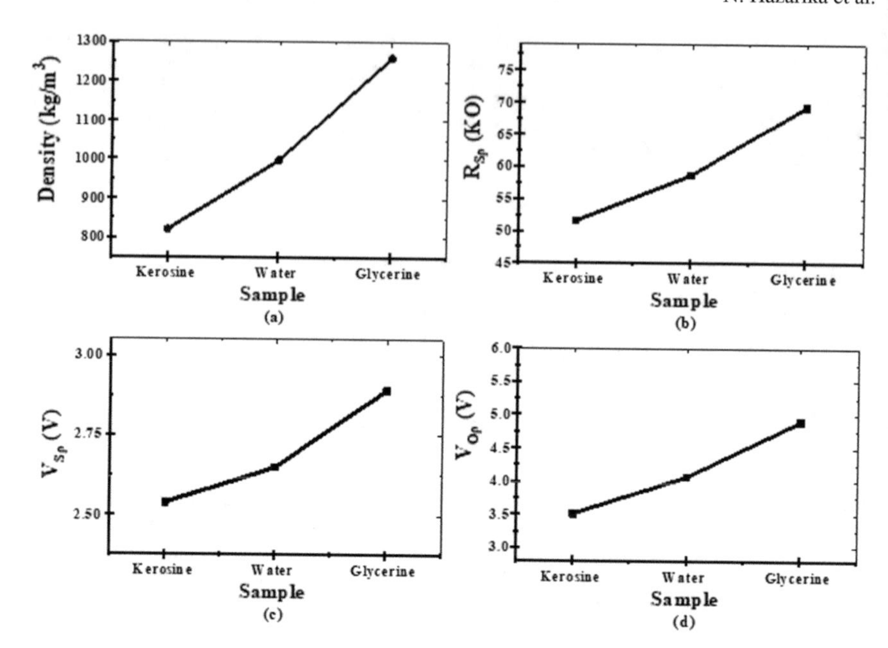

Fig. 5 Scattered plot showing: **a** standard density versus three samples, **b** volume of 50 mL for three samples versus resistance $R_{S\rho}$ of LDR_S, **c** volume of 50 mL for three samples versus voltage $V_{S\rho}$ across LDR_S, **d** volume of 50 mL for three samples versus output voltage $V_{O\rho}$

constant liquid volume considered for density measurement is 50 mL at a level of 35 mm. The obtained responses have been plotted in the Fig. 5a–d, respectively.

6 Results and Discussion

In Fig. 3a, e, it is shown that longitudinal distance between the light source and LDR_S is found to be increased linearly with the increase in liquid volume and level for three different liquid samples, respectively. Again, Fig. 3b, f shows that elongation in length of the spring increases linearly with the increase in liquid volume and level, respectively, for three different liquid samples. With the increase in liquid volume or level or density, the weight of the liquid also increases linearly. Therefore, elongation in length of the spring is directly proportional to the liquid volume or level, which clearly depicts the validity of Hooke's law. Resistance of the LDR_S increases nonlinearly with the increase in liquid volume and level as shown in the Fig. 3c, g, respectively. The plot in the Fig. 3d, h, shows the variation of V_S with liquid volume and level, respectively, for each liquid sample. The V_S varies slight nonlinearly with the increase in volume and level. The variation of output voltage V_O with the change in volume and level of the three liquid samples is shown in the Fig. 4a, b, respectively. The response of the two graphs depicts that the output voltage

V_O varies slight nonlinearly with the increase in volume and level. Figure 5a–d shows the variation in density, R_S, V_S, and V_O of the liquid samples used for recording the observations. From the plots, it is observed that R_S, V_S, and V_O vary nonlinearly with density.

7 Conclusion

A detailed experimental study of an opto-mechanical method for measuring liquid volume, level, and density by using linear extension of spring is presented. The most important aspect of the proposed system is that the single system is capable of measuring liquid volume, level, and density. The system has the novelty in its simple design and inexpensive instrumentation for measurement. The measurements are independent of color or refractive index of the liquids. The method is also suitable to measure the volume, level, and density of highly inflammable, toxic, alkaline, or volatile liquids.

References

1. Sisk HB (1996) Fundamental instrument design review for liquid fuel volume measurement. In: IEEE IMTC and IMEKO Tec. pp 798–803
2. Singh HK et al. A new non-intrusive optical technique to measure transparent liquid level and volume. IEEE Sens J 11(2)
3. Rosolem JB et al. Fiber optic bending sensor for water level monitoring: development and field test: a review. IEEE Sens J 13(11):4113–4120
4. Lin X et al. Low-cost multipoint liquid-level sensor with plastic optical fiber. IEEE Photon Technol Lett 26(16):1613–1616
5. Khaliq S, James SW, Tatam RP. Fiber-optic liquid-level sensor using a long-period grating. Opt Lett 26:1224–1226
6. Liu Y, Li Y, Yan X, Li W. High refractive index liquid level measurement via coreless multimode fiber. IEEE Photon Technol Lett 27(20):2111–2114
7. Hazarika N, Roy RK, Bezboruah T (2019) A non-intrusive Opto-mechanical technique for the measurement of liquid level. In: International conference on innovations in electronics, signal processing and communication (IESC). India, pp 269–272
8. Roy RK, Hazarika N, Singh HK, Bezboruah T. A novel optical method to measure weight by using linear extension Spring. IEEE Sens Lett 2(3)

An Analysis of Sentiment Using Aspect-Based Perspective

F. Arockia Rexi and N. Preethi

Abstract Opinions play a major role in almost every human practice. Finding product and service reviews is made easy online. Product reviews are readily available in huge quantities. Considering each review and making a concise decision about a product is not feasible or even possible. Aspect-based sentiment analysis (ABSA) is one of the best solutions to this problem. Summary and online review's analysis is delivered in this paper. ABSA has made extensive use of machine learning techniques. Recent years have seen deep learning take off due to the growth of computer processing power and digitalization. When applied to various deep learning techniques, numerous NLP tasks produced futuristic results. An overview of various deep learning models used in the field of ABSA is presented in this chapter after an introduction to ABSA.

Keywords Aspect-based sentiment analysis · Convolutional neural network · Natural language processing · Deep learning

1 Introduction

People often turn to Web portals to formulate an opinion about a certain product, topic or service because of the simple openness of the Web. Reviewers expressed opinions on these sites. Making decisions using online opinions is a valuable resource. In fact, the sheer volume of reviews does not ease our task, but rather makes it more difficult since one cannot read each and every review. A thoughtful analysis of the sentiment associated with a product's reviews is crucial to forming an opinion about it. Opinion-based sentiment analysis can be done at various levels. Document level, sentence level and aspect level are a few. While the document level reveals the overall opinion, the

F. A. Rexi · N. Preethi (✉)
Christ University, Bengaluru, India
e-mail: preethi.n@christuniversity.in

F. A. Rexi
e-mail: arockia.f@science.christuniversity.in

A. Kumar et al. (eds.), *Proceedings of the International Conference on Cognitive and Intelligent Computing*, Cognitive Science and Technology,
https://doi.org/10.1007/978-981-19-2350-0_76

sentence level mines the sentence level opinions. Documents and sentences can speak about an entity's different features (aspects), but these levels ignore these differences. When extracted, information must have its equalling sentiment polarities. The process of analysing sentiment based on aspect is called aspect-based sentiment analysis (ABSA). There are many subtasks within the ABCA task. Listed below are some examples of subtasks [1] 1. Identify the concept of aspects 2. Term identification 3. Extraction of category aspects 4. Identifying sentiments (polarity) 5. Identification of the level of sentiment 6. Identification of mood shifters 7. Name and time of opinion holder 8. Generation of an opinion tuple 9. Summarizing opinions.

The expression or understanding of an opinion can be very different. In contrast, according to [2], a definition of opinion is objectively defined as follows:

Defining a quintuple (opinion)

$$(Cij, Yij, Uij, Vij, Tij) \tag{1}$$

Sample reviews (Table 1).

For a sentence *Cij*, its respective entity *Yij*, aspect term *Uij* and aspect category *Vij* and the time at which it was expressed. In ABSA for identifying the sentiment *Bij*, *Tij* is vital. Aspect term is the hyponym of aspect category. The aspect categories may not always be direct in the context (example reviews 2 and 4). At times they have to be understood by the context or their relationship with the aspect terms (example reviews 1, 3 and 5). Understanding elements like entity, aspect term and aspect category is crucial for getting started with ABSA. The opinions that we extract from the sentences need not apply to a single category but maybe for multiple categories as well (example review 4). ABSA becomes significantly more tough and intriguing as a result of this. Unlike previous methods including tedious chores such as deep feature extraction, multiple models and so on, the rapid exponential expansion of

Table 1 Sample reviews

Review	Entity	Aspect term	Aspect category	Opinion word
The earphones are amazing. product delivered faster than expected	Online shopping	Earphones	Electronics	Amazing
The movie was spectacular, especially on the big screens	Theatre	Movie	Movie	Spectacular
The burger was great would definitely recommend	Restaurant	Burger	Food	Great
Was able to find all the things I looked for, the service was good and things were worth the price	Supermarket	Service	Service, price	Good
Much variety was not there. The shirts were not organized well. Overall bad experience	Clothing store	Shirts	Clothing	Bad

data and resources has enabled us to grasp data and go beyond with analysis. Deep learning is a fascinating area of artificial intelligence that allows us to do a variety of activities considerably more quickly. We were able to capture features and classify them using various CNN, RNN, LSTM and other models, allowing us to apply multiple architectures.

ABSA's deep learning contributes to and faces challenges in this chapter. Each chapter is related in a complementary way. Part 2 describes ABSA's problem formulation. Part 3 will highlight some useful observations/assumptions by researchers in the field. Part 4 examines deep learning input representation. In Part 5, various deep learning methods are introduced. Part 6 highlights the different deep learning architectures used in ABSA. Finally, Sect. 7 finishes up the chapter.

2 Defining the Problem

The problem of extraction of aspect terms and the detection of aspect categories must be conceptualized in view of the subjectivity that is involved in opinion mining and ABSA. In this section, we discuss these two issues.

2.1 *Identifying Aspects and Terms*

The goal of this aspect-term extraction: The ATE problem is investigated as a sequence labelling task, in which each token is labelled to indicate whether or not it is a piece of aspect term. To put this labelling in place, the popular BIO tagging technique [3] is typically used, with B representing the start of a word, I representing within a term and O representing others not included in the term.

The result is a list of labels $y = (y{*}1 >, y{*}2 >, \ldots, y)$, where w is the word position in the review phrase and y is the individual label, such as B, I or O. The task can be thought of as a T-class multi-class classification problem.

2.2 *Identifying Aspect Categories*

$C = [c1, c2, c3, \ldots, ck]$ represents the category label space, which contains k categories, and $R = [r1, r2, \ldots, rn]$ indicates a review dataset containing n review sentences. Aspect category detection can be expressed as h: $R \rightarrow 2C$ from a multi-category training set $D = (ri, Yi) \mid 1 \leq$ in, $Yi \subseteq C$ is a set of category labels associated with ri. The aspect category prediction function $h(*)$ predicts the proper category label for a review based on the unseen review $r * R$.

3 Observations/Assumptions Regarding ABSA

Based on an analysis of various articles regarding ABSA, we have come up with the following conclusions/assumptions made by researchers. The consideration of these assumptions can help to reduce the complexity of ABSA by tackling the problem in a very efficient way.

1. Using a bag-of-words model can never be adequate because the bag-of-words model loses all context information [4]. There is a great deal of significance in a word's location and the way it interacts with another word. Generally, words of negative sentiment are negated by positive sentiment words.
2. A reasonable price is more important than a poor price in the context of the aspect of price. The most important aspect of 'service' is poor, instead of reasonable. Certain words from the review can only be interpreted using commentary, such as context words [5].
3. Aspect terms must be accompanied by opinion words, which are indicative of sentiment polarity [6].
4. Many sentiment words have aspect-specific meanings, such as the adjectives 'delicious' and 'tasty' which are used exclusively for the aspect food, and 'cheap' and 'costly' which are used exclusively for the aspect price. In dependence parsing, the connection between aspect terms and sentiment words (opinions) can be captured [7].
5. Sequential labelling reduces the error space of the current prediction by considering the predictions at previous time steps. A previous prediction of O cannot be used in the current prediction of I in the B-I-O tagging [6].
6. A deep neural network such as the LSTM cannot produce usable results on a sentiment classification task.
7. As for a sequence labelling task, CNN should only be applied after the convolution layers since sequence labelling consumes a representation for every position, whereas the max-pooling process mixes up representations for different positions [8].

4 Representation of Inputs

1. In a neural network, input and output should be matched [9, 10]. The input must match the desired output of the network. Text can be handled by a deep neural network by embedding words, similar to natural language processing. A sentence, word, or part of the text can be represented in a variety of ways.
2. After convolution layers for aspect-term extraction, the max-pooling operation should not be used since a sequence labelling model requires accurate representations of all positions, and max-pooling mixes the representations of different positions, which is undesirable [8]. Nevertheless, for multi-word aspect terms, the average of each word can be used as a representation. An aspect term that consists of only one word can be represented only by its word embedding [5].

3. In some cases, it is better to display words in terms of word embeddings or character embeddings [11] to illustrate the effect of word morphology
4. The six-dimensional vector can be represented in the case of six tags (noun, verb, adjective, adverb, preposition and conjunction). The word + POS features dimension will be 306 [12] if the word embedding dimension is 300.
5. Aspect information can be taken into account within ABSA [13] by concatenating aspect vectors into sentence hidden representations and attaching the aspect vector to the input word vectors.

5 Benefits of Deep Learning

5.1 *Embeddable Word Documents*

In Word2Vec (word embeddings model), words are represented in vector space, enabling similarities to be calculated by calculating the vector distance. A symbolic representation, WordNet, is a hierarchical representation used to compute similar words, whereas word embeddings are based on learned context words derived from large corpora. Using this architecture, word vectors are first learned by the neural network language model, and then the n-grams neural network language model is trained using these distributed representations. Based on the current word, the Skip-gram model, out of two models suggested by [9, 10, 14], anticipates the context based on the bag-of-words model. To generate semantic associations, various methods can be used: Linked Statistics Data (LSD), WordNet, Word Embedding, etc.

5.2 *Long Short-Term Memory (LSTM)*

As part of the hidden layer of LSTM, the inputs are generated by the past state $a*t*1 >$ and the present input $x*t >$. This memory is known as a cell. Input gates determine which new input information in the cell state has to be changed. For its state dynamics, LSTM uses the concept hof gates to tackle the vanishing gradient problem. Input I forget (f) and output (o) gates are commonly used in LSTM (o). Based on the input and the content of the memory cell, the output gate determines what to output. The LSTM can add new information to the cell state, update current information and delete data [15] (Fig. 1).

5.3 *Bidirectional Long Short-Term Memory (Bi-LSTM)*

Both networks' outputs are pooled to forecast the final outcome. Since two different networks are used (one for each direction), all available information will be captured.

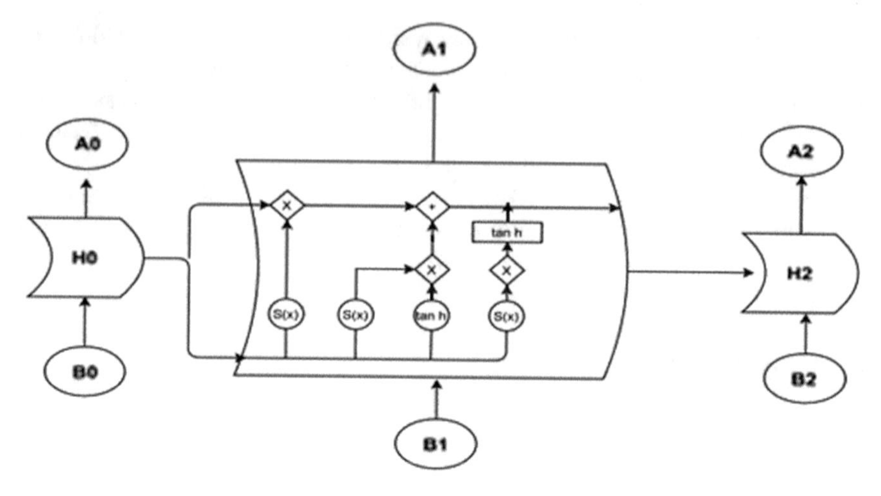

Fig. 1 Long short-term memory (LSTM) model

The purpose of Bi-LSTMs is to have the output come from the previous and future elements as well as the current one [16].

5.4 Convolution Neural Network (CNN)

The image classification task was performed using CNN. CNN is currently producing top-of-the-line results when it comes to performing various NLP tasks as well [12, 17]. The word embeddings tools (word2vec, glove, etc.) can be used to convert each sentence into a sentencing matrix. Given d word-embedding dimensions, sentence matrix S dimension becomes $d * s$. Then, a CNN can be applied to the sentence. CNN produces a feature map by convoluting a filter over an image. Feature maps will be $(n * f + 1) * (n * f + 1)$ if the image size is $n * n$ and the filter size is $f * f$. The same procedure applies to convolution with text: a filter of w * IRhd is applied to the window of h words to produce a new feature [12, 17] (Fig. 2).

6 Architectures for Deep Learning at ABSA

This section summarizes the various applications of deep learning in sub-tasks of ABSA.

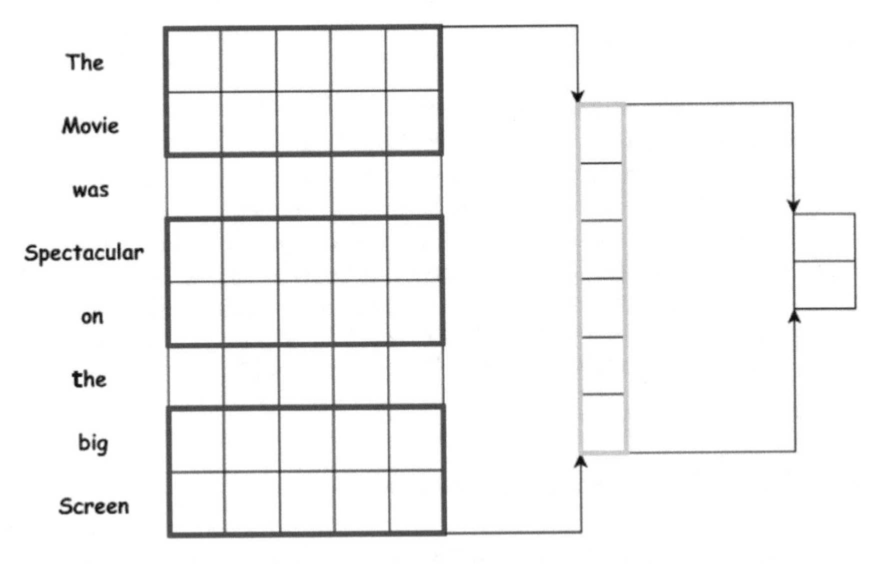

Fig. 2 Convolutional neural network architecture on text

6.1 Sentiment Analysis

A convolutional neural network-based sentence-level sentiment analysis is done by [18]. The author records notable improvement when categorizing the sentences into different groups. While CNN architecture is used for its model, [4] uses basic sentiment classification. In CNN, the operation is done by moving a filter within a window of size h on the words. In sentences with uni-sentiment, this approach is successful.

6.2 Extraction of Aspect Terms

Character-level features were used in this model word-level features are incorporated in the form of embeddings while predicting aspect terms. A similar architecture could be used to extract aspect terms by using the Bi-LSTM-CRF. Al-Smadi et al. [18] used a model based on bidirectional LSTM. Extracting term aspects from reviews with CRF (Bi-LSTM-CRF). The sentences are below. Initialized characters and words randomly an LSTM layer received the inputs. An indirect feature vector was derived by combining the inputs of the word embeddings provided by fastText and a character embedding for each token in a sentence. The resultant vector is fed as input to the main algorithm. For the CRF feature extraction, a LSTM bi-layer is used.

6.3 Aspect Category Extraction

Using a two-layer neural network, we classified the aspect categories, and this network had a hidden layer dedicated to sharing. The second neural networks were trained for only one category (different neural networks were trained), the categories are as follows [4]. The probability distribution for the word vectors and their outputs were derived. The inputs were generated from the average of the distribution. As an entity and attribute pair (E#A pair), aspects refer to the entity and attribute pairs. Each pair has its own name. Aspect-specific and shared features are learned by the neural network during training. Two layers were used. The detection of aspects (aspect categories) is handled by a fully connected neural network [7].

6.4 Aspect-Based Sentiment Detection

In aspect-based sentiment analysis (ABSA), text data is categorized by aspects and emotions are assigned to every one of them. ABSA can be used for customer feedback analysis by assigning a sentiment to all aspect of the product or service.

The elements or characteristics of a product or a service like 'user experience with interface', 'user complaint on delivery or response time' or 'new product review' are referred to as 'aspects' of the product or service.

Aspect-based sentiment analysis reveals:

- Positive or negative opinions about a particular aspect.
- There are several aspects: the feature, the category, or the topic discussed (Table 2).

7 Conclusion

Here are a few introductions to the different types of deep learning in ABSA. Various ABSA tasks and the methodology used for them are discussed in this chapter. A common method used to label sequences, such as aspect-term extraction, is deep learning models based on Bi-LSTM-CRF. In terms of sentiment classification, vanilla neural networks and CNN show outstanding results. Exploiting the relationship between this aspect and the opinion of the aspect, and the word in context is useful when detecting aspect sentiment. Many diverse approaches were used to extract the relationships and were fed into the deep neural networks. The aspect detection and sentiment analysis tasks are also a pair of tasks that can be combined to perform together. It is a great starting point to dive into ABSA using a deep learning approach.

Table 2 Summary of state-of-the-art ABSA methods

References	Dataset	Domain	Embeddings	Performance	Method
[19]	SemEval-16 Arabic	Hotel	Word2vec + FastText	F1-score: 69.98 Accuracy: 82.70	Bi-LSTM-CRF LSTM + Attention
[18]	SemEval-16 Task-5	Restaurant	Word2vec	F1-score: 72.44	Bi-LSTM-CRF
[20]	SemEval-14 Task-4	Restaurant Laptop	FastText	F1-score: 84.12 F1-score: 77.96	2layer Bi-LSTM-CRF 2layer Bi-LSTM-CRF
[7]	SemEval-14 Task-4	Restaurant	–	F1-score: 90.10	2-layer NN + Logistic R
[21]	Dong et al. 2014 [22]	Twitter	SSWE	F1-score: 69.00 F1-score: 69.50	TD-LSTM TC-LSTM
[13]	SemEval-14 Task-4	Restaurant Laptop Restaurant	Glove	Accuracy: 76.60 Accuracy: 77.20 Accuracy: 68.90 Accuracy: 68.70 Accuracy: 82.50 Accuracy: 83.10 Accuracy: 84.10	AE-LSTM AT-LSTM AE-LSTM AT-LSTM AE-LSTM ATAE-LSTM ATAE-LSTM
[22]	SemEval-14	Restaurant Laptop	Glove	Accuracy: 81.16 Accuracy: 74.12	Bi-GRU Attention
[23]	SemEval-14 Task-4 SemEval-14 Task-4 SemEval-15 Task-12 SemEval-16 Task-5	Restaurant Laptop Restaurant Restaurant	Glove	Accuracy: 80.36 Accuracy: 71.94 Accuracy: 81.67 Accuracy: 84.64	LSTM + Attention
[4]	SemEval-15 Task-12	Laptop	Word2vec	F1-score: 51.30 Accuracy: 78.30	2-layer NN Parser + CNN

References

1. Feng J, Cai S, Ma X (2018) Enhanced sentiment labeling and implicit aspect identification by integration of deep convolution neural network and sequential algorithm. Cluster Comput 1–19
2. Liu B (2012) Sentiment analysis and opinion mining. Synth Lect Hum Lang Technol 5(1):1–167
3. Ramshaw LA, Marcus MP (1999) Text chunking using transformation-based learning. In: Natural language processing using very large corpora. Springer, Berlin, pp 157–176
4. Wang B, Liu M (2015) Deep learning for aspect-based sentiment analysis. Stanford University report
5. Tang D, Qin B, Liu T (2016) Aspect level sentiment classification with deep memory network. arXiv preprint arXiv:1605.08900

6. Li X, Bing L, Li P, Lam W, Yang Z (2018) Aspect term extraction with history attention and selective transformation. arXiv preprint arXiv:1805.00760
7. Zhou X, Wan X, Xiao J (2015) Representation learning for aspect category detection in online reviews. In: Twenty-Ninth AAAI conference on artificial intelligence. AAAI Press, pp 417–424
8. Xu H, Liu B, Shu L, Yu PS (2018) Double embeddings and cnn-based sequence labeling for aspect extraction. arXiv preprint arXiv:1805.04601
9. Mikolov T, Chen K, Corrado G, Dean J (2013) Efficient estimation of word representations in vector space. arXiv preprint arXiv:1301.3781
10. Pennington J, Socher R, Manning CD (2014) Glove: global vectors for word representation. In: Proceedings of the 2014 conference on empirical methods in natural language processing (EMNLP). ACL, pp 1532–1543
11. Güngör O, Güngör T, Üsküdarli S (2019) The effect of morphology in named entity recognition with sequence tagging. Nat Lang Eng 25(1):147–169
12. Poria S, Cambria E, Gelbukh A (2016) Aspect extraction for opinion mining with a deep convolutional neural network. Knowl-Based Syst 108:42–49
13. Wang Y, Huang M, Zhao L et al (2016) Attention-based LSTM for aspect-level sentiment classification. In: Proceedings of the 2016 conference on empirical methods in natural language processing. pp 606–615
14. Mikolov T, Sutskever I, Chen K, Corrado GS, Dean J (2013) Distributed representations of words and phrases and their compositionality. Advances in neural information processing systems. pp 3111–3119
15. Hochreiter S, Schmidhuber J (1997) Long short-term memory. Neural Comput 9(8):1735–1780
16. Schuster M, Paliwal KK (1997) Bidirectional recurrent neural networks. IEEE Trans Signal Proc 45(11):2673–2681
17. Zhang Y, Wallace B (2015) A sensitivity analysis of (and practitioners' guide to) convolutional neural networks for sentence classification. arXiv preprint arXiv:1510.03820
18. Chen T, Ruifeng X, He Y, Wang X (2017) Improving sentiment analysis via sentence type classification using BiLSTM-CRF and CNN. Expert Syst Appl 72:221–230
19. Al-Smadi M, Talafha B, Al-Ayyoub M, Jararweh Y (2018) Using long short-term memory deep neural networks for aspect-based sentiment analysis of Arabic reviews. Int J Mach Learn Cybern 1–13
20. Giannakopoulos A, Musat C, Hossmann A, Baeriswyl M (2017) Unsupervised aspect term extraction with b-lstm & crf using automatically labelled datasets. arXiv preprint arXiv:1709.05094
21. Tang D, Qin B, Feng X, Liu T (2015) Target-dependent sentiment classification with long short-term memory. CoRR, arXiv:1512.01100
22. Gu S, Zhang L, Hou Y, Song Y (2018) A position-aware bidirectional attention network for aspect-level sentiment analysis. In: Proceedings of the 27th international conference on computational linguistics. pp 774–784
23. He R, Lee WS, Ng HT, Dahlmeier D (2018) Effective attention modeling for aspect-level sentiment classification. In: Proceedings of the 27th international conference on computational linguistics. pp 1121–1131

Vaccine Hesitancy to Vaccine Hope: Comparison of MR Vaccine and COVID Vaccine Trends in India

Jayan Vasudevan and Sreejith Alathur

Abstract Social media played a major role during the distress in the era of Web 3.0 technologies. The use of social media for relief and rescue operations is common nowadays. But the Web 3.0 and its technologies have made the situation worse sometimes, especially in the healthcare sector. The measles-rubella (MR) vaccine campaign in India had a huge setback due to the social media. The World Health Organization (WHO) has observed that the misinformation in social media is one of the ten reasons for vaccine hesitancy. In the COVID-19 situation, the misinformation has increased tremendously but at the same time people were expecting a vaccine. The vaccine hesitancy depends on the severity of the cause of the disease and the cure of the disease. There were trends of vaccination hesitancy during the MR vaccination campaign and that has changed to vaccine hope during the COVID-19 in the social media.

Keywords Vaccine hesitancy · Vaccine hope · Misinformation · Agenda setting theory · Measles-rubella vaccine

1 Introduction

The literature shows that social media and vaccination hesitancy have close relationships. The engagements for the articles with negative sentiments toward the vaccine show that any propaganda related to the anti-vaccination campaign can get much attention. This influences the psychological behavior of an individual. The classification of articles shared in the social media is based on the 5C psychological antecedents of vaccination. The 5C antecedents are confidence, constraints, complacency, calculation, and collective responsibility [1].

J. Vasudevan (✉)
Centre for Development of Advanced Computing (C-DAC), Trivandrum, Kerala 695582, India
e-mail: jayan@cdac.in

S. Alathur
National Institute Technology Karnataka, Surathkal, Karnataka, India

A. Kumar et al. (eds.), *Proceedings of the International Conference on Cognitive and Intelligent Computing*, Cognitive Science and Technology,
https://doi.org/10.1007/978-981-19-2350-0_77

The situation of MR vaccination and COVID-19 vaccination is different. There were enough studies and results available with the MR vaccine. On the contrary, the corona disease emerged with a panic and stress among the people globally. There was no vaccine, and people were really wanting for a cure due to the severity of the disease. The hope and hype about the new vaccine in the media are the major engagements in social media [2]. The drugs like hydroxychloroquine and remdesivir were advocated based on some contradictory and inconclusive studies [3]. The non-availability of the efficacy report among the public is the main concern, and it may result in refusal of COVID vaccines [4]. Many healthcare professionals themselves raised concerns over some vaccines developed in India.

There are many factors behind vaccine hesitancy across the globe ranging from religious sentiments to psychological factors [5]. The emergence of Web 3.0 technologies enabled the anti-vaccine campaign easier and faster. The SNS are the main source of misinformation that is affecting the public health programs. There is a significant contribution on misinformation in the social media leading to vaccine hesitancy [6]. The rumors and misinformation about the corona vaccine have already started circulating in the SNS, and it is the collective responsibility of all individuals to tackle such a situation [7].

Misinformation is getting more attention than the factual information due to the false narrative in the content and its presentation [8]. Misinformation in social media has become a great concern in the healthcare domain globally [9]. The scientifically valid health information should be made available with the support of social media platforms for the successful healthcare programs [10]. Caregivers seek information from the social media, and if they are not having adequate skills and expertise in the caregiving added with the low health literacy, it will lead to the improper usage of medication. The impact of the misinformation will be severe in that case [11–13].

Hope is having multiple definitions like "pleasure of anticipation," "self-soothing response to anxiety," "appetite with an opinion of obtaining," "vulnerable to fear and superstition," and so on [14]. A progressive disease will make the people despair especially when there is no cure available as on date. So this despair out of the deadly disease like COVID will generate hope for a cure. Miceli and Castelfranchi [15] define "hope as an anticipatory emotion based on three components: the assumption that the occurrence of an event is possible; the desire for the event to occur; and the assumption that one has no or only little influence on the occurrence."

2 Methods

The study was mainly focused on the assessment of the engagement of the news shared in the social media during the MR vaccination campaign and the COVID-19 situation. The Web site https://app.buzzsumo.com was used for retrieving the news articles. The news articles are then classified manually according to the variables identified. The application retrieved 185 articles using the keywords "MR Vaccination" and "measles-rubella". The duplicate results were removed, and finally, we got 179 articles for the analysis. At least 5 engagements were considered for the shortlisting of the articles. The more engagements mean more reach in the social media platform and more impact on the user. Facebook and Twitter are having more engagements compared to other Social Network Sites (SNS).

A similar assessment was conducted during the COVID-19 disease period. The keyword used was "vaccin." The geographic location selected was India for our study. The SNS engagement was considered during the period February 14, 2020 and February 12, 2021. There were 61,474 articles extracted during this period and a total engagement of 13.8 million. More engagement was seen on Facebook. Twitter is not much used by the common people due to its framework. Facebook is more common in India due to the functionalities which suit the common man as it is not having restrictions in text length and sharing images.

3 Results

The articles belonging to General Information and Awareness were shared more during the MR vaccination campaign in social media. Next to that, more cases are reported on Adverse Events Following Immunization (AEFI) and have more articles shared in social media. When considering social media engagement and articles shared, social engagement and the number of articles are not related much. The accessibility has two news links but the social engagement is about 19%. Articles related with the religious aspects are 9% but the social media engagement is 5%. Social media engagement is more for YouTube. Figure 1a shows the share of articles based on the engagements as per the classification. Figure 1b shows the share of articles related with the corona vaccine. The hope is having more engagements which amount to nearly 83%. The General Information (GI) share stood at the second most engagements with a contribution of 4.53%. Though the percentage of the engagement is very low in the case of corona vaccine, the absolute number is much higher than MR vaccine engagements.

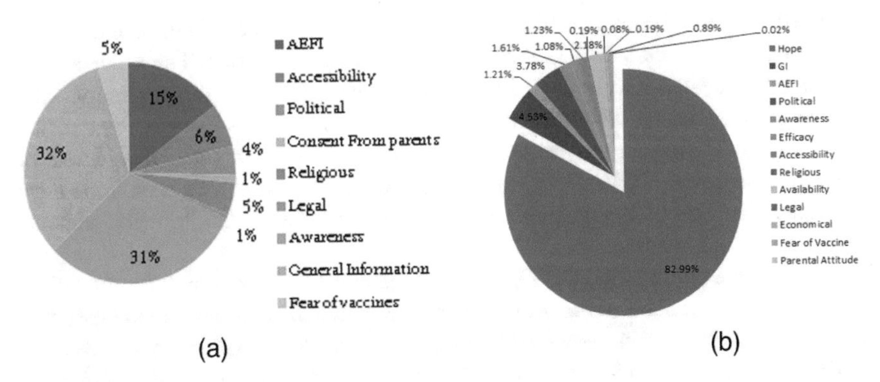

Fig. 1 The article shared in social media and their categories: **a** MR vaccine **b** COVID vaccine

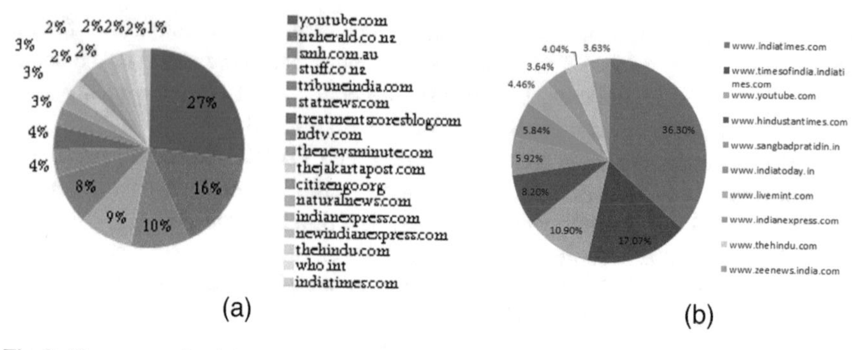

Fig. 2 The source of articles shared in social media: **a** MR vaccine **b** COVID vaccine

Only the top 17 sources of articles based on social media engagements are considered for comparison. 27% share social media engagement containing the video contents. There were 86 links that contained articles having more engagements. The sharing of images will have more engagements than the texts. The India Times got the maximum number of engagements which amounts up to 36.30%. In MR vaccine, YouTube was the leading source which was sharing the MR vaccine information. But in the corona vaccine, the YouTube engagement is nearly one by third of the India Times engagement. Here, only the leading 10 sources of information sharing links for the analysis were considered (Fig. 2b).

The articles related with the COVID vaccine keep on increasing right from the beginning of the coronavirus infection. The engagements were also increasing substantially. There are many types of articles shared during this period. As per WHO, the COVID time is the real challenge for the information exchange and information retrieval for the health information processing. A new terminology, Infodemics, has come up during this period to indicate the misinformation. The Infodemics made the health information seeking people more difficult to segregate it from facts and fakes. Figure 3 shows the trends of the number of articles shared on a monthly basis starting from the month of February 2020. There is a dip in the month of February 2021 as the data collection period is up to February 12, 2021.

Apart from the factors considered for the MR vaccine, a new factor is introduced for the corona vaccine which is much relevant during this period. The articles were continuously dealt with the hope for a new vaccine right from the beginning. The articles on hope during this period came out of despair, and people are really worried out of the uncertainty in the progress of the disease. Figure 4 is the engagement of articles related with COVID. The articles shared in the social media on hope have the highest number of engagements. There were more than 10 million engagements on hope in social media.

In the initial phase of the COVID, the engagements were very high for the article related with hope. The major trends were the availability of new medicines and the expected outcome in the vaccine research. Millions of engagements were there during the initial period for the articles related with hope, and the engagement had a

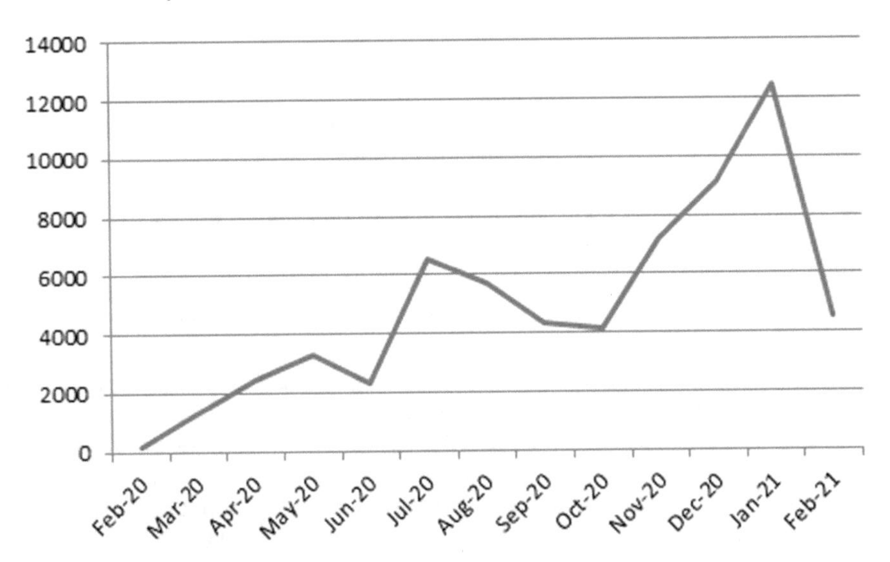

Fig. 3 Total engagement of articles related with vaccine during COVID

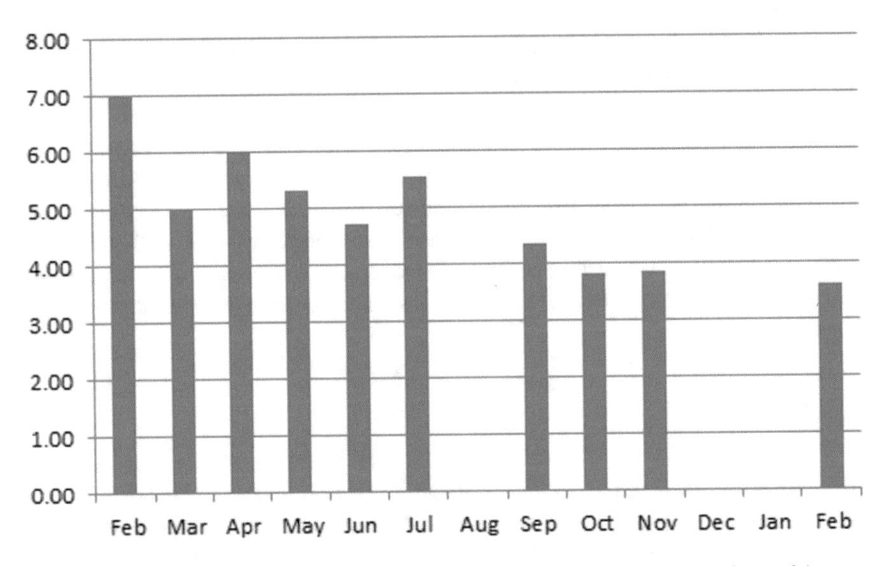

Fig. 4 The engagement of articles related with hope during coronavirus disease (log scale)

decreasing tendency in the subsequent months. The confirmation of the availability of new vaccines and the news about the clinical trials were given some confidence in the social media user. On the other hand, the article count increased in the subsequent months.

4 Discussion

The vaccine hesitancy and the vaccine hope are the psychological intent of an individual. Depending on the necessity and the situation, that can be either hesitancy or hope. The MR vaccine got many negative feedbacks in the form of misinformation in social media, and people were reluctant to take the vaccination. Measles and rubella were not a common term among India, and people were not aware about the disease though it took the lives of many children in India. At the same time, coronavirus was a common term all over the world and created a fear among the people globally. That was the main reason behind the vaccine hope.

People are following Twitter and other SNS to get awareness about the vaccines. During the MR vaccine campaign, Facebook and Twitter got 72,326 engagements from 179 articles. This data was extracted using the Web site https://app.buzzsumo.com. The engagement in Facebook is more compared to Twitter. Similarly during the corona infection, the engagements of vaccine-related articles were extracted. There were more engagements during this time due to the deadliness of the disease and the non-availability of the cure for the disease.

The motivation behind the vaccine hesitancy is initiated by many factors, and at the same time, vaccination hope is arising out of the despair to get the cure for a disease in this situation. The fear of the deadly disease like COVID-19 and the uncertainty in the situation made the people to hope for the new vaccine.

The situation of the vaccine hate and denial of vaccine during the MR vaccination campaign was turned into vaccination hope during the COVID situation. The uncertainty in the cure for the coronavirus disease, the restrictions in movement, and several economic and social restrictions made the situation worse. The lockdown and the shutdown of all the leisure activity centers made the people despair to get something as a cure. The proportion of article engagements concerned with the vaccine denial and the vaccine hesitancy were very less in the case of corona vaccine.

5 Conclusion

The main observation of the study is the healthcare program can be made effective if the stakeholders give adequate awareness to the beneficiaries. The awareness about the MR vaccine was limited, and the public was not educated with the importance of getting their children vaccinated. Anti-vaccine campaigners used this opportunity and were able to misguide a large number of populations to sometime. Government was forced to extend the campaign, and the people vaccinated their children in the extended schedule. Parents should be informed about the seriousness of the Measles Rullella and the government should share the actual scenario of the death due to the same. So the resistance against vaccination will get suppressed. Any healthcare program needs a collaborative effort to implement it successfully. At the same time, the public also needs to know the importance and seriousness of the same. The

COVID situation was an eye-opening case where people voluntarily looked for a solution. They understood the seriousness of the situation. So the vaccine hope comes in place instead of vaccine hesitancy.

References

1. Betsch C, Schmid P, Heinemeier D, Korn L, Holtmann C, Böhm R (2018) Beyond confidence: Development of a measure assessing the 5C psychological antecedents of vaccination. PLoS ONE 13(12):e0208601. https://doi.org/10.1371/journal.pone.0208601
2. Forni G, Mantovani A (2021) On behalf of the COVID-19 commission of Accademia Nazionale dei Lincei, Rome et al COVID-19 vaccines: where we stand and challenges ahead. Cell Death Differ 28:626–639. https://doi.org/10.1038/s41418-020-00720-9
3. Raoult D, Hsueh PR, Stefani S, Rolain JM (2020) COVID-19 therapeutic and prevention. Int J Antimicrob Agents 105937
4. Calnan M, Douglass T (2020) Hopes, hesitancy and the risky business of vaccine development. Health Risk Soc 22(5–6):291–304. https://doi.org/10.1080/13698575.2020.1846687
5. Salmon DA, Dudley MZ, Glanz JM, Omer SB (2015) Vaccine hesitancy: causes, consequences, and a call to action. Vaccine 33(Suppl 4):D66–D71. https://doi.org/10.1016/j.vaccine.2015.09.035
6. Wilson SL, Wiysonge C (2020) Social media and vaccine hesitancy. BMJ Glob Health 5(10):e004206. https://doi.org/10.1136/bmjgh-2020-004206
7. Donovan J (2020) Social-media companies must flatten the curve of misinformation. Nature. https://doi.org/10.1038/d41586-020-01107-z
8. Wang Y, McKee M, Torbica A, Stuckler D (2019) Systematic literature review on the spread of health-related misinformation on social media. Soc Sci Med 240:112552
9. Larson HJ (2018) The biggest pandemic risk? Viral misinformation. Nature 562:309
10. Hill JA, Agewall S, Baranchuk A et al (2019) Medical misinformation: vet the message! Circulation 139:571–572
11. Arnett JJ (2003) Conceptions of the transition to adulthood among emerging adults in American ethnic groups. New Dir Child Adolesc Dev 100:63–75
12. Rikard RV, Thompson MS, McKinney J, Beauchamp A (2016) Examining health literacy disparities in the United States: a third look at the national assessment of adult literacy (NAAL). BMC Public Health 16:975
13. Bangerter LR, Griffin J, Harden K, Rutten LJ (2019) Health information seeking behaviors of family caregivers: analysis of the health information national trends survey. JMIR Aging. 2:e11237
14. Kube T, Blease C, Ballou SK, Kaptchuk TJ (2019) Hope in medicine: applying multidisciplinary insights. Perspect Biol Med 62(4):591–616. https://doi.org/10.1353/pbm.2019.0035
15. Miceli M, Castelfranchi C (2010) Hope: the power of wish and possibility. Theor Psychol 20(2):251–276

IoT-Based Home Intrusion Detection System for Enhanced Smart Environment

Aman Shah, Rukmani Panjanathan, Graceline Jasmine, C. Rajathi, Modigari Narendra, and Benson Edwin Raj

Abstract An interconnected collection of smart devices which is expanding at a very rapid pace as many devices and sensors are being added to the inter or getting connected to it is referred to as Internet of Things or IoT for short. One of the most widely used part of IoT is integrating it in home security. The IoT increases the ability to control and monitor all the operations occur at home. This work aims at using the IoT features in creating an inexpensive home security system which tracks the digital footprint of all the visitors. Home security is an essential part of today's world, especially with all the technological advancements. Traditional systems detect the intrusion but not the intruder. The systems can be easily fooled by using physical props. Objective of this work is to enhance the current security system to detect the identity of the intruders digitally which can later be used by authorities to trace them easily.

Keywords Intrusion detection · Smart environment · IoT · Security

A. Shah · R. Panjanathan (✉) · G. Jasmine · C. Rajathi
School of Computer Science and Engineering, Vellore Institute of Technology, Chennai, India
e-mail: rukmani.p@vit.ac.in

A. Shah
e-mail: aman.shah2017@vitstudent.ac.in

G. Jasmine
e-mail: graceline.jasmine@vit.ac.in

C. Rajathi
e-mail: rajathi.c2019@vitstudent.ac.in

M. Narendra
Department of Computer Science and Engineering, Vignan Deemed to Be University, Andhra Pradesh, India

B. E. Raj
Higher Colleges of Technology, Fujairah Women's Campus, Fujairah, UAE
e-mail: braj@hct.ac.ae

A. Kumar et al. (eds.), *Proceedings of the International Conference on Cognitive and Intelligent Computing*, Cognitive Science and Technology,
https://doi.org/10.1007/978-981-19-2350-0_78

1 Introduction

A home security system device protects both the expensive things and people's personal safety. Given the fact that IoT allows us to gather data from various types of sources including appliances, cars, humans, pets, etc., it is described as "Infrastructure of Information Society." In the world, all the objects which equipped with a physical IP, which allow exchange of information, can be able to be part of IoT system. IoT connections use the World Wide Web to allow items to communicate with one another using their logic and circuits; it is not the same as the Internet. In past few years, the number of gadgets being connected to World Wide Web has seen a tremendous boom. All these gadgets integrated with the Internet are associated with IoT and its infrastructure. This allows exchange of useful data and information between one another, and this is exactly why it is useful to use such preexisting hardware infrastructure which is our advantage for the proposed security system.

Utilizing Wi-Fi module as a test sniffer to gather all virtual impressions close by and classified into individual and non-individual. Individual gains admittance to house and others not. The logs of the occasion are dispatched to the proprietor through e-mail, and it is caught utilizing PyShark. JSON is used to impart among module and the server.

Wi-Fi module works by sniffing the incoming network packets to get the MAC address of devices in its vicinity. Face detection algorithm with OpenCV is used to distinguish faces of known family members quickly and accurately from possible intruders or outsiders. The details will be stored for future reference. The dual nature of the system allows us to first get the MAC address of the intruders even when their Wi-Fi [1] in the phones is off and even if they are covering their faces.

2 Related Work

IEEE 802.11 standard defines Wi-Fi probe requests as an active mechanism with which cellular gadgets can request data from access points and speed up the Wi-Fi connection process [2]. Researcher recognized the privacy risk which the previous label access point is not limited to the previous person. Wireless Sensors Networks (WSNs) with Internet of Things (IoT) and increasing home security ideas and solutions and also packages have been addressed. The researchers show that the concept of exploiting Wi-Fi probe request in which smartphones sends de-anonymize the starting place of the user [3]. Home Area Network [4] offers the function to continue connection for communication and data transfer. In [5], a method is recommended for estimating the range of cellular gadgets present at a positive vicinity and time through Wi-Fi probe. The study [6, 7] enforces real-time surveillance, tracking development of the house protection based on ZigBee era and GSM/GPRS network [3, 7]. Smart telephones have emerged as an essential part of our everyday lives because of their competencies of having access to the net in the use of Wi-Fi and cellular record networks. These Wi-Fi devices continuously give out data packets known as probe requests, which may be traced in the use of Wi-Fi sniffers. In this thesis, first we investigate taking pictures of Wi-Fi probe request packets in the use of the assist of Wi-Fi Pineapple gadgets and examine how we will use sign electricity statistics of probe request records for indoor occupancy tracking by PyShark [1]. To enforce actual house protection, the shrewd faraway tracking machine turned into advanced for domestic protection primarily based totally on "ZigBee" era and GSM/GPRS community [6]. Wi-Fi probe requests present a unique privacy risk which researchers have recognized; this includes and is not limited to leaking a person's previous access point labels and their movements. Even though this shows us the privacy risks of using Wi-Fi probe requests, it is still in widespread use and other alternative technologies have not been used over this. In this work, the author quantifies Wi-Fi probe requests' threat to privacy by conducting an experimental study of the most popular smartphones in different settings.

3 Proposed System

The traditional tracking system is slow and difficult to get accuracy. The proposed system provides the advancement by using Wi-Fi module. Wi-Fi module tracks the footprints of the intruders by the devices. The Wi-Fi module present inside the house captures the probe requests sent by the devices to connect to any wireless network. All the MAC addresses of the members living in the house are stored in a list. The non-members like guests or intruders' MAC addresses will be identified and emailed to the owner so that he can keep a track of the people entering his house. Wi-Fi module is used to sniff the Wi-Fi probe requests from the visitor. Our system will identify the members and non-members from the retrieved MAC addresses. After differentiating them, the list will be shared via JSON and will be updated accordingly in the UI.

3.1 Architecture of the Home Intrusion Detection System for Enhanced Smart Environment

Wi-Fi module is utilized as a probe sniffer which collects the virtual footprints of all the close by gadgets and sends them to the station to get processed which categorizes them as individuals or non-individuals of the house. The individuals are granted get entry to get into the house, while the non-individuals are not. The log of these types of events is dispatched to the proprietor through e-mail. Alongside the Wi-Fi module which works by sniffing the incoming network packets to get the MAC address of devices in its vicinity, the system also utilizes a fast and efficient face detection algorithm with OpenCV using a model shipped with the library in order to detect faces using deep learning. It allows to quickly and accurately distinguish faces of

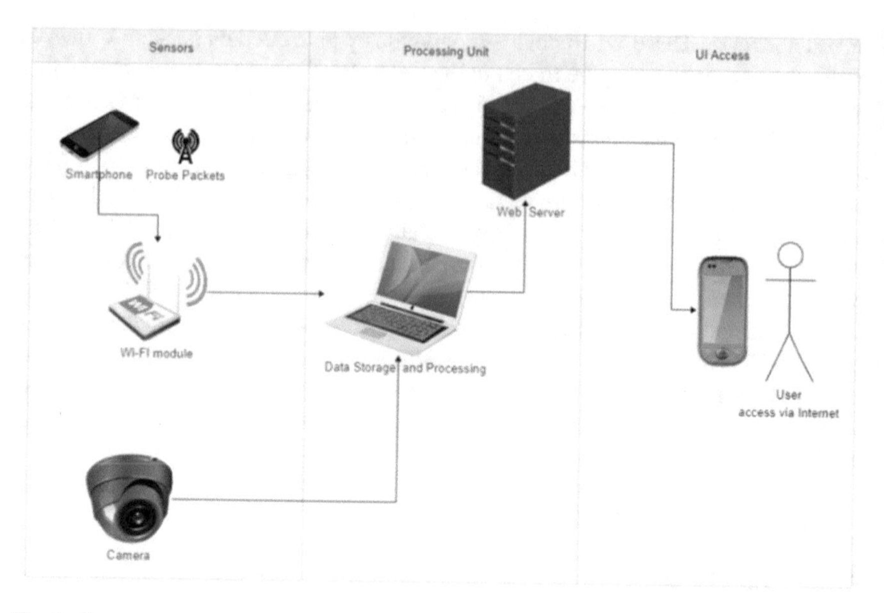

Fig. 1 System overview

known family members from possible intruders or outsiders. The system would then store the faces of all unknown individuals in a database for future use in order to help identify and apprehend the intruders. Wi-Fi module is used to sniff the Wi-Fi probe requests from the visitor. Our system will identify the members and non-members from the retrieved MAC addresses. After differentiating them, the list will be shared via JSON and will be updated accordingly in the UI. The overall flow is detailed in Fig. 1.

If the intruder is not carrying a smartphone, the system also has a video surveillance module with facial recognition model using OpenFace deep neural network to accurately recognize or identify intruders from family members. This dual nature of our system allows to detect any possible intrusions in the house in an intuitive but not intrusive manner. The system can be further upgraded to include additional modules like adding additional sensors like motion detectors, glass-break sensors, infrared sensors, etc. on an Arduino Uno microcontroller or a Raspberry Pi controller to allow installation of multiple sensors and to trigger an alarm or a notification using wireless connection via Wi-Fi or local Internet (Fig. 2).

4 Experimental Results

The MAC address is detected to sniff the packets detected to be intruders. The possible intruders are detected and are stored in the database. The webpage is rendered to find

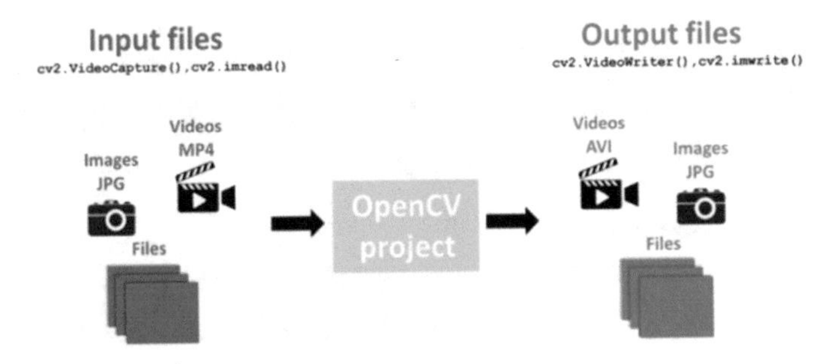

Fig. 2 Workflow for identifying intruders

the members and non-members using the MAC address. The output of the MAC detection is shown in Fig. 3.

Each image is iterated over and over in the training dataset to extract the face from the image using openCV face detection model and localize the faces in the training dataset. A blob is created for the face from the image and then passes it through the embedding model to create a 128-d quantification for it and store the embedding and name of the person to a list. The serialised face detector and the face embedding model are loaded in order to recognise and produce facial embedding for faces that appear in the video frames that the camera is recording. Create the webcam video stream and extract each frame for facial recognition. The recognized face and the accuracy are shown in the Fig. 4a. If an unknown face is seen by the model which does not match the faces listed in the training dataset, it saves it in memory as a possible intruder along with the date and time for future identification and is shown in Figs. 4b and 5.

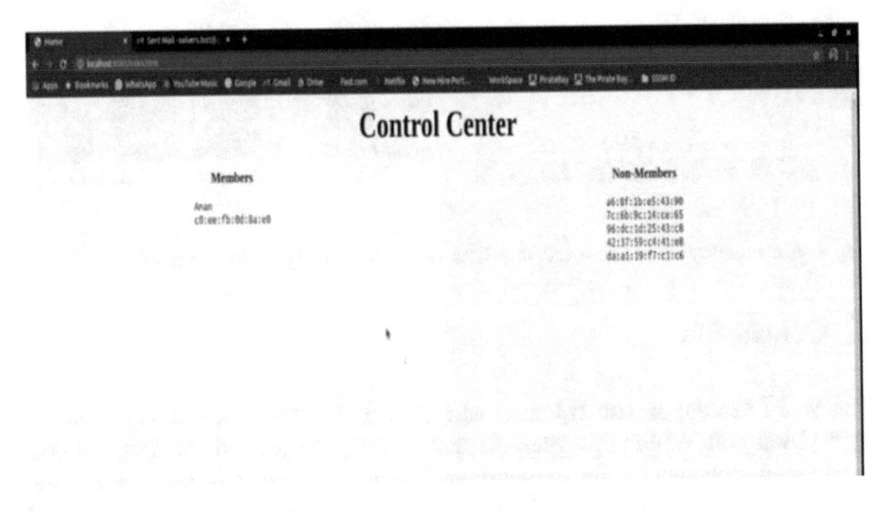

Fig. 3 Output of MAC detection

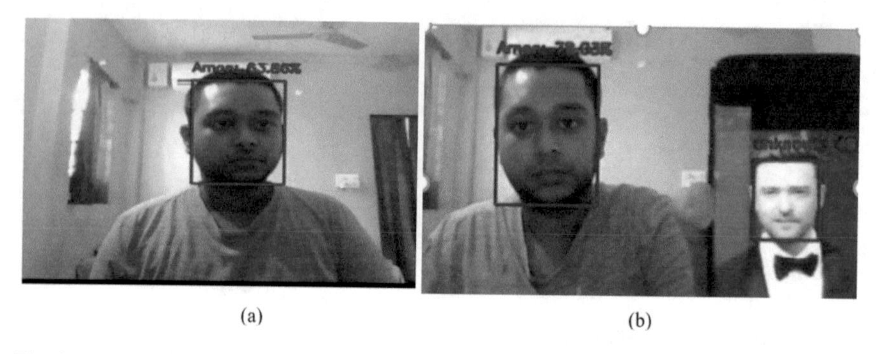

(a) (b)

Fig. 4 **a, b** Recognized face and accuracy

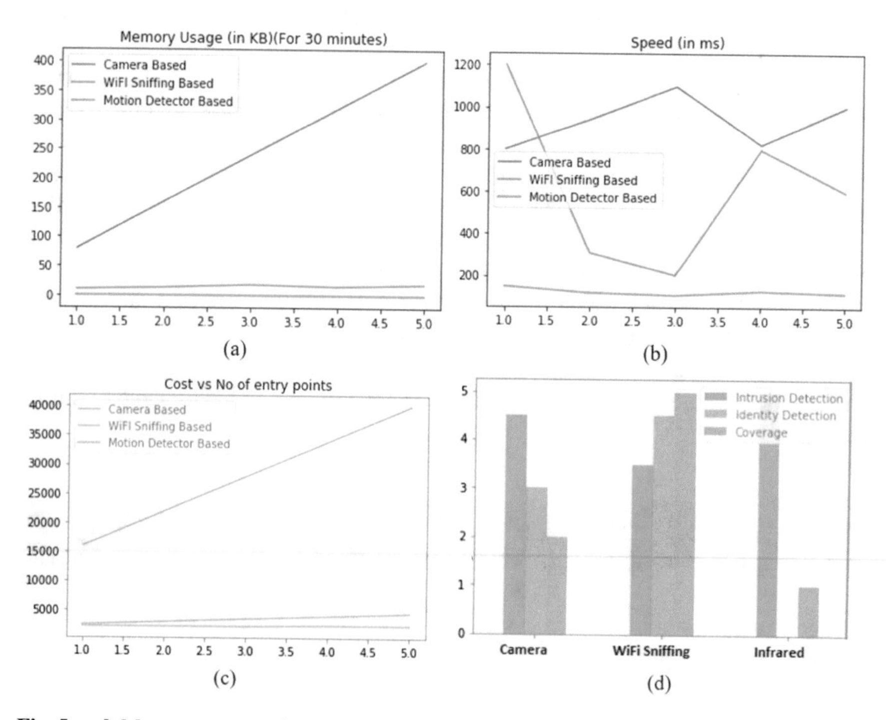

(a) (b) (c) (d)

Fig. 5 **a, b** Memory usage and speed, **c, d**. Cost versus No of entry points and coverage

5 Conclusion

The Wi-Fi module present at homes will identify the members and non-members by capturing the Wi-Fi probe requests sent by their devices and checking with the database. By tracking the digital footprint of the people visiting the house, it is easier to track the identity of the intruder in case of any criminal activity like theft. The facial recognition model is working as intended such that it detects and recognizes faces in indoor settings efficiently and stores any unidentified faces in its database. The video feed from the cameras can also be stored but currently we are storing only images for efficient storage management.

The created framework additionally can be used in business and business programs which incorporate workplaces, stockrooms, and various areas wherein a couple of locales are held for legitimate representatives just or various areas wherein insurance and safety measures are of number one concerns which incorporate net server room of a huge MNC from wherein organization records might be taken.

References

1. Anbarasi J, Mala A (2015) Verifiable multi secret sharing scheme for 3D models. Int Arab J Inf Technol (IAJIT) 12
2. Jadhav L, Pai V (2018) Smart home automation and security using internet of things
3. Ketchen DJ, Shook CL (2006) The application of cluster analysis in strategic management research: an analysis and critique. Strateg Manag J 17(6):441–458
4. Basem B, Ghalwash AZ, Sadek RA (2015) Multilayer secured SIP based VoIP architecture. Int J Comput Theory Eng 7(6):453
5. Ding C, Tao D (2017) Pose-invariant face recognition with homography-based normalization. Pattern Recogn 66:144–152
6. Julien F, Raya M, Felegyhazi M, Papadimitratos P (2007) Mixzones for location privacy in vehicular networks. In: Association for computing machinery (ACM) workshop on wireless networking for intelligent transportation systems (WiN-ITS)
7. Kerns AJ, Shepard DP, Bhatti JA, Humphreys TE (2014) Unmanned aircraft capture and control via GPS spoofing. J Field Robot 31(4):617–636

SMART—Stockpile Management with Analytical Regulation Technology

L. Maria Michael Visuwasam, K. Dhinakaran, G. Kalpana, Ashwin Balakrishna, V. Kowsalyaa, and S. R. Nikitha Keerthana

Abstract Stockpile Management with Analytical Regulation Technology (SMART) is an advanced inventory management system, specifically curated for warehouse management, large retail stores, and the export/import industry. This system is developed using Pega which is a business process management (BPM) platform and also supports customer relationship management (CRM). This keeps track of the current stock details of the products, checks the quantity, automatically generates bills, and will calculate the number of items in the stock. When the stock quantity crosses the minimum threshold value or when the stock has reached its expiry date, an automatic alert will indicate the warehouse to restock the goods. The data pages (DP) contain all the information required for the working of the application and also use data transform (DT) for accessing the fields across the case types (CT). A predictive system will generate reports based on the export quantities in a particular area and the sale of each product in the past, present, and future. Finally, the system will send an automated e-mail containing the business intelligence (BI) report which would contain insights on possible future merchandising for better returns.

Keywords SMART · Inventory · Management · Stock · Threshold · Analysis · Prediction · Expiry · Reports · BI · PCA · PEGA · BPM · CRM · DP · DT · CT

L. M. M. Visuwasam (✉)
Department of Artificial Intelligence and Data Science, R.M.K. College of Engineering and Technology, Chennai, India
e-mail: mariamichaelads@ritchennai.edu.in

K. Dhinakaran
Department of Artificial Intelligence and Data Science, Dhanalakshmi College of Engineering, Chennai, India

G. Kalpana
Department of Computer Science and Engineering, Rajalakshmi Institute of Technology, Chennai, India

A. Balakrishna · V. Kowsalyaa · S. R. Nikitha Keerthana
Department of Computer Science and Engineering, Rajalakshmi Institute of Technology, Chennai, Tamil Nadu, India

A. Kumar et al. (eds.), *Proceedings of the International Conference on Cognitive and Intelligent Computing*, Cognitive Science and Technology,
https://doi.org/10.1007/978-981-19-2350-0_79

1 Introduction

Stockpile management in warehouses is a crucial factor in the production and distribution industry [1, 2]. The safe storage of all the goods is mandatory, which requires continuous monitoring and a standard outlook for resources. Our system, developed using Pega, is a platform for faster access to these resources through data pages and various case types.

These resources are updated constantly, and instant reports are available for download. The generated business intelligence report provides insights on future merchandising and investments for sales and purchases of all the products dispatched from the warehouse [3, 4]. The principal motivation of SMART is to develop a system that advances the features and functionalities of any existing inventory management system(s). This new framework provides a low-cost, accessible, and efficient structure with exclusive features such as an automated alert generation functionality for a stock reorder, a business intelligence report using prediction and data analysis, identification of expiry date, bill generation, and bar code generator and scanner. The utilization of modern technologies such as Pega system, business intelligence and analysis, and DevOps has embellished all existing inventory management systems [5, 6].

Working of the system
Pega Systems
Pega is a platform to build and develop process-oriented, rule-based solutions and applications. The features provided by Pega are business process management, customer relationship management, situational layer cake (SLC), and case management. The development process incorporates stage, process, steps, and rules.

Pega provides an App Studio that acts as a development environment to build applications and a Dev Studio, a web-based application development platform where the developer can work and create solutions remotely. Pega is a recently developed rules engine from Pegasystems. System architects are the developers who build the Pega instances while business analysts gather requirement details and have the option of changing workflow rules during runtime.

Pega is a business process management tool developed on Java and uses OOPS concepts and Java programming. People with a development background get the required inputs and work with business analysts to make changes or add new functionalities if required. The Pega's user interface saves a lot of time and effort.

2 Business Intelligence

Business intelligence is a procedure that is used to examine data and provide actionable instructions which help chief, supervisor, and employees to take knowledgeable business conclusion. Organizations gather data, draw up it for analysis, run queries for these data, and generate visualizations, BI dashboards, and reports to make the outcomes of these analytics, which is made available to business users for making operational decision-making.

The goal of business intelligence is to make business choices that would enable a company to increase its assets, improve operational profitability, and gain competitive improvements over business rivals. To achieve this, BI engulfs an aggregate of analytics, data management, and reporting tool methodologies for managing and analyzing data.

The structure includes data that must be typically stored as a data warehouse built for an entire organization or in the form of small data marts which hold subsets of business data for specific departments and business segments.

The data used would usually include past or previous information and real-time data collected from the node systems from which are generated, enabling BI tools to support both strategic and tactical decision-making processes. To use the data present in BI applications, raw data from various source systems must be integrated, combined, and cleansed using the data integration and data quality management techniques to ensure the data is accurate and consistent.

3 Stockpile Management

Stockpile management system is a software system for tracking commodities levels, sales, orders, and deliveries [7, 8]. All organizations own an inventory. It is also used in the manufacturing industry to create production-related documents. Any enterprise that handles goods will need a system to accurately control and track the stock. Without a stockpile management system, the venture will be working on an entirely offhand basis and firm will quickly run into plight where the market is overstocked or understocked. Stockpile systems lets you know the number of ingredients that an enterprise will need to create or assemble the output. Without this knowledge the firm may end up with excess stock or with insufficient stock to meet customer demand.

4 Proposed Model

It is necessary to resolve each query of the users and provide them the required solution. In many situations, it is difficult to achieve the results. The proposed model overcomes the problem and improves the search results. It retrieves the search results of the users by analyzing the search engine logs containing click through data [9–11].

5 Problem Statement

SMART is into grocery business by running chain of stores and does quite a lot of retail and bulk sales in a given day. The customer is facing challenge in terms of managing the inventory its gets on the daily or regular basis.

On one side, the products arrive from the distributors or direct sellers into the customer ware house, and on the other side, the same product is being stacked in the shelf for the people to buy.

The challenges the customer face are:

- At any point of time what is the stock available in the inventory (i.e., in warehouse), and available in the shelf.
- The edible stocks, that do not move too fast, are tend of get spoiled too frequently.
- The system should give warning 2–3 days in advance when that situation arises.
- Invoices to the suppliers to be managed for every stock that moves in, and reverse invoices for the goods that move out.
- Alert should be made to the administrator, when some stocks go below the threshold.
- To categorize and identify the damaged goods, these factors are considered:
- Whether during transport.
- During shipping.
- By expiry dates.

6 Overview

The application overcomes the challenges faced by the customer and provides an efficient solution by implementing the following features:

- Login to the application using secure credentials.
- User able to enter or upload data pertaining to incoming inventories.
- Rules and category management (RCM), so that the system generates needed alerts. The alerts are generated if the goods are reaching low in inventory and if the goods are getting damaged.
- On daily basis, when goods move from warehouse to the shop, the system has to update the inventory details accordingly.
- Bar-code generation is done for identifying the products accordingly.
- Invoice management and bill generation.
- Reports are generated for business insights.

Figure 1 explains the working process of SMART. The whole flow of the application is simple with very few steps as mentioned.

There will be a bar-code checker role within the shop, where inventory entered by a bar code, would be verified by the checker.

Once approved by checker, the data will not get tampered with.

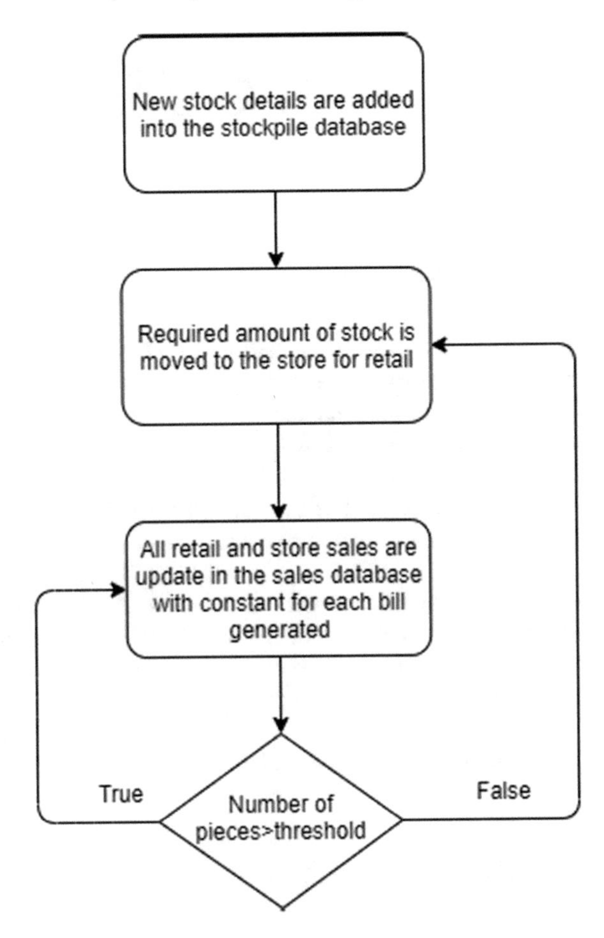

Fig. 1 Representation of the system

Invoice would be generated as PDF files and sent through e-mail to the corresponding distributor/seller.

Alerts are sent as e-mail and SMS to the corresponding distributor/seller.

7 Implementation

The implementation process involves the need to check the required software and hardware for smooth functioning of this system.

Fig. 2 Representation of the input in the form of data provided to the system and output generated as reports after updating the retail stock sale

7.1 Requirements

This system requires the stipulations to get the stock by examining all potential user needs. The analysis on performing the task involves a couple of requirements as follows. Input Data: Input data consists of product details, sales details, staff details, member details, purchase details, etc., which is fed into the stockpile management system. This data is processed for checking the stock, expiry dates, alert generation, and billing purposes. The data is also processed for report generation which is further utilized in business intelligence and analysis.

Output Data: Output data consists of business intelligence reports which is sent to the respective dealers and sellers. E-mails are generated for alerting the dealers on expiry dates and low stock threshold, and bills that are generated for the purchases. The sellers receive e-mails containing business intelligence reports and bills generated on the sales.

Figure 2 explains the flow of stock input data which is processed in updating retail sales and given as output data, i.e., sales report generation.

7.2 Resources Required

It consists of list of hardware and software components needed to develop the product.

Hardware

Processor	Intel Core
RAM	2 GB or above
Hard Disk	8 GB or above

Software
Windows OS
Pega Platform Personal Edition
Java Runtime Environment
PostgreSQL
Apache Tomcat
Python
Seaborn
Pandas
Matplotlib

DESIGN:
The design of the system addresses the aspects of its architecture, functions, and also the modules connected. The design of our proposed model is as follows,

8 System Architecture

System architecture explains the process of the product in detail and also provides a summary of the functions implemented by that product. The architecture is specified in Fig. 4.2.
Create Account
Stock Update
Purchase Order Update
Data Table
Business Intelligence Report

In Fig. 3, the overall diagrammatic view of the stockpile management system is given. The stock is first added by the staff after verifying with the bar-code scanner. The expiry date is updated in the database for alert generation. When the alert is generated before the expiry date, manager proceeds to reorder the stock and take action on the expired products [12, 13]. The staff will also check for damaged goods

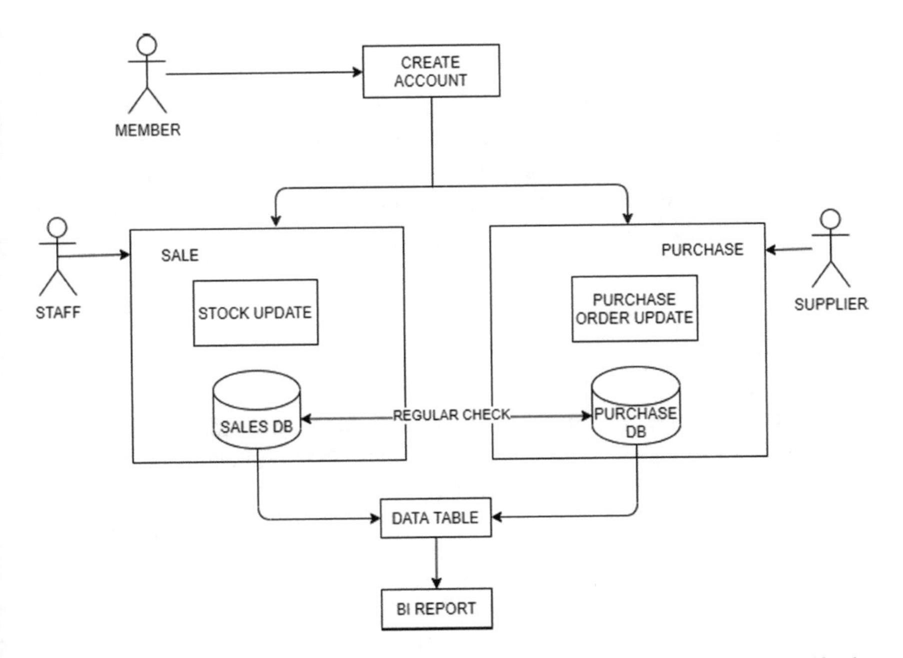

Fig. 3 Architecture of the designed system. The designed system can be generally categorized as stock storage and stock sale update which once action is completed updated in database

which are loaded in the warehouse. If any such goods are found, they immediately intimate the manager [14, 15]. The manager will either request for a return or refund for the product.

After performing the tasks, a final table of data can be viewed. The report generated is used for business intelligence and analysis. The BI report is sent to the sellers and dealers, respectively, for future merchandising and better returns.

The system consists of 5 interfaces and 3 significant features.

9 Interfaces

9.1 *Supplier*

This interface includes the party that is responsible for manufacturing and distributing products to various docks or inventories. The user will receive e-mails from the member (manager of store) regarding the supply they require for the store. Upon receiving these orders, they place the order for the respective stores and make sure the products reach the members store.

9.2 *Member*

This interface includes the party that is responsible for the functioning of the store and is usually the store manager. They are responsible for various activities like placing order for goods into the store and inventory when in demand and also decide on what proceeding decisions have to be taken on items that are nearing expiry date and various other such activities.

9.3 *Staff*

This interface includes the party that is responsible for storing and moving products in and out of the storage unit. They are also responsible for intimating the managers about expiry date arrivals and damaged good quantities.

9.4 *Purchase*

This interface includes the purchase of products from either the supplier to the inventory or from the inventory to the store or shop. This interface can only be accessed by either the member or the staff account. These include auto updating of database based on the purchase made by the store.

9.5 Sales

This interface includes the sales of various goods from the store. These are recorded by the system, and these real-time values are used to update the database automatically along with the bill generated and then these values are used to further carry out various analysis and business insights.

10 Features

10.1 Expiry Date Alert System

This feature helps us keep track of the products in the stockpile and can help reduce loss incurred due to wastage or disposal of products. Usually, there are lots of products in the inventory that get disposed yearly due to poor management of products by not managing goods in a timely manner. This feature lets us know whether any of the products are nearing its expiration date and if the condition states true, then the MEMBER user can decide what has to be done to these products. Either sell them to smaller vendors who sell on a daily basis or give these items away on a sale.

10.2 Damaged Good Regulation

This feature includes returning and requesting refund for damaged goods which either happened while shipping or docking process. This responsibility is usually done by staff users who are involved in the stocking and regulating phase of the goods.

10.3 BI Report Generation

This feature includes the generation of various graph from the previous month sales and gives you a monthly report on the number of products that have been sold, the highest bought items, the product with the highest revenue returns and many other insights to develop your business investment and help gain higher returns [16].

10.4 Working

The process starts with the member (manager) user requesting goods from the supplier. The supplier acknowledges the request and makes necessary arrangements for the doc items to be shipped and loaded into the storage unit. Once every good reaches, it is all scanned and important fields such as expiry date and other product info are logged into the system. The member user then generates a request to the inventory management, about the need and supply of selective goods from the storage unit. Once the product makes it to the store, all purchases made by customers are recorded through every bill generated and copies are sent to the Member user for tally purposes. If any product is nearing its expiration date, the system would generate an alert to the member user asking him how he would like to proceed. The member can choose to whether or not sell these goods to local vendors who sell on a daily basis or to sell those items on a sale for faster regulation of the product. Any damages incurred to these products during shipping or docking or any other stage of the product life span, the products are moved to a separate unit and is intimated to the member user by the staff user. The member user would send a request to replace or refund the product with an e-mail which is pre-defined by the system to be sent to the supplier. The last phase includes the generation of monthly BI report, which provides business insights on the various possible improvements that can be made based on the graphs and charts provided.

11 Result and Inference

Stockpile management is vital for any manufacturing establishment. It helps the company in reducing the cost of managing the warehouse.

The implementation of the stockpile management system must be supported by skilled personnel. Training and hiring qualified staff are necessary. It is also essential to have inputs like demand prediction and other expense rates to be practical to make productive use of the stockpile system.

The system is a solution and an improvement for stockpile management factoring in various aspects of warehouses, stocks, distribution, and purchasing. As demonstrated, the proposed model improves the components of existing systems, each with one prominent feature. This model allows the end-user to access multiple resources in a single application. Once a staff on boarded to the SMART application, they can update the data on new sales and purchases, intimate their managers on the expiry dates and damaged goods. The update is saved in the database to use in business intelligence reports. These reports are insightful on product sales and distribution.

Acknowledgements We would like to thank Mr. L. Maria Michael Visuwasam for guiding and helping us to finish the paper successfully.

References

1. Ray BR, Chowdhury M, Abawajy J (2013) Critical analysis and comparative study of security for networked RFID systems. In: 2013 14th ACIS international conference on software engineering, artificial intelligence, networking and parallel/distributed computing (SNPD). Honolulu, HI, pp 197–202
2. EPCglobe. EPC Radio-frequency identify protocols generation-2 UHF RFID specification for RFID air interface protocol for communication at 860 MHz–960 MHz version 2.0.0 ratified, November 2013. Available online at http://www.gs1.org
3. Boneh D, Lynn B, Shacham H (2001) Short signature from the weil pairing. Proc Asia Crypt, LNCS 2248:514–532
4. Elkiyaoui K, Blass E-O, Molva R (2012) CHECKER: on-site checking in RFID-based supply chains, paper presented to proceedings of the fifth ACM conference on security and privacy in wireless and mobile networks. Tucson, Arizona, USA
5. Pasic F, Wohlers B, Dziwok S, Becker M, Heinrich M (2019) A KPI-based condition monitoring system for the beer brewing process. In: 2019 24th IEEE international conference on emerging technologies and factory automation (ETFA). Zaragoza, Spain, pp 1469–14
6. Graham I, Goodall P, Peng Y, Plamer C, West A, Conway P, Mascolo JE, Dettmer FU (2015) Performance measurement and KPIs for remanufacturing. J Remanufacturing 5(10):1–17. ISSN 2210-4690, Springer, Heidelberg
7. ISO-international organization of standardization. Referenced 2015 at http://www.iso.org/
8. Kapoor G, Zhou W, Piramuthu S (2011) Multitag and multi-owner RFID ownership transfer in supply chain. Decis Support Syst 52(1):258–270
9. Samelin K, Pohls HC, Bilzhause A, Posegga J, de Meer H (2012) Redactable signature for independent removal of structure and content. Inf Sec, Pract, Experience, LNCS 7232:17–33
10. Cretu M, Ceclan A, Czumbil L, Şteţ D, Bârgăuan B, Micu DD (2019) Key performance indicators (KPIs) for the evaluation of the demand response in the technical university of Cluj-Napoca buildings. IEEE 8th international conference on modern power systems (MPS). pp 1–4
11. Rahman M, Soshi M, Miyaji A (March 2009) A secure RFID authentication protocol with low communication cost. pp 559564
12. Ouafi K, Vaudenay S (2009) Pathchecker: an RFID application for tracing products in supply-chains, paper presented to information security group (GSI)
13. de Andrade PRM, Sadaoui S (2017) Improving business decision making based on KPI management system. In: Proceedings of IEEE international conference on systems, man, and cybernetics (SMC). pp 1280–1285
14. Yang MH (2012) Secure multiple group ownership transfer protocol for mobile RFID, electronic commerce research and applications
15. Trujillo-Rasua R, Solanas A, Prez-Martnez PA, Domingo-Ferre J (2012) Predictive protocol for the scalable identification of RFID tags through collaborative readers. Comput Ind 63(6):557–573
16. Trujillo-Rasua R, Solanas A (2011) Efficient probabilistic communication protocol for the private identification of RFID tags by means of collaborative readers. Comput Netw 55(15):3211–3223

Implementation of Unusual Human Activity Detection in Warehouse Using SSD

L. Maria Michael Visuwasam, G. Kalpana, K. Dhinakaran, N. Kallish Kumar, and V. Manigandan

Abstract In this emerging world, technology and crimes are increasing correspondingly. Unusual activity may be anything that seems illegal or abnormal which is not performed by every normal person. Unusual activities that can happen in warehouse are theft and damaging of warehouse products, unnecessary fights among employees and accidents that occur during work which cause heavy loss of products and wealth. It is a tedious task to control those unusual activities since the working area is very large which involves many number of workers. Even though implementation of CCTV cameras helps in security, it requires a well-trained professional to handle, which can be further automated by machine learning and deep learning models. It will reduce the use of manpower and improve the security by using modern solutions. To provide security to warehouse, we developed an unusual human activity detection (UHAD) model which is trained through SSD model that detects unusual activities.

Keywords CCTV · Deep learning · Machine learning · SSD · Unusual activity · UHAD · Warehouse

L. M. M. Visuwasam (✉)
Department of Artificial Intelligence and Data Science, R.M.K. College of Engineering and Technology, Chennai, Tamilnadu, India
e-mail: mariamichaelads@rmkcet.ac.in

G. Kalpana
Research Scholar, Saveetha School of Engineering, Saveetha Institute of Medical and Technical Sciences, Chennai, India

K. Dhinakaran
Department of Artificial Intelligence and Data Science, Dhanalakshmi College of Engineering, Chennai, India

N. K. Kumar · V. Manigandan
Rajalakshmi Institute of Technology, Chennai, Tamilnadu, India

A. Kumar et al. (eds.), *Proceedings of the International Conference on Cognitive and Intelligent Computing*, Cognitive Science and Technology,
https://doi.org/10.1007/978-981-19-2350-0_80

1 Introduction

The world is developing at a faster rate where privacy and security threats are the major risk. Organizations require a large manpower to run a successful management to reduce and safeguard from security threats. Many technologies have been evolved, and society has started adopting it for better performance. Many domains have been developed a large set of technologies to improve safety among people and to reduce security threats that cause heavy loss of wealth and damages to human lives and products. Security threats are caused due to unusual activities that occur among normal peoples and environment which leads to dangerous activities like murder, theft, riots, and other life-killing actions. The increase in crime rate is also due to the occurrence of unusual activities which can be reduced before occurring by proper supervision. As per the past records, supervision also known as surveillance is performed by humans named security guards who work for more time than normal hours to prevent any unusual activities to happen.

Later, manual supervision by security guards was replaced by closed circuit television (CCTV) cameras installed in an organization to record the activities that are happening around. Surveillance using CCTV also requires some manpower to monitor and manage the recorded videos from the CCTV. A person with technical skills is employed to maintain the CCTV and also to monitor the recorded videos. This process is similar to the traditional method of supervision by security guards with some additional technology to help but did not reduce the manpower involved in the process of surveillance. A warehouse is place where products from different organization are stored to be get packed and distributed among customers. Since many different organizations involved in warehouse, the amount of cost involved is also higher than any other domain of industry. Many small organizations including subdealers, contractors, manufacturers, and distributers were also considered to be the stakeholders of warehouse organization. Warehouse securities rely mostly on CCTV to monitor suspicious activity before it turns into something harm to the warehouse.

Figure 1 describes the view of warehouse which stores products belonging to various organizations. Storage of many organization's products is clearly shown, and the maintenance of those products is very difficult.

Security threats in warehouse can happen like warehouse theft, break-ins, devastation, and warehouse fires. Also, accidents while using machineries like fork lifter, wheel chocks, and truck restraints can cause harm to the employees and also for the products that have been stored in the warehouse. Those threats have to be supervised, and an additional support has to be provided to monitor the suspicious activities that happen in the warehouse. Figure 2 describes the ratio of threats that happen in warehouse which causes harm to employees, products, and goods that stored in the warehouse.

According to Transported Assets Protection Association (TAPA), five of 46 countries recorded cargo theft in the half of 2020 accounted for 87% of loses of products. United Kingdom (UK) stands first with 50.9% thefts of products from the warehouse; Germany stands second with 25.2% theft rate, and Netherland stand third

Fig. 1 Warehouse

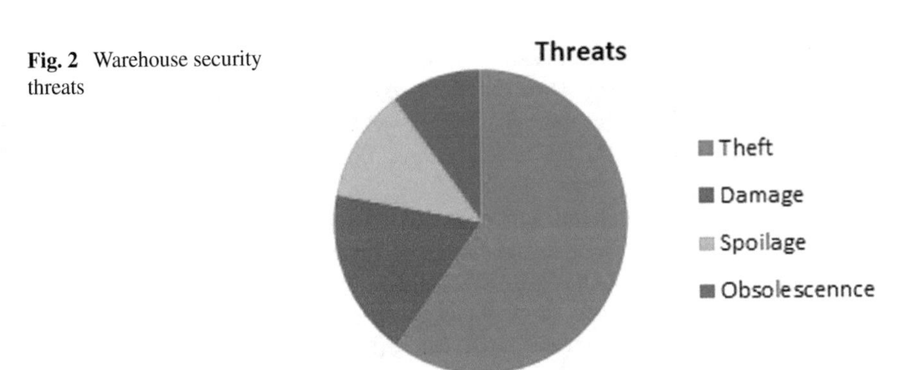

Fig. 2 Warehouse security threats

with 5.1% of theft rate. Products like computers, mobiles, clothing, foods and drinks, and cosmetics are some of the major target products for theft.

Many incidents like warehouse devastation have been increased in the past 10 years which causes damages to the products worth of many lakhs in many countries including India. Incidents like riots have been planned near the location of warehouse and that lets the people inside the warehouse by break-ins and cause damaging and theft of warehouse products. Devastation further leads to warehouse fires which causes injuries to the employees and even loosing of their lives, damage to products, and ultimately ends in explosion of warehouse. Surveillance has to be improved at a rigorously to provide safety to the products and for the lives of employees working in the warehouse.

Manual supervision in warehouse does not give hope for security since many other reasons may affect it, and there is a lot of chance for human error. Supervision is further automated by using the technologies like machine learning and deep learning which can help to automate the process of finding unusual activities from the video recorded by the CCTV. The major advantage of using machine learning and deep learning in surveillance will yield a solution of high accuracy and helps in reducing manpower involved in security purpose. Machine learning and deep learning can be used to identify patterns and make decisions from the provided set of information.

References [1–3] training an object detection model with images of unusual activities can help to detect unusual activities that occur in real time. Once an unusual activity is detected by the object detection model in the CCTV video, an alarm sound is produced to alert the employees and the control center of warehouse.

This will reduce the amount of cost spent to implement various security tools that require man power to maintain and reduce the manpower involved. Since no manpower or very low manpower is required for this model to operate, this can provide solution with high accuracy and efficiency.

2 Related Work

In this section, the related works so far in the field of security by object detection algorithm and its various domains are discussed. There is lot of technologies and many object detection algorithms that were used earlier will be discussed briefly. Some steps are common in all the detection algorithms like preprocessing and feature extraction.

[4–6] Using long short-term memory neural network (LSTM) algorithm to detect abnormal vehicle traffic in highways and local roads. LSTM belongs to recurrent neural network (RNN) that can learn order dependence in sequence prediction issue. LSTM holds feedback connections that can process entire data sequence as a whole. A LSTM consists of a cell followed input, output, and a forget gate [7]. LSTM works primarily for classification which can be followed by data processing, analyzing, and to take decisions. Then, the frames are analyzed, and the abnormal actions from the frames of CCTV are detected by one-class support vector machine (OC-SVM).

[8] This model detects suspicious activity in general places using You Only Look Once (YOLO) model. The images from the videos of CCTV are extracted and are processed by a YOLO pretrained model on COCO dataset to detect each and individual objects from the video. The frames of the images are then processed by residual network model (ResNet) which is trained with interaction dataset to classify the actions in the input frames. YOLO helps to detect objects from frames of the image from CCTV camera video by applying a single CNN [9] and then divides the frames into grid which is evaluated with the confidence scores of bounding boxes that covers an object against the threshold scores. ResNet makes use of skip connection that skips training from few layers and connects directly to the output that is expected, which represents the residual mapping that fill the network. YOLO algorithm helps to identify and detects the suspicious human activities in a very fast fashion with high accuracy.

[2] The model to recognize anomalous activity from CCTV uses neural networks composed of convolutional and recurrent neural networks. The images from the CCTV video are extracted and processed where high-level feature maps are obtained

from convolutional neural network which will reduce the difficulty for the input of next level of recurrent neural network. Images of various classes from ImageNet are then retrained to acquire new classes [10]. The inception model output is fed into the convolutional neural network to extract the last pooling layer called high-level feature map which is a vector containing features to be feed as input to recurrent neural network. In order to reduce the training complexity of recurrent neural network, the input is the feature map instead of images from the frame of the video. RNN helps in extracting the actions portrayed in the fixed time duration and is used to classify the actions as normal or anomaly activity.

[1] This proposed model is to identify suspicious activity using deep learning technologies that can learn data in an unsupervised manner. This model will help to detect unusual human activities from the videos recorded from the CCTV and will notify the authority if any unusual activity occurs. The video available from the CCTV is converted to frames before processing and is preprocessed using OpenCv Python library. VGG-16 which is already trained neural network is used for feature extraction which is trained on ImageNet. It uses three dimensional filters with one convolutional layer and is a 16-layer deep learning architecture that includes rectified linear unit (ReLU) which is an activation function. VGG-16 belongs to a classification model through which the activities are classified as normal or suspicious from the video of CCTV.

[3] the model detects human-based suspicious activity using CNN for feature extraction and deep belief neural network (DBNN) for action classification from the surveillance video obtained. CNN is a branch of DNN that is used mainly for processing image-based data and to obtain a decision. The pre-processing procedure is made less which consists of input layers followed by hidden layers that perform convolution and an output layer [11, 12]. ReLU is made available to process that followed by multiple pooling layers and normalized layers to reduce the dimension of the data provided with the cluster from one layer onto a single neuron in the next layer. DBNN follows greedy learning algorithms since they are fast and efficient to detect normal and unusual activities.

[13] it detects gun-based crimes using TensorFlow implementation of Faster R-CNN and abandoned bags detection using ResNet model. Faster R-CNN belongs to a model which is convolution-based network consists of region proposal network (RPN) for creating region proposals and a network using these proposals to detect object in the frame. Faster R-CNN practices selective search technique that helps to generate proposals containing objects which will be examined by a classifier and regress or to find the object in the frame of the image. A residual neural network belongs to an artificial neural network (ANN) which uses skip connection that skips training from few layers and connects directly to the output instead learning the underlying mapping; the network itself fits the residual mapping. Residual mapping supports in training the DNN without causing vanishing gradient by skipping some layers in the middle and connecting to the output layer directly. The crimes which are based on gun and deserted objects are detected using Faster R-CNN and ResNet with reasonable accuracy and speed.

[14] it proposed a model to recognize human activity which is unusual by using convolutional 3D network. The video from the CCTV is converted to frames that are intended to detect objects in the frame, and a visual interest map is developed from the object detected on the frame. The model is trained using the visual interest map that is created earlier, and the frames are made to undergo a smoothening process that provides an adaptive compression. Feature extraction will help to reduce the dimension of the frames that is sliced from the video [15]. Adaptive compression is followed since it compresses the video without losing the significant information through which it is used for classification. The features that were extracted are used by 3D convolutional networks by following the respective features for classification of anomalous activity and normal activity. The features are provided to the fully connected neural network that gives the score of the anomalous activity where the frame with higher score is identified using multiple-instance learning to classify unusual and normal activity. Adaptive compression is combined with activity recognition model to improve efficiency and speed.

3 Proposed Model

The proposed model—unusual human activity detection (UHAD) is a general object detection model trained through single-shot detector (SSD) using a large set of images of unusual activities as dataset. Figure 3 shows the architecture diagram of the proposed model which is used to detect unusual human activities that occur in warehouse and generate alarm sound when detected.

4 Single-Shot Detector

Single-shot detector is an object detection algorithm that detects an object from an image at a single shot of view. Unlike any other object detection algorithm, the time required to process an image is less when compared, and the efficiency is correspondingly improved. SSD consists of a backbone component which is an image classification network that acts as the feature extractor. The result is a deep neural network that helps to extract the root meaning from the input image provided. Another component is SSD head which is a convolutional layer that is added to the backbone component, and the solution is a bounding box and class of the object that appears on the image. SSD divides an image into multiple grids, and the detection is made by identifying objects in each cell by predicting the bounding box and class of the object.

Each grid cell is assigned with a set of anchor boxes which are predefined and is meant for size, shape, color of the object. The object from the image is detected by matching the anchor box that contains the details like aspect ratio and zoom level while training and testing the model. SSD eliminates the process of proposal

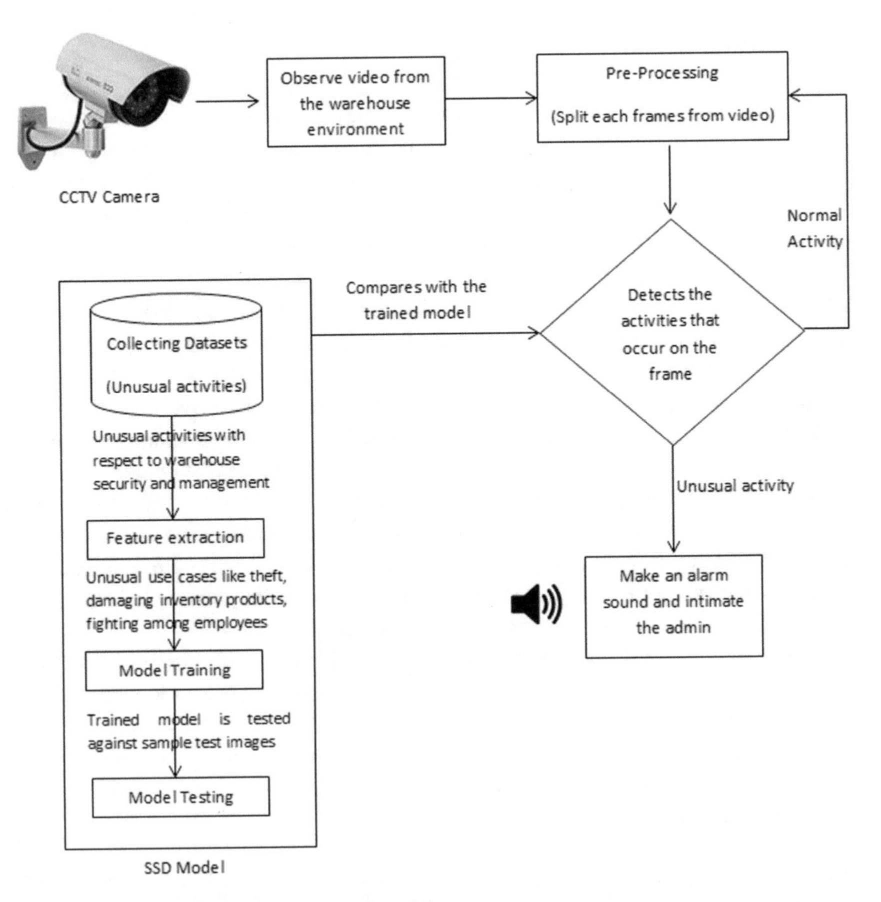

Fig. 3 Architecture diagram of proposed model

generation and feature sampling and thus combine as a single phase. Thus, SSD is easy to train and integrate to detect objects of various domains. Figure 3 shows the architecture and working of unusual human activity detection (UHAD) model that detects and produces alarm in case of unusual activity occurs.

SSD model is trained with a set of images of unusual activities as dataset. The feature extraction process is done by the SSD backbone and is preceded to model training. The bounding boxes and the anchor boxes of the images are matched and trained to the model. The model is tested against a set of unusual activity images for calculating the efficiency and accuracy of the model.

5 Working Procedure

The images of unusual activities are collected from Internet on various sources and is annotated using LabelImg. LabelImg is an annotation tool that helps to point the bounding boxes and to name the classes that we are expecting to detect by training the model. The result is an XML file that contains the annotation details like maximum and minimum coordinates for pointing the bounding box along with the class of the object belongs. Once the annotation processed is completed, the XML files are processed, and it is converted as a single CSV file for generating TF-Record which store sequence of binary record. Then, the batch size is fixed for the model and is preceded for training with the available data.

The trained model can be installed in a system which is used to monitor the CCTV camera videos so that detection can be done from live video input source, and alarm is made immediately when unusual activity occurred. The CCTV camera installed in a warehouse is used for monitoring, and the video is stored for a specific period of time for future reference. The video is processed, and the frames are extracted which is reduced to a common size through which SSD detection takes place. The image is compared with the provided dataset images and the bounding boxes, anchor boxes, and classes of the objects are identified by matching them.

This is an iterative process which takes place on the entire frames extracted from the video. Once an unusual activity is detected, an alarm sound is produced, and the control center of the warehouse is alerted, and the detection process continues with the next frame which is extracted. If the frame extracted from the video doesn't have any unusual activities, the process of detection continues to proceed with the next frame that is extracted.

```
Algorithm UHAD (source_video)
{
Unusual = { }//Model is trained with set of images
Loop:
If (video.frame == unusual) then
      Message ("Alert!! An unusual activity has been occurred!!");
        Go to: Loop
Else
        Go to: Loop
End if
}
```

6 Result and Discussion

Unusual human activity detection model is trained and tested against a set of around 250 unusual activity images as dataset. The process of collecting dataset in this case is difficult, as unusual activities that occur in warehouse cannot be captured outside the organization.

Figures 4, 5, and 6 shows the UHAD model detects an accident that occurs in a warehouse as unusual activity. It will raise an alarm sound which will notify other employees and the control center for immediate rescue and other helps.

SSD achieves 72.1% mAP on VOC2007 test at 58 FPS on an Nvidia Titan X and for 500 × 500 input; SSD achieves 75.1% mAP, outperforming a comparable state

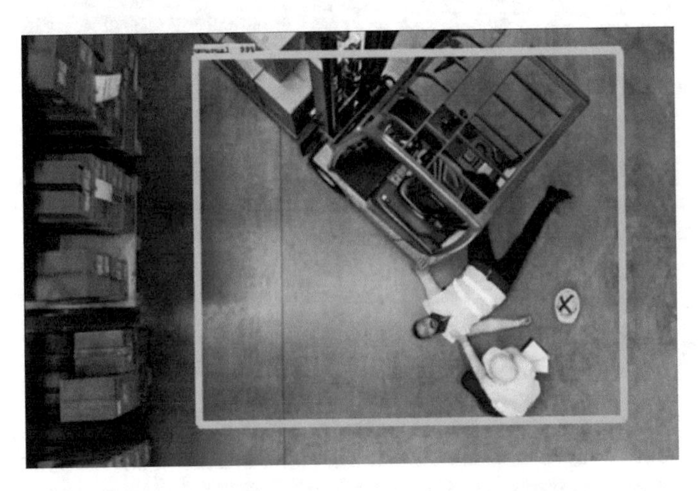

Fig. 4 Detecting and Locating an accident as unusual activity-1

Fig. 5 Detecting an accident as unusual activity-2

Fig. 6 Detecting an accident as unusual activity-3

of the art Faster R-CNN model. Since SSD model follows single-shot detection, the response time that any other object detection model is low, which means it can react quicker and detect even fast-moving objects.

7 Conclusion

Warehouse security is one of the major things which have to be focused a lot since the cost and lives involved is higher than any other business domain. Since unusual human activity detection (UHAD) model follows single-shot detector algorithm, it is better to implement in warehouse for security on considering the efficiency and response time. Many activities that have been classified as unusual can be easily detected using SSD even in fast-moving video. Many object detection algorithms have been evolved and been used for a while, but using SSD for warehouse security is new and useful. The model is limited to a set of use cases like detecting accidents, theft, and falling as unusual activities that occur in warehouse. Extending the use cases wider to find more possible unusual activities in warehouse and training the model to detect those will be the future enhancement of this model. This model will be highly useful for warehouse management which will reduce the cost spent for security tools and devices and man power involved.

References

1. Amrutha C, Jyotsna C, Amudha J (2020) Deep learning approach for suspicious activity detection from surveillance video. In: 2020 2nd international conference on innovative mechanisms for industry applications (ICIMIA). https://doi.org/10.1109/icimia48430.2020.9074920
2. Singh V, Singh S, Gupta P (2020) Real-time anomaly recognition through CCTV using neural networks. Procedia Comput Sci 173:254–263. https://doi.org/10.1016/j.procs.2020.06.030
3. Basha A, Parthasarathy P, Vivekanandan S (2019) Detection of suspicious human activity based on CNN-DBNN algorithm for video surveillance applications. In: 2019 innovations in power and advanced computing technologies (i-PACT). https://doi.org/10.1109/i-pact44901.2019.8960085
4. Kong X, Gao H, Alfarraj O, Ni Q, Zheng C, Shen G (2020) HUAD: hierarchical urban anomaly detection based on spatio-temporal data. IEEE Access 8:26573–26582. https://doi.org/10.1109/access.2020.2971341
5. Kapoor A, Biswas KK, Hanmandlu M (2017) Unusual human activity detection using Markov logic networks. In: 2017 IEEE international conference on identity, security and behavior analysis (ISBA). https://doi.org/10.1109/isba.2017.7947700
6. Sandhya Devi G, Suvarna Kumar G, Chandini S (2017) Automated video surveillance system for detection of suspicious activities during academic offline examination, world academy of science, engineering and technology. Int J Comput Inf Eng 11(12)
7. Samir H, Abd El Munim HE, Aly G (2018) Suspicious human activity recognition using statistical features. In: 2018 13th international conference on computer engineering and systems (ICCES). https://doi.org/10.1109/icces.2018.8639457
8. S. Adarsh, Giridhar Kannan, S. Poorvaja, B. S. Vidhyasagar, J. Arunnehru (May 2020) Suspicious activity detection and tracking in surveillance videos. Int J Emerg Technol Innovative Res 7(5):75–79. www.jetir.org. ISSN:2349-5162
9. Chaudhary S, Khan MA, Bhatnagar C (2018) Multiple anomalous activity detection in videos. Procedia Comput Sci 125:336–345. https://doi.org/10.1016/j.procs.2017.12.045
10. Albukhary N, Mustafah YM (2017) Real-time human activity recognition. IOP Conf Ser: Mater Sci Eng 260:012017. https://doi.org/10.1088/1757-899x/260/1/012017
11. Mohite A, Sangale D, Oza P, Parekar T, Navale M (Jan 2020) Unusual human activity detection using open CV python with machine learning. Int J Adv Res Comput Commun Eng 9(1). https://doi.org/10.17148/IJARCCE.2020.9109
12. Senthilkumar T, Narmatha G (2015) Suspicious human activity detection in classroom examination, computational intelligence. Cyber security and computational models. pp 99–108. https://doi.org/10.1007/978-981-10-0251-9_11
13. Loganathan S, Kariyawasam G, Sumathipala P (2019) Suspicious activity detection in surveillance footage. In: 2019 international conference on electrical and computing technologies and applications (ICECTA). https://doi.org/10.1109/icecta48151.2019.8959600
14. Shreyas DG, Raksha S, Prasad BG (2020) Implementation of an anomalous human activity recognition system. SN Comput Sci 1(3). https://doi.org/10.1007/s42979-020-00169-0
15. Lee D-G, Suk H-I, Park S-K, Lee S-W (2015) Motion influence map for unusual human activity detection and localization in crowded scenes. IEEE Trans Circuits Syst Video Technol 25(10):1612–1623. https://doi.org/10.1109/tcsvt.2015.2395752

Attacks and Security Schemes in Cloud Computing: A Survey

Priyanka Jain, Raju Barskar, and Uday Chourasia

Abstract Cloud computing provides the virtual environment for users to access the hardware resources and services. The resources such as storage, applications, services, networks and servers that quickly prerequisite and remain unhindered with insignificant management effort. Security is one of the major factors in cloud computing, and many users are paid for on-demand cloud resources and services. Some of the users are not authorized to access, misappropriate or change data that are called as attackers. The cloud is committed to preserving the integrity and dependability of the data, it contains by the cloud service provider. The attackers are degrading the performance of services as well as make the legitimate users unable to receive response on time. This paper presents an overview on cloud computing, attacks on cloud and their possible security solutions with merits and demerits. There are many security schemes proposed previously, but each of them had some constrained related to configuration, computing and storage. The researchers are continuously giving the contribution in the field of cloud security to improve communication and the performance of the network. It is very typical for the security scheme to identify the normal or a malicious user on the network. Hence, there is demand of strong security schemes that can easily identify the malicious users as well as can handle the adverse effect of attackers in cloud.

Keywords Cloud computing · Cloud security · Attacks · Security schemes · Malicious users

---

P. Jain (✉) · R. Barskar · U. Chourasia
University Institute of Technology RGPV, Bhopal, Madhya Pradesh, India
e-mail: Priyanka.jain0111@gmail.com

R. Barskar
e-mail: rajubaraskar@rgtu.net

U. Chourasia
e-mail: udaychourasia@rgtu.net

A. Kumar et al. (eds.), *Proceedings of the International Conference on Cognitive and Intelligent Computing*, Cognitive Science and Technology,
https://doi.org/10.1007/978-981-19-2350-0_81

1 Introduction

Cloud computing (CC) is characterized as the collection and availability of data and computing resources over the Internet. No data is saved on your personal computer. The on-demand provision of computing resources is servers, data storage, networking, databases, etc. The key aim of CC is to allow multiple users to access the data centres [1, 2]. Users can also view data on a remote server. The opportunity for cost savings is the key explanation for the implementation of CC by many organizations. Many of the devices or uses are fulfil their requirement by using CC mention in Fig. 1.

CC allows you the right to access resources as necessary and to pay only for what you use [3]. Cloud infrastructure has made it easier to manage IT activities as an outsourced unit without a lot of in-house personnel. Cloud computing has evolved as a cutting-edge technology that provides developers with a variety of computing resources. IaaS, PaaS, SaaS and are all examples of CC resources [1].

The introduction of the cloud represents a watershed moment in technical development for quick information processing. When a new computing system is released, academics and researchers place a high focus on its security. The success of any information processing system depends on the security of data processing across all systems [3]. CC enables data processing to be done quickly and without regard to location. As a result, one of the most significant issues that cloud customers confront when using their resources is trust [2]. The articles [4] provide a comprehensive picture of the security problems associated with the cloud environment and also discuss about the high data rate communication in network [5].

Cloud is vulnerable to a variety of adversaries such as DDoS attack, malware injection attack, spoofing attack, flooding attack, DoS and DDoS attack. Detecting and filtering of attack packets to protect cloud users are the one of difficult task. There are four cloud models presented below that provide services according to the company requirements as mentioned in Fig. 2.

Fig. 1 An architecture of cloud computing

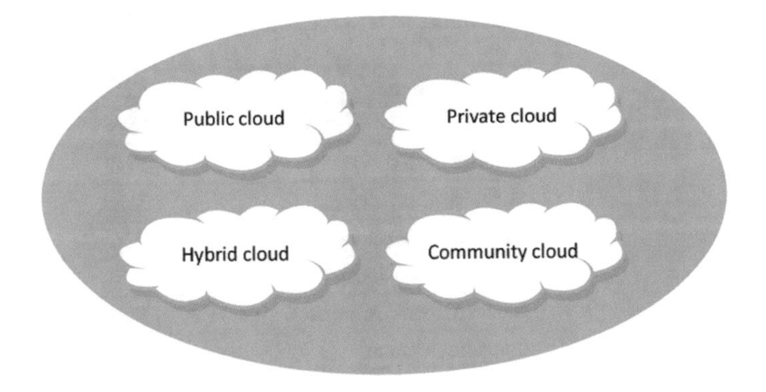

Fig. 2 Types of cloud model

1.1 Private Cloud

This is where computer services are deployed by a single entity. This approach is most commonly used for intra-business encounters. Where computing services can /be managed, controlled and run by that company only.

1.2 Community Cloud

The community and organizations are equipped with computational services.

1.3 Public Cloud

This cloud is typically used for B2C-type experiences. Here, the computing resource is owned, managed and controlled by govt., academia or a business entity.

1.4 Hybrid Cloud

It can be used for all types of interactions B2B or B2C. This implementation approach is known as the hybrid cloud since computing services are joined together by various clouds (Table 1).

Table 1 List of acronyms used in paper

Full forms	Acronym
Cloud computing	CC
Infrastructure as a service	IaaS
Platform as a service	PaaS
Software as a service	SaaS
Business to consumer	B2C
Business to business	B2B
Media authentication code	MAC
Operating system	OS

2 Features of Cloud Computing

It provides a range of appealing features for both companies and customers. Some of the features mentioned below are [1, 2].

2.1 Device and Location Independence

Users can connect to the cloud network from any location and device, e.g. computers or smartphones, since they can be reached via the Internet and servers that are centrally located (off-site and service provider maintained) independent of their particular location.

2.2 Pay Per-Use

Users need to only pay for tools they have used out of the pool of software and facilities available and do not need to pay for the entire infrastructure.

2.3 Multi-tenancy

It provides the sharing of resources, software applications, networks and their costs by large users. The same physical infrastructure, such as servers and hardware, is shared between multiple users, but all of them maintain privacy and data protection.

2.4 *Stability*

The reliability of the infrastructure is enhanced with the use of several redundant servers for databases and data storage, so that data can be recovered quickly in case of failure.

2.5 *Efficiency and Performance*

The productivity of projects using cloud networks with applications running improves as many users operate on the same database and program concurrently. Thus, by evaluating in their way at the same time, it will provide better performance.

3 Cloud Computing Advantages

The CC is really important for accessing information and resources. The some of the advantages [1, 4] are as follows.

3.1 *Cost-Effective*

The usage of cloud technology in networking and storage can minimize the total cost of buying and maintaining hardware and technological resources for the organization's mission. This would make the project cost-effective as companies do not need to spend money on constructing data centres, buying infrastructure, technical updates and other tools required to operate the project, as all of these facilities are handled by the cloud service provider. Moreover, the expense of renting these cloud services to businesses is very economic relative to managing them rather than buying them.

3.2 *Accessibility*

The usage of CC technology can offer simplicity and mobility for end-users to extract, store and exchange data from anywhere, at any moment, only by providing an Internet connection. This ensures that consumers are not forced to hold hard discs and CDs to carry their data from one location to another. They can only store their data on Google Drive or Dropbox, and then view it from anywhere on the Internet.

3.3 Simple Data and Programme Management

Because companies do not need to customize the software and hardware of systems and programmers they are operating on, they can also concentrate a lot on designing software applications quickly. In addition, all data is stored on a single server such that it is easier to monitor the data and track who is accessing whatever form of data at that location through the management.

3.4 Platform Flexibility

In CC, the same data and software can be downloaded on multiple platforms such as tablets, desktop PCs and iPads.

3.5 Increased Storage Space

The capacity of servers to store data is much greater than the storage capacity of the user computer. Cloud storage can make it possible for consumers and organizations to maintain their large units of personal and project-related data on cloud data servers.

3.6 Automation of Software Upgrade

Cloud infrastructure can deliver automated updates to all systems and software programmes operating on its network on a timely basis.

4 Disadvantages in Cloud Computing

The some of the drawbacks [1, 4] of cloud computing are as follows.

4.1 Need for Reliable and High Speed Internet Access

The entire definition of cloud infrastructure relies on the provision of a permanent Internet connection. If the customer has no bandwidth or low network communication level, he or she would not be able to access his or her data and other software on the cloud service.

4.2 Protection Issues

Security and secrecy of resources and software is a major problem in CC, since more than one company share the same computer space on public cloud networks to store and access their data and applications. As a result, there has always been a security issue concerning the protection of users' data and details in the minds of organizations, as they do not have much exposure at all.

To fix this issue, cloud service providers have instaled firewalls to protect unauthorized access to the network and have issued passwords to customers so that they can only have restricted access privileges to their individual accounts.

4.3 Relocation Issue

If, in any event, the services of a service provider are interrupted for any reason, it is very difficult for customers to transfer a massive data and application device to another cloud network. It would exhaust a lot of time and resources, and there is still no guarantee of full data migration, some of which could be missed due to connectivity problems.

5 Data Security Issue in Cloud

The Internet is the main source to transfer the data on cloud. There are many users use the high speed and personal network for send data on the cloud as well as receive from cloud. MAC or content-level authentication are examples of streaming-level authentication mechanisms to safeguard data which is multimedia in a wireless network, methods such as digital watermarking can be utilized [6]. As a result, digital watermarking is employed to provide security for user data in the cloud environment in this study. The following are the models of cloud service [7].

5.1 Software as a Service (SaaS)

Customer has access to provider's apps, which are hosted on cloud architecture. A thin client interface, such as a web browser or a programme interface can be used to access the apps from a variety of devices which are clients. The cloud infrastructure includes servers, network, OS, specific application capabilities, storage and even is not managed by the consumer, with the possible exception of restricted user-specific application configuration options.

5.2 Platform as a Service (PaaS)

The capacity to instal consumer made or acquired apps produced with programming onto the cloud infrastructure is the capability supplied to the consumer. The consumer does not have control over the underlying cloud infrastructure, such as the network, servers, operating systems or storage, but does have control over the instaled apps and maybe the application-hosting environment's configuration settings.

5.3 Infrastructure as a Service (IaaS)

The customer's capacity is defined as the ability to provide processing, storage, networks and other critical computer resources on which the customer can deploy and run arbitrary software such as operating systems and applications. The consumer does not manage or control the core cloud infrastructure, but he or she does have control over operating systems, storage, and instaled applications, as well as perhaps limited control over specific networking components (e.g., host firewalls).

6 Data Storage Issue in Cloud

The cloud's distributed structure allows large volumes of data to be stored with minimal effort. The demand for video/picture-based information exchange has increased dramatically. As a result, a large amount of data in the form of content (multimedia), structure (links) and usage is generated and analysed on a daily basis (logs). Microsoft claims that in a single year, its cloud storage service housed around 11 billion photographs. Facebook has also reported the removal of 220 billion photographs, with 300 million new photographs being added every day [8]. The location information can also maintained for reduce the extra overhead on search destination [9]. A good compression approach would not only lighten the load on cloud storage, but also on the application devices used to analyse picture data.

7 Attacks in Cloud

An intruder's main purpose is to crack a cryptosystem and decipher the plaintext from the cypher text. The following are some examples of cryptanalysis attacks.

7.1 Brute Force Attack

In a brute force attack, the attacker exhausts all possible options in order to figure out the key. The number of possible keys is 216 = 65,536 when the key is 16 bits long. If there is a long key [10], this method will take a long time to complete the attack.

7.2 Dictionary Attack

A dictionary of words is kept in a dictionary attack. The network is used to send the encryption text. The intruder uses a dictionary to attack the cypher text; the cypher text is searched in the dictionary, and if the cypher text contains one or more dictionary words, the assault is successful [11].

7.3 Denial of Service Attack (DoS)

Generally, DoS attackers have the purpose to assault on the server. Many consumers use the services provided by the server. DoS attackers act as if they are a legitimate client, with the goal of making services unavailable by generating a large number of pending requests and overflowing the service queue [12].

7.4 Distributed Denial of Service (DDoS) Attack

Network attackers are now posing a threat to interconnected systems like database servers, web servers, CC servers and so on. A DDoS attack consists of multiple connected online devices that overwhelm an online service with bogus traffic. Traffic assaults, bandwidth attacks and application attacks are some of the most popular DDoS attacks [13].

7.5 SQL Injection Attacks

That is a sort of code injection attack in which malicious code is injected into a programme by inserting malicious code strings into the program's instructions. Users' information is stolen via this attack. When malicious code is inserted into a valid SQL query, the nature of the original query is changed.

7.6 *Spoofing Attack*

The one of the security risks in wireless networks is the spoofing attacks. Spoofing attacks occur when an intruder or malicious programme successfully impersonates other network devices, allowing the attacker to launch attacks against network hosts or steal data. IP address spoofing, ARP spoofing and DNS spoofing are some examples of common spoofing attacks [14].

7.7 *Jamming Attack*

Jamming is a type of DoS attack that aims to block other nodes or users from receiving information in a timely and adequate manner. It works by blocking the channel and obstructing the traffic on the network. The attacker is only active for a short period of time in this attack, and the important information is targeted [15].

7.8 *Man-In-The-Middle Attack*

Between two trusted parties, data is exchanged via the network (e.g. sender and receiver). The attacker attempts to establish contact between the two parties in this attack. Attackers gain access to the communications and attempt to impersonate a sender and a recipient. As a result, they alter traffic patterns and focus on information secrecy and integrity [16].

7.9 *Phishing Attack*

Phishing is a form of social engineering. The phishing email message is contained a link to a faked website, which is sent to a large number of people by the attackers. Attackers persuade users to reveal personal information such as usernames and passwords, as well as other sensitive information. When a user clicks on the provided link, he is redirected to a faked website, where he is requested to provide sensitive information [17].

8 Literature Survey

In this section presents the previous work that has already done in field security in cloud. The detailed description of work with drawback is mentioned in Table 2.

Table 2 Comparative study among previous work on the basis of different parameters and performances

S. No.	Author/s	Worked done	Overhead	Infection	Loss	Efficiency	Technique used	Result
1	Singh et al. [18]	They compare AES, RSA-and SHA security algorithms	High	Not evaluated	Not evaluated	Good	The network security algorithms are used for secure communication	Only concept of security has explained
2	Ranjan et al. [19]	They proposed a scheme to secure from malware injection attack	High	Not evaluated	Not evaluated	Good	SQLIA for stop the injection of packets on server	Measure the speed of server
3	Bin et al. [20]	They proposed an approach to control data loss or stolen at the terminal	Not evaluated	Not evaluated	Not evaluated	Not evaluated	Smart saver grid discussions are there	Only concept of cloud has explained in smart grid
4	Thangavel et al. [21]	Focused on the hash tree-based approach	High	Negligible	Low	Good	Third party auditor (TPA) and Ternary hash tree (THT)	Measures the storage of replica blocks and normal storage blocks
5	Salunke et al. [22]	They proposed approach is granting the users read and write access control	Not evaluated	Not evaluated	Not evaluated	Not evaluated	KDC technique is used for enhanced distributed control access	Only scenario is given in paper
6	Selvanayagam et al. [23]	Focused on cloud service provider (CSP) security using encryption and decryption technique	Not evaluated	Not evaluated	Not evaluated	Not evaluated	Encryption and decryption techniques are used	Only concept of security has explained

(continued)

Table 2 (continued)

S. No.	Author/s	Worked done	Overhead	Infection	Loss	Efficiency	Technique used	Result
7	Sinchana [24]	They are discuss about the factors affected the cloud performance	Not evaluated	Not evaluated	Not evaluated	Not evaluated	Security schemes overview and flaws	Only the description of techniques are discussed
8	Dhakrey [25]	Proposed a hybrid encryption technique for encrypted data	Low	–	–	Good	Use the combination of AES, DES and blowfish or	Only the encryption and decryption time comparison evaluated
9	Shaohua et al. [26]	Proposed a security scheme to detect the abnormal flow of data in the cloud	Low	–	Low	Good	Distinct feature number (DFN) for detecting traffic	Detect the unwanted flow of packets by measuring the entropy
10	Sengupta et al. [27]	Proposed a data storage scheme for arrange and handle data efficiently	High	–	–	Good	Cloud storage protocols for dynamic data (DSCSI and DSCSII)	Calculate the storage overhead, communica-tion cost in different file size
11	Huang [28]	Proposed a scheme to calculate the credibility of loss and apply the security after the judgement of loss	High	Based on credibility	Low	Good	Only focus on the security factors not affect on cloud or not able to provides better security	Measure the vulnerability of confidentiality, integrity, availability and reliability

(continued)

Table 2 (continued)

S. No.	Author/s	Worked done	Overhead	Infection	Loss	Efficiency	Technique used	Result
12	An et al. [29]	Proposed a security scheme to secure gathered data of cloud and collect security information from cloud	–	High	–	Good	Apply modified hierarchical attack representation model (HARM) for secure gathered data	Evaluate sum risk, maximum risk, attacker risk, attacker success and loss probability

9 Conclusion and Future Scope

Data or knowledge is relevant today, its ideals are unrecognized. It has a number of advantages, indicating that you should only implement CC services by reviewing all the big security problems in CC. If an attacker uses all available resources, others cannot use those resources which contributes to the denial of service which may slow down access to those resources. Customers who use cloud services and who are affected by caps could also work to impact the availability of other providers. Many companies feel more comfortable placing important data on their platform than on the cloud. Consumers have no knowledge about the location of records, the transition of dates, cloud operations, etc. In this paper, the survey of attacks and security schemes is mentioned. The previous work mention the effective security scheme for different attacks. The basic knowledge of CC is also mentioned in different sections. A reliable security scheme definitely improves the performance and stops malicious activities in the network.

Holding this information on an open network automatically raises questions regarding security, availability and abuse of information. We tested the technique on simulation on a network deployed on OMNET++ with over 50 users. The experimental findings demonstrate that the proposed technique is successful and can detect dropping attacks with a high detection rate and low data packets received on the network.

Acknowledgements I take this opportunity to gratefully acknowledge the assistance of my college faculty and guides Prof. Uday Chourasia and Dr. Raju Baraskar, who had faith in me. I am also very much grateful to UIT RGPV, Dept. of CSE for motivating the students.

References

1. Mell P, Grance T (7 Oct 2009) The NIST definition of cloud computing, version 15, national institute of standards and technology
2. Zhao G, Liu, Tang Y, Sun W, Zhang F, Ye X, Tang N (2009) Cloud computing: a statistics aspect of users. In: First international conference on CC (CloudCom). Beijing, Springer, Berlin, Hiedelberg, pp 347–358
3. Mishra H, Makad J, Pathak N, Soni G (April 2021) Cloud based career guidance system. Int J Sci Res Comput Sci, Eng Inf Technol (IJSRCSEIT) ISSN No. 2456-330
4. Yu S, Guo S, Stojmenovic I (2012) Can we beat legitimate cyber behavior mimicking attacks from botnets? In: Proceedings INFOCOM. pp 2851–2855
5. Soni G, Chandravanshi K (26–27 Aug 2021) Security scheme to identify malicious maneuver of flooding attack for WSN in 6G. In: 8th international conference on signal processing and integrated networks
6. Podilchuk C, Delp E (2001) Digital watermarking algorithms and its applications. IEEE Sig Process Mag 18(4):33–46
7. Armbrust M, Fox A, Griffith R, Joseph AD, Katz R, Konwinski A, Lee G, Patterson D, Rabkin A, Stoica I, Zaharia M (2010) A view of CC. ACM Commun 53(4):50–58

8. Ong Z. Picture this: chinese internet giant Tencents Qzone social network now hosts over 150 billion photos 2012. Available at: http://thenextweb.com/asia/2012/08/09/picture-thischinese-internet-giant-tencents-qzone-social-network-now-hosts-over-150-billion-photos
9. Soni KG, Chandravanshi K, Jhariya MK, Tomar DS (Feb 2020) A multipath location based hybrid DMR protocol in MANET. IEEE international conference on emerging technologies in computer engineering (ICETCE)
10. Gautam T, Jain A (2015) Analysis of brute force attack using TG–dataset. IEEE SAI intelligent systems conference (IntelliSys)
11. Jose J, Tomy TT, Karunakaran V, Krishna VA, Varkey A, Nisha CA (2016) Securing passwords from dictionary attack with character-tree. IEEE international conference on wireless communications, signal processing and networking (WiSPNET)
12. Somani G, Gaur MS, Sanghi D, Conti M, Buyya R (2017) DDoS attacks in CC: issues, taxonomy and future directions. Comput Commun 107:30–48
13. Khadka B, Withana C, Alsadoon A, Elchouemi A (2015) Distributed denial of service attack on cloud: detection and prevention. IEEE International Conference And Workshop On Computing And Communication
14. Jindal K, Dalal S, Sharma KK (2014) Analyzing spoofing attacks in wireless networks. IEEE fourth international conference on advanced computing and communication technologies
15. Sowmya S, Malarchelvi PDSK (2014) A survey of jamming attack prevention techniques in wireless networks. IEEE international conference on information communication and embedded systems
16. Eigner O, Kreimel P, Tavolato P (2016) Detection of man-in the- middle attacks on industrial control networks. IEEE international conference on software security and assurance
17. Andri J, Oreski D, Kisasondi T (2016) Analysis of phishing attacks against students. MIPRO, May 30–June 3
18. Singh P, Saroj SK (2020) A secure data dynamics and public auditing scheme for cloud storage. IEEE 6th international conference on advanced computing and communication systems (ICACCS)
19. Ranjan I, Agnihotri RB (2019) Ambiguity in cloud security with malware-injection attack. Proceedings of the third international conference on electronics communication and aerospace technology (ICECA)
20. Bin C, Xiaoyi Y (2019) Research on the application and security of CC in smart power grids. IEEE 4th international conference on mechanical, control and computer engineering (ICMCCE)
21. Thangavel M, Varalakshmi (2020) Enabling ternary hash tree based integrity verification for secure cloud data storage. IEEE computer society
22. Salunke PM, Mahale VV (2018) Secure data sharing in distributed cloud environment. Proceedings of the second international conference on I-SMAC (IoT in Social, Mobile, Analytics and Cloud) (I-SMAC 2018)
23. Selvanayagam J, Singh A, Michael J, Jeswani J (March 2018) Secure file storage on cloud using cryptography. Int Res J Eng Technol (IRJET) 5(3) e-ISSN: 2395-0056
24. Sinchana MK, Savithramma RM, Liu DQ (2020) Survey on CC security. Innovations in computer science and engineering. pp 1–6
25. Dhakrey H (2018) Secure file storage in cloud computing, School of computer science engineering. Galgotias University
26. Shaohua H, Nanfeng X (2020) Abnormal traffic monitoring methods based on a cloud computing platform. IEEE 5th international conference on cloud computing and big data analytics
27. Sengupta B, Dixit A, Ruj S (2020) Secure cloud storage with data dynamics using secure network coding techniques. IEEE transactions on cloud computing

28. Huang M (2021) Design of basic process of information security risk assessment in cloud computing environment. IEEE international conference on consumer electronics and computer engineering (ICCECE 2021)
29. An S, Eom T, Park JS, Hong JB (2019) CloudSafe: a tool for an automated security analysis for cloud computing. 18th IEEE international conference on trust, security and privacy in computing and communications/13th IEEE international conference on big data science and engineering

Studying the Effectiveness of Community Detection Algorithms Using Social Networks

R. Kiruthika and M. S. Vijaya

Abstract Social network analysis is a significant area of research for analyzing the interconnection between the people within network. Community detection is one of the most important applications in SNA. The main motive of CD is to discover the collection of node that are tightly correlated within the network and weakly correlated to another network for partitioning the network to form the group of communities. The aim of this work is to detect communities from undirected disjoint social networks in which it is implemented on lesmis and email-Eu-core-department-labels networks. Effective partitioning and detection of the network are the primary factors for implementing this work by using Girvan–Newman, greedy modularity maximization, and Kernighan–Lin bipartition CD algorithms. The effectiveness of these CD algorithms is analyzed with respect to ground-truth communities based on measures such as recall, normalized mutual information score, precision, and F1-score. Experimental results show that the greedy modularity maximization algorithm provides best results for CD on email-Eu-core-department-labels network with respect to corresponding ground-truth communities.

Keywords Social network analysis (SNA) · Community detection (CD) · Ground-truth communities (GT)

1 Introduction

Extracting and analyzing the data from social network have become a trending research problem in real-time networks analysis [1]. Social network analysis is the solution for this problem. SNA is the collection of methods and methodologies for analyzing the communications between the people from social network. The social

R. Kiruthika (✉) · M. S. Vijaya
Department of Computer Science, PSGR Krishnammal College for Women, Coimbatore, India
e-mail: kirthikamole@gmail.com

M. S. Vijaya
e-mail: msvijaya@psgrkcw.ac.in

A. Kumar et al. (eds.), *Proceedings of the International Conference on Cognitive and Intelligent Computing*, Cognitive Science and Technology,
https://doi.org/10.1007/978-981-19-2350-0_82

network gets partition into group of clusters to form the communities [2]. Communities in network are cluster of nodes that are interconnected with more edges within the group than between groups. The method of detecting cluster of communities from the networks is known as community detection [3]. The aim of community detection is to identify communities based on the network structure [4]. CD is an unsupervised learning method [5]. The applications of CD are graph compression, identifying online clients with mutual traits, track COVID affected patients, identifying individuals who are geographically close to each other, vertex classification, identifying customers with similar interests, etc.

Edge betweenness is an important element in Girvan–Newman algorithm [6] which works to find the number of shortest paths between the edges that connect communities in network. The key work of Kernighan–Lin bipartition algorithms is to partition the network into set of segments to detect the communities from the network [7]. Greedy modularity maximization algorithm works basis of the modularity score in which it is an important measure to computes the maximum modularity score from the network.

This paper mainly focuses on detecting communities from the benchmark [8] network and analyzing their structural properties of the network using SNA techniques. The effectiveness of the algorithms is evaluated by comparing the results of the retrieved communities with ground-truth communities [9] by using performances evaluation metrics like recall, normalized mutual information score, precision, and F1-score measures.

2 Literature Review

The significant number of studies has been done on community detection using complex network systems. The majority of previous related work is focused on directed networks in order to improve the efficiency of detecting communities in complex directed networks. But, the current related works are mainly focused on community detection problem on undirected networks and many solutions have been presented. The most important research findings on the community detection problems are detailed reviewed below.

Newman et al. [10] suggested an algorithm for discovering community structure in graph. This algorithm sequentially removes the edges from the network to segment it into communities where the eliminated edge has been recognized using betweenness measure. The recommended work was focused on the divisive and agglomerate algorithm, but the implementation was based on shortest-path betweenness, resistor networks, and random walks visualized using dendrogram. The suggested algorithms are enormously beneficial at discovering the structure of communities in both computer-produced networks and actual networks, namely tests on computer-generated networks, karate club network, and collaboration network. The results of the suggested work show that the Zachary karate club network got a higher modularity score using a random-walk algorithm.

Liu and Ma et al. [11] suggested the novelty in similarity-constrained Adamic and Adar (AA) index which extracts both national and global data to provide the best possible community identification. AA index was one of the similarity index suggested by Adamic and Adar. This paper recommended a divisive and agglomerate algorithm with other six methods: fast greedy, info map, leading Eigen, label propagation algorithm (LPA), walk trap, Louvain algorithms were used. Social networks used for detecting the communities include Zackary's karate club, email, dolphin social network, US college football, Netscientist, GC_Hep_PH (Arxiv HEP-TH collaborators), poll books (books about US politics), GR_QC (Arxiv GR-QC collaborators), Facebook, GC_Hep_TH (Arxiv HEP-TH collaborators), and Co-authors network. This proposed work provided the best result for Netscientist network in which the modularity value comprises 0.96 using fast greedy, walk trap, and Louvain methods.

Chen et al. [12] conducted the research which aimed to identify social relations using a novel non-overlapping algorithm. The main aim behind this work was to calculate the trust model using the trust-based local community detection algorithm (TLCDA). This method was compared with the most popular community detection algorithms: GN algorithm, Clauset, Newman, and Moore (CNM), eigenvector, info map algorithms using Lesmis network and Gemo network. The best modularity score achieved for the Lesmis network was 0.56 using the infomap algorithm. The modularity score for the Girvan–Newman algorithm was 0.78 achieved using the Gemo network.

Sathiyakumari et al. [13] proposed the method of community detection based totally. This analysis aimed to use association rule mining to identify communities of users that had common goals and values. The work on this paper was proposed on clique percolation method (CPM), hybrid clique percolation (HCPM), and extended clique percolation (ECPM). The work focused on frequent pattern mining using the Apriori algorithm. Community detection was implemented on the real time of the directed cricket player network crawled from twitter network datasets. The result of this work was based on modularity score for ECPM was high as 0.87 compared with other methods such as CPM and HCPM.

Based on the preceding literature review, all community detection techniques differ based on the type of network. There are hardly few metrics or methodologies for determining accuracy of the algorithm in community detection. The modularity score is the effective metric used by the most of researchers to determine accuracy of the algorithm. The majority of researchers uses the most effective method such as clique percolation method to detect the community from the network [13]. This research is proposed to measure the effectiveness of greedy modularity maximization [14], Girvan–Newman and Kernighan–Lin bipartition algorithms [15] on lesmis and email-Eu-core-department-labels social networks.

3 Methodology

The key problem in community detection is to identifying the best method for finding the communities accurately from the social network. Girvan–Newman, greedy modularity maximization, and Kernighan–Lin bipartition are community discovery techniques implemented in this study on lesmis and email-Eu-core-department-labels social networks. Ground-truth communities are the most effective approach to evaluate the partitioned communities from the networks using algorithms. This work is implemented using Python on Google Colab for community detection in social network. The main purpose of this work is to analysis the three major detection methods based on their performance with respect to the benchmark values known as ground-truth communities. Shortest path eigenvector centralities, density, betweenness centrality, transitivity, and closeness centrality are the basic elements of the network structure analysis measures which affect the performance of these algorithms. Normalized mutual information score, precision, recall, and F1-score are the key performance evaluation metrics which applied for analysis the performance of CD algorithms.

3.1 Greedy Modularity Maximization Algorithm

Greedy modularity maximizes algorithm is intended to detect the communities from the network by maximizing the modularity at every phase. Measuring the intensity of relationships within a community is evaluated by modularity. Network with a high modularity score has maximum number of links inside a community. The procedure of greedy modularity maximization algorithm is given below.

- Initially, every vertex belongs to a specific community.
- Compute the modularity score by combining nodes from the communities.
- Repeat the step 2 until only one community left in the network.
- Select the partitioned network with the highest modularity.

All vertices in the networks are allocated to a specific community at the initial stage of the process. The modularity of the entire network is determined by combining the nodes in the network in which it should increase the modularity. So that node became the part of the same community. The step 2 in the process continues until there is only one community left in the network. The network partition is with the highest modularity is consider to be selected as the community.

3.2 Girvan–Newman Algorithm

The Girvan–Newman algorithm is one of the finest divisive hierarchical clustering algorithms. Girvan–Newman algorithm is based on the edge-betweenness approach in which it is calculated by calculating the number of shortest routes between the single vertexes to all other vertices in the network. The procedure of Girvan–Newman algorithm is given below.

- Calculate betweenness centrality at each edge of the network.
- Eliminate the edge that has the greatest score.
- Recalculate all of the scores.
- Repeat the second step.

For each and every edge, the betweenness centrality of the network is calculated. Eliminate the edge that has the greatest score. Recalculate all of the scores to find the greatest betweenness centrality score. The second step gets repeats until the process gets completed. Finally, connected components are considered as communities with respect to edge-betweenness.

3.3 Kernighan–Lin Bipartition Algorithm

The Kernighan–Lin algorithm is a network partitioning iterative technique that produces very closely related outcomes. The simplest form of the method divides the input graph into two blocks of equal size. This algorithm is based on modularity score which is donated as ΔQ. The procedure of Kernighan–Lin bipartition algorithm is given below.

- Initially, partition the vertices in the network into two equivalent groups as A and B randomly.
- Compute the ΔQ for every distinct adjacent vertex between A and B.
- Switch the two pairs of vertices with the highest Q gain.
- Repeat steps 3 and 4 until all vertices have been swapped once
- Return to the series of swapping and locate the swap with the maximum Q.

3.4 Datasets

Network datasets are complex graph model which are utilized for analysis the large collection of interlinked information. The main aim of the network datasets is to understand the connectivity between the edges within a connectivity group of the network. Lesmis and the email-Eu-core-department-labels social networks are used to detect communities in this study, which is detailed below.

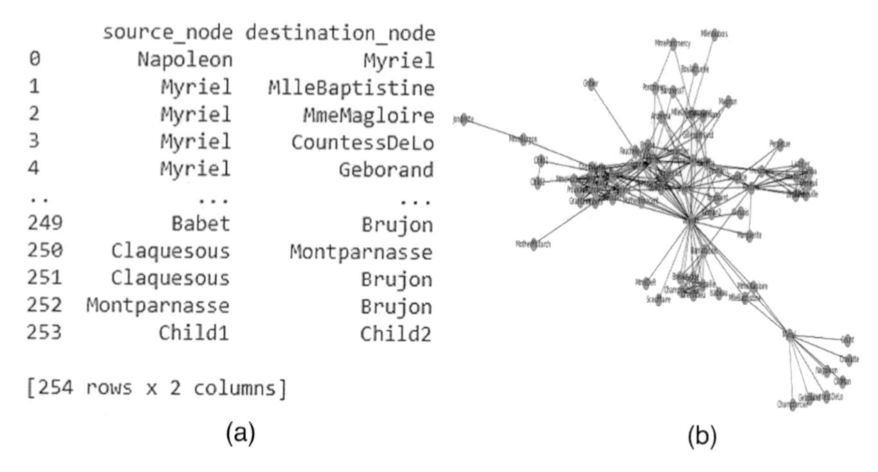

	source_node	destination_node
0	Napoleon	Myriel
1	Myriel	MlleBaptistine
2	Myriel	MmeMagloire
3	Myriel	CountessDeLo
4	Myriel	Geborand
..	...	...
249	Babet	Brujon
250	Claquesous	Montparnasse
251	Claquesous	Brujon
252	Montparnasse	Brujon
253	Child1	Child2

[254 rows x 2 columns]

(a)

(b)

Fig. 1 **a** Sample network data and **b** Initial network diagram for Les Miserable network (Lesmis)

Les Miserable graph network data. Professor Donald Knuth was originally developed this lesmis network by collected characters as novel named Victor Hugo's Les Misérables published in French in the year 1862 as part of the Stanford Graph Base. Les Miserable graph network data is the social network graph which represent interlink between characters and their interactions within the novel Les Misérables [9]. This network also called as lesmis network which is an undirected weighted disjoint network. The total number of node in the network is 77 which indicate the characters in the novel. The total number of connection within the networks is 254 which indicate the co-occurrence of characters between them. The ground-truth communities of the lesmis graph network are six. The sample network data and initial network diagram for lesmis network are shown in Fig. 1.

Email-Eu-core-department-labels network data. Email-Eu-core-department-labels network data [16] is collected from a major European research organization. This network contains only receiving and sending email among professional members of the organization is analyzed. But this network does not contain organizational member's personal messages. The research members in the organizational is 1005 are consider as the node of the network. The total number of relations between the members in the organizational through email is 25571 in which it donated as edge between the networks. The ground-truth communities for the email network are 42. The sample network data and initial network diagram for email-Eu-core-department-labels network are shown in Fig. 2.

3.5 Network Analysis

The interacting behavior of nodes within the network structure is captured by using different network measures [17]. The main motive of the network measures is

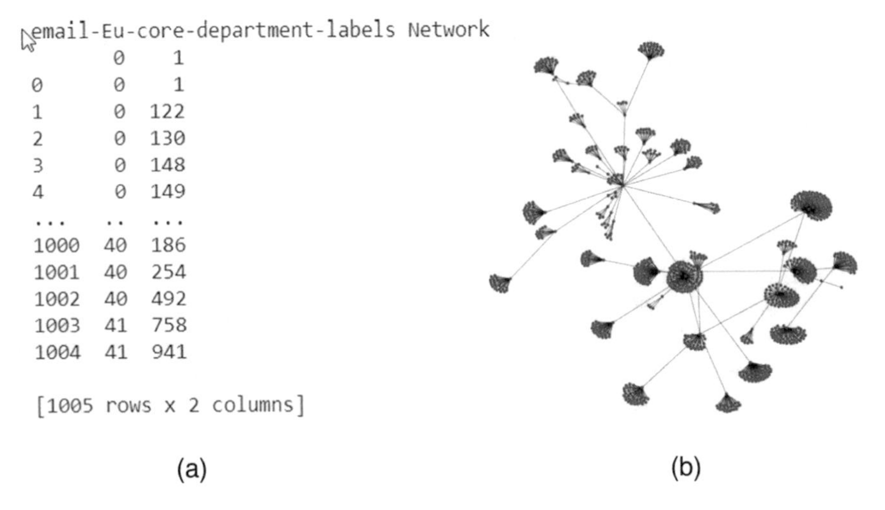

Fig. 2 **a** Sample network data and **b** Initial network diagram for email-Eu-core-department-labels network

to understand the natural structural behavior properties of the network. Network structural metrics are classified as local and global metrics based on the quantifying structural properties of a graph. Local metrics are applied to a single node in a graph, whereas global metrics apply to the whole network. The following network analysis metrics are used to analysis the network on the lesmis and email-Eu-core-department-labels networks.

Degree Centrality. The most basic centrality measure is a known as degree centrality in which it counts the number of connections between elements. The number of edges occurring to a node of a network is called its degree, and loops are counted twice in general. This is the local types of structural metrics which are applied to each and every node in a network, and it is based on node centrality measure. This degree is a measure which varies from the types of network which have been used. If the network is a directed network, then the degree is classified as in-degree and out-degree. Since in this work undirected type of the network is used, so the degree of the node is twice the amount of edges. Degree centrality is important simple measure in social network analysis to visualize the connection between node and edges in the network.

$$C_D(n) = \deg(n) \tag{1}$$

where n represents the node in the network and deg represents degree of the node.

Eccentricity. Eccentricity is a vertex centrality index measure which is calculated by computing the maximum distance between the nodes in the network. This eccentricity measure is based on node centrality measure of the network.

$$EC(V) = \max d(V, E) \tag{2}$$

where $EC(V)$ denotes eccentricity of the vertex, max d denotes the maximum number of distance, and the V and E denotes vertexes and edges in the graph.

Density. Density of the network is defined as the ratio between definite edges over all probable links, and V is the number of nodes in the network.

$$\text{Density} = \frac{V(V-1)}{2} \tag{3}$$

where E is number of edges in the graph and A_{ij} represents adjacency matrix.

Closeness Centrality. The closeness connection between each entity is from all other elements are measured by closeness. Closeness centrality metric indicates that how a node is adjacent to the remaining of the nodes in the network. This is also the local types of structural metrics that are used on every node in a network. The reciprocal of average of all minimum shortest paths between each node is used to calculate closeness centrality (C_c).

$$C_c = \frac{1}{\sum_h d(h, g)} \tag{4}$$

where $d(h, g)$ is the distance between the node g and h.

Betweenness Centrality. The significance of a node in communicating with other sets of nodes in a network is measured by its betweenness centrality. Communication flow is the foundation for betweenness centrality. The degree through which a node appears on trails between other vertices is measured. Betweenness centrality is computed by calculating the number of shortest paths in the graph that pass through the node divided by the total number of shortest paths.

$$B_c = \sum_{j<k} \frac{g_{jk}(n)}{g_{jk}} \tag{5}$$

where g_{jk} denotes the number of geodesics connecting jk, and $g_{jk}(i)$ denotes the number of geodesics that node n in the network. The minimum number of links in between any intermediate vertices is known as the geodesic distance.

The various network properties and their interacting behavior are investigated for two sample networks such as lesmis and email-Eu-core-department-labels networks using the above measures. The obtained values for triadic closure as 0.49 using lesmis network which is comparatively very less than 0.01 for email network. The email network has a density of 0.002, but the lesmis network has a density of 0.08. The reason behind the drastic variation in density values is that the lesmis is a small network, whereas email is a larger network. The minimum degree value for both the network is same as value one. The maximum and average degree values for the lesmis are 36 and 6.6. But for email network, the average degree is two and the maximum degree value as 623, which is higher than lesmis network. This deviation in this degree value affects the detecting of communities from the network. This variation indicates that lesmis network provides better results than email network. The summarization of the network analysis measure for the lesmis and email-Eu-core-department-labels networks is tabulated in Table 1.

4 Experiment and Results

The experimental analysis of community detection methods is achieved by using Google Colab with Python. Community detection is the technique for extracting the communities from networks in which it used to study and analysis the structural properties of the social network. This structural analysis of the network is accomplished by using different network analysis measures. The extraction of communities from large network is an exhausting task for determining the exact communities by using community detection algorithms with respect to ground-truth communities. A complex network is being partitioned in a variety of methods. This work is carried out through Kernighan–Lin bipartition, Girvan–Newman, and greedy modularity maximization on the lesmis and email-Eu-core-department-labels networks. The effectiveness of the community detection algorithm on these networks is assessed by using performance evaluation metrics with respect to ground-truth communities of the network. The assessment of this experiment is carried out using different performance evaluation metrics like normalized mutual information (NMI), precision, recall, and F1-measure.

The Girvan–Newman method removes the edges repetitively with the highest group of shortest routes by connecting vertices crossing the network. The Girvan–Newman methods implement to detect communities on the lesmis and email-Eu-core-department-labels networks. The lesmis network has six ground-truth communities, whereas the email network has 42 ground-truth communities. Girvan–Newman method detects five communities for lesmis network and seven communities for email network. Communities detected using Girvan–Newman algorithm from the lesmis and email-Eu-core-department-labels networks with respect to ground-truth values are visualized in Fig. 3.

The performance analysis of this work studies the effectiveness of the Girvan–Newman method on undirected disjoint networks like the lesmis and email-Eu-core-department-labels networks. The performance evaluation measure includes normalized mutual information (NMI), precision, recall, and F1-measure and are used to evaluate this observation. The Girvan–Newman algorithm got the precision value of 0.67 for the lesmis network and 0.86 for the email-Eu-core-department-labels network. This variation in higher precision values illustrates that email network work better than lesmis network based on the detected communities. The recall and F1-score value for lesmis network are 0.57 and 0.62. But when comparing, the lager email network values of recall and F1-score measures as 0.014 and 0.25 which are comparatively less than lesmis network. Both the network obtains the same NMI score values as one with respect to the ground-truth communities. The comprehensive results of performance analysis measures for Girvan–Newman algorithm illustrate that lesmis network provides the better results ground-truth communities.

Table 1 Network analysis of lesmis and email-Eu-core-department-labels networks

S. No.	Network analysis	Average clustering coefficient	Average degree	Density	Maximum degree	Minimum degree	Transitivity
1	Lesmis network	0.57	6.6	0.08	36	1	0.49
2	Email-Eu-core-department-labels network	0.4	2	0.002	623	1	0.01

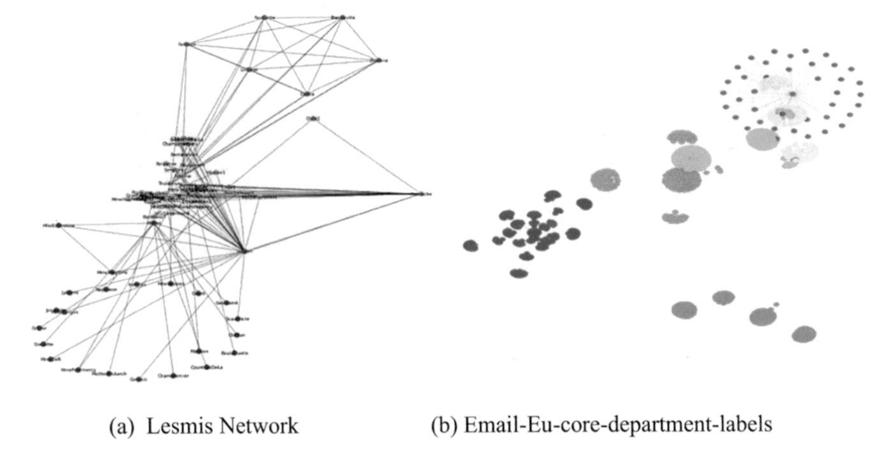

(a) Lesmis Network (b) Email-Eu-core-department-labels

Fig. 3 Community detection using Girvan–Newman algorithm

Performance analysis metrics for the Girvan–Newman algorithm yields best results on lesmis network which are despite in Table 2 and graphically visualized in Fig. 4.

Kernighan–Lin bipartition algorithm is a heuristic optimization approach for dividing a network into two communities under certain criteria. Kernighan–Lin bipartition algorithm works by interchanging set of vertices from distinct partitions. The Kernighan–Lin bipartition algorithm implements to detect communities on the lesmis and email-Eu-core-department-labels networks. The lesmis network has six ground-truth communities, whereas the email network has 42 ground-truth communities. Kernighan–Lin bisection method detects five communities for lesmis network and seven communities for email network. Communities detected using Kernighan–Lin

Table 2 Performance evaluation of Girvan–Newman community detection algorithm

S. No.	Metrics	Precision	Recall	F1-score	NMI score
1	Lesmis network	0.67	0.57	0.62	1
2	Email-Eu-core-department-labels network	0.86	0.14	0.25	1

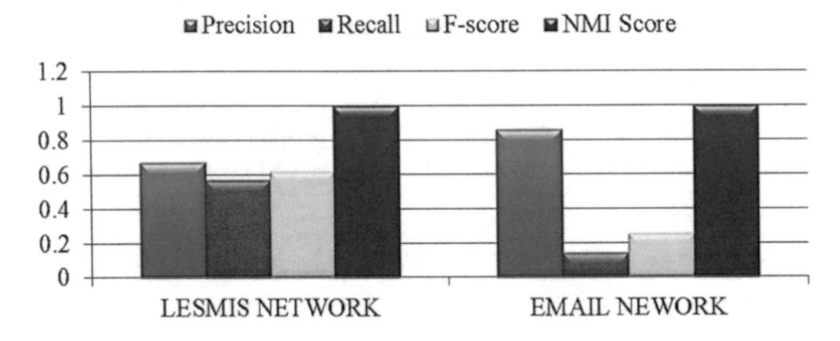

Fig. 4 Performance analysis for Girvan–Newman community detection algorithm

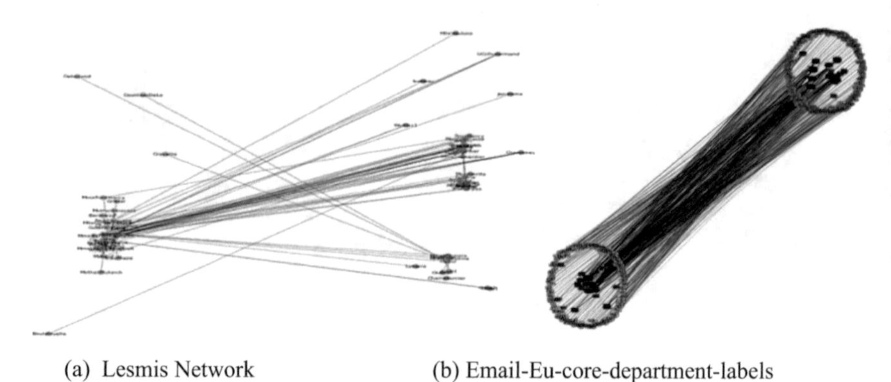

(a) Lesmis Network (b) Email-Eu-core-department-labels

Fig. 5 Community detection using Kernighan–Lin bisection algorithm

bisection algorithm from the lesmis and email-Eu-core-department-labels networks with respect to ground-truth values are visualized in Fig. 5.

The performance analysis of this work studies the effectiveness of the Kernighan–Lin bisection algorithm in undirected disjoint networks like the lesmis and email-Eu-core-department-labels networks. Kernighan–Lin bipartition algorithm has a precision value of one for both lesmis and email-Eu-core-department-labels networks. The recall and F1-score value for lesmis network are 0.5 and 0.67. But when comparing, the lager email network values of recall and F1-score measures as 0.05 and 0.09 which are comparatively less than lesmis network. This huge variation in higher precision values illustrates that lesmis network work better than email network based on the detected communities. Both the networks obtain the same NMI score values as one with respect to the ground-truth communities. The comprehensive results of performance analysis measures for Kernighan–Lin bipartition algorithm illustrate that lesmis network provides the better results ground-truth communities. Performance analysis metrics for the Kernighan–Lin bipartition algorithm yields best results on lesmis network which are despite in Table 3 and graphically visualized in Fig. 6.

Greedy modularity maximization algorithms merge the communities with highest modularity. The main objective of this optimization algorithm is to reduce the disparity between the number of connections between communities and the number of connections within communities. Greedy modularity maximization methods are

Table 3 Performance evaluation of Kernighan–Lin bisection algorithm

S. No.	Metrics	Precision	Recall	F1-score	NMI score
1	Lesmis network	1	0.5	0.67	1
2	Email-Eu-core-department-labels network	1	0.05	0.09	1

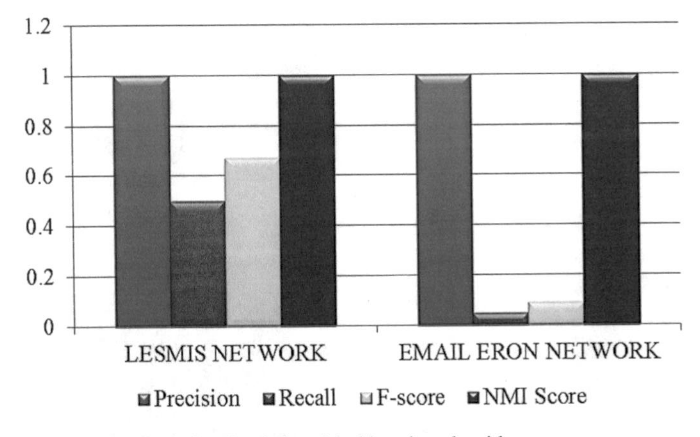

Fig. 6 Performance analysis for Kernighan–Lin bisection algorithm

used to implement for detecting the communities on the lesmis and email-Eu-core-department-labels networks. The lesmis network has six ground-truth communities, whereas the email network has 42 ground-truth communities. Greedy modularity maximization algorithm detects five communities for lesmis network and seven communities for email network. Communities detected using greedy modularity maximization algorithm from the lesmis and email-Eu-core-department-labels networks with respect to ground-truth values are visualized in Fig. 7.

The performance analysis of this work studies the effectiveness of the greedy modularity maximization algorithm on undirected disjoint networks like the lesmis and email-Eu-core-department-labels networks. The greedy modularity maximization algorithm has a precision value of one for email-Eu-core-department-labels

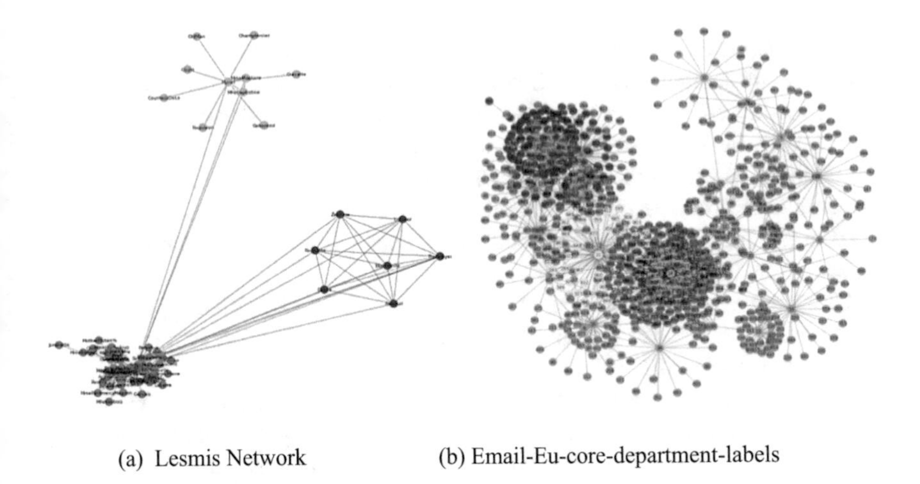

Fig. 7 Community detection using greedy modularity maximization algorithm

Table 4 Performance evaluation for greedy modularity maximization algorithm

S. No.	Metrics	Precision	Recall	F1-score	NMI score
1	Lesmis network	0.67	0.29	0.4	1
2	Email-Eu-core-department-labels network	1	0.6	0.78	1

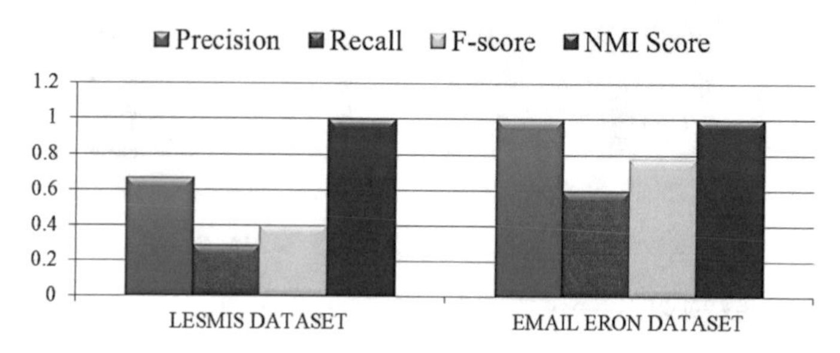

Fig. 8 Performance analysis for greedy modularity maximization algorithm

networks and 0.67 lesmis network. The recall and F1-score value for email network are 0.6 and 0.78. But when comparing, the lager email network values of recall and F1-score measures as 0.29 and 0.4 which are comparatively less than lesmis network. This huge variation in higher precision values illustrates that email network work better than lesmis network based on the detected communities. Both the networks obtain the same NMI score values as one with respect to the ground-truth communities. The comprehensive results of performance analysis measures for greedy modularity maximization algorithm illustrate that lesmis network provides the better results ground-truth communities. Performance analysis metrics for the greedy modularity maximization algorithm yields best results on email network which are despite in Table 4 and graphically visualized in Fig. 8.

This study on community identification is based on the most popular community detection methods such as Girvan–Newman, Kernighan–Lin bipartition, and greedy modularity maximization algorithms through lesmis network and email-Eu-core-department-labels social networks with respect to ground-truth communities. Comprehensive comparison of community detection approaches for both networks based on their ground-truth communities in addition to efficiency parameters such as accuracy, recall, F1-score, and normalized mutual information score which are shown in Fig. 9.

The summarization of the comparison results is achieved by utilizing performance assessment measures for Girvan–Newman, greedy modularity maximization, and Kernighan–Lin bipartition algorithms are tabulated in Table 5.

The observational results indicate that the greedy modularity maximization method has a higher precision value as one for detecting communities on email eron network. The recall and F1-score values are also high as 0.6 and 0.78 on the email eron network. This outcome specifies that greedy approach works better on larger email network. When comparing the performance evaluation results of the

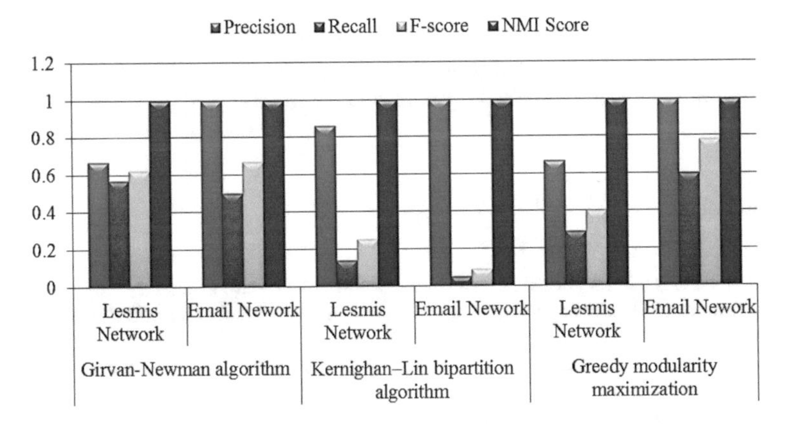

Fig. 9 Performance evaluation comparison of three community detection algorithms

Table 5 Analysis of performance evaluation using three community detection algorithms

S. No.	Metrics	Girvan–Newman algorithm		Kernighan–Lin bipartition algorithm		Greedy modularity maximization algorithm	
		Lesmis network	Email network	Lesmis network	Email network	Lesmis network	Email network
1	Precision	0.67	1	0.86	1	0.67	1
2	Recall	0.57	0.5	0.14	0.05	0.29	0.6
3	F1-score	0.62	0.67	0.25	0.09	0.4	0.78
4	NMI score	1	1	1	1	1	1

greedy modularity maximization algorithm with Girvan–Newman and Kernighan–Lin Bisection algorithms, the greedy modularity maximization technique gives best results with respect to ground-truth communities.

5 Conclusion

This work studied the effectiveness of the community detection approaches in social networks. The greedy modularity maximization, Girvan–Newman and Kernighan–Lin bipartition approaches were implemented to discover communities from the two social networks like lesmis and email-Eu-core-department-labels networks. This study was implemented by using Python in Google Colab. Precision, recall, F1-measure, and normalized mutual information score were the performance analysis measures used to evaluate the outcomes. It was observed that greedy modularity maximization provides the better results for community detection on lesmis networks based on ground-truth communities. In the future, the suggested approach might be used to discover communities by executing it on an artificially formed social network with minimal computing time. Artificial networks may provide a significant solution to community identification challenges in social networks.

References

1. Cherifi H, Gaito S, Mendes JF, Moro E, Rocha LM (eds) (2020) Complex networks and their applications {VIII}. Springer International Publishing. https://doi.org/10.1007/978-3-030-36687-2
2. (2010) Handbook of social network technologies and applications. https://doi.org/10.1007/978-1-4419-7142-5
3. Javed MA, Younis MS, Latif S, Qadir J, Baig A (2018) Community detection in networks: a multidisciplinary review. J Netw Comput Appl 108:87–111. https://doi.org/10.1016/j.jnca.2018.02.011
4. Ghoshal AK, Das N, Das S (2021) Influence of community structure on misinformation containment in online social networks. Knowl-Based Syst 213:106693. https://doi.org/10.1016/j.knosys.2020.106693
5. Wang X, Li J, Yang L, Mi H (2021) Unsupervised learning for community detection in attributed networks based on graph convolutional network. Neurocomputing 456:147–155. https://doi.org/10.1016/j.neucom.2021.05.058
6. Girvan M, Newman MEJ (2002) Community structure in social and biological networks. Proc Natl Acad Sci USA 99:7821–7826. https://doi.org/10.1073/pnas.122653799
7. Rafique W, Khan M, Sarwar N, Dou W (2019) SocioRank*: a community and role detection method in social networks. Comput Electr Eng 76:122–132. https://doi.org/10.1016/j.compeleceng.2019.03.010
8. Inuwa-Dutse I, Liptrott M, Korkontzelos I (2021) A multilevel clustering technique for community detection. Neurocomputing 441:64–78. https://doi.org/10.1016/j.neucom.2021.01.059
9. Rossi RA, Ahmed NK (2015) The network data repository with interactive graph analytics and visualization. In: AAAI
10. Newman MEJ, Girvan M (2004) Finding and evaluating community structure in networks. Phys Rev E—Stat Nonlinear Soft Matter Phys 69:1–15. https://doi.org/10.1103/PhysRevE.69.026113
11. Liu Z, Ma Y (2019) A divide and agglomerate algorithm for community detection in social networks. Inf Sci (Ny) 482:321–333. https://doi.org/10.1016/j.ins.2019.01.028
12. Chen X, Xia C, Wang J (2018) A novel trust-based community detection algorithm used in social networks. Chaos, Solitons Fractals 108:57–65. https://doi.org/10.1016/j.chaos.2018.01.025
13. Sathiyakumari K, Vijaya MS (2020) Association rule mining for clique percolation on community detection. Int J Adv Sci Technol 29:1492–1503
14. Savić M, Ivanović M, Jain LC (2019) Complex networks in software, knowledge, and social systems. Springer International Publishing. https://doi.org/10.1007/978-3-319-91196-0
15. Kernighan BW, Lin S (1970) An efficient heuristic procedure for partitioning graphs. Bell Syst Tech J 49:291–307. https://doi.org/10.1002/j.1538-7305.1970.tb01770.x
16. Leskovec J, Krevl A (2014) {SNAP Datasets}: {Stanford} large network dataset collection
17. Sathiyakumari K, Vijaya MS (2016) Community detection based on Girvan-Newman algorithm and link analysis of social media. Commun Comput Inf Sci 679:223–234. https://doi.org/10.1007/978-981-10-3274-5_18

Segmentation of Handwritten Characters from Answer Scripts

M. Ravikumar and S. Sampathkumar

Abstract In this work, we propose a method for segmentation of handwritten Kannada character from answer scripts. Character segmentation plays an important role in Kannada optical character recognition (OCR) system, because characters incorrectly segmented perform to unrecognized character. This paper provides improved segmentation algorithm based on contour and bounding box method. To improve segmentation accuracy, built system works on two stages: preprocessing and segmentation stage. The modified abovementioned method has been successfully tested on 100 real-time documents of Kannada handwritten scripts collected from different schools. The results are very promising, indicating the efficiency of the suggested approach.

Keywords Answer scripts · Kannada · Preprocessing · Segmentation · OCR · Augmentation

1 Introduction

Segmentation of characters is a very challenging and important step for any OCR (optical character recognition) system. It segregates the handwritten text into lines, words, and characters. To build high accuracy OCR or to yield good results, segmentation of characters is needed. Usually, handwritten word or text written by person varies from person to person, as the nature of writing style of persons' changes depending on font styles, cursive, and touching characters. It is complicated to segment these types of handwritten text. Cursive handwriting character segmentation is very difficult task for many real-time applications. From the last few decades, to achieve high recognition accuracy, many segmentation techniques are proposed

M. Ravikumar (✉) · S. Sampathkumar
Department of Computer Science, Kuvempu University, Jnanasahyadri, Shimoga, Karnataka, India
e-mail: ravi2142@yahoo.co.in

S. Sampathkumar
e-mail: sampath1447@gmail.com

A. Kumar et al. (eds.), *Proceedings of the International Conference on Cognitive and Intelligent Computing*, Cognitive Science and Technology,
https://doi.org/10.1007/978-981-19-2350-0_83

but till now there is no universal segmentation method proposed for segmentation of handwritten text. Kannada is the regional language of Karnataka. Segmentation of Kannada handwritten word into character is also one of the most tedious and ongoing challenging tasks from few decades. As it contains vowels and consonants and apparent mixture of overlapped ottaksharas, vyanjanas, and its structured complexity as compared with Latin-based languages, Kannada handwritten text is difficult to segment. Moreover, Kannada language has distinct characters [1], and segmentation of handwritten Kannada script into lines, words, and character is of great importance and much demanded for some specific applications. Segmentation of handwritten Kannada script poses major challenges due to additional modifier characters.

Some of the major difficulties while segmentation of handwritten text are as follows:

There can be variation in shapes and font styles of different writers connected or overlapped nature of kannada handwriting containing ligatures. Kannada language contains ottaksharas as it is overlapped with other characters. Some characters have similar contours. Kannada scripts contain semi-cursive characters which is difficult to segment.

To get a desirable result in a recognition phase, a robust and efficient character segmentation is needed. Only proper segmentation will improve the performance of OCR. Without segmentation of characters, we cannot directly build an efficient OCR and it is needless to state decomposition of individual characters of a word reveals that much more effort is required to develop a complete optical character recognition (OCR). To segment the characters individually, directly characters cannot be segmented, so before that, some segmentation steps have to be done, i.e., line segmentation and word segmentation because to get a high accuracy OCR. Recognition can be carried either for a single character or the whole word. For any recognition process either character or word, fragmentation of single character in a word is necessary. To achieve best results, three stages of segmentation are to be performed, they are as follows.

1.1 Text Segmentation

From the document, entire lines are initiated from top to bottom. It represents a key step in recognition of many optical character recognition systems [2]. Handwritten text line segmentation still remains an open research field. Hough transform method, smearing method, and touching method are used to segment text.

1.2 Word Segmentation

Word segmentation is one of the preprocessing steps for any handwritten document. The main aim of word segmentation is to extract whole word image in a handwritten document. It is necessary that word segmentation is very important in a retrieval system. To build a proper character segmentation and to get a good result, line and word segmentation is necessary. There are several segmentation methods available in literature such as projection profile method and connected component analysis.

1.3 Character Segmentation

Several segmentation techniques are proposed for achieving high recognition accuracy. Recognition accuracy depends on the correct segmentation. In this paper, methods have been proposed to extract answer and to segment an answer scripts into sentences, words, and then finally characters for isolated character. Figure 1 shows the various steps used as a flowchart. Further methods have been used to preprocess the image in such a way that reduces the memory needed by the models to process the characters.

The paper is organized in the following sections: Sect. 2 contains the review of related work. Section 3 gives brief idea of proposed method used. Section 4 contains results and discussion, and Sect. 5 explains the conclusion of this research work.

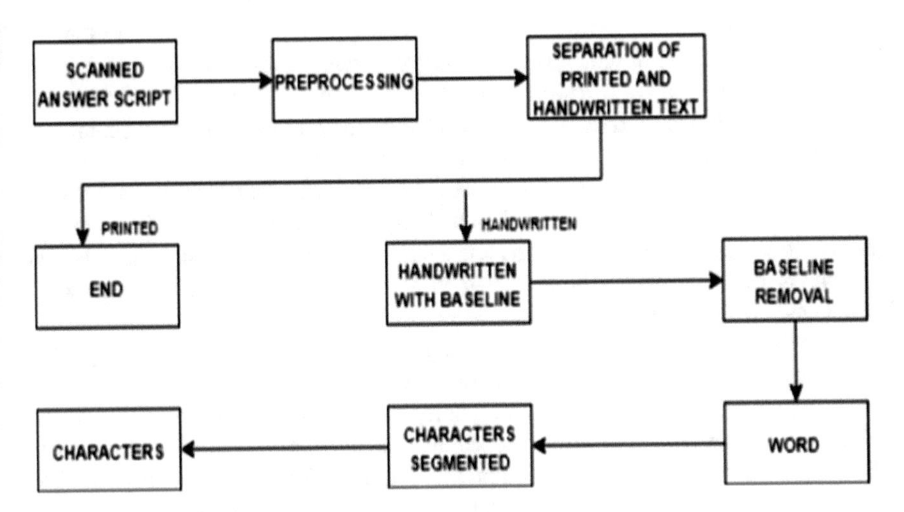

Fig. 1 Various stages of proposed method

2 Related Work

In this review paper [3], Kannada character segmentation algorithm is presented. The proposed method is based on the thinning, branch points, and mixture models. From touching lines, characters are segmented, the proposed method works good for handwritten Kannada words and has shown very well results up to 85.5, connected component analysis algorithm is proposed [1], and segmentation of handwritten characters and feature extraction are carried out by using feature extraction algorithms. For recognition and classification of characters, support vector machine classifier is used. Experimentation explains that an improvement in accuracy and validity for segmentation of characters.

Projection profiles and morphological operation methods are proposed [4], and Kannada handwritten scripts are segmented into lines, words, and typescript. In preprocessing phase, organizing and disintegration are done. The main goal is to concentrate on singular content line, and the experiments are done on absolutely unconstrained Kannada handwritten scripts. Recognition based on zoning is proposed [5], feature extraction method is used for character recognition, and support vector machine classifier is used for classification. The algorithm used for training and testing are used to compare handwritten characters. In the preprocessing stage, cropping, resizing, and noise removal are done. Median filter is used to enhance the image. In the segmentation stage, ON pixel and OFF pixel are taken as X and X1 and X and X2 to find all the characters in a word. At last, classifier is applied which increases the speed of the accuracy and which requires less memory to store training samples.

A twelve-directional feature extraction technique is proposed [6], and data is collected from 25 different writers having different handwritten styles. A feed forward back propagation neural network is used for classification and recognition of Kannada handwritten characters. Morphology and projection technique is to segment the characters. Experimentation is carried out, and feature extraction method and classifier give better accuracy rate. Hidden Markov model method is proposed [7], and implicit segmentation technique is used to avoid explicit segmentation. To segment characters and to reduce number of classes in a characters, HMM method is used. Experiment is conducted on 74 k char dataset, and to train character, gradient-based feature is extracted.

Water reservoir and thinning concept is proposed [8], and connected component analysis (CCA) algorithm is to detect the components in a word. Based on touching position, close loop positions and morphological structure of touching region, the character cutting path is generated to segment characters. Proposed concept holds best results. Skew detection, correction, and segmentation method is proposed [9], and bounding box technique is applied to skew angle detection and correction. It is a way to find the extreme corners of text image. Skew correction is done by rotating the document through an angle with respect to horizontal line. Unconstrained handwritten Kannada script is segmented into lines and words. For line segmentation, Sobel edge detector is applied which lists the points in an image. Experiment is conducted on 40 document images; an average segmentation rate 0f 91% and 70% for lines and words is obtained. Horizontal projection profile and windowing method is proposed [10], and to segment and recognize, feedback-based approach is applied. The gap between segmentation and recognition phase is fixed by attempting a proposed method. KHTD standard dataset is used and got recognition accuracy of 95.02%. From the literature, some more good works on character segmentation are mentioned [11–15].

3 Proposed Method

In this section, we discuss the proposed method for handwritten answer extraction at word level and segmentation of characters.

Scanned Kannada handwritten answer scripts are collected, and dataset is shown in Fig. 1. In the step 2, preprocessing techniques contain removing of shadow or dark spots, ink smudges, printer scales while scanning, etc. It is necessary to process the answer script to get necessary data. It helps in saving of memory required to store the answer script data. To perform further analysis on the image is drastically reduced. In the step 3, printed and handwritten text is separated from answer script because we are concentrating only on the segmentation of handwritten answer by using baseline method. In the step 4, horizontal line detection is applied to extract exact handwritten answers and word is segmented with baseline from answer script. In the step 5, subtraction of baseline is applied, so that word is passed to further stage. The step 6 is a segmentation stage; here, character is segmented. Based on contours and bounding box method, characters are segmented individually. The bounding box method chosen for character segmentation provides considerable results.

In the abovementioned equation, the P determines pixel strength of the text image. To segment a line, the coordinates x, y are the function to calculate the line where $y = mx + c$. l is the length of the line. For baseline.

The above equation represents the criteria for assigning a baseline to a particular text line, where P is the particular length of baseline present in the answer. In the next section, experimentation is discussed.

4 Results and Discussion

For the purpose of experimentation, we have created our own dataset. The dataset is created in such a way that initially we have formulated 10 questions that is in printed form and all questions are one-word answers. In that, we have provided a baseline to write answers. The student has to answer above that baseline. We conducted experimentation on 3 cases. Case 1 contains dataset with both question and answer written in handwritten form. Case 2 contains combination of printed and handwritten text without ottaksharas dataset, and Case 3 contains printed and handwritten text

Table 1 Test accuracy of combination handwritten with and without ottakshara characters

Answer script	Total characters in extracted answer	Correctly segmented characters	% of characters correctly segmented
1	50	40	80.0
2	50	49	98.0
3	45	42	93.3
4	60	53	88.3
3	40	38	95.0

with ottaksharas. This type of dataset is complicated in case of answer detection. In this way, we have created 100 documents. All the answer scripts are written by 100 different students.

The results of the aforementioned character segmentation techniques and results are displayed.

CASE 1: Both question and answer are in the handwritten text

Table 1 illustrates the answer scripts with both question and answer written in handwritten. Second column explains the total characters extracted from answer, the third column shows the result of correctly segmented characters, and finally, percentage of segmented characters is shown. By observing from table, we came to know that highest efficiency gained is 98.0% for the script 2. The least accuracy gained is 80.0% for script 1, because of too many variations present in the characters.

Fig. 3 explains the graphical representation of both question and answer which are in the handwritten text. The results of the answer script vary from one to other as it contains unconstrained characters. Better results are achieved for segmentation of characters.

CASE 2: Combinations of printed and handwritten text (without ottakshara)

Table 2 illustrates the answer scripts without ottakshara present in words. Second column explains the total characters extracted from answer, the third column shows the result of correctly segmented characters, and finally, percentage of segmented characters is shown. By observing from table, we came to know that highest efficiency achieved is 92.7% for the script 3. The least accuracy achieved is 84.0% for script 1, because of too many variations present in the character.

Fig. 4 explains the graphical representation of handwritten answers without ottakshara characters. The highest result achieved is 92.7%. In the bar graph, we have shown results for 10 answer scripts but in the tables for 3 cases, we mentioned results for only 5 answers (Fig. 5).

CASE 3: Combinations of printed and handwritten text (with ottakshara)

Table 3 illustrates the answer scripts with ottaksharas present in words. Totally there are 5 answer scripts. Each row in the second column of the table represents the answers extracted from 5 scripts. In the third column, results of correctly segmented

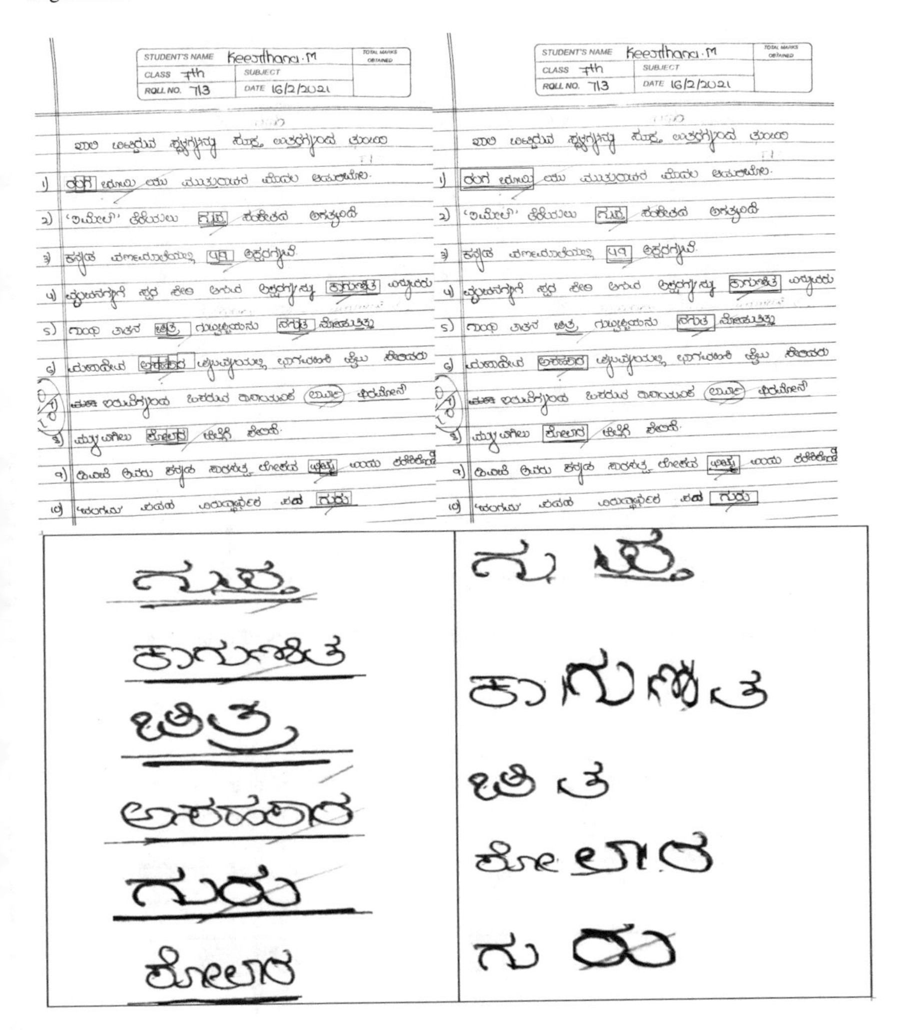

Fig. 2 **a** Original script. **b** Script with answer detection. **c** Words with baseline. **d** Segmented characters without baseline

characters are mentioned. Finally, percentage of each answer script segmentation results is shown. By observing from table, we came to know that highest efficiency achieved is 92.7% of script 3, because the characters present in the script have least number of characters without ottaksharas, and font style of the character is clear as compared with other answers. The least accuracy achieved is 84.0% for script 1, because of invariant styles of character present in the answer.

To the best of our knowledge, no work has been done on Kannada answer scripts, and from the survey, we came to know that many researchers have worked on Kannada word segmentation with different dataset but our research work aims on answer scripts. Most of the words are presented with ottaksharas with one or two ottaksharas

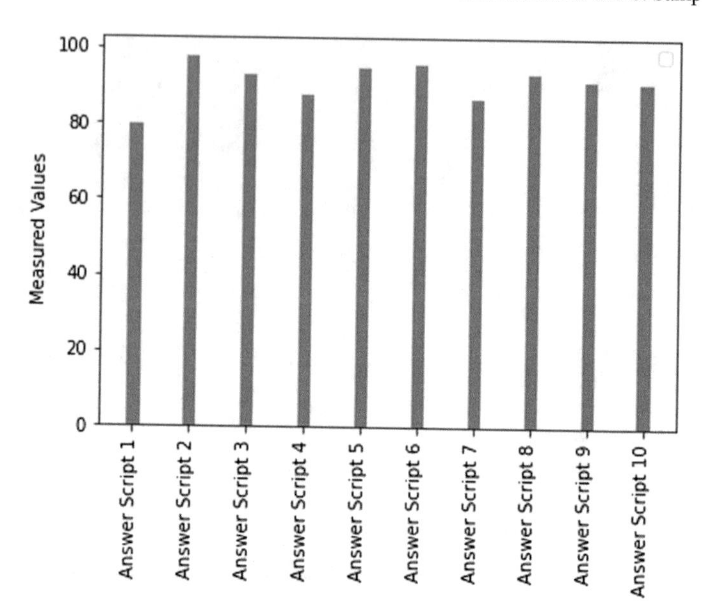

Fig. 3 Graphical representation of measured values for combinations of ottakshara and without ottakshara character segmentation

Table 2 Test accuracy of handwritten Kannada without ottakshara characters

Answer script	Total characters in extracted answer	Correctly segmented characters	% of characters correctly segmented
1	50	42	84.0
2	40	29	72.5
3	55	51	92.7
4	60	53	88.3
3	25	23	92.0

present below the characters. To segment a characters into ideal parts of individual characters, bounding box and contour method is used. The obtained segmentation accuracy of the proposed method is 86.29%. The successful segmentation results are shown in fig (D) of every cases. The least accuracy obtained is 84.0% for without ottakshara, and for without ottakshara, accuracy is 92.07%. It contains complicated characters with touching consonants. Our dataset consists of unconstrained words with touching characters. If the dataset has constrained characters which is separated and untouching baseline, we can achieve 100% results. The main drawback of the proposed system is wherever one or more ottaksharas are present below the character, the accuracy will be less (Fig. 6).

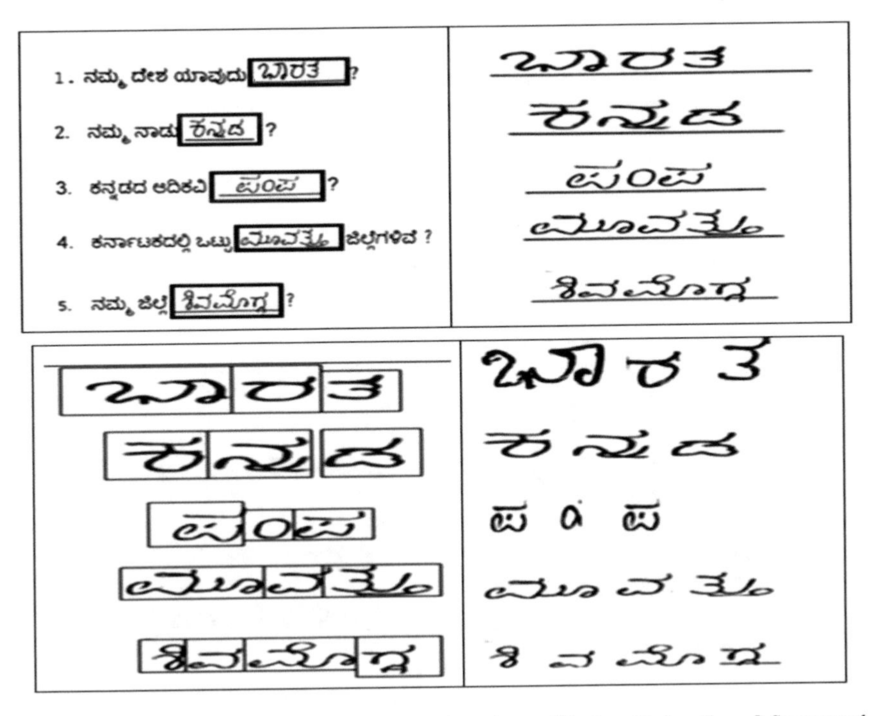

Fig. 4 **a** Original script. **b** Script with answer detection. **c** Words with baseline. **d** Segmented characters without baseline

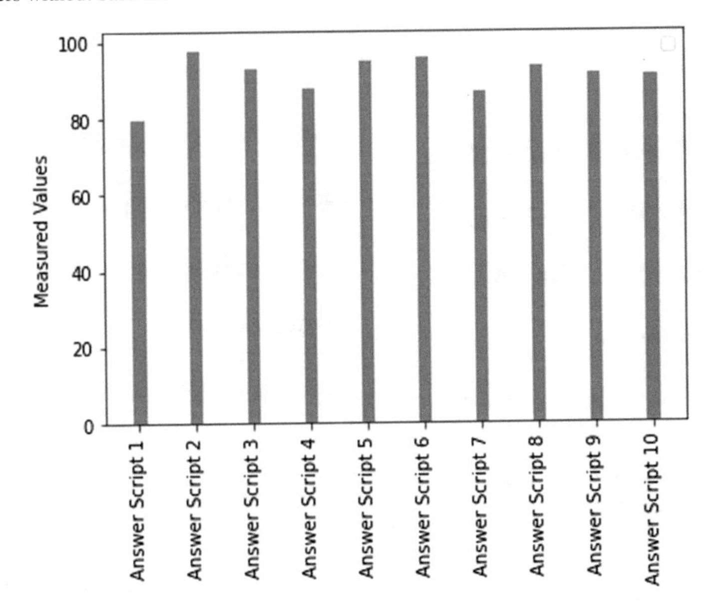

Fig. 5 Graphical representation of measured values for combinations of ottakshara and without ottakshara character segmentation

Table 3 Test accuracy of handwritten Kannada with ottakshara characters

Answer script	Total characters in extracted answer	Correctly segmented characters	% of characters correctly segmented
1	50	42	84.0
2	40	29	72.5
3	55	51	92.7
4	60	53	88.3
3	25	23	92.0

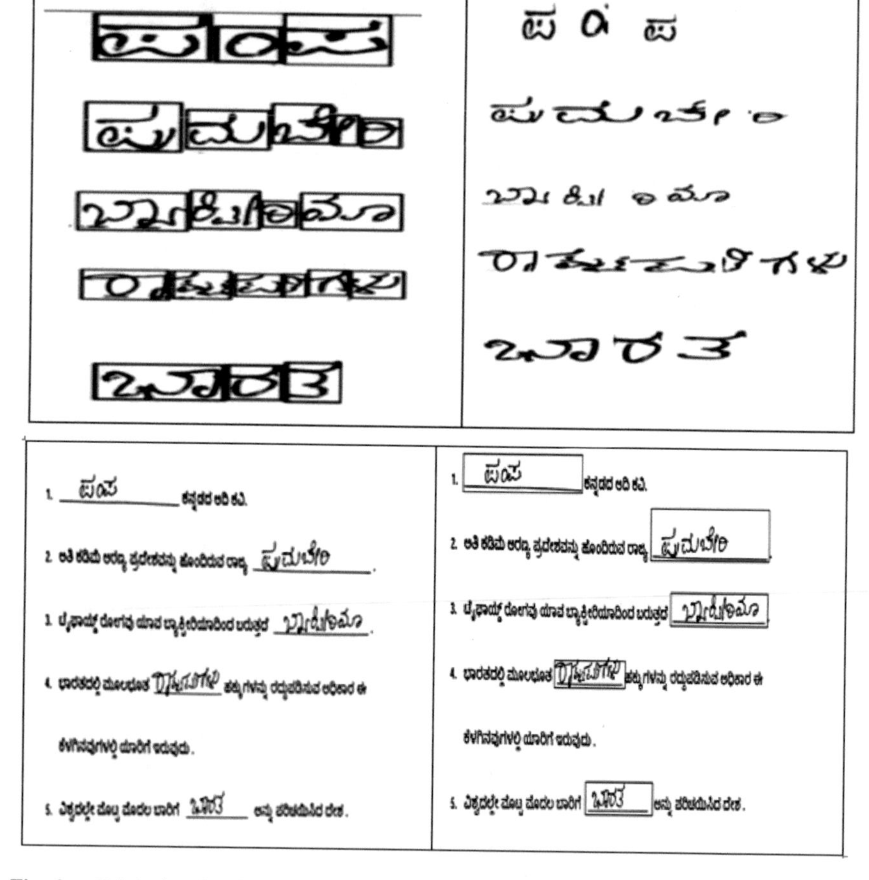

Fig. 6 **a** Original script. **b** Script with answer detection. **c** Words with baseline. **d** Segmented characters without baseline

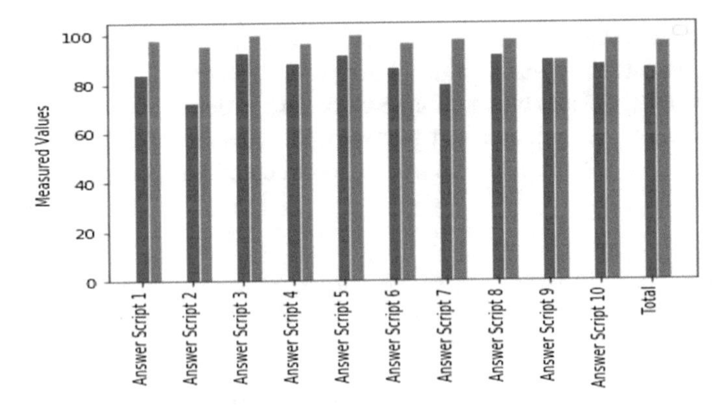

Fig. 7 Graphical representation of measured values for ottakshara and without ottakshara character segmentation

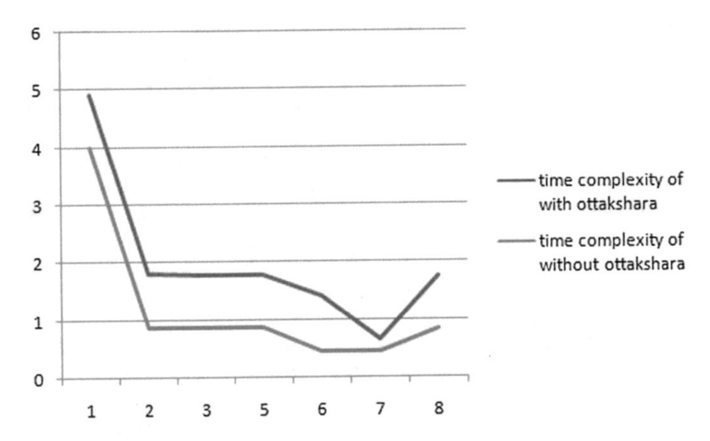

Fig. 8 Graphical representation of measured values for ottakshara and without ottakshara character segmentation

Fig. 7 represents the time complexity of above 3 cases. The x-axis shows the result of measured values in minutes for with and without ottakshara and combination of printed and handwritten dataset (Fig. 8).

5 Conclusion

In this paper, techniques have been proposed to extract handwritten answer from answer script, and segmentation of handwritten answer is shown. The proposed work explains the usage of openCV, contour, horizontal line detection, and bounding box method. From the Tables 1, 2, and 3, it can be observed that the abovementioned method proves to be good model for answer detection and segmentation of characters. Experiments show that significant improvements are achieved in both answer extraction and segmentation. Our future work aims to develop still better character segmentation and recognition tool, which provides better result for answer evaluation system. Also, we plan to extend our work to more complicated Kannada handwritten dataset [16–20].

References

1. Mr. Chethan Kumar G.S, Dr. Y.C Kiran (2016) Kannada handwritten character segmentation using curved line and recognition using support vector machine. Int J Combined Res Dev (IJCRD) 5:641–645
2. Razak Z, Zulkiflee K et al (12 June 2018) Off-line handwriting textline segmentation: a review. Int J Comput Sci Netw Sec
3. Naveena C, Manjunath Aradhya VN (2012) Handwritten character segmentation for Kannada scripts. World congress on information and communication technologies. pp 144–149
4. Mamatha HR, Srikantamurthy K (Oct 2012) Morphological operations and projection profiles based segmentation of handwritten Kannada document. Int J Appl Inf Syst 4(5) ISSN: 2249-0868
5. Tonashyal R, Kiran YC (June 2015) Offline handwritten Kannada character segmentation and recognition based on Zoning. Int J Comput Sci Inf Technol Res 3(2):1107–1114
6. Venkatesh M, Majjagi V, Vijayasenan D (2014) Implicit segmentation of Kannada characters in offline handwriting recognition using hidden Markov models. 1–6
7. Chethan Kumar GS, Kiran YC (April 2016) Kannada handwritten character segmentation using curved line and recognition using support vector machine. Int J Combined Res Dev (IJCRD) 5(4)
8. Ramappa MH, Krishnamurthy S (Nov 2012) Skew detection, correction and segmentation of handwritten Kannada document. Int J Adv Sci Technol 48
9. Karthik S, Murthy KS (2016) Segmentation and recognition of handwritten Kannada text using relevance feedback and histogram of oriented gradients—a novel approach. Int J Adv Comput Sci Appl 7(1)
10. Alaei A, Nagabhushan P, Pal U (2011) A Benchmark Kannada handwritten document dataset and its segmentation. International conference on document analysis and recognition. pp 141–145
11. Alaei A, Pal U, Nagabhushan P (2011) A new scheme for unconstrained handwritten text-line segmentation. Pattern Recogn 44(4):917–928
12. Smith LN (2017) Cyclical learning rates for training neural networks WACV
13. Muller AC, Guido S (2016) Introduction to machine learning with Python—a guide for data scientists. Sebastopol, CA, O'Reilly Media, Inc
14. Goodfellow I, Bengio Y, Courville A (3 Jan 2017) Deep learning. MIT Press
15. Klep D, Schoenmakers GS, Van M (3 June 2016) Data augmentation of a handwritten character dataset for a convolutional neural network and integration into a Bayesian linear framework
16. Azhaguvarthani G, Ramani B (2017) Handwritten text recognition system for English. Int J Pure Appl Math 117(16). ISSN: 257-263
17. Gatos B, Stamatopoulos N, Louloudis G (2009) ICDAR 2009 handwriting segmentation contest, proceedings of 10th ICDAR. pp 1393–1397
18. Aradhya VNM, Naveena C (2011) Text line segmentation of unconstrained handwritten Kannada script. ICCCS'11, pp 231–234

19. Sen S, Prabhu SV, Jerold S, Pradeep JS, Choudhary S (2018) Comparative study and implementation of supervised and unsupervised models for recognizing handwritten Kannada characters, 3rd IEEE conference on recent trends in electronics, information international and communication technology (RTEICT-2018). pp 774–778
20. Chollet F and others (2015) Keras. https://keras.io

A Survey for the Early Detection and Classification of Malignant Skin Cancer Using Various Techniques

Jinu P. Sainudeen and N. Meenakshi

Abstract The unrepaired DNA in skin cells causes skin cancer disease which leads to mutations or genetic defects in the skin. This disease will be widely spread in other parts of the body which can be cured in the initial stages. So, the early deterrence of this disease is a vital factor. This skin cancer has increased the mortality rate, and the treatment is highly expensive. Researchers have undergone several techniques for skin lesion detection based upon parameters of skin such as symmetry, color, size, and shape. These parameters are useful for distinguishing non-melanoma cancer from melanoma cancer. This paper gives a detailed study of the early detection of skin cancer with the aid of image processing, machine learning, and deep learning techniques.

Keywords Skin cancer detection · Convolutional neural networks · Deep learning · Transfer learning

1 Introduction

Skin cancer is the abnormal growth of cells in the skin which may invade into other parts of the body. Majority of the disease is caused due to the exposure of UV radiation. Other causes include sunburn, previous history of melanoma, people living in geographic locations, eczema, psoriasis, xeroderma pigmentation, and over exposure to particular chemicals. The unrepaired DNA in skin cells causes skin cancer which leads to the mutations or genetic defects in the skin. Mainly, the arrival of skin cancer is from the moles on the skin which leads to the inflammation around epidermal layer. It will cause drastic increase in the temperature.

J. P. Sainudeen (✉) · N. Meenakshi
Information Technology, Hindustan University, Chennai, India
e-mail: jinups@gmail.com

N. Meenakshi
e-mail: nmeenakshi@hindustanuniv.ac.in

A. Kumar et al. (eds.), *Proceedings of the International Conference on Cognitive and Intelligent Computing*, Cognitive Science and Technology,
https://doi.org/10.1007/978-981-19-2350-0_84

Basal cell carcinoma, squamous cell carcinoma, and malignant melanoma are some of the categorization of the skin cancer. Former two are non-malignant, and later one is malignant.

Melanoma and non-melanoma are the most dangerous and widespread skin cancer. Of this most dangerous one is melanoma which have raised the mortality. This cancer has been spread over 5.4 million [1] years in USA. Based upon the annual calculation, there is around 52% increase in melanoma. In the coming decades, there will be an increase in mortality. If diagnosis of the skin cancer is delayed, the survival rate will be less than 15%. By the early detection, we can raise the survival rate. This raises the necessity of skin cancer deterrence system. For the detection of skin cancer, a well-experienced dermatologist has to inspect the lesions, undergoes some microscopical process, and has to undergo biopsy. This time-consuming process led the patients to the last stage of cancer. A perfect accurate result is obtained based upon the skill of the doctors which is subjective. The above said has only around 80–85% of accuracy in proper diagnosis of the skin disease. In addition to this, there is a scarcity in skilled dermatologists. For the easy diagnose of skin cancer in a very fast manner at early stages and a solution for the previously said problems, there are vast research solutions by developing models using deep convolutional neural networks. So this paper deals with the review based on the early deterrence of skin cancer using different non-invasive techniques.

2 State of the Art

2.1 *Literature Survey*

This section deals with a review on the researches done latest for the detecting skin cancer in the earlier stages and classification by implementing image processing, deep learning, and machine learning concepts. At current scenario, researchers across the world are conducting researches in machine and deep learning concepts in a wide range on medical, industrial, and agricultural image analyses/object recognition/understanding tasks and have acquired significant results in this area. Many researches have been undergone for the deterrence of skin cancer, which is a complex area. Early detection in this field will cure the disease or else this cancer will pervade to other parts of body which will lead to the death of that patient.

Reference [2] proposes a system which is integrated with a software developed with the aid of deep learning concept (CNN). Here, the classification of skin lesion is done using DEAN. RGB color space, LBP, and GLCM methods are used to bring out the features of the skin cancer image. These techniques have played vital role in classification process of skin diseases. Automated segmentation of images is done with the help of fuzzy C-means clustering. Taking the above said dataset, it has achieved an accuracy of 97.4%. This work has been compared with other traditional classifiers such as GA-ANN, CNN, and SVM and concluded that this work has achieved higher performance. This system will be extended for evaluating the relation between skin burn and also for detecting other types of skin cancer diseases.

Hasan et al. [3] demonstrated a peer-to-peer system using a deep learning technique for the deterrence of skin cancer. It uses CNN which have high accuracy in image processing. Here, a three-layered system is used. Input layer gets data from ISIC dermoscopic image set. A weight is added into it and given to the hidden layer. For detecting patterns, neurons separate the features. From the patterns obtained, it is given to output layer which performs binary classification as class 1 and class 0. Class 1 is harmful malignant cells, and class 0 is harmless cells. A ReLu activation function is applied on output layer for obtaining proper classification. Here, the results are evaluated with parameters such as accuracy, precision, recall, and f1score. 89.5% accuracy and 93.7% training accuracy have been achieved.

Reference [4] proposes a computer-aided model which classifies skin images obtained from ISIC 2019 into eight different groups of skin areas. The model is implemented with the help of DCN and transfer learning mechanism. Google Net architecture is modified to enhance the features and remove the noise by adding the layers. This is done by restoring the last three layers with new fully connected SoftMax layer and output classification layers and also by renouncing the last two layers and keeping the original fully connected layer of Google Net as features extractors. Classification purpose bootstrap multi-class support vector machine is used. This change mainly focused for detecting outliers. Performance metric is evaluated, and accuracy obtained is 94.92%.

Krishna Monika [5] proposes a deterrence and analysis system for different types of cancerous cells in skin with the help of machine learning and image processing techniques. During the initial processing, input is taken from dermoscopic images. For removing noises in the image system, Dull razor method is used on the skin lesion, and for image smoothing, Gaussian filter is used. Median filter is considered for filtering the noise and for preserving the edges of the lesion. Color-related clustering using k-means is performed in segmentation stage for analyzing the type of cancer. ABCD and GLCM are used for extracting the statistical and texture features. Work is done on ISIC 2019 dataset which contains eight variants of dermoscopic images. MSVM was used for classification purpose, and 96.25% accuracy is obtained on this system.

In the research [6], network model is designed with the help of meta-heuristic enhanced CNN classifier which is already pre-trained. The weight and biases of the CNN model are improved with the help of whale optimization algorithm. Comparative study of this method is done with different classifiers on DermIS Digital Database and Dermquest Database. By evaluating the performance, they obtained an accuracy 91%, NPV 95%, PPV 84%, specification 92%, and sensitivity 95%. Feature extraction should be extended to get more features so as to achieve more accuracy to improve the efficiency of the system.

Research work conducted on [7] used a triplet loss function for the classification purpose with the help of deep CNN model. Differential features of skin disease were learned from this model. Here, each layer is fine-tuned instead of considering blockwise for improving the performance of the peer-to-peer learning method.

Kaur et al. [8] propose a deterrence system for skin cancer (BCC) which is designed in MATLAB. First, the image should undergo initial processing; then, features have to be extracted; and the finally, classification will be done. For initial processing of the images, K-mean clustering is implemented for identifying both the foreground and background of an image and checks the background after k-means. This issue can be solved using particle swarm optimization algorithm. Feature extraction is implemented using Speed Up Robust Features. Both segmented image features are extracted which can be used for enhancing the quality of the image. Based upon the extracted features, ANN is trained. For determining the efficiency of the system, true positive rate, false positive rate, precision, and accuracy are evaluated. Obtained parameter values are as follows: Accuracy is about 98.75, TPR is 94.9, FPR is 71, and precision is 97.1.

Reference [9] analyzes skin images, which chooses the best mathematical classifier model for classification purpose for easy diagnosis of melanoma. Here they have concluded that the logistic regression model was the most efficient model. Detection of melanoma using logistic regression was concluded as the best model with 97.0%. Classification based on gradient boosting was found as the second-best method.

In [10], a system was developed which classifies skin lesion on the basis of transfer learning, deep learning, and IoT system. Affected skin areas are classified based on transfer learning and CNN. They have combined different CNN models with different classifiers and the accuracy value obtained is 96.805% while using ISBI-ISIC and 93.167% is accuracy while using PH 2. Accuracy value obtained is 96,805% while using ISBI-ISIC and 93.167% while using PH 2.

In another research paper [11], pre-processing is done on input image for noise removal, and then, for segmentation, ROI based-features are extracted with K-means. Features are augmented; then, transfer learning with CNN is applied for learning; and a model is created for testing purpose.79.5% accuracy is obtained.

Li et al. in [12] developed a Lesion Indexing Network for image segmentation and classification and formed a Lesion Feature Network for lesion feature extraction. While using the Lesion Indexing Network they have obtained 75.3% of the Jaccard index and an accuracy of 91.2%. In LFN, it achieves the best average precision and sensitivity of 42.2% and 69.3%.

Hosny [13] proposed a model on the basis of deep learning and transfer learning mechanisms. Here, input images are fine-tuned and augmented by AlexNet. A transfer learning mechanism is applied by SoftMax, which is done by replacing last layer of the architecture. Obtained parameter values after the classification of skin lesion are accuracy 98.6, sensitivity 98.33, specificity 98.93, and precision 97.73.

A review paper [14] on skin cancer detection reviewed various artificial neural network techniques based on the classification and detection of cancerous cells in skin. Reviews on various non-invasive cancer detection method are described in this paper. Detection system is not an easy task which requires several stages which includes pre-processing and segmentation of the images. Here, they reviewed papers on artificial NNs, convolutional NNs, KNNs, and RBFNs for classifying skin lesion images. All the above said techniques have got several pros and cons. For getting good outcome, we have to choose proper classification algorithms. Finally, they came into a conclusion that convolutional neural network has put forward the good outcomes in the classification purpose.

Mahbod et al. [15] propose a model that evaluates a fusion approach (a multi-scale multi-CNN). Here, it utilizes a three-layered network architecture which undergoes training and cropping of dermoscopic images of variant scales. They actually tried the skin lesion classification based on different pre-trained networks of CNN which used ISIC challenge dataset. The dataset has been resized and cropped which resulted in three CNNs, namely EfficientNetB0, EfficientNetB1, and SeReNeXt-50. They concluded that image cropping can be selected as efficient strategy by comparing with image rescaling. Merging of the above said networks with the help of cropped images in all the six scaling improved the performance of the system as compared to the single network. Applying this model in ISIC 2018 classification, it yields 86.2% accuracy.

Hosny et al. [16] have used color images for classifying skin lesion images. They have created a model which undergoes three subsections. First section undergoes image pre-processing; second section used transfer learning mechanism; and in third section, they have replaced the last three layers of pre-trained network model with new layers. Transfer learning mechanism used the concepts AlexNet, ResNet101, and Google Net architectures for classifying tumorous cells. Here, they obtained a best performance from modified Google Net and accuracy is high. Here, ensemble of advanced deep neural architecture can be implemented for medical image diagnostics. From the study, they say that ensemble advanced deep architectures will lead to successful diagnosis.

Hameed et al. [17] used two approaches for the differentiation of skin lesion images. Initially, they used algorithms of machine learning for classification purpose, and later, they used deep learning approach for the same. First approach has undergone initial processing, segmentation, feature extraction, and classification. Second approach is good in object recognition and classification purpose. Peer-to-peer training is done from pixels of raw images and labeled images using deep learning algorithm. Here, CNN model is used for image classification purpose. Transfer learning mechanism is used. Comparison between proposed algorithm and multi-class single-level classification algorithm has taken place and has achieved high accuracy in both the approaches.

Thurnhofer et al. [18] have developed a framework in deep learning for the deterrence of skin cancer. They adapted a transfer learning mechanism for creating a plain and two class hierarchical classifiers which are capable of differentiating seven types of moles. To increase the dermoscopic image size, they have used HAM10000 dataset, and for improving performance, data augmentation has taken place. They have analyzed the model with six other deep learning networks for performance evaluation. Of that, DenseNet201 has shown good performance using transfer learning techniques. Classification accuracy is high while using Google Net network. First model, a plain model, has good performance than hierarchical model.

Thanh et al. [19] proposed a method for the deterrence of malicious cancer using automatic image processing techniques which undergoes pre-processing, segmentation, and feature extraction. These steps are done by adaptive principal curvature, color normalization, and ABCD rule. They have chosen International Skin Imaging Collaboration (ISIC) skin lesion dataset. This method has achieved high accuracy. Even though they have achieved a good result, subset of images which they have taken varies in resolution. To solve this, they concluded in taking images of same resolution and fuse it with the help of deep learning approach.

Qin et al. [20] propose the model which analyzes medical lesion images of skin using modified GAN-based data augmentation technology. This has been done for the improvement of performance of the classifier which is been developed. For the synthesis of the image, a style-based GAN model has been implemented for lesion images on the basis of original architecture of style-based GANs. Here, the system has been redesigned and came to a conclusion that redesigned structure is superior than the original GAN structure.

Manzo et al. [21] propose a pre-trained deep convolutional neural network which was useful for the representation of the lesion images and classification purpose. This concept used transfer learning approaches for feature extraction and ensemble method for classification purpose. They have shown a good performance compared to competitors.

Sertan et al. [22] propose a Gabor wavelet-based deep convolutional neural network for the early deterrence of cancerous cells. This system classified lesion images into seven directional sub-bands which can be used as inputs for the eight possible predictions by giving it to eight parallel CNNs. Based on the sum rule of decision fusion, classification is done.

Shahin et al. [23] propose a model based on deep convolutional neural network (DCNN) which purely depends on deep learning approach. Initial stage is pre-processing where filters or kernel is applied for removing the unwanted things of the image such as noise and artifacts. Then, normalization of image is done for the feature extraction for the exact classification. Data augmentation increased the size of images for improving the accuracy. This system is compared with other transfer learning mechanisms to evaluate the performance of the model. Model used HAM10000 dataset and has acquired an accuracy level above 90%.

Jain et al. [24] proposed a melanoma skin cancer detection methodology using image processing techniques. Here the input image of the skin lesion is pre-processed for enhancing the image quality by gamma correction technique. Here image segmentation is done with the help of an automatic thresholding process and masking operation in edge detection. The segmented image has undergone feature extraction by standard geographic features. The extracted features are then given for classification process (Table 1).

3 Conclusion

The paper deals with the survey of the state-of-the-art methods for performing analysis of skin lesion images. In this paper, review is based upon the techniques and algorithms which is implemented in the processing skin lesion images in detecting

Table 1 Comparison study

Work	Techniques/dataset	Accuracy
[2]	DE-ANN, fuzzy clustering, image segmentation, HAM10000	97.4
[3]	CNN	89.5
[4]	Bootstrap multi-class SVM, convolution neural network, ISIC 2019	94.2
[5]	ABCD method, Dull razor method, GLCM method, ISIC 2019	96.25
[6]	CNN, whale optimization algorithm, DermIS, Dermquest	91
[7]	CNN	87.42
[9]	GLCM matrix machine learning, ISIC 2017	97
[10]	CNN, transfer learning, ISBI-ISIC, PH2	96.86
[11]	CNN, transfer learning, DermIS	97.2
[12]	Deep neural network, fully convolutional residual network, ISIC 2017	91.2
[15]	Deep learning, transfer learning MSM-CNN, ISIC	86.2
[17]	MCML, PH2	96.47
[18]	Deep learning, HAM10000	96
[19]	ABCD rule, ISIC	96.6

malicious cancerous disease of skin. Techniques implemented include pre-processing methods, methods for feature extractions, algorithms used for segmentation process, and classification techniques.

Several deep learning techniques which researchers included in their work are fully convolution neural network, transfer learning mechanism, pre-trained model, and ensemble methods for the detection of skin cancer. From the works, it has been noted that deep learning techniques do not require more complicated and complex techniques for pre-processing of the images. Datasets like PH2, ISBI data obtained from different challenges, DermIS, Dermquest, HAM10000, Mednode, and open-access datasets has been used for the evaluation of researchers' work. From the literature, the image processing-related papers deal with variant data and human interaction is high. Here, manually tuned parameters are taken for getting correct results and better accuracy. In machine learning-related works, they have taken different features and have identified the cancerous cells but have shortcomings that features are not properly identified so results are not much accurate. So, by considering both of these approaches, it is found that the performance of deep neural networks is found to be satisfactory. The evaluation metric of accuracy, sensitivity, and specificity is up to the mark. So, in the future work we can go for hybrid and ensembling techniques for acquiring a good result.

References

1. Kadampur MA, Al Riyaee S (2020) Skin cancer detection: applying a deep learning based model driven architecture in the cloud for classifying dermal cell images. Inf Med. Unlocked 18(8 2019):100282. https://doi.org/10.1016/j.imu.2019.100282
2. Kumar M, Alshehri M, AlGhamdi R, Sharma P, Deep V (2020) A DE-ANN inspired skin cancer detection approach using fuzzy C-means clustering. Mob Networks Appl 25(4):1319–1329. https://doi.org/10.1007/s11036-020-01550-2
3. Hasan M, Das Barman S, Islam S, Reza AW (2019) Skin cancer detection using convolutional neural network. ACM international conference proceeding series. pp 254–258. https://doi.org/10.1145/3330482.3330525
4. Kassem MA, Hosny KM, Fouad MM (2020) Skin lesions classification into eight classes for ISIC 2019 using deep convolutional neural network and transfer learning. IEEE Access 8:114822–114832. https://doi.org/10.1109/ACCESS.2020.3003890
5. Krishna Monika M, Arun Vignesh N, Usha Kumari C, Kumar MNVSS, Laxmi Lydia E. Skin cancer detection and classification using machine learning. Mater Today Proc 33(xxxx):4266–4270. https://doi.org/10.1016/j.matpr.2020.07.366
6. Zhang L, Gao HJ, Zhang J, Badami B (2020) Optimization of the convolutional neural networks for automatic detection of skin cancer. Open Med 15(1):27–37. https://doi.org/10.1515/med-2020-0006
7. Ahmad B, Usama M, Huang CM, Hwang K, Hossain MS, Muhammad G (2020) Discriminative feature learning for skin disease classification using deep convolutional neural network. IEEE Access 8:39025–39033. https://doi.org/10.1109/ACCESS.2020.2975198
8. Kaur R, Kumar GP, Babbar G (2019) An enhanced and automatic skin cancer detection using K-mean and PSO technique. Int J Innov Technol Explor Eng 8(9):634–639. https://doi.org/10.35940/ijitee.I1101.0789S19
9. Almeida MAM, Santos IAX (2020) Classification models for skin tumor detection using texture analysis in medical images. J Imaging 6(6):1–15. https://doi.org/10.3390/JIMAGING6060051
10. de Rodrigues DA, Ivo RF, Satapathy SC, Wang S, Hemanth J, Filho PPR (2020) A new approach for classification skin lesion based on transfer learning, deep learning, and IoT system. Pattern Recogn Lett 136:8–15. https://doi.org/10.1016/j.patrec.2020.05.019
11. Ashraf R et al (2020) Region-of-interest based transfer learning assisted framework for skin cancer detection. IEEE Access 8:147858–147871. https://doi.org/10.1109/ACCESS.2020.3014701
12. Li Y, Shen L (2018) Skin lesion analysis towards melanoma detection using deep learning network. Sensors (Switzerland) 18(2):1–16. https://doi.org/10.3390/s18020556
13. Hosny KM, Kassem MA, Foaud MM (2019) Skin cancer classification using deep learning and transfer learning. In: 2018 9th Cairo international biomedical engineering conference CIBEC 2018—proceedings no. December, pp 90–93. https://doi.org/10.1109/CIBEC.2018.8641762
14. Dildar M et al (2021) Skin cancer detection: a review using deep learning techniques. Int J Environ Res Public Health 18(10). https://doi.org/10.3390/ijerph18105479
15. Mahbod A, Schaefer G, Wang C, Dorffner G, Ecker R, Ellinger I (2020) Transfer learning using a multi-scale and multi-network ensemble for skin lesion classification. Comput Methods Programs Biomed 193:105475. https://doi.org/10.1016/j.cmpb.2020.105475
16. Hosny KM, Kassem MA, Foaud MM (2020) Skin melanoma classification using ROI and data augmentation with deep convolutional neural networks. Multimed Tools Appl 79(33–34):24029–24055. https://doi.org/10.1007/s11042-020-09067-2
17. Hameed N, Shabut AM, Ghosh MK, Hossain MA (2020) Multi-class multi-level classification algorithm for skin lesions classification using machine learning techniques. Expert Syst Appl 141:112961. https://doi.org/10.1016/j.eswa.2019.112961
18. Thurnhofer-Hemsi K, Domínguez E (2020) A convolutional neural network framework for accurate skin cancer detection. Neural Process Lett. https://doi.org/10.1007/s11063-020-10364-y

19. Thanh DNH, Prasath VBS, Hieu LM, Hien NN (2020) Melanoma skin cancer detection method based on adaptive principal curvature, colour normalisation and feature extraction with the ABCD rule. J Digit Imaging 33(3):574–585. https://doi.org/10.1007/s10278-019-00316-x
20. Qin Z, Liu Z, Zhu P, Xue Y (2020) A GAN-based image synthesis method for skin lesion classification. Comput Methods Programs Biomed 195:105568. https://doi.org/10.1016/j.cmpb.2020.105568
21. Manzo M (2020) Bucket of deep transfer learning features and classification models for melanoma detection. https://doi.org/10.3390/jimaging6120129
22. Serte S, Demirel H (2019) Gabor wavelet-based deep learning for skin lesion classification. Comput Biol Med 113(8):103423. https://doi.org/10.1016/j.compbiomed.2019.103423
23. Ali MS, Miah MS, Haque J, Rahman MM, Islam MK (2021) An enhanced technique of skin cancer classification using deep convolutional neural network with transfer learning models. Mach Learn Appl 5(4):100036. https://doi.org/10.1016/j.mlwa.2021.100036
24. Jain S, Jagtap V, Pise N (2015) Computer aided melanoma skin cancer detection using image processing. Procedia Comput Sci 48(C):735–740. https://doi.org/10.1016/j.procs.2015.04.209

Adoption of Green Cloud Computing—A Review

Kirti Baghele, Raju Barskar, and Uday Chourasia

Abstract With the advent of cloud computing, the IT industry has picked a pace in case of development and new technologies. With the emergence of technology, users are also increasing at a rapid speed. If development is increasing, then at the same time we are setting back in terms of the environment by degrading it at a very high pace. Initially, any degradation is not visible but when use increases, degradation appears, and at that time, research is conducted to minimize or neutralize the demerits. Same with cloud computing, as it has been adopted by every second person in a crowd of thousands, much degradation is being caused, and to stop or to minimize green cloud computing has opted. The computer industry is contributing to global warming with high pollution, high energy, and high water enterprises. Green computing is in the process to minimize power consumption, energy consumption, and carbon dioxide emission.

Keywords Green cloud computing · Cloud computing · Virtualization

1 Introduction

Green cloud computing is becoming a hot research topic among Information and Communication Technologies (ICT). Continuation of technology requires taking care of the environment and not degrading it at such a high pace. In today's scenario, many government organizations and societies are opting for greener technology [1]. Many IT companies have opted for greener computing to slow down the increase in carbon footprint [2]. There are many organizations and consumers which are

K. Baghele (✉) · R. Barskar · U. Chourasia
University Institute of Technology RGPV, Bhopal, Madhya Pradesh, India
e-mail: Kirtibaghele01@gmail.com

R. Barskar
e-mail: rajubaraskar@rgtu.net

U. Chourasia
e-mail: udaychourasia@rgtu.net

A. Kumar et al. (eds.), *Proceedings of the International Conference on Cognitive and Intelligent Computing*, Cognitive Science and Technology,
https://doi.org/10.1007/978-981-19-2350-0_85

concerned for the environmental degradation. With the economical point of view also, they have joined green cloud computing which will make individual the brand ambassador of product; and by this, profit is caused to their organizations. To save energy, Green IT is the solution [3].

Now that ICT has taken certain steps to minimize or nullify the environmental degradation, there is increased pressure on producers to produce such products and provide such services that cause less influence on the ecosystem [4]. Factors contributing to global warming by IT companies are energy consumption, power consumption, and carbon dioxide emission.

2 Figures Influencing for the Adoption of Green Computing

Certain data has been collected on the basis of which, importance to opt for green cloud computing can be understood in better ways:

- A system not in use but still power on consumes 70% of energy [1].
- Greenpeace Guide to Greener Electronics in their May 2010 edition had listed only 2 PCs and other devices out of 18 who obtained green rating. Those 2 PCs were from Nokia and Sony Ericsson [5].
- About Rs. 1500–4500 can be saved per PC, if they are shut down when not in use [5].
- A report from McAfee (an antivirus company) revealed that 2 million houses in United States can be powered in the amount of electricity used in receiving everyday trillions of spam emails [2].
- Around 35 million tons of exhaust emission is produced by IT devices, which matches the carbon dioxide emissions observed in an aviation industry [6].
- At present, with the emergence of new devices every day, 66% of organizations are still preferring to use desktops over laptops which have the in-built feature of switching to sleep mode when the machine is not in use [3].
- It is predicted in case of carbon emission from different data centers that around 11% increase will be there but now it is estimated to be 440 metric megatons by 2025 [7].
- Making use of cloud computing can save around 30–60% of carbon emission in large organizations. In case of medium-sized enterprises, saving of carbon emission can be done up to 60% to 90%, whereas in case of small businesses, around 90% can be saved only by using cloud technology [7].
- On an average, around 3 billion people use electricity on a daily basis. In 2010, a report was published in which the electricity consumption is 1.5% of the world's electricity, only by the computers [8].
- Reports suggests that in 2014, data centers consumed power around 42 TWh which will result in CO_2 emission of 670 million tons annually by 2020 [8].
- Till now after going through the data statistically, it was found out that computers and data centers with cloud technology are the biggest consumers of energy [9].

- Around 30% of the power is consumed by the networking infrastructure used in data centers [10].

3 Cloud Computing

In the early 1950s, the concept of cloud computing had emerged with an identity of time sharing system [11]. Salesforce.com for the very first time gave a kick start through the Web site. The term cloud computing was introduced by Amazon in 2002, after which in 2009 Google and Yahoo have also included their presence. Cloud computing is considered as the subset of ICT [4].

Some examples of cloud computing being used widely are Gmail, Google Drive, Facebook, Instagram, and many more.

3.1 Data Centers

Like any other technology, the cloud also has physical infrastructure called a data center. Data center consists of hardware and network infrastructure. Physical hardware contains all the computer hardware required in the functioning of cloud computing, and the network provides two aims: 1. Network helps to build the connectivity between data center hardware and cloud and 2. facility to maintain connectivity between user and cloud.

Some progress has been made in the traditional data centers to reduce their energy consumption up to 30% with the use of a distributed computing platform and is named as Nano Data centers (NaDa) [12].

4 Virtualization

4.1 Origin

In 1964, in an IBM project the concept of virtualization originated [13].

4.2 Goal

To combine IT resources such as processor, storage, and network to utilize them to their fullest which will reduce the costing by combining all the ideal resources into a shared pool and then virtual machines are created to perform different tasks [14].

4.3 Standard Definitions

- Virtualization: "It is the application of the layering principle through enforced modularity, whereby the exposed virtual resource is identical to the underlying physical resource being virtualized." [15]. Two terms need to be explained in the above definition: Layering—single abstraction is presented with the addition of indirection level, when the indirection relies on a single lower layer, and to expose the abstraction, a well-defined namespace is used. Enforced modularity—it makes sure that the abstraction layer is not bypassed by the clients of that layer [15].

In simple words, virtualization uses a physical server also called a host, which is shared among different virtual machine monitors and hypervisors which also increases utilization of resources [16].

- Virtual machines: "It is an abstraction of a complete compute environment through the combined virtualization of the processor, memory, and I/O components of a computer." [15].
- Hypervisor: "It is a specialized piece of system software that manages and runs virtual machines." [15].
- Virtual machine monitor (VMM): "It refers to the portion of the hypervisor that focuses on the CPU and memory virtualization." [15].
- Host computer: The computer on which virtual machine is operated/run/installed.

4.4 Need for Virtualization

Opting for virtualization techniques under cloud computing, organizations will be benefited by saving the management, hardware, and power utilization costs. It also facilitates improved mobile applications, administration, and deployment simplifications. Use of virtualization can be beneficial in many fields such as data centers, education, finance, and government [13]. DVMC (dynamic virtual machine consolidation) is used to reduce the energy consumption in data centers which is obtained by virtualization [16].

4.5 Techniques to Maximize Benefits in Virtualization

- Bin packing.
- Greedy methods.
- Graph theory.
- Meta-heuristic approaches (genetic algorithm, ant colony system) [9].

4.6 Emulation

It is a technique in which the system hardware working is converted into a software program and works over the operating system. This is done to provide the flexibility to guest OS, and drawback is that the translation speed is slowed down [14].

5 Green Cloud Computing

5.1 Origin

The idea of green cloud computing is initiated with a simple idea of Sustainable Development [12]. This idea was initiated when an U.S.-based energy protection agency had launched energy star rating on monitors and various electronic devices in 1992 [5].

5.2 Principle

Sustainability undergoes—reduce, reuse, and recycle [17].

5.3 Goal

Green chemistry and green computing are similar if we study at their goals which says:

- Hazardous material use should be reduced.
- In a product's lifespan, energy efficiency should be high.
- Waste from factories and products with non-operationally behavior should be recycled or biodegradable.

5.4 Steps Taken by ICT for Greener IT

- Standards for display power consumption: American Video Electronics Standards Association.
- TCO certification: Swedish Confederation of Professional Employees had developed for display environmental requirements. Ecology, energy, emissions, and ergonomics are four main criteria.

- EPEAT certification: Standard No. 1680 of the American Association of Electrical and Electronic Engineers (IEEE) was opted in 2006 by the American National Standards Institute (ANSI) under United States Green Electronics Council. This certification is used on all the electronic products throughout their life cycle [6].

6 Literature Review

Author analyzed some surveys performed on both academic and non-academic studies to understand in brief about green cloud computing. This survey is important to understand the view of organizations, enterprises, and government institutes toward green cloud computing. Involvement of green cloud computing is not limited to only energy consumption from computer devices but also emphasizes emission of carbon dioxide, e-waste, and natural resources consumption. Author suggested that replacement of high power computers with low powered computers will bring positive change in energy efficiency. As compared to traditional cloud, there is a decrease of 27% in energy consumption by using heuristic algorithms in which optimization of virtual machine placement is proposed and was based on dynamic migration technology supporting virtual machines, and this was done in 2009 [4].

Author has presented some experiments explaining their findings on virtualization to attain green cloud computing and also environmental sustainability. Some energy usage methods had been discussed with the help of server virtualization. As data centers cause much more power consumption, I am looking forward to some non-profit group named Green Grid, which was set up in February 2007 by some IT industries. An experiment was conducted in between Xen and KVM, and the conclusion turns out to be in favor of virtualization techniques for the decrease in consumption of power in data centers. There are various technologies like application development, migration metrics and algorithms, CPU design, etc. which can be used to reduce energy [1].

In comparison with the trends in 2012 with the ongoing trends in 2019, there had been so much evolution in the solution and with that problems had also evolved differently. In 2010, researchers had designed a simulator based on NS-2 which is capable of measuring power consumed at data centers with green computing. Equipments used were Intel Xeon 4-core processor, 8 MB DDR3 RAM, and 3.33 GHz cache [14].

Author has presented research to increase the efficiency of data centers and server virtualization. With the increment in the use of cloud computing, data center setup also increases which proportionally increases the consumption of energy. The main cause of the heat production is through the processor chips present in data centers. For working, these chips need to be cooled down, which also needs electricity. Techniques opted for achieving green cloud computing are virtual machine (VM) consolidation, server virtualization, dynamic voltage and frequency scaling (DVFS), and live migration [18].

In this paper, researchers find most significant ways to obtain green cloud computing such as virtualization, scheduling, clustering, and proportional computing and some more energy conservation techniques are studied. Author suggested that cloud computing is all about sharing of computing resources and does not include any activity of maintenance and construction of local servers. Due to increase in emission of carbon dioxide and energy consumption day by day, there occurred need to opt for some techniques like scheduling of workloads, clustering, data virtualization, proportional computing, renewable energy, and power consumption [2].

A brief introduction to cloud computing is presented in [19] with models: deployment and service, characteristics and technologies including virtualization, service-oriented architecture (SOA), grid computing, and utility computing. Grid computing is called distributed computing in which a problem is divided into various problems and is then solved by various computers which are at different locations, and this is used in performing large computational problems. Utility computing is all about managing pay per use services.

Reference [20] presents the benchmarks with challenges in research for green cloud computing. To accomplish the requirements of IT companies for a very long time, this has caused environmental degradation and has influenced the use of green cloud computing. Factors leading to degradation are data centers increase with a high pace, CPU advancement, cost for energy consumption, and utilization of servers (Fig. 1).

Some challenges are presented which are obstacles in opting for green cloud computing. Some of them are new optimizing technique requirements, architecture complexity minimization, data center efficiency requirements, and data center cooling, etc.

Author in [12] had used data mining and engineering approaches driven by a model. This is estimated in the reduction of carbon footprints and operational cost being optimized. Carbon footprint calculating formula is used. Also, it is suggested to use autoscaling feature to be applied on cloud computing. Performance evaluation formula is used for utilized resources.

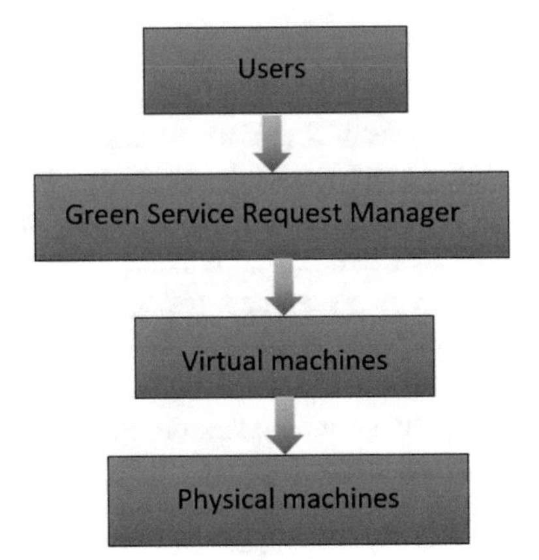

Fig. 1 Green computing mechanism

The paper [8] consists of a resource management solution named dynamic and energy-aware in which consolidation of virtual machines is obtained which is responsible for energy efficiency and utilization. Degradation in performance of host and virtual machine (VM) is caused due to aggressive VM consolidation. For this, researchers have proposed efficient energy enhancing and QoS dynamic resource management methods. Some constraints are affinity, security, and migration constraints (Table 1).

7 Challenges

- Power consumption: Due to increase in the use of computers, power consumption is increasing at a rapid pace. Many technologies are suggested like virtualization which consumes less power as compared to use of traditional servers.
- Energy consumption: As increase in computer so does increase in processors. For working and cooling of processors, there is need of energy in both the scenarios. By developing green computers, energy consumption is being minimized with each advent.
- Carbon emission: Data centers are the main cause for the increase in the carbon emission. Constructing green data centers called NaDa is helpful but also some improvement in terms of carbon emission is to be made.
- Resource management: With the adoption of virtualization technology, resources are shared in virtual machine of a single physical server. To obtain better virtualization, virtual machine consolidation is opted but some constraints are observed which are listed in [8].

8 Conclusion

After reviewing various research papers, many methods had been proposed by various researchers to obtain green cloud computing. Various factors causing degradation to the environment are studied, and their solutions need to be improved with more techniques. ICT (Information and Communication Technology) is paying much more efforts to minimize the effect of computer industry on environment.

Various fields which are causing degradation to the environment do not have much limelight like resource management, data mining, etc. One such technology which need to be researched in deeper is containerization.

This paper is all about being aware of the amount of degradation being caused all over the world, a brief introduction to cloud computing, virtualization, green cloud computing—understanding their origins, goals, principles, and measures taken by some special group.

Table 1 Summary of existing review papers

S. No.	Paper title	Reference	Year	Work details	Technology used
1	Green cloud computing: a literature survey	[4]	2017	Academic and non-academic studies. Virtual machine optimization used. Figures and facts about degradation of environment	Dynamic migration technology
2	An analysis report on green cloud computing current trends and future research challenges	[14]	2019	Simulator designed which can measure power consumption at data centers	Power management methods
3	Green computing strategies and challenges	[5]	2015	Improvement in the concept of reuse, repair, and recycling	Energy-efficient coding
4	Green cloud computing: a review on efficiency of data centers and virtualization of servers	[18]	2017	Elimination of the heat produced by processor chips	Dynamic voltage and frequency scaling (DVFS)
5	Green cloud computing: a review	[2]	2017	Lack in security and connectivity areas. Many techniques suggested to opt green cloud computing	Queuing theory principle
6	Power saving strategies in green cloud computing systems	[10]	2015	Various techniques opted by cloud service providers to minimize the energy consumption	Nano Data centers
7	Cloud computing technologies	[19]	2018	Benefits to use cloud computing. Explanation to grid computing, utility computing	Utility computing
8	A brief survey on benchmarks and research challenges for green cloud computing	[20]	2016	Some factors which are not studied very often for environmental degradation like data center increment and many more	Taxonomy of benchmarks

(continued)

Table 1 (continued)

S. No.	Paper title	Reference	Year	Work details	Technology used
9	Green cloud computing solution for operational cost efficiency and environmental impact reduction	[12]	2019	Various techniques including data mining and engineering are used to calculate the carbon emission. Virtual machine migration and evolution of Nano Data centers	Virtualization
10	An emerging technology: green computing	[17]	2015	Various technologies which are using green cloud computing concepts and the invention of new modes say: sleep and hibernate	Green cloud computing
11	Green cloud computing: efficiency energy—aware and dynamic resources management in data centers	[8]	2018	Resource management techniques following some constraints	Virtual machine consolidation

Acknowledgements I take this opportunity to gratefully acknowledge the assistance of my college faculty and guides Dr. Raju Baraskar and Prof. Uday Chourasia, who had faith in me. I am also very much grateful to UIT RGPV, DoCSE for motivating the students.

References

1. Motochi V, Barasa S, Owoche P, Wabwoba F (2017) The role of virtualization towards green computing and environmental sustainability. Int J Adv Res Comput Eng Technol (IJARCET) 6(6):851–858
2. Masoud RI, AlShamrani RS, AlGhamdi FS, AlRefai SA, Hemalatha M (2017) Green cloud computing: a review. Int J Comput Appl 167(9):5–7
3. Agarwal S, Ghosh A, Nath A (2016) Green enterprise computing-approaches towards a greener IT. Int J Innov Res Adv Eng (IJIRAE) 1–5
4. Radu L-D (2017) Green cloud computing: a literature survey. Symmetry 9(12):1–20
5. Singh S (2015) Green computing strategies and challenges. In: 2015 international conference on green computing and internet of things (ICGCIoT). IEEE, pp 758–760
6. Zhang X, Gong L, Li J (2012) Research on green computing evaluation system and method. In: 2012 7th IEEE conference on industrial electronics and applications (ICIEA). IEEE pp 1177–1182
7. Kaur P, Sachdeva M (2016) A review on green cloud computing, journal. Int J Comput Technol, ISSN 2277–3061:6686–6691
8. Diouani S, Medromi H (2018) Green cloud computing: efficient energy-aware and dynamic resources management in data centers. Int J Adv Comput Sci Appl 9(7):124–127

9. Fathi MH, Khanli LM (2018) Consolidating VMs in green cloud computing using harmony search algorithm. International conference on internet and e-business. pp 146–151
10. Borah AD, Muchahary D, Singh SK, Borah J (2015) Power saving strategies in green cloud computing systems. Int J Grid Distrib Comput 8(1):299–306
11. Atrey A, Jain N, Iyengar N (2013) A study on green cloud computing. Int J Grid Distrib Comput 6(6):93–102
12. Bindhu V, Joe M (2019) Green cloud computing solution for operational cost efficiency and environmental impact reduction. J ISMAC 1(02):120–128
13. Ding WM, Ghansah B, Wu YY (2016) Research on the virtualization technology in cloud computing environment. In: International journal of engineering research in Africa, vol 21. Trans Tech Publications Ltd, pp 191–196
14. Durairaj M, Kannan P (2014) A study on virtualization techniques and challenges in cloud computing. Int J Sci Technol Re 1, no 1. 3(11):147–151
15. Bugnion E, Nieh J, Tsafrir D (2017) Hardware and software support for virtualization. Synth Lect Comput Archit 12(1):1–206
16. Farahnakian F, Pahikkala T, Liljeberg P, Plosila J, Tenhunen H (2015) Utilization prediction aware VM consolidation approach for green cloud computing. In: 2015 IEEE 8th international conference on cloud computing. IEEE, pp 381–388
17. Mishra R, Jain S, Kurmi N (2015) An emerging technology: green computing. Int Res J Eng Technol 2(2):175–179
18. Shakeel F, Sharma S (2017) Green cloud computing: a review on efficiency of data centres and virtualization of servers. In: 2017 international conference on computing, communication and automation (ICCCA). IEEE, pp 1264–1267
19. Malik MI, Wani SH, Rashid A (2018) Cloud computing-technologies. Int J Adv Res Comput Sci 9(2):379–384
20. Patel YS, Jain K, Shukla SK (2016) A brief survey on Benchmarks and research challenges for green cloud computing. Int J Comput Appl 975–8887

Printed in the United States
by Baker & Taylor Publisher Services